Nanosafe 2012: International Conferences on Safe Production and Use of Nanomaterials

Journal of Physics: Conference Series Volume 429

Grenoble, France
13-15 November 2012

ISBN: 978-1-62748-306-3
ISSN: 1742-6588

Printed from e-media with permission by:

Curran Associates, Inc.
57 Morehouse Lane
Red Hook, NY 12571

Some format issues inherent in the e-media version may also appear in this print version.

Printed by Curran Associates, Inc. (2013)

For permission requests, please contact the Institute of Physics
at the address below.

Institute of Physics
Dirac House, Temple Back
Bristol BS1 6BE UK

Phone: 44 1 17 929 7481
Fax: 44 1 17 920 0979

techtracking@iop.org

Additional copies of this publication are available from:

Curran Associates, Inc.
57 Morehouse Lane
Red Hook, NY 12571 USA
Phone: 845-758-0400
Fax: 845-758-2634
Email: curran@proceedings.com
Web: www.proceedings.com

TABLE OF CONTENTS

EXPOSURE ASSESSMENT

DETECTION AND IDENTIFICATION

TOXICOLOGY

ENVIRONMENTAL INTERACTIONS

NANOMATERIALS RELEASE

INDUSTRIAL PRODUCTION

ETHICS AND SOCIETAL ISSUES

COMMERCIAL EQUIPMENT

RISK MANAGEMENT FOR OHS EXPERTS

REGULATION AND STANDARDIZATION FOR NANO PRODUCTS

GRENOBLE-ALPES MÉTROPOLE

VILLE DE GRENOBLE

DE LA RECHERCHE À L'INDUSTRIE cea

liten

MINATEC®

DEKATI Excellence in Particle Measure

TSI® UNDERSTANDING, ACCELERATED

® PARTICLE MEASURING SYSTEMS

CytoViva®

GRIMM

CORDOUAN Technologies

HORIBA Scientific

ECOMESURE

NANOBADGE ALCEN

iZON

trimatec Une technologie maîtrisée, un avenir préservé

IOP Conference Series conferenceseries.iop.org

INERIS maîtriser le risque pour un développement durable

European Technology Platform on Industrial Safety

NANOSIGHT

Welcome from the organizers

The rapidly developing field of nanotechnologies opens up many exciting opportunities and benefits for new materials with significantly improved properties as well as some revolutionary applications in the fields of energy, environment, medicine, etc.
These new materials potentially pave the way to considerable innovations in many industries of the 21st century strictly only if associated risks are perfectly under control for workers, consumers and the environment.
So, one can understand easily why the so-called Nanosafety is now considered as a specific new scientific area, gaining in importance and maturity each days.

Following the successful outcome of the two past **international conferences on Safe production and use of nanomaterials: Nanosafe 2008 and 2010,** the organizing committee has the pleasure to welcoming you again in Grenoble with some of the most famous specialists of the field in the world.

In addition of the standard issues addressed in previous Nanosafe conferences such as Toxicology, Eco-toxicology, Expology, Detection, Life Cycle Analysis, Regulation and Standardization, new topics of great interest will be dealt with this year, concerning Governance plus practical Risk Management for OSH experts and Societal issues.
In order to enhance the exchanges and conviviality three debates will be organized around Nano Governance, Toxicology and Ethics.

We hope that you will appreciate this new Nanosafe edition like the previous ones.

This is your Nanosafe conference, please enjoy.

The Nanosafe conference organisers

François Tardif

Frédéric Shuster

Management

Vanessa Gaultier

NANOSAFE 2012
Conference Organization

Local Organizing Committee

Vanessa GAULTIER (CEA)
François TARDIF (CEA)
Dominique BAGUET (CEA)
Frédédric SHUSTER (CEA)
Yves SICARD (CEA)
Catherine DURAND (CEA)
Didier MOLKO (MINATEC)

International Scientific Committee

Chair: Frédéric SCHUSTER (CEA, FR)
François TARDIF (CEA, FR)
Co-chair: Georgios KATALAGARIANAKIS (EC, BE)

Mélanie AUFFAN (CEREGE, FR)
Jorge BOCZKOWSKI (INSERM, FR)
Jean-Yves BOTTERO (CEREGE, FR)
Jacques BOUILLARD (INERIS, FR)
Derk BROUWER (TNO, NL)
Marie CARRIERE (CEA, FR)
Claude EMOND (U. MONTREAL, CA)
Cassandra ENGEMAN (UCSB, USA)
Eric GAFFET (CNRS, FR)
François GENSDARMES (IRSN, FR)
Alexei GRINBAUM (CEA, FR)
Antje GROBE (U.STUTTGART, DE)
Peter HOET (KUL, BE)
Eric QUEMENEUR (CEA, FR)
Olivier LE BIHAN (INERIS, FR)
Robert MUIR (NANEUM, UK)
Tinh NGUYEN (NIST, USA)
Bernd NOWACK (EMPA, CH)
Günter OBERDÖRSTER (U. ROCHESTER, USA)
David PUI (U. MINNESOTA, USA)
Michael RIEDIKER (IST, CH)
Yves SAMSON (CEA, FR)
Ken TAKEDA (U. TOKYO, JP)
Olivier WITSCHGER (INRS, FR)

NAME	FIRST NAME	COMPANY	COUNTRY
AIT- OUMEGHAR	Stephane	CORDOUAN TECHNOLOGIES	FRANCE
AL-KATTAN	Ahmed	EMPA	SWITZERLAND
AMBLARD	Frédéric	CEA/DRT/DSP/GMR	FRANCE
AMBLARD	Frédéric	CEA	FRANCE
AMDAOUD	Malika	NANOBADGE - ALCEN	FRANCE
AMODEO	Tanguy	INERIS	FRANCE
ANSTEINSSON	Vibeke	UNIVERSITY OF BERGEN FACULTY OF MEDICINE AND DENTISTRY, BIOMATERIALS	NORWAY
ARTOUS	Sébastien	CEA	FRANCE
ASCHBERGER	Karin	NANOBIOSCIENCES UNIT, JOINT RESEARCH CENTRE ISPRA, EUROPEAN COMMISSION	ITALY
ASIMAKOPOULOU	Akrivi	CERTH/CPERI/APTL	GREECE
AUBERT	Dimitri	IZON SCIENCE LTD	GREAT BRITAIN
AUBLANT	Jean-Marc	LNE	FRANCE
AUFFAN	Melanie	CEREGE CNRS	FRANCE
AUGER	Aurélien	CEA GRENOBLE	FRANCE
BABIN	Mar	INIA	SPAIN
BACHLER	Gerald	ETH ZÜRICH	SWITZERLAND
BAE	Gwi-Nam	KOREA INSTITUTE OF SCIENCE AND TECHNOLOGY	SOUTH KOREA
BAEZA	Armelle	UNIVERSITY PARIS DIDEROT, UNITÉ DE BIOLOGIE FONCTIONNELLE ET ADAPTATIVE	FRANCE
BAGUET	Dominique	CEA GRENOBLE	FRANCE
BAHRI	Susana	CEA	FRANCE
BALASUBRAMANIAN	Sivakumar	BIO-NANO ELECTRONICS RESEARCH CENTER, GRADUATE SCHOOL OF INTERDISCIPLINARY NEW SCIENCE, TOYO UNIVERSITY	JAPAN
BARTCZAK	Dorota	LGC	UNITED KINGDOM
BAU	Sébastien	INSTITUT NATIONAL DE RECHERCHE ET DE SÉCURITÉ	FRANCE
BAUBLYTE	Lijana	UNIVERISTY OF LIMERICK	IRELAND
BAUMERTE	Anda	UNIVERSITY OF LATVIA	LATVIA
BELOKRYLOVA	Ekaterina	INSTITUTE OF LAW, UDMURT STATE UNIVERSITY	RUSSIA
BELOT	Jean-Marc	CETIM	FRANCE
BELUT	Emmanuel	INRS	FRANCE
BENCSIK	Anna	ANSES	FRANCE
BERNARD	Daniel	ARKEMA	FRANCE
BERNARD-BRUNEL	Aude	ENTREPRISES POUR L'ENVIRONNEMENT (EPE)	FRANCE
BERNE	Philippe	CEA GRENOBLE	FRANCE
BEUZELIN OLLIVIER	Marie-Gabrielle	LTP EPFL	SWITZERLAND
BILLARD	Alain	LRC CEA IRTES-LERMPS	FRANCE
BILLAT	Bernard	PARTICLE MEASURING SYSTEMS	UNITED STATES
BLOCH	Daniel	CEA/DRT	FRANCE
BLYUM	Yaroslav	INSTITUTE OF CELL BIOLOGY AND GENETIC ENGINEERING OF NATIONAL ADADEMY OF SCIENCES OF UKRAINE	UKRAINE
BOCZKOWSKI	Jorge	INSERM U955, TEAM 4	FRANCE
BÖHME	Steffi	HELMHOLTZ-CENTRE FOR ENVIRONMENTAL RESEARCH / DEP. BIOANALYTICAL ECOTOXICOLOGY	GERMANY
BOIS	Frédéric	INERIS	FRANCE
BOMBARDIER	Pierre	FAURE QEI	FRANCE
BOSSA	Nathan	CEREGE	FRANCE
BOTTERO	Jean-Yves	CEREGE-CNRS-AMU	FRANCE
BOUILLARD	Jacques	INERIS	FRANCE
BOULANGER	Pascal	CEA	FRANCE
BOURGUINON	Philippe	CEA	FRANCE
BOUTOU-KEMPF	Odile	INSTITUT DE VEILLE SANITAIRE - DÉPARTEMENT SANTÉ TRAVAIL	FRANCE
BRAINI	Céline	UJF	FRANCE
BREGOLI	Lisa	ECSIN, VENETO NANOTECH	ITALY
BRESSOT	Christophe	INERIS	FRANCE
BROUARD	Christophe	CEA GRENOBLE	FRANCE
BROUWER	Derk	TNO	NETHERLANDS
BRZICOVÁ	TáňA	VŠB – TECHNICAL UNIVERSITY OF OSTRAVA, FACULTY OF SAFETY ENGINEERING	CZECH REPUBLIC
BUSQUETS-FITE	Marti	CATALAN INSTITUTE OF NANOTECHNOLOGY (ICN)	SPAIN
BUSSY	Cyrill	NANOMEDICINE LAB, CDDR, UCL SCHOOL OF PHARMACY, UNIVERSITY COLLEGE LONDON	UNITED KINGDOM

CALZOLAI	Luigi	EUROPEAN COMMISSION. JRC\	ITALY
CANCINO BERNARDI	Juliana	UNIVERSITY OF SÃO PAULO / PHYSICS INSTITUTE OF SÃO CARLOS	BRAZIL
CAPRON	Philippe	CEA GRENOBLE	FRANCE
CARRIERE	Marie	CEA	FRANCE
CASSETTE	Simone	THALES RESEARCH & TECHNOLOGY FRANCE	FRANCE
CHARLES	Sandrine	ANSES	FRANCE
CHEN	Lie	HEALTH CANADA	CANADA
CHENTOUF	Aouatif	ROQUETTE FRERES	FRANCE
CHEVALLET	Mireille	CEA-GRENOBLE / IRTSV / LCBM	FRANCE
CHUEN-JINN	Tsai	NATIONAL CHIAO TUNG UNIVERSITY/INSTITUTE OF ENVIRONMENTAL ENGINEERING	TAIWAN (ROC)
CIMPAN	Mihaela Roxana	BIOMATERIALS - FACULTY OF MEDCINE AND DENTISTRY	NORWAY
CLANCY	Catherine	BLUE SCOPE STEEL LIMITED	AUSTRALIA
CLAVAGUERA	Simon	CEA GRENOBLE	FRANCE
CLERC	Frédéric	INRS	FRANCE
COCCINI	Teresa	MAUGERI FOUNDATION IRCCS - INSTITUTE OF PAVIA	ITALY
COHIGNAC	Vanessa	INSERM U955	FRANCE
COLIN	Claudine	TRIMATEC	FRANCE
COSSON	Céline	AIRBUS HEALTH AND SAFETY TEAM	FRANCE
COSTA	Carla	INSTITUTE OF PUBLIC HEALTH OF PORTO UNIVERSITY (ISPUP)	PORTUGAL
DANROC	Joel	CEA	FRANCE
DE MERINGO	Alain	COMPAGNIE DE SAINT-GOBAIN	FRANCE
DECOM	Anne-Laure	AIRBUS	FRANCE
DEKALI	Samir	INERIS	FRANCE
DELABY	Stephane	CSTB	FRANCE
DELAITRE	Caroline	MICHELIN	FRANCE
DELAVAL	Mathilde	UNIV PARIS DIDEROT, (SORBONNE PARIS CITÉ), UNIT OF FUNCTIONAL AND ADAPTIVE BIOLOGY (BFA) LAB OF MOLECULAR AND CE	FRANCE
DENOO	Koen	SOLAE BELGIUM NV	BELGIUM
DERROUGH	Samir	CEA	FRANCE
DIAZ-TORRES	Roberto	NATIONAL UNIVERSITY OF MEXICO	MEXICO
DILABIO	Gino	NATIONAL INSTITUTE FOR NANOTECHNOLOGY	CANADA
DRAIS	Eric	INRS	FRANCE
DUBOSSON	Muriel	CEA/DRT/DSP/SMR/LABM	FRANCE
DURAN	Nelson	INSTITUTO DE QUÍMICA, UNIVERSIDADE ESTADUAL DE CAMPINAS	BRAZIL
DURAND	Catherine	CEA GRENOBLE	FRANCE
DUTOUQUET	Christophe	INERIS	FRANCE
DUVAL-ARNOULD	Guy	SAINT-GOBAIN	FRANCE
DZANTIEV	Boris	A.N. BACH INSTITUTE OF BIOCHEMISTRY, RUSSIAN ACADEMY OF SCIENCES	RUSSIA
EINIG	Antinéa	ECOLE CENTRALE PARIS	FRANCE
EMOND	Claude	U. MONTREAL	CANADA
ENDES	Carola	ADOLPHE MERKLE INSTITUTE, UNIVERSITY OF FRIBOURG	SWITZERLAND
ENGEMAN	Cassandra	UCSB	UNITED STATES
EZAN	Eric	CEA/DSV/IBEB/SBTN - CENTRE DE MARCOULE	FRANCE
FALK	Andreas	BIONANONET FORSCHUNGSGESELLSCHAFT MBH	AUSTRIA
FAUQUANT	Caroline	CEA, LABORATOIRE DE CHIMIE ET BIOLOGIE DES MÉTAUX, UMR 5249 CEA-CNRS-UJF	FRANCE
FAURE	Bertrand	STOCKHOLM UNIVERSITY	SWEDEN
FAYE	Benoît	INERIS	FRANCE
FAYE	Laurent	UKTI - UK TRADE & INVESTMENT - BRITISH TRADE & INVESTMENT DELEGATION - LYON	FRANCE
FEDOTOVA	Nataliya	ETH ZURICH	SWITZERLAND
FERREIRA	Paula Silvia Hadd	FEDERAL UNIVERSITY OF SÃO PAULO - UNIFESP	BRAZIL
FIGAROL	Agathe	ECOLE NATIONALE SUPÉRIEURE DES MINES DE SAINT-ETIENNE	FRANCE
FILIMUNDI	Eric	TSI GMBH	GERMANY
FILIPPONI	Luisa	NANOPINION	DENMARK
FLEURY	Dominique	INERIS	FRANCE
FLORES-CERVANTES	Deborah Xanat	EAWAG	SWITZERLAND
FRANCHI	Leonardo	UNIVERSITY OF SÃO PAULO	BRAZIL

FREYER	Annette	LEIBNIZ INSTITUTE FOR SURFACE MODIFICATION	GERMANY
FRIJNS	Evelien	VITO	BELGIUM
FUCHS	Dietmar	MEDICAL UNIVERSITY	AUSTRIA
FUJIOKA	Kouki	JIKEI UNIVERSITY SCHOOL OF MEDICINE	JAPAN
GAMBARDELLA	Chiara	DISTAV, UNIVERSITY OF GENOVA	ITALY
GARDINI	Davide	CNR-ISTEC	ITALY
GAULTIER	Vanessa	CEA GRENOBLE	FRANCE
GAUTHIER	Laury	ECOLAB	FRANCE
GEORGE	Isabelle	UNIVERSITÉ PARIS DIDEROT / UNITÉ DE BIOLOGIE FONCTIONNELLE ET ADAPTATIVE	FRANCE
GHARBI	Aïcha	CNRS - OMNT	FRANCE
GHIAZZA	Mara	DEPARTMENT OF CHEMISTRY, UNIVERSITY OF TURIN	ITALY
GLOMSTAD	Berit	NORWEGIAN UNIVERSITY OF SCIENCE AND TECHNOLOGY	NORWAY
GÖHLER	Daniel	RESEARCH GROUP MECHANICAL PROCESS ENGINEERING, INSTITUTE OF PROCESS ENGINEERING AND ENVIRONMENTAL TECHNC	GERMANY
GOLANSKI	Luana	CEA	FRANCE
GÓMEZ	Virginia	INSITUTO DE NANOCIENCIA DE ARAGÓN/INSTITUTE OF NANOSCIENCE OF ARAGON	SPAIN
GONZALES	Nicolas	CYTOVIVA	UNITED STATES
GRIDELET	Laurent	SOPROREAL	FRANCE
GRINBAUM	Alexei	CEA SACLAY	FRANCE
GROBE	Antje	RISIKO -DIALOGU	SWITZERLAND
GROSO	Amela	EPFL	SWITZERLAND
GROSSEAU	Philippe	ECOLE NATIONALE SUPÉRIEURE DES MINES	FRANCE
GUIOT	Arnaud	CEA GRENOBLE	FRANCE
GUIZARD	Benoit	CEA	FRANCE
GUO	Muyao	THE UNIVERSITY OF HONG KONG	CHINA
GUSEVA CANU	Irina	FRENCH INSTITUTE FOR PUBLIC HEALTH SURVEILLANCE (INVS)	FRANCE
HAAS	Karl-Heinz	FRAUNHOFER-INSTITUT FÜR SILICATFORSCHUNG ISCÜ	GERMANY
HANADA	Sanshiro	NATIONAL CENTER FOR GLOBAL HEALTH AND MEDICINE	JAPAN
HANSEN	Steffen Foss	DTU ENVIRONMENT	DENMARK
HARTMANN	Georg	TECHNISCHE UNIVERSITÄT MÜNCHEN	GERMANY
HAUPT	Mélanie	ETH ZURICH, INSTITUTE FOR ENVIRONMENTAL ENGINEERING	SWITZERLAND
HEIDMANN	Ilona	UNIVERSITY OF KOBLENZ-LANDAU / INTERDISCIPLINARY RESEARCH GROUP FOR ENVIRONMENTAL STUDIES	GERMANY
HEJAZY	Marzie	TOXICOLOGY DEPARTMENT, FACULTY OF VETERINARY MEDICINE, UNIVERSITY OF TEHRAN, IRAN	IRAN
HENDRIKSON	Olga	A.N. BACH INSTITUTE OF BIOCHEMISTRY, RUSSIAN ACADEMY OF SCIENCES	RUSSIA
HENRY	François	INERIS	FRANCE
HERLIN	Nathalie	CEA	FRANCE
HERTZ	Audrey	CEA DEN MARCOULE	FRANCE
HERTZ	Audrey	CEA	FRANCE
HILDEBRAND	Heike	HELMHOLTZ-ZENTRUM DRESDEN-ROSSENDORF	GERMANY
HIRSCH	Vera	ADOLPHE MERKLE INSTITUTE	SWITZERLAND
HISCHIER	Roland	EMPA / TECHNOLOGY & SOCIETY LAB	SWITZERLAND
HOET	Peter	KU LEUVEN	BELGIUM
HOOL	Alessandra	MATSEARCH COSULTING HOFMANN	SWITZERLAND
HUBER	Stephany	UNCAMP/ HEMOCENTRO	BRAZIL
ICHIHARA	Gaku	NAGOYA UNIVERSITY GRADUATE SCHOOL OF MEDICINE	JAPAN
INOUE	Yuriko	JAPAN	JAPAN
IRIE	Masaru	WASEDA UNIVERSUTY	JAPAN
IRIZAR	Amaia	FORECOM BIOSCIENCE LTD	UNITED KINGDOM
ISAACSON	Carl	EAWAG (SWISS FEDERAL INSTITUTE OF AQUATIC SCIENCE)	SWITZERLAND
ISAXON	Christina	LUND UNIVERSITY	SWEDEN
IZAK	Emilia	BAYER TECHNOLOGY SERVICES GMBH	GERMANY
JAENSSON	Alia	STORA ENSO AB	SWEDEN
JAFFRES	Cristina	L'ORÉAL RECHERCHE	FRANCE
JAHNEL	Jutta	KARLSRUHE INSTITUTE OF TECHNOLOGY, INSTITUTE FOR TECHNOLOGY ASSESSMENT AND SYSTEMS ANALYSIS	GERMANY
JAMIER	Vincent	CATALAN INSTITUTE OF NANOTECHNOLOGY	SPAIN
JENSEN	Keld Alstrup	NATIONAL RESEARCH CENTRE FOR THE WORKING ENVIRONMENT	DENMARK
JOHANSON	Gunnar	KAROLINSKA INSTITUTET	SWEDEN

JORDAN	Robert	CHUBB INSURANCE COMPANY OF EUROPE SE	NETHERLANDS
JUNG	Younjung	KIST EUROPE	GERMANY
KAISER	Jean-Pierre	EMPA, SWISS FEDERAL INSTITUTE FOR MATERIALS SCIENCE AND TECHNOLOGY	SWITZERLAND
KALIčÁKOVÁ	ZdeňKa	VŠB- TECHNICAL UNIVERSITY OF OSTRAVA, FACULTY OF SAFETY ENGINEERING	CZECH REPUBLIC
KANAYA	Fumihide	NATIONAL CENTER FOR GLOBAL HEALTH AND MEDICINE	JAPAN
KANNOSTO	Jonna	DEKATI LTD.	FINLAND
KATALAGARIANAKIS	Georgios	EUROPEAN COMMISSION	BELGIUM
KATALAGARIANAKIS	Georgios	EUROPEAN COMMISSION	BELGIUM
KIM	Young-Rok	KYUNG HEE UNIVERSITY	SOUTH KOREA
KINSNER-OVASKAINEN	Agnieszka	EUROPEAN COMMISSION, JOINT RESEARCH CENTRE, NANOBIOSCIENCES UNIT	ITALY
KOETH	Anja	FEDERAL INSTITUTE FOR RISK ASSESSMENT/ DEPARTMENT SAFETY OF CONSUMER PRODUCTS	GERMANY
KOLAROV	Felix	TÜBINGEN UNIVERSITY	GERMANY
KOWAL	Sophie	INERIS	FRANCE
KRISTENSEN	Henrik Vejen	DANISH TECHNOLOGICAL INSTITUTE	DENMARK
KUBO-IRIE	Miyoko	DEPT. OF HYGIENE CHEM., FAC. OF PHARM. SCI., TOKYO UNIV. OF SCI.	JAPAN
KUKUTSCHOVA	Jana	VSB - TECHNICAL UNIVERSITY OF OSTRAVA	CZECH REPUBLIC
KUPER	Frieke	TNO	NETHERLANDS
KURANO	Norikazu	SHIGEMATSU WORKS CO., LTD.	JAPAN
L'ALLAIN	Catherine	PACTE-CNRS	FRANCE
LABILLE	Jerome	AIX-MARSEILLE UNIVERSITÉ, CNRS, CEREGE UMR 7330	FRANCE
LACH	Karel	CZECH TECHNOLOGY PLATFORM ON INDUSTRIAL SAFETY	CZECH REPUBLIC
LACROIX	Genevieve	TMMX - HERAKLES - SAFRAN	FRANCE
LACROIX	Ghislaine	INERIS	FRANCE
LALOY	Julie	UNIVERSITY OF NAMUR - DEPARTMENT OF PHARMACY	BELGIUM
LANDRY	Marion	INSERM	FRANCE
LANONE	Sophie	INSERM	FRANCE
LARUE	Camille	ISTERRE	FRANCE
LAWRENCE	Sam	CYTOVIVA	UNITED STATES
LE BIHAN	Olivier	INERIS	FRANCE
LE DUR	David	ECOMESURE	FRANCE
LE LOUREC	Patrick	NANOBADGE - ALCEN	FRANCE
LE ROUX	Gaël	INSERM U1066 MINT	FRANCE
LECONTE	Yann	CEA-DSM/IRAMIS/SPAM	FRANCE
LECOQ	Pierre	ANSES	FRANCE
LEGROS	Samuel	CEA GRENOBLE	FRANCE
LEMEUNIER	Marie	APAVE SUDEUROPE SAS	FRANCE
LEVARD	Clement	CEREGE-CNRS	FRANCE
LEWINSKI	Nastassja	INSTITUT DE SANTÉ AU TRAVAIL	SWITZERLAND
LI	Lingxiangyu	TECHNICAL UNIVERSITY OF MUNICH	GERMANY
LIGUORI	Biase	TECHNICAL UNIVERSITY OF DENMARK	DENMARK
LIU	Wei	CEREGE - UMR7330 CNRS- AIX MARSEILLE UNIV	FRANCE
LOMBARD	Alain	ALLOTOXCONSULTING	FRANCE
LOMELLO	Fernando	LRC CEA/LERMPS - PERSEE	FRANCE
LOPEZ DE IPIÑA	Jesús M.	TECNALIA RESEARCH AND INNOVATION	SPAIN
LOSERT	Sabrina	EMPA	SWITZERLAND
LOUNSBURY	Michael	UNIVERSITY OF ALBERTA	CANADA
LOZANO GARCIA	Omar	UNIVERSITY OF NAMUR / NAMUR NANOSAFETY CENTER	BELGIUM
LUCAS	Stéphane	UNIVERSITY OF NAMUR (FUNDP)	BELGIUM
MAJEDI	Seyed Mohamma	NATIONAL UNIVERSITY OF SINGAPORE/ DEPARTMENT OF CHEMISTRY	SINGAPORE
MALARD	Véronique	CEA	FRANCE
MANTOVANI	Elvio	AIRI/NANOTEC IT	ITALY
MARCHETTO	Alain	CEA/GRENOBLE	FRANCE
MARQUARDT	Clarissa	KARLSRUHE INSTITUTE OF TECHNOLOGY (KIT), INSTITUTE OF APPLIED COMPUTER SCIENCES (IAI)	GERMANY
MARTINEZ	Diego Stéfani Tec	INSTITUTE OF CHEMISTRY / STATE UNIVERSITY OF CAMPINAS - UNICAMP	BRAZIL
MARTIROSYAN	Alina	UNIVERSITÉ CATHOLIQUE DE LOUVAIN, INSTITUTE OF LIFE SCIENCES	BELGIUM
MARUCCO	Arianna	UNIVERSITÀ DEGLI STUDI DI TORINO	ITALY

MASSON	Christian	ONERA	FRANCE
MASUDA	Kensuke	INDUSTRIAL MARKETING CONSULTANTS CO.,LTD.	JAPAN
MAURIZI	Lionel	EPFL/ POWDER TECHNOLOGY LABORATORY (LTP)	SWITZERLAND
MAZEAU	Pascal	MICHELIN	FRANCE
MEJIA	Jorge	UNIVERSITY OF NAMUR / NAMUR NANOSAFETY CENTER	BELGIUM
MENA	Maria	DESARROLLOS Y ACABADOS EN POLIURETANO SA DE CV	MEXICO
MENEGHETTI	Fabio	MATECH	ITALY
MEUNIER	Thierry	TREDI	FRANCE
MHADHBI	Mohsen	NATIONAL INSTITUTE OF RESEARCH AND PHYSICOCHEMICAL ANALYSIS	TUNISIA
MIčKA	Vladimír	CZECH TECHNOLOGY PLATFORM ON INDUSTRIAL SAFETY	CZECH REPUBLIC
MICHAUD-SORET	Isabelle	LCBM /IRTSV / CEA	FRANCE
MONCOQ	Damien	CNRS	FRANCE
MONTIGEL	Esther	BASLER VERSICHERUNGEN	SWITZERLAND
MOONS	Miranda	EXXONMOBIL	BELGIUM
MORIN	Fabrice	FUNDACION TECNALIA	SPAIN
MOTELLIER	Sylvie	CEA GRENOBLE	FRANCE
MOTZKUS	Charles	LABORATOIRE NATIONAL DE MÉTROLOGIE ET D'ESSAIS	FRANCE
MOUCHET	Florence	ECOLAB	FRANCE
MUIR	Robert	NANEUM LIMITED	UNITED KINGDOM
NEUMEISTER	Lothar	BG ETEM	GERMANY
NGUYEN	Tinh	NIST, US	UNITED STATES
NGUYEN	Kathy	HEALTH CANADA	CANADA
NICOLAS	Melanie	CSTB	FRANCE
NICOLINO	Philippe	CEA-SACLAY	FRANCE
NIEMINEN	Elina	DEKATI LTD.	FINLAND
NOWACK	Bernd	EMPA, SWITZERLAND	SWITZERLAND
OBERDÖRSTER	Günter	U. ROCHESTER, US	UNITED STATES
OGURA	Isamu	NATIONAL INSTITUTE OF ADVANCED INDUSTRIAL SCIENCE AND TECHNOLOGY (AIST)	JAPAN
OJEA	Isaac	CENTRE OF NANOBIOSAFETY AND SUSTAINABILITY	SPAIN
ONO-OGASAWARA	Mariko	JAPAN NATIONAL INSTITUTE OF OCCUPATIONAL SAFETY AND HEALTH	JAPAN
OTSUKA	Kenichi	JFE TECHNO-RESEARCH CORPORATION	JAPAN
ØVERJORDET	Ida Beathe	SINTEF MATERIALS AND CHEMISTRY	NORWAY
OZIL	Samuel	HONEYWELL HPC	FRANCE
PAGET	Vincent	INERIS	FRANCE
PAKRASHI	Sunandan	VIT UNIVERSITY	INDIA
PAPAIOANNOU	Eleni	CERTH/CPERI/APTL	GREECE
PAPILLON	Francine	CEA GRENOBLE	FRANCE
PATEL	Stephanie	CEA	FRANCE
PEIKERTOVÁ	Pavlína	VŠB - TECHNICAL UNIVERSITY OF OSTRAVA, NANOTECHNOLOGY CENTRE	CZECH REPUBLIC
PEOTTA	Pierre	NANOSIGHT	UNITED KINGDOM
PERREIN-ETTAJANI	Hanane	LUNAM UNIVERSITÉ, UCO, MMS, ANGERS	FRANCE
PESCH, DR.	Markus	GRIMM AEROSOL TECHNIK GMBH & CO. KG	GERMANY
PIELLA BAGARIA	Jordi	INSTITUT CATALA DE NANOTECNOLOGÍA	SPAIN
PINI	Martina	UNIVERSITY OF MODENA AND REGGIO EMILIA	ITALY
PIRET	Jean-Pascal	UNIVERSITY OF NAMUR	BELGIUM
PLACHÁ	Daniela	VŠB-TECHNICAL UNIVERSITY OF OSTRAVA, NANOTECHNOLOGY CENTRE	CZECH REPUBLIC
PLANQUE	Alizée	CEA	FRANCE
PONCE DEL CASTILLO	Aida	ETUI	BELGIUM
PONCELET	Olivier	CEA	FRANCE
POURCHEZ	Jérémie	ECOLE DES MINES DE SAINT-ETIENNE, LINA	FRANCE
PRAETORIUS	Antonia	ETH ZURICH	SWITZERLAND
PRAS	Maxime	INSA LYON	FRANCE
PRAT	Odette	CEA/DSV/IBEB/SBTN	FRANCE
PRESUME	Mirlande	INSERM U955	FRANCE
PUI	David	U. MINNESOTA, US	UNITED STATES
PUNTES	Víctor	CATALAN INSTITUTE OF NANOTECHNOLOGY	SPAIN

PUOLAMAA	Maila	EUROPEAN COMMISSION, BRUSSEL	BELGIUM
RABILLOUD	Thierry	IRTSV/LCBM, UMR 5249	FRANCE
RAFFIN	Guy	CNRS SCA	FRANCE
RAMIREZ-NOGUERA	Patricia	NATIONAL UNIVERSITY OF MEXICO	MEXICO
RASMUSSEN	Susanne	SOLAE DENMARK A/S	DENMARK
RASMUSSEN	Pat	HEALTH CANADA	CANADA
RAVINDRAN GIRIJA	Aswathy	BIO-NANO ELECTRONICS RESEARCH CENTER, GRADUATE SCHOOL OF INTERDISCIPLINARY NEW SCIENCE, TOYO UNIVERSITY	JAPAN
RELIER	Charlène	INERIS	FRANCE
REYNES	Mathias	CEA	FRANCE
REYNIER	Martine	INRS	FRANCE
RIEDIKER	Mickael	INSTITUTE FOR WORK AND HEALTH	SWITZERLAND
ROES	Lex	UTRECHT UNIVERSITY	NETHERLANDS
ROLLIN	Francoise	CEA/DSV/SBTN/LEPC	FRANCE
ROSE	Jérôme	CEREGE	FRANCE
ROSSANO	Manon	UNIVERSITÉ DU HAVRE	FRANCE
ROURE	Francoise	MINISTRY FOR ECONOMY, FINANCE AND INDUSTRY, FRANCE	FRANCE
SACHSE	Sophia	CRANFIELD UNIVERSITY	UNITED KINGDOM
SAFARIKOVA	Miroslava	INSTITUTE OF NANOBIOLOGY AND STRUCTURAL BIOLOGY OF GCRC	CZECH REPUBLIC
SAINT-ANTONIN	François	CEA	FRANCE
SAKULKHU	Usawadee	EPFL/ POWDER TECHNOLOGY LABORATORY (LTP)	SWITZERLAND
SALVI	Olivier	EU-VRI/	GERMANY
SAMUELSEN	Jan T.	NIOM	NORWAY
SANTAELLA	Catherine	CNRS	FRANCE
SANTOS	Sandra	CENTER FOR NEUROSCIENCE AND CELL BIOLOGY	PORTUGAL
SARRET	Geraldine	ISTERRE	FRANCE
SAVEL	Jean Rémy	CEA	FRANCE
SCHLAGENHAUF	Lukas	EMPA	SWITZERLAND
SCHMITT	Franck	BUREAU VERITAS	FRANCE
SCHUBERT	Chloe	CEA/LCNS	
SCHULTE	Paul	CDC/NIOSH, US	UNITED STATES
SCHUSTER	Frédéric	CEA SACLAY	FRANCE
SCHYMURA	Stefan	HELMHOLTZ-ZENTRUM DRESDEN ROSSENDORF - INSTITUTE OF RESOURCE ECOLOGY	GERMANY
SCIFO	Lorette	TECNALIA-FRANCE	FRANCE
SEABRA	Amedea	UNIVERSIDADE FEDERAL DE SÃO PAULO	BRAZIL
SEBASTIEN	Patrick	SAINT GOBAIN CREE	FRANCE
SEITZ	Stefanie	KIT, ITAS	GERMANY
SENE	Christophe	STEPAN EUROPE	FRANCE
SHINOHARA	Naohide	NATIONAL INSTITUTE OF ADVANCED INDUSTRIAL SCIENCE AND TECHNOLOGY	JAPAN
SICARD	Yves	CEA GRENOBLE	FRANCE
SIGWARD	Estelle	FACULTÉ DE PHARMACIE - PARIS DESCARTES - LABORATOIRE DE GALÉNIQUE	FRANCE
SIMON-DECKERS	Angélique	INSERM U955 EQ04	FRANCE
SIMONELLI	Federica	EUROPEAN COMMISSION - JOINT RESEARCH CENTRE	ITALY
SIMONIN	Marie	LABORATOIRE D'ECOLOGIE MICROBIENNE UMR5557/ UNIVERSITÉ LYON1	FRANCE
SIRVEN	Jean-Baptiste	CEA SACLAY	FRANCE
SMITH	Janet	EXXONMOBIL CORPORATION	UNITED STATES
SOUZA	Tiago	UNIVERSITY OF SÃO PAULO/ FACULTY OF MEDICINE	BRAZIL
STAVES	Malcolm	L'OREAL	FRANCE
STEINFELDT	Michael	UNIVERSITY OF BREMEN, FG 10 TECHNOLOGICAL DESIGN AND DEVELOPMENT, FB 4 PRODUCTION ENGINEERING	GERMANY
STINTZ	Michael	TU DRESDEN, INSTITUTE OF PROCESS ENGINEERING AND ENVIRONMENTAL TECHNOLOGY	GERMANY
STOIKA	Rostyslav	INSTITUTE OF CELL BIOLOGY OF NATIONAL ACADEMY OF SCIENCES OF UKRAINE	UKRAINE
SULTANA	Sadequa	CNRS - CSPBAT	FRANCE
SUN	Tianyin	EMPA/ ETH ZURICH	SWITZERLAND
SZEWCZYńSKA	MałGorzat	CENTRAL INSTITUTE FOR LABOUR PROTECTION- NATIONAL RESEARCH INSTITUTE	POLAND
TABET	Lyes	HÔPITAL DU SACRÉ-COEUR DE MONTRÉAL, CENTRE DE RECHERCHE (J3115), UNIVERSITÉ DE MONTRÉAL	CANADA
TAMURA	Hisaaki	INDUSTRIAL MARKETING CONSULTANTS CO.,LTD.	JAPAN
TANTRA	Ratna	NATIONAL PHYSICAL LABORATORY, UK	UNITED KINGDOM

TARANTINI	Adeline	ANSES - LABORATOIRE DE FOUGÈRES	FRANCE
TARDIF	Francois	CEA	FRANCE
THIERIET	Nathalie	ANSES	FRANCE
THOMASSON	Alexandre	AIR RHÔNE-ALPES	FRANCE
TOMIC	Marija	FACULTY OF MECHANICAL ENGINEERING	SERBIA
TRAN	Dinh-Trinh	ARMINES	FRANCE
TRITSCHER	Torsten	TSI GMBH	GERMANY
TROUILLER	Bénédicte	INERIS	FRANCE
TSURUOKA	Shuji	SHINSHU UNIVERSITY	JAPAN
UANG	Shi-Nian	INSTITUTE OF OCCUPATIONAL SAFETY AND HEALTH, COUNCIL OF LABOR AFFAIRS	TAIWAN (ROC)
UMBUZEIRO	Gisela	UNICAMP - TECHNOLOGY FACULTY	BRAZIL
US	Svitlana	SCIENCE AND TECHNOLOGY CENTER IN UKRAINE	UKRAINE
USUI	Yuki	SHINSHU UNIVERSITY	JAPAN
VALLEN	Jochen	GRIMM AEROSOL TECHNIK GMBH & CO. KG	GERMANY
VAN DER MEEREN	Anne	CEA	FRANCE
VÅZQUEZ-CAMPOS	Socorro	ACONDICIONAMIENTO TARRASENSE - LEITAT TECHNOLOGICAL CENTER	SPAIN
VIDAUD	Claude	CEA/DSV/SBTN/LEPC	FRANCE
VILAR PALOS	Gemma	ACONDICIONAMIENTO TARRASENSE - LEITAT TECHNOLOGICAL CENTER	SPAIN
VINCENT	Phil	NANOSIGHT	UNITED KINGDOM
VINCHES	Ludwig	ÉCOLE DE TECHNOLOGIE SUPÉRIEURE	CANADA
VOLLMER	Nathalie	HORIBA SCIENTIFIC	FRANCE
VON DER KAMMER	Frank	UNIVERSITY OF VIENNA/ DEPARTMENT OF ENVIRONMENTAL GEOSCIENCES	AUSTRIA
WAGNER	Stephan	UNIVERSITY OF VIENNA/ DEPARTMENT OF ENVIRONMENTAL GEOSCIENCE	AUSTRIA
WALSER	Tobias	INSTITUTE OF ENVIRONMENTAL ENGINEERING	SWITZERLAND
WEBER	Philippe	ANSES	FRANCE
WEDIN	Irene	STORA ENSO AB	SWEDEN
WEI	Taotao	INSTITUTE OF BIOPHYSICS, CHINESE ACADEMY OF SCIENCES	CHINA
WITSCHGER	Olivier	INRS	FRANCE
WOHLLEBEN	Wendel	BASF SE	GERMANY
YANAGIUCHI	Mamoru	SHIGEMATSU WORKS CO., LTD.	JAPAN
ZAPÓR	Lidia	CENTRAL INSTITUTE FOR LABOUR PROTECTION - NATIONAL RESEARCH INSTITUTE	POLAND
ZIMMERMANN	Eric	CEA/PNS	FRANCE
ZUIN	Stefano	CONSORZIO VENEZIA RICERCHE	ITALY

Monday 12/11/12	Tuesday 13/11/12	Wednesday 14/11/12	Thursday 15/11/12
	07:30. Registration 08:30 **Welcome from the organizing committee** **Conference opening. F. Tardif** Nanogovernance, trends and limitations **Chairman: G. Katalagarianakis** **F. Schuster- F. Roure - C. Engeman - D. Bernard - A. Grobe** **Auditorium**	08:00 Session 3 Toxicology limited to: nano-bio-interactions modelling, nanoparticles biotransformation and degradation in cells, tissues and environment, nanoparticles translocation across barriers G. Oberdörster Auditorium 08:45 Toxicology Auditorium 08:45 Detection Room B 08:45. Secure industrial production Room C	08:00 Session 4 Environmental interactions of nanomaterials J. Rose Auditorium 08:45 Environmental interactions Auditorium 08:00 Toxicology Room B
	10:30-11:00: coffee break	10:00-10:30: coffee break	10:00-10:30: coffee break
	11:00 Session 1 Exposure: assessment, modelling and case studies D. Brouwer Auditorium	10:30 Session 12 Regulation and standardization for nano products M. Puolamaa/O.Salvi Auditorium 10:30 Detection Room B 10:30 Session 11 Risk management for OSH expert P. Schulte Room C	10:30 Session 5 Nanomaterials release from: ageing, mechanical constraints and sollicitation, incineration, etc. T. Nguyen Auditorium 10:30 Session 9 Ethics and societal issues A. Grinbaum Room C
	11:45 Exposure assessment Auditorium 11:45 SCAFFOLD Room C	11:15 Regulation and standardization Auditorium 11:45 Toxicology Room B 11:15 Risk management for OSH experts. P. Schulte Room C	11:15 Nanomaterial release Auditorium 11:15 Toxicology Room B 11:15 Ethics and societal issues Room C
	13:00-14:00: lunch	12:30-13:30: lunch	12:30-13:30: lunch
	14:00 Session 2 Detection and identification of engineered nanoparticles D. Pui Auditorium 11:45 TRIMATEC Room C	13:30 Session 8 Life cycle analysis, recycling, waste management and disposal B. Nowack Auditorium 13:30 Toxicology Room B 13:30 Risk management for OSH Room C	13:30 Environmental interactions Auditorium 13:30 Panel discussion The opinion of the civil society A. Grinbaum Room B 13:30 Nanomaterial release Room C
	14:45 Detection and identification Auditorium 14:45 Exposure assessment Room B	14:15 Life Cycle Analysis Auditorium 14:30 Panel discussion Toxicology C. Emond Room B 14:30 Regulation and standardization Room C	14:30 Toxicology Room B
	16:00-16:30: coffee break	15:30-16:00: coffee break	15:30-16:00: coffee break
	16:30 Session 6 et 7 Secure industrial production and protection technology Luana Golanski Auditorium	16:00-18:00 Session 10 Commercial equipment Robert Muir Auditorium 16:00-18:00 Toxicology Room B 16:00-17:15 Life Cycle Analysis Room C	16:00-17:00 Environmental interactions Auditorium 16:00-19:00 NANOHOUSE Room B 16:00-18:00 Nanomaterial release Room C
16:30-19:30 Registration	17:15 Secure industrial production Auditorium 17:15 Exposure assessment Room B 17:15 Detection Room C	17:15. Secure industrial production Room C	17:00 -18:00 Toxicology Auditorium
	18:30-19:30 Panel discussion Governance F. Roure Auditorium 18:30-21:00 Poster exhibition/Evening event Ballroom 1st floor	20:00-23:30: cocktail party Transport by cabble car to La Bastille	Conclusions: V. Gaultier; F. Schuster; F. Tardif. **End of the conference: 18:15** **Next NanoSafe Conference: November 2014**

Observation of nanoaerosol release from a hand dryer

G N Bae, S H Park, and S B Lee

Center for Environment, Health and Welfare Research, Korea Institute of Science and Technology, Seoul 136-791, Republic of Korea

E-mail: gnbae@kist.re.kr

Abstract. Consumer exposure is one of exposures in the life cycle perspective to risk assessment. Many nanoproducts have been commercialized in the fields of health, electronics, home, food etc. Hand dryers are widely used at public rest rooms. In this study, we selected the hand dryer having a filter coated with silver nanoparticles as a popular electronic nanoproduct. A test chamber system was developed to detect nanoaerosols released from the test nanoproduct during the operation. The particle number concentration of the chamber background was confirmed to be <1 particles/cm^3 by using a condensation particle counter with a detection limit of 10 nm. When the hand dryer was turn on for 1 min, the particle number concentration in the chamber was almost linearly increased up to the peak concentration (<20 particles/cm^3). It was slowly decreased after the hand dryer was turn off. When the operation was extended to 10 min, similar release pattern of nanoaerosols was observed. Therefore, it was concluded that the nanoaerosols was emitted from the hand dryer during the operation. However, the amount of released nanoaerosols was negligible, compared to indoor nanoaerosol concentration (~10,000 particles/cm^3). Also, the released nanoaerosols could not be analyzed by filter sampling due to small quantity of mass.

1. Introduction

As nanotechnology is developed, new nanoproducts using nanomaterials are commercialized. Due to the dramatic increase of surface area, the nanomaterials are of concern in adverse human health effect, compared to well-known particulate matter of PM_{10} or $PM_{2.5}$. Health risks are considered in the three aspects such as worker exposure, consumer exposure, and human population and ecological exposure. To assess worker exposure, many field studies have been conducted at workplaces. In the workplace, local exhaust ventilation systems or personal protection equipments could be applied to protect exposure of nanomaterials to workers [1-3]. However, a few investigations on consumer exposure were reported [4, 5]. In addition, proper measures are not available to protect nanomaterial exposure to consumers who use nanoproducts. To avoid consumer exposure to significant amount of nanomaterials, extensive studies on nanomaterial emission from nanoproducts are urgently needed. Accumulation of emission test data could make a standard emission test method, which enables to certify safe nanoproducts according to the emission level.

In this study, we investigated nanoaerosol release pattern of an electronic nanoproduct. The commercial hand dryer having a filter coated with silver nanoparticles was selected as a test nanoproduct and the experiment was carried out in a test chamber.

2. Experimental

To quantify the amount of nanoaerosol release, a test chamber system was designed as shown in figure 1. It consisted of a cube-shape chamber, a pure air generator, and a real-time aerosol monitoring instruments. The volume of the chamber was 93 L (41 × 35 × 65 cm) and it was enough to install a full-scale test product such as a hand dryer. The chamber was made of stainless steel except a small acrylic window for monitoring by eyes. An electric power socket attached on the wall was used to operate the electronic nanoproduct under the test.

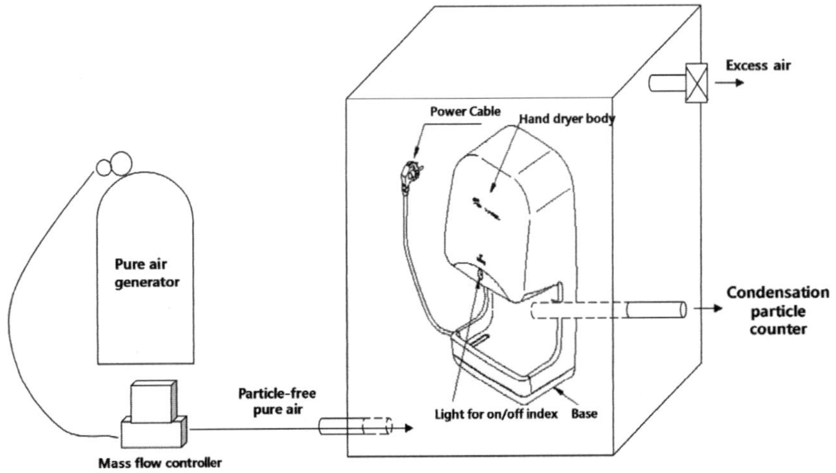

Figure 1. Experimental schematic for testing the release of nanoaerosols from a hand dryer.

The total particle number concentration inside the chamber was monitored every second by using a condensation particle counter (CPC, TSI model 3010) with a detection limit of 10 nm. As shown in figure 1, a hand dryer was selected as a test product. It had a filter coated with silver nanoparticles for antibacterial function.

The test was conducted as follows. First, a test product was set in the test chamber. Prior to main test, the chamber was purged with a particle-free pure air of 10 L/min under no operation of the hand dryer to minimize a background particle concentration. Initial particle concentration in the chamber was higher than 3,000 particles/cm^3 and the background level of <1 particles/cm^3 was achieved after 90-min purging. After purging, the air flow rate of the pure air was reduced to 5 L/min for the main test.

The release experiment was carried out twice. At the first test, the hand dryer was turned on for 1 min and then turned off for 30 min four times cyclically. In the case of second test, the operation of the hand dryer lasted 10 min to investigate the effect of operating time on the release of nanoaerosols.

3. Results

3.1. Background level of nanoaerosols in the test chamber

As shown in figure 2, the initial particle number concentrations in the chamber ranged 4,000~11,000 particles/cm^3 because indoor aerosols entered during installation of the hand dryer. As soon as purging air was supplied at 0 min, the particle number concentration started to decrease linearly in its logarithm plot versus time because of constant air exchanging effect. The particle number concentrations became below 10 particles/cm^3 after 60-min purging, and then the background level of <1 particles/cm^3 was achieved after 90-min purging regardless of initial concentrations. Based on this background level measurement, it was possible to start the main release test after 90-min purging with 10 L/min of pure air under the indoor contaminant conditions of this study.

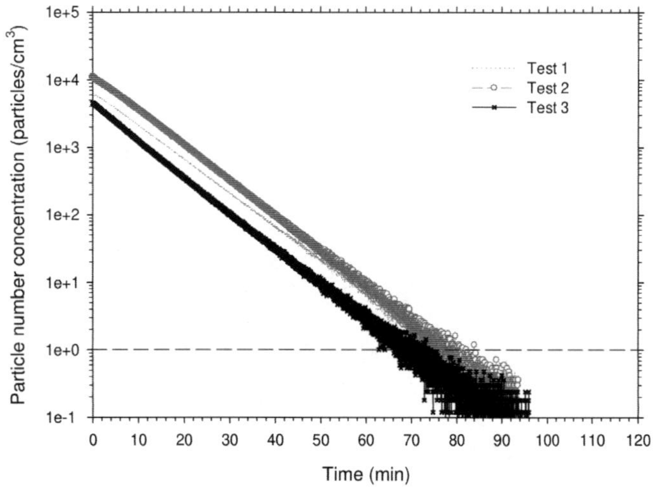

Figure 2. Variation of particle number concentration during the background level test.

3.2. Release of nanoaerosols from a hand dryer

When the test hand dryer was turn on at 0 min after purging, the particle number concentration abruptly increased up to a peak value as shown in figure 3. And then it exponentially decreased down to the background level after its operation stopped. The peak particle concentrations ranged 10-20 and 150-200 particles/cm^3 for the 1-min and 10-min operation, respectively, which are significantly lower than that in indoor environment (~5,000 to 10,000 particles/cm^3). The almost linear increasing pattern of the particle number concentrations was observed for each 10-min operation period, which implies that nanoaerosols were released continuously at a similar pattern. Repetition of operation of the test electronic hand dryer showed a periodic particle release, implying that the test hand dryer generates a small amount of nanoaerosols during the use period. It is not confirmed to be silver nanoparticles due to low quantity of emission.

4. Conclusions

The release of nanoaerosols from the commercial hand dryer having a filter coated with silver nanoparticles was evaluated during the operation using the test chamber equipped with a CPC. The particle number concentration in the chamber started to almost linearly increase from <1 particles/cm^3 as a background level up to 20 and 200 particles/cm^3 for 1-min and 10-min operations, respectively. The particle number concentration exponentially decreased just after the hand dryer was turned off because of air exchange. Based on the similar increasing patterns of the particle number concentration during each operation, it is concluded that a small amount of nanoaerosols was released from the hand dryer. The chemical composition of the released nanoaerosols was not identified yet, because the amount of released nanoaerosols was negligible. More tests during the longer operation of the hand dryer are necessary for obtaining more information such as particle size distribution and chemical composition of the released nanoaerosols.

Further studies are needed to establish the standard test method for evaluating the exposure level to consumers of nanomaterials released from nanoproducts.

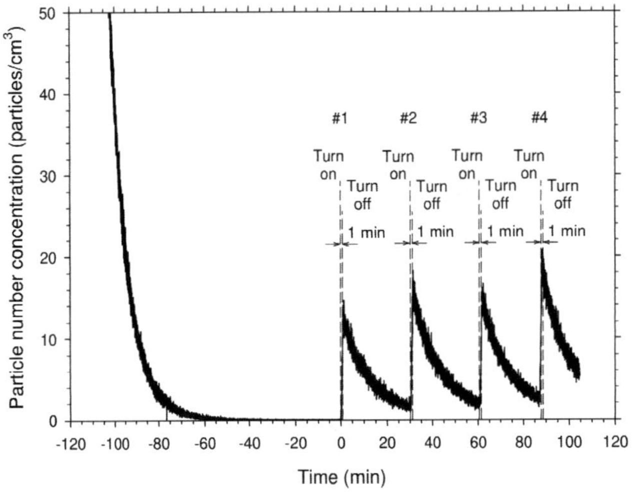

(a) 1-min operation

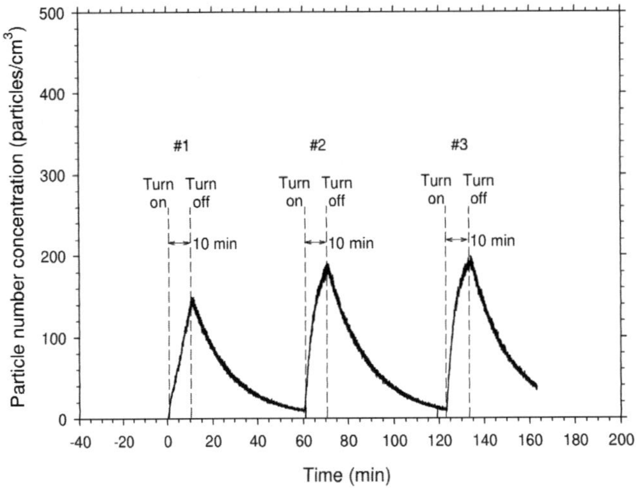

(b) 10-min operation

Figure 3. Variation of particle number concentration during the cyclic operation of the hand dryer.

References

[1] Lee M, McClellan W J, Candela J, Andrews D and Biswas P 2007 *J. Nanopart. Res.* **9** 127
[2] Methner M 2008 *J. Occup. Environ. Hyg.* **5** D63
[3] Old L, Dunn K H, Garcia A and Echt A 2008 *J. Occup. Environ. Hyg.* **5** D103
[4] Hsu L Y and Chein H M 2007 *J. Nanopart. Res.* **9** 157
[5] NIOSH 2011 *Current Intelligence Bulletin 63: Occupational Exposure to Titanium Dioxide* DHHS (NIOSH) Publication No. 2011160

A modular tool for analyzing cascade impactors data to improve exposure assessment to airborne nanomaterials

Sébastien Bau and Olivier Witschger

Institut National de Recherche et de Sécurité (INRS), Laboratoire de Métrologie des Aérosols, Rue du Morvan, CS 60027, 54519 Vandoeuvre Cedex, France

E-mail: sebastien.bau@inrs.fr

Abstract. Cascade impactors are widely used to provide particle size distributions for the study of aerosols in workplaces and ambient air. In the frame of exposure assessment to airborne particles, one of their main advantages is the possibility to perform further off-line analysis (e.g. electron microscopy, physical-chemical characterization by XRD, ICP-MS, etc.) on the collected samples according to particle size. However, the large channel width makes the particle size distributions not enough size-resolved. Furthermore, in spite of the sharpness of the collection efficiency curves, the existence of an overlap between stages renders data interpretation difficult. This work aim was to develop a modular program allowing the inversion of data stemming from cascade impactors based on the mass (or any quantity) collected on each impaction stage. Through a precise description of the collection efficiency curves of the different stages, the software provides a continuous curve (from 100 to 1000 points) using the Markowski method, and more particularly the Twomey iterative algorithm, according to several publications about inverse problems in cascade impactors. An additional option consists in determining the experimental error at each point of the inverse curve, performed by realizing several consecutive inversions. The inversion procedure was first tested and optimized for the case of the SIOUTAS personal sampler. Validation of the calculation was performed considering theoretical aerosols. Then, the software was used for two sets of data obtained during field measurement campaigns.

1. Introduction

Nanomaterials offer applications in fields as wide-ranging as health, food and agriculture, energy, materials, and transport. In parallel, their unique properties raise also questions about their potential related health effects. Due to their rising use, exposure of people working in research laboratories, staff in state-run or public industry is probably increasing. Consequently, assessing inhalation exposure to airborne nanoparticles constitutes an important challenge [1].

Conventionally, occupational exposure to particulate compounds is characterized by the mass concentration combined with the size range of particles penetrating different regions of the respiratory system (inhalable, thoracic and alveolar fractions, as defined by [2]). This approach to occupational exposure is applied to all chemical substances (except for fibers where the concentration is expressed in terms of number of fibers per unit of air volume) found as aerosols, whatever the particle size. To ensure continuance in occupational exposure assessment and in absence of specific regulation for nanoparticles, it seems necessary to pursue measuring particle mass size distribution in workplace atmospheres.

In addition, there are already several published studies that reveal that exposure (*via* the inhalation route) could be characterized by nano-sized and respirable particles during production and handling of manufactured nanomaterials as well as the use of products that contain manufactured nanomaterials [3, 4]. This is one more reason why, in the few strategies for assessing workplace exposure to airborne nanomaterials already proposed, there are the cascade impactors [5, 6].

Cascade impactors have been studied and modeled extensively; they constitute the instrument of choice for the determination of aerosol mass size distributions [7]. As a consequence, there is a wide variety of cascade impactors commercially available, with various cutoff diameters, flow rates and sampling substrates. Associated to analyses (gravimetric, physical-chemical, etc.) performed on each impaction stage, they provide the mass size distribution of the aerosol averaged over the sampling period.

Presenting the cascade impactor data in a histogram ignores information on particle size distribution within each impactor stage and does not take into account the possible overlap between them, *i.e.* that the particles of a given diameter may deposit over several stages [8].

The objective of this study was to develop and test a generic inversion tool devoted to post-analyzing data from cascade impactor measurements. According to several publications about inverse problems in cascade impactors, the software realizes the data inversion using the Markowski method, and more particularly the Twomey iterative algorithm. The results provide a continuous curve defined by 100 to 1000 data points.

From this inverted curve, several particulate concentrations can be determined, such as those corresponding to:

- The conventional health-related fractions [2] used in the field of occupational health;
- The Particulate Matter fractions (PM_{10}, $PM_{2.5}$, PM_1, and even $PM_{0.1}$) used in the field of ambient air quality (non-occupational);
- The lung-deposited fractions [9] as recently recommended by ISO [10] for better estimation of exposure, especially when submicrometer particles are significant.

Furthermore, these data can be converted into any other metric related to the initial one (*e.g.* from mass to number or surface area concentration) that might be more relevant regarding potential health effects (see *e.g.* [11, 12]).

2. Inversion software

2.1. Mathematical description

Within cascade impactors, particles are separated according to their aerodynamic properties, related to their inertia. The parameter governing impaction is the Stokes number (*Stk*), defined by the ratio of the stopping distance of a particle to a characteristic dimension of the obstacle, corresponding here to the radius of a circular nozzle [13]:

$$Stk(d_p) = \frac{\tau \cdot U}{\frac{d_j}{2}} = \frac{\rho_p \cdot d_p^2 \cdot Cu(d_p) \cdot U}{9 \cdot \eta_g \cdot d_j}$$

where ρ_p is the particle density, Cu the slip correction coefficient, U the particle velocity, η_g the gas viscosity and d_j the nozzle diameter. The Stokes number can be used to predict whether or not a particle will impact ($Stk \gg 1$) or not ($Stk \ll 1$) on an impaction plate.

Although collection efficiency (E) depends on particle size, it is not easily related to the Stokes number [14]. Thus, the actual collection efficiency curves are usually determined experimentally by measuring responses of monodisperse calibration aerosols [15-26]. In this work, the following general expression was used:

$$E(d_p) = \left[1 + \left(\frac{d_{p,50}}{d_p}\right)^{2s}\right]^{-1}$$

where $d_{p,50}$ corresponds to the cutoff diameter and s represents the stiffness of the collection efficiency curve. Their values are specific to an impactor model, and mainly depend on its nominal flow rate and stage geometry. Each stage of a cascade impactor will consequently be described by a set of ($d_{p,50}$, s)-values.

For a cascade impactor made of *N* stages, the probability to collect a particle of diameter d_p on the ith stage is given by the Kernel function:

$$k_i(d_p) = E_i(d_p) \cdot \prod_{j=i+1}^{N} [1 - E_j(d_p)] \qquad i = 1, \ldots, N$$

The determination of a continuous size distribution of the aerosol sampled, notated $f(x)$, based on the *N* observations, notated m_i, consists in solving the Fredholm integral equations:

$$m_i = \int_0^{\infty} k_i(x) \cdot f(x) \cdot dx + \epsilon_i, \qquad i = 1, \ldots, N$$

where ϵ corresponds to the uncertainty associated to the observation for each stage. The last equation constitutes the common start point of all inversion algorithms, commonly used under discrete form:

$$m_i = \sum_{j=1}^{M} k_i(x_j) \cdot f(x_j) \cdot \Delta x_j + \epsilon_i, \qquad i = 1, \ldots, N\,; j = 1, \ldots, M$$

where *M* is the number of points where the function f is defined in discrete form.

The latter equation, called Fredholm equation in the remainder of this paper, constitutes the condition to be fulfilled to ensure that the inversion procedure has completed. Indeed, when Fredholm equation is satisfied, the difference between observed (m_i) and calculated quantity for stage i ($\sum_{j=1}^{M} k_i(x_j) \cdot f(x_j) \cdot \Delta x_j$) is less or equal to the tolerated error (ϵ_i).

The aim of the inversion procedure is to determine the values of the aerosol size distribution $f(x_j)$ allowing Fredholm equation to be satisfied on each stage. Because the number of points expected is higher than the number of stages ($M \gg N$), the inversion problem is ill-posed and does not admit a unique solution based on a finite set of observations.

2.2. Short description of the inversion method implemented within the software

A deterministic method mainly based on the work of Twomey [27] was chosen to invert data from cascade impactors. Among its main advantages, this non linear iterative method does not assume a mathematical model for the aerosol size distribution resulting from the inversion procedure.

Initiated by Chahine [28], the algorithm proposed consists in multiplying the function at p^{th} iteration by a weighting factor taking into account the Kernel function:

$$f^{(p+1)}(x_k) = f^{(p)}(x_k) \cdot \prod_{i=1}^{N} \left[1 + \left(R_i^{(p)} - 1\right) \cdot k_i(x_k)\right]$$

with:

$$R_i^{(p)} = \frac{m_i}{\sum_{j=1}^{M} k_i(x_j) \cdot f^{(p)}(x_j) \cdot \Delta x_j}$$

Practically, when the resulting function converges to the solution, then:

$$R_i^{(p)} \rightarrow 1 \Rightarrow f^{(p+1)}(x_k) \rightarrow f^{(p)}(x_k)$$

Because the function resulting from Twomey's algorithm is particularly dependent on initial supposition and conserves its roughness, the calculation procedure was improved by smoothing the resulting function $f(x_j)$ [29] and by enlarging the weighting functions to avoid important variations due to the narrowness of cascade impactors Kernel functions [30].
Figure 1 is a schematic of the resolution algorithm.

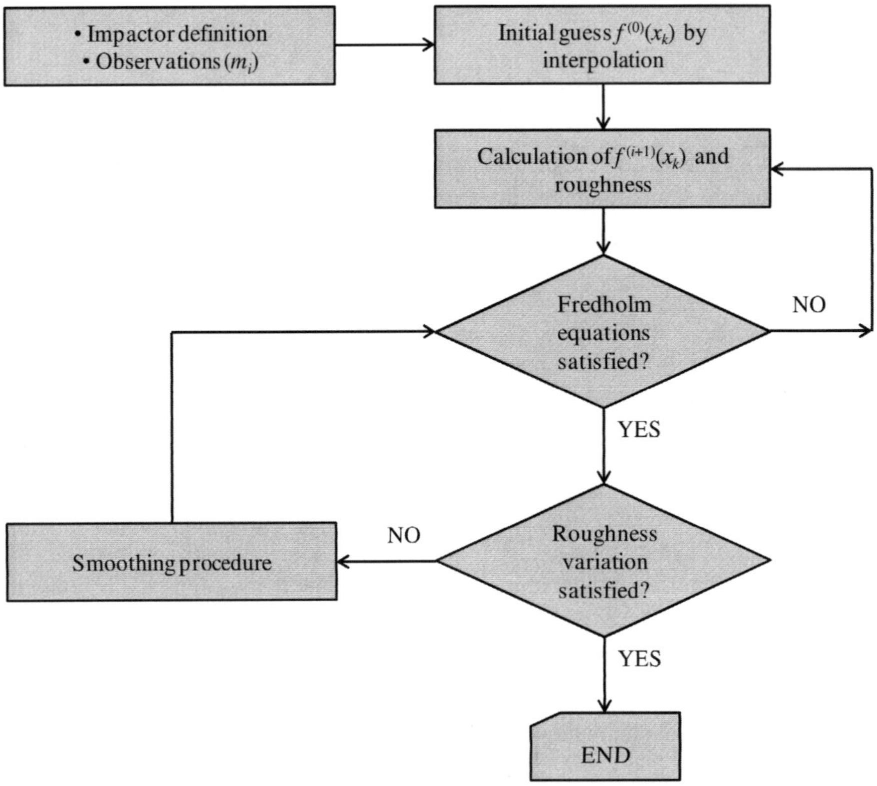

Figure 1. General resolution scheme implemented within the inversion software.

2.3. Software features

Following items describe how the sequence is processed within the modular tool:

1. Impactor selection or definition: if the cascade impactor used is not already present in the list of pre-defined impactors, the user is asked to input the ($d_{p,50}$, s)-values for the N stages. Optionally, the collection efficiency and deposition probabilities can be displayed to check whether the cascade impactor chosen is well defined;
2. Data entry: the user is asked to enter the data (*i.e.* masses, mass concentrations, or any quantifiable parameter) measured for each stage of the cascade impactor. The uncertainty associated to each of the data, notated σ_i, can be added in this window. In this case, the error accepted for each stage (ϵ_i) is taken equal to σ_i, otherwise it can be acquainted later in the inversion window;
3. The histogram corresponding to the aerosol size distribution is displayed;
4. The inversion window appears where inversion parameters (number of consecutive smoothing loops, roughness variation condition, error accepted ϵ_i) are automatically pre-filled with a set

of "universal" values. The user is allowed to modify all of them if necessary;

5. Once realized, the inversion result corresponds to a continuous curve having from 100 to 1000 points and can be directly exported to an Excel Worksheet for further analysis.

When uncertainties (σ_i) have been informed, an additional option consists in determining the experimental error at each point of the inverse curve. Several consecutive inversions are performed, each of them being realized on a random-generated size distribution within each channel confidence interval, leading to statistical variations for each point of the inverse curve.

3. Application to the SIOUTAS personal cascade impactor

3.1. Introduction

This software was originally developed for the case of the SIOUTAS personal cascade impactor (SKC inc, USA). This impactor contains four impaction stages where particles are accelerated in rectangular-shaped nozzles and collected on 0.5 µm-PTFE 25mm filters and a terminal filter consisting in a 2 µm-PTFE 37 mm filter. The SIOUTAS personal impactor has a nominal flow rate of 9 L.min^{-1}.

Besides the collection efficiency curves described by [23], the aspiration efficiency was calculated for thin-walled probe and upwards sampling according to Su and Vincent [31]. Terminal filter efficiency is assumed to be 100% whatever particle size. This could be improved by implementing filtration models related to the filter characteristics (filter solid fraction, gas velocity, etc.) as described for example by Raynor *et al.* [32].

We performed mathematical adjustment based on the experimental data from [23] with substrates used without any coating or adhesive material; this led to the stage parameters gathered in Table 1.

Table 1. Stage characteristics of the SIOUTAS personal impactor.

stage	$d_{p,50}$ (µm)	s (-)
4	2.50	2.79
3	1.00	2.51
2	0.50	1.85
1	0.25	1.88

The corresponding deposition probabilities are displayed in Figure 2.

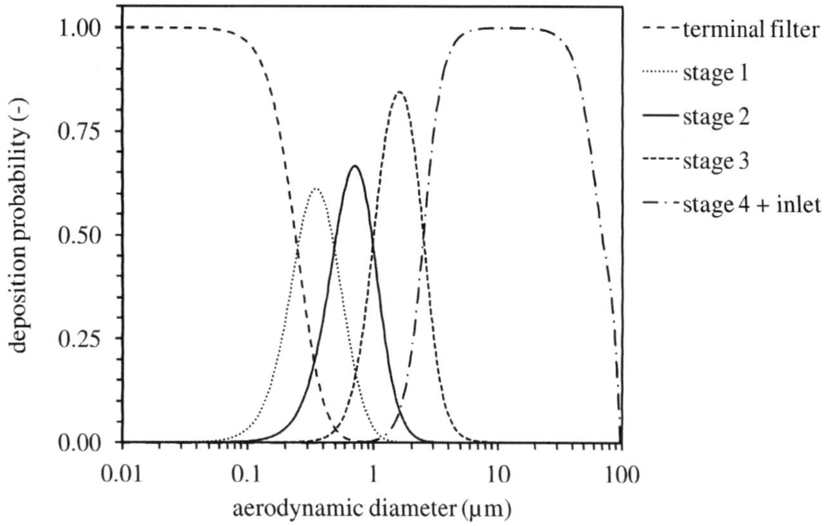

Figure 2. Particle deposition probabilities within the SIOUTAS personal impactor.

It can also be highlighted from Figure 2 that particle overlapping cannot be ignored. Let us take the example of 0.3 µm-particles, attributed to stage 1 according to Table 1. In reality, only 58% of them will deposit on stage 1, while 28% will be found on the terminal filter and 14% collected on stage 2. It is even worth for particles of 1 µm in diameter, presenting 47% deposition probability on stage 2 and 49% on stage 3. As a consequence, it is crucial to take into account the collection efficiency curves when inverting cascade impactors data.

3.2. Theoretical inversions (validation of the calculation procedure)

To validate the inversion method, a series of theoretical data corresponding to the particle mass on each stage was calculated according to the following procedure: (1) definition of the "input" particle size distribution, (2) deposition on the different stages based on collection efficiency curves, and (3) calculation of the quantity of particles collected on each impaction stage. After inversion, the resulting curve ("output") was compared to the original one.

As an example, Figure 3 presents input and output distributions for the case of a bimodal aerosol ($\text{mode}_1 = 0.5\ \mu\text{m}, \sigma_{g,1} = 2, \alpha_1 = 40\%, \text{mode}_2 = 3\ \mu\text{m}, \sigma_{g,2} = 2$), as well as the deposition of particles on the different stages and the corresponding histogram.

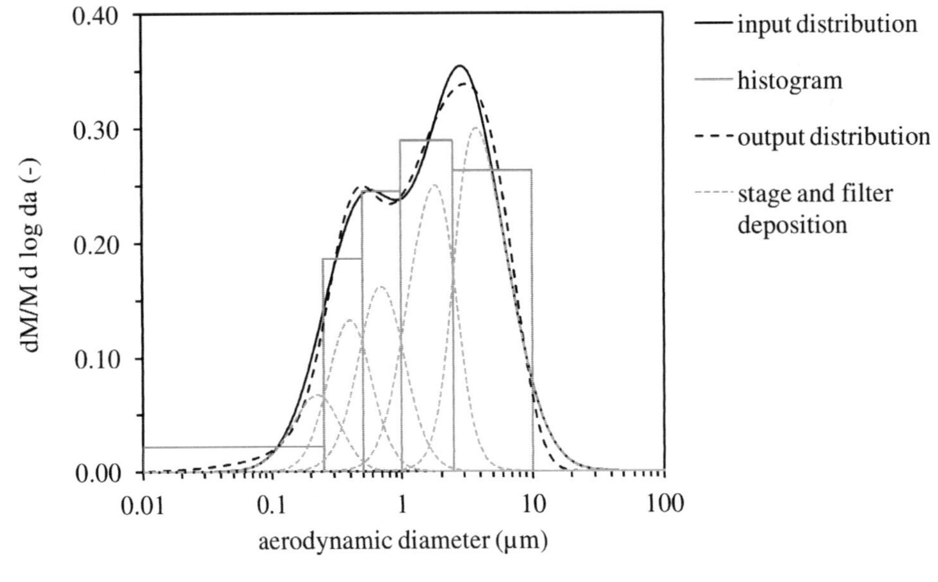

Figure 3. Particle deposition within the SIOUTAS personal impactor for the considered theoretical aerosol, input and output distributions.

Although the histogram (in grey) does not indicate the presence of two populations, the inverse curve resulting from the computation is found to be bimodal and in good agreement with the input size distribution. In such case, it is probable that an adjustment procedure would not bring out the presence of two populations.

For the case corresponding to Figure 3, Fredholm equations are satisfied within $\epsilon_i = 1\ \mu\text{g}$, as shown in Table 2.

Table 2. Comparison of input and output masses (theoretical case).

Stage	terminal	1	2	3	4
Input mass (mg)	0.072	0.129	0.170	0.265	0.364
Output calculated mass (mg)	0.072	0.128	0.171	0.264	0.364

A set of 18 theoretical cases was studied to validate the inversion procedure. Among them, 12 cases correspond to monomodal lognormal aerosols (mode between 0.2 µm and 10 µm, geometric standard deviation between 1.1 and 3) and 6 to bimodal lognormal aerosols. In 65% of the cases, the masses stemming from the inverse curve agreed the initial ones within ± 1 µg, 94% within ± 2 µg and 100% within ± 5 µg, as shown in Figure 4.

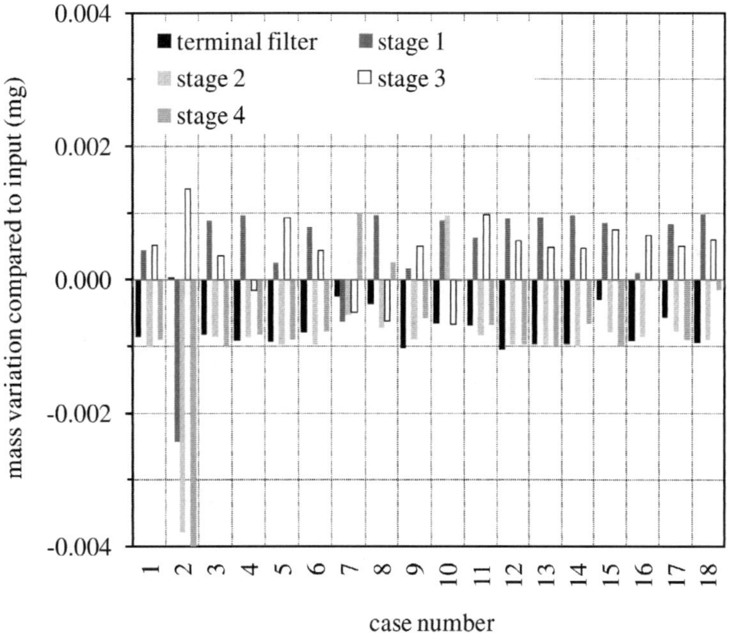

Figure 4. Mass variations between input and calculated from inverse curve for the different stages of the SIOUTAS personal impactor for theoretical cases.

Furthermore, it should be reminded that an inversion problem does not accept a unique solution. Consequently, in addition to the masses per stage, the shape of the curve was examined. In particular, the comparison between input and output distributions was based on observing the concordance of the modal diameter(s), the standard deviation(s) and the height of the curve – without quantified criteria. These observations indicate a good agreement between input and output distributions in most of the cases. Although not universal, it can be concluded that the inversion procedure is valid.

3.3. Example of application of the inversion tool to real measurement data

The first set of data results from a study performed by Witschger *et al.* [33] investigating the potential for exposure from cleanout operations by sandpapering of a reactor producing nanocomposite thin films embedded with silver nanoparticles (~ 8 nm).

Close to the source of sandpapering operation, the total mass concentration was 459 µg.m$^{-3,}$ while the silver mass concentration was 89 µg/m^3, with a bimodal mass size distribution confirmed by electron microscope observations (modal sizes of ~200 nm and 4 µm). Figure 5 displays the mass distribution measured by ICP-MS and the corresponding inverse curve. If the finer mode is in agreement with microscope observations, the coarse one is found around 9 µm in the inverse curve.

Based on ICP-MS elemental composition, the density of the collected particles – mainly containing silver, copper, nickel and iron – was estimated to be 8.5 g.cm^{-3}. The particle dynamic shape factor was taken equal to 1.3. From these parameters and assuming that the particle size stemming from electron microscope pictures is the particle volume equivalent diameter, the corresponding aerodynamic diameter was found to be 10 µm, which is in agreement with the mode observed in the inverse curve.

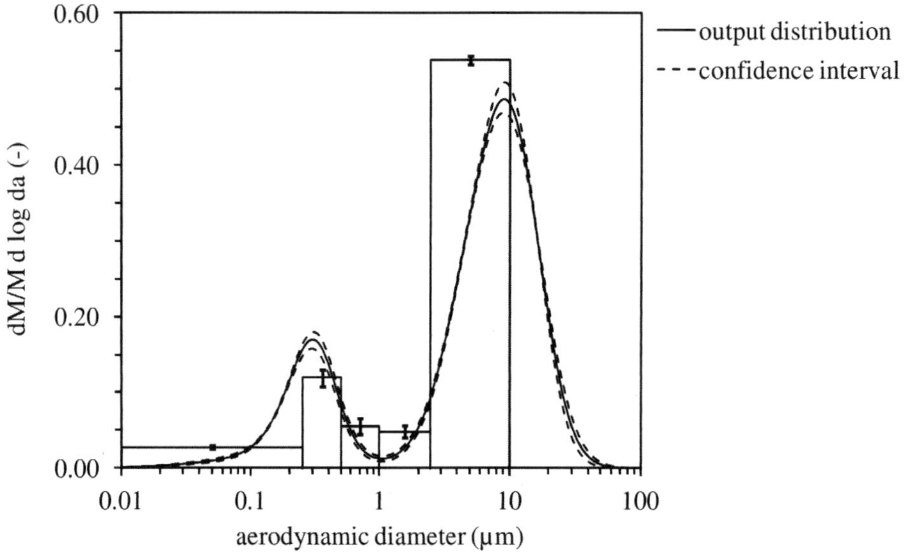

Figure 5. Output distribution and associated confidence interval from airborne silver particles emitted by sandpapering operation – data from [33].

Moreover, the inverted data from total mass collected on each stage (data not shown) was used to calculate the respirable fraction of the aerosol according to CEN [2]. This fraction was found to be 42% of the total mass concentration when the calculation is applied to the inverted data, and 55% when performed on the raw data.

A second set of data was provided by Simon *et al.* [34] in the frame of characterizing airborne actinomycetes (*Thermoactinomyces vulgaris*) produced by means of a liquid sparging generator described in [35]. In this case, the quantities measured on each of the impaction stages correspond to the amount of cultivable bacteria (notated C_B), as shown in Figure 6.

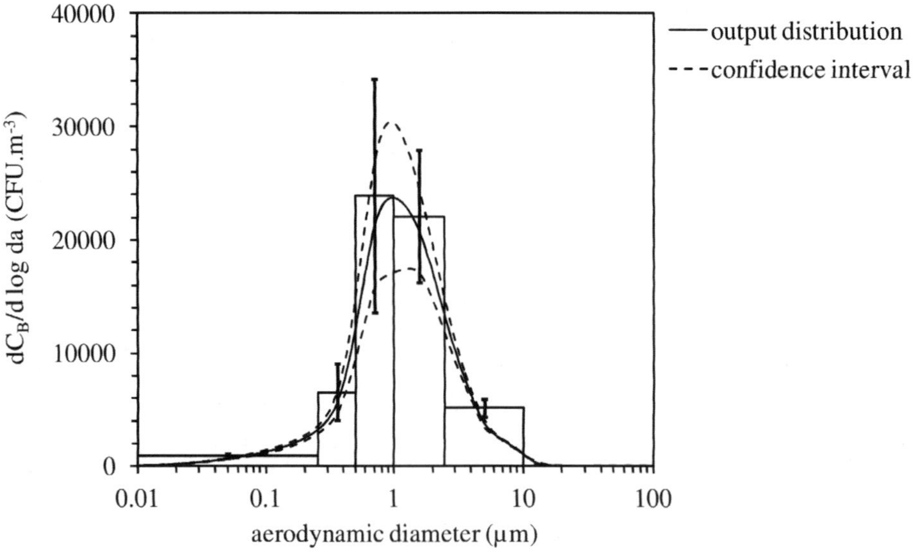

Figure 6. Output distribution and associated confidence interval from cultivable bacteria resulting from the generation of *T. vulgaris* – data from Simon *et al.* [34].

Although Reponen *et al.* [36, 37] measured airborne spores' aerodynamic diameters of 0.57-0.60 µm for *Thermoactinomyces vulgaris*, the inverse curve shows a widespread distribution with a modal size near 1 µm. This indicates the presence of cultivable species other than isolated spores in each of the impaction stages, which can be attributed to mycelium, fragments and agglomerates of various sizes.

4. Discussion and conclusions

A modular tool was developed to inverse data from cascade impactors based on a precise description of the collection efficiency curves. A series of theoretical tests has led to satisfying agreement between input and output distributions for the case of the SIOUTAS personal cascade impactor.

Before being applied to any other widespread cascade impactor (e.g. DLPI — Differential Low Pressure Impactor —, MARPLE, etc.), the inversion software has to be theoretically tested by means of a similar set of data to ensure relevance of the results.

This program is perfectible, and further developments are still needed. Among them:

- diffusion losses are not taken into account in the actual version of the software;
- the aspiration efficiency is only valid for the case of the SIOUTAS impactor;
- calculation of lung-deposited or conventional fractions could be automated;
- terminal filter efficiency has to be improved by the implementation of a filtration model;
- for the cases where extreme stages reveal the highest amount, fictive stages can be required to avoid divergence of the computation. This feature already exists in the program but a series of specific test cases has to be proceeded for validation.

Acknowledgments

Authors would like to thank Noël Castignolles from ESSTIN Services for developing the code and Xavier Simon from INRS for providing experimental data of Figure 6.

References

[1] Maynard A. D., Aitken R. J., Butz T., Colvin V., Donaldson K., Oberdörster G., Philbert M. A., Ryan J., Seaton A., Stone V., Tinkle S. S., Tran L., Walker H. J. and Warheit D. 2006. *Nature* **444**: 267-269.

[2] Comité Européen de Normalisation (CEN) 1993. *Workplace Atmospheres: Size Fraction Definitions for Measurement of Airborne Particles in the Workplace*, CEN standard EN 481.

[3] Brouwer D. 2010. *Toxicology* **269**: 120-127.

[4] Kuhlbusch T., Asbach C., Fissan H., Göhler D. and Stintz M. 2011. *Particle and Fiber Toxicology* **8**: 22.

[5] Ramachandran G., Ostraatb M., Evans D. E., Methner M. M., O'Shaughnessy P., D'Arcy J., Geraci C. L., Stevenson E., Maynard A. D. and Rickabaugh K. 2011. *Journal of Occupational and Environmental Hygiene* **8**: 673-685.

[6] Witschger O., LeBihan O., Reynier M., Durand C., Marchetto A., Zimmermann E. and Charpentier D. 2012. *Hygiène et Sécurité au Travail* **226**: 41-55.

[7] ISO 2008. *Nanotechnologies — Health and safety practices in occupational settings relevant to nanotechnologies*, ISO/TR 12885.

[8] Dong Y., Hays M. D., Smith N. D. and Kinsey J. S. 2004. *Journal of Aerosol Science* **35**: 1497-1512.

[9] ICRP 1994. *Publication 66. Human respiratory tract model for radiological protection*, Oxford: Pergamon.

[10] ISO 2012. *Air quality — Sampling conventions for airborne particle deposition in the human respiratory system*, (EN) ISO 13138.

[11] Heitbrink W. A., Evans D. E., Ku B. K., Maynard A. D., Slavin T. and Peters T. 2009. *Journal of Occupational and Environmental Hygiene* **6**: 19-31.

[12] Maynard A. D. and Aitken R. J. 2007. *Nanotoxicology* **1**: 26.

[13] Marple V. A. and Olson B. A. 2011. *Sampling and measurement using inertial, gravitational, centrifugal, and thermal techniques*, in: Kulkarni P., Baron P. A. and Willeke K., *Aerosol measurement. Principles, techniques and applications*, Third edition, John Wiley & Sons, Inc., 129-151.

[14] Hinds W. C. 1999. *Aerosol Technology. Properties, behavior, and measurement of airborne particles*, Second edition, John Wiley & Sons, Inc., 483 p.

[15] Arffman A., Marjamäki M. and Keskinen J. 2011. *Journal of Aerosol Science* **42**: 329-340.

[16] Demokritou P., Lee S. J., Ferguson S. T. and Koutrakis P. 2004. *Journal of Aerosol Science* **35**: 281-299.

[17] Hillamo R., Mäkelä T., Schwarz J. and Smolik J. 1999. *Journal of Aerosol Science* **33**: S901-S902.

[18] Hillamo R. E. and Kauppinen E. I. 1991. *Aerosol Science and Technology* **14**: 33-47.

[19] Keskinen J., Pietarinen K. and Lehtimäki M. 1992. *Journal of Aerosol Science* **23**: 353-360.

[20] Kwon S. B., Lim K. S., Jung J. S., Bae G. N. and Lee K. W. 2003. *Journal of Aerosol Science* **34**: 289-300.

[21] Marjamäki M., Keskinen J., Chen D. R. and Pui D. Y. H. 2000. *Journal of Aerosol Science* **31**: 249-261.

[22] Marjamäki M., Lemmetty M. and Keskinen J. 2005. *Aerosol Science and Technology* **39**: 575-582.

[23] Misra C., Singh M., Shen S., Sioutas C. and Hall P. M. 2002. *Journal of Aerosol Science* **33**: 1027-1047.

[24] Rubow K. L., Marple V. A., Olin J. and McCawley M. A. 1987. *American Industrial Hygiene Association Journal* **48**: 532-538.

[25] Vaughan N. P. 1989. *Journal of Aerosol Science* **20**: 67-90.

[26] Virtanen A., Marjamäki M., Ristimäki J. and Keskinen J. 2001. *Journal of Aerosol Science* **32**: 389-401.

[27] Twomey S. 1975. *Journal of Computational Physics* **18**: 188-200.

[28] Chahine M. T. 1968. *Journal of the Optical Society of America* **12**: 1634-1637.

[29] Markowski G. R. 1987. *Aerosol Science and Technology* **7**: 127-141.

[30] Winklmayr W., Wang H.-C. and John W. 1990. *Aerosol Science and Technology* **13**: 322-331.

[31] Su W.-C. and Vincent J. H. 2004. *Journal of Aerosol Science* **35**: 1119-1134.

[32] Raynor P. C., Leith D., Lee K. W. and Mukund R. 2011. *Sampling and analysis using filters*, in: Kulkarni P., Baron P. A. and Willeke K., *Aerosol measurement. Principles, techniques and applications*, Third edition, John Wiley & Sons, Inc., 107-128.

[33] Witschger O., Bau S., Bianchi B., Wrobel R. and Matera V. 2013. *Journal of Physics: Conference Series* **This issue**:

[34] Simon X., Betelli L., Koehler V., Coulais C. and Duquenne P. 2013. *Aerosol Science and Technology* **47**: 158-168.

[35] Simon X., Duquenne P., Koehler V., Piernot C., Coulais C. and Faure M. 2011. *Journal of Aerosol Science* **42**: 517-531.

[36] Reponen T., Gazenko S. V., Grinshpun S. A., Willeke K. and Cole E. C. 1998. *Applied and Environmental Microbiology* **64**: 3807-3812.

[37] Reponen T., Grinshpun S. A., Conwell K. L., Wiest J. and Anderson M. 2001. *Grana* **40**: 119-125.

EXPLORATORY STUDY ON A STATISTICAL METHOD TO ANALYSE TIME RESOLVED DATA OBTAINED DURING NANOMATERIAL EXPOSURE MEASUREMENTS

F. Clerc[1], G-H Njiki-Menga, O.Witschger
INRS, Département Métrologie des Polluants, 1 rue du Morvan, 54500 Vandoeuvre les Nancy, France

E-mail: frederic.clerc@inrs.fr

Abstract. Most of the measurement strategies that are suggested at the international level to assess workplace exposure to nanomaterials rely on devices measuring, in real time, airborne particles concentrations (according different metrics). Since none of the instruments to measure aerosols can distinguish a particle of interest to the background aerosol, the statistical analysis of time resolved data requires special attention. So far, very few approaches have been used for statistical analysis in the literature. This ranges from simple qualitative analysis of graphs to the implementation of more complex statistical models. To date, there is still no consensus on a particular approach and the current period is always looking for an appropriate and robust method. In this context, this exploratory study investigates a statistical method to analyse time resolved data based on a Bayesian probabilistic approach. To investigate and illustrate the use of the this statistical method, particle number concentration data from a workplace study that investigated the potential for exposure via inhalation from cleanout operations by sandpapering of a reactor producing nanocomposite thin films have been used. In this workplace study, the background issue has been addressed through the near-field and far-field approaches and several size integrated and time resolved devices have been used. The analysis of the results presented here focuses only on data obtained with two handheld condensation particle counters. While one was measuring at the source of the released particles, the other one was measuring in parallel far-field.. The Bayesian probabilistic approach allows a probabilistic modelling of data series, and the observed task is modelled in the form of probability distributions. The probability distributions issuing from time resolved data obtained at the source can be compared with the probability distributions issuing from the time resolved data obtained far-field, leading in a quantitative estimation of the airborne particles released at the source when the task is performed. Beyond obtained results, this exploratory study indicates that the analysis of the results requires specific experience in statistics.

[1] To whom any correspondence should be addressed.

1. Introduction

Manufactured nanomaterials are becoming more and more common in a wide range of domains: material, food industry, healthcare, energy or transportation industry [1]. The manufacturing of such materials is realized by workers who might be exposed to aerosols containing manufactured nanoparticles (MNP). Even if the danger related with MNP is not always known, the risk always exists and consequently, the reduction of exposures should be an objective.

Three routes of exposure exist: skin permeation, ingestion and inhalation. Concerning MNP, the inhalation is considered as the main route of exposure. For reducing this exposure, a relevant and reliable quantitative characterization of exposure at workplaces has to be developed.

Background aerosols from natural and incidental sources in or outside the work environment are ubiquitous and present a major issue to overcome when using time resolved instruments that are not specific for MNP; as indeed any other aerosol instrument used in industrial hygiene to characterize larger airborne particles (i.e. photometers or optical particle counters). Adequate characterization of exposures to aerosols of MNP cannot be accomplished without successfully distinguishing them from background aerosols. One of the approaches that can be used when using time-resolved instruments is based on a combined analysis of time series with and without activity and spatial analysis.
With respect to aerosol instrumentation, almost all the studies used the commonly type of instrument for counting airborne particles: condensation particle counter (CPC). The CPC is a time-resolved size-integrated instrument which detection range varies with the specific type but usually goes from about 2 nm up to about 1 μm. With this instrument, so called time series of particle number concentrations can be subject to statistical studies.
In other fields of science and technology, time resolved data is a common tool used for monitoring activity. Beyond signal processing, focused on mathematical methods, the field of economic sciences makes an extensive use of statistical methods used for analysing time resolved data, in particular ARIMA methods [2]. In most cases, the aim is to forecast the (near) future of time series. The input time series is often long term data possibly presenting cycles or seasonality. Apart exceptions, these requirements do not match with occupational exposure to chemical substances objectives. Indeed, is often consists of making a reliable comparison with a limit value; while in most cases, the duration of the measurement is a shift or less. This is why novel methods should be developed.

In most case studies, graphical representations of time series are provided and commented, with the support of common statistical indicators [3, 4, 5]. Recently, advanced statistical methodology for estimating exposure to MNP has been presented; it uses ARIMA time series analysis [6]. In the present study, a different statistical approach is proposed for assessing potential exposure based on time-resolved particle concentration measurements.

2. Methodology

2.1. Time-resolved data used

The time-resolved data used in this exploratory work are from a recent workplace related exposure study that investigates the potential for exposure via inhalation from cleanout operations by sandpapering of a reactor producing nanocomposite thin films embedded with silver nanoparticles [7]. In this academic research laboratory, small stainless steel samples were coated with nanocomposite thin films using a plasma deposition process reactor. During the experimental phases which could last several consecutive days, the cleaning of the reactor was a daily operation. This operation was done by hand with sandpaper; the objective being to polish all the internal walls of the reactor to recover the original color of stainless steel. As all sanding operations of this type, emissions of airborne particles could be suspected but almost no emission or exposure data are available in the literature so far. Therefore, the objective was to characterize the emitted aerosol in terms of particle sizes, particle concentrations and morphology, and also to look how the element silver, signature of the embedded nanoparticles in the thin films produced, could be found in the workplace atmosphere in the near field or far field environment of the reactor. The measurement strategy was based on the recent one developed in France [8]. Among aerosol instruments used were two handheld CPCs. These are the data from these two CPCs that are used in this exploratory study. One of the two was located close to the source; the distance to the source was about 10 to 20 cm. The second was located at 5 to 6 meters away in the same room (see Fig. 1). The only person performing a task was the one that worked on the cleaning of the reactor. No other activity or task was running in the laboratory room during the measurement period. Prior, during and after the cleaning operation, no incidental sources of nano-sized airborne particles in or outside the work environment has been identified. Also, the doors were left open, and the room was not equipped with special built ventilation compartment or local exhaust ventilation.

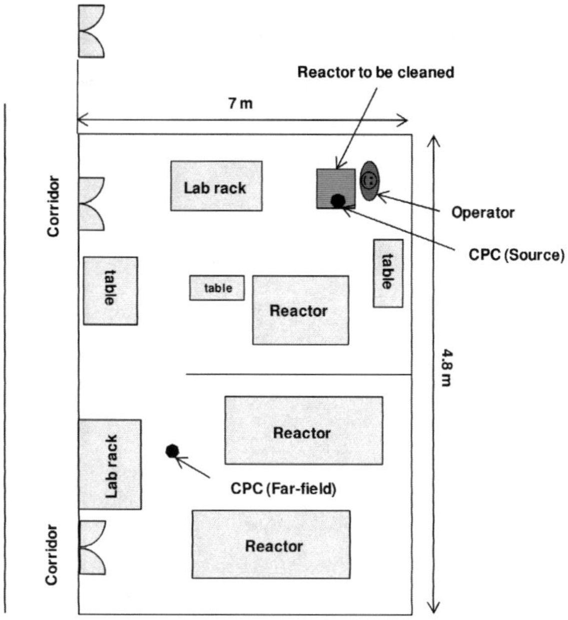

Figure 1. Floor plan of the laboratory room where the measurements took place.

Both CPCs recorded the particle number concentration during a total time of 68 minutes. The different tasks of the cleaning operation are labeled A to F; these correspond to different parts that were sanded inside the reactor like the walls, the silver target etc.

In the initial strategy, operators in charge of the measurements have used two different frequencies for each of the two CPCs: the "far field" CPC was recording a value every 30 seconds while the "source" CPC was recording the number concentration every 5 seconds. For the purposes of statistical analysis, a frequency identical to the two instruments is required. This is why the first action was to equalize the two time series at the same frequency of 30 seconds Thus, each set of source values corresponding to a single far field value are identified. The average of these values is then computed.

Finally, the time series analyzed are composed of 51 values as shown on Fig. 2.

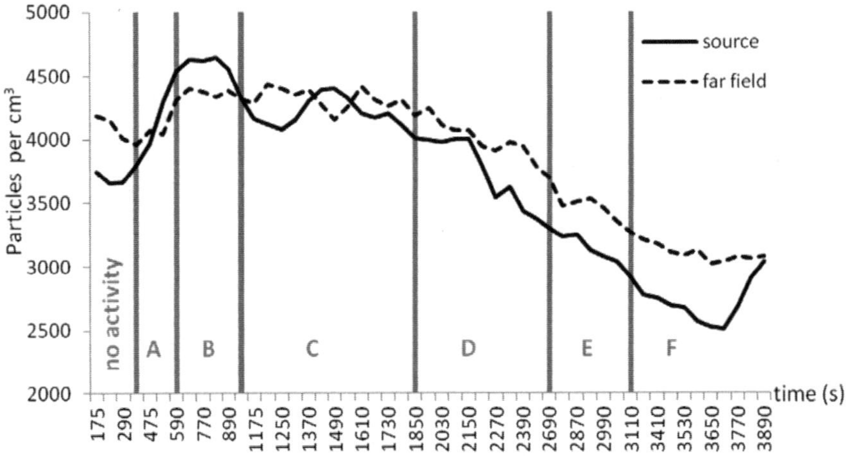

Figure 2. Particle number concentration profiles over time measured at the source and far field during a cleaning operation. The letters above the X-axis correspond to different parts that were sanded inside the reactor. During the time period 0-350 s there was no activity or task performed.

2.2. *Statistical analysis*

Three hypothesis are assumed

(1) Both source (black curve) and far field (dotted curve) CPCs record the same aerosol when there is no activity. The measured values include a systematic difference due to the devices characteristics.

(2) When the cleaning operation is realized, the CPC close to the source of emission also record the airborne particles released by the hand sandpapering.

(3) There was no specific ventilation in the laboratory room during the activity: the background aerosol behaves similarly around both CPCs, while the released airborne particles remain in the near field of the source.

Based on these assumptions, the principle is composed of four steps (see Fig. 3)

(1) Computing the systematic difference between measurement devices. The systematic difference is measured when no activity is realized. It is represented as a probability distribution of possible values.

(2) Including the systematic difference in the "source" time series thus computing a corrected source time series. This corrected source time series is comparable with the "far field" time series. Since the systematic difference is a probability distribution, each point of the corrected source time series comes with uncertainty.

(3) Computing the time series of emitted particles. For each data point, the far field value is subtracted from the source value with uncertainty. The result is the time series of extra

particles measured at the source of emission, with a global uncertainty coming from corrected source time series.

(4) Computing the quantity of particles counted during each task, in the form of probabilistic distributions taking into account the uncertainty related with the systematic difference and the time variation.

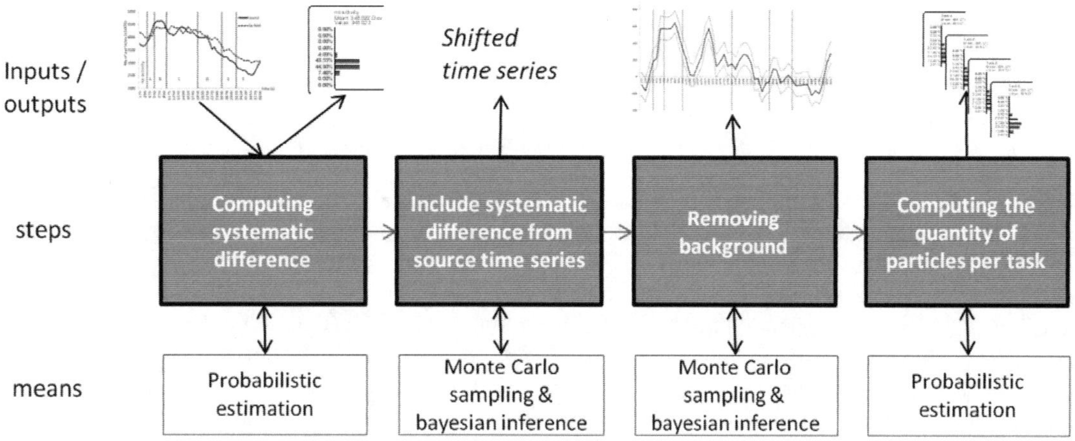

Figure 3. Overview of statistical analysis realized in this study.

3. Results

3.1. Computing the systematic difference

The systematic difference can be considered as a single value such as the difference of averages. Nevertheless, the number of values supporting the calculation of this difference is very small (4 data points). Consequently, it is more relevant to use a probability distribution as a systematic difference. In the same manner, the number of particles emitted by each task is presented as a probability distribution. This type of probabilistic computation can be done by sampling through Markov Chain Monte Carlo methods [9] (supported for instance by BUGS software [10]). It can also be done by exact computation through the use of a Bayesian network [11].

The aim of a Bayesian network is to provide a formal and graphical framework for representing knowledge and to support probabilistic calculations: Bayesian inference. A Bayesian network is an oriented graph whose nodes represent variables and whose arcs represent relationships between variables. The relationships between variables are quantified with conditional probabilities. The software used for Bayesian network design and inference is BayesiaLab [12]. If there are two arcs from variables "far field" and "source" to variable "no activity", the conditional probability table of "no activity" represent the distribution of values of "no activity" as a function of "far field" and "source". In the example (see Fig. 4), the function is "far field"-"source", the conditional probabilities of "no activity" are computed by sampling.

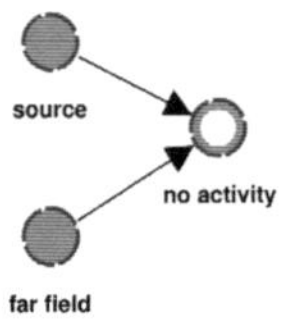

source	<=3705	<=3757	>3757
	50%	25%	25%

far field	<=4033	<=4107	>4107
	50%	0%	50%

No activity											
Far field	source	<=-360	<=-220	<=-80	<=60	<=200	<=340	<=480	<=620	<=760	>760
<=4033	<=3705	0	0	0	0	0	79%	21%	0	0	0
<=4033	<=3757	0	0	0	0	0	1	0	0	0	0
<=4033	>3757	0	0	0	0	33%	67%	0	0	0	0
<=4107	<=3705	0	0	0	0	0	2%	98%	0	0	0
<=4107	<=3757	0	0	0	0	0	52%	48%	0	0	0
<=4107	>3757	0	0	0	0	0	99%	1%	0	0	0
>4107	<=3705	0	0	0	0	0	0	70%	30%	0	0
>4107	<=3757	0	0	0	0	0	0	100%	0	0	0
>4107	>3757	0	0	0	0	0	23%	77%	0	0	0
Sum		0	0	0	0	0.33	4.21	4.16	0.3	0	0

Figure 4. Bayesian network used in the study (left), Conditional Probability Tables of "source" and "far field" nodes (middle), Conditional Probability Table of "no activity" node: systematic difference.

The nodes representing "far field" and "source" are darker because their conditional probabilities distributions come from the data; the difference "no activity" has a lighter patch because it is computed. The computation or *Bayesian inference* consists of transforming the "no activity" conditional probability table into a so call *marginal* distribution. In this simple case, it is the sum of all probabilities per column, which is normalized afterwards for obtaining the marginal probability distribution of differences when there is no activity (see Fig. 5).

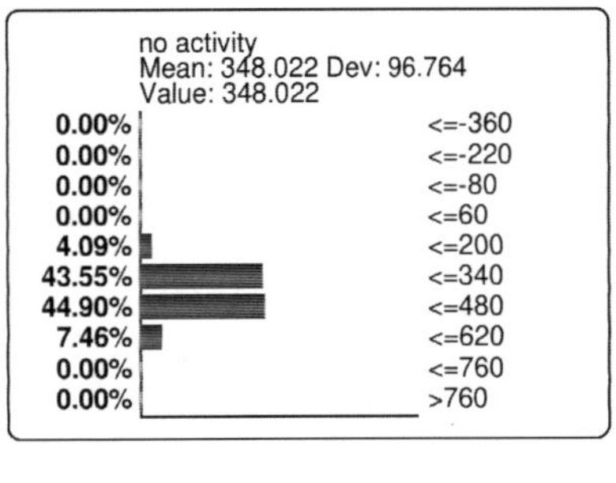

Figure 5. Marginal probability distribution of "no activity" node. This is the distribution of systematic difference between "source" and "far field" observed when no activity is realized.
When there is no activity, the concentration measured at far field is higher than at source, which fits with the graphical observation that can be made from Fig. 2.

3.2. Removing systematic difference from source time series & computing time series of emitted particles

During the "no activity" period, the number concentration at the source is greater than that obtained in the far field of 348 $1/cm^3$ on average (mean indicator in Fig. 5). The probability of observing a number concentration within the range]200; 340] is 43,38%. This can be explained by that fact that the particles size ranges of the CPCs used were not exactly the same. Based on the "no activity" distribution i.e. the systematic difference between devices, the background aerosol part of the source time series can be removed. Graphically, with regards to Fig. 2, this means shifting the black curve upwards. With respect to the hypothesis, the number concentration C due to the activity is the number concentration measured at the source C_{source}, where the number concentration measured at far field $C_{far\ field}$ is removed and the systematic difference "no activity" is added. For adding the "no activity" systematic difference, 200 values are sampled out of the distribution for each time step. The real "far field" and "source" time values are used. A total of 51 time steps are defined, which means that 51*200=10200 data points are sampled in total. The following algorithm describes this process.

For each time step t
  For i=1 to 200
  Sample d_i out of "no activity" distribution
  $C_i(t) = C_{source}(t) - C_{far\ field}(t) + d_i$
  End for
End for

The result C provided by using this approach is in the form of probability distributions (one per time step) based of the 200 C_i values. With this, it is possible to draw the time series of number concentration of released particles, considering the median and 5th percentile and 95th percentile as confidence interval for each time step as shown on the Fig. 6.

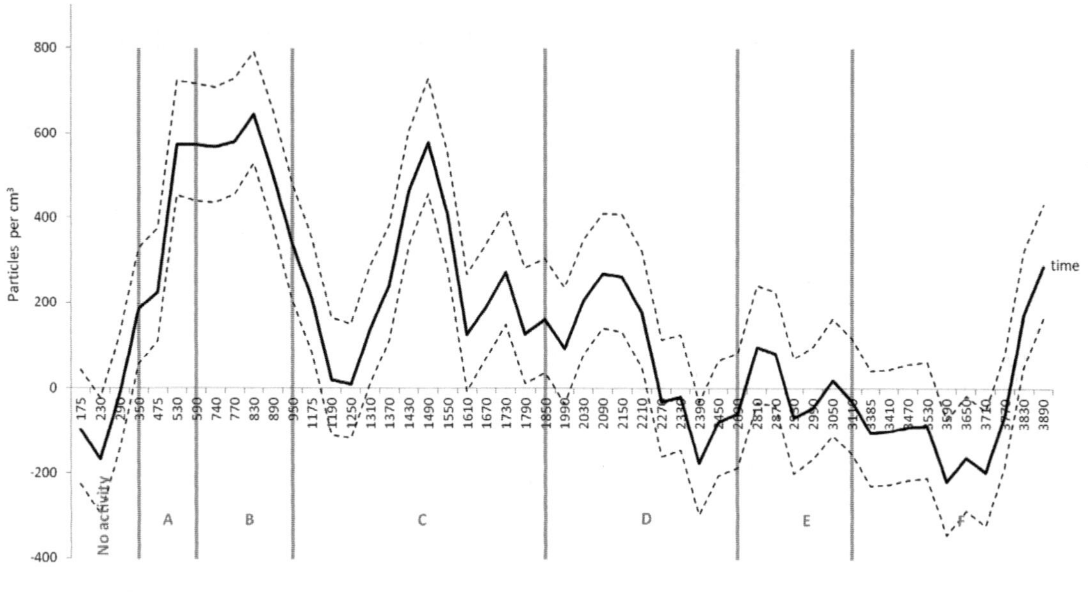

Figure 6. Number concentration of particles emitted at the source after the procedure of removal of the aerosol background as median value with 95%-confidence intervals.

On the basis of this time series, it is possible to quantify the number concentration of released airborne particles during each task through the probability distributions. By observing the graph, it seems that tasks A and B are generating a surplus of airborne particles, while the peak observed during task C may also create an average surplus of airborne particles. On the contrary, tasks D, E and F seem to have no impact. For quantifying these assumptions from graph data, the same methodology allowing building probability distributions of particle counts is applied. For each task, the number concentration of released particles at the source is depicted in Fig. 7. The statistical analysis shows that the mean number concentrations of released airborne particles are about 460 1/cm^3 and 540 1/cm^3 for task A and task B respectively. For the task C it is about 250 1/cm^3, while for the tasks D, E and F our analysis shows that no airborne particles was released.

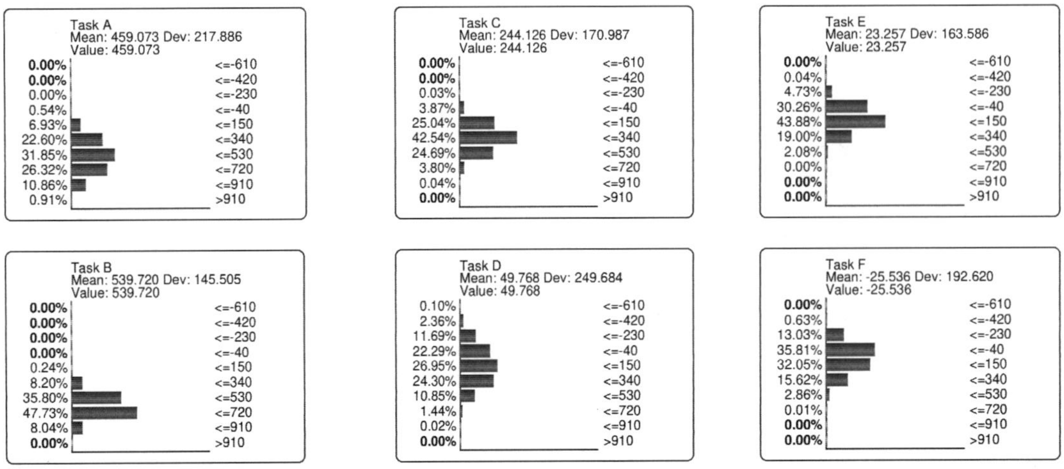

Figure 7. Probability distribution of the number concentration of release airborne particles for tasks A and B, C, D, E and F.

4. Discussion

It is important to note first of all that the measurement strategy used for collecting data was not established for this type of statistical analysis. This is why data present some specificity that complicates here the statistical analysis. For example, because of the difference in sampling frequency of CPCs, the data had to be combined, which dramatically reduced the number of concentration data on which the statistical analysis could be performed.

The results of the statistical analysis suggests that, overall, there is a positive effect of the cleanout operation by sandpapering of the reactor on the number concentration of particles measured by the CPC close to the source. However this effect is not positive or significant for all tasks, which correspond to sanding different parts of the reactor. For example, task A shows a mean number concentration of released airborne particles of about 460 $1/cm^3$, with a peak over 910 $1/cm^3$ (p=0.91%). For task C, the mean number concentration of released airborne particles is about 250 1/cm3 with peaks superior to 720 $1/cm^3$ (p=0.04%). The amount of airborne particles counted (and therefore released) during the following D, E and F tasks is around zero.

It is important to remember that the statistical analysis focused on measurement data using CPCs which are time resolved instruments giving no information on the nature of the detected particles.

5. Conclusion

Since measurements strategies [8, 12, 13, 14] published so far to characterize workplace exposure to manufactured nanoparticles put emphasis on the use real-time instrumentation for task-based evaluation of (size-resolved) aerosol concentrations, appropriate analysis of the results obtained from these instrumentation is a key factor.

In this exploratory study, it is assumed that the so-called background concentration is similar for source and far field sampling location: the far field concentration is not affected by the released aerosol at source. A statistical method based on a Bayesian probabilistic approach was developed to analyze time resolved data. The Bayesian probabilistic approach allows a probabilistic modeling of data series, and the observed task is modeled in the form of probability distributions. The probability distributions issuing from time resolved data obtained at the source can be compared with the probability distributions issuing from the time resolved data obtained far-field, leading in a quantitative estimation of the airborne particles released at the source when the specific task is performed.

To illustrate the use of this statistical method, particle number concentration data from a workplace study that investigated the potential for exposure via inhalation from cleanout operations by sandpapering of a reactor were used. The results suggests that, overall, there is a positive effect of the cleanout operation by sandpapering of the reactor on the number concentration of particles measured by the CPC close to the source.

What is interesting in this approach is that, in the perspective of future limit values based on measurement results obtained in real time, it becomes possible to use a probabilistic indicator as the 95th percentile.

Beyond obtained results, this exploratory study indicates that the analysis of the results requires specific experience in statistics, and therefore highlights the need for an association between statisticians and specialists in aerosol metrology.

6. References

[1] *(In french)* M. Lahmani and C. Brechignac 2006 *Les nanosciences. Tome 2 : Nanomateriaux et nanochimie* (Paris, FR:Belin)

[2] G. Box and G. Jenkins 1976 *Time series analysis: forecasting and control* (San Francisco: Holden-Day)

[3] D.H. Brouwer, J.H.J. Gijsbers and M.V.M. Lurvink 2004 *Ann Occup Hyg.* **48** 439

[4] E. Demou, P. Peter and S. Hellweg 2008 *Ann Occup Hyg.* **52** 695

[5] D.E. Evans, B. Ki Ku, M.E. Birch and K.H. Dunn 2010 *Ann Occup Hyg.* **54** 514

[6] R.H. Klein Entink and W. Fransman 2009 *J Nanopart Res.* **13** 6991

[7] Witschger O., Bau S., Bianchi B., Wrobel R. and Matera V. 2013. Journal of Physics: Conference Series This issue:

[8] *(In french)* O. Witschger, S. Le Bihan, M. Reynier, C. Durand, A. Marchetto, E. Zimmermann and D. Charpentier 2012 *Hygiène et Sécurité du Travail* **226** 41

[9] W. Gilks, S. Richardson and D. Spieghalter 1996 *Markov Chain Monte Carlo in Practice* (Chapman & Hall/CRC)

[10] D. Lunn, D. Spieghalter, A. Thomas and N. Best 2009 *Statistics in Medicine* **28** 3049

[11] D. Koller and N. Friedman 2009 *Probabilistic Graphical Models* (MIT Press)

[12] BayesiaLab *www.bayesia.com* [Online].

[13] D. Brouwer, B. van Duuren-Stuurman, M. Berges, E. Jankowska, D. Bard and D. Mark 2009 *J Nanopart Res.* **11**, 1867

[14] G. Ramachandran, M. Ostraat, D. E. Evans, M. M. Methner, P. O'Shaughnessy, J. D'arcy, C. L. Geraci, E. Stevenson, A. Maynard and K. Rickabaugh 2011 *J. Occup Environ Hyg* **8** 673

Approach to the Exposure Assessment of MWCNT by Considering Size Distribution and Oxidation Temperature of Elemental Carbon

M Ono-Ogasawara[1], M Takaya[1], H Kubota[1], Y Shinohara[1], S Koda[1], E Akiba[2], S Tsuruoka[3] and T Myojo[4]

[1] Japan National Institute of Occupational Safety and Health, 6-21-1, Nagao, Tama-ku, Kawasaki 214-8585, Japan
[2] Kuraray Living Co., Ltd. Umeda Hankyu Building Office Tower, 8-1, Kakudacho, Kita-ku, Osaka 530-8611, Japan
[3] ENCs, Shinshu University, 4-17-1, Wakasato, Nagano, 380-8553, Japan
[4] University of Occupational and Environmental Health, Kitakyushu 807- 8555, Japan

E-mail: ono@h.jniosh.go.jp

Abstract. Multi-walled carbon nanotubes (MWCNTs) have many beneficial characteristics, but it is concerned that exposure to MWCNTs may pose health risks. As an approach to the quantitative exposure assessment of MWCNTs, we have characterized and determined MWCNTs by elemental carbon (EC) using an aerosol carbon monitor. The EC fractions oxidized at different temperatures correspond sample characteristic such as diameter of MWCNT or origins of particles. As MWCNTs aggregate/agglomerate easily, they are usually observed as micron-size particles. Whereas, EC contained in ambient particulate matter (APM) is mainly observed in fine particles. Therefore, the size of airborne particles is a good parameter to distinguish MWCNT from other carbonaceous particles, especially APM. The size and oxidized temperature of EC suggested the origin of the carbonaceous aerosol samples. Exposure assessment of MWCNT was conducted by utilizing the size distribution of EC in the environment where particulate MWCNT or MWCNT-containing composite was handled. A procedure for exposure assessment of MWCNT-related workplace is proposed.

1. Introduction

Novel carbon nanotubes (CNTs) have many beneficial characteristics, but exposure to them may pose health risks. At present, some OELs (Occupational Exposure Limits) are proposed as follows; 0.0025 mg/m^3[1], 0.007 mg/m^3[2], 0.05 mg/m^3[3], and 0.03 mg/m^3 [4]. The OEL proposed by the US NIOSH [2] is derived as elemental carbon (EC) concentration, not mass concentration. Therefore, exposure measurement of CNTs requires a method capable of determining CNTs at least a few to 0.01 mg/m^3.

EC is considered as an index of CNT because EC is the main component of CNT. Thermal-optical carbon analysis, which is a sensitive method to measure carbon, has been used to assess diesel exhaust particles (DEPs) in occupational [5,6] and ambient environments [7]. This method has also been applied to monitor CNT concentrations in an inhalation exposure study [8] and to control the environment of a CNT laboratory [9]. EC has been used as an index of CNT and carbon nanofiber by the US NIOSH [10–12]. However, in an environment where concentration of background carbon is relatively high, the total EC may not represent exposure amount of CNT well. The key issue for exposure assessment of CNTs is the distinction between EC from MWCNT and ambient particulate matter (APM). Therefore, we have applied a modified IMPROVE protocol [13–15] to determine multi-walled carbon nanotube (MWCNT) by carbon analysis. This method can determine MWCNT as graphitic EC separated from other graphitic ECs in APM. As MWCNTs easily aggregate/agglomerate to micron-size particles, by taking into account the size distribution of graphitic ECs, micron-size aerosols of MWCNTs can be determined separately from submicron-size ECs originating from fuel combustion.

Not only for the process in which particulate MWCNT is handled, but also those in which MWCNT-containing composites are used, MWCNTs and/or fragments of those composites can be generated as micron-size particles in the working environment. At present, although the health effects of such fragments are not well known, it is important to collect the exposure data for such particles proactively in the view of an occupational health and life-cycle assessment.

In this paper, we conducted exposure assessment for particulate MWCNT and fragment of MWCNT-containing composite by carbon analysis of size-segregated particles, and propose a procedure to conduct exposure assessments of such particles.

2. Method

2.1. Carbon analysis

There are many variations of MWCNTs having differences of numbers of graphene layers, crystalline structures, and impurity contents. The temperature and rate of oxidation of MWCNTs reflect certain characteristics of individual MWCNT. Therefore, we characterized MWCNTs according to their oxidized temperature. Different types of MWCNTs and other carbonaceous particles were analyzed by a Carbon Aerosol Monitor (Sunset laboratory Inc.)

We employed a modified IMPROVE protocol (**table 1**) to characterize and determine MWCNT: 1) particles collected on quartz fiber filters were heated under a helium atmosphere to evolve organic carbons (OC); 2) ECs were oxidized at 550°C, 700°C and 920°C under an oxygen/helium atmosphere. The evolved carbon was oxidized to carbon dioxide and then reduced to methane to be monitored using a flame ionized detector. The evolved carbons denoted EC1, EC2 and EC3 are ECs oxidized at different temperatures. EC2 and EC3 are attributed to graphitic carbon [16]. MWCNTs with diameters greater than 150 nm were not oxidized below 900°C.

Table 1. Modified IMPROVE protocol.

	Time (sec)	Oven Temperature (°C)	Oven gas
OC1	180	120	He
OC2	180	250	
OC3	300	450	
OC4	300	550	
EC1	360	550	2% O_2/He
EC2	600	700[a] or 600[b]	
EC3	360	920[c]	

[a] for particulate MWCNT
[b] for MWCNT-containing composite, to distingush MWCNT from base material
[c] changed from 800°C in the original IMPROVE protocol

For exposure assessment of fragments of MWCNT-containing composites, we conducted exposure assessment in an environment where MWCNT-containing yarn was woven (see 2.3.2) [17]. The yarn was produced from coating MWCNT with binders on polyester yarn [18,19]. An appropriate protocol was set to distinguish the original polyester, the MWCNT (Baytube C150P) that was coated on the yarn, and the yarn coated with MWCNT. The Baytube was oxidized at 700°C and the polyester yarn and binder were oxidized at 600°C. We applied another modified protocol for this case by setting the temperature of EC2 to 600°C.

2.2. EM observation

For scanning electron microscopic (SEM) observation, aerosol was collected on a polycarbonate filter (pore size 100 nm; Millipore). A field emission SEM (Hitachi S-4700, Tokyo) was used to observe the samples after coating with platinum. Samples for SEM analysis of different types of MWCNTs were prepared by dispersing MWCNTs in 2-propanol with sonication followed by filtration

MWCNT-coated yarn was observed by transmitting electron microscopy (TEM). A yarn sample cut to ca. 5 mm was embedded in epoxy resin. The embedded resin block was sliced into 0.001 mm-thick sections on a microtome for the optical microscopic observation. Sections in which the MWCNT-coated yarn was detected were sliced into ultra-thin sections of 50–80 nm thick. The ultra-thin sections were examined by TEM (FEI Tecnai G2 sprit, FEI) at an acceleration voltage of 120 kV. The oxidation residue of the carbon analysis, which contains only EC3, was also observed using the same procedure.

2.3. Sampling

We conducted exposure assessments for two different processes: handling particulate MWCNT and applying MWCNT-containing composites. In both processes, the size-segregated aerosol was sampled, and then the aerosol was analyzed using a carbon aerosol monitor. A Sioutas Cascade Impactor (SKC Inc.) was used to collect particles with a quartz fiber filter (2500-QAT, Pall). The impactor can classify particles into five fractions with a flow rate of 9 L/min. The size of each fraction in terms of their aerodynamic particle diameters is: >2500 nm, 2500–1000 nm, 1000–500 nm, 500–250 nm, and back-up stage < 250 nm. Experimentally determined D_{50}s of each impaction stage for PTFE substrate are 2600 nm, 950 nm, 520 nm and 230 nm [20].

2.3.1. Handling of particulate MWCNTs. The process monitored was an automated packaging of MWCNT. MWCNT was automatically fed from a reservoir into a plastic bag in an enclosed system installed in a facility. Since the enclosed system had a small opening to check operation, small amounts of MWCNT can leak into the facility. The sampling point was ca. 2 m from the opening. The sampling duration was the same as the packaging process of about one hour. The background sample was collected at the same point when no operation was being conducted (at night).

2.3.2. Handling of MWCNT-containing composite. In the facility where MWCNT-coated yarn was being woven, exposure assessment of fragments of the yarn was conducted [17]. Since there were one loom for MWCNT-coated yarn and over twenty other looms for polyester yarn, fragments of polyester yarn could affect the monitoring of EC. We collected size-segregated samples around the MWCNT-coated yarn loom. Sampling was conducted while the loom for MWCNT-coated yarn was in operation. Background samples were collected at the same point outside the hours of operation (at night) while other polyester looms were operated.

3. Results

3.1. MWCNT oxidation pattern from carbon analysis.

Typical results of carbon analysis and SEM images are shown in **figure 1**. Thin MWCNTs with diameters less than 10 nm were oxidized as EC2. However, the MWCNTs with diameters exceeding 20 nm were detected mainly as EC3. The temperature at which MWCNTs were oxidized was affected by the diameter of the MWCNT, but not by its length [14] and the size of agglomerate [13]. Although the ratio EC3/EC2 of MWCNTs varies depending on crystallinity, metal content and the amount of sample, EC3 can be used as a good indicator of MWCNTs.

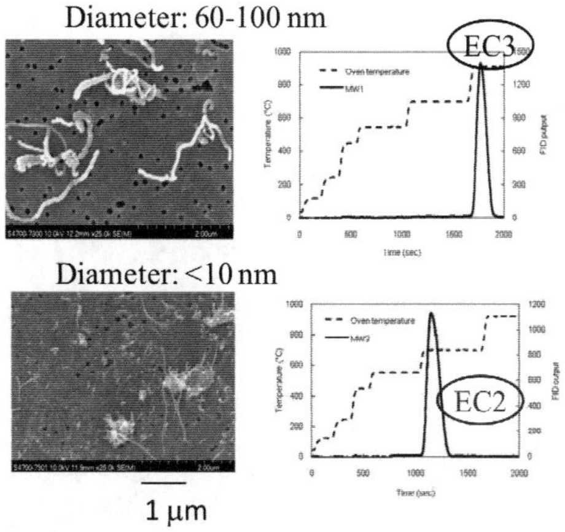

Figure 1. SEM images and EC oxidation pattern of typical MWCNTs.

3.2. MWCNT in factories

3.2.1. Packaging process of particulate MWCNTs. We characterized the airborne MWCNT particles by taking into account the oxidation temperature and size distribution of EC2 and EC3. From previous research, MWCNTs, whether simulated or in the real environment, are aerosolized as micron-size particles arising from aggregate/agglomerate [13,14]. When the packaging process was in operation, micron-size EC3 was dominant. At night in the same facility, the amount of EC2 arising primarily from APM was larger than EC3. There were trace amounts of MWCNTs because micron-size EC3 was observed and SEM observation confirmed this result. Most ECs in APM, which is considered as background EC, is characterized as EC2 and dominant in particles collected as fraction smaller than 250 nm. We determine EC3 in all particles collected by SCI as MWCNT by separating from APM.

EC3 concentration was about one tenth of mass concentration. SEM observation showed the presence of small number of single or aggregate/agglomerate MWCNTs.

3.2.2. Weaving process of MWCNT-coated yarn. SEM observation showed the presence of micron-size particles containing MWCNT as fragments of the MWCNT-coated yarn (**figure 2**). We could not detect single MWCNT fibers in this field survey.

In the workplace, there were a large numbers of polyester fragments from polyester yarn and small numbers of fragments from MWCNT-coated yarn. In order to distinguish between particles with/without MWCNTs, we adjusted the protocol of carbon analysis. The concentration of MWCNTs contained in the micron-size particles did not exceed 0.0053 mg/m^3. The respirable mass concentration was about ten times higher than the EC3 concentration.

To confirm that the EC3 was attributed to MWCNTs in the MWCNT-coated yarn, we conducted TEM observations of both MWCNT-coated yarn and the oxidation residue. As shown in **figure 3**, the polyester disappeared from the yarn after oxidation at 600°C. Therefore, it is thought that MWCNT was detected as EC3.

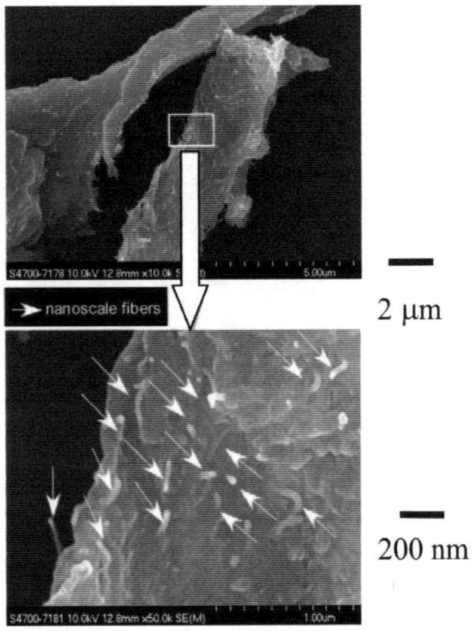

Figure 2. SEM images of fragments of MWCNT-coated yarn collected near the loom. (from Takaya et al. 2012 Ind Health [15])

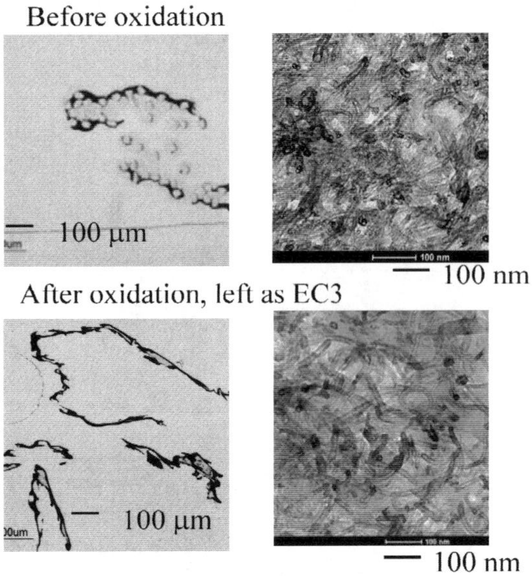

Figure 3. TEM images of MWCNT-coated yarn before after the oxidation step of carbon analysis.

4. Discussion

A method for measuring exposure to MWCNTs is required to conduct risk management of MWCNTs in the workplace. The analysis of MWCNTs separated from other carbonaceous substances was conducted using a carbon aerosol monitor. MWCNTs were determined by monitoring graphitic carbon. For the carbon monitor measurements, the limit of quantitation was around 0.002 mg. If the sample volume is 1 m^3 and the entire area of a single filter is analyzed, the quantitation limit for airborne MWCNT is expected to be 0.002 mg/m^3.

Since MWCNTs are oxidized and evolved as EC2 and EC3, EC2 and EC3 are used as quantitative indices of MWCNT.

The size dependency of EC2 and EC3 can be directly applied to an exposure assessment of MWCNTs in the workplace. By analyzing various carbonaceous particles, we presented two-dimensional mapping of size distribution of EC1 (less graphitic carbon), EC2 and EC3 [14]. Generally, MWCNTs are observed as EC2 and EC3 in micron-size particles because they easily aggregate/agglomerate. ECs in APM originating from combustion are observed as EC2 in fine particles. Carbon black (CB) is observed as EC2 and EC3 mainly in micron-size particles. Single-walled carbon nanotubes (SWCNTs) and fullerenes (C_{60}) are easily oxidized to evolve as EC1. Interference from these carbonaceous materials should be taken into account.

In addition, this method is applicable for the measurement of MWCNTs in composites by adjusting the protocol to separate MWCNTs from resins or other base materials of the composite.

In **figure 4,** we propose MWCNT exposure assessment procedure employing this approach. The first step of the procedure is different for particulate MWCNTs and MWCNT-containing composites, because the protocol for the carbon analysis must be changed to correctly measure the target MWCNT. The background concentration of EC2 and EC3 should be measured simultaneously outside the workplace. If outside air does not represent a background of the environment, for example an environment in which no direct intrusion of outside air or other carbonaceous particles are continuously present, the background sample can be sampled in the same place during non-working hours. When the background air contains a high concentration of ECs, assessment of MWCNTs in the workplace should be carefully conducted by analyzing background data.

It is preferable that a person with sufficient knowledge of the processes and of occupational hygiene checks the workplace and decides where and how exposure should be measured. When the gravimetric mass concentration of respirable particles including background is lower than a tentative OEL, for example 0.03 mg/m^3 proposed by AIST [4], there may not be needed to conduct carbon analysis. Even if the mass concentration is lower than the OEL, SEM observation of airborne sample or wipe sample is recommended to confirm the effectiveness of the applied control measures.

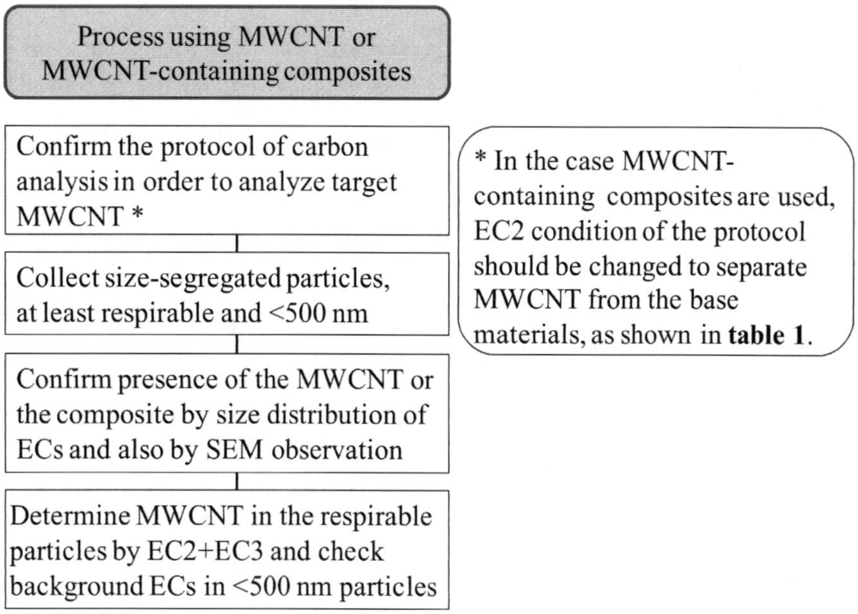

Figure 4. MWCNT exposure assessment procedure by using carbon analysis.

5. Conclusion

Exposure measurement of MWCNTs was conducted for the processes of handling particulate MWCNTs and weaving MWCNT-containing yarn by using graphitic carbon as an index of MWCNTs. MWCNTs were observed in micron-size particles as EC oxidized at 700°C and/or 920°C with carbon analysis. Since EC contained in APM is usually observed as EC oxidized at 700°C in fine particles, MWCNTs can be determined separately from background by considering size distribution of ECs. By adjusting the protocol for carbon analysis, MWCNT in MWCNT-containing composite can be analyzed by this approach.

We propose MWCNT exposure assessment procedure employing this approach. The first step of the procedure is different for particulate MWCNT and MWCNT-containing composites, because the protocol for the carbon analysis must be changed to measure correctly the target MWCNT. The background EC should be measured simultaneously. When the background air contains a high concentration of EC, assessment of MWCNTs in the workplace should be carefully conducted by analyzing background data.

6. Acknowledgement

This research was supported in part by a Grant-in-Aid for Scientific Research (No. 23615014) of the Japan Society for the Promotion of Science. We are deeply grateful to Matsubun Textile Co. Ltd for giving us an opportunity to conduct a field survey in a MWCNT-related textile factory.

The authors declare no conflicts of interest.

References

[1] Ma-Hock L, Treumann S, Strauss V, Brill S, Luizi F, Mertler M, Wiench K, Gamer A O, van Ravenzwaay B, Landsiedel R 2009 *Tox Sci* **112** 468–81

[2] National Institute for Occupational Safety and Health (NIOSH) 2010 niosh current intelligence bulletin: Occupational Exposure to Carbon Nanotubes and Nanofibers. available at http://www.cdc.gov/niosh/docket/review/docket161a/pdfs/carbonNanotubeCIB_PublicReviewOfDraft.pdf

[3] Pauluhn J 2010 *Tox Sci* **113** 226–42

[4] Nakanishi J 2011 Risk Assessment of Manufactured Nanomaterials:"Approach"-Overview of Approaches and Results- Final report issued on Aug 17, 2011. NEDO project (P06041) "Research and development of Nanoparticle characterization method." available at https://www.aist-riss.jp/projects/nedo-nanorisk/nano_rad2/download/download.cgi?ES_e

[5] Birch M E, Cary R A. 1996 *Aerosol Sci Technol* **25** 221–4

[6] Birch M E 2002 *Appl Occup Environ Hyg* **17** 400–5

[7] Chow J C, Watson J G, Pritchett L C, Pierson W R, Frazier C A, Purcell R G 1993 *Atmos Environ* **27A** 1185–1201

[8] Myojo T, Oyabu T, Nishi K, Kadoya C, Tanaka I, Ono-Ogasawara M, Sakae H, Shirai T 2009 *J Nanoparticle Res* **11** 91–9

[9] Han J H, Lee E J, Lee J H, So K P, Lee Y H, Bae G N, Lee S-B, Ji J H, Cho M H, Yu I J 2008 *Inhal. Toxicol* **20** 741–9

[10] Maynard A D, Baron P A, Foley M, Shvedova A A, Kisin E R, Castranova V 2004 *J Toxicol Environ Health,* Part A **67** 87–107

[11] Methner M, Hodson L, Dames A, Geraci C 2010 *J Occup Environ Hyg* **7** 163–76

[12] Birch M E, Ku B K, Evans D E, Ruda-Eberenz T A 2011 *Ann Occup Hyg* **55** 1016–36

[13] Ono-Ogasawara M, Myojo T 2011 *Ind Health* **49** 726–34

[14] Ono-Ogasawara M, Myojo T 2013 *Adv Powder Technol* **24** 263–269

[15] Takaya M, Serita F, Ono-Ogasawara M, Shinohara Y, Koda S 2010 *San Ei Shi* **52** 182–8 (in Japanese)

[16] Han Y M, Cao J J, Chow J C, Watson J G, An Z S, Jin Z D, Fung K, Liu S X 2007 *Chemosphere* **69** 569–74

[17] Takaya M, Ono-Ogasawara M, Shinohara Y, Kubota H, Tsuruoka S, Koda S 2012 *Ind Health* **50** 147–55

[18] Fugetsu B, Han W, Endo N, Kamiyama Y, Okuhara T 2005 *Chem Lett* **34** 1218–9

[19] Patent application 2007-225966, 2008-224821

Misra C, Singh M, Shen S, Sioutas C, Hall P M 2002 *J Aerosol Sci* **33** 1027–47

Urban Air Pollution by Nanoparticles in Ostrava Region

Authors: Kaličáková Zdeňka[2] , Míčka Vladimír[1], Lach Karel[1], Danihelka Pavel[2]

(1) Institute of Public Health Ostrava, Partyzánské náměstí 7, 702 00 Ostrava, Czech Republic

(2) VŠB–Technical University of Ostrava, Faculty of Safety Engineering, Lumírova 13, 700 30 Ostrava Výškovice, Czech Republic

E-mail: zdenka.kalicakova@centrum.cz

Abstract. Air pollution harms human health and the environment. Ostrava's agglomeration and its immediate vicinity suffer regular exceeding of air pollution limits due to its geomorphologic location and present heavy industry. Maximum exceedances of air quality standards and especially PM10 which 24 hour limit value is in EU 50µg.m-3, must not be exceeded more than 35 days per year. This limit is being still often exceeded. In the year 2011 such as situation occurred 126 times. It is very important then for identify sources of air pollution to find out maximum information about air borne dust, like size distribution, chemical composition of individual size fractions, morphology of particulate matter together with other parameters like meteorological conditions, year season etc. Our measurement started two years ago. We focus on the critical situation when there are values of PM10 over a long period above the limit. In winter season it is so called inversion. By default, during the campaign it is measured size distribution of air born dust in range 5.6 nm - 560nm by FMPS and using the sampler NanoId are collected samples in range 1nm - 35µm in 12 size fractions for chemical analysis and morphological observations. This contribution deals with results of size distributions only.

1. Introduction

The town of Ostrava is the third largest town in the Czech Republic. The region is the most air-polluted in the country. As for geomorphology, Ostrava is located in the Ostrava basin [1]. Ostrava's geomorphology and poor dispersion conditions both contribute to the air pollution and cause the pollution to concentrate here, especially in winter.

Although in the 1990´s the air quality in the Czech Republic improved considerably, industrial growth and transport expansion in the 2000´s made the situation worse again.

Industrial sources, emissions from road vehicles and local heating are the main causes of degraded air quality. Due to the prevailing north-west direction of airflow also come into consideration sources in the nearby industrial agglomeration and possibly also from large residential area in the Polish Lower Silesia region. There are freely available sources of information to confirm this hypothesis [2]. To evaluate the formation of fine and ultrafine

Published under licence by IOP Publishing Ltd

aerosol was used reverse trajectory model HYSPLIT [3]. Estimate of the impact of potential sources of pollution by PM10 was carried out for 22 days in which it was in years 2006-2011 one of the selected stations the daily (24-hour) average concentrations of PM10 at least 200µg.m-3. Air quality was monitored on nine locations. All episodes studied occurred in winter season. Only in four cases it was possible to reject the hypothesis of fine aerosol transport from the Polish territory where the main sources are rather local heating than metallurgical industry[2] [3].

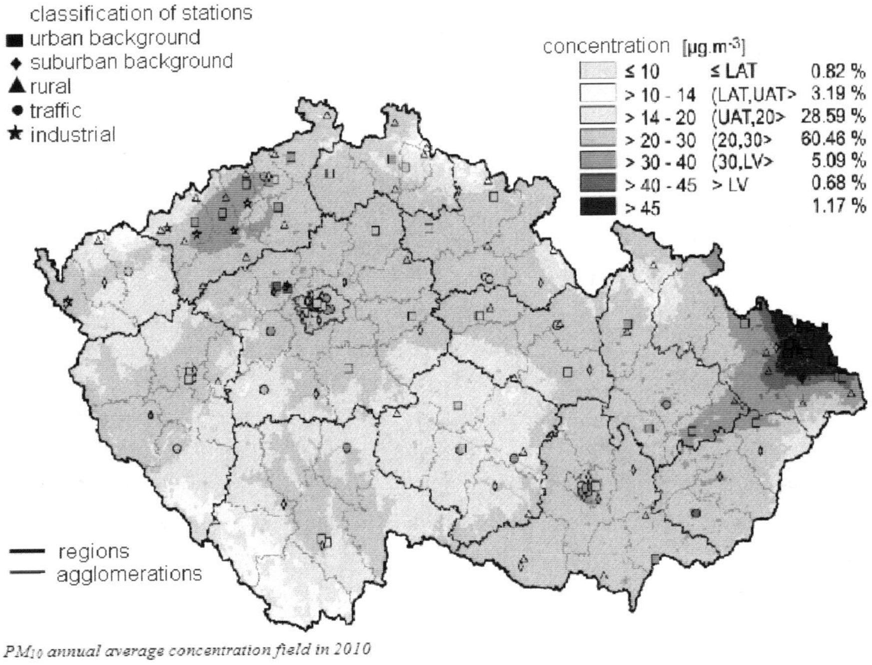

Fig. 1: Map of the Czech Republic – air pollution concentration by PM_{10} [4]

Limit values for particulate matter and other pollutants such as arsenic and benzo(a)pyrene are exceeded on a regular basis in Ostrava. These dangerous substances are, beside of particles, a major threat for people living in industrial areas. The exceeding of limit values for carcinogenic benzene also occurs in the Ostrava region. In warm seasons, problems arrive in the Czech Republic with tropospheric ozone, presence of which is related to the formation of photochemical smog [5] [6].

The range of monitored pollutants grew over the years. Initially, only sulfur dioxide and suspended particulates were measured in most monitoring stations; a few years later nitrogen oxides monitoring was introduced. The monitoring was later extended by measurement of mass fraction of selected metals in particulate matter samples and carbon dioxide monitoring. The boom in automatic analyzers made possible independent monitoring of nitric oxide and nitrogen dioxide and, recently, monitoring of ozone, selected organic substances and dust aerosol PM_{10} fraction. The PM_{10} daily limit value of 50 $\mu g.m^{-3}$ is exceeded repeatedly and often by several times in some regions, especially the Ostrava´s one. The limit should not be exceeded more than 35 times a year; however, in some parts of the

Czech Republic it is exceeded more than 100 times a year (source of information: Czech Hydro-meteorological Institute). With time, $PM_{2,5}$ began to be measured and nowadays PM_1 is measured at some stations, too. The smaller the particle, the more easily it enters the human body, in addition, were also expect some of the negative health effects of nanoparticles in the air. Therefore, in the future nanoparticles should be monitored. Another reason to monitor nanoparticles is their properties: they work as carriers of organic pollutants and they may carry information about their origin and, thus, their source. This is the reason why this pilot study was performed [7] [8] [9].

In Ostrava, an increasing number of people suffer from upper and lower respiratory tract inflammations, asthma, nasopharyngitis, allergic rhinitis and atopic eczema. Long-term exposure to particles aerosol can lead to chronic bronchitis or shorten life expectancy. The inhabitants of Ostrava suffer much more often from cardiovascular diseases and the overall illness rates are higher than in other parts of the country. The most affected groups are children, senior citizens and people with respiratory and cardiovascular diseases [10] [11].

Objective of the measurement was to determine whether nanoparticles and microparticles under 560 nm contribute significantly to the contamination of air in Ostrava and whether their composition and distribution allow drawing any conclusions about their origin.

2. Methods

The area of investigation is located in the Moravian-Silesian Region in Ostrava-Mariánské Hory and Ostrava-Radvanice town districts. The first sampling location in Ostrava-Radvanice was located close to an industrial zone which was 1.5 to 3 kilometers from Mittal Steel iron and steel works, one from the main industrial polluters. The character of Ostrava-Radvanice is individual houses settlement with additional air pollution sources from local heating. The second sampling location in Ostrava-Mariánské Hory was chosen as it was situated in a densely populated area – a preschool and vegetation were close and the nearby traffic was high.

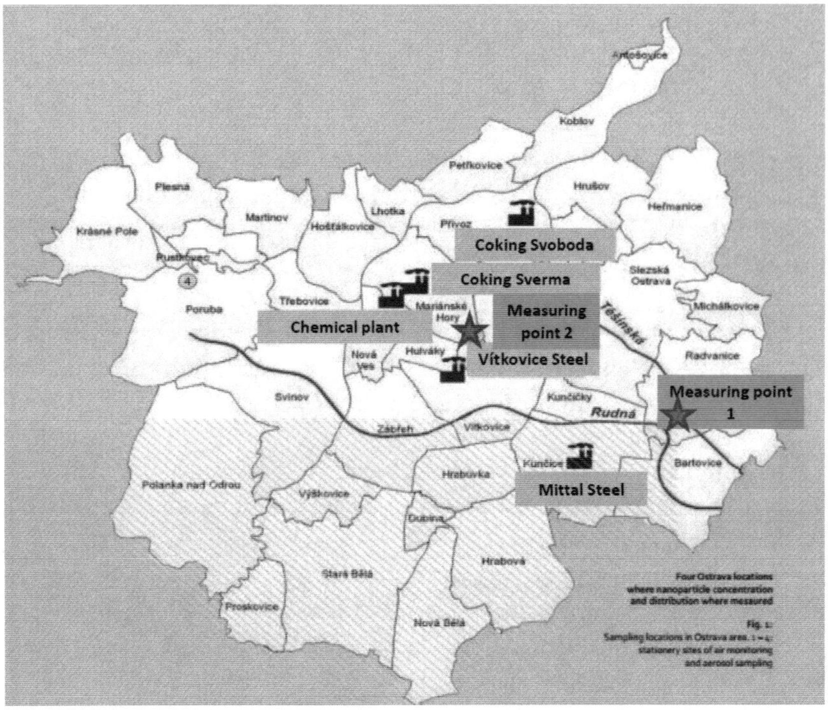

Fig. 2: The map of Ostrava with industrial air pollution sources and measuring locations. [12]

The measurements took place in four different days in years 2011-2012. Two consecutive measurements in one day were performed at each sampling location.

The distribution was measured with Fast Mobility Particle Sizer Spectometer Model 3091 (FMPS ™, ITS Inc., USA), a 5.6 – 560 nm range particle spectrometer. FMPS offering a total 32 channels of resolution. Due to its unique desing, the FMPS spectrometer is capable of making particle size distribution measurement with one-second resolution, enabling you to visualize particle size distributions and events in real time.

Nano-ID SelectTM (Naneum Ltd, UK) was used for sampling the entire aerosol ranging from 0,001 nm to 30 μm in 12 stages, 5 of which are in the 0.001-250 nm rage. The wide range capabilities of this sampler are particularly useful when the area of interest extends into the nano and micro regions. The retrieved samples were analyzed in a laboratory using a scanning electron microscope (SEM) and an inductively coupled plasma mass spectrometer (ICP-MS) model THERMO X-Series2 [13] [14]. Results concerning chemical analysis and electron microscope observations will be subject of a separate contribution in a professional journal or conference [13] [14] [15].

3. Results and Discussion

The first measurement campaign started on 1st February 2011. That day the Czech Hydrometeorological Institute announced regulatory signal as the nitrogen dioxide or sulfur dioxide or PM_{10} special limit values had been exceeded. It means, according to Czech legislature, that the PM_{10} concentration exceeded the 24-hour average concentration of 150 $\mu g.m^{-3}$ or the sulfur dioxide concentration exceeded 500 $\mu g.m^{-3}$ in three successive hours or the nitrogen dioxide concentration exceeded 400 $\mu g.m^{-3}$ in three successive hours [16]. The average daily temperature on that day was minus 6.3 °C. The wind direction was southwest and the average wind speed was 3 m/s. The first two measurements in Ostrava-Mariánské Hory on 1st February 2011 started in the afternoon. Distant local heating sources may have been, together with nearby industries, the sources of air pollution in this winter season. Transport may also have affected the measurement results as the measurement was carried out during rush hour. Two busy intersections were 400 m and 500 m far from the sampling location. Another two measurements started on 1st February 2011 in Ostrava-Radvanice. The wind was blowing southwest. One of sources that may have contributed to the air pollution was Mittal Steel iron and steel works. In immediate vicinity of the sampling site there were local air pollution sources from solid fuel local heating.

The second measurement started on 30th March 2011. The air dispersion conditions were good. The average daily temperature was 9.5 °C. The wind direction was east southeast and the average wind speed was 1.7 m/s. Two measurements took place in Ostrava-Mariánské hory on 30th March 2011 before noon. The measurement results may have been affected, among other things, by local heating sources. Sources located in Poland may also have contributed to the air pollution. Another two measurements started on 30th March 2011 in Ostrava-Radvanice. The measurement was carried out during rush hour. The measurement results may have been affected by local heating air pollution sources. The wind was blowing from Poland to the sampling site, which could influence the measured values.

The third measurement started on 7th September 2011. The dispersion conditions were good. The average daily temperature was 14.7 °C. The wind direction was southwest and the average wind speed was 2.7 m/s. In Ostrava-Mariánské Hory two measurements took place. The wind was blowing southwest from the air pollution sources. Local heating air pollution sources may have contributed to the pollution of the air. Two measurements took place on 7th September 2011 in Ostrava-Radvanice. The wind was blowing southwest from the air pollution sources.

The fourth measurement started on 22nd March 2012. The dispersion conditions were good. The average daily temperature was 8.2 °C. The wind direction was northwest to west and the average wind speed was 2 m/s. Two measurements took place in Ostrava-Radvanice on 22nd March 2012 at noon. At this measurement, particles of 6 nm to 12 nm were absent. The wind was blowing northwest from the air pollution sources. Another two measurements took place in Ostrava-Mariánské Hory on 22nd March 2012 before noon. The buildings of the preschool were being under reconstruction. The wind was blowing northwest. A coking plant, a steel mill and a chemical plant were among the sources that may have contributed to the air pollution. The local air pollution sources from local heating may also have contributed to the air pollution due to the cold weather.

The total number concentrations of particles and particle size distributions show remarkable differences in both locations. The differences may have been caused by different daily temperatures. On cold days, increased local heating may contribute to the air pollution more significantly. In individual houses areas, residents themselves pollute the air through local heating. The problem is that they frequently use low quality solid fuels and even household waste burning is not excluded. The overall air pollution in Ostrava-Radvanice is comparable to that in industrial Chinese cities of Guiyu, Taizhou, Tianjin [7]. Transport is another air pollution source. During rush hours the concentration of particles were increased. The wind direction had a measurable effect on the distribution of particles. When the wind blew from large industrial air pollution sources to the sampling locations, the concentration of particles was elevated. Air dispersion conditions may also have an effect on the air contamination.

In various cases, one, two or three from maxima of size distribution at ~10 nm, ~40 nm, ~100 nm were observed (see Fig. 3), which implies that various, but limited sources combines to form overall aerosol composition. Small number of measurements does not allow making deeper conclusions concerning sources.
In some cases, the air pollution by nanoparticles is not homogenous in time and local maxima at different size distribution lasting for about minutes were frequently observed, which can be seen in Figure 4. We can explain this phenomenon by changing the direction and velocity of air flow making it difficult to assign a particular distribution of aerosols to a specific source. The fact is that the suspended aerosol can be transported at very large distances and thus resources can be tens of kilometers away from the measuring point.
On Fig. 5 is demonstrated the opposite situation when the distribution and concentration rather constant which implies the hypothesis that pollution is already well mixed and stabilized in air and so, the source is rather significant and far away from place of measurement, because small local sources would not cause, due to the turbulence of air, stable concentration and distribution of fine aerosol. Generally it is very important complex statistics evaluation of all data measured with chemical composition and morphology of individual fractions collected by Nano-ID. It was not the intention of this brief study to detect all links leading to the identification of all sources of such as high PM10 pollution. Generally, higher concentrations were detected in the vicinity of industrial areas. In three out of four days of measurement, the pollution in Ostrava-Radvanice was higher than in Ostrava-Mariánské Hory.

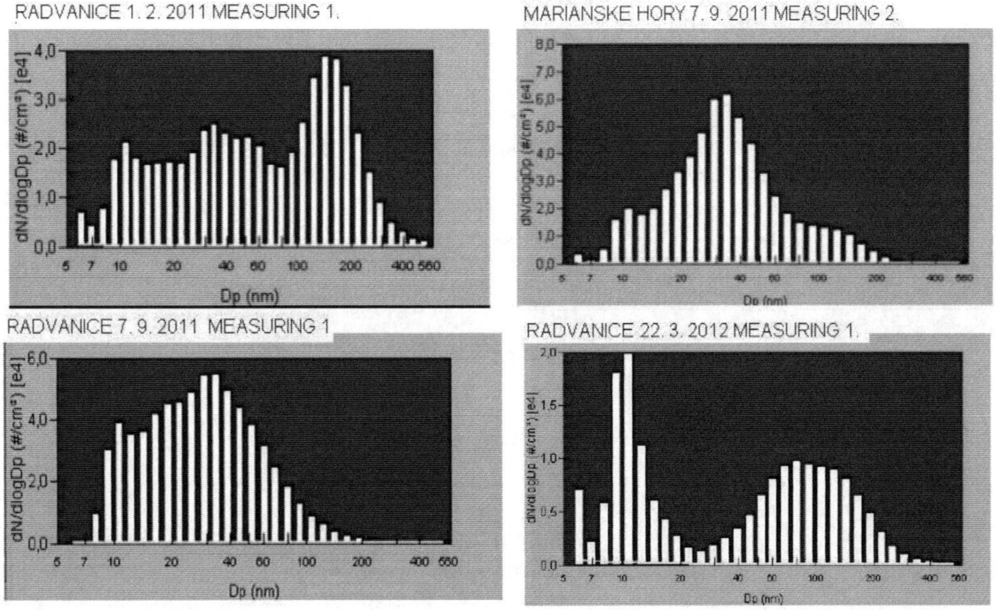

Fig. 3a),b),c),d): Demonstration of results of size distribution of fine and ultrafine aerosol on two localities in Ostrava after 5s integration time measured by FMPS.

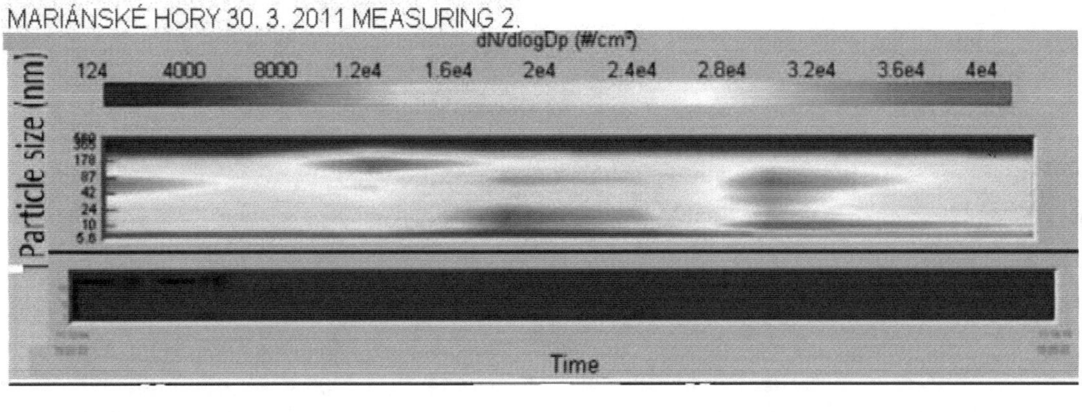

Fig. 4: Particle size distribution in time showing results of long term (1 hour) measurement of fine aerosol in Ostrava's district Marianske Hory . The red color indicates the size area with the highest concentration of particles. There were very unstable conditions as for fine aerosol size distribution. The probable reasons for it were the changes in the air flow directions.

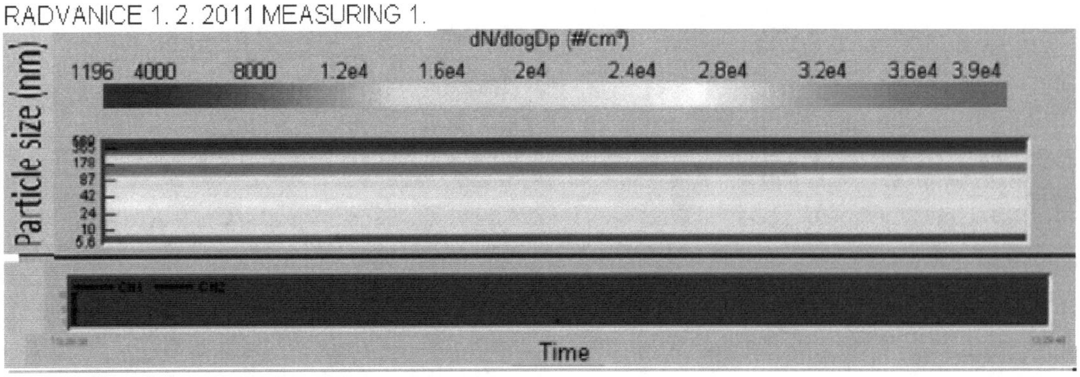

Fig. 5: Demonstration of very steady particle size distribution in locality Ostrava-Radvanice with the mode 150nm.

Another interesting fact is that the same maximum in the distribution of particles size was observed at both locations on the same day. On 7th September 2011, both in Ostrava-Mariánské Hory and Ostrava-Radvanice particles of 34 nm prevailed. Similarly, on 22nd March 2012, in one measurement in Ostrava-Radvanice and in one measurement in Ostrava-Mariánské Hory there were present particles that occurred in the largest distribution of particles of 81 nm.

The total mass concentration of nanoparticles and ultra-fine particles (5,6 - 560 nm) is relatively high, varying from few percent (w/w) of PM_{10} at good dispersion conditions to more than 50 percent in the case of smog situation. While the number concentration is much higher for small particles, the distribution of measured particles represented at Fig. 3 gives the good picture describing overall particle number concentration distribution.

4. Conclusions

Fine and ultrafine aerosol participate significantly to the air pollution in the Ostrava region and during bad meteorological condition, the fraction 5.6 – 560 nm composes even more than 50% of total particulate matter mass in ambient air. Generally, higher concentration fine and ultrafine aerosol was detected in the vicinity of industrial areas. In three out of four days of measurement, the pollution in Ostrava-Radvanice near the big metallurgical plant ArcelorMitall Ostrava a.s. was higher than in Ostrava-Mariánské Hory in the centrum of the city Ostrava.

Various types of sources probably contribute to pollution by nanoparticles and all suspended particulate matter. An important source of variability in the distribution of ultrafine aerosol is the weather where the major factor for determining sources of nanoparticles is air flow direction. Typically were observed maxima at three sizes regions ~10 nm, ~40 nm, ~100 nm. This corresponds well to uni- bi- and trimodal distribution. These findings imply opinion that there is a complex system of different industrial sources, local heating emission and partly also road transport in this region which emit significant quantities of nanoparticles.

Further, systematic study of Ostrava region air pollution by fine and ultrafine particles is needed and therefore another project plans are preparing now, which will study this problem in more detailed way, involving chemical analyzes, not only air pollution but also emissions directly at the expected sources, etc. It is assumed also closer co-operation with Polish state authorities and relevant environmental organizations.

Acknowledgment

Authors thank to projects SAFETY AGENT CZ.1.07/2.4.00/31.0049 and INEF CZ.1.05/2.1.00/01.0036. for support.

Bibliography

[1] DEMEK, Jan a Jaromír. *Z nížin do hor*: geomorfologické jednotky České republiky. Vyd. 1. Praha: Academia, 2012, 343 p. ISBN 978-802-0020-260

[2] *Śląski monitoring powietrza* [online]. 2013 [cit. 2013-01-06]. Available from: http://stacje.katowice.pios.gov.pl/monitoring/

[3] BÍLEK, Jiří. *Hodnocení znečištění ovzduší v Ostravě za roky 2006 – 2011 – Zpětné trajektorie*.Dychamproostravu [online]. 2012 [cit. 2012-10-21]. Available from: Hodnocení znečištění ovzduší v Ostravě za roky 2006 – 2011 – Zpětné trajektorie

[4] *Map of the Czech Republic - PM_{10} pollution.* CHIM [online]. 2010 [cit. 2012-10-21]. Available from: http://portal.chmi.cz/files/portal/docs/uoco/isko/grafroc/groc/gr10cz/png/oIl42x8PM10rp.png

[5] *Informace o vyhodnocení výsledků imisního monitoringu.* Ministry of Environment of the Czech Republic [online]. 2009 [cit. 2012-10-21]. Available from: http://www.mzp.cz/C1257458002F0DC7/cz/kvalita_ovzdusi/$FILE/OOO-Monitoring_2007-20090112.pdf

[6] KAŠPAR, Jakub. *Největším českým ekologickým problémem zůstává kvalita ovzduší.* Ministry of Environment of the Czech Republic [online]. 2007 [cit. 2012-10-21]. Available from: www.mzp.cz/cz/news_tz071217ovzdusi

[7] KAZMAROVÁ, Helena. *Vývoj znečištění ovzduší v ČR.* Centrum pro otázky životního prostředí [online]. 2011 [cit. 2012-10-21]. Available from: http://www.czp.cuni.cz/projekty/sdcz/moduly/2A/2A2/kazmarova.pdf

[8] *Imisní limity.* Czech Hydrometeorological Institute [online]. 2012 [cit. 2012-10-21]. Available from: http://www.chmi.cz/files/portal/docs/uoco/isko/info/limity_CZ.html

[9] LACH, Karel. *Nanočástice v ovzduší Ostravska.* Monitoring nanočástic v životním prostředí [online]. 2012 [cit. 2012-10-21]. Available from: http://iszp.kr-moravskoslezsky.cz/assets/nanocastice-v-ovzdusi-ostravska.pdf

[10] *Ovlivňuje znečištěné ovzduší náš zdravotní stav?*. Institute of Experimental Medicine AS CR, v. v. i. [online]. 2012 [cit. 2012-10-21]. Available from: http://www.iem.cas.cz/press/press/120612-ovlivnuje-znecistene-ovzdusi-nas-stav.html

[11] *ZNEČIŠTĚNÍ OVZDUŠÍ NA ÚZEMÍ ČESKÉ REPUBLIKY V ROCE 2010.* Czech

Hydrometeorological Institute [online]. 2010 [cit. 2012-10-21]. Available from: http://portal.chmi.cz/files/portal/docs/uoco/isko/grafroc/groc/gr10cz/kap2421.html

[12] DANIHELKA, Pavel. *Nanoparticles in Ostrava urban Atmosphere*. [online]. 2011 [cit.2012-03-08]. Available from: http://www.zuova.cz/informace/poster_nanoparticles_ in_the_ostrava_urban_atmosphere.pdf.

[13] Model 3091 Fast Mobility Particle Sizer TM Spectromete [online]. 2004 [cit. 2012-03-27]. Available from: http://www.kenelec.com.au/sitebuilder/products/files/116/3091.pdf.

[14] Wide Rande Aerosol Sampling System: Nano ID Select 005. Cantebery, 2009

[15] Model 3091 Fast Mobility Particle Sizer TM Spectrometer. TSI Incorporated [online]. 2004[cit.2013-01-23].Dostupnéz: http://www.tsi.com/uploadedFiles/Product_Information/Literature/Spec_Sheets/3091FMPS.pdf

[16] *Analýza dopadů návrhu zákona o předcházení a nápravě ekologické újmy*. Ministry of Environment of the Czech Republic [online]. 2006 [cit. 2012-10-21]. Available from: www.cemc.cz/clen_www/legislativa/prip_rizeni/.../ostatni/11_05.doc

Detection and Identification: Instrumentation and Calibration for Air/Liquid/Surface-borne Nanoscale Particles

Tsz Yan Ling, Zhili Zuo and David Y.H. Pui

Particle Technology Laboratory, Department of Mechanical Engineering, University of Minnesota, Minneapolis, Minnesota, 55455, USA.

Email: dyhpui@umn.edu

Abstract. Nanoscale particles can be found in the air-borne, liquid-borne and surface-borne dispersed phases. Measurement techniques for nanoscale particles in all three dispersed phases are needed for the environmental, health and safety studies of nanomaterials. We present our studies on connecting the nanoparticle measurements in different phases to enhance the characterization capability. Microscopy analysis for particle morphology can be performed by depositing air-borne or liquid-borne nanoparticles on surfaces. Detection limit and measurement resolution of the liquid-borne nanoparticles can be enhanced by aerosolizing them and taking advantage of the well-developed air-borne particle analyzers. Sampling electrically classified air-borne virus particles with a gelatin filter provides higher collection efficiency than a liquid impinger.

1. Introduction

Nanoscale particles, e.g., engineered nanoparticles, soot agglomerates from engines, and viruses, exist in the air-borne, liquid-borne and surface-borne dispersed phases [1]. It is necessary to develop instrumentation and measurement techniques for all three dispersed phases in order to gain better understanding of the life cycle of these nanoscale particles. For example, engineered nanoparticles often are produced in flame synthesis in the air-borne phase. They are coated on various surfaces to enhance performance of nano-products in the surface-borne phase. Upon disposal of these nano-products, the coated nanoparticles may be released in the waste streams in the liquid-borne phase. Physico-chemical characterization of nanoscale particles in these phases are therefore important in the environmental, health and safety studies of nanomaterials in support of sustainable nanotechnology [2, 3].

Each characterization method for the nanoscale particles in different phrases has its own applicable size and concentration ranges. Decision on the use of a specific method will depend on the properties of the particles of interest. Particles are best characterized at their original phase, with minimum handling or perturbation. For example, nanoscale particles in liquid-borne phase will form aggregates when dried or pH changes. However, transferring the particles from one phase to another, sometimes, allows additional information of the particles to be obtained. In this paper, we will describe some of the experimental techniques we have developed to connect nanoscale particle measurements in three dispersed phases. By aerosolizing liquid-borne nanoparticles into the air-borne phase, the particle sizing resolution improves significantly because the air-borne size analyzer, the Scanning Mobility Particle Sizer (SMPS), has sizing resolution significantly better than the liquid-borne Nanoparticle

Tracking Analysis (NTA) or the Dynamic Light Scattering (DLS) techniques. By depositing air-borne or liquid-borne nanoparticles on surfaces, they can be inspected by the Scanning Electron Microscopy (SEM), Transmission Electron Microscopy (TEM), Energy Dispersive X-ray Spectrometry (EDS) and Atomic Force Microscopy (AFM) for particle morphology and chemical compositions. While viruses are often transmitted in the air-borne phase, most of the virological tools are based on liquid-phase analysis. It is therefore particularly important to develop tools that can disperse and deposit nanoparticles in all three phases. We will demonstrate that by correlating nanoscale particles dispersed in different phases, it is possible to obtain accurate calibration of instruments for particle size and concentration.

2. Liquid-borne/surface-borne correlation

Liquid-borne nanoscale particles are encountered when they are manufactured and stored in liquid phase, e.g. CMP slurry, or released to the water stream in the environment. Common commercialized methods for sizing liquid-borne nanoparticles include: Field-Flow Fractionation (FFF), Dynamic and Static Light Scattering (DLS and SLS), Laser Particle Sizing (LPS) and Nanoparticle Tracking Analysis (NTA) [4-6]. It should be noted that these methods use different principles, namely the particle diffusion and light scattering properties, to measure the equivalent particle size distribution. One should be careful when comparing the size measured by different methods.

In addition to various liquid-borne nanoparticle characterization techniques, microscopy methods can further provide valuable information of the nanoparticles. In order to perform microscopy analysis, nanoscale particles in liquid-borne phase need to be transferred onto a surface. Common microscopy techniques, having different resolutions, include: SEM, Environmental-SEM, which allows the sample to be measured under variable pressure and humidity, TEM and AFM. From microscopy analysis, the shape, structure and area-equivalent size of the particles can be determined [4]. Additionally, electron microscopy methods allow the determination of the elemental composition of the particles if an EDS is included. Nanoscale particles in liquid-borne phase are often sensitive to physical (e.g., mixing) or chemical (e.g., pH, ionic strength) disturbances. The primary disadvantage of microscopy analysis is that the sample preparation procedures require drying the nanoparticles and transferring them to a surface. This may cause particle aggregation and change in particle size distribution [4]. The solute in the liquid phase may also precipitate onto the particle surface, affecting the image quality. Secondly, microscopy analysis usually has a limited sample size. It is time consuming to obtain and analyze sufficient images in order to have a statistically significant analysis [7].

The nanoscale particles do not always exist in spherical form. As a single value of particle size cannot completely represent the non-spherical particles, any equivalent particle size can be different from another. Researchers have been studying the correlations between various particle size representations [8-10]. Different particle size representations may have different significance. For example, the hydrodynamic size, which relates the drag experienced by the particles, will determine the inertial and diffusion behaviors. On the other hand, the geometric size may be more relevant for particle filtration by interception mechanism [11].

Beads mill has been used in industrial processing for grinding and dispersing agglomerated particles into primary particles in submicron size range [12]. We used a beads mill to disperse liquid-borne Titanium oxide (TiO_2) aggregated particles. By controlling the milling time, particles of different degree of agglomeration can be obtained. Figure 1 shows example images of the aggregated particles collected onto a 50 nm-rated Nuclepore filter for milling times at 0, 1 10 and 60 min. We first characterized the geometric properties of the aggregated particles by analyzing the SEM images. The image analysis software, ImageJ, was used to determine the projected-area equivalent diameter of the particles. We also used the NTA, (LM10, NanoSight Ltd.) to characterize the hydrodynamic properties of the particles. Table 1 lists the statistics of the analysis. Figure 2 shows the comparison of hydrodynamic and area equivalent size distributions of the TiO_2 aggregates.

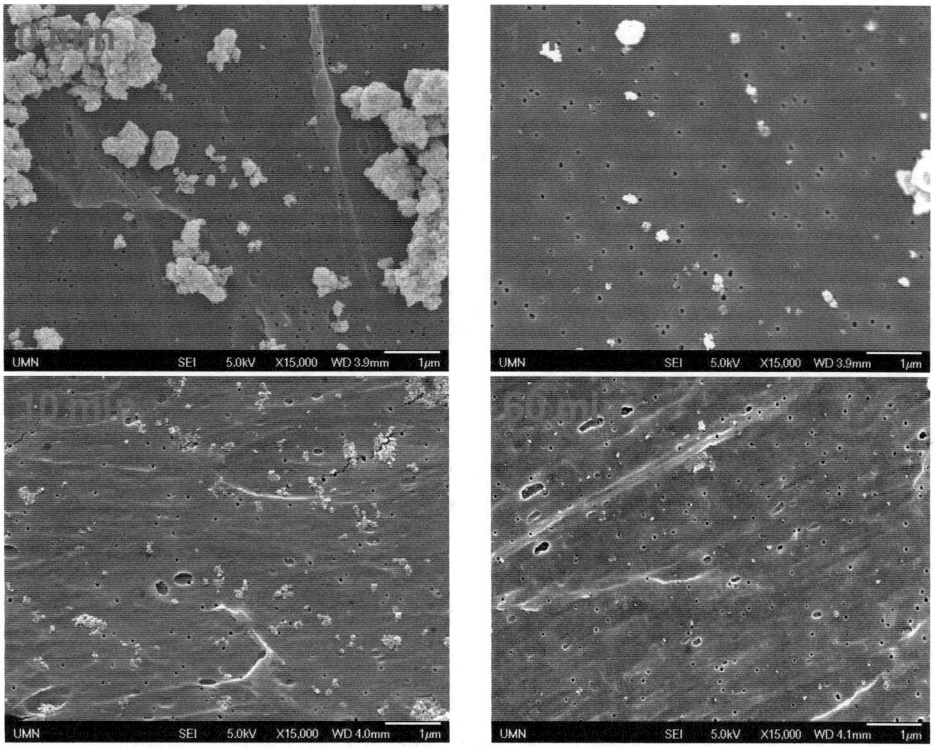

Figure 1. TiO_2 aggregate particles collected at different milling times.

Table 1. Statistics of the Nanoparticle Tracking Analysis and microscopy analysis for TiO_2 aggregate particles.

Time	50th percentile diameter (nm)		# of particles analyzed from SEM
	Hydrodynamic	Area equivalent	
0 min	220	180	650
1 min	149	170	901
10 min	65	58	509
60 min	44	35	80

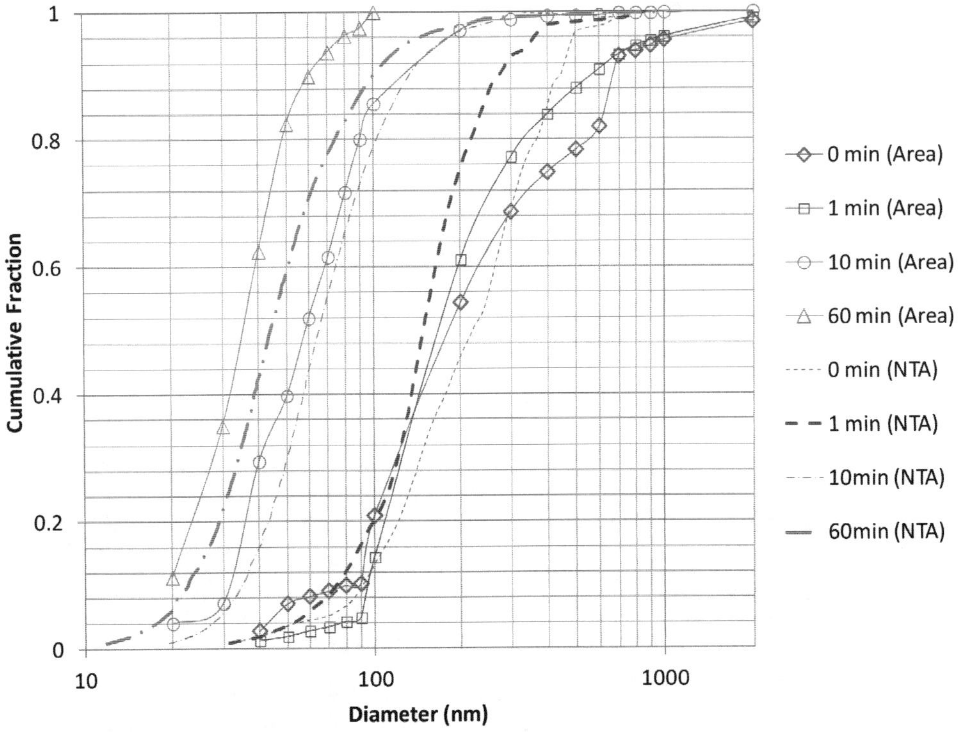

Figure 2. Cumulative size distribution of TiO_2 aggregates based on projected-area equivalent diameter and hydrodynamic diameter

It has been shown that the drag of fractal cluster scales with the radius of gyration and orientationally averaged projected area-equivalent size in the continuum and free molecular regimes, respectively [8]. Zhang et al. [10] also suggested the Smoluchowski radius can represent the drag of any non-spherical particles in the continuum regime. Binder et al. [9] reported that the mean drag on aggregates (with D_f=1.85) can be related to the projected area in the continuum regime with a power law with 15% variation, thus giving a rough estimate of the hydrodynamic properties from projected area. In the example shown, we have demonstrated a combined liquid and surface characterization method for determining the hydrodynamic size and projected area of aggregate particles and their correlation. Overall, surface-borne microscopy characterization provides supplementary information on particle properties, namely projected area, shape and structure, chemical composition etc, which enhance the understanding on the nanoscale particles under study.

3. Air-borne/liquid-borne correlation

In the previous section, the use of various liquid-borne particle instruments and microsocpy methods to analyzed nanoscale particles in liquid-borne phase have been discussed. Since liquid particle instruments, primarily rely on the light scattering from nanoparticles, they are limited due to the low scattered light intensity of the particles and small refractive-index contrast between particles and the liquid medium [13]. While microscopy analysis can offer additional information of the particles, the technique is not able to provide fast-respond measurements. Aerosol instruments, on the other hand, are considered to be better developed compared to those for liquid-borne particles [14]. Therefore, researchers have been exploring aerosol techniques for characterizing nanoscale particles in liquid-borne phase. Aerosolizing liquid-borne nanoparticles with an atomizer or nebulizer is relatively simple and inexpensive. After aerosolization and subsequent evaporation of solvent in the droplets, we can size and count the air-borne particles by a SMPS [15].

The challenge for this method is to preserving the original size distribution as in the liquid-borne phase. It needs to produce single particles which are agglomeration-free and avoids the influence from the impurities (soluble or insoluble) present in the liquid phase. The nonvolatile surfactant, sometimes necessary to stabilize the nanoparticles in liquid-borne phase, is also considered as part of the impurities. After evaporation of the volatile solvent, the nonvolatile impurities can either form smaller residue particles, which interfere with the particles of interest on the size distribution plot, or coat onto the particles of interest, increasing the size and changing the chemical nature of them. To overcome these problems, Lenggoro et al. [16] applied electrospray operating in the cone-jet mode, instead of an atomizer, to disperse liquid-borne nanoparticles into air-borne phase. Since the electrospray is capable of generating finer droplets compared to conventional atomizers, one can ensure that only one nanoscale particle is contained in a droplet, avoiding particle aggregation. The small droplets also minimize the effects of nonvolatile impurities after solvent evaporation. Lenggoro et al. [17] further demonstrated that a tubular furnace can evaporate residue particles and coatings on the nanoparticles of interest, leading to size measurement with better accuracy. Park et al. [18] also suggested that the effect of impurities can be suppressed by reverse osmosis process under high pressure. The aerosolization method to analyze nanoscale particles in liquid-borne phase has been applied to monitoring water quality and primary marine aerosol formation in seawater [13] and studying filtration of liquid-borne nanoparticles [14, 19]. Reversing the aerosolization process, it is sometimes advantageous to transfer air-borne particles to liquid-borne phase. Greenwald et al. [20] reported a technique to measure the insoluble fraction of ambient air-borne particles, by sending them to an aqueous medium and measuring them with a liquid particle counter. Also, air-borne viruses are usually transferred to liquid medium for virological analysis.

The nanoparticle nebulizer (TSI Inc.) has an improved design for dispersing droplets smaller than the conventional atomizers or nebulizers, making it suitable for the purpose of characterizing nanoscale particles by the aerosolization method. We have compared the size distribution of gold nanoparticles measured by 1) a nanoparticle nebulizer-SMPS system and 2) NTA (NanoSight LM10). The samples include monodisperse 80, 50 and 30 nm gold particles (Ted Pella, Inc.) and a mixture of them with equal number concentration. The concentration limit for the two methods should be noted. SMPS gives results with good statistics with particle concentration > 1E9 #/ml, while the NTA has a concentration window between 1E8 and 1E9 #/ml. Therefore, proper dilution was needed.

Figure 3 shows the size distribution of 30 nm gold particles measured with SMPS and NTA. On the SMPS size distribution, the first and second peaks correspond to the residue and gold particles, respectively. Residue particles have significantly higher concentration compared to the gold particles. However, since the residue particles are small (mode diameter at 10 nm), they do not interfere with the gold particles on the size distribution. In many cases, they can be removed by using a diffusion screen or by electrostatic classification using a Differential Mobility Analyzer (DMA). Figure 4 shows the size distributions of mixed 80, 50 and 30 nm gold particles with equal number concentration. On the SMPS plot, while the residue particles are present in high concentration, they are kept in small size, not interfering with the particles of interest. The peaks at 30, 50, 80 nm can be distinctly resolved. The NTA plot, which bases on particle light scattering, shows a biased measurement towards the larger particles. From the comparison on the two measurements, it is obvious that SMPS gives more accurate representation for nanoparticles with polydisperse distribution. However, one drawback is that it usually requires higher particle concentration to give results with good statistics.

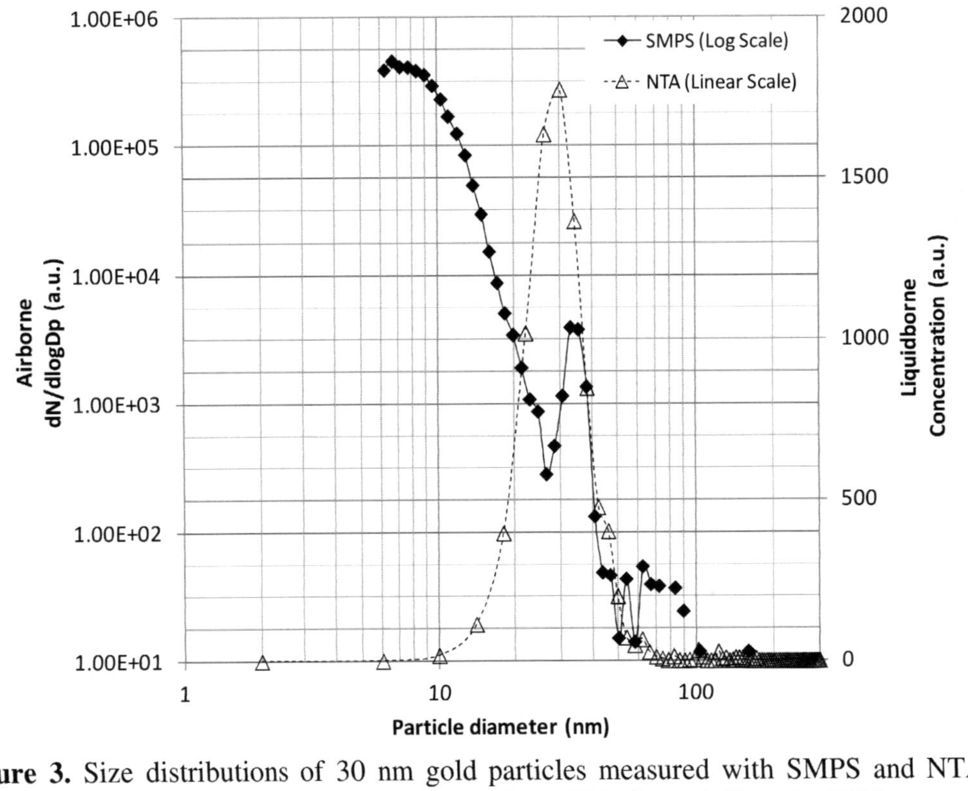

Figure 3. Size distributions of 30 nm gold particles measured with SMPS and NTA. The SMPS measurement is represented with resolution of 32 channels/decade. NTA measurement is represented with resolution of 4nm/channel.

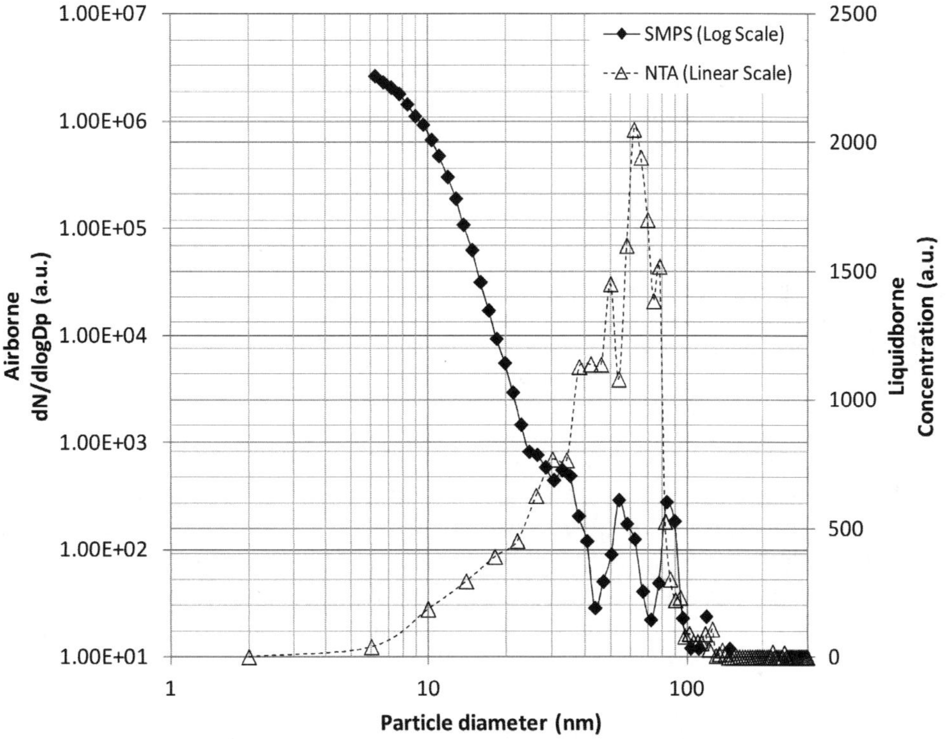

Figure 4. Size distributions of mixed 80, 50 and 30 nm gold particles with equal number concentration measured with SMPS and NTA. The SMPS measurement is represented with resolution of 32 channels/decade. NTA measurement is represented with resolution of 4nm/channel.

4. Air-borne/surface-borne correlation

As in the case for liquid-borne nanoscale particles, additional analyses can be performed by depositing nanoparticles from air-borne phase to surface-borne phase, or vice versa. Researchers have been using microscopy tools for characterizing air-borne nanoparticles collected onto a surface [11, 21-22]. Another example is the calibration of Condensation Particle Counters (CPC). Liu and Pui [23] introduced singly charged air-borne particles into an electrometer current sensor, where particles are removed by a filter, and related the particle concentration to the electrical current reading, air flowrate and elementary unit of charge. Tapered Element Oscillating Microbalance (TEOM) is a device for near real-time particle mass measurement by depositing the particles onto a surface and relating the mass to the change in vibrating frequency [24]. Yook et al. [25] reported a method to deposit the monodisperse air-borne particles onto a surface with controlled number and spot size. Their work has found application in surface inspection tool characterization. In the remaining section, we describe how depositing air-borne virus particles onto a gelatin filter can enhance collection efficiency with easy recovery process.

Virus aerosols, also known as air-borne viruses, are assemblies of solid particles or liquid droplets suspended in a gaseous medium that carry viruses. They are raising more and more public concern due to their involvement in infectious disease transmission [26]. For example, respiratory viruses such as influenza virus have been detected in ambient aerosols [27] as well as human expiratory particles [28]. The physical sizes of most viruses typically range from 25 to 300 nm. Liquid impingers are commonly used for virus aerosol sampling, which transfer air-borne particles to liquid-borne phase. However, they have $d_{50\%}$ (particle size with 50 % collection efficiency) at around 300 nm and their physical collection efficiency decreases rapidly to 10% at 100 nm [29]. Therefore, they have low physical collection efficiency for nanoscale particles. In addition, liquid impingers provide little information on the particle size of the collected virus aerosol samples, which governs virus transport in air, its deposition in the respiratory tract, and its control by filtration [30]. Although size-segregated samplers such as cascade impactors can also be used [31], they provide poor size resolution for submicron particles.

Recently, we developed a DMA and gelatin filter based sampling system to characterized laboratory-generated virus aerosols [32]. A schematic diagram of the experiment is shown in Figure 5. Virus aerosol particles generated using a Collison nebulizer were first size-classified (100-450 nm) by a DMA and then collected onto a piece of gelatin filter. The use of a DMA enables the measurement of submicron airborne virus with high size resolution. Gelatin filters have excellent physical collection efficiency, e.g. >96% for airborne MS2 [33] and high biological collection efficiency, e.g. help preserve virus infectivity of the collected virus aerosol [34-35]. More importantly, after virus aerosol sampling, the collected virus can be recovered easily by dissolving the gelatin filter in a liquid medium for subsequent virus infectivity assay. Therefore, there is no additional step needed to extract the collected virus from filters. The virus extraction efficiency from filters depends on virus-fiber surface interaction [36] and may be difficult to quantify. Compared with traditional virus aerosol sampling methods, the novel DMA-gelatin filter sampling method transfers airborne particles to surface-borne particles with high collection efficiency, high size resolution, and easy virus recovery.

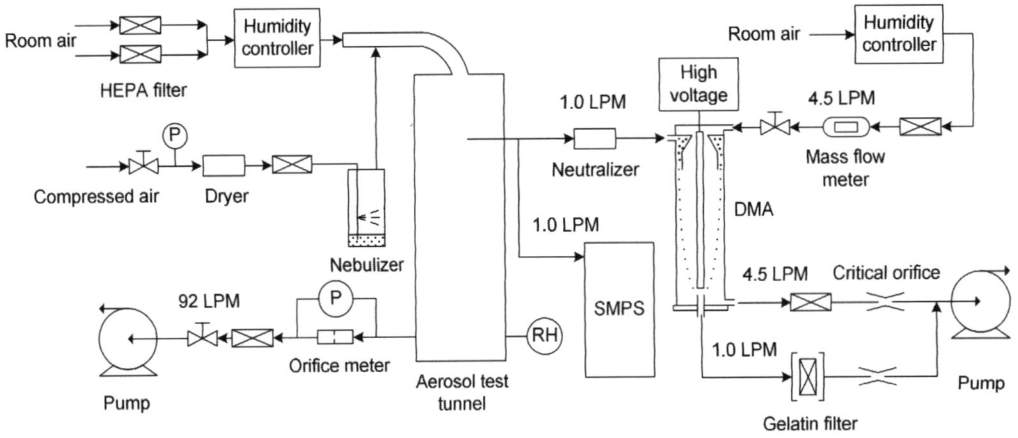

Figure 5. Schematic diagram of the experimental setup for the characterization of virus aerosols.

We used the DMA and gelatin filter based sampling system to measure four types of viruses, namely MS2 bacteriophage (MS2), transmissible gastroenteritis virus (TGEV), swine influenza virus (SIV), and avian influenza virus (AIV). Virus aerosol size distribution measured by real-time instruments such as SMPS does not necessarily represent virus infectivity distribution among polydisperse particles. We defined a particle size distribution particularly for particles carrying infectious virus, which was known as infectious virus size distribution, $IV(D_p)$ and was calculated as [32]:

$$IV(D_p) \equiv \frac{dC_{IV}(D_p)}{d\log_{10} D_p} \approx \frac{C_{IV,gel}(D_p)V_{gel}}{Q_{gel}t[\Delta \log_{10} D_p]} \frac{1}{f_{+1}(D_p)P(D_p)} \quad (1)$$

where C_{IV} is the concentration of airborne infectious virus, $C_{IV,gel}$ is the concentration of infectious virus recovered in the effluent from the gelatin filter, V_{gel} is the volume of eluent used to dissolve the gelatin filter, Q_{gel} is the air sampling flow rate through the gelatin filter, t is the sampling time, $\Delta log_{10}D_p$ is the logarithm of the width of the size interval of the DMA, f_{+1} is the fraction of singly positively charged particles, and P is penetration of particles through the DMA. For the four types of viruses tested, the infectious virus size distributions generally followed particle volume distribution rather than particle number distribution (Figure 6). The results show that there is strong correlation between airborne virus infectivity and particle size, suggesting the importance of particle size in viral disease transmission via the aerosol route.

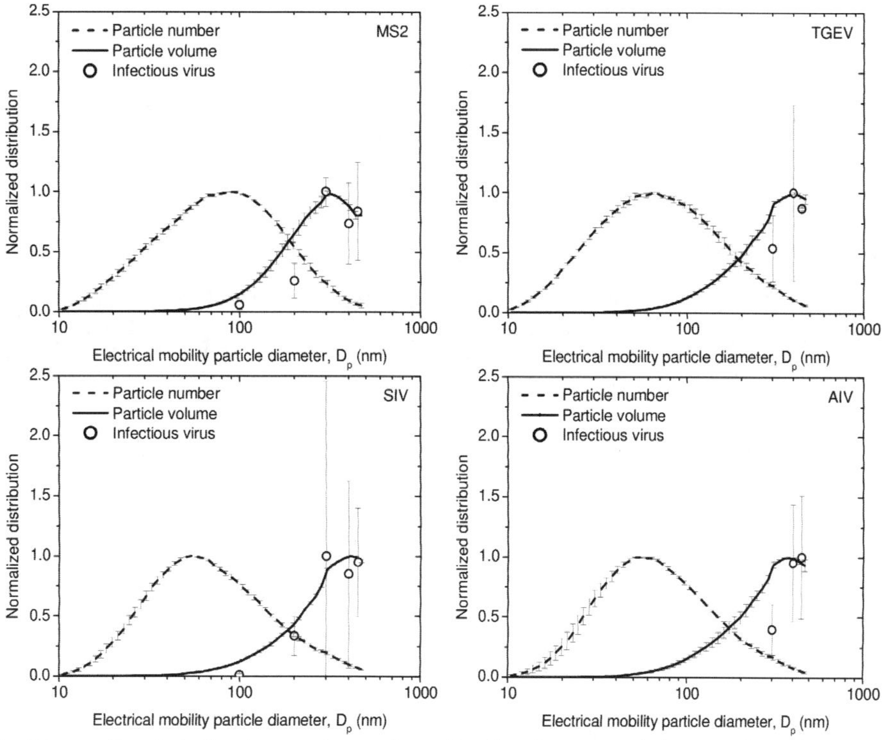

Figure 6. Normalized infectious virus size distribution, particle number distribution, and particle volume distribution for airborne MS2, TGEV, SIV, and AIV.

5. Summary

Nanoscale particles exist in air or liquid phases or on surfaces by deposition. Techniques for physico-chemical characterization of nanoparticles in these phases are therefore important in the environmental, health and safety studies of nanomaterials in support of sustainable nanotechnology. We describe experimental techniques that enhance characterization capability by transferring the nanoparticles from one dispersed phase to another. Figure 7 summarizes our work in bridging nanoscale particle measurements in different dispersed phases. Aerosolizing liquid-borne nanoparticles and measuring them with a SMPS improves the sizing resolution of liquid-borne particle measurements. By correlating particle concentrations in the air-borne and surface-borne phases, accurate calibration of both air-borne and liquid-borne instruments over a wide range of particle concentration may be achieved. Depositing air-borne and liquid-borne nanoparitcles on a surface for microscopy analysis enables more complete physical morphology and chemical analysis. We also demonstrated that the DMA-gelatin filter system is a sampling method that has high collection efficiency and allows easy recovery for air-borne viruses.

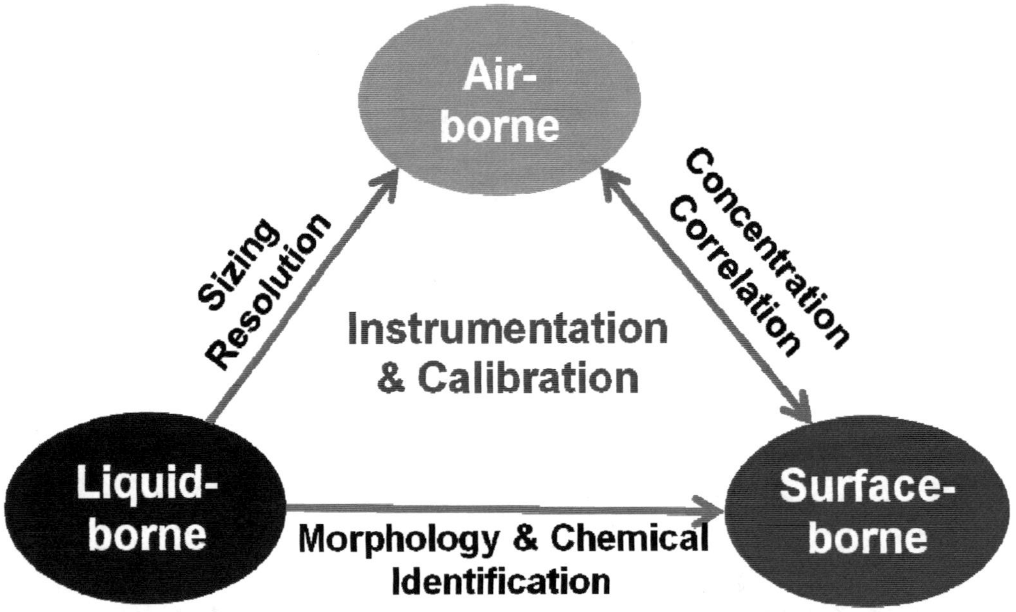

Figure 7. Relation of the measurements of nanoscale particles in air-borne, liquid-borne and surface-borne dispersed phases.

Acknowledgements

This work was supported by the NSF Grant (Award ID: 1236107) on "GOALIE: Unipolar Diffusion Charging of Spherical and Agglomerated Nanoparticles and its Application toward Surface-Area Measurement". Parts of this work were carried out in the Characterization Facility, University of Minnesota, a member of the NSF-funded Materials Research Facilities Network (www.mrfn.org) via the MRSEC program. The authors would like to thank Prof.

Chris Hogan's group for assisting us in beads milling experiments and Dr. Axel Zerrath of TSI Inc. for assisting in the experiments with the nanoparticle nebulizer.

References

[1] Roco MC 2005. *J Nanopart Res* **7** 707-712
[2] Colvin VL 2003. *Nat Biotechnol* **21** 1166-1170
[3] Maynard A, Aitken RJ, Butz T, Colvin V, Donaldson K, Oberdörster G, Philbert MA, Ryan J, Seaton A and Stone V 2006. *NATURE-LONDON-* **444** 267
[4] Hassellöv M, Readman JW, Ranville JF and Tiede K 2008. *Ecotoxicology* **17** 344-361
[5] Lee HS, Chae SK and Liu BYH 1989. *Particle & Particle Systems Characterization* **6** 93-99
[6] Filipe V, Hawe A and Jiskoot W 2010. *Pharm Res* **27** 796-810
[7] Mavrocordatos D, Perret D and Leppard GG 2007. *IUPAC Seriese on Analytical and Physical Chemistry of Environmental Systems* **10** 345
[8] Rogak SN, Flagan RC and Nguyen HV 1993. *Aerosol Sci Tech* **18** 25-47
[9] Binder C, Feichtinger C, Schmid HJ, Thürey N, Peukert W and Rüde U 2006. *J Colloid Interface Sci* **301** 155-167
[10] Zhang C, Thajudeen T, Larriba C, Schwartzentruber TE and Hogan Jr CJ 2012. *Aerosol Sci Tech* **46** 1065-1078
[11] Wang J, Kim SC and Pui DYH 2011. *J Nanopart Res* **13** 4565-4573
[12] Inkyo M, Tokunaga Y, Tahara T, Iwaki T, Iskandar F, Hogan Jr CJ and Okuyama K 2008. *Ind Eng Chem Res* **47** 2597-2604
[13] Park JY, Lim S, Han S and Park K 2011. *Aerosol Sci Tech* **45** 1010-1018
[14] Ling TY, Wang J and Pui DYH 2009. *Environ Sci Technol* **44** 774-779
[15] Wiedensohler A, Krämer M and Hansson HC 1991. *J Aerosol Sci* **22** S329-S330
[16] Lenggoro IW, Xia B, Okuyama K and de la Mora JF 2002. *Langmuir* **18** 4584-4591
[17] Lenggoro IW, Widiyandari H, Hogan CJ, Biswas P and Okuyama K 2007. *Anal Chim Acta* **585** 193-201
[18] Park JY, McMurry PH and Park K 2012. *Aerosol Sci Tech* **46** 354-360
[19] Ling TY, Wang J and Pui DYH 2011. *J Nanopart Res* **13** 5415-5424
[20] Greenwald R, Bergin MH, Carrico CM and Grant D 2005. *Environ Sci Technol* **39** 4967-4973
[21] Kim SC, Wang J, Shin WG, Scheckman JH and Pui DYH 2009. *Aerosol Sci Tech* **43** 1033-1041
[22] Shin WG, Wang J, Mertler M, Sachweh B, Fissan H and Pui DYH 2009. *J Nanopart Res* **11** 163-173
[23] Liu BYH and Pui DYH 1974. *J Colloid Interface Sci* **47** 155-171
[24] Pui D and Chen D 2001. *Air sampling instruments* 377-414
[25] Yook SJ, Fissan H, Engelke T, Asbach C, van der Zwaag T, Kim JH, Wang J and Pui DYH 2008. *J Aerosol Sci* **39** 537-548
[26] Tellier R 2009. *J R Soc Interface* **6** S783-S790
[27] Yang W, Elankumaran S and Marr LC 2011. *J R Soc Interface* **8** 1176-1184
[28] Fabian P, McDevitt JJ, DeHaan WH, Fung ROP, Cowling BJ, Chan KH, Leung GM and Milton DK 2008. *PLoS One* **3** e2691

[29] Hogan Jr C, Kettleson E, Lee MH, Ramaswami B, Angenent L and Biswas P 2005. *J Appl Microbiol* **99** 1422-1434
[30] Zuo Z, Kuehn TH and Pui DYH 2012 *Am J Infect Control*, in press. doi:10.1016/j.ajic.2012.01.010
[31] Appert J, Raynor PC, Abin M, Chander Y, Guarino H, Goyal SM, Zuo Z, Ge S and Kuehn TH 2012. *Aerosol Sci Tech* **46** 249-257
[32] Zuo Z, Kuehn TH, Verma H, Kumar S, Goyal SM, Appert J, Raynor PC, Ge S and Pui DYH 2012. *Aerosol Sci Tech* **47** 373-382
[33] Burton NC, Grinshpun SA and Reponen T 2007. *Ann Occup Hyg* **51** 143-151
[34] Grinshpun SA, Adhikari A, Honda T, Kim KY, Toivola M, Rao KSR and Reponen T 2007. *Environ Sci Technol* **41** 606-612
[35] Verreault D, Moineau S and Duchaine C 2008. *Microbiol Mol Biol Rev* **72** 413-444
[36] Zuo Z, Abin M, Chander Y, Kuehn TH, Goyal SM, Appert J, Raynor PC, and Pui DHY 2012. Comparison of influenza virus recovery from non-woven fabrics after spike tests vs. aerosol challenge tests. *Influenza Other Respi Viruses*, under review

Metal impurities provide useful tracers for identifying exposures to airborne single-wall carbon nanotubes released from work-related processes

Pat E Rasmussen[1,2], Innocent Jayawardene[1], H David Gardner[2], Marc Chénier[1,2], Christine Levesque[1] and Jianjun Niu[1]

[1]Health Canada, 50 Colombine Driveway, Ottawa, ON Canada, K1A 0K9
[2]University of Ottawa, 140 Louis-Pasteur, Ottawa, ON Canada, K1N 6N5

E-mail: pat.rasmussen@hc-sc.gc.ca

Abstract. This study investigated the use of metal impurities in single-wall carbon nanotubes (SWCNT) as potential tracers to distinguish engineered nanomaterials from background aerosols. TEM and SEM were used to characterize parent material and aerosolized agglomerates collected on PTFE filters using a cascade impactor. SEM image analysis indicated that the SWCNT agglomerates contained about 45% amorphous carbon and backscatter electron analysis indicated that metal impurities were concentrated within the amorphous carbon component. Two elements present as impurities (Y and Ni) were selected as appropriate tracers in this case as their concentrations were found to be highly elevated in the SWCNT parent material (% range) compared to ambient air particles (μg/g range), and background air concentrations were below detection limits for both elements. Bioaccessibility was also determined using physiologically-based extractions at pH conditions relevant to both ingestion and inhalation pathways. A portable wet electrostatic precipitation system effectively captured airborne Y and Ni released during sieving processes, in proportions similar to the bulk sample. These observations support the potential for catalysts and other metal impurities in carbon nanotubes to serve as tracers that uniquely identify emissions at source, after an initial analysis to select appropriate tracers.

1. Introduction

There is an urgent need for practical and cost-effective methods to distinguish process-related nanoparticle emissions from background aerosols in the workplace.[1] While direct readings of particle number, mass, size distribution, and surface area concentrations are recommended as a first step in identifying workplace exposures, these instruments do not provide the specific information (e.g. morphology or chemical composition) needed to distinguish nanomaterials from background[2,3]. Wide spatial and temporal variability of background aerosols make it especially difficult and time consuming to identify process-related emissions using direct-reading instruments.[4] Elemental carbon methods have been recommended[5] for measurement of occupational exposures to carbon nanotubes and carbon nanofibers (using NIOSH Method 5040 or an equivalent method), but such methods have potential limitations in situations where the background aerosol is also largely composed of elemental carbon (e.g., diesel soot).

The present study investigated the use of metal impurities in single-wall carbon nanotubes (SWCNT) as airborne tracers to distinguish process-related emissions from background. Electron microscopy and elemental analyses provided morphological and chemical information about the bulk parent material and aerosolized samples, which supported the concept of using catalysts and other impurities as airborne tracers of carbon nanotubes for workplace monitoring. Electrostatic precipitation and filter-based approaches were used for sampling aerosolized SWCNT particles. Estimates of metal bioaccessibility, a relevant parameter in risk assessment, were calculated for both inhalation and ingestion pathways.

2. Materials and Methods

The SWCNT material used in this study (Aldrich no. 698695; batch MKBB3788) was purchased from Sigma-Aldrich Co. (Gillingham, UK). This product was manufactured by arc discharge, with 40-60 wt% carbon content, 30-35% metal content, average diameter 1.4 nm, and bundled dimensions of 2 to 10 nm in diameter and 1 to 5 μm in length (information provided by Sigma-Aldrich).

2.1. Physical characterization using TEM and SEM

Test samples of SWCNT dispersed in 99% ethanol were placed on 300 mesh copper grip coated with holey carbon film, then dried for 12 hr under vacuum (10^{-8} torr) generated by a turbo pump (Varian V-81M Turbo Station). Diameter and cell wall thickness of aerosolized and parent SWCNT samples were measured using a JEM-2100F Transmission Electron Microscope (TEM; JEOL Industry) at high resolution (± 0.2 nm.) and length was measured using a JSM-7500F Scanning Electron Microscope (SEM; JEOL Industry). Secondary electron imaging (SEI) was used to examine morphology of SWCNT agglomerates loaded on filter substrates (described below). Transmission electron diffraction (TED) mode was used for image analysis. Backscatter electron (BSE) mode was used to locate elements heavier than carbon (catalysts) in the amorphous carbon component of the agglomerates and x-ray energy dispersive spectrometry (EDS) was used to identify these heavier elements (>1%). SWCNT particles in an oil emersion were examined using a Carl Zeiss polarizing microscope equipped with a 100 x 10 objective.

2.2. Collection of SWCNT on filters using cascade impactor

A dry powder aerosol generator (TSI Model 3400A) was employed to aerosolize the SWCNT for subsequent size fractionation and collection on filters using a cascade impactor. Approximately 10 cc of SWCNT were required to fill the reservoir. Samples of SWCNT that had passed through the aerosolizer were collected from the inside wall of the elutriator for morphological analysis. A micro-orifice uniform deposit impactor (NanoMoudi-II, Model 125B, MSP Co., MN, USA), which uses inertial impaction to separate particles into discrete fractions according to their aerodynamic diameter, was equipped with pre-weighed PTFE filters (47 mm Teflo® filter with PMP support ring, Pall Córp.). The aerosolized SWCNT entered the impactor at a flow rate of 10 L/min, for three 2-hr runs and one 3.5-hr run, using 14 filters per run (13 stages plus an after filter) at nominal cut sizes ranging from 10 micron to 10 nm. PTFE filters were conditioned before pre-weighing and post-weighing for 24 hr in a controlled environmental chamber at a constant air temperature of 21°C (± 0.5°C) and constant relative humidity (RH) of 40% (± 0.2%), using Health Canada's buoyancy-corrected gravimetric analysis facility[6] (Archimedes M3™). The filters were weighed using a Mettler UMX2 microbalance (readability of ± 0.1 μg) after removing static electricity by passing each filter between two polonium-210 strip deionizers.

2.3. Collection of SWCNT using wet electrostatic precipitation

An Aerosol-to-Liquid Particle Extraction System (ALPXS; Meinhard Glass Products, Golden, CO) was used to monitor airborne SWCNT released from a sieve shaker. After entering the ALPXS air intake (~300 L/min), particles are first negatively charged by a fine ionization wire (-8,000V DC) and then collected and concentrated in de-ionized water that continuously re-circulates around a counter electrode (+8,000V DC).[7] A mechanical sieve shaker (Ro-Tap RX-94; W.S.Tyler) and the ALPXS were placed inside a self-ventilating fume hood, with the ALPXS air sampling intake located about 25 cm from the source. A mass of about 10 mg SWCNT was maintained on the sieve pan throughout the shaking tests. The rate of shaking was increased incrementally, with 30 min monitoring at each increment, for a total of 4 monitoring tests. The ALPXS aqueous samples were then prepared for elemental analysis (described below). Before sieving was initiated, background air concentrations of tracer elements in the laboratory and other areas of the building were below limits of quantification (≤ 12 ng/m^3 for Ni and ≤ 7 ng/m^3 for Y)

2.4. Elemental Analysis using ICP-MS

Elemental concentrations were determined in the sample digests using a Perkin Elmer Sciex Elan DRC II ICP-MS operated in standard mode and equipped with a Meinhard Concentric Quartz Nebuliser, cyclonic spray chamber, and platinum skimmer/sampler cones. Internal standards were Germanium, Indium and Rhenium. Total elements were determined in bulk parent material using two approaches: microwave digestion using 20 mg test samples in nitric acid[8] and 2-hr ultrasonic digestion using 2 mg test samples in nitric/hydrofluoric acid mixture based on a method previously developed for air filter samples.[9] Bioaccessibility was estimated using two physiologically-based extraction techniques (PBET): a weak hydrochloric acid extraction (0.07 M; pH 1.5) to simulate gastric acid,[10] and an ammonium acetate buffer (0.01 M; pH 7) to simulate the neutral lung environment.[11] The extraction tubes were placed in a covered hot water bath at body temperature (37°C) for 2 hr (1 hr with agitation and 1 hr without agitation) for both PBET methods.[10,11] Samples collected by ALPXS were pre-concentrated by evaporation on a hot block, and then digested using nitric acid and ultrasonic digestion. High purity acids (Seastar Chemical Inc, Sidney BC, Canada) and high purity standard stock solutions (High-Purity Standards, SC, USA) were used for all stages of sample preparation and analysis.

3. Results and Discussion

All samples of SWCNT examined by SEM/TEM showed a mixture of bundled nanotubes (individual nanotubes were rare) and irregularly-shaped agglomerates consisting of nanotube structures tangled with amorphous carbon/catalyst (figure 1). Both SEM micrographs in figure 1 cover the same area of dry-mounted sample. BSE mode (right image) was used to investigate the location of elements heavier than carbon within the agglomerates. Elemental impurities (in this case Y and Ni, as identified by EDS) appeared as bright spots on the BSE micrograph, and were associated almost exclusively with the amorphous carbon component (figure 1).

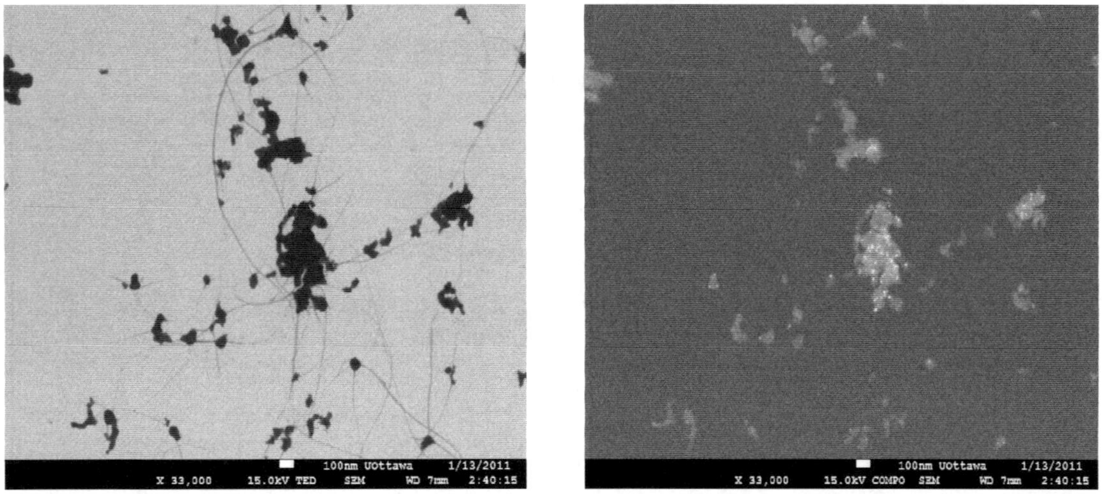

Figure 1. Two SEM micrographs of same area of SWCNT sample (Aldrich). Left image in TED mode shows distribution of amorphous carbon/catalyst and bundled SWCNT. Right image in backscatter mode shows location of metal catalysts (Ni, Y) as bright spots associated with amorphous carbon. Scale bar = 100 nm.

3.1. Selection of Y and Ni as suitable tracers

Analysis of bulk SWCNT (Aldrich 698695) by ultrasonic digestion and ICP-MS confirmed elevated concentrations of Y (25,916 µg/g = 2.6%) and Ni (132,745 µg/g = 13%), as listed in table 1. The next highest concentrations of elemental impurities were Al (492 µg/g) and Fe (475 µg/g).The presence of Y and Ni at concentrations that were orders-of-magnitude higher other elemental impurities indicated that Y and Ni were used as catalysts in the manufacture of the Aldrich SWCNT.

Concentrations of the two potential tracer elements (Y and Ni) and other impurities (Al and Fe) were analyzed in bulk samples of outdoor air particles that had been collected in a previous study.[11] Concentrations of Y (6.6 µg/g) and Ni (40 µg/g) in ambient air particles were thousands of times lower than in the SWCNT. In contrast, concentrations of Al (8500 µg/g) and Fe (17,700 µg/g) in ambient air particles were several times higher than in the SWCNT. Thus, elemental ratios in SWCNT, such as Y/Al (52), Y/Fe (54), Ni/Al (270) or Ni/Fe (279), provided unambiguous discrimination from the ambient air particles (in which these ratios are <0.01). Moreover, Y and Ni were undetectable in indoor air monitored in the sieving laboratory and two other locations within the building. Therefore, Y and Ni qualified as appropriate tracers to identify releases of Aldrich SWCNT for this monitoring situation.

Metallic components of inhaled particles have the capacity to generate an increase in reactive oxygen species in the respiratory system.[12] As the agglomerated forms of SWCNT, containing nanotube bundles tangled with amorphous carbon and metal impurities (figure 1), are likely candidates for inhalation and/or ingestion exposure, estimates of the bioaccessibility of metal impurities in SWCNT are relevant for risk assessments. Bioaccessibilities of the catalyst elements were determined using physiologically-based extractions at pH conditions relevant to inhalation and ingestion pathways (table 1).

Table 1. Concentrations (mg/g) of yttrium (Y) and nickel (Ni) in Aldrich SWCNT determined using ICP-MS and three extraction techniques. Concentrations shown are the mean of three replicates for each test, expressed as dry wt. PBET = physiologically based extraction technique.

Element	Total Extraction	Neutral PBET	Gastric PBET	Bioaccessibility	
				Inhalation Pathway [a]	Ingestion Pathway [b]
^{89}Y	25.9	4.50	12.7	17%	49%
^{60}Ni	133	14.4	51.7	11%	39%

[a] neutral PBET/total x 100
[b] gastric PBET/total x 100

The microwave digestion method resulted in an incomplete extraction (not included in table 1), attributed to inefficient contact between particles and acid caused by the higher solid:liquid ratio and absence of agitation in the microwave method.

3.2. Characterization of aerosolized agglomerates

SWCNT particles that had passed through the aerosolizer (collected before entering the MOUDI) were characterized for morphology by SEM/TEM to determine if aerosolization had caused a change in morphology compared to the parent material. Post-aerosolization measurements (provided in table 2) indicated that aerosolization had no observable impact on morphology. Nanotube diameter of the parent material (mean=1.9 ± 0.6 nm; n=15 measurements) was not significantly different from the aerosolized sample (mean=1.7 ± 0.2 nm; n=27 measurements). Likewise, there was no difference in wall thickness between the two samples (0.4 ± 0.1 nm for both aerosolized and non-aerosolized). Bundles of both parent material and aerosolized SWCNT ranged from 5-10 nm in diameter and were well-formed, typically containing from 3 to 20 individual nanotubes. All nanostructure measurements were normally distributed (both Kolmogorov-Smirnov and Shapiro-Wilk normality tests $p > 0.2$).

Table 2. Physical characterization of SWCNT after aerosolization, using TEM to determine diameter and wall width and SEM to determine length (n= number of measurements).

	Diameter (nm) n=27	Wall Width (nm) n=27	Length SEM (μm) n=31
Mean	1.7 (sd 0.2)	0.4 (sd 0.1)	2.13 (sd 1.08)
Median	1.8	0.4	2.15
Range	1.3 – 2.0	0.3 – 0.5	0.32 – 4.70

SEM micrographs of aerosolized SWCNT agglomerates collected on PTFE filter samples collected on two MOUDI stages (nominal cut sizes 1 μm and 3.2 μm) are shown in figures 2 and 3. These agglomerates appear as scattered clusters of bundled nanotubes and amorphous carbon/catalyst material against the backdrop of PTFE fibers. These micrographs confirm that aerosolized SWCNT agglomerates include amorphous carbon/catalyst material. Such evidence of the presence of amorphous carbon/catalyst material in aerosolized nanomaterials substantiates the concept of using catalyst elements as airborne tracers.

Overall, the aerosol generator-MOUDI system yielded very lightly loaded filter samples, required labour-intensive cleaning between runs to minimize carry-over, and resulted in inconsistent particle size distributions between runs. Gravimetric analysis detection limits (≈ 2 μg) were generally exceeded in filter samples from the 32 nm to 3.2 μm range, but many samples were below detection limits and between-run reproducibility was poor. These factors, combined with the impact of agglomeration on particle size distribution, suggested little advantage in using a MOUDI for monitoring SWCNT in the workplace, compared to simpler 37 mm cassettes recommended previously.[2,3] However, the aerosol generator-MOUDI system did prove to be useful for the preparation of filter samples of SWCNT agglomerates suitable for examination by SEM (i.e., minimal particle overlap) as shown in figures 2 and 3.

Figure 2. Clusters of aerosolized SWCNT agglomerates collected on PTFE filter (MOUDI cut size 1 μm); scale bar = 10 μm; particle mass 51 μg.

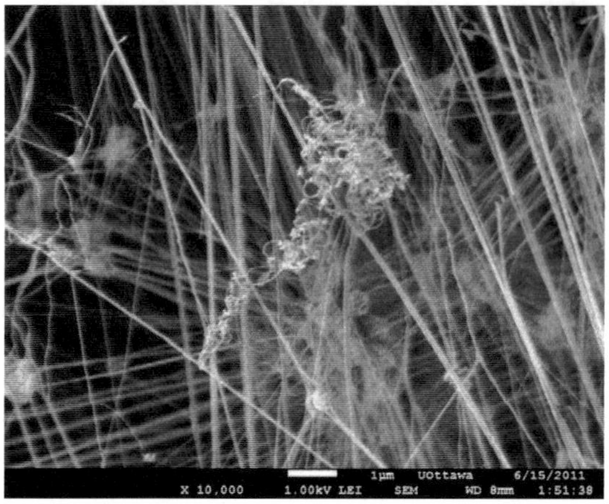

Figure 3. Close-up of SWCNT agglomerate impacted on PTFE filter fibres (MOUDI cut size 3.2 μm); scale bar = 1 μm, particle mass 22 μg.

3.3. Image analysis

Significant amorphous carbon/catalyst material was observed in all SWCNT agglomerates in all size ranges. Nano- to micro-scale SWCNT agglomerates were observed using SEM/TEM (figures 1 to 3), and larger agglomerates in the 10 to 100 μm size range (along the long axis) were observed using a light microscope. In order to quantify the proportion of amorphous carbon/catalyst compared to nanotube structures in the agglomerates, SEM images were classified into intervals corresponding to the grey-scale pixel value of individual points. Figure 4 shows the original SEM micrograph of an area of dry-mounted SWCNT sample, plus the classified image distinguishing nanotube structures, amorphous carbon / catalyst material, and background. The relative proportion of nanotube structures to amorphous carbon/catalyst was calculated by completing an area analysis for each of 19 images (including figure 4) randomly selected along transects of the mounted sample. Excluding background material, the average proportions were: 55% nanotube structures to 45% amorphous carbon/catalyst (in two-dimensional view).

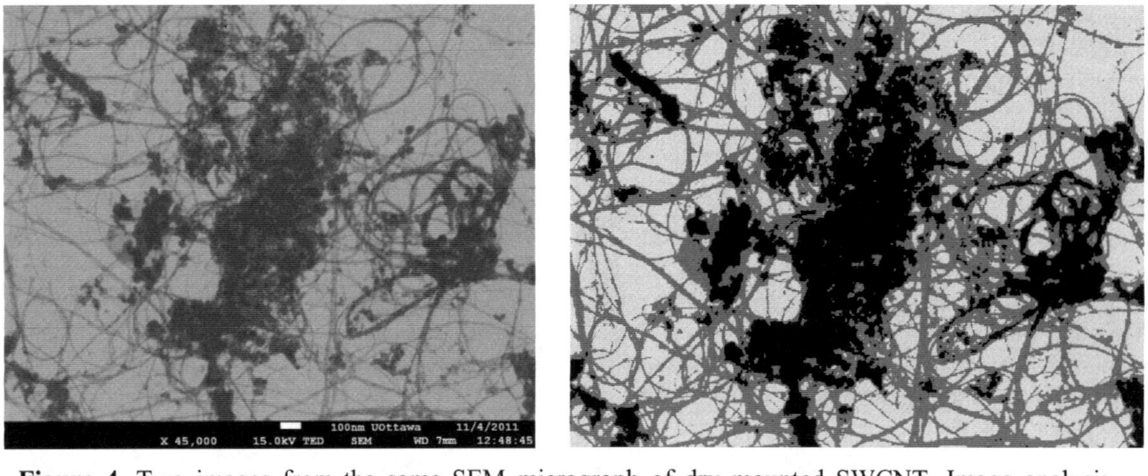

Figure 4. Two images from the same SEM micrograph of dry mounted SWCNT. Image analysis (right) shows how the TED micrograph (left) is classified as SWCNT (grey) and amorphous carbon/catalyst (black). Scale bar = 100 nm

3.4. Monitoring airborne tracers using electrostatic precipitation

After electrostatic precipitation, SWCNT particles were visible as black supermicrometre clusters in the aqueous ALPXS sample. Figure 5 shows results of monitoring Y and Ni tracers of SWCNT released from the sieve shaker as the rate of shaking was increased incrementally. The particle release rate increased as the shaking rate increased, indicated by an increase in airborne Y concentration from 27 ng/m^3 to 1839 ng/m^3 (figure 5a), and in airborne nickel concentration from 151 ng/m^3 to 8794 ng/m^3 (figure 5b). These air concentrations are consistent with table 1 which indicates that Y and Ni occur as impurities in the SWCNT parent material in approximately the same proportion (total concentration Y:Ni ratio $\approx$ 1:5). It was established (figures 1 to 4) that a large percentage of aerosolized agglomerate is amorphous carbon/catalyst material. The observed increase in airborne Y and Ni concentrations (figure 5) is consistent with the release of amorphous carbon/catalyst material as the SWCNT agglomerates disaggregate.

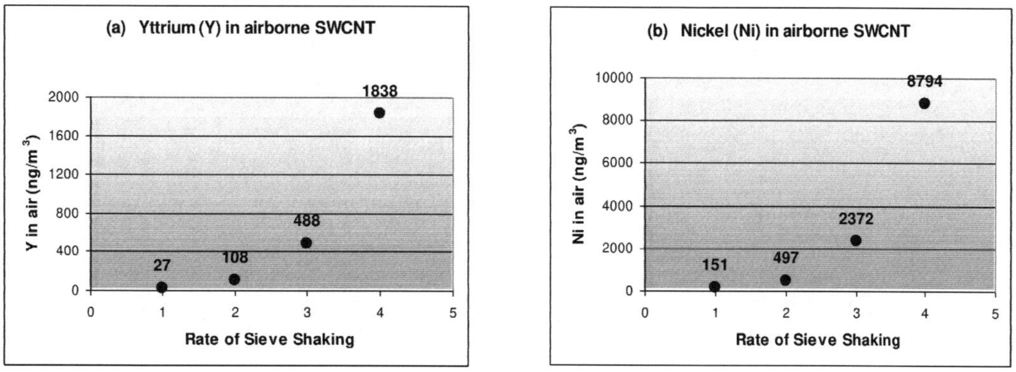

Figure 5. Yttrium (5a) and nickel (5b) tracers of airborne SWCNT released in an R&D lab during sieving. Rate of shaking was increased incrementally with 30 min monitoring at each increment, using wet electrostatic precipitation and ICP-MS determination. Background air concentrations of both elements were below limits of quantification.

Determination of metal impurities in SWCNT (as shown in figure 5) as an alternative to elemental carbon (e.g. NIOSH Method 5040) could improve detection sensitivity significantly. Carbon content of the SWCNT parent material used in this study ranges from 40 to 60% and elemental impurities are present at 13% for Ni and 2.6% for Y. Based on upper limits of quantification (LOQ) for detecting airborne SWCNT using the above electrostatic precipitation method (12 ng/m^3 for Ni and 7 ng/m^3 for Y), the equivalent LOQ estimated for elemental carbon ranges from 0.04 to 0.16 μg/m^3. This is one- to two- orders of magnitude lower than that of NIOSH Method 5040 (7 μg/m^3).[5] Note that LOQs for elemental tracer methods will vary depending on the particle sampling approach (e.g. filter media), sample handling, ICP-MS equipment and ICP-MS personnel.

4. Summary and Conclusions

The goal of this research is to develop practical monitoring techniques within the scope and budget of typical occupational health and safety investigations[4]. A promising strategy to identify process-related releases of carbon nanotubes is to monitor elemental tracers using electrostatic precipitation, filter cassettes, or possibly surface wipes, followed by elemental analysis of the collected samples by ICP-MS. Selection of appropriate tracers requires an initial analysis of (1) the bulk parent nanomaterial, as different elements (e.g. Co, Cu, Fe, W, Ta, Mo) are used as catalysts and anodes and these are unique to each manufacturing process, and (2) the background environment to ensure that the proposed tracer does not arise from local sources. In this study, Y and Ni proved to be suitable tracers because they were undetectable in background air, and their concentrations in SWCNT were orders of magnitude higher than in background particles. Elemental ratios such as Y/Al, Y/Ni, Ni/Al, or Ni/Fe unambiguously distinguished SWCNT from ambient aerosols. Wet electrostatic precipitation was an effective method for capturing airborne Y and Ni released during sieving processes. SEM analysis indicated that metal catalysts occur predominantly in the amorphous carbon component, which constitutes about 45% of SWCNT agglomerates. Examination of SWCNT agglomerates captured on PTFE filters demonstrated that amorphous carbon/catalyst material remains a significant component after aerosolization. The above lines of evidence support the concept of using metal impurities as airborne tracers of carbon nanotubes for workplace monitoring.

Acknowledgments
This study was funded by the Chemicals Management Plan Nanotechnology Initiative (Health Canada). SEM and TEM measurements were made at the Centre for Catalysis Research and Innovation (CCRI) at University of Ottawa. Sincere thanks go to Dr. Yun Liu (CCRI) for valuable assistance, and to Drs. Yong-Lai Feng and Xinghua Fan (Health Canada), and members of the Nanosafe International Scientific Committee, for helpful review comments.

References
[1] Kuhlbusch T A J, Asbach C, Fissan H, Göhler D and Stintz M 2011 Nanoparticle exposure at nanotechnology workplaces: A review *Part. Fibre Toxicol.* **8** 22
[2] OECD 2009 Emission assessment for the identification of sources and release of airborne manufactured nanomaterials in the workplace: compilation of existing guidance ENV/JM/MONO 16
[3] Methner M, Hodson L and Geraci C 2010 Nanoparticle emission assessment technique (NEAT) for the identification and measurement of potential inhalation exposure to engineered nanomaterials — Part A. *J. Occup. Environ. Hyg.* **7** 127-32
[4] CSA 2012 Nanotechnologies — Exposure control program for engineered nanomaterials in occupational settings. Canadian Standards Association Z12885 Mississauga Canada 122p
[5] NIOSH Current Intelligence Bulletin Occupational Exposure to Carbon Nanotubes and Nanofibers. November 10 Draft. National Institute for Occupational Safety and Health. National Institute for Occupational Safety and Health, 149p
[6] Rasmussen P E, Gardner H D and Niu J 2010 Buoyancy-corrected gravimetric analysis of lightly loaded filters *J. Air Waste Manage.* **60** 1065-77
[7] Wickersham Jr C E Air Quality Analyzer USA Patent US2012/0048748-A1 http://www.freepatentsonline.com/20120048748.pdf (Accessed Oct 11, 2012)
[8] Decker J E et al, 2009 Sample preparation protocols for realization of reproducible characterization of single-wall carbon nanotubes *Metrologia* **46** 682-92
[9] Niu J, Rasmussen P E, and Chénier M. 2012 Ultrasonic dissolution for ICP-MS determination of trace elements in lightly loaded airborne PM filters *Int. J. Envron. Anal. Chem.* DOI:10.1080/03067319 672980
[10] Dodd, M, Rasmussen, P, and Chénier, M 2012 Comparison of two *in vitro* extraction protocols for assessing metals' bioaccessibility using dust and soil reference materials *Human Ecol. Risk Assess.* DOI:10.1080/10807039.2012.719381
[11] Niu J, Rasmussen P E, Hassan N, Vincent R. 2010 Concentration distribution and bioaccessibility of trace elements in nano and fine urban airborne particulate matter: influence of particle size *Water Air and Soil Pollut.* **213** 211-25
[12] Pacurari M, Castranova V, and Vallyathan V 2010 Single- and multi-wall carbon nanotubes versus asbestos: are the carbon nanotubes a new health risk to humans? *J. Toxicol. Environ. Health Part A: Current Issues* **73** 378-95

Risk assessment of released cellulose nanocrystals – mimicking inhalatory exposure

C Endes[1], S Müller[2], O Schmid[3], D Vanhecke[1], EJ Foster[2], A Petri-Fink[1], B Rothen-Rutishauser[1], C Weder[2] and MJD Clift[1]

[1] Bionanomaterials, Adolphe Merkle Institute, University of Fribourg, Route de l'Ancienne Papeterie CP 209, 1723 Marly, Switzerland.

[2] Polymer Chemistry and Materials, Adolphe Merkle Institute, University of Fribourg, Route de l'Ancienne Papeterie CP 209, 1723 Marly, Switzerland.

[3] Comprehensive Pneumology Centre, Institute of Lung Biology and Disease, Helmholtz Zentrum Muenchen, Neuherberg, Germany.

Email: carola.endes@unifr.ch

Abstract. Cellulose nanocrystals (CNCs) exhibit advantageous chemical and mechanical properties that render them attractive for a wide range of applications. During the life-cycle of CNC containing materials the nanocrystals could be released and become airborne, posing a potential inhalatory exposure risk towards humans. Absent reliable and dose-controlled models that mimic this exposure *in situ* is a central issue in gaining an insight into the CNC-lung interaction. Here, an Air Liquid Interface Cell Exposure system (ALICE), previously designed for studies of spherical nanoparticles, was used for the first time to establish a realistic physiological exposure test method for inhaled fiber shaped nano-objects; in this case, CNCs isolated from cotton. Applying a microscopy based approach the spatially homogenous deposition of CNCs was demonstrated as a prerequisite of the functioning of the ALICE. Furthermore, reliability and controllability of the system to nebulise high aspect ratio nanomaterials (HARN, *e.g.* CNCs) was shown. This opens the potential to thoroughly investigate the inhalatory risk of CNCs *in vitro* using a realistic exposure system.

1. Cellulose nanocrystals: composition and applications

Cellulose is the most abundant organic polymer on earth [1]. Individual cellulose molecules consist of linear chains, each of about 10,000 linked D-glucose units, and are the main component produced by condensation reactions of glucose during photosynthesis within the primary cell wall of green plants [1, 2]. Cellulose microfibrils contain both a crystalline and an amorphous fraction. The repeated crystalline domains that are formed via hydrogen bonds are disrupted by disordered regions of cellulose chains, which is the amorphous part. The ratio between the two fractions varies by species specificity, although in each case forms a natural composite material that provides a highly rigid structure (*e.g.* plant cell walls) [1]. The amorphous part can be degraded via controlled hydrolysis with mineral acids creating high aspect ratio cellulose nanocrystals (CNCs). Depending upon the source (*e.g.* cotton, soft wood pulp, rice and banana husks, bacteria, tunicates) rod shaped crystals with typical diameters of 5-50 nm and a length of 100-2,000 nm can be obtained [3-7]. Thus, CNCs represent high aspect ratio nanoparticles (HARNs) with an aspect ratio typically between 10 and 100 [8].

These characteristics, together with outstanding material properties, *i.e.* high strength and stiffness (elastic modulus ~80-150 GPa) give rise to the potential use of CNCs in many applications, including for example high-performance polymer nanocomposites [9], mechanically adaptive materials [10], membranes for water purification [11] and photonic films [12]. Overall CNCs may be viewed as an oil-independent, lightweight and inexpensive alternative to carbon nanotubes (CNTs) [8].

2. Opportunities and risks of CNCs

Due to their low cost and ease of availability, combined with their properties and the possibility to simply modify the surface, CNCs have gathered increased attention within the field of material science in the past decades [2]. Taking into consideration the high potential for their industrial exploitation, it is essential to look into the potential risk posed by this nanomaterial [13].

The life-cycle of CNCs that are produced from different sources and embedded into polymers to form composites for specific applications includes the isolation of CNCs, compounding with polymers, incorporation into a product, processing and eventually their disposal. During these steps the materials undergo mechanical mixing processes as well as possible finishing steps such as sanding, drilling, abrasion and degradation. Thus, contained CNCs may be released as single nanocrystals, small agglomerates or even composite dust into the air and therefore potentially reach the human body via inhalation [14, 15]. There are three main exposure routes for nano-objects into the human body that highlight occupational, consumer related and environmental exposure: via the skin, the gastrointestinal tract and the respiratory tract. The lung, due to its large surface area is widely accepted as the primary entry route for ambient particulate materials into the human body [16]. Therefore, CNCs pose not only an occupational but also a consumer related and an environmental exposure risk [17]. One reason for a thorough investigation into the potential inhalatory risk of CNCs is due to their similarity to CNTs, especially in terms of their physico-chemical characteristics. Previously it has been reported that long, straight and stiff CNTs that were injected into the abdomen of mice caused severe adverse effects (*i.e.* inflammatory granulomas), which highlighted the discussion as to whether or not the 'asbestos-like effects' reported should be considered for any fiber shaped particle in the nano-dimension [18]. In accordance with previous fiber toxicology research that formed the 'fiber paradigm' (*i.e.* fibers must be long, thin and biopersistent) [19, 20], it has emerged in recent years that to fully understand the potential advantages of nanofibers, the interaction of HARN with biological systems must be taken into consideration.

In addition, it should also be highlighted that there is considerable investigation into the adverse health effects with humans following exposure to cotton dust. Workers of cotton mills, weaving and soft tissue paper-producing industries have been shown to suffer from reduced lung function, airway obstruction, 'flu-like' symptoms, also called byssinosis (brown lung disease), eye and skin problems due to exposure to cotton dust when compared with silk mill workers [21, 22]. A distinction of the observed health effects within occupational settings induced from the different fractions of cotton dust, including cotton fibers, is currently limited. Furthermore the role of bacteria, fungi [23] and related contents such as endotoxin [24, 25] and spores existing in the air due to growing on the cotton material is under debate for the determined occupational related health problems. These aspects indicate that not only the shape and size of CNCs must be taken into consideration but also the material itself (cotton), its composition and purity, when assessing the potential associated adverse effects of CNCs.

Studies for the release kinetics of CNCs from nanocomposites do currently not exist, since these materials are not yet produced at a large scale. In the case of CNTs, however, which are at a state of mass production, the potential aerosol exposure to carbon nanofibers and multiwalled carbon nanotubes (MWCNTs) during production was shown in several cases due to poorly controlled transfer and bagging of the material or within a blending laboratory [26, 27]. Wohlleben et al. investigated the potential release of nanofillers such as CNTs and silica from different polymer matrices [28]. The authors were unable to show, however, that a countable fraction of single CNTs are released into the air following different abrasion scenarios. In contrast, Schlagenhauf et al. showed the release of CNTs in an abrasion scenario with CNT nanocomposites, freeing single CNTs as well as agglomerates in the size range of 13 nm up to 20 µm [29].

Therefore, according to the potential heightened use and broad interest in the application of CNCs in nanocomposites, due to their outstanding characteristics and renewable nature, their potential health risk must be investigated. The comparability of CNTs and CNCs in terms of dimensions, stiffness and application justifies the need for a thorough investigation of the potential impact of CNCs on biological systems in general and on the human respiratory system in particular.

2.1. In vitro and in vivo interactions of CNCs: an evaluation of recent findings

Up to now scientists have mainly focussed upon the improvement of protocols for CNCs in terms of their mechanical, thermal or chemical characteristics [5, 30-34]. Risk assessment of any CNC-based nanomaterial or CNCs themselves has so far been limited. Despite this, due to the plethora of applications associated with CNCs, research is beginning to focus upon the biological interaction of CNCs.

One of the first studies to address the potential risk upon inhalation of CNCs was performed using CNCs derived from cotton (220±67 x 15±5 nm) and assessed their effects with a novel 3D triple cell co-culture model of the human epithelial airway barrier [35]. This model consists of a layer of a human bronchiolar epithelial cell line (16HBE14o$^-$) forming a polarised cell layer and specific characteristics such as tight junctions after differentiation [36]. Human peripheral blood monocyte derived macrophages (MDMs) and dendritic cells (MDDCs) are added to form a stratified, *in vivo* like multicellular layer, enabling the interaction between the different cell types [37-39]. Applying the different test materials via suspensions (0.005-0.03 mg/mL) to the apical side (MDM and epithelial cell side) of the cell model it was shown that upon interaction with the MDM contained in the multicellular system, MWCNTs and crocidolite asbestos fibers (CAFs) induced a significantly higher (pro-)inflammatory response when compared to the cotton derived CNCs used in the study. In contrast to CNCs, CAFs showed classical frustrated phagocytosis, a phenomenon commonly associated with asbestos fibers due to the inability of the leukocyte (*e.g.* macrophage) to adequately engulf a long, stiff fiber [40]. In addition to this, other studies have also shown induction of cell alterations and genotoxicity by nanofibers derived from cotton and curaua (0.01-1 %; 130-180 x 6-14 nm) [41]. Jeong and co-workers could not detect any adverse effects upon the exposure to bacterial cellulose under the investigated conditions *in vitro* (0.1-1.0 mg/mL) nor *in vivo* (0.5-5 mg/mL) [42]. A set of classical toxicological tests conducted with nanocrystalline cellulose (NCC, typically 200 x 10 x 5 nm; at a range of concentrations from 0.03 to 10 g/L) by Kovacs et al. revealed no concerning effects with organisms representing different trophic levels of the aquatic ecosystem, representing the receiving environment for NCC production site effluents [43].

In addition, several research groups performed animal experiments with cellulosic fibers, more precisely microfibrillated cellulose, of different dimensions (mass median aerodynamic diameter 4.8 µm up to 15 µm), origin (cotton dust, commercial cellulose powder, mechanical wood pulp dusts), doses (0.75 mg/100 g animal - twice per week - 6 weeks, 15 mg single dose, 300-757 fibers/cm^3 within 2 weeks, 10^6-10^9 fibers) and application mode (inhalation, intratracheal instillation, intraperitoneal injection) [44-51]. The various exposures showed that instillation into the lung, pleura or peritoneal cavity or inhalation of the cellulosic materials can lead to inflammation, fibrosis, granulomata, sarcomas as well as fibrosing bronchio-/alveolitis under the employed conditions with rats and hamsters [45-52]. For all detected effects the authors associated the biopersistence of the cotton fibers within the lungs. *In vivo* for example, Muhle et al. reported that cellulose fibers (4.2 µm x 0.87 µm, instillation dose 2 mg) were present in the lung after one year, with a calculated half-time of fiber clearance of about 1,000 days [52]. Compared to the existing number of publications concerning the risk of CNTs and their biological interactions, knowledge as to their potential interactions and effects of CNCs is severely limited. As previously highlighted, although CNCs could pose (potential) risk as CNTs, cellulose and its derivatives are perceived as benign materials, which is subverted by the findings of the research discussed above.

Despite such results, however, these studies [35, 41-52] have to be carefully evaluated. Due to different production protocols (milling vs. acid hydrolysis) or sources (cotton, wood pulp, paper, plants, bacteria) the dimensions and composition (surface charge, surface groups) of the CNCs vary dramatically and were often not reported in detail. This lack of characterisation and the application of the potentially inappropriate techniques can have a significant influence on the evaluation of resultant biological effects following exposure. When characterising high aspect ratio nanomaterials there is controversial debate as to the gold standard methods that should be used, such as employing measurements of zeta potential for the materials' surface charge, or dynamic light scattering for its dimensions (underlying algorithms for these methods are generated on spherical particle behaviour in an electric field and the diffusion motion in liquid, respectively). Generally, the importance of close material characterisation has been highlighted and a minimum set of physico-chemical characteristics, such as dimensions, surface area, chemical composition (production protocols, purity) or aggregation behaviour have been widely accepted in order to properly conduct a risk assessment of any nanomaterial [53]. Furthermore, instillation of extremely high doses, so-called 'overload situations', might not necessarily reflect realistic exposure scenarios, but on the other hand might reveal the risk potential in a worst case scenario or highlight possible translocation (if exposed to a material over an elongated period of time), as shown with CNTs [18]. Furthermore, detailed knowledge is lacking as to whether observed effects are provoked solely from the fiber fraction of the investigated substances or are due to a combination with impurities (*e.g.* bacteria, fungi, pesticides, wood matter) contained within the sample [50, 51]. It is also important to highlight that results from *in vivo* and *in vitro* studies are also not directly comparable, taking into consideration the close interplay of different cell types within the exposed organ. In the process of understanding a potential risk in occupational, consumer related or environmental exposure scenarios towards nanoparticles the route of inhalation should be especially focussed upon as previously highlighted. Therefore, with regards to assessment of the exposure risk of CNCs, a relevant *in situ* exposure system, realistic doses and systematic characterisation of the material, as well as a suitable *in vitro* model of the investigated organ (*e.g.* the human lung) are of prime importance when investigating their biological impact.

3. *In situ* scenarios and the 3R`s concept

With emerging nanotechnology products there are more and more materials to be tested for their potential risks towards human health [54]. Efficient and reliable tests are necessary to evaluate in which context these materials can be used or should be modified in order to avoid any harmful effects to humans and the environment [55]. Alternatives to animal testing are, therefore, a required prerequisite not only from an ethical perspective but also when considering significant differences when comparing reactions of the rodent and human immune system (*i.e.* inflammatory reactions) [56, 57]. Within the aim of the three R's concept (refinement, reduction and replacement) [58] it is therefore important to develop, validate and use *in vitro* models. In addition, high throughput test strategies are of highest importance regarding the growing field of nanotechnology and the emerging need for risk assessment [55].

In conjunction, several *in vitro* systems have been under investigation to copy the *in vivo* situation (*e.g.* in the lung). Researchers simulated the alveolar-capillary barrier *in vitro* as principle biological barrier in the lower respiratory tract [59]. Rothen-Rutishauser et al. described a triple cell co-culture model forming the defined architecture of the human epithelial airway barrier (section 2.1) [37]. Other approaches focus on the interplay of different cell types with little regard to the *in vivo* structure of the lung [60]. Investigation of such systems highlight the differences in the biological response when comparing mono- and co-culture techniques (two, three or four cell types) and show different reactions toward nanoparticle exposure which have to be closely discussed and compared as is widely highlighted in literature [35, 36, 57, 61]. An additional aim when using optimal *in vitro* systems is to realistically investigate the potential biological impact of nanoparticle exposure. This can be achieved by the simulation of the *in situ* application of nanoparticles. For studying the potential risks of inhalation of nanoparticles many systems have already been established for spherical or atmospheric (nano)particles including a glove-box system for occupational exposure [62] and air–liquid interface cell exposure systems for exposure of engineered nanoparticles [63], as well as to study the environmental exposure to both exhaust emissions and brake wear particles [64-66]. Other possibilities to generate nanoparticle aerosols and facilitate the investigation of inhaled nanoparticles either in the field of basic research or for risk oriented purposes were highlighted by Creutzenberg et al., including spark generators, dry dispersion techniques, using pressurised air or aerosolising particle suspensions to form droplet aerosols [67]. These systems have the advantage to overcome artefacts derived from suspension related exposure set-ups that include agglomeration and sedimentation of nanoparticles or cell medium interactions (protein interactions), which can influence observed effects as shown in different studies [68, 69]. Additionally, the direct comparison of aerosol and suspension exposure *in vitro* revealed considerably different outcome concerning the dose-response relationship and effective moment of maximum development of the investigated parameters following exposure to zinc oxide nanoparticles [70]. On the other hand, the comparison offers the possibility to distinguish exposure to nanoparticle containing aerosols composed of particulate and gaseous phase versus exposure in suspension to solely the particle-associated effects [70]. These studies contribute to the trend aiming at the most realistic exposure scenario depending on the research focus (occupational exposure, basic research) overcoming experimental artefacts (aggregation, sedimentation, interaction with cell medium), together with sophisticated *in vitro* systems, although, as of yet none have been used with HARN.

4. The Air Liquid Interface Cell Exposure system (ALICE) for HARN

As previously mentioned, the use of realistic *in vitro* models in combination with a relevant exposure system is of the utmost importance when evaluating the effects of the nanoparticle-cell interaction. In order to mimic the exposure to airborne nanoparticles via inhalation, an Air Liquid Interface Cell Exposure system (ALICE) was used [63]. The ALICE consists of an exposure chamber in which a perforated vibrating membrane nebuliser (customised eFlow, PARI Pharma GmbH, Germany) equipped with a class 45 nominal size aerosol head which produces a dense cloud of 6.00 µm (mass median diameter) droplets that fills the chamber and settles gently onto the cells. The chamber is connected to a continuous air-flow system that is generated by a pump. Humidity and temperature are controlled to create a stable physiological environment for the cells. Filters present at the inlet and outlet permit a 'semi-sterile' environment. Additionally, the deposition via ALICE nebulisation of any material can be monitored by a quartz crystal microbalance (detection limit 90 ng/cm^2) which is inserted into the chamber prior to the nebulisation process. This system has previously been described in accordance with studies of *in vitro* exposure to spherical particles (*e.g.* gold nanoparticles [71]) where it has been shown to provide an efficient, repeatable, dose controlled and spatially uniform deposition [63].

For the first time, the ALICE was applied for the nebulisation of suspended HARN, specifically CNCs derived from cotton (168±72 nm x 19±7 nm; Figure 1). CNCs were obtained from Whatman No.1 filter paper via hydrolysis with concentrated sulphuric acid following the protocol of Dong et al. [72]. Subsequent sonication for 4 h led to a well dispersed CNC suspension. Since anisotropic particles differ much in their physical properties from spherical particles, a close investigation of the nebulisation process is indispensable to ensure the previously described functioning of the ALICE [64]. A schematic drawing of the system is given in Figure 2.

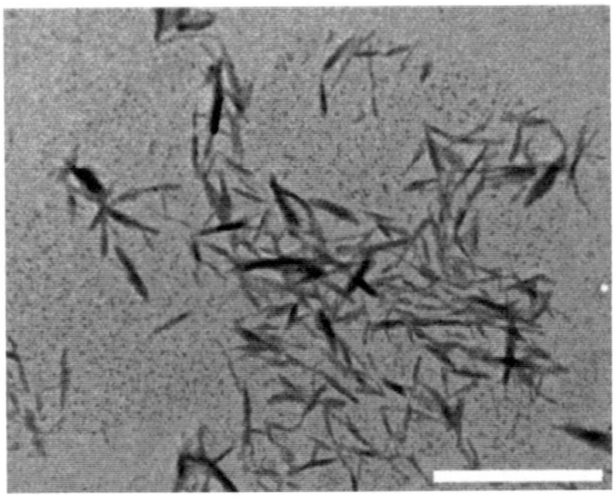

Figure 1: Transmission electron microscopy (TEM) image of 1 mg/mL cotton cellulose nanocrystals after 4 h sonication in ultrapure water. Scale bar represents 1 µm.

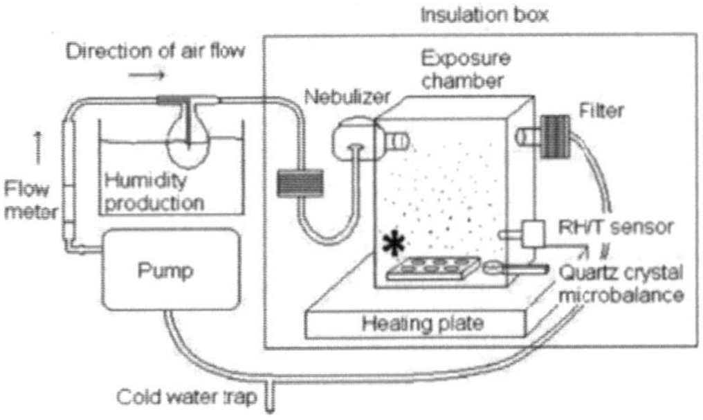

Figure 2: Schematic depiction of the Air Liquid Interface Cell Exposure system (ALICE), consisting of a closed air flow system including humidity production, an exposure chamber with aerosol generator, heating plate and a quartz crystal microbalance. * Equates to the position of 6-well plates with inserts containing glass slides, TEM grids or cells. Image is adapted under open access availability and with the permission of the authors of reference [64].

To document the spatially uniform deposition of CNCs after nebulisation as an important prerequisite for reliability and cornerstone of the ALICE, a microscopic analytical approach was used. Initially, 1 mL of an aqueous suspension of cotton CNCs (1 mg/mL) was nebulised via the nebuliser of the ALICE. After complete nebulisation onto glass slides (Ø 20 mm, Fisher Scientific AG, Switzerland) contained within 6-well plate inserts (BD Falcon™ Cell Culture Inserts, 3 µm pores, BD USA), the dried droplets comprising CNCs were analysed using confocal laser scanning microscopy (Zeiss LSM 710, Carl Zeiss AG, Germany). The dried CNC residuals were stained with a 0.2 % aqueous Calcofluor solution (Fluorescent Brightener 28, Sigma-Aldrich, USA). This fluorescent label is widely used and can be easily observed at an emission of 435 nm (excitation wavelength of 345 nm) [75-77]. Homogeneously distributed dry residuals of deposited CNCs with a bimodal size distribution were found arranged in a 'halo' shape (Figure 3 (a)).

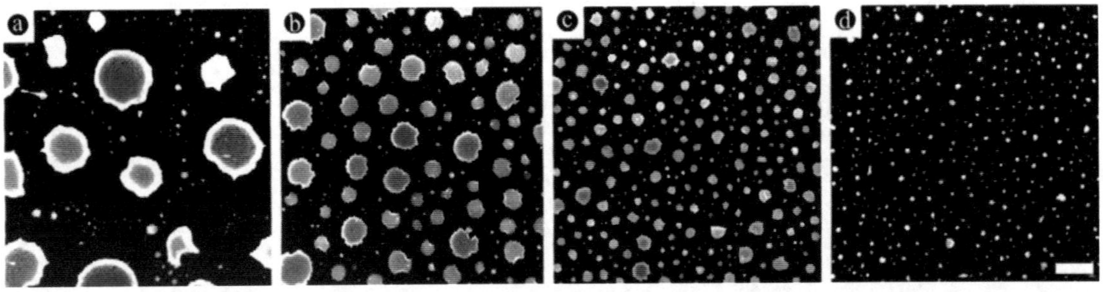

Figure 3: Confocal laser scanning microscopy images showing dried residuals of the nebulised 1 mg/mL cotton cellulose nanocrystal (CNC) suspension on glass slides after nebulisation times of (a) 80 s, (b) 40 s, (c) 20 s, (d) 10 s using the Air Liquid Interface Cell Exposure system. All images were taken at 20x magnification and the scale bar represents 100 µm.

To further investigate this observation time resolved nebulisation of the 1 mg/mL suspension of cotton CNCs was conducted. The nebulisation process was stopped after 40, 20 and 10 seconds respectively, and compared to a complete nebulisation (80 s). Figure 3 (a-d) shows that the deposition pattern and the size of dried residuals on the exposed glass slides changes as a function of exposure time. All residual particles appear as irregular shaped compact particles due to surface tension effects during the drying process. It is evident that the deposition pattern is spatially uniform for all nebulisation times. However, the size of the residual particles increases substantially from ~3 µm up to ~150 µm with increasing nebulisation time. Due to the continuing deposition process the dried residual particles are much larger than the CNC residuals expected from a single droplet. With decreasing nebulisation time, *i.e.* decreasing deposition of droplets, the fraction of single droplets increases which explains the reduction in the size of the residual particles reaching a minimum of ~3 µm for 10 s. The drying process also explains why surrounding each large droplet a 'halo' of very small droplets is visible (Figure 3 (a-b)). During drying, evaporating water pulls the included CNCs to a central point leaving an empty space between the 'halo' and the central, larger, dried residual of deposited CNCs. With less time of nebulisation this effect dissipates (Figure 3 (c-d)). It should be noted that the highly uniform distribution of droplets disregarding the nebulisation time highlights the efficient function of the ALICE, even with HARN. The hypothesis of an observed drying effect was supported by exposing copper grids (Plano AG, Germany) for transmission electron microscopy (TEM; Hitachi H-7100, Hitachi, Japan, at 75 kV equipped with a Morada 11 MPix digital CCD camera, Olympus, Japan) in the same manner as the glass slides for confocal microscopy. After complete nebulisation of 1 mg/mL cotton CNCs this reveals droplets in the nanometer size regime (Figure 4), pointing at the production of small droplets.

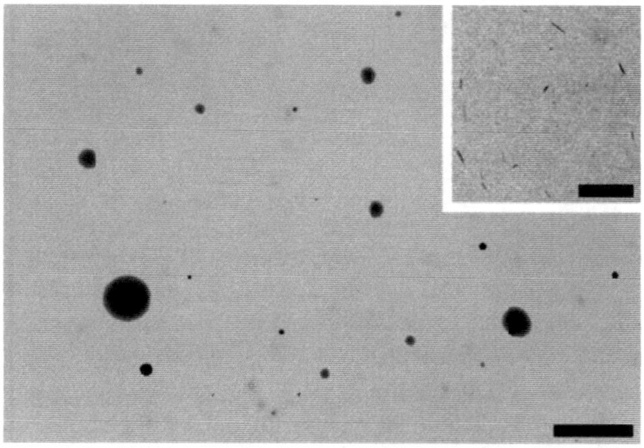

Figure 4: Transmission electron microscopy (TEM) image of 1 mg/mL cotton cellulose nanocrystals after being nebulised using the Air Liquid Interface Cell Exposure system onto a copper grid for TEM (Plano GmbH, Germany). Scale bar represents 10 µm. Inset shows single deposited crystals after exposure, scale bar represents 500 nm.

In summary, it has been shown with a microscopic approach that the ALICE provides spatially uniform deposition of aerosolised HARN (CNCs) in transwell inserts for cell culturing. An adverse effect of fiber shape on the spatial uniformity of particle deposition can be ruled out. Also it was shown that independent of concentration or exposure time (leading to varying aerosolised volumes of CNC suspensions) uniform distribution can be achieved as proven with repeated nebulisations of the HARN (n•3, data not shown). Due to merging of droplets on the glass slides measurements of the exact size of the produced aerosol droplets from imaging the dried residual particles on the slide cannot be achieved with this method (see Figure 3 and 4). Imaged dry residuals after deposition suffer from a variety of artefacts (*e.g.* drying, surface tension and contact angle). It is important to note that Figure 3 and Figure 4 are not representative for the CNC particles deposited onto the surface of a cellular *in vitro* model especially since cellular membranes, even exposed to the air liquid interface, never pose a dry surface, *i.e.* the compaction due to the drying process will not occur on the cells. The fluidic model of the double phospholipid layer of cell membranes with liquid coating will, therefore, encounter the entity produced by the nebuliser. Detailed analysis will elucidate if these are still fiber-shaped or already compacted in spherical shaped droplets both observed by TEM after exposure (Figure 4, inset).

Nevertheless, the proof of spatially uniform deposition of aerosolised CNCs with the ALICE is of greatest importance when subsequently analysing deposited mass with the integrated quartz crystal microbalance and also the exposure of *in vitro* systems to study the biological impact of CNC-interactions in a controlled and dose-dependent manner.

5. Conclusion

The emerging field of nanotechnology gathers much attention due to the outstanding properties of nanoparticles that can be used in many different applications. CNCs in particular propose high chances for industrial exploitation due to their advantageous physico-chemical characteristics. In the field it is widely accepted that new materials are required to be closely investigated in regard to their potential adverse health effects prior to application. Only with the combination of novel, sophisticated cell culture models and realistic exposure systems together with close characterisation of the material and appropriate material doses close to the reality of occupational settings can the potential adverse effects of CNCs be determined. In the case of CNCs derived from cotton, the basis for an in-depth risk assessment of inhalatory exposure is now possible due to the establishment of nebulising these high aspect ratio nanomaterials with the existing ALICE system. For the first time the fulfilment of functioning cornerstones of the ALICE such as spatially uniform deposition, as well as reliability and controllability of the deposited dose have been experimentally verified, which facilitates the investigation of the potential adverse effects of inhaled CNCs with *in vitro* assays.

6. Acknowledgments

The authors wish to thank the Swiss National Science Foundation (National Research Programme 64, Project #406440_131264/1) and the Adolphe Merkle Foundation for financial support. Many thanks also to Fabian Herzog, Isabelle Wey and Sandra Camarero Espinosa for their contributions to this work.

References

[1] Eichhorn SJ 2011 *Soft Matter* **7** 303-15
[2] Habibi Y, Lucia LA, Rojas OJ 2010 *Chem Rev* **110** 3479-500
[3] Bolio-Lopez GI, Valadez-Gonzalez A, Veleva L, Andreeva A 2011 *Rev Mex Ing Quím* **10** 291-9
[4] van den Berg O, Capadona JR, Weder C 2007 *Biomacromolecules* **8** 1353-7
[5] Sani A, Dahman Y 2009 *J Chem Technol Biotechnol* **85** 151–64
[6] Johar N, Ahmad I, Dufresne A 2012 *Ind Crop Prod* **37** 93-9
[7] Mueller S, Foster EJ, Weder C manuscript in preparation
[8] Eichhorn SJ, Dufresne A, Aranguren M, Marcovich NE, Capadona JR, Rowan SJ, et al. 2010 *J Mater Sci Mater* **45** 1-33
[9] Tang L, Weder C 2010 *ACS Applied Materials & Interfaces* **2** 1073-80
[10] Shanmuganathan K, Capadona JR, Rowan SJ, Weder C 2010 *Progress in Polymer Science* **35** 212-22
[11] Ma H, Hsiao BS, Chu B 2012 *ACS Macro Letters* **1** 213-6
[12] Shopsowitz KE, Qi H, Hamad WY, MacLachlan MJ 2010 *Nature* **468** 422-5
[13] Borm PJA, Robbins D, Haubold S, Kuhlbusch T, Fissan H, Donaldson K, et al. 2006 *Part Fibre Toxicol* **3** 11
[14] Hagens WI, Oomen AG, de Jong WH, Cassee FR, Sips AJAM 2007 *Regul Toxicol Pharmacol* **49** 217-29
[15] Maynard AD, Kuempel ED 2005 *Journal of Nanoparticle Research* **7** 587-614
[16] Arora S, Rajwade JM, Paknikar KM 2012 *Toxicol Appl Pharmacol* **258** 151-65
[17] Nowack B, Brouwer C, Geertsma RE, Heugens EHW, Ross BL, Toufektsian M-C, et al. 2012 *Nanotoxicology* 1-5

[18] Poland CA, Duffin R, Kinloch I, Maynard A, Wallace WAH, Seaton A, et al. 2008 *Nat Nanotechnol* **3** 423-8
[19] Donaldson K 2009 *Crit Rev Toxicol* **39** 487-500
[20] Donaldson K, Murphy FA, Duffin R, Poland CA 2010 *Part Fibre Toxicol* **7**
[21] Kraus T, Pfahlberg A, Zobelein P, Gefeller O, Raithel HJ 2004 *Chest* **125** 731-6
[22] Tahir MW, Mumtaz MW, Tauseef S, Sajjad M, Nazeer A, Farheen N, et al. 2011 *Environ Monit Assess* **184** 4879-88
[23] Furness G, H.B. M 1952 *Brit J industr Med* **9** 138-45
[24] Mayan O, Da Costa JT, Neves P, Capela F, Pinto AS 2002 *Ann Occup Hyg* **46** 277-80
[25] Paudyal P, Semple S, Niven R, Tavernier G, Ayres JG 2011 *Ann Occup Hyg* **55** 403-9
[26] Evans DE, Ku BK, Birch ME, Dunn KH 2010 *Ann Occup Hyg* **54** 514-31
[27] Han JH, Lee EJ, Lee JH, So KP, Lee YH, Bae GN, et al. 2008 *Inhal Toxicol* **20** 741-9
[28] Wohlleben W, Brill S, Meier MW, Mertler M, Cox G, Hirth S, et al. 2011 *Small* **7** 2384-95
[29] Schlagenhauf L, Chu BTT, Buha J, Nüesch F, Wang J 2012 *Environ Sci Technol* **46** 7366-72
[30] Capadona JR, Shanmuganathan K, Tyler DJ, Rowan SJ, Weder C 2008 *Science* **319** 1370-4
[31] Capadona JR, Tyler DJ, Zorman CA, Rowan SJ, Weder C 2012 *MRS Bulletin* **37** 581-9
[32] Heath L, Thielemans W 2010 *Green Chem* **12** 1448–53
[33] Hoeger I, Rojas OJ, Efimenko K, Velev OD, Kelley SS 2011 *Soft Matter* **7** 1957-67
[34] Moon RJ, Martini A, Nairn J, Simonsen J, Youngblood J 2011 *Chem Soc Rev* **40** 3941–94
[35] Clift MJD, Foster EJ, Vanhecke D, Studer D, Wick P, Gehr P, et al. 2011 *Biomacromolecules* **12** 3666-73
[36] Lehmann AD, Blank F, Baum O, Gehr P, Rothen-Rutishauser BM 2009 *Part Fibre Toxicol* **6** 26
[37] Rothen-Rutishauser BM, Kiama SG, Gehr P 2005 *Am J Respir Cell Mol Biol* **32** 281-9
[38] Rothen-Rutishauser B, Blank F, Mühlfeld C, Gehr P 2008 *Expert Opin Drug Metab Toxicol* **4** 1075-89
[39] Rothen-Rutishauser B, Müller L, Blank F, Brandenberger C, Mühlfeld C, Gehr P 2008 *ALTEX* **25** 191-6
[40] Brown D, Kinloch I, Bangert U, Windle A, Walter D, Walker G, et al. 2007 *Carbon* **45** 1743-56
[41] de Lima R, Mattoso, Feitosa, Maruyama, Barga, Yamawaki, et al. 2012 *Int J Nanomed* 3555-65
[42] Jeong SI, Lee SE, Yang H, Jin YH, Park CS, Park YS 2010 *Mol Cell Toxicol* **6** 373-80
[43] Kovacs T, Naish V, O'Connor B, Blaise C, Gagne F, Hall L, et al. 2010 *Nanotoxicology* **4** 255-70
[44] Davis JM 1993 *Br J Ind Med* **50** 187-90
[45] Milton DK, Godleski JJ, Feldman HA, Greaves IA 1990 *Am Rev Respir Dis* **142** 184-92
[46] Tatrai E, Adamis Z, Bohm U, Meretey K, Ungvary G 1995 *J Appl Toxicol* **15** 45-8
[47] Tatrai E, Brozik M, Adamis Z, Meretey K, Ungvary G 1996 *J Appl Toxicol* **16** 129-35
[48] Warheit DB, Snajdr SI, Hartsky MA, Frame SR 1998 *Two-week inhalation study in rats with cellulose fibers* In: Advances in the prevention of occupational respiratory diseases. ed Chiyotani K, Hosoda Y, Aizawa Y (Amsterdam: Elsevier Science) p. 579-82
[49] Adamis Z, Tatrai E, Honma K, Ungvary G 1997 *J Appl Toxicol* **17** 137-41
[50] Cullen RT, Miller BG, Clark S, Davis JM 2002 *Inhal Toxicol* **14** 685-703
[51] Cullen RT, Miller BG, Jones AD, Davos JMG 2002 *Ann Occup Hyg* **46** 81-4
[52] Muhle H, Ernst H, Bellmann B 1997 *Ann Occup Hyg* **41** Suppl. 1 184-8
[53] Warheit DB 2008 *Toxicol Sci* **101** 183-5
[54] Nel A, Xia T, Madler L, Li N 2006 *Science* **311** 622-7
[55] Damoiseaux R, George S, Li M, Pokhrel S, Ji Z, France B, et al. 2011 *Nanoscale* **3** 1345-60
[56] Mestas J, Hughes CCW 2004 *J Immunol* **172** 2731-8
[57] Clift MJD, Gehr P, Rothen-Rutishauser B 2011 *Arch Toxicol* **85** 723-31
[58] Russell WMS 1959 *The principles of humane experimental technique.* ed Russell W, Burch R (London: Methuen)

[59] Kasper J, Hermanns MI, Bantz C, Maskos M, Stauber R, Pohl C, et al. 2011 *Part Fibre Toxicol* **8** 6
[60] Alfaro-Moreno E, Nawrot TS, Vanaudenaerde BM, Hoylaerts MF, Vanoirbeek JA, Nemery B, et al. 2008 *Eur Respir J* **32** 1184-94
[61] Lehmann AD, Parak WJ, Zhang F, Ali Z, Rocker C, Nienhaus GU, et al. 2010 *Small* **6** 753-62
[62] Rothen-Rutishauser B, Grass RN, Blank F, Limbach LK, Mu hlfeld C, Brandenberger C, et al. 2009 *Environ Sci Technol* **43** 2634-40
[63] Lenz AG, Karg E, Lentner B, Dittrich V, Brandenberger C, Rothen-Rutishauser B, et al. 2009 *Part Fibre Toxicol* **6** 32
[64] Müller L, Gasser M, Raemy DO, Herzog F, Brandenberger C, Schmid O, et al. 2011 *Insciences Journal* **1** 30-64
[65] Müller L, Compte P, Czerwinski J, Kasper M, A.C.R. M, Gehr P, et al. 2010 *Environ Sci Technol* **44** 2632-8
[66] Perrenoud A, Gasser M, Rothen-Rutishauser B, Gehr P, Riediker M 2010 *IJBNN* **1** 17-33
[67] Creutzenberg O 2012 *Arch Toxicol* **86** 1117-22
[68] Maiorano G, Sabella S, Sorce B, Brunetti V, Malvindi MA, Cingolani R, et al. 2010 *Acs Nano* **4** 7481-91
[69] Wick P, Manser P, Limbach LK, Dettlaff-Weglikowska U, Krumeich F, Roth S, et al. 2007 *Toxicol Lett* **168** 121-31
[70] Raemy D, Grass R, Stark W, Schumacher C, Clift M, Gehr P, et al. 2012 *Part Fibre Toxicol* **9** 33
[71] Brandenberger C, Rothen-Rutishauser B, Muhlfeld C, Schmid O, Ferron GA, Maier KL, et al. 2010 *Toxicol Appl Pharmacol* **242** 56-65
[72] Dong XM, Kimura T, Revol J-F, Gray DG 1996 *Langmuir* **12** 2076-82

Stability of polymer encapsulated quantum dots in cell culture media

I. Ojea-Jiménez[1],* J. Piella[1], T.-L. Nguyen[2], A. Bestetti[2], A. D. Ryan[2], V. Puntes[1,3]
(1) Catalan Institute of Nanotechnology (ICN), Universitat Autònoma de Barcelona (UAB), 08205, Cerdanyola del Vallés, Barcelona, (Spain) (2) School of Chemistry and Bio21 Institute, University of Melbourne, Parkville, Vic 3010 (Australia) (3) Institució Catalana de Recerca i Estudis Avançats (ICREA), 08093, Barcelona, (Spain).

E-mail: Isaac.ojea.jimenez@gmail.com

Abstract. The unique optical properties of Quantum Dots have attracted a great interest to use these nanomaterials in diverse biological applications. The synthesis of QDs by methods from the literature permits one to obtain nanocrystals coated by hydrophobic alkyl coordinating ligands and soluble in most of the cases in organic solvents. The ideal biocompatible QD must be homogeneously dispersed and colloidally stable in aqueous solvents, exhibit pH and salt stability, show low levels of nonspecific binding to biological components, maintain a high quantum yield, and have a small hydrodynamic diameter. Polymer encapsulation represents an excellent scaffold on which to build additional biological function, allowing for a wide range of grafting approaches for biological ligands. As these QD are functionalized with poly(ethylene)glycol (PEG) derivatives on their surface, they show long term stability without any significant change in the optical properties, and they are also highly stable in the most common buffer solutions such as Phosphate Buffer Saline (PBS) or borate. However, as biological studies are normally done in more complex biological media which contain a mixture of amino acids, salts, glucose and vitamins, it is essential to determine the stability of our synthesized QDs under these conditions before tackling biological studies.

1. Introduction

The unique optical properties of Quantum Dots (QDs) have attracted a great interest to use these nanomaterials in diverse biological applications as fluorescent probes [1-2]. In contrast to common organic fluorophores, QDs have broad absorption spectra enabling excitation by a wide range of wavelengths, and their emission spectra are symmetric and narrow. The most widely used and best characterized QDs are CdX/ZnS (X = Se, Te) nanocrystals, where the CdX core is overcoated with a few atomic layers of a material with a larger bandgap such as ZnS, which improves its luminescence properties while protecting surface atoms from oxidation and preventing chemical degradation [3-4]. The synthesis of QDs by methods from the literature permits one to obtain gram quantities of nanocrystals with a narrow size distribution (Standard deviation < 5 %), coated by hydrophobic alkyl coordinating ligands but soluble only in organic solvents [5-6]. The ideal biocompatible QD must be homogeneously dispersed and colloidally stable in aqueous solvents, exhibit pH and salt stability, show low levels of nonspecific binding to biological components, maintain a high quantum yield, and have a small hydrodynamic diameter. To disperse QDs in aqueous solvents it is necessary to manipulate the surface chemistry, and three main strategies have evolved during the past few years: i) ligand exchange, ii) silica encapsulation and iii) polymer encapsulation. Given that these different capping strategies address the requirements of various potential applications, the advantages of each method have to be careful weighted against the drawbacks. For instance, our investigations suggest that with ligand exchange methods the QD photoluminescence is often compromised and that the growth of a thin, uniform silica shell that encapsulates a single QD is at present nontrivial. By contrast, polymer encapsulation represents an excellent scaffold on which to build additional biological function, allowing for a wide range of grafting approaches for biological ligands. We have shown that the commonly synthesized QD in organic solvents transfer well using a poly(stryrene-*co*-maleic anhydride) polymer and that the resulting biocompatible QD is homogeneously dispersed and colloidaly stable in aqueous solvents, exhibits pH and salt stability and maintains a high quantum yield [7].

As these QD are functionalized with poly(ethylene)glycol (PEG) derivatives on their surface, they show long term stability for more than 6 months without any significant change in the optical properties (absorbance and fluorescence spectra), and they are also highly stable in the most common buffer solutions such as PBS or borate. However, as biological studies are normally done in more complex biological media which contain a mixture of aminoacids, salts, glucose and vitamins, it is mandatory to determine the stability of our synthesized QDs under these conditions before embarking into any *in vitro* or *in vivo* studies.

2. Stability tests

We tested a sample of CdSe/ZnS Core/Shell QDs with first absorption maximum at 535 nm. These particles have 50 % of azide groups on their surface, QD_{50} indicates coating of Jeffamine M1000 and $H_2N\_PEG\_N_3$ in a 50:50 ratio. Samples were incubated at two different concentrations 1 μM and 10 μM of QDs, and stock solutions were previously prepared in order to keep constant the added volume to the different media. Sample were incubated at 37 ºC in four different media: water, Dulbecco's modified Eagle's medium (DMEM), DMEM supplemented with 10 % fetal bovine serum (FBS) and piperazine-N,N′-bis(2-ethanesulfonic acid) (PIPES). The stability was monitored by Ultraviolet-visible (UV-vis) and fluorescence spectroscopies, fluorescence images, Dynamic Light Scattering (DLS) and Transmission Electron Microscopy (TEM) for periods of time up to 48 hr.

The absorption spectra of QD_{50} at a concentration of 1 μM showed no significant variation in most of the cases except in DMEM, indicating in this case a major nanoscale aggregation or surface degradation after 48 hr (Figure 1A). By contrast, for samples of 10 μM concentration of QDs incubation in all the media showed a great stability, including the case of DMEM (Figure 1B). This result is somewhat surprising since concentration determines the chances of two particles interacting with each other and subsequently aggregate. The higher stability at higher concentrations of QDs is likely to occur because the stability is majorly due to steric (stabilized by PEG chains) instead of electrostatic interactions. For a given concentration of salt at low concentration of QDs the screening of the fixed surface charges is more effective, but with high concentration of QDs there is also a greater number of PEG chains in solution which interact more efficiently preventing the formation of a diffuse double layer around the particles.

Figure 2A shows fluorescence images of 547 nm emitting QD_{50} samples incubated in the four different media and taken after certain periods of time. After 48 hr in DMEM for a 1 μM concentration of QDs there is evidence of colloidal instability seen by the aggregation of the QDs, which crashed out from solution. The photoluminescence intensities for the cases of 1 μM concentration of QDs were also lower after incubation in DMEM + 10% FBS and PIPES for 48 hr when compared to water, and this was drastically reduced for the case of DMEM (Figure 2B). In comparison with its analogue sample incubated in DMEM, the higher colloidal stability observed by UV-vis and fluorescence spectroscopies of QD_{50} samples at 1 μM concentration incubated in DMEM + 10% FBS after 48 hr is likely due to the serum proteins sticking on the surface of the QDs and forming a protein corona which stabilizes the particles.

Journal of Physics: Conference Series **429** (2013) 012009 doi:10.1088/1742-6596/429/1/012009

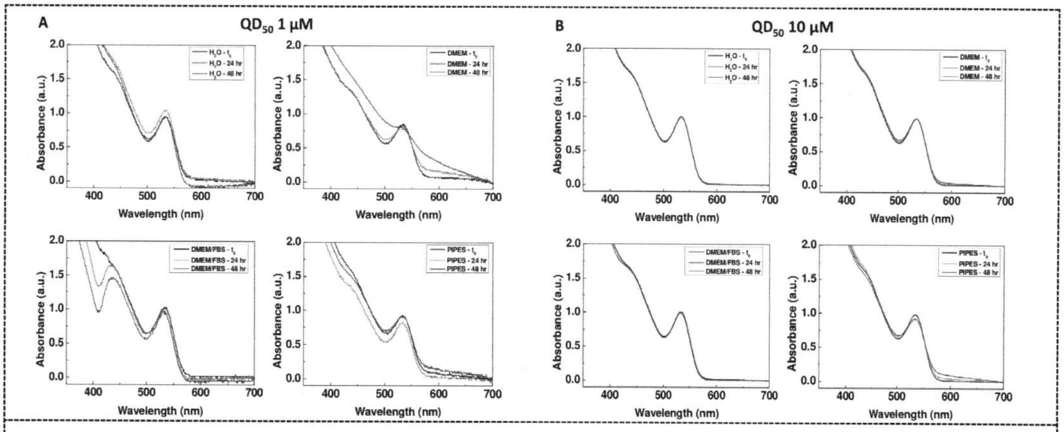

Figure 1: Normalized absorbance of two different concentrations 1 µM (A) and 10 µM (B) of QD_{50} in H_2O, DMEM, DMEM + 10% FBS and PIPES.

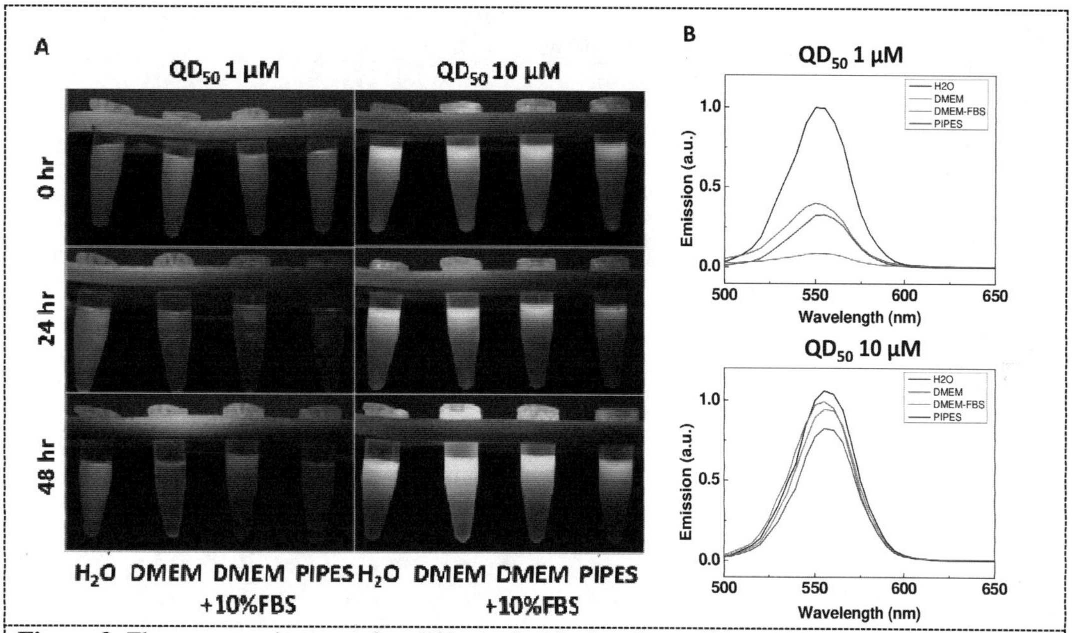

Figure 2: Fluorescence images after different incubation times (A) and fluorescence spectra after 48 hr (B) of 1 µM and 10 µM samples of QD_{50} in H_2O, DMEM, DMEM + 10% FBS and PIPES.

Most of the QD_{50} samples incubated in the different media exhibit low variations of the hydrodynamic diameter obtained from DLS measurements (~ 15-20 nm) (Figure 3). However, incubation in DMEM at a concentration of 1 µM drastically increased the Hydrodynamic Diameter (HD) to ~ 270 nm, corroborating the aggregation effect. TEM also confirmed these data, as images taken from the QD_{50} samples incubated in DMEM showed major signs of aggregation (Figure 4).

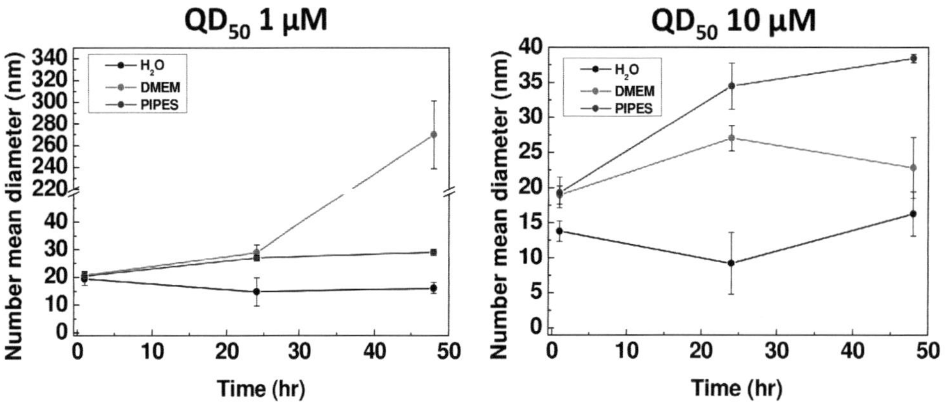

Figure 3: Size evolution by DLS (number mean diameter) at different times of 1 μM and 10 μM samples of QD_{50} in H_2O, DMEM and PIPES. Samples incubated in DMEM + 10% FBS were not analyzed by this technique as the abundance of proteins in the medium would overpower the signal of QDs present at much lower concentrations.

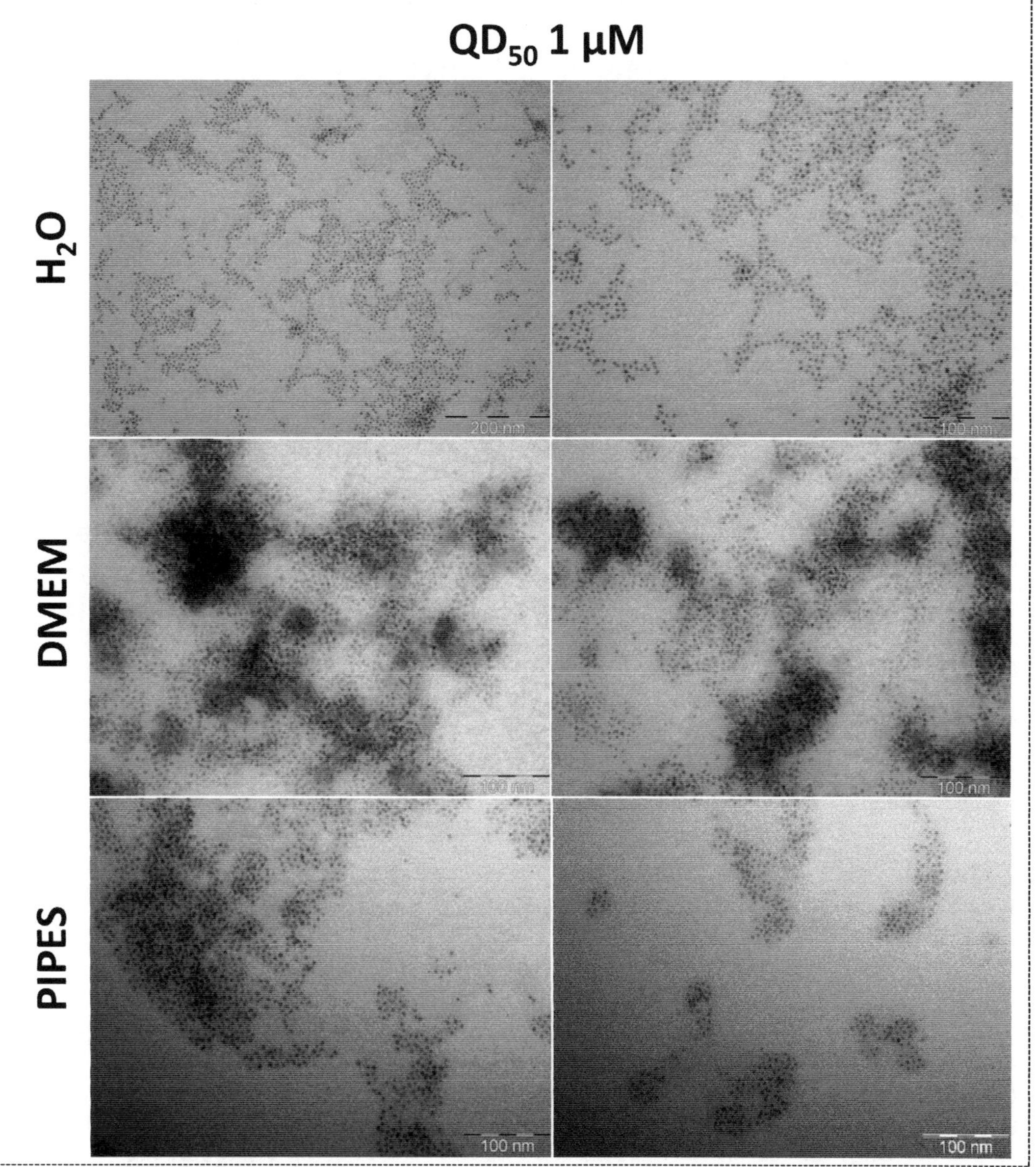

Figure 4: TEM images of 1 µM QD_{50} samples after 48 hr in H_2O, DMEM and PIPES.

3. References

[1] Bruchez M, Moronne M, Gin P, Weiss S and Alivisatos A P 1998 *Science* **281** 2013
[2] Chan W C W and Nie S 1998 *Science* **281** 2016
[3] Dabbousi B O, Rodriguez-Viejo J, Mikulec F V, Heine J R, Mattoussi H, Ober R, Jensen K F and Bawendi M G 1997 *J. Phys. Chem. B* **101** 9463
[4] Hines M A, and Guyot-Sionnest P 1996 *J. Phys. Chem.* **100** 468
[5] Murray C B, Norris D J and Bawendi M G 1993 *J. Am. Chem. Soc.* **115** 8706
[6] Peng X, Wickham J and Alivisatos A P 1998 *J. Am. Chem. Soc.* **120**, 5343
[7] Lees E E, Nguyen T-L, Clayton A H A and Mulvaney P. 2009 *ACS Nano* **3** 1121

The use of PIXE for engineered nanomaterials quantification in complex matrices

O Lozano[1,2,*], O Toussaint[1,3], J-M Dogné[1,4], and S Lucas[1,2]

[1] Namur Nanosafety Centre (NNC), NAmur Research Institute for LIfe Sciences (NARILIS), University of Namur (UNamur), Rue de Bruxelles 61, B-5000 Namur, Belgium
[2] Research Centre for the Physics of Matter and Radiation (PMR-LARN), University of Namur (UNamur), Rue de Bruxelles 61, B-5000 Namur, Belgium
[3] Department of Pharmacy, NAMEDIC, Namur Thrombosis and Hemostasis Centre (NTHC), University of Namur (UNamur), Rue de Bruxelles 61, B-5000 Namur, Belgium
[4] Laboratory of Biochemistry and Cellular Biology (URBC), University of Namur (UNamur), Rue de Bruxelles 61, B-5000 Namur, Belgium
[*] Corresponding author e-mail: omar.lozanogarcia@fundp.ac.be

Abstract. Engineered nanomaterials (ENMs) quantification in complex media is an area under development, much demanded by stakeholders due to the introduction of a myriad of consumer products containing ENMs. In this work, Particle-Induced X-ray Emission (PIXE) will be shown as capable of quantifying ENMs in complex media for both in vivo and in vitro assessments using minimal conditioning. ENM quantifications (SiC, TiC, SiO_2) were performed on complex media (rat feces, rat lungs and cell culture), with applied ENM concentrations corresponding to ranges of interest for in vivo or in vitro assessments. The case studies presented in this work show the capability and versatility of PIXE measurements for biopersistence, biodistribution and dose assessment studies.

1. Introduction

With the advent of engineered nanomaterials (ENMs) discovery and their fundamental properties different to those of bulk materials, there has been a rush in the industry to produce and use these new materials in different products, touching wide and diverse areas such as semiconductors [1], medicine [2], cosmetics [3], and food [4]. Correspondingly, stakeholders have raised health concerns due to the unprecedented progress uncoupled with a proper understanding of the possible associated risks. Reports point out that ENMs may be more toxic than their larger counterparts [5]. An important area is ENM quantification, not only in ideal conditions (i.e. ENMs dispersed in ultrapure water) but rather in complex media, should it be for biopersistence, biodistribution or dose assessment purposes.

The current recommendations about the definition of a nanomaterial [6] considers them to be anything with a dimension lower than 100 nm and with a greater than 50% ratio in number in a product. A current challenge in the nanosafety field is the quantification of ENMs, especially when reaching small sizes such as 20 nm or less [7]. In this regard, ion beam techniques, particularly Particle-Induced X-ray Emission (PIXE), can overcome this challenge given that an ion beam interacts on an atomic scale basis. PIXE requires a minimal sample conditioning and yields a detection limit down to a few ppm [8-10]. In addition, special set-ups allow to probe the local concentrations on the micron range, and therefore build spatial elemental composition maps with micrometre lateral resolution [9, 11]. Non-vacuum PIXE is also possible, making this technique suitable to analyse the contents of NPs in liquids [12, 13].

The aim of this work is to present case studies of ENM quantification by PIXE in complex media: rat feces and lungs exposed to ENMs, and ENMs dispersed in cell media and ultrapure water. The presented case studies are part of larger studies focusing on the nanotoxicology or fate of ENMs.

2. Materials and methods

2.1. *Engineered Nanomaterials*

SiC and TiC ENMs were purchased from Ionic Liquids Technologies GmbH & Co. KG, Germany [14, 15] and used as received. Mesoporous SiO_2 ENMs synthesized via sol-gel were supplied by Nanologica™ through the NanoValid project (FP7 2007-2013).

2.2. *Sample preparation*

2.2.1. Rat feces. Female Sprague–Dawley rats (Charles River, France) were orally administered with 50 mg/kg of SiC or TiC ENMs in an acute assessment. The administration was done following the OECD (Organization for Economic Cooperation and Development) 420 guideline for acute toxicity study of chemicals [16] using tap water as delivery media. Feces were collected 24 hours later. The procedure of feces preparation for PIXE analysis has been described previously [10]. Briefly: feces were dried in a 60 °C oven for 24 hours; chromium nitride (Cr_2N) powder (Goodfellow, 99% purity, 45 μm particle size) was added as 7-10 % of the feces weight. Both feces and Cr_2N powder were ball milled to produce a homogeneous powder mixture. This mixture was hard pressed into a 2 cm diameter, 1-2 mm thick pellet. Chromium was added as an internal standard; it is not present in the biological matter and does not interfere in the PIXE measurements.

2.2.2. Rat lungs. Female Sprague–Dawley rats (Charles River, France) were exposed to a nanoaerosol of SiC ENMs in an acute assessment during 6 hours and sacrificed 6 hours later. The exposure was done according to the OECD 403 guideline for chemicals [17]. The lungs were prepared into pellets following the previously described procedure [10]. Briefly: lungs were dried in a 37°C oven for 24 hours, then they were froze dried [18] and chromium nitride (Cr_2N) powder (Goodfellow, 99% purity, 45 μm particle size) was added as 7-10 % of the feces weight. Both lungs and Cr_2N powder were ball milled to produce a homogeneous powder mixture. This mixture was hard pressed into a 2 cm diameter, 1-2 mm thick pellet. Chromium was added as an internal standard; it is not present in the biological matter and does not interfere in the PIXE measurements.

2.2.3. Cell media and ultrapure water. SiO_2 ENMs, at a 20 μg/mL concentration, were dispersed in minimum essential medium (MEM, Gibco), containing 10% fetal bovine serum (Gibco) and 1% penicillin –streptomycin (BioWhittaker). These ENMs were also dispersed in ultrapure water at the same concentration. A drying method was used to prepare the samples and has been described elsewhere [19]. Briefly: ten 20 μL droplets were placed in a sample holder and each droplet was left to dry during one hour in a 60 °C oven before placing the next droplet.

2.3. *Particle-Induced X-ray Emission (PIXE)*

PIXE measurements were performed with the University of Namur ALTAÏS accelerator. PIXE is an ion beam technique, its physical principles can be found extensively explained elsewhere [9]. Briefly: ion-matter interactions occur on an atomic scale basis, meaning that, ion beam techniques like PIXE are not limited by particle size in an ENM quantification context. At the same time, ion beam techniques cannot separate particles sizes, but can be coupled with other size distribution techniques like Centrifuge Liquid Sedimentation (CLS) to provide precise quantification and size distribution in toxicological assessments [10].

Ion-matter interactions can produce several phenomena, amongst which characteristic X-ray emissions for each element present, due to the disturbance in its electronic shell. In this disturbance, an electron emitted from a shell in the atom is then replaced by an electron from an upper shell, During this transition, the difference in energy between the shells is emitted as an X-ray. The X-ray yield of a specific Z element is described by the following equation [9]:

$$Y(Z) = QH\varepsilon^{i}{}_{Z} t_{Z} C_{Z} Y_{t}(Z) \qquad (1)$$

where Q is the accumulated charge from the ion beam, H is the solid angle, ε^{i}_{Z} is the detector intrinsic efficiency, t_Z the transmission through any absorber, C_Z is the concentration of element Z, and $Y_t(Z)$ is the theoretical X-ray yield production of element Z. The GUPIXWIN software was used to analyze the PIXE spectra [20]. The measurements calibration was validated using two standards from the International Atomic Energy Agency [21, 22] and a standard from the Institute of Reference Materials and Measurements [23] following an H-value adjustment methodology [24].

PIXE, as an ENM quantification tool, offers several advantages: concurrent multi-element acquisition in a single measurement, minimal sample preparation when compared to other techniques, parts per million (ppm) levels sensitivity, fast measurements (a few minutes per measurement). Also, when compared to similar techniques like energy dispersive X-ray (EDX) spectroscopy, PIXE has a higher sensitivity due to lower background noise [10]. In addition, a high volume sampling for solid samples is achieved by the use of a rotating stage providing a total scan area of 140.5 mm^2 [10].

In order to have fully quantitative PIXE measurements, a precise knowledge of the major elements present in any sample is needed. This means the knowledge of the carbon (C), oxygen (O), and nitrogen (N) contents for biological samples. This matrix of elements acts as an attenuator for X-ray emission. In our setup, all PIXE measurements are recorded simultaneously using Rutherford Backscattering Spectrometry (RBS), a technique that relies on backscattered ion beam energy loss due to its interaction with the samples. RBS can precisely quantify light elements like C, O, and N, whose values are then used as the main matrix for PIXE analysis. In the case of feces and organs, the amount of Cr_2N is added as part of the matrix and represents 7-10 % of the biological matter weight. This kind of methodology is known as 'Total Ion Beam Analysis' [25].

The geometry setup for the reported measurement was the following: with respect to the beam direction, the samples are tilted at 45°, a Canberra LEGe (Low Energy Germanium) detector is located at 90° for PIXE measurements, and a Canberra PIPS detector is positioned at 145° for RBS measurements. The current PIXE detector setup is suitable for X-ray measurement on elements of $Z \geq 13$ (sodium and heavier elements). The PIXE detector is calibrated with a ^{57}Co source and the RBS detector is calibrated with a SnO_2 thin film on glass. The used ion beam was composed of protons, although it is possible to work with alpha particles [9]. Two sets of parameters were used: Protons of 2 MeV and an aluminum collimator for the LEGe detector of 3 mm in aperture and 1 cm thick, or protons of 2.5 MeV and an aluminum collimator for the LEGe detector of 1 mm in aperture and 0.2 mm thick.

3. Results and discussion

3.1. Oral administration of SiC or TiC to rats

Within the framework of possible effects to SiC or TiC ENMs exposure, the oral exposure route was carried on. One of the purposes was to assess the ENMs biodistribution in feces and the possibility of crossing the intestinal barrier [10]. Figure 1 shows the PIXE spectra of feces from control rats and rats administered with 50 mg/kg of either SiC or TiC ENMs. Table 1 summarizes the ENM quantification in both cases and their expulsion rate on day 1. Figure 1 highlights the ability of detecting the Si-Kα or Ti-K X-rays coming from the SiC or TiC ENMs respectively. In a single measurement, all the elements present in the feces were measured, a feature which allows matrix changes evaluation due to the presence of ENMs. PIXE showed that feces composition is altered due to orally administered ENM [10].

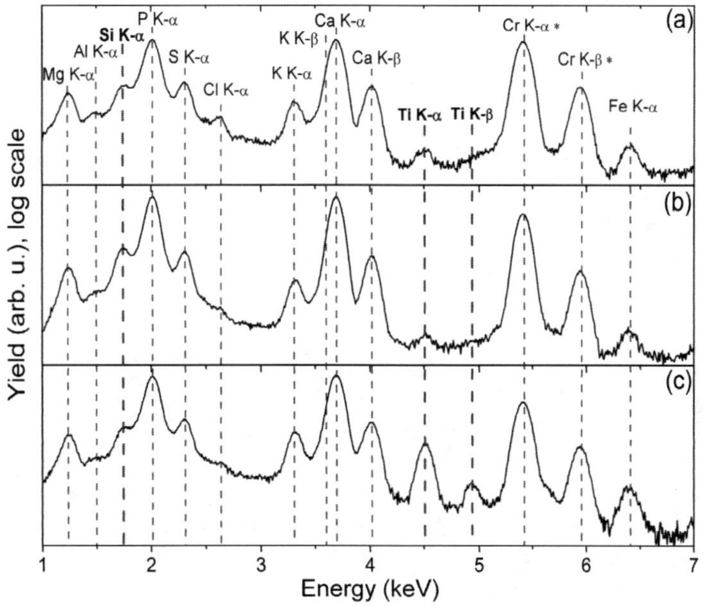

Figure 1. PIXE rat feces spectra from oral administration studies: (a) control, (b) SiC administered at 50 mg/kg, (c) TiC administered at 50 mg/kg. Chromium (Cr) is due to the standard added to the feces. It is clear to the eye that the increase on Si-Kα and Ti-K X-rays yield for the feces where rats were instilled with SiC or TiC respectively.

Table 1. Administered ENMs, quantity found in feces and expulsed amount. The administered quantity is based on rat weight average (n=3). The amount found in feces subtracted the background Si or Ti signal found in control feces. The expulsed amount is the average ENM ratio found in feces with respect to the average administered quantity.

ENM administered	Quantity administered (ppm)	Amount found in feces 1 day later (ppm)	Expulsed amount on day 1 (%)
SiC	13000 ± 209	9652 ± 604	74.25
TiC	10250 ± 700	5932 ± 35	57.87

3.2. SiC nanoaerosol exposure to rats

The respiratory pathway is one of the most important routes of exposure to ENMs [26]. A whole body exposure system for rodents has been developed and validated at the University of Namur, using SiC ENMs as the produced nanoaerosol [27]. Lungs from control and exposed rats are shown in Figure 2. SiC ENMs presence in the exposed lungs was clearly seen as a raise in the Si-Kα X-ray yield, showing a load of 1413 ± 477 ppm (n=3) of SiC ENMs in the lung 6h after exposure. The elemental lung composition was not affected by the presence of SiC ENMs.

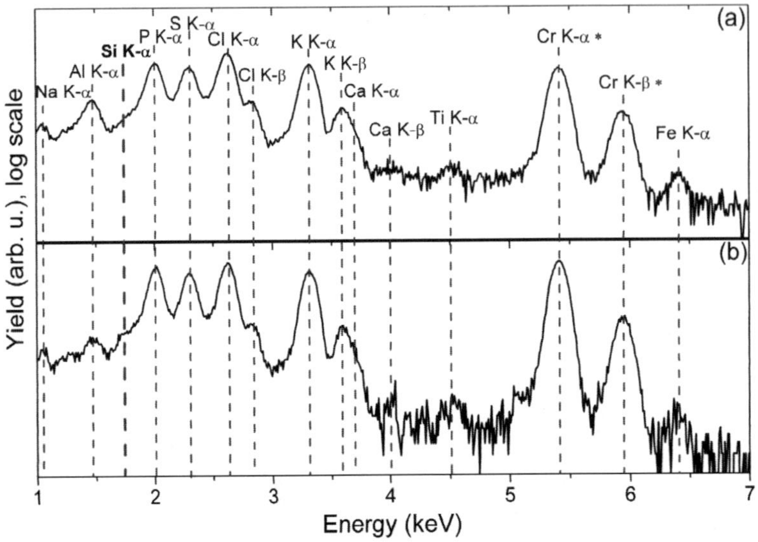

Figure 2. PIXE spectra of rat lungs from inhalation studies: (a) control, (b) exposed to SiC nanoaerosol. Chromium (Cr) is due to the standard added to the lungs. The presence of Si-Kα X-rays is seen only for the exposed lungs.

3.3. ENM dispersion into cell media and ultrapure water

ENM quantification in cell media is important as a mean to understand cell-nanoparticle dynamics, especially with incubation times where the quantity of ENMs present in the media can vary with time [28]. Firstly, the drying method route was preferred to a droplet measurement, given that the detection limits can be greatly improved. For example, taking the case of SiO_2 ENMs dispersed in water, the limit of detection (LOD) was estimated in 500 µg/mL using a droplet measurement methodology [13]. In a recent set of studies 20 µg/mL of SiO_2 were dispersed in ultrapure water and was measured by PIXE using the drying method (see Figure 3). The estimated LOD was 435 ng/ml, i.e. ~3 orders of magnitude lower than using the droplet methodology. This dramatic improvement in the LOD is due to 2 factors: the lack of other elements generating a high background noise like argon in air, and the fact that the matrix is evaporated using the drying method and thus leaves only ENMs for measurement. Given the ion beam penetration depth , a proton beam of 2 MeV penetrates 76.92 ± 2.98 µm of water [29], the drying method is preferred when going to low concentrations. The droplet measurement methodology still remains relevant when droplets cannot be dried (i.e. oils) or when the drying method may to aerosolize the ENMs (i.e. a very volatile matrix). While the matrix effect is not important when analyzing ENMs on droplets of different matrices [13], it is an important issue in the drying method for complex matrices which partially evaporate. For example, this is the case for cell media where proteins remain when dried. In a recent study of the dynamics of ENMs in cell media, it was found that 20 µg/mL dispersed in MEM present a LOD of 16.2 µg/mL [19]. While this result was clearly not achievable using the droplet method, it is still two orders of magnitude higher from the same concentration measurement in a fully evaporable matrix. Table 2 summarizes the SiO_2 ENM value measured in water using both methods and in using the drying method in MEM.

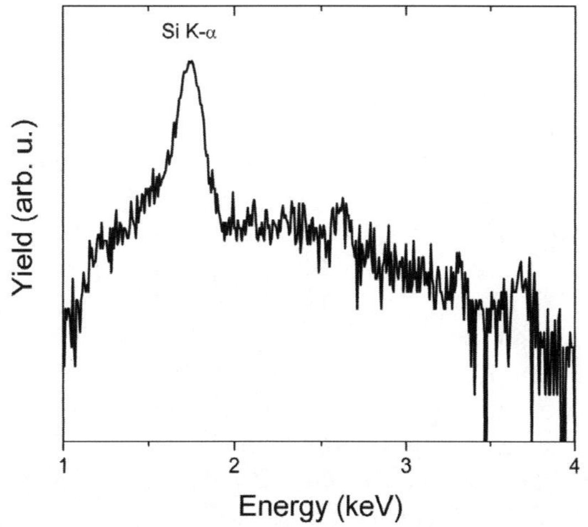

Figure 3. PIXE spectrum of SiO_2 ENMs measured using the drying method. Dispersion was done in H_2O at a 20 µg/mL concentration.

Table 2. Summary of SiO_2 ENM detection using the droplet or drying method dispersed on H_2O or MEM. The detection limit (LOD) of each measurement is included.

Preparation method	Matrix	Measured concentration (µg/mL)	LOD (µg/mL)
Droplet	H_2O	750	500
Drying	H_2O	20	0.435
Drying	Mem	20	16.2

In conclusion, case studies of ENM quantification in complex media by PIXE were presented, where minimal conditioning for either solid or liquid sample was required. These examples show the capability and versatility of PIXE measurements for biopersistence, biodistribution and dose assessment studies.

Acknowledgements. This work was supported by the Service Public de Wallonie (SPW) – Direction générale opérationnelle – Economie, Emploi et Recherche (DGO6), Département des Programmes de Recherche (Nanotoxico Project, SPW/FUNDP research convention No. 516252). The research leading to these results has received funding from the European Community's Seventh Framework Programme (FP7/2007-2013) under grant agreement n° 263147 (NanoValid - Development of reference methods for hazard identification, risk assessment and LCA of engineered nanomaterials). O. Toussaint is a Belgian FNR/ FNRS Research Associate.

References

[1] Piccione B, Cho C-H, van Vugt L K and Agarwal R 2012 All-optical active switching in individual semiconductor nanowires *Nat Nano* **7** 640-645.
[2] Wang W, Zhang S, Chinwangso P, Advincula R C and Lee T R 2009 Electric potential stability and ionic permeability of SAMs on gold derived from bidentate and tridentate chelating alkanethiols *J Phys Chem C* **113** 3717-3725.
[3] Calzolai L, Gilliand D and Rossi F 2012 Measuring nanoparticles size distribution in food and consumer products: a review *Food Addit Contam A*

[4] Napierska D, Thomassen L C, Lison D, Martens J A and Hoet P H 2012 The nanosilica hazard: another variable entity *Part Fibre Toxicol* **7**

[5] Oberdörster G 2001 Pulmonary effects of inhaled ultrafine particles *Int Arch Occup Environ Health* **74** 1-8.

[6] European Comission 2011 *Nanomaterials*. In http://ec.europa.eu/environment/chemicals/nanotech/index.htm. Accessed: November 2011.

[7] Mitrano D M, Barber A, Bednar A, Westerhoff P, Higgins C P and Ranville J F 2012 Silver nanoparticle characterization using single particle ICP-MS (SP-ICP-MS) and asymmetrical flow field flow fractionation ICP-MS (AF4-ICP-MS) *J Anal At Spectrom* **27** 1131-1142.

[8] Demortier G 1988 Application of nuclear microprobes to material of archaeological interest *Nuclear Instruments and Methods in Physics Research Section B: Beam Interactions with Materials and Atoms* **30** 434-443.

[9] Johansson S A E, Campbell J L and Malmqvist K G 1995 *Particle- Induced X-Ray Emission Spectrometry (PIXE)* (Place: John Wiley and Sons Inc) pp

[10] Lozano O, Mejia J, Masereel B, Toussaint O, Lison D and Lucas S 2012 Development of a PIXE analysis method for the determination of the biopersistence of SiC and TiC nanoparticles in rat lungs *Nanotoxicology* **6** 263-271.

[11] Deconnick G 1973 Analyse des microquantities d'elements traces dans des substances solides par bombardement d'ions *J Radioanal Chem* **17** 29-43.

[12] Deconninck G 1977 Trace element analysis in liquids by proton induced X-ray emission *Nuclear Instruments and Methods* **142** 275-284.

[13] Lozano O, Mejia J, Tabarrant T, Masereel B, Dogné J-M, Toussaint O and Lucas S 2012 Quantification of nanoparticles in aqueous food matrices using Particle-Induced X-ray Emission *Anal Bioanal Chem* **403** 2835–2841.

[14] Iolitec 2007 *Silicon(IV)carbide - technical data sheet*. In http://www.nanomaterials.iolitec.de/en/Nanomaterials/carbide.html. Accessed: October 2012.

[15] Iolitec 2007 *Titanium(IV)carbide - technical data sheet*. In http://www.nanomaterials.iolitec.de/en/Nanomaterials/carbide.html. Accessed: October 2012.

[16] OECD 2001 *OECD guideline for testing of chemicals. Acute oral toxicity – fixed dose procedure. Guideline 420*. In http://iccvam.niehs.nih.gov/SuppDocs/FedDocs/OECD/OECD_GL420.pdf. Accessed: October 2012.

[17] OECD 2009 *OECD guideline for testing of chemicals. Acute Inhalation Toxicity. Guideline 403*. In http://browse.oecdbookshop.org/oecd/pdfs/free/9740301e.pdf. Accessed: Ocotober 2012.

[18] Galuszka J, Jarczyk L, Rokita E, Strzalkowski A and Sych M 1984 The influence of target preparation and mode of irradiation on PIXE analysis of biological samples *Nucl Inst Meth Phys Res B* **3** 141-146.

[19] Lozano O, Mejia J, Piret J-P, Jacques D, Saout C, Dogné J-M, Toussaint O and Lucas S 2012 How do oxide nanomaterial dispersions evolve in an in vitro assessment? *J Phys Conf Series*

[20] Campbell J L 2005 *GUPIX and GUPIXWIN homepage [Internet]*. In http://pixe.physics.uoguelph.ca/gupix/main/. Accessed: Ocotober 2012.

[21] IAEA 1989 *IAEA-153 , Milk Powder*. In http://nucleus.iaea.org/rpst/ReferenceProducts/ReferenceMaterials/Trace_Elements_Methylmercury/IAEA-153.htm. Accessed: Ocotber 2012.

[22] IAEA 1990 *IAEA-155 , Whey Powder*. In http://nucleus.iaea.org/rpst/ReferenceProducts/ReferenceMaterials/Trace_Elements_Methylmercury/IAEA-155.htm. Accessed: October 2012.

[23] IRMM 2010 *Certified reference material BCR - 126A*. In https://irmm.jrc.ec.europa.eu/html/reference_materials_catalogue/catalogue/attachements/BCR-126A_cert.pdf. Accessed: July 2012.

[24] Gama S, Volfinger M, Ramboz C and Rouer O 2001 Accuracy of PIXE analyses using a funny filter *Nucl Inst Meth Phys B* **181** 150-156.

[25] Jeynes C, Bailey M J, Bright N J, Christopher M E, Grime G W, Jones B N, Palitsin V V and Webb R P 2012 ''Total IBA'' – Where are we? *Nucl Inst Meth B* **271** 107-118.

[26] Oberdörster G, Maynard A, Donaldson K, Castranova V, Fitzpatrick J, Ausman K, Carter J, Karn B, Kreyling W, Lai D, Olin S, Monteiro-Riviere N, Warheit D and Yang H 2005 Principles for characterizing the potential human health effects from exposure to nanomaterials: elements of a screening strategy *Part Fibre Toxicol* **2** 8.

[27] Laloy J, Lozano O, Nonet A, Demortier A, Toussaint O, Rolin S, Masereel B, Dogné J and Lucas S 2012 Development of a nanoparticle whole-body exposure model for rats using silicon carbide nanoparticles *Particle and Fibre Toxicology* **(submitted)**

[28] Mejia J, Toussaint O, Masereel B and Lucas S 2012 Fate of SiC and TiC nanoparticles dispersions in simulated gastric fluid *Int J Nano Biomat* **(accepted)**

[29] Mayer M 2008 *SIMNRA homepage [Internet]*. In http://www.rzg.mpg.de/~mam/. Accessed: October 2012.

µTAS (micro total analysis systems) for the high-throughput measurement of nanomaterial solubility

R Tantra and J Jarman
National Physical Laboratory, Hampton Road, Teddington, Middlesex, TW11 0LW, United Kingdom

E-mail: ratna.tantra@npl.co.uk

Abstract. There is a consensus in the nanoecotoxicology community that better analytical tools i.e. faster and more accurate ones, are needed for the physicochemical characterisation of nanomaterials in environmentally/biologically relevant media. In this study, we introduce the concept of µTAS (Micro Total Analysis Systems), which was a term coined to encapsulate the integration of laboratory processes on a single microchip. Our focus here is on the use of a capillary electrophoresis (CE) with conductivity detection microchip and how this may be used for the measurement of dissolution of metal oxide nanomaterials. Our preliminary results clearly show promise in that the device is able to: a) measure ionic zinc in various ecotox media with high selectivity b) track the dynamic dissolution events of zinc oxide (ZnO) nanomaterial when dispersed in fish medium.

Keywords: µTAS, nanoecotoxicology, dissolution, nanomaterial, zinc oxide

1. Introduction

Under the guidelines given by the Organisation for Economic Co-operation and Development (OECD) [1] and International Organization for Standardization (ISO) [2], solubility is an important physicochemical property to measure for nanomaterials. In particular there is a need to understand the extent of toxicity arising from the ions released as result of the dissolution process. In the past, it is the dissolution of metal oxides nanomaterial such as ZnO that has been of most concern and it is zinc in the ionic rather than the particulate form that has been blamed for potential toxicity [3]. Current protocols to measure solubility often involve a separate sample preparation step prior to analysis; this is often required to distinguish particulate zinc from ionic zinc. In the past this has been achieved either through the use of a dialysis membrane [4] or the extraction of a supernatant using a pipette after centrifugation [5]. In either case, the extracted sample can then be analysed using elemental analysis based techniques such as: Inductively coupled plasma mass spectrometry (ICP-MS) [5] or Inductively coupled plasma atomic emission spectroscopy (ICP-AES) [6]. The reliability of solubility data acquired from such methods hinges largely on the proper sample preparation step e.g. on the careful extraction of the supernatant without disturbing the resulting pellet after centrifugation, a highly operator-dependent step. In the case of the dialysis membrane, an assumption is made that any nanosize zinc particulates have smaller size than that of the pore size of the membrane, which may not be the case. In addition to the potential unreliability of the data, the methods proposed are laborious and not suitable for rapid routine analysis. What is needed is a highly selective device capable of measuring ionic zinc (as opposed to particulate zinc) that is potentially suitable for mass screening.

In the early 1990s, the concept of μTAS (Micro Total Analysis Systems) was put forward, which builds on the idea that a multifunctional chemical laboratory process can be integrated on to a single microfabricated device. The advantages of such devices in comparison to their "bulkier" counterparts are clear and have been well documented to include: faster analysis, lower cost, less sample consumption, and suitability for high-throughput analysis [7]. One such device is the CE(capillary electrophoresis)-conductivity microchip [8]. This is a μTAS device capable of performing: subsampling, sample separation and detection, all achieved by applying high voltages to microchannel networks. The microchip has a capillary electrophoretic column for the separation of ions and a contactless capacitive coupled conductivity detection (C4D) system at the end of the separation column to quantify the concentration of the separated ionic species. The theory of the CE-conductivity microchip has been well documented and was recently shown to be suitable for the quantification of ions in highly complex medium such as blood [9]. Its ability to quantify ion concentration in complex media suggests its potential to be used for the measurement of nanomaterial dissolution in nanoecotoxicological applications.

In this study, the feasibility of the CE-conductivity microchip in nanoecotoxicological applications is investigated. Our approach is to conduct a preliminary investigation that involves the use of $ZnCl_2$; this was chosen as our model for analyte dissolution due to the fact that $ZnCl_2$ is highly soluble in aqueous solutions. $ZnCl_2$ is dissolved in two separate ecotox media i.e. fish and Daphnia, in order to establish the performance of the microchip for analysis of ionic zinc. Results presented show, for the first time, the feasibility of the CE-conductivity microchip for the analysis of ZnO nanomaterial dissolution when dispersed in fish medium. Its usefulness in monitoring the dynamics of ZnO dissolution for a period of several weeks is assessed. As suggested by Tanyanyiwa *et al.*, the choice of background electrolyte (BGE) buffer mixture is important as it contributes to both selectivity and efficiency of separation; for zinc the BGE of a hydroxyisobutyric acid (HIBA)/histidine (His) combination has been recommended [10].

2. Experimental

2.1 Materials

Unless stated otherwise, all chemicals used in the study were of analytical grade or better, purchased from Sigma Aldrich UK and used as received. De-ionised water was used throughout and had a resistivity value of 18.2 MOhm/cm. Standard test solutions of: a) 1 mM LiCl, KNO_3 and Na_2SO_4 in deionised water and b) BGE of 0.5 M acetic acid, needed to qualify the CE-conductivity microchip, were purchased from eDAQ Europe; these were supplied with the associated certified weight reports provided by Absolute Standards, Inc. Nanomaterial Z-Cote Zinc Oxide was supplied by BASF SE through an OECD global initiative for nanomaterial testing [1]. The physicochemical property of this material has been reported elsewhere and it has been revealed by scanning electron microscopy that the powdered form has a primary particle size of 151.0 ± 55.6 nm [5] ; the range reported here represented the broadness of the size distribution.

2.2 Sample Preparation

Stock solutions of $ZnCl_2$ (100 mM) were dissolved either in fish or Daphnia medium; the appropriate dilutions were made from this stock for subsequent studies. The fish and Daphnia media were prepared based on the recommendations of the OECD guidelines and the detailed protocol for this has been covered elsewhere [11]. The main components of the ecotox media and final pH are summarised in Table 1.

Table 1. Components of Daphnia and fish ecotox medium

	Chemical formula	Fish medium concentration (mM) ; final pH~7	Daphnia medium concentration (mM) ; final pH~8
Calcium Chloride	$CaCl_2$	4.87	1.08
Magnesium Sulphate	$MgSO_4$	1.92	0.43
Sodium Bicarbonate	$NaHCO_3$	1.54	0.51
Potassium chloride	KCl	0.15	0.05
Sodium selenite	Na_2SeO_3	N/A	Trace i.e. 0.004

The protocol employed for the dispersion of the ZnO nanomaterial was based on the recommended protocol used for recent global OECD nanomaterial testing, as previously described [5]. In summary, this involved mixing the nanomaterial powder into a paste before adding more liquid and sonicating using a sonicating probe (Cole Palmer ® 130 - Watt Ultrasonic Processors (50/60 Hz, VAC 220)) for 20 s. A final dispersion of 500 mg/L concentration (0.5 L total volume) was prepared and stored in a 1L storage media bottle. A corresponding "control" bottle containing just the fish medium i.e. no ZnO was also prepared. Throughout a 3 week period, 2 mL samples were aliquoted from each bottle; prior to extracting the sample each bottle was shaken gently by hand to aid homogeneity. Each 2 mL aliquot was passed through a 20 nm Anatop syringe filter syringe (Fisher Scientific UK) and the filtrate was collected for further analysis.

For the microchip experiments, a BGE solution of 5 mM His and 3 mM hydroxyisobutyric acid (HIBA) was prepared (the final pH adjusted to 4.5 with acetic acid). All solutions were filtered through a 20 nm syringe filter prior to introduction into the microchip to prevent potential blockages in the microchannels.

2.3 The CE-conductivity microchip

Borosilicate glass microchips were purchased from Micronit, Netherlands (model ET145-4). The fabricated microchip contained a manifold in a double-T geometry and integrated contactless conductivity detection, as schematically shown in Figure 1. The chip has a separation channel ~33 mm long, with a semi-circular cross section of ~100 μm width and ~10 μm depth. The arrangement of the detector consisted of two parallel strips of ~ 200 μm width, ~500 μm height and ~200 nm thick, with the electrodes separated by ~2.9 mm.

The CE-conductivity instrument was purchased from eDAQ Europe and consisted of the following units:

a) Two high voltage sequencers (HVS, Model ER230). Each HVS unit has a dual channel high voltage power supply especially designed for electroosmotic flow applications.
b) Capacitively Coupled Contactless Conductivity Detection (C4D) data system (ER225); this comprised of a conductivity detector and signal recording functions.

c) C4D Micronit chip electrophoresis platform (ET225). This unit has high voltage cables that allow the microchip to be connected to the HVS units; there are four reservoirs that connect the microchip to outside platinum electrodes. The unit includes a safety interlock, to switch off the HVS if the cover plate is lifted during the experiment. The platform also has a ground connector to minimise noise in the C4D signal.

2.4 Electrophoretic analysis on a microchip

The capillary electrophoresis measurement was performed using a BGE that is optimal for the said analyte. The instrumental setting is governed by the type of BGE employed and this was established by following the manufacturer's instructions [12]. The instrument settings for the test standard BGE (0.5 M acetic acid) were: frequency 800 kHz, amplitude 100%, high gain off, 5 Hz low pass. The instrument settings for the BGE employed for zinc analysis i.e. 5 mM His, 3 mM HIBA, pH 4.5 were: frequency 1150 kHz, amplitude 20%, high gain on, 5 Hz low pass).

For the electrophoretic analysis, the microchip was pre-conditioned prior to use (on a day-to-day basis). This was achieved by flushing the microchannels with the following solutions under vacuum: 0.5 M NaOH for 2 minutes, DI water for 10 minutes and finally BGE for 5 minutes. The microchip was mounted on the C4D Micronit chip electrophoresis platform and electrophoresis was carried out by applying the appropriate voltages to the four reservoirs (see Figure 1b). The voltage programme was set so as to achieve: sample plug formation, sample injection, separation and detection [12]. After use at the end of the day, the microchannels were flushed with deionised water, placed on a thermal hotplate (100°C) for a few minutes and stored dry in a petri dish. PowerChrom ® software purchased from eDAQ Europe (ES280) was used to collect, display and analyse data.

3. Results and Interpretation

Figure 1c. shows a typical electropherogram for the standard test sample containing cations of lithium, potassium and sodium. The three peaks are attributed to the elution of the three cations, with retention times between 1 and 1.4 minutes. Taking into account that sample injection occurred after 47 s, the net elution time of the cations are: K^+ 22s, Na^+ 29s and Li^+ 38s; these results coincided with findings previously reported [12]. Results show that the peaks are negative, which has been previously associated with the fact that the cations analysed are less conductive than that of the H^+ cation that is being displacing in the acetic acid BGE [12, 13]. The expected retention time of the ions is correlated to their corresponding electrophoretic mobility, which in turn is a function of *its charge and size that is modulated by the structure and dynamics of the hydration shells and the solvent* [14]. Hence, high mobility cations are expected to be eluted first i.e. K^+ in this case. The lower electrophoretic mobility of Li^+, compared to the other two cations, is attributed to the large solvation shell around Li^+ [14].

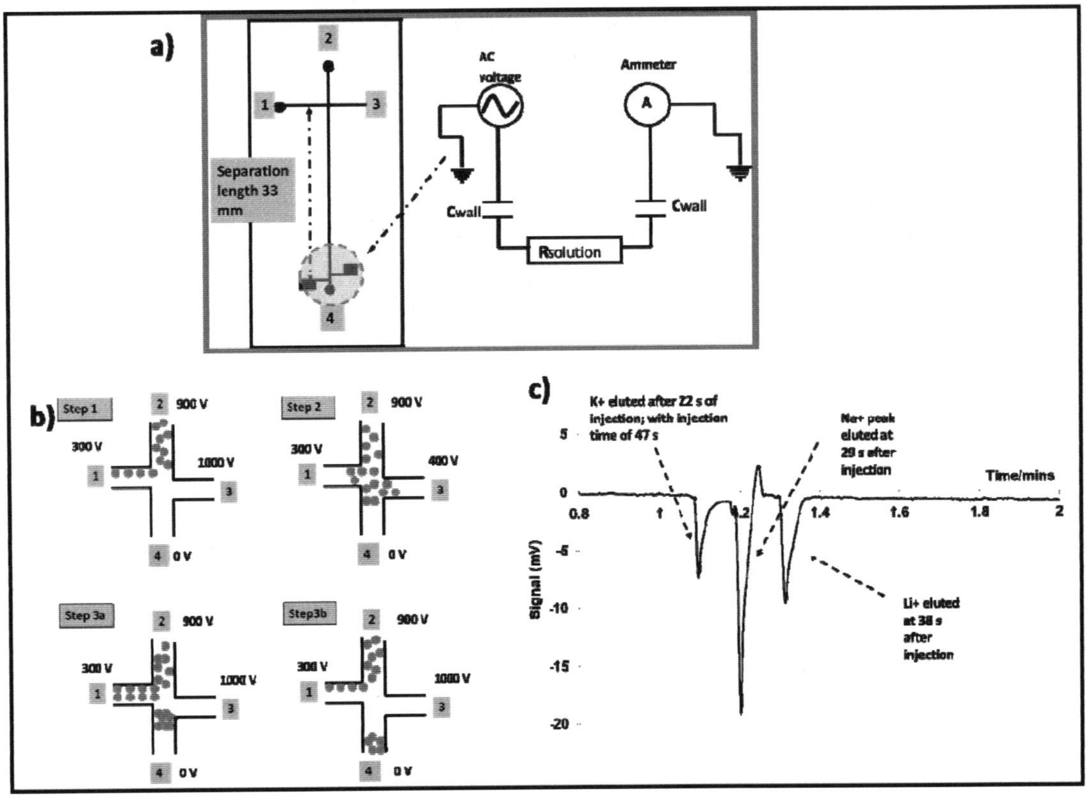

Figure 1. CE-conductivity microchip analysis. The Figure illustrates: a) a schematic of the microchip set up detailing the equivalent circuit model of the C4D b) a schematic of sample flow depicting plug formation and subsequent injection at the double cross T-junction; the Figure also details the corresponding voltage programme applied on the four reservoirs c) a typical electropherogram of the cations eluted as a result of analysing the test solution at 100 microMolar concentration (with an acetic acid buffer BGE).

Figure 2 shows the electrophoregrams of $ZnCl_2$ (at concentrations of 0, 1 and 5 mM) when dissolved in Daphnia and fish medium. The electropherograms clearly show a peak on top of a flat baseline that is attributed to zinc cations. At 0 mM, both electropherograms show a small peak eluted at 1.3 minutes; this peak is due to cations present in the ecotox medium and shall be referred to as the blank peak for the remainder of the paper. It is likely that the broad peak in the blank is attributed to a combination of: Na^+, Mg^{2+} and Ca^{2+} that exists at the few mM concentrations levels found in the blank ecotox medium. However, in order to verify this, there is a need to spike the blank sample with the appropriate cations to see what their corresponding elution times. It is important to note that the elutions times for the cations observed in the electrophoregram Figure 1c. (this is the reference test samples (containing only Li+, Na+ and K+))cannot be employed to elucidate the elution times in Figure 2, simply because different BGEs are being employed in each case. The choice of BGE for the analysis of a particular ion is important as they affect separation selectivity.

In the presence of $ZnCl_2$, another peak elutes straight after the blank peak (at 1.4 minutes), and the size of the peak increases with increasing $ZnCl_2$ concentration. It is clear that this peak, which will be referred to as the zinc peak, is attributable to the concentration of the ionic zinc present in the media.

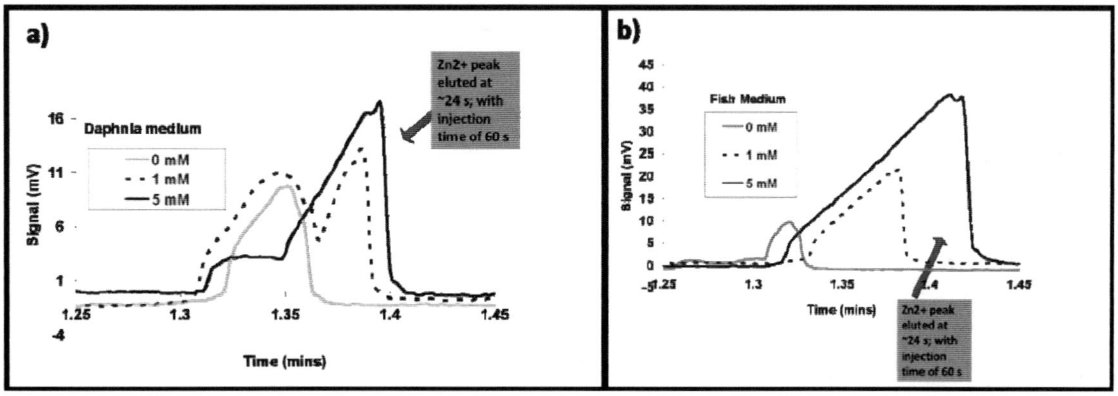

Figure 2. Electropherogram for $ZnCl_2$ (0 mM, 1 mM, 5 mM) when dissolved in: a) Daphnia b) fish media. Experimental conditions of: BGE of 5 mM His, 3 mM HIBA, pH 4.5; injection time 60 s; frequency 1150 kHz; amplitude 100%; headstage gain on.

Interestingly, there are distinct differences in the profile of the zinc peak between the two media, in how much the blank peak interferes with the zinc peak. When in Daphnia medium, the blank peak is present in all concentrations and at 5 mM the blank peak exists as a shoulder peak on the left of the zinc peak. In the corresponding fish medium, the blank peak does not interfere with the zinc peak and at 1 mM concentration the blank peak disappears from the electropherogram. Another apparent difference between the two media is the size of the zinc peak, with this being much larger in the case of the fish medium compared to the corresponding Daphnia. At 5 mM, the peak height for the fish is ~40 mV whereas this is ~16 mV in the case of Daphnia. The much higher signal in the fish medium can be attributed to the actual concentration of free zinc at the different pH in the two different media. It is well known that zinc dissolution is not straightforward and the zinc can exists in various forms depending on factors such as temperature, the presence of complexing ions, and pH [15, 16, 17, 18]. According to Reichle and co-workers [15], the lower the pH the more likely it is that the zinc will exist as a free ion. In this case, the lower pH of fish medium at pH ~ 7 will mean that more of the zinc will exist as the free ion compared to the Daphnia medium at pH ~ 8, thus explaining the larger signal associated with the fish medium. As suggested by past workers [16], at the higher pH of 8 the following reaction can occur, in which the ion forms a protective, water insoluble zinc hydroxide ($Zn(OH)_2$):

$$Zn^{2+} + 2OH^- \rightarrow Zn(OH)_2(s)$$

Interestingly, this interpretation can also explain findings related with the previous OECD study, in which the solubility measurements of ZnO dispersion showed that measured zinc concentrations in fish medium was always higher in comparison to the corresponding medium. The reader should refer to the relevant NPL's open access report for further details [5]. Another interesting observation from Figure 2 is the shape of the peaks, which is not Gaussian in shape. This can be attributed to the effects of electromigration dispersion, in which diffusional and dispersive forces can arise inside the channels resulting in wedge shaped peaks; these often appear when the sample concentrations are too high [13].

Figure 3 shows the performance of the microchip in the measurement of free zinc in fish medium at various concentrations of $ZnCl_2$, as assessed through its retention time and peak area; the plots show the mean value of 3 replicates with +/-1SD. Figure 3a shows that the eluted zinc peak at t_R of ~ 1.4 minutes (with a 60s sample injection time) is consistent and as expected not affected with change in $ZnCl_2$ concentration. The RSD associated with the retention time is good, being less than 4% in all cases.

a)

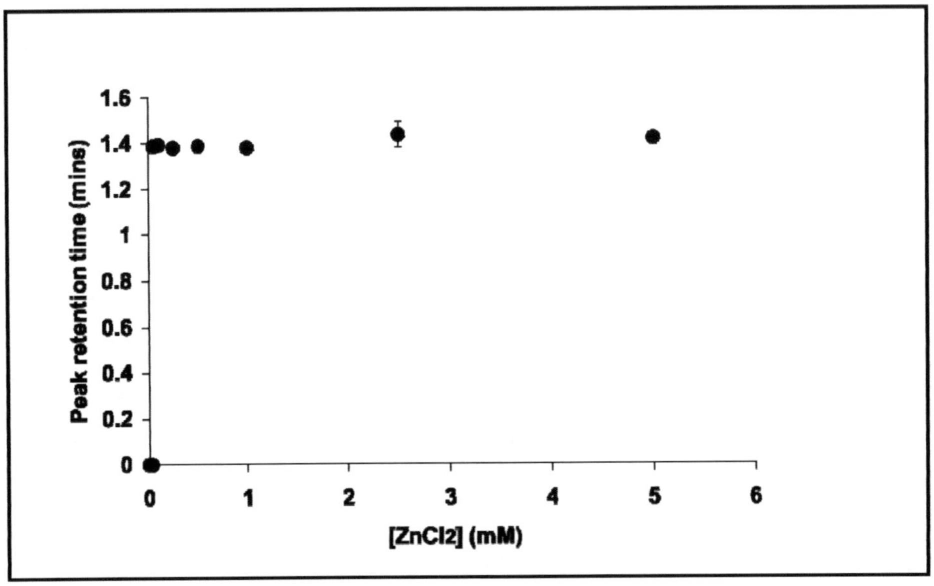

b)

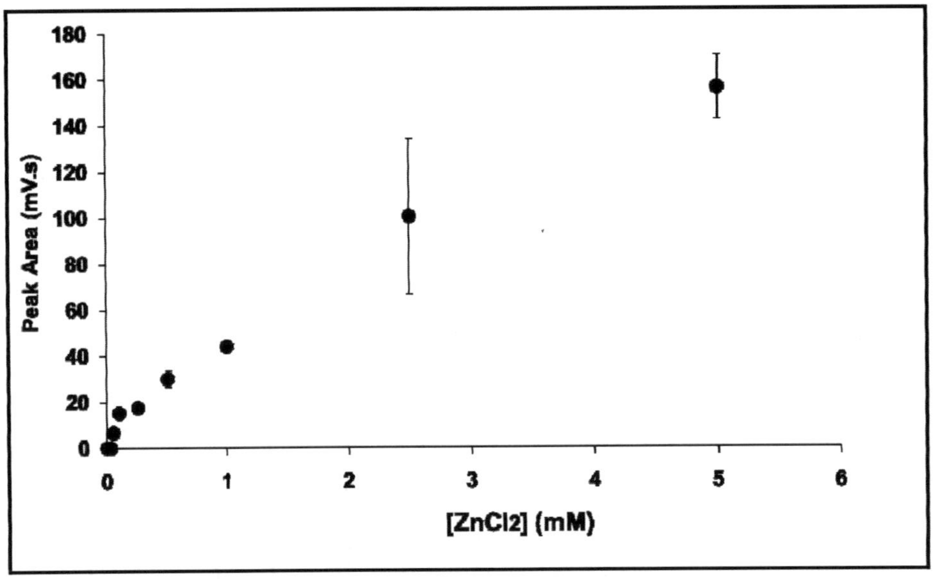

Figure 3. The effect upon increasing [$ZnCl_2$] when in fish medium as assessed by the analysis of eluted zinc peak: (a) retention time, t_R and (b) peak area. Electrophoretic separation and detection were acquired as in Figure 2b. The plot shows mean of 3 replicates with +/-1SD.

Figure 3b shows the corresponding zinc peak area w.r.t $ZnCl_2$ (when in fish medium) concentration, indicating a strong linear relationship. Repeatability for peak area is less than 10%; this is true apart from concentrations of 0.05 and 2.5 mM, with RSD of 37 and 34%, respectively. The high RSD associated with the low concentration of 0.05 mM can be attributed to the fact that the measurements are close to limit of quantification. The poor RSD associated with 2.5 mM, however, is puzzling, with no real concrete explanation put forward at this stage. Future repeatability study will need to be conducted in order to establish if this is a real effect and if there is a need to further develop the method further to ensure better repeatability.

Figure 4 shows results from the ZnO dissolution study. Figures 4a and 4b show the electropherograms of the blank i.e. fish medium only, and the 500 mg/L ZnO dispersion in fish medium, respectively; both were taken at the end of the 3 week study. It is clear from Figure 4b that there is a second peak eluted after the blank peak; the blank peak is apparent in both Figures. The second peak can be attributed to dissolved zinc from the ZnO nanomaterial dispersion; this result is similar to before (Figure 2b), in that the zinc peak elutes soon after the blank peak. The size of the zinc peak observed here, however, is very small with peak height < 5 mV. This zinc peak was monitored throughout the 3 weeks study and it first emerged at the end of the first week. After this, the zinc peak seems to appear and disappear within the 3 weeks timeframe. Even though the zinc peak is not always present after week 1, this peak was never present in the corresponding blank samples throughout the study.

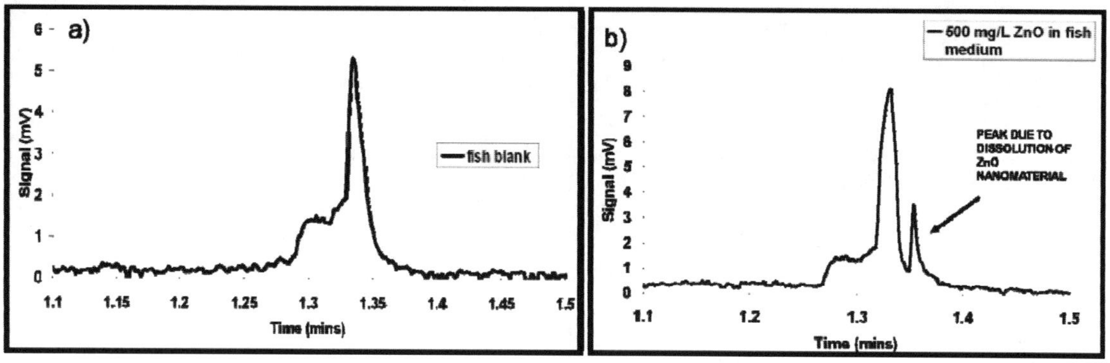

Figure 4. Electropherograms from the ZnO dissolution study of: a) blank b) 500 mg/L ZnO dispersion at the end of the 3 weeks study. Electrophoretic separation and detection were acquired under the same experimental conditions as in Figure 2b.

The appearance/disappearance of the zinc peak in the case of the ZnO dispersion can only be explained by the fact that, as stated before, zinc dissolution from nanomaterial is not straightforward and there is a possibility that free zinc re-precipitates in the solid form. As indicated by past workers, several factors can affect the kinetics associated with this re-precipitation event including: pH, concentration of other dissolved ions/organic compounds in the liquid medium. [16, 17,18].

4. Conclusion

In this study, a μTAS device was shown to be useful in the measurement of dissolution of metal oxide nanomaterial. In particular, we had shown the feasibility of the CE-conductivity microchip to investigate the solubility ZnO nanomaterial in fish medium. The device was used to highlight the complexity the dissolution events and that the re-precipitation of ionic zinc (and its kinetics) must be considered and further investigated in order to elucidate the mechanism of nanomaterial toxicity.

References

[1] OECD Series on the Safety of Manufactured Nanomaterials, *ENV/JM/MONO(2010)46*, No. 27 - 46, 2010 List of manufactured nanomaterials and list of endpoints for phase one of the sponsorship programme for the testing of manufactured nanomaterials:revision. http://www.oecd.org/document/53/0,3746,en_2649_37015404_37760309_1_1_1_1,00.html

[2] *ISO, ISO/TR 13014* 2012 Nanotechnologies -- Guidance on physico-chemical characterization of engineered nanoscale materials for toxicologic assessment.

[3] Gilbert B, Fakra S C, Xia T, Pokhrel S, Mädler L, Nel A E, 2012 The fate of ZnO nanoparticles administered to human bronchial epithelial cells *ACS Nano*. **6**(6):4921-30.

[4] Fabrega J, Tantra R, Amer A, Stolpe B, Tomkins J, Fry T, Lead J R, Tyler C R and Galloway, T S 2012 Sequestration of zinc from zinc oxide nanoparticles and life cycle effects in the sediment dweller amphipod corophium volutator *Environ. Sci. Technol* **46**(2) 1128-35.

[5] Tantra R, Boyd R, Cackett A, Fry A T, Gohil D D, Goldberg S, Lee J L S, Minelli C, Peck R, Quincey P, Smith S, Snowden J, Spencer S, Tompkins J, Wang J, Yang L 2012 NPL report: final report on the physico-chemical characterisation of PROSPEcT engineered nanomaterials: http://publications.npl.co.uk/dbtw-wpd/exec/dbtwpub.dll?&QB0=AND&QF0=ID&QI0=%206281%20&TN=NPLPUBS&RF=WFullRecordDetails&DL=0&RL=0&NP=4&AC=QBE_QUERY.

[6] Pasqui D, Golini L, Della Giovampaola C, Atrei A and Barbucci R 2011 Chemical and biological properties of polysaccharide-coated titania nanoparticles: the key role of proteins. Biomacromolecules *Biomacromolecules* **12**(4) 1243- 49.

[7] Weigl B H, Bardell R L and Cabrera C R 2003 Lab-on-a-chip for drug development *Adv Drug Deliver Rev* **55**(3) 349-377.

[8] Elbashir A A and Aboul-Enein H Y 2012 Recent advances in applications of capillary electrophoresis with capacitively coupled contactless conductivity detection (CE-C4D): an update *Biomed. Chromatogr.* **26**(8) 990-1000.

[9] Vrouwe E X, Luttge R, Vermes I and van den Berg A 2007 A Microchip capillary electrophoresis for point-of-care analysis of lithium *Clin. Chem.* **53**(1) 117-123.

[10] Tanyanyiwa J and Hauser P C 2002 High-voltage capacitively coupled contactless conductivity detection for microchip capillary electrophoresis *Anal. Chem.* **74**(24) 6378- 82.

[11] Tantra R, Jing S and Gohil D 2010 Technical issues surrounding the preparation, characterisation and testing of nanoparticles for ecotoxicological studies, V.B. Popov (Editor) WIT Press, *Environmental Toxicology 3* 165-176.

[12] eDAQ, Application Note C4D016. Procedure for microchip electrophoresis with C4D. http://www.edaq.com/index.php.

[13] Brito-Neto J G A, da Silva J A F, Blanes L and do Lago C L 2005 Understanding capacitively coupled contactless conductivity detection in capillary and microchip electrophoresis. part 1. fundamentals *Electroanal.* **17** (13) 1198-1206.

[14] Koneshan S, Rasaiah J C, Lynden-Bell R M and Lee S H 1998 Solvent structure, dynamics, and ion mobility in aqueous solutions at 25 degrees C *J. Phys. Chem. B* **102**(21) 4193-4204.

[15] Reichle R A, McCurdy K G and Hepler L G 1975 Zinc hydroxide: solubility product and hydroxy-complex stability constants from 12.5 - 75 degrees C *Can. J. Chem.* **53** 3841- 45.

[16] Degen A and Kosec M 2000 Effect of pH and impurities on the surface charge of zinc oxide in aqueous solution *J Eur Ceram Soc* **20**(6) 667-673.

[17] Lv J, Zhang S, Luo L, Han W, Zhang J, Yang K and Christie P 2012 Dissolution and microstructural transformation of ZnO nanoparticles under the influence of phosphate *Environ Sci. Technol.* **46**(13) 7215- 21.

[18] Mudunkotuwa I A, Rupasinghe T, Wu C M and Grassian V H 2012 Dissolution of ZnO nanoparticles at circumneutral pH: a study of size effects in the presence and absence of citric acid *Langmuir* **28**(1) 396-403.

Acknowledgments

The authors gratefully acknowledge contributions from Drs. Robert Boyd, Paul Quincy, Richard Brown and Mr. Kenny Robinson at NPL. The work was conducted initially as part of PROSPEcT, which is a public-private partnership between DEFRA, EPSRC and TSB and the Nanotechnology Industries Association (NIA Ltd.) and its members, and was administered by the DEFRA LINK Programme. The authors gratefully acknowledge funding from the National Measurement Office (NMO) through the Chem-Bio programme.

Characterization of manufactured TiO_2 nanoparticles

C Motzkus[1], T Macé[1], S. Vaslin-Reimann[1], P. Ausset[2] and M. Maillé[2]
[1] Laboratoire National de métrologie et d'Essais (LNE), 1 rue Gaston Boissier, 75724 Paris Cedex 15, France
[2]°Laboratoire Interuniversitaire des Systèmes Atmosphériques (LISA), UMR CNRS 7583, Université Paris-Est Créteil et Université Paris-Diderot, 61 Avenue du Général de Gaulle, 94010 Créteil.

E-mail: charles.motzkus@lne.fr

Abstract.
Technological advances in nanomaterials have allowed the development of new applications in industry, increasing the probability of finding airborne manufactured and engineered nano-objects in the workplace, as well as in ambient air. Scientific studies on health and environmental risks have indicated that airborne nano-objects in ambient air have potential adverse effects on the health of exposed workers and the general population. For regulatory purposes, ambient measurements of particulate matter are based on the determination of mass concentrations for PM10 and PM2.5, as regulated in the European Directive 2008/50/EC. However, this legislation is not suitable for airborne manufactured and engineered nano-objects. Parameters characterising ultrafine particles, such as particle number concentration and size distribution, are under consideration for future health-based legislation, to monitor workplaces and to control industrial processes. Currently, there are no existing regulations covering manufactured airborne nano-objects. There is therefore a clear, unaddressed need to focus on the toxicology and exposure assessment of nano-objects such as titanium dioxide (TiO_2), which are manufactured and engineered in large quantities in industry. To perform reliable toxicology studies it is necessary to determine the relevant characteristics of nano-objects, such as morphology, surface area, agglomeration, chemical composition, particle size and concentration, by applying traceable methods. Manufacturing of nanomaterials, and their use in industrial applications, also require traceable characterisation of the nanomaterials, particularly for quality control of the process. The present study arises from the OECD WPMN sponsorship programme, supported by the French Agency for Environmental and Occupational Health Safety (ANSES), in order to develop analytical methods for the characterization of TiO_2 nanoparticles in size and count size distribution, based on different techniques to characterize five different manufactured TiO_2 nanoparticles. In this study, different measurement techniques have been implemented: Transmission Electron Microscopy (TEM), Scanning Mobility Particle Sizer (SMPS) and Aerodynamic Particle Sizer (APS).
The TEM results lead to a relatively good agreement between data from the manufacturer and our characterizations of primary particle size. With regard to the dustiness, the results show a strong presence of agglomerates / aggregates of primary particles and a significant presence of emitted airborne nanoparticles with a diameter below 100 nm (composed of isolated primary particles and small aggregates / agglomerates formed from a few primary particles): the number proportion of these particles varies from 0 to 44 % in the measurement range 14-360 nm depending on the types of powders and corrections of measurements.

1. Introduction

The technological advances in nanomaterials during the last twenty years have allowed the development of new applications in various fields such as medicine (for therapeutic and diagnostic purposes), environment (water treatment), food (membranes for purification), industry (electronics, energy), etc.

The scientific knowledge on health and environmental risks associated to nanoparticles shows that airborne nanoparticles in ambient air have potential adverse effects on the health of exposed workers and on the general population (Witschger *et al.*, 2005 [1]). Nowadays, only particles smaller than 2.5 microns are considered and measured by air quality monitoring networks in France. However, studies have shown that excessive amounts of solid and poorly soluble aerosols are able to penetrate deeply into the lungs by inhalation, to stay there permanently and to create a pulmonary overload weakening the ability of the body's defense (Witschger *et al.*, 2005 [1]; Hervé-Bazin B, 2007 [2]). These effects are all the more important since the particles are fine.

In the context of exposed workers, a report by the French Agency for Environmental and Occupational Health Safety (ANSES), entitled "Nanomaterials and safety" [3], indicates a production of several thousands tons of manufactured TiO_2 nanoparticles in french industry. Therefore, in order to fulfil its aim of protecting human health, ANSES began to focus on the toxicology of nanoparticles such as titanium dioxide (Vincent, 2008 [4]) in order to protect the health of the workers involved in their production process. However, to perform reliable toxicology studies, it is necessary to determine the relevant characteristics of TiO_2 nanoparticles given by the International Organisation of Standardardization TC 229 "Nanotechnologies" and the OECD, such as the particle size/size distribution, shape, purity/impurity, solubility, surface chemistry, surface charge, crystallinity, surface area, chemical composition, aggregation/agglomeration state, concentration.

Consequently, we have developed an analytical strategy to characterize some key parameters of TiO_2 nanoparticles such as size and count size distribution, based on different techniques to characterize five different manufactured TiO_2 nanoparticles. The characterization results have been obtained after having developed traceable measurement methods (to the international system) and measurement uncertainty estimation (repeatability and reproducibility) to ensure reliability of the characterization of nanoparticles. In this study, different measurement techniques have been implemented: Transmission Electron Microscopy (TEM), Scanning Mobility Particle Sizer (SMPS) and Aerodynamic Particle Sizer (APS). The two measuring instruments SMPS (ISO 15900 [5]) and APS allow measuring and comparing the count size distribution, over a large range between 15 nm to 20 µm on the airborne particles generated with different manufactured TiO_2 powders in the same conditions of dispersion ; this work allows to determine different dustiness of the nano-powders (Hamelmann et Schmidt, 2003 [6]). This study deals with the physico-chemical characterization of different manufactured TiO_2 powders containing primary nanoparticles: P25 (Evonik), PC105 and Tiona AT1 (Cristal global – Milennium), and UV TITAN M262 and M212 (Sachtleben). In this paper, we present only the results for the P25 (Evonik) taking as an example.

This work arises from the OECD WPMN sponsorship programme.

2. Experimental methods

2.1. TEM measurement method

The objective of this protocol is to set up the size distribution of TiO_2 particles from powder by Transmission Electron Microscopy (TEM) after their deposition on a TEM grids.

For each type of TiO_2 powder, a very small amount (2 mg) was introduced in 6 ml of filtered ethanol, then the mixture was stirred and sonicated for 1 min (ultrasonic bath of BIOBLOCK 88155 type 460 with Freq 35 KHz).

A drop of this suspension was deposited on copper TEM grids, and dried within a laminar flow in a fume cupboard and then observed. Two hundred mesh copper grids covered with Formvar/Carbon film (reference S162 of Agar Scientific) were used for transmission electron microscope measurements. A Jeol 100 CXII with magnifications between x100,000 and x320,000 and an acceleration voltage of 100 kV was coupled to a CCD camera (Dualvision 3000W model 780, Gatan Inc.) allowing digital images acquisition for the particles size and morphology determinations. The camera calibration was performed, at low magnification, with a Carbon grating replica grid (type 10020, 2160 lines/mm = 0.463 micron /line, E.F. FULLAM) and at high magnification, with Polystyrene Latex Spheres of (102 ± 3 nm with a standard deviation of 7.6 nm from the particle size distribution calibrated, Duke Scientific Corporation) certified by National Institute of Standards and Technology (NIST). The images were taken at different magnifications, depending on the size of the particles encountered: x 270 000.

For each powder, 20 to 30 images were performed at different random positions on the grid (in different meshes and different positions in the mesh): a maximum of five images were taken in the same mesh.

The obtained images are mostly composed of aggregates of primary particles with a size between 10 and 50 nm excepted for TIONA AT1. The primary particle size distribution was determined from 15 images of aggregates measured by TEM in order to obtain at least 100 primary particles measured.

A recent work of Song et al. (2009 [7]) confirms the importance of a statistical point of view to have at least 100 particles to correctly describe the size distribution when the expected size distribution is either normal or lognormal. In the case of a population with a log-normal or normal distribution, taking into account the measurement of 200, 500 or 1000 particles doesn't give a significant contribution to the statistical level, see negligible ([7] and [8]).

The average primary particle size was calculated according the formula:

$$\bar{d} = \frac{\sum (d_l + d_s)}{2n}$$

Where,

- d_l is the long diameter which is the length of the longest axis of the counted particle,

- d_s is the short diameter which is the length of the axis perpendicular to and through the center of the longest axis of each counted particle,

- n is the total number of particles.

The diameter of primary particles is determined by the average length (l) and width (s) of the ellipse circumscribed with the particles using the software Image J. This software is currently used to process images produced by TEM (Song et al. 2009 [7]).

2.2. Dustiness measurement method

In the first time, a bibliographic research has been performed to determine the actual methods of the nano-powder dispersion and shows that there is a few industrial systems that are able to disperse the nanoparticles starting from the nano-powders (Ma-Hock *et al.*, 2007 [9]).

Although the most used and reproducible technique is atomisation (using a liquid suspension containing the powder), we decided to produce an aerosol with a powder disperser based on the vortex principle. This disperser has the advantage to produce an aerosol with a size distribution of the potentially particles close to the original powder (aggregates/agglomerates and primary particles) and the chemical nature unmodified by the use of a liquid suspension, ultrasonic protocols or thermic treatment. Furthermore, dispersion directly made by a vortex air allows to determine the potential or the maximum emission particles during an hypotethic accident or incident during industrial process and storage.

2.2.1. Generation method

In a first step, we selected a commercial generator to study the dustiness of the nano-powders of TiO_2 nanoparticles. The chosen generator was the commercial disperser "model PA100" of Naneum based on the principle of a vortex. This system generates aerosol from powder by concentrating high velocity vibrating jets of clean air at the powder surface, disturbing and separating particles from the powder to create an aerosol.

Vortices are formed by the turbulent air flow inside the aerosolisation chamber which acts to further break apart the aerosol particles. The aerosol is then passed through a 10 liter gravitational chamber which allows to homogenize the concentration and to remove the larger aerodynamic particles sizes, above a few ten micrometers on the basis of gravitational settling. The concentration of aerosol particles and the separation of the particles constituting an agglomerate can be controlled by adjusting the position of the jet nozzle in relation to the surface of the aerosolising powder.

A generation protocol was developed to optimize the protocol for determining the dustiness of nanoparticles from the primary particles constituting the agglomerates/aggregates of nanostructured powders.

The goal of this protocol is to achieve the most stable and reproducible aerosol in size and number concentration including the following steps.

- The aerosol dispersion state from powders; dispersion of the powder in aerosol phase by injection of filtered air (choice of the jet nozzle, distance between the surface of the powder and the nozzle, flow rate of the disperser and powder mass used).
- The aerolic conditions of the dispersed material; this addresses mainly the dilution step with clean dry air.
- The change of some plastic parts with stainless steel parts to reduce the loss of particles by electrostatic effects.

In addition, a cleaning protocol to be implemented before each measurement has been developed and validated. With our experimental conditions (geometric dimensions, flow rates and concentrations), this system allows to homogenize the aerosol while minimizing losses by diffusion and sedimentation of the generated aerosol in the measurement range based on calculations of deposit particles in the settling and homogenization chamber (Hinds, 1999 [10]).

2.2.2. Experimental facilities

Figure 1 presents the facilities developed to measure the count size distribution of an aerosol containing the nanometer primary particles and aggregates/agglomerates over a large range from 15 nm to 20 µm. This set-up is composed of an air filtration system, a generator of aerosol from a powder, a flow splitter, a diluter and two systems of parallel measurements: Scanning Mobility Particle Sizer (SMPS) and Aerodynamic Particle Sizer (APS). In this study, the results are expressed in aerodynamic diameter (d_{ae}) and electrical mobility diameter (d_m) respectively for the results obtained by the APS and SMPS.

We observed that the concentration was above the APS concentration limit (>> at 1000 particles /cm^3), thus it was necessary to use a diluter (VKL of PALAS) with clean dry air in order to measure the aerosol emitted by the generator. The used SMPS is composed of the impactor to remove the larger particles above 1 µm and thus measure the submicrometer particles, an aerosol neutralizer with a Kr85 source (model TSI 3077), a Long Differential Mobility Analyzer (model TSI 3081) and a Condensation Nuclei Counter (CNC). The used CNC (TSI 3022) detects particles from 7 nm (minimum particle size corresponding to an efficiency of 50%) at concentrations up to 10^7 particles/cm^3 (Sem, 2002 [11]).

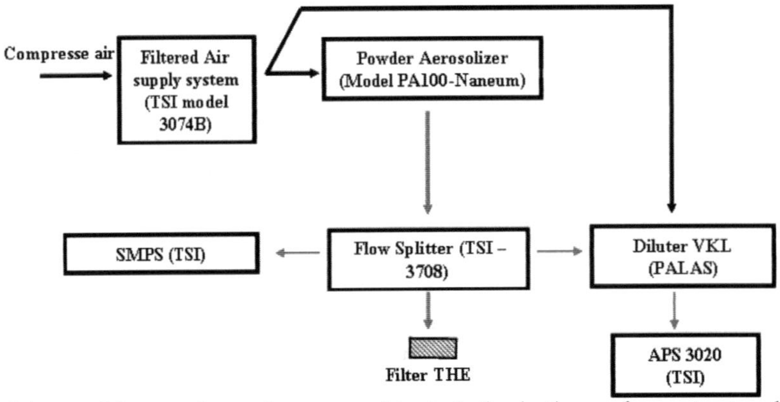

Figure 1. Scheme of the experimental set-up used to study the dustiness of nanostructured powders

Experimental conditions for SMPS

The flow rate of the polydisperse aerosol inlet was 0.3 L/ min. The sheath flow rate used in the Electrostatic Classifier (DMA 3080) was equal to 3 L/min. In these conditions, the electrical mobility diameter d_m range was between 14 and 673 nm. The cut-off sizes (d_{50}) of the impactor were respectively of 812 and 360 nm for the particles with a density of 1 and 4.2 g/cm^3 (TiO_2 density). It is very important to underline that the cut-off size of 360 nm was lower than the maximum electrical mobility diameter d_m of 673 nm. Thus, the larger particles above the d_{50} , with a strong inertia, will be removed by the impactor. We decided to take into account only the particles with a d_m below 360 nm. We used the Aerosol Instrument Manager Software of TSI to perform the diffusion and charge corrections for the following densities: 4.2 g/cm^3 for the particle and 0.0012 g/cm^3 for gas.

Experimental conditions for APS and the dilutor

To compare the APS and SMPS measurements, it was decided to use the results of the APS on the same scanning time as the SMPS. With regard to the diluter VKL, its flow rate input was checked before each measurement to ensure that it was not clogged by a too high concentration of particles. The results presented in this paper take into account the dilution factor.

Experimental conditions for the generator AERO PA100

The flow rate of dispersion is 6 L/ min and the mass of powder in the sample holder was fixed at 2.5 g. However, it was necessary to reduce this amount to 1.5 g when the dustiness was too high for some nanostructured powders, and thus to produce a "puff" of aerosol that can block up the jet nozzle. Two distances between the orifice and the surface of the powder were used in the experiments.

For each distance, the results of the count size distribution of twenty scans have been taken into account to determine the average and the standard deviation of the following parameters: mean diameter, median diameter, geometric diameter, modal diameter and geometric standard deviation.

3. Experimental results for the powder AEROXIDE P25 (Evonik)

3.1. Dustiness results

Preliminary work has been performed to develop measurement methods to ensure the traceability of measurements (connecting instruments to the international system) and to estimate the uncertainties of measurements (repeatability and reproducibility) of the nanoparticles characterization.

On the same batch of powder, repeatability (several experiments in one day) and reproducibility (experiments over several days) experiments have been performed to have a better understanding of the count size distribution and the associated uncertainties.

A slow decrease in number concentrations measured by the APS and the SMPS during the time dispersion has been observed. But this decrease did not affect the stability of particles size distribution. Figure 2 shows the count size distribution for a measurement realized by the APS in the case of the powder P25 of Evonik taking as an example.

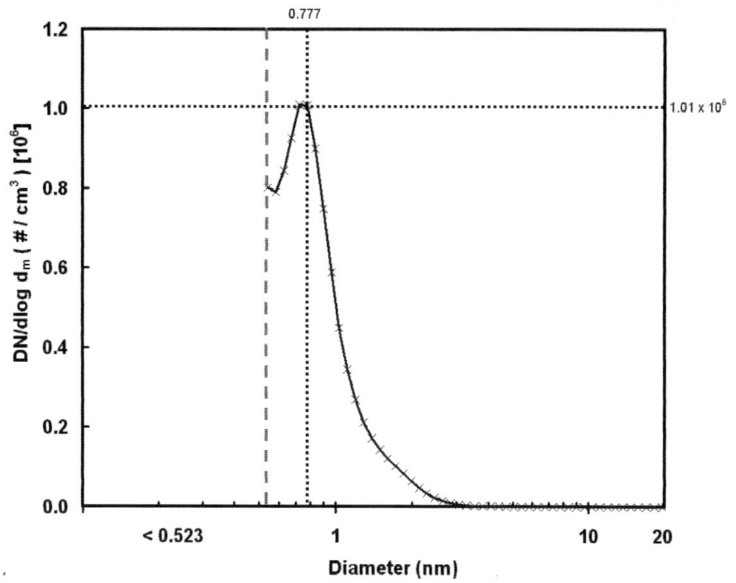

Figure 2. Count size distribution of the aerosol obtained for one sample with APS in the case of the powder AEROXIDE P25

Table 1. Parameters characteristics of size distribution obtained by APS in the case of the powder AEROXIDE P25

Statistical diameters	Experiment 1st day	Experiment 2nd day	Experiment 3rd day			
	Average on 20 samples	Average on 20 samples	Average on 20 samples	Total Average	Repetability Standard deviation	Reproducibility Standard deviation
Median (µm)	0.77	0.78	0.76	0.77	0.03	0.03
Mean (µm)	0.87	0.89	0.92	0.90	0.04	0.05
Geometric Mean (µm)	0.81	0.83	0.82	0.82	0.03	0.03
modal (µm)	0.65	0.64	0.57	0.62	0.10	0.11

The average parameters characteristics of size distribution obtained by APS were determined from several experiments (table 1) and twenty samples (scan measurements).

The observed size distributions have statistical diameters (median, arithmetic mean, geometric mean and modal) which are repeatable and reproducible (table 1). We obtained low repeatability standard deviations (<0.10 µm) and low reproducibility standard deviations (<0.11 µm). The obtained statistical average diameters are in a range between 0.6 and 0.9 µm. It is important to highlight the low presence of particles (agglomerates) above a few microns (see Figure 2). Most of the particles measured in the range 0.5 to 20 µm have an aerodynamic diameter between 0.5 µm and 1 µm.

Figure 3 presents respectively the count size distributions for a measurement performed with SMPS by taking into account both the diffusion and charge corrections (a) and with only the diffusion correction (b) in the case of the powder P25 (Evonik). This figure show that size distributions are influenced by taking into account the charge correction, as demonstrated by Ogura et al. [12] which studied NTC, fullerenes, ZnO and also TiO_2 dustiness powders.

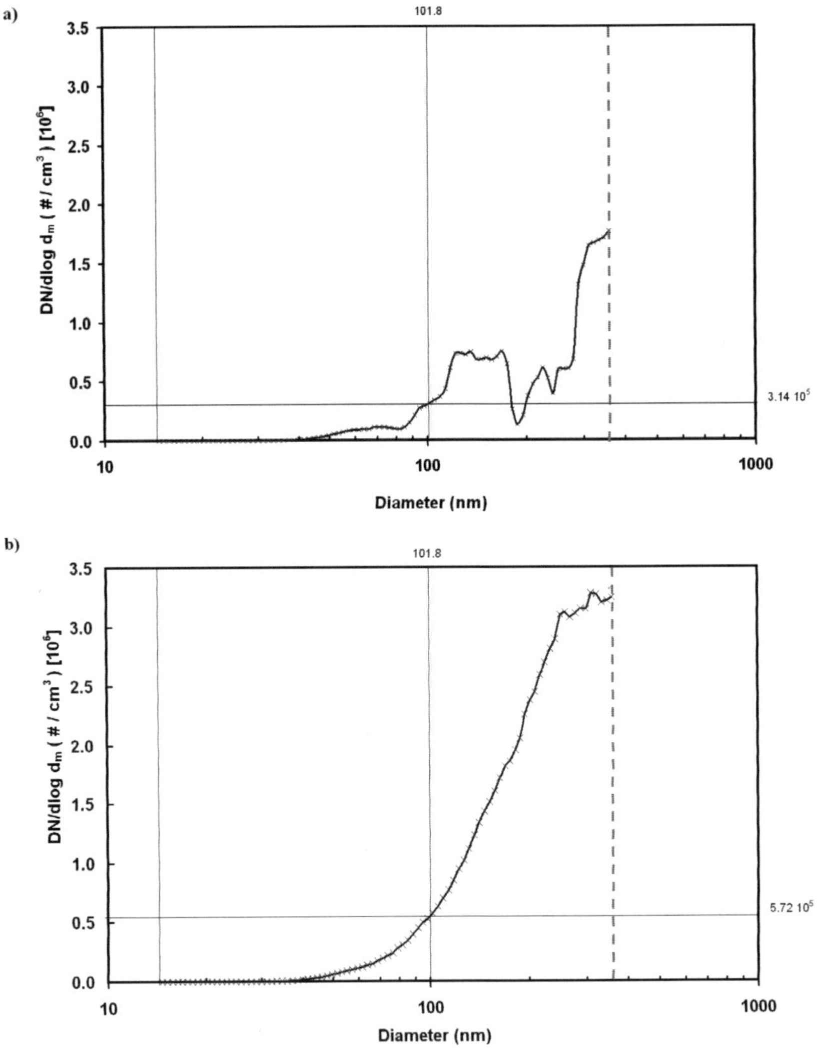

Figure 3. Count size distribution of the aerosol obtained for one sample with SMPS in the case of the powder Aeroxide P25: (a) both the diffusion and charge corrections and (b)with only the diffusion correction

Table 2. Parameters characteristics of size distribution obtained by SMPS with both diffusion and charge corrections in the case of the powder AEROXIDE P25

Statistical diameters	Experiment 1st day	Experiment 2nd day	Experiment 3rd day			
	Average on 20 samples	Average on 20 samples	Average on 20 samples	Total Average	Repetability Standard deviation	Reproducibility Standard deviation
Median (nm)	303	300	287	297	30	31
Mean (nm)	254	256	244	251	21	21
modal (nm)	357	354	351	354	9	9

Table 3. Parameters characteristics of size distribution obtained by SMPS with both with only the diffusion correction in the case of the powder AEROXIDE P25

Statistical diameters	Experiment 1st day	Experiment 2nd day	Experiment 3rd day			
	Average on 20 samples	Average on 20 samples	Average on 20 samples	Total Average	Repetability Standard deviation	Reproducibility Standard deviation
Median (nm)	257	254	253	255	11	11
Mean (nm)	249	248	246	248	10	10
modal (nm)	351	352	343	349	14	14

We observed that the parameters of the size distribution (tables 2 and 3) are repeatable and reproducible (low standard deviations of repeatability and reproducibility less than 14 and 31 nm with only the diffusion correction and with both the diffusion and charge corrections respectively). The average statistical diameters (mean, median and mode) obtained are between 240 nm and 360 nm. In the case of charge correction, we observed that the total average of these statistical diameters were slightly higher until 14 % compared to the case without charge correction. It is important to notice that a high presence of agglomerates / aggregates of primary particles of a few hundred of nanometers is observed. Most of the particles measured in the range 14-360 nm have electric mobility diameter between 100 and 360 nm.

Nevertheless, the results show a significant presence of emitted airborne nanoparticles with a diameter below 100 nm. Taking into account only the diffusion correction, we obtained a concentration of ultra fine particles (< 100nm) that could vary between 1 and 12% compared to the total concentration in the measurement range 14 -360 nm. In contrast, with both the diffusion and charge corrections, this concentration can vary between 1 and 20%, involving a greater proportion of emitted particles with a diameter below 100 nm. It is therefore very important to turn attention to the used measurement methods and the employed corrections because they can lead to different conclusions.

3.2. Size distribution of primary particles

Figure 4 shows aggregates of TiO_2 nanoparticles from powder AEROXIDE P25 (Evonik).

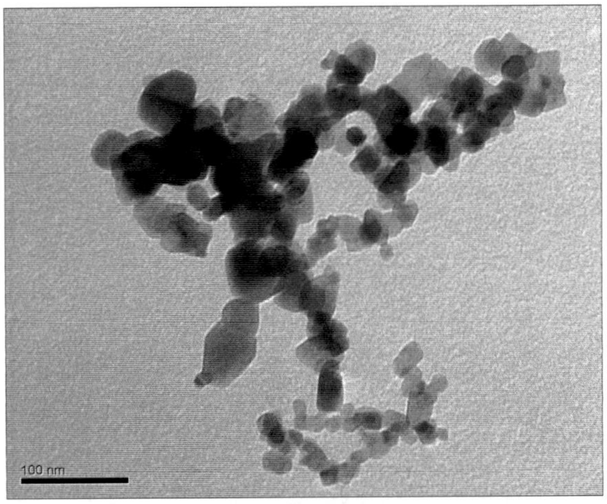

Figure 4. TEM image of TiO_2 particles from powder Aeroxide P25

The aggregates are composed of primary particles with diameters between 9 and 53 nm. The primary particle size distribution was performed from 13 images of aggregates measured by TEM (see Figure 5). This size distribution was established with measurements of 138 primary particles. We found a size distribution with a modal diameter of 20 nm and a mean diameter of 23 nm with a standard deviation of 7 nm.

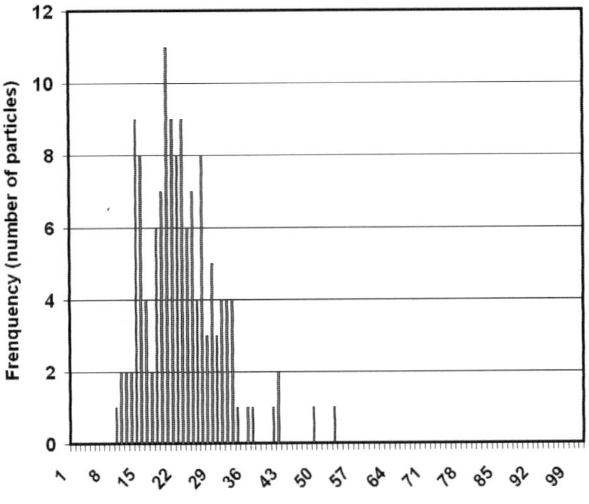

Figure 5. Count size distribution of the primary particles of TiO_2 from powder AEROXIDE P25 determined by TEM

We observed that the obtained standard deviations are very high compared to the measurement uncertainty associated with the microscopy method (1 pixel corresponds to 0.42 nm for magnifications of x 270 000 and 0.78 nm for x 140 000).

4. Conclusion

This study allows us to develop analytical methods for the characterization of TiO_2 nanoparticles in size and count size distribution, based on different techniques. The characterization results include the development of traceable measurement methods (to the international system) and measurement uncertainty estimation (repeatability and reproducibility) to ensure the reliability of the characterization of nanoparticles. In this study, different measurement techniques have been implemented: Transmission Electron Microscopy (TEM), Scanning Mobility Particle Sizer (SMPS) and Aerodynamic Particle Sizer (APS). The two measuring instruments SMPS (ISO 15900) and APS allow the measurement and the comparison of count size distribution, over a large range from 15 nm to 20 µm. This work has been focused on airborne particles generated with different manufactured TiO_2 powders in the same conditions of dispersion.

This study was performed on different TiO_2 manufactured powders: P25 of Evonik, PC105 and Tiona AT1 of Cristal global – Milennium, and UV TITAN M262 and M212 of Sachtleben. But we have presented in this paper only the results for the P25 (Evonik) taking as an example.

The TEM results allow to observe a good agreement between data coming from the manufacturer and our characterizations of primary particle size. With regard to the dustiness, the results show a strong presence of agglomerates / aggregates of primary particles and show a significant presence of emitted airborne nanoparticles with a diameter below 100 nm (composed of isolated primary particles and small aggregates / agglomerates formed from a few primary particles): the number proportion of these particles varies from 0 to 44 % in the measurement range 14-360 nm depending on the types of powders and corrections of measurements.

5. References

[1] Witschger O and Fabriès J F 2005 " Ultra-fine particles and occupational health. 1 - Characterization of the potential effects on health " INRS, Hygiène et sécurité au travail *ND 2227* **199** 21-35
[2] Hervé-Bazin B, 2007 Nanoparticles: a major challenge for occupational health *?* 1ère édition Les Ulis EDP Sciences
[3] Agence Française de Sécurité Sanitaire de l'Environnement et du Travail (AFSSET) 2008 *Nanomaterials - work safety*
[4] Vincent J M, 2008 Toxicology of titanium dioxide nanoparticles, INERIS, report DRC-08-94464-11281A
[5] ISO 15900, 2009 "Determination of particle size distribution-Differential electrical mobility analysis for aerosol particles" 57p
[6] Hamelmann F, Schmidt E, 2003 Methods for Estimating the Dustiness of Industrial Powders- A review. *KONA 21* pp7-18
[7] Song N W, Park K M, Lee I-H and Huh H, 2009 Uncertainty estimation of nanoparticle size distribution from a finite number of data obtained by microscopic analysis *Metrologia* **46** 480-488
[8] ISO 13322-1, 2004 Particle Size Analysis - Image Analysis Methods - Part 1: Static Image Analysis Methods
[9] Ma-Hock L, Gamer A O, Landsiedel R, Leibold E, Frechen T, Sens B, Linsenbuehler M et van Ravenzwaay B, 2007 Generation and characterization of test atmospheres with nanomaterials. *Inhalation Toxicology* **19** no 10 833-848
[10] Hinds W 1999 *Aerosol Technology- Properties, behavior, and measurement of airborne particles* (Wiley-interscience, second edition)
[11] Sem GJ, 2002 Design and performance characteristics of three continuous-flow condensation particle counters: a summary. *Atmos. Res.* **62** 267–294
[12] Ogura I, Sakurai H and Gamo M, 2009 Dustiness testing of engineered nanomaterials *Journal of Physics: Conference Series* **170** p. 012003

How does the deposited dose of oxide nanomaterials evolve in an *in vitro* assay?

O Lozano[1,2,*,†], J Mejia[1,2,*], J-P Piret[1,3], C Saout[1,3], J-M Dogné[1,4], O Toussaint[1,3], and S Lucas[1,2]

[1] Namur Nanosafety Centre (NNC), NAmur Research Institute for LIfe Sciences (NARILIS), University of Namur (UNamur), Rue de Bruxelles 61, B-5000 Namur, Belgium
[2] Research Centre for the Physics of Matter and Radiation (PMR-LARN), University of Namur (UNamur), Rue de Bruxelles 61, B-5000 Namur, Belgium
[3] Laboratory of Biochemistry and Cellular Biology (URBC), University of Namur (UNamur), Rue de Bruxelles 61, B-5000 Namur, Belgium
[4] Department of Pharmacy, NAMEDIC, Namur Thrombosis and Hemostasis Centre (NTHC), University of Namur (UNamur), Rue de Bruxelles 61, B-5000 Namur, Belgium
[*] Both authors have participated equally in the study reported here.
[†] Corresponding author e-mail: omar.lozanogarcia@fundp.ac.be

Abstract. In this work, the evolution of some key physicochemical parameters of oxide engineered nanomaterial (ENM) dispersions was studied during an *in vitro* biological assessment. Commercial oxide ENMs, SiO_2 and TiO_2, were dispersed in aqueous solutions (20 μg/mL) to A549 cells and N-hTERT keratinocytes and were assessed at several incubation times: 6, 24, 48, and 72 hours. The ENMs deposited dose and its particle size distribution (PSD) were followed each time. Centrifuge Liquid Sedimentation (CLS) measured the PSD and the ENMs deposited dose from a particle entity perspective, while Particle-Induced X-ray Emission (PIXE) measured the ENMs deposited dose from an elemental mass perspective. No significant variations in PSD were observed for SiO_2 ENMs during incubation in A549 cells and TiO_2 ENMs in both cell lines, while a continuous evolution of the PSD is observed for SiO_2 in N-hTERT keratinocytes. The deposited dose for TiO_2 ENMs remained stable and similar in both cell lines due to a smaller specific surface area and a higher quantity of primary particles present during incubation. It is concluded that the observed differences in the deposited dose are related to an interaction between the proteins present in the media and the ENMs specific surface.

1. Introduction

Engineered nanomaterials (ENMs) exhibit remarkable properties, which are fundamentally different to those of bulk materials. These properties induced a rush in industry to produce and use these new materials in different products, touching wide and diverse areas such as semiconductors [1], medicine [2], cosmetics [3], and food [4, 5]. However, stakeholders have raised health concerns due to the unprecedented progress uncoupled with a proper understanding of the possible associated risks. Indeed, several reports have indicated that ENMs may be more toxic than their larger counterparts [6-9].

The field of nanotoxicology, and more broadly nanosafety, was then established with the aim to assess the potential hazard of nanomaterials [10]. Among the different efforts in the field, the *in vitro* approach allows to investigate the intracellular pathways induced by ENMs [8] and their consequences [11-13]. A relatively recent acknowledgement is the need of ENMs physicochemical characterization [11, 14, 15], not only as dry powders or aqueous predispersions but also as suspensions in cell culture medium added to their "behaviour" evaluation during *in vitro* exposure. Commercial ENMs are not always only composed of primary particles (PPs) but can also contain agglomerates and/or aggregates [10]. Once dispersed in liquid milieu the PPs/agglomerates-aggregates ratio could evolve and be influenced by sedimentation (sub-micron particles) and diffusion (nanometer-sized particles) [16]. In this respect there have been efforts to understand ENMs dynamics in cell culture media, both experimentally [17] and computationally [18]. Even if the general principles are understood in cell culture media, other parameters should play a role during cell lines incubation, such as the cell culture media consumption as a function of incubation time and cell growth. There is not any generic information about the ENMs deposited dose during a specific cell line incubation nor a description of the key parameters that play a role in predicting this deposited dose.

Oxides are a class of ENMs that are currently used for several commercial applications. For example, silica (SiO_2) is used as a food additive [4], and titania (TiO_2) is used in cosmetic products such as sunscreens [3]. These ENMs, which are part of the OECD ENM list [19], were chosen as relevant model oxides due to their current uses. Human alveolar epithelial A549 and Human telomerase-immortalized (N-hTERT) keratinocytes [20] were chosen to represent two of the most important exposure routes to ENMs [21].

The aim of this work is to study the evolution, with two complementary techniques for PSD and weight determination, of the SiO_2 and TiO_2 ENMs deposited dose at 20 µg/mL after increasing incubation times (6, 24, 48, and 72 hours), on A549 cell lines and N-hTERT keratinocytes.

2. Materials and methods

2.1. ENMs

SiO_2 ENMs dry powders were supplied by Nanologica™ through the NanoValid project (FP7 2007-2013). This is a mesoporous ENM synthesized via sol-gel. TiO_2 ENMs dry powders were purchased from Sigma Aldrich (CAS number 13463-67-7). Both ENMs were employed as received.

2.2. ENM stock dispersions

A stock dispersion of each ENM was prepared at a 1 mg/mL (w/v) concentration in ultrapure water and magnetically stirred during 15 min at 400 rpm. A 48 µL volume of the stock solution was added to each cell culture box containing each of the two culture media, specific to each cell line, to achieve a final ENM 20 µg/mL concentration.

2.3. Cell lines

Human alveolar epithelial A549 cells were purchased from the American Type Culture Collection (ATCC). A549 cells were cultured in minimum essential medium (MEM, Gibco) plus serum. MEM contains 10% fetal bovine serum (Gibco) and 1% penicillin–streptomycin (BioWhittaker). This culture medium will be referred throughout this paper either as MEM plus serum or A549 culture medium.

Human telomerase-immortalized keratinocytes (N-hTERT) [20] were kind gifts from Dr. J.G. Rheinwald (Department of Medicine and Harvard Skin Disease Research Center, Boston, MA, USA). They were maintained in Epilife medium with human keratinocyte growth supplement (Cascade Biologics) and antibiotics (penicillin/streptomycin, BioWhittaker).

Both cells were grown in a 5% CO_2 incubator in a humidified 37°C atmosphere.

2.4. In vitro assay

995,000 A549 cells or 900,000 N-hTERT keratinocytes were seeded in 25 cm^2 polystyrene flasks (T25, Corning) and incubated with 20 µg/mL (6 mL of culture medium/T25 was used) of either SiO_2 or TiO_2 ENMs for 6, 24, 48 and 72 hours (one flask was incubated for each incubation time). After incubation, the cell culture medium was collected by tilting the cell boxes towards one of the bottom corners and then stored in glass containers.

2.5. ENMs characterization

2.5.1. Dry powders. The average pristine ENMs diameter was obtained with a Field Emission Gun Scanning Electron Microscope (FEG-SEM) Jeol JSM 7500 operated at 20 keV. Both ENMs bulk composition was measured with an EDX (Energy Dispersive X-ray) detector coupled to the FEG-SEM. Each powder was deposited on a double face carbon tape.

Specific Surface Area (SSA) measurements were obtained by the BET method with an Accelerated Surface Area and Porosimetry System (ASAP, Micromeritics 2010).

Surface composition was analysed with an X-ray Photoelectron Spectroscopy (XPS) system. The used apparatus is a K-Alpha (Thermo Scientific) system using Al K-α X-rays, with spectra recorded at 90° take-off angle. The XPS analysis depth is around 5 nm. Core-level lines (C1s, Si2p) were calibrated to the C1s peak (284.6 eV). The spectra were analysed, fitting the Gaussian function to the experimental curve, with a non-linear least squares scheme, and using a Shirley background. Nominal resolution was measured as full width at half maximum of 1.0 eV (core-level spectrum) to 1.5 eV (survey spectrum). Each powder was deposited on a double face carbon tape.

2.5.2. ENM characterization during in vitro *incubation.* After incubation, the culture medium was collected and two 0.5 mL aliquots were taken: one for Centrifuge Liquid Sedimentations (CLS) and one for Particle-Induced X-ray Emission (PIXE) measurements. They were analysed within an hour in the case of CLS. Three independent experiments were conducted and the average and standard deviations were calculated.

Particle size distributions (PSD) (in number and weight) were measured with a DC24000 system Disc Centrifuge (CPS Instruments Inc.). This measurement is based on the CLS method according to Stokes' law using a 405 nm wavelength laser. This method is also known as Differential Centrifugal Sedimentation (DCS). The diameters measured are hydrodynamic diameters, which provide the real diameter (sphericity factor equal to 1) for spherical particles. A certified PVC microparticle calibration standard (226 nm), provided by the instrument supplier, was used to calibrate all measurements. Each measurement was done by injecting a 0.5 mL volume.

PIXE was used for ENM quantification in cell medium. The principles of PIXE have been previously reported [22, 23]. For this study, a proton ion beam s with a2.5 MeV energy was used at low currents (< 1 nA) in order to avoid changes due to heating in the matrix. Samples were prepared in the following way: a truncated conical shape sample holder was used, with the following dimensions: 2 cm in diameter, 4 mm of depth and 2 mm in the truncation. This holder was designed as to hold a droplet for drying and preventing horizontal spread. Double face carbon tape was placed on top of the holder, providing a conductive surface and almost no background noise for the PIXE measurements. A total volume of 200 µL was deposited as 10 droplets of 20 µL, leaving each droplet to dry in a 60 °C oven during 1 hour before placing the next. Before taking each droplet, the glass vials were vortexed for 5 seconds.

The deposited dose is defined as the ENM dose that reaches the cells by either sedimentation (sub-micron particles) or diffusion (nanometer-sized particles). It was calculated, independently with CLS and PIXE techniques, from ENM content measurements in the cell culture medium. For CLS, the estimation was the difference between the integrated particle weight distributions at each incubation time compared to the control. For PIXE, the estimation was the difference between the measured ENM elemental concentrations at each incubation time with respect to a control with the applied dose (20 µg/mL).

3. Results and discussion

3.1. ENMs characterization (dry powders)

The average diameter of both dry ENMs, calculated from SEM images, was 197.5 ± 40.5 and 25.6 ± 4.9 nm for SiO_2 and TiO_2, respectively. In the case of SiO2 ENMs, the average diameter was calculated from the particles with more than 100 nm, and not from the small particles observed at the bottom (cf. CLS results in section 3.2.1). A sample SEM image of each ENM is presented in Figure 1.

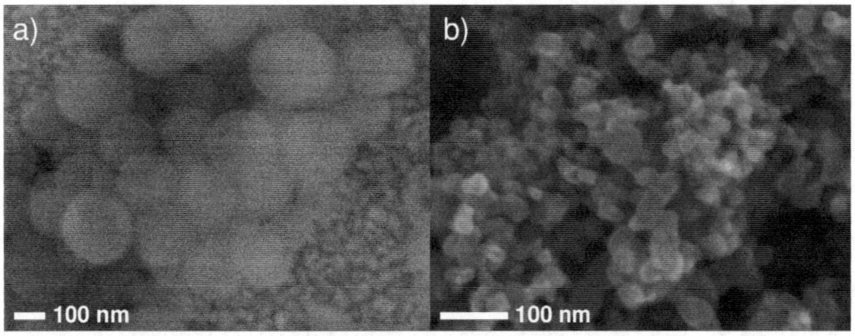

Figure 1. SEM micrographs of a) SiO_2 NPs and b) TiO_2 NPs.

The XPS surface composition spectra for both ENMs are presented in Figure 2, and summarized along with other ENMs physicochemical properties in Table 1. It is noticeable that the SiO_2 SSA was more than one order of magnitude greater than that of TiO_2. In terms of composition, both ENMs at the surface are close to their stoichiometric 1:2 metal ratio (Si or Ti) to O, while the C content values are assumed to be due to environmental and synthesis contaminations. At bulk composition, SiO_2 contains more Si than O, a 1.5:1 ratio Si to O, as well as a considerable amount of C (26.3 %). The TiO_2 bulk composition kept a 1:2 stoichiometric ratio, highlighting that the C content is mostly located on the surface.

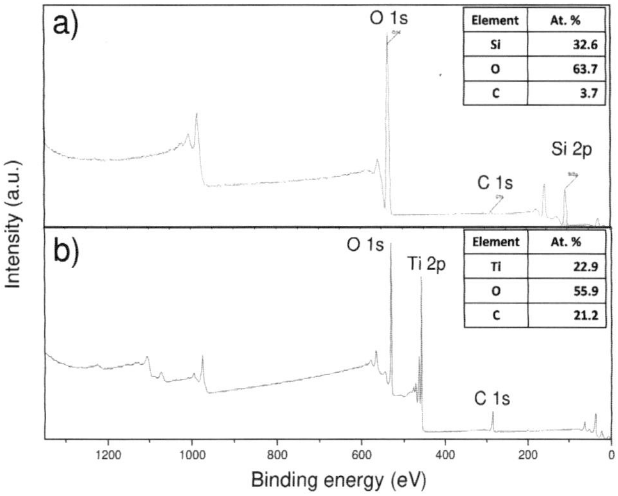

Figure 2. XPS surface composition spectra of a) SiO_2 NPs and b) TiO_2 NPs. The constituent elements atom percentage is presented.

Table 1. Physicochemical characterization of pristine SiO_2 and TiO_2 ENMs.

Physicochemical property	SiO_2						TiO_2					
	Measured			Supplier[1]			Measured			Supplier[1]		
Average diameter (SEM) nm	197.5 ± 40.5			100			25.6 ± 4.9			21		
Specific surface area m^2/g	962			-			32			35-65		
at. %	Si	O	C	Si	O	C	Ti	O	C	Ti	O	C
Bulk composition (EDX)	43.7	30.0	26.3	-			27.4	65.3	7.36	99.5%*		
Surface composition (XPS)	32.6	63.7	3.7	-			22.9	55.9	21.2			

* Product purity reported as TiO_2 only.

3.2. ENMs size distribution

3.2.1. Size distribution in ENMs stock dispersions. The stock ENMs dispersions PSD is presented in Figure 3. The PSD was evaluated in terms of weight or volume distribution (emphasis on agglomerates/aggregates) and in number distribution (emphasis on PPs). Both ENMs weight distribution , Figure 3a and 3c, shows the presence of agglomerates / aggregates and sub-micron particles. In the case of SiO_2, the particles around 150 nm correspond to the agglomerates observed by SEM, see Figure 3a. For TiO_2 the weight distribution shows an even distribution between PPs and agglomerates / aggregates, see Figure 3c. The PPs for SiO_2 have an average hydrodynamic diameter of 20 nm and 18 nm for TiO_2, according to the number distributions in Figures 3b and 3d, respectively. Both of these ENMs, even though containing significant agglomerates / aggregates when dispersed, are classified as nanomaterials in accordance to the recommendation of the European Commission for the definition of a nanomaterial: more than 50 % of particles in number being less than 100 nm in one of their dimensions [24].

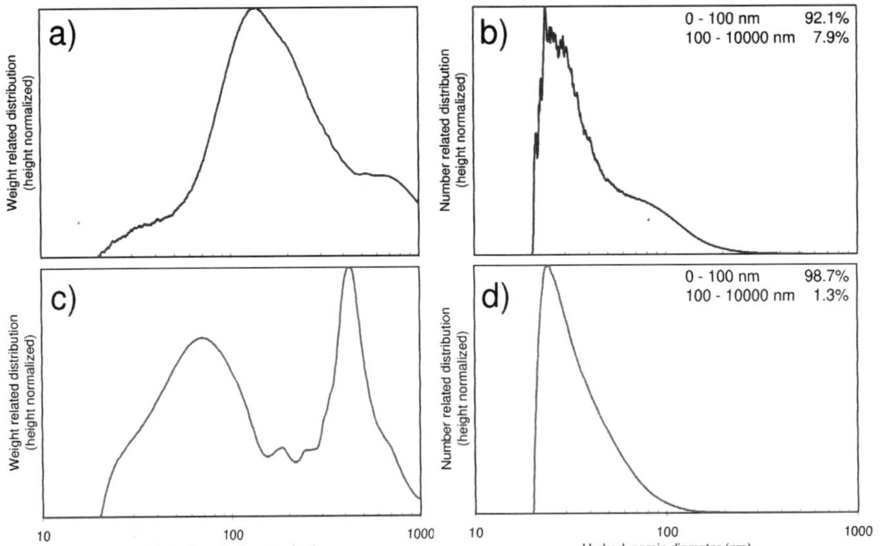

Figure 3. CLS PSD of the ENMs in the stock dispersion: a) SiO_2 weight related distribution, b) SiO_2 number related distribution, c) TiO_2 weight related distribution, d) TiO_2 number related distribution. The percentages of particles smaller and bigger than 100 nm are indicated in the number related distributions.

3.2.2. Size distribution on cell culture medium. The PSD of SiO_2 and TiO_2 ENMs in MEM plus serum and Epilife cell culture media at several incubation times are presented in Figures 4 and 5. The cut-off size for PPs was 60 nm SiO_2 and 40 nm TiO_2, values that were determined from the CLS number related distribution (see Figures 3b and 3d, respectively). In the A549 culture medium, SiO_2 ENMs PSD stabilized with time. Such a behaviour is associated to the progressive filling of the open pores on the surface and to the presence of proteins in the medium, helping in the stabilization and de-agglomeration processes [25]. The TiO_2 ENMs PSD in the A549 culture medium was nearly stable at all incubation times.

In Epilife cell medium, the SiO_2 ENMs PSD presents a de-agglomeration with time. Such a de-agglomeration is associated to the culture medium type and to SiO_2 ENMs physicochemical properties. The SiO_2 ENMs high SSA and porosity and the higher content of Si in the core are signatures of a core-shell structure. Given that Epilife is a protein free medium (no serum), this medium likely fills the SiO_2 ENMs open pores, inducing a de-agglomeration process. In the case of TiO_2 ENMs in Epilife, its PSD remains unchanged with time. This result is due to the TiO_2 ENMs hard core structure, which is not strongly affected by the de-agglomeration process.

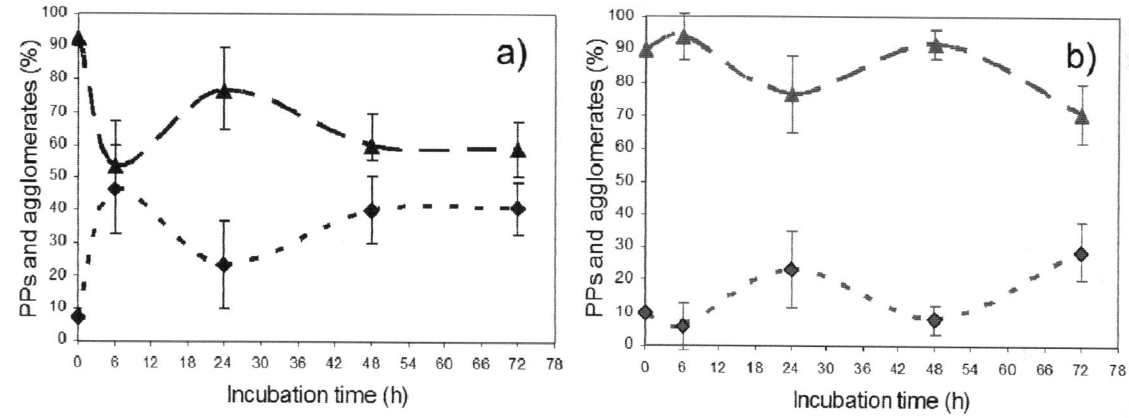

Figure 4. Percentage of PPs (---) and agglomerates (— —) from CLS measurements of a) SiO_2 ENMs and b) TiO_2 ENMs dispersed in MEM plus serum culture medium at a 20 μg/mL nominal dose during increasing incubation times with A549 cells. Values for incubation time of "0 h" correspond to the water pre-dispersion before contact with the culture medium. Both dashed lines are guides to the eye.

3.3. Deposited dose

The deposited dose (see last paragraph of Materials and Methods section) of SiO_2 and TiO_2 ENMs in both cell lines was calculated by CLS from the PSD measurements in cell culture media, see Figure 6. The CLS technique provides a weight value that corresponds to the total particle weight contained in the analysed sample, irrespective of their size. The PSD obtained from the incubation control without ENMs was subtracted from each particle weight distribution. In addition, the measurements were corroborated with respect to a control containing the applied dose without incubation (to avoid medium contributions). The results show that the SiO_2 ENMs deposited dose in the A549 culture medium was lower at 24 and 48 hours in comparison to the 6 and 72 hours where the deposited dose was similar. The progressive deposited dose decrease was less marked in the Epilife culture medium. For TiO_2 ENMS, their behaviour is quite similar in both cell culture media.

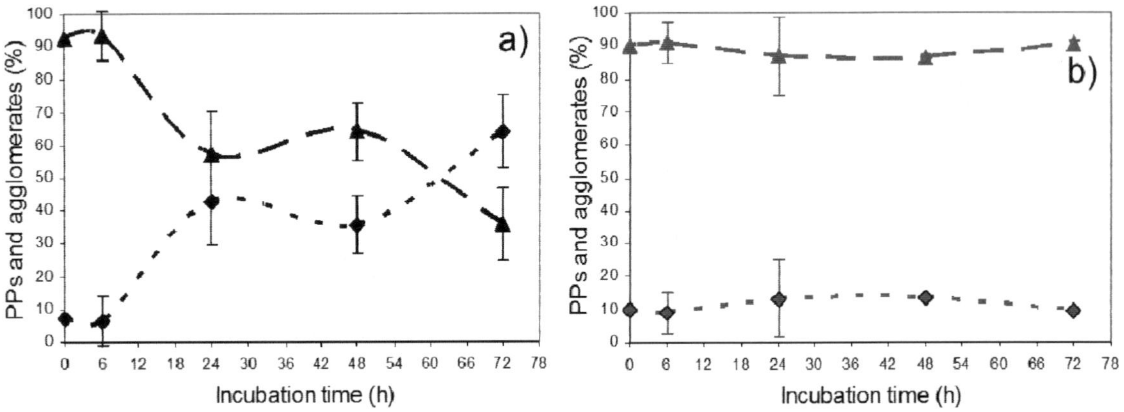

Figure 5. Percentage of PPs (---) and agglomerates (— —) from CLS measurements of a) SiO_2 ENMs and b) TiO_2 ENMs dispersed in Epilife culture medium at a 20 µg/mL nominal dose during increasing incubation times N-hTERT keratinocytes. Values for incubation time of "0 h" correspond to the water pre-dispersion before contact with the culture medium. Both dashed lines are guides to the eye.

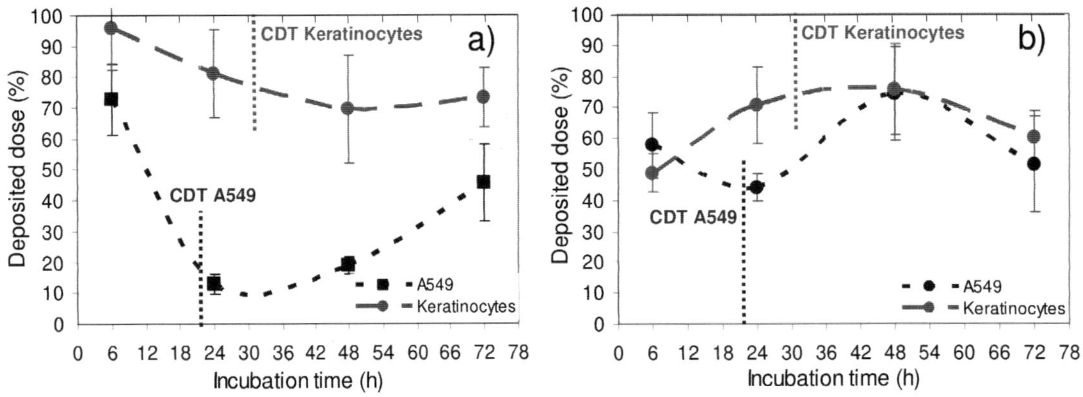

Figure 6. Deposited dose calculated from CLS measurements of a) SiO_2 ENM and b) TiO_2 ENMs dispersions during *in vitro* incubations in two cell culture media: MEM plus serum for A549 cell line (---) and Epilife for N-hTERT keratinocytes (— —) for a 20 µg/mL nominal dose. CDT stands for 'cell doubling time' which is 22 h for A549 cells and 31 h for n-HTERT keratinocytes, marked by a dotted line (···). Both dashed lines are guides to the eye.

The SiO_2 and TiO_2 ENMs deposited dose in both cell lines was also calculated by PIXE from the elemental weight measurement, see Figure 7. The SiO_2 ENMs deposited dose in the A549 culture medium increased at 24 and 48 hours, and then it went back at 72 hours to levels similar to those at 6 hours. In the case of Epilife culture medium, the SiO_2 ENMs deposited dose decreased with time. The TiO_2 ENMs deposited dose behaviour was mostly constant, except for the 6h value in the A549 culture medium. The trend observed indicates a slight reduction with time. The deposited dose calculated by PIXE represents the elemental weight measurement including contributions from particles (ENMs included), particle surface constituents released in the culture medium (wetting and dissolution), ionic species, proteins and nutrients (from the culture medium) and dead cells (if any). Also, another factor governing detection is the sample preparation: the ion beam used in PIXE has an average penetration depth of 63.18 ± 2.66 µm for MEM plus serum or Epilife, thus anything that has dried below is not measured by PIXE. This last point is especially important in the presence of high amounts of ENMs agglomerates, when the sample from the culture medium is being dried as multilayer.

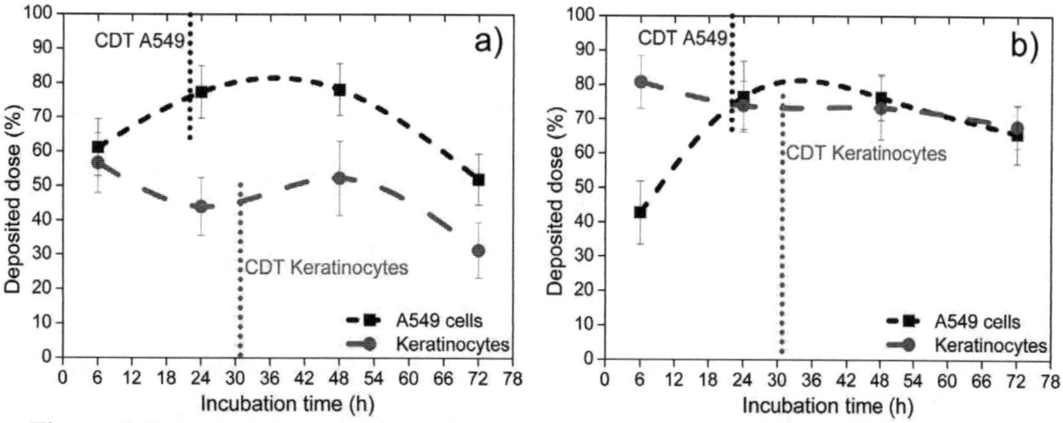

Figure 7. Deposited dose calculated from PIXE measurements of the a) SiO_2 and b) TiO_2 dispersions during *in vitro* assessment in two culture media: MEM plus serum for A549 cell line (---) and Epilife for N-hTERT keratinocytes (— —) for a 20 µg/mL nominal dose. CDT stands for 'cell doubling time' which is 22 h for A549 cells and 31 h for n-HTERT keratinocytes, marked by a dotted line (···). Both dashed lines are guides to the eye.

The calculated deposited doses by CLS and PIXE seem to indicate different results; nevertheless, these differences can be explained from the basis of each determination technique. In CLS, the nature of measurements is particle-based (entity with a defined and measurable volume by extinction of an optical detector and independent of the chemical signature), thus anything outside the range of the hydrodynamic measurement (like ionic species) is not measured. The nature of PIXE, an elemental weight measurement, has already been explained. For example, let us take the case of TiO2 in both cell lines, see Figures 6b and 7b. Both CLS and PIXE measurements show nearly the same deposited dose at all times. The ratio of TiO_2 agglomerates versus SiO_2 is 2 at all times (see Figure 4) and therefore, the sedimentation of agglomerates / aggregates does not play an important role in the differences between CLS and PIXE measurements. The deposited dose of SiO_2 in keratinocytes, compare Figures 6a and 7a, is lower for PIXE. This difference may be associated to the SiO_2 ENMs open pore filling by the Epilife medium causing a de-agglomeration process. This process not only induced more PPs, see Figure 5, but also a detachment or ion release only measured by PIXE. Therefore, the de-agglomeration process, related to the ENM SSA and the specific medium, is one key factor in the apparent discrepancy of the obtained deposited doses.

The influence of the CDT for each cell line is evidenced by the trend variations observed for the deposited dose, more marked in the case of SiO_2 in presence of A549 cell lines. These observations open the debate about the use of a stationary model, where the culture medium is not continuously supplied and the specific cell growth rate influence is highlighted.

While the general conclusion goes to the presence of agglomerates and the different SSA, more ENMs should be tested, and this will be part of a future set of studies in regard to the cell deposited dose and the dynamics of ENMs in *in vitro* assays.

Acknowledgements. The research leading to these results has received funding from the European Community's Seventh Framework Programme (FP7/2007-2013) under grant agreement n° 263147 (NanoValid - Development of reference methods for hazard identification, risk assessment and LCA of engineered nanomaterials). O. Toussaint is a Belgian FNR/ FNRS Research Associate.

References

[1] Piccione B, Cho C-H, van Vugt L K and Agarwal R 2012 All-optical active switching in individual semiconductor nanowires *Nat Nano* **7** 640-645.

[2] Wang W, Zhang S, Chinwangso P, Advincula R C and Lee T R 2009 Electric potential stability and ionic permeability of SAMs on gold derived from bidentate and tridentate chelating alkanethiols *J Phys Chem C* **113** 3717-3725.

[3] Calzolai L, Gilliand D and Rossi F 2012 Measuring nanoparticles size distribution in food and consumer products: a review *Food Addit Contam A*

[4] Dekkers S, Krystek P, Peters R J B, Lankveld D P K, Bokkers B G H, van Hoeven-Arentzen P H, Bouwmeester H and Oomen A G 2011 Presence and risks of nanosilica in food products *Nanotoxicology* **5** 393-405.

[5] Napierska D, Thomassen L C, Lison D, Martens J A and Hoet P H 2012 The nanosilica hazard: another variable entity *Part Fibre Toxicol* **7**

[6] Brayner R 2008 The toxicological impact of nanoparticles *Nano Today* **3** 48-55.

[7] Dhawan A and Sharma V 2010 Toxicity assessment of nanomaterials: methods and challenges *Analytical and Bioanalytical Chemistry* **398** 589-605.

[8] Jones C F and Grainger D W 2009 In vitro assessments of nanomaterial toxicity *Advanced Drug Delivery Reviews* **61** 438-456.

[9] Nel A, Xia T, Madler L and Li N 2006 Toxic Potential of Materials at the Nanolevel *Science* **311** 622-627.

[10] Oberdörster G, Stone V and Donaldson K 2007 Toxicology of nanoparticles: A historical perspective *Nanotoxicology* **1** 2-25.

[11] Hristozov D R, Gottardo S, Critto A and Marcomini A 2012 Risk assessment of engineered nanomaterials: a review of available data and approaches from a regulatory perspective *Nanotoxicology* **0** 1-19.

[12] Krug H F and Wick P 2011 Nanotoxicology: an interdisciplinary challenge *Angew Chem Int Ed Engl* **42** 1260-1278.

[13] Ostiguy C, Brigitte R, Woods C and Soucy B, Engineered Nanoparticles, Current Knowledge about OHS Risks and Prevention Measures, in: Studies and Research Projects, IRSST, Quebec, Canada, 2010, pp. 119.

[14] Bouwmeester H, Lynch I, Marvin H J P, Dawson K A, Berges M, Braguer D, Byrne H J, Casey A, Chambers G, Clift M J D, Elia G, Fernandes T F, FjellsbÃ¸ L B, Hatto P, Juillerat L, Klein C, Kreyling W G, Nickel C, Riediker M and Stone V 2010 Minimal analytical characterization of engineered nanomaterials needed for hazard assessment in biological matrices *Nanotoxicology* **doi: 10.3109/17435391003775266**

[15] Monteiro-Riviere N A and Tran C L 2007 *Nanotoxicology : characterization, dosing and health effects* (Place: Informa Healthcare) pp

[16] Cohen J, DeLoid G, Pyrgiotakis G and Demokritou P 2012 Interactions of engineered nanomaterials in physiological media and implications for in vitro dosimetry *Nanotoxicology* **0** 1-15.

[17] Kato H, Fujita K, Horie M, Suzuki M, Nakamura A, Endoh S, Yoshida Y, Iwahashi H, Takahashi K and Kinugasa S 2010 Dispersion characteristics of various metal oxide secondary nanoparticles in culture medium for in vitro toxicology assessment *Toxicology in Vitro* **24** 1009-1018.

[18] Hinderliter P, Minard K, Orr G, Chrisler W, Thrall B, Pounds J and Teeguarden J 2010 ISDD: A computational model of particle sedimentation, diffusion and target cell dosimetry for in vitro toxicity studies *Particle and Fibre Toxicology* **7** 36.

[19] OECD, LIST OF MANUFACTURED NANOMATERIALS AND LIST OF ENDPOINTS FOR PHASE ONE OF THE SPONSORSHIP PROGRAMME FOR THE TESTING OF MANUFACTURED NANOMATERIALS: REVISION, in: OECD (Ed.) Series on the Safety

of Manufactured Nanomaterials, Organisation for Economic Co-operation and Development, Paris, 2010.

[20] Dickson M, Hahn W, Ino Y, Ronfard V, Wu J and Weinberg R 2000 Human keratinocytes that express hTERT and also bypass a p16(INK4a)-enforced mechanism that limits life span become immortal yet retain normal growth and differentiation characteristics *Mol Cell Biol* **20** 1436-1447.

[21] Oberdörster G, Maynard A, Donaldson K, Castranova V, Fitzpatrick J, Ausman K, Carter J, Karn B, Kreyling W, Lai D, Olin S, Monteiro-Riviere N, Warheit D and Yang H 2005 Principles for characterizing the potential human health effects from exposure to nanomaterials: elements of a screening strategy *Part Fibre Toxicol* **2** 8.

[22] Lozano O, Mejia J, Masereel B, Toussaint O, Lison D and Lucas S 2012 Development of a PIXE analysis method for the determination of the biopersistence of SiC and TiC nanoparticles in rat lungs *Nanotoxicology* **6** 263-271.

[23] Lozano O, Mejia J, Tabarrant T, Masereel B, Dogné J-M, Toussaint O and Lucas S 2012 Quantification of nanoparticles in aqueous food matrices using Particle-Induced X-ray Emission *Anal Bioanal Chem* **403** 2835–2841.

[24] European Comission 2011 *Nanomaterials*. In http://ec.europa.eu/environment/chemicals/nanotech/index.htm. Accessed: November 2011.

[25] Valenti L E, Fiorito P A, García C D and Giacomelli C E 2007 The adsorption–desorption process of bovine serum albumin on carbon nanotubes *Journal of Colloid and Interface Science* **307** 349-356.

Interaction of fibrinogen and albumin with titanium dioxide nanoparticles of different crystalline phases

Arianna Marucco, Ivana Fenoglio, Francesco Turci and Bice Fubini

University of Torino, Dept. of Chemistry, Interdepartmental Centre "G. Scansetti" for Studies on Asbestos and Other Toxic Particulates and Interdepartmental Center for Nanostructured Interfaces and Surfaces, Via P. Giuria 7, 10125 Torino, Italy

ariannamaria.marucco@unito.it

Abstract. TiO_2 nanoparticles (NPs) are contained in different kinds of industrial products including paints, self-cleaning glasses, sunscreens. TiO_2 is also employed in photocatalysis and it has been proposed for waste water treatment. Micrometric TiO_2 is generally considered a safe material, while there is concern on the possible health effects of nanometric titania. Due to their small size NPs may migrate within the human body possibly entering in the blood stream. Therefore studies on the interaction of NPs with plasma proteins are needed. In fact, the interaction with proteins is believed to ultimately influences the NPs biological fate. Fibrinogen and albumin are two of the most abundant plasma proteins. They are involved in several important physiological functions. Furthermore, fibrinogen is known to trigger platelet adhesion and inflammation. For these reasons the study of the interaction between these protein and nanoparticles is an important step toward the understanding of the behavior of NPs in the body. In this study we investigated the interaction of albumin and fibrinogen with TiO_2 nanoparticles of different crystal phases (rutile and anatase) using an integrated set of techniques. The amount of adsorbed fibrinogen and albumin for each TiO_2 surface was investigated by using the bicinchoninic acid assay (BCA). The variation of the surface charge of the NP-protein conjugates respect to the naked NPs was used to indirectly estimate both surface coverage and reversibility of the adsorption upon dilution. Surface charge was monitored by measuring the ζ potential with a conventional electrophoretic light scattering (ELS) system. The extent of protein deformation was evaluated by Raman Spectroscopy. We found that both proteins adsorb irreversibly against electrostatic repulsion, likely undergoing conformational changes or selective orientation upon adsorption. The size of primary particles and the particles aggregation rather than the crystal phase modulate the affinity of fibrinogen for the TiO_2 surfaces.

1. Introduction

The rapid diffusion in the market of nanotechnological products opens new concerns on the possible adverse effects following the direct or indirect exposure of humans to nanoparticles (NPs) [1-4]. The hazard associated to nanomaterials is often difficult to be defined, because of knowledge of the interaction of nanoscale objects with living matter is still incomplete [5-7].

When a foreign body comes in contact with a biological fluid its surface is quickly covered by the components of the fluid, particularly proteins. A layer of proteins, defined as "protein corona" in the case of nanoparticles [8], is rapidly formed. Such layer is a dual system, composed by a core of strongly bound proteins and an outer layer of fast exchanging molecules [9, 10]. The composition of the protein corona depends upon a series of competitive processes driven by both thermodynamic and kinetic factors [9]. Interaction with blood proteins is considered the most critical step among the events determining the fate of NPs in the body. At the same time the interaction with blood proteins may trigger adverse responses like thrombosis and inflammation [11]. Plasma is a very complex system composed by thousand of different proteins, the two most abundant being albumin (35-50 mg/ml) and fibrinogen (1.8-3.5 mg/ml). Albumin is a soft protein of 67 kDa exhibiting a globular shape (9 × 6 × 5 nm). At physiological pH it has a net negative charge the isoelectric point (IP) being at pH 4.8. This protein is synthetized in the liver and its main physiological role is the transport of solutes in the blood stream to their target organs. It also contributes to maintain the pH and the osmotic pressure of plasma [12].

Fibrinogen is a large and complex glycoprotein (340 kDa). It has a rod-like shape with dimensions of 9 × 47.5 × 6 nm and, like albumin, it shows a negative net charge at physiological pH (IP at pH 5.2) [13]. It is synthesized in the liver and it is involved in several different biological functions such as blood clotting and stabilization of thrombi following vessel injury [14]. It also triggers foreign body reactions following implantation of prosthesis [15] and inflammatory responses to NPs [16]. These effects are mediated by surface-driven conformational changes leading to the exposure of specific epitopes when fibrinogen is adsorbed [15,16]. It is generally accepted that the physico-chemical properties of the surface play a crucial role in the interaction of a solid with living matter [15, 17, 18, 19, 20] but there is not yet a clear definition of the physico-chemical determinants of such interactions.

Here we report about the interaction of albumin and fibrinogen with three different samples of TiO_2 having different crystal phase (rutile and anatase). The study aims to elucidate the possible influence of the crystal phase on the behaviour of these two proteins at the titania surface.

Rutile and anatase nanometric titanium dioxide have been chosen for this study since they are largely used in industry mainly as UV blockers in sunscreen and plastics. TiO_2 finds also application in catalysis and has been proposed for water and air remediation thanks to its high photo-reactivity. Nanometric TiO_2 is also a promising material in nanomedicine [21]. Exposure of workers and users to this kind of nanomaterial is therefore expected to increase in the next years.

The interaction of nano- TiO_2 with the two proteins was investigated by evaluating the amount of protein interacting with the powder extent of surface coverage and irreversibility of the process (spectrophotometry and electrophoretic light scattering). The occurrence of conformational changes was also investigated by Raman spectroscopy.

2. Materials and Methods

2.1 Materials

Pyrogenic nanometric anatase/rutile powder (Aeroxide P25) ($T_{A/R}$) was purchased from Degussa-Evonik (Germany) while rutile (MT500B) (T_R) was from LCM-Trading (Italy)

Nanometric anatase powder (T_A) was synthesized by a sol gel process as described in previous work [22].

Bovine Plasma Fibrinogen (BPF) and Bovine Serum Albumin (BSA) were chosen as model for human fibrinogen and albumin and purchased from Sigma Aldrich (Germany) and used without further purification.

In all experiments, ultrapure Milli Q (Millipore, Billerca, MA) water was used. All other reagents were from Sigma Aldrich (Germany).

2.2 Methods

2.2.1 X ray diffraction

XRD spectra were collected on a diffractometer (PW1830, Philips) using CoKα radiation, in the (20–90) 2θ range, with step width 2θ = 0.05, and diffraction peaks have been indexed according to the ICDD database (International Centre for Diffraction Data). The spectra have been elaborated (X'-pert Highscore 1.0c, PANalytical B.V.) in order to assess the primary particle mean diameter of the different specimens.

2.2.2 HR-TEM investigation

The morphology of rutile nanoparticles were investigated by JEOL 3010-UHR HR-TEM equipped with a LaB6 filament operated at 300 kV, beam current 114 μA and equipped with a 2k x 2k pixels Gatan US1000 CCD camera. The nanoparticles were dispersed in ultrapure water (MilliQ system, Millipore), sonicated for 20 minutes and a droplet was deposited on lacey carbon Cu grids.

2.2.3 X ray fluorescence diffraction

The TiO_2 samples were analyzed using an EDAX Eagle III energy dispersive micro-XRF spectrometer equipped with a Rh X-ray tube and a polycapillary exciting a circular area of nominally 30 μm diameter. Data collection occurred at each point for 200 s detector live time, with X-ray tube settings adjusted for 30% dead time. About 1×10^6 Cps were counted per scan. At least 4 points were collected for each sample.

2.2.4Surface Area Measurements

The surface area of the particles was measured by means of the Brunauer, Emmett, and Teller (BET) method based on N_2 adsorption at 77 K (Micrometrics ASAP 2020).

2.2.5 ζ Potential.

The ζ Potential was evaluated by means of electrophoretic light scattering (ELS) (Zetasizer Nano-ZS, Malvern Instruments, Worcestershire, U.K.)

In this technique the velocity of particles in an oscillating electric field, which is proportional to their ζ Potential, was measured by light scattering. TiO_2 particles were suspended in ultrapure water and then sonicated for 2 min with a probe sonicator (100 W, 60 kHz, Sonoplus, Bandelin, Berlin, Germany).

The ζ Potential in function of pH was measured for all the samples; in water and in 10 mM phosphate buffer (PBS) for monitoring nanoparticles surface charge. The ζ potential at pH 7.4 was obtained by interpolation of a curve obtained by measuring the ζ potential at different pH (2–9) by adding 0.1 M HCl or NaOH to the suspension

The measurements were also repeated on titania powders incubated with protein solution and on titania powders incubated with protein and then washed three times in PBS.

2.2.6 Evaluation of the quantity of protein adsorbed on nanoparticles

Titania powders were suspended (33 mg/ml) in buffered protein solution (0.01M phosphate buffer: pH 7.4, 0.138 M NaCl and 2.7 mM KCl) at different concentrations (100-10000 mg/L).

The suspension were stirred in a thermostatic stirrer at 37°C for 1h and centrifuged at 11.000 RPM, then filtered through a membrane filter (cellulose acetate, pore diameter 0.45 μm) and the concentration of protein in the supernatant was determined spectrophotometrically (562 nm) by using the bicinchoninic acid (BCA) assay.

The amount of protein adsorbed was calculated as a difference between the final and the initial concentration of protein in the supernatant. The results are reported as the mean value of at the least three separate determinations ± standard error. The data are expressed as nano moli of protein adsorbed on m^2 as a function of the protein concentration present in solution after the incubation.

The theoretical monolayers were calculated by protein sizes as the amount needed for obtain a side-on protein monolayer on a totally available flat surface.

2.2.7 Raman Spectroscopy.

Raman spectra were acquired using an integrated confocal Raman system that includes a Horiba Jobin-Yvon HR 800 microspectrometer, an Olympus BX41 microscope, and a CCD air-cooled detector operating at -70°C.

A polarized solid-state Nd laser operating at wavelength of 532.11nm and a power of 80 mW was used as excitation source.

Calibration of the instrument was performed by measuring the Stokes and the anti-Stokes bands of the Si band at 520.7 cm^{-1} .

For the powder samples the measure were made placing the sample on a polished stainless steel slide, and a 50X objective delivering a power of ca. 15 mW to the sample was used.

Spectra were acquired with a spectral resolution of ca. 2 cm^{-1} and an integration time spanning from 100 to 400 s for the spectral region of 400-1800 cm^{-1}.

3. Results and discussion

3.1 Physico-chemical characteristic of nanoparticles

The three different samples of TiO_2, designated as T_R (rutile), $T_{A/R}$ (anatase/rutile), T_A (anatase) having different proportion of crystalline phase were chosen and thoroughly characterized (Table 1).

Table 1. Physico-chemical characterization

Sample	Crystalline Phase	Purity	Surface area BET (m^2/g)	Primary particles dimension (nm)	Hydrodynamic size (nm)	PCZ	ζ Potential (pH 7.4)
T_R	100% Rutile	>99.5	39	Lenght 40.5 ± 12.1 [a] Diameter 27.4 ± 6.5 [a]	156 PDI 0.12	5.5	-35.7
$T_{A/R}$	77% Anatase 23% Rutile	>99.6	53	19±3 [b] 45±7 [b]	214 PDI 0.18	6.9	-13.2
T_A	100% Anatase	>99.6	83	11±2 [b]	428 PDI 0.38	6.7	-17.5

a. Evaluated by TEM
b. Evaluated by XRD in according to the Scherrer's equation

The degree of purity of the TiO_2 samples evaluated by X-rays fluorescence (XRF) was found to be very high (>99%). The samples showed a different specific surface area which correspond to a different nanostructure (Table 1). In fact, the estimated crystallites size of anatase was lower than in the other two samples, in agreement with to the higher SSA value obtained.

The size analysis of the NP aggregates has been obtained by DLS in water at pH 10. At this pH value the particles exhibit a highly negative surface charge which minimize the formation of agglomerates because of electrostatic repulsions among particles. All samples exhibited aggregates having a wide range of diameters as suggested by the high polydispersity index (PDI) values (Table 1). The anatase sample appears to be composed by aggregates larger than the other two samples.

In water the TiO_2 NP surface is charged due to the the presence of dissociated hydroxyl groups and partially uncoordinated $Ti^{(IV)}$ ions [23]. The charge depends upon the pH value of the suspension and may be estimated by the evaluation of the ζ potential, i.e. the potential across the double layer of ions around the particles.

The variation of the ζ potential of the samples as a function of pH in deionised water is reported in Figure 1A. Under these experimental conditions, the point of zero charge (PZC) of T_R (5.5) resulted to be different from the PZC of the other two samples (6.7 and 6.9 respectively) suggesting a higher acidity of the hydroxyl groups in rutile. The other two samples exhibit a similar behavior according with the prevalence of the anatase phase in the $T_{A/R}$ sample.

The ζ potential was also measured by suspending the powders in phosphate buffer (PBS) (Figure 1B) as the interaction with proteins has been measured in PBS to simulate physiological conditions (see below). A decrease in ζ potential values was observed in the whole pH range as a consequence of the high ionic strength of the suspension. Furthermore, a dramatic shift of the PZC was observed for all samples. This effect may be due to a strong adsorption of phosphate ions at the surface. Interestingly the ζ potential at physiological pH appears nearly the same for the three samples, suggesting that, in physiological conditions, the differences observed in water among rutile and anatase disappear thus the electrostatic interactions with proteins are expected to be similar for the two phases.

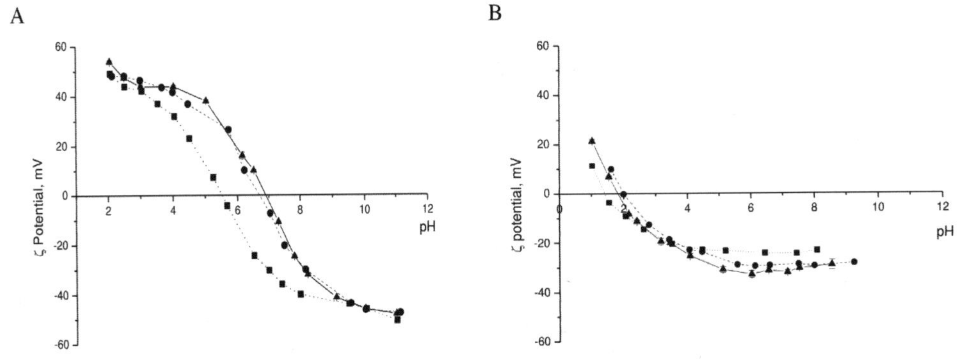

Figure 1 Variation of ζ Potential of TiO_2 samples measured in different dispersant. The curves are measured after the suspension of samples nanoparticles T_R (dot), $T_{A/R}$ (solid), T_A (dash) in water (A) and PBS (B) by sonication. The acquisition of these curve is repeated for three times.

3.2 Evaluation of the amount of protein interacting with nanoparticles

To investigate the affinity of the proteins for the TiO_2 buffered solutions of BSA or BPF at increasing concentration (500-10000 mg/L) were incubated for 1 hour with the powders. The amount of protein interacting with the particles was calculated by the decrease of their concentration in the supernatant after incubation. The results are reported in Figure 2 and are expressed as amount of protein adsorbed on TiO_2 per unit surface area *vs* the final concentration of the protein in solution after adsorption. The amount of adsorbed protein corresponding to a theoretical monolayer on an ideal flat surface, considering a side-on adsorption, is also reported in the graph (dotted line).

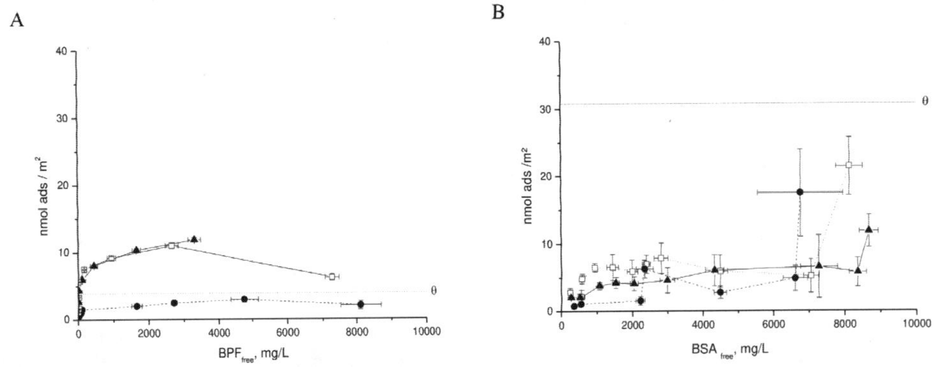

Figure 2 Comparisons of protein adsorption curves. The quantity of adsorbed fibrinogen (A) and albumin (B) on titania samples T_R squares), $T_{A/R}$ (triangles), T_A (circles) is show. The theoretical monolayer (θ) is not reached for all titania samples after the adsorption of BSA, while in the case of BPF the adsorption on $T_{A/R}$ and T_R largely exceeds of the θ.

BSA (Figure 2B) interacted with the three samples in a similar manner since the three adsorption curves appear close to each other. The amount of interacting protein increased with the protein concentration and reached a plateau at values lower than the theoretical monolayer. A different behavior was observed for BPF (Figure 2A). A rapid initial growth of the amount of protein adsorbed with a very low residual protein concentration was observed. The amount of protein adsorbed increased and overcame at high coverage, the theoretical monolayer.

The amount of fibrinogen adsorbed on T_R nanoparticles results to be similar to $T_{A/R}$ sample while a different behavior was observed for the anatase which accommodates a lower amount of proteins on its surface.

The data obtained suggest that fibrinogen has an affinity for TiO_2 surface higher than albumin. Differences in the properties of the two proteins may account for such differentiation. Both proteins have a negative net charge at pH 7.4 and therefore the adsorption of the proteins occurs against the repulsive electrostatic forces with the negative TiO_2 surface. However, BSA exposes at the solvent in the native form a largely uniform negative charge. Repulsive forces act both against the adsorption onto the surface but also laterally among proteins which are therefore unable to adsorb close to each other to form a compact monolayer. Conversely fibrinogen has a negatively charge domain in the central region, but the αC regions, exhibit a positive net charge due to the presence of arginine and lysine [13]. The presence of these positive domains possibly allows orientating the protein thus facilitating the adsorption onto negative surfaces and the formation of multi-layers.

The anatase sample exhibits a lower capacity to absorb proteins. This effect was much more evident for BPF than for BSA. Since in PBS the ζ potential of the three samples was similar such differences are not expected to be due to a different contribution of the electrostatic forces. On the other hand the anatase sample exhibits primary particles smaller than the other two samples, which are organized in large aggregates. The presence of pores among particles accessible to nitrogen but not to fibrinogen may be one reason of the lower amount of protein interacting with this sample. On the other hand, the lower primary particle size and, consequently, the higher curvature of each single primary particle makes rougher the surface of the aggregates possibly lowering the number of possible points of interaction with fibrinogen. In the case of BSA, this effect is expected to be less important being BSA a much smaller molecule.

3.3 Evaluation of the extent of coverage of the NPs surface.

To get insight on the extent of the coverage of the surface the shift of ζ potential following protein adsorption was also evaluated.

One of the most relevant effects of the formation of the protein corona on NPs is the consequent masking of the NP surface. Being charged entities proteins are expected to modify the ζ potential of the NPs. We therefore measured the ζ potential shift of the titania powders, $T_{A/R}$, previously incubated with the fibrinogen and albumin at different concentrations (Figure 3A and B respectively). Similar results were obtained with the other two powders (data not shown).

The adsorption of BPF led to a shift of the PZC toward the isoelectric point of the proteins, suggesting that the coverage of the surface increased proportionally with protein concentration (Figure 3A) in agreement with the adsorption curve. However, the isoelectric point (pI) of the protein was not reached even at high coverage suggesting that the particle surface was still exposed at the solvent. This was expected for BSA, since the amount absorbed was not enough to fully cover the surface. However a PZC lower than pI was also observed for BPF. A preferential orientation of the protein at the surface with the positive side toward the surface may account for this result.

A similar masking effect is expected to occur with the protein corona is formed by both proteins reversibly or irreversibly interacting with the surface. Therefore the data obtained did not give any information on the reversibility of the process. This information is important to predict the behavior of proteins in competitive conditions.

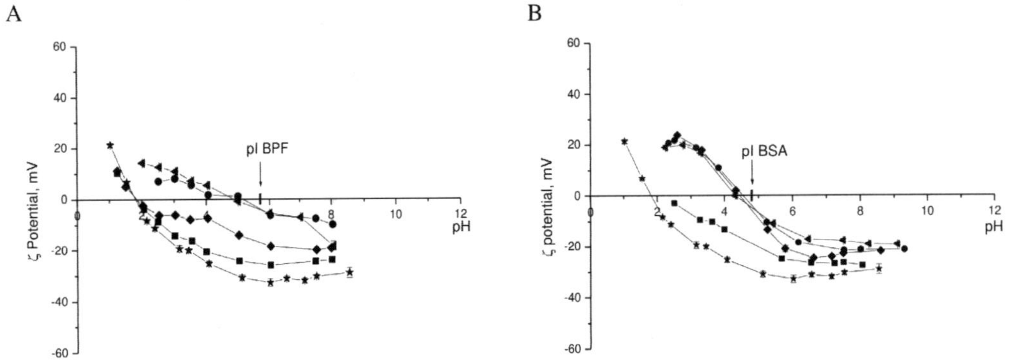

Figure 3 Variation of ζ potential after protein adsorption. Increasing concentration of fibrinogen (A) and albumin (B) is associated with the shift of the nanoparticles point of zero charge(PZC) toward the protein isoelectric point (pI). The curves are relative to $T_{A/R}$ sample before (stars) and after incubation with different protein concentration 10 (triangles), 5 (circles), 2.5 (diamonds) and 0.5 (squares) g/L.

3.4 Evaluation of reversibility

The reversibly-bound fraction of the protein corona is expected to be removed by applying a negative protein concentration gradient, i.e. re-suspending the protein –NP conjugates in a fresh protein-free buffer. If the protein corona is totally reversible, the ζ potential of the NPs is expected to shift back to the values observed for the free NP surface. On the contrary, if a totally irreversible corona is formed upon NP surface subsequent washing will not modify the conjugate ζ potential. In Figure 4 the ζ potential curves measured after suspending $T_{A/R}$ previously incubated with the proteins in PBS are reported. BPF appeared to interact irreversibly with TiO_2 nanoparticles since no shift of the ζ potential curves was observed (Figure 4A). Similar results were obtained with the other two powders (data not shown). Also for albumin the interaction appeared almost irreversible (Figure 4B).

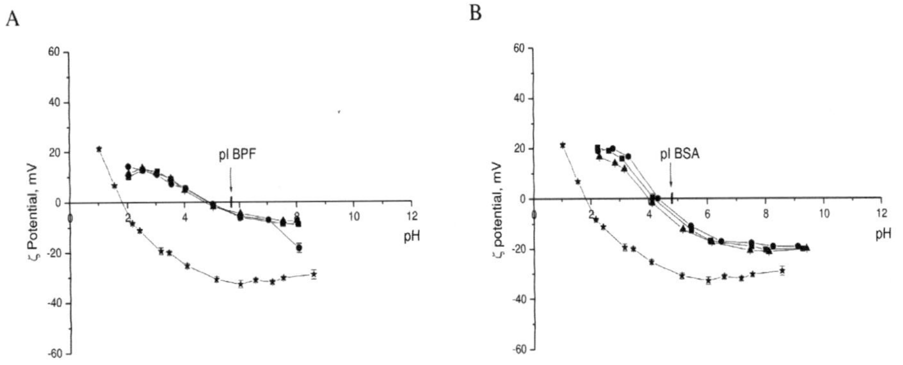

Figure 4 Variation of ζ potential after washing. The curves are acquired for the sample before the adsorption (stars), after the incubation with a high (10000 mg/L) protein concentration (circles), one cycle (squares) and three cycles of washing (triangles). This experiment has been carried out also for the other nanoparticles and for two other protein concentrations (5000 and 500 mg/L) and in all cases the adsorption resulted to be irreversible.

3.5 Evaluation of surface-driven protein conformational changes with Raman Spectroscopy

Albumin and fibrinogen are both soft proteins and their adsorption on nanoparticles is irreversible. It is therefore to be expected that BSA and BPF interacting with TiO_2 NPs, undergo some kind of conformational change. The evaluation of such conformational changes may be studied with micro-Raman spectroscopy. Raman spectroscopy can be used to evaluated the protein secondary structure by analyzing the vibrational frequencies of the bands associated with the backbone amide bonds that are dependent upon the secondary arrangement of the polypeptide chain [24].

The amide I band corresponds to the sum of coupled modes of the polypeptide backbone. A major contribution to the amide I modes comes from the C-O stretching of the peptide carbonyl groups [25]. It is possible to evaluate the surface-driven conformational modifications by comparing the Raman spectral pattern of the native folded protein with that of the same protein adsorbed on the titania surface. Note that the Raman modes of both anatase and rutile TiO_2 are located in the region between 200 and 800 cm^{-1} and therefore no interference by these picks is expected. The spectra of BSA are acquired in the native form or adsorbed on the titania nanoparticles; the spectra relative to the interaction with the $T_{A/R}$ sample are reported in Figure 5. The amide I Raman bands for the intensity maximum of each Raman spectrum was normalized to 1. The amide I band of native BSA (Figure 5, black spectrum, 1657 cm^{-1}) is slightly shifted toward higher wavenumbers (1660 cm^{-1}) when BSA is adsorbed on $T_{A/R}$, and the amide III band (1278 cm^{-1}) is slightly shifted toward lower wavenumbers (1267 cm^{-1}) after the adsorption. These shifts are typical of a protein conformational change; it is possible that these modifications in protein structure are due to oxidative damage linked with the reactivity (photocatalitic activity) of titania samples [26]. The three peaks of aromatic residues (1557, 1588, and 1618 cm^{-1}) are assigned according to previous data. [24, 25].

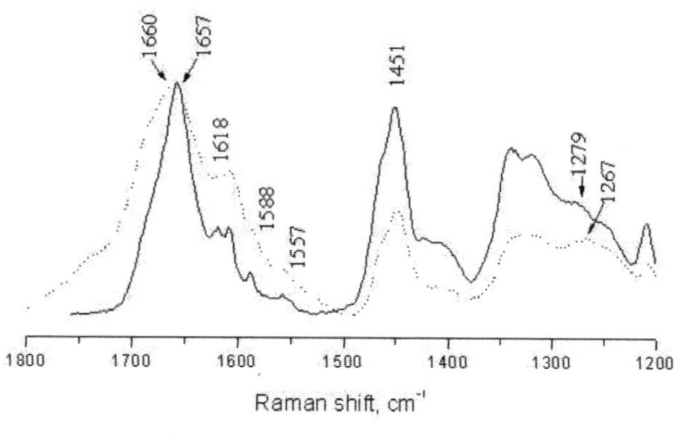

Figure 5 Evaluation of protein conformational change after adsorption on $T_{A/R}$ Raman spectra of BSA native (solid spectra) and adsorbed on nanoparticles (dot spectra).These data are the average of ten acquisition. The arrows point the peak of amide I (1657 cm^{-1}) and amide III (1279 cm^{-1}) bands. The shift of these bands is typical of a conformational change. Spectra intensities were normalized to amide I peak.

4. Conclusions

In this study, the interaction of albumin and fibrinogen with three samples of TiO_2 having distinct crystal phases and different primary particle size, has been evaluated.

We found that both albumin and fibrinogen interact irreversibly with the negative surface of TiO_2 NPs. The interaction with albumin appears to be driven by conformational changes of the protein at the surface. This was expected since BSA has a low resistance to deformation. In the case of BPF both conformational changes and electrostatic forces between TiO_2 surface and the positive domains of the proteins my account for such interaction. Fibrinogen interacts in large quantity with T_{AR} and T_R while a lower amount was observed in the case of the anatase sample. This effect was likely due to a lower availability of the surface to the large fibrinogen molecule and to the higher roughness of this sample as a consequence of the lower dimension of the primary particles, while the differences in surface properties related to the crystal phase seemed to play a minor role. A smaller effect was observed for BSA because this protein may closely contact the particles surface while BPF is expected to "see" the TiO_2 powders as rough aggregates (Figure 6).

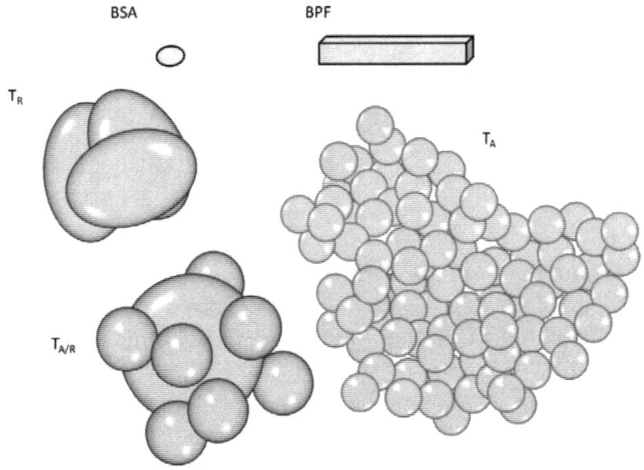

Figure 6 Proteins and titania nanoparticles. Schematic representation

It is well known that small proteins are faster to diffuse and then arrive at the surface before large proteins. The early formed protein corona is however expected to evolve: in fact proteins having higher affinity for surfaces like fibrinogen may replace smaller proteins through a mechanism known as Vroman's effect [17, 27, 20]. The present data suggest that the surface curvature of particles may modulate the amount of fibrinogen interacting with the nanoparticles and possibly its affinity for the surface. Since the interaction of fibrinogen with NPs has been shown to trigger inflammatory responses [16] the lower affinity of fibrinogen for the sample having smaller primary particles may correspond to a possible higher biocompatibility of small TiO_2 nanoparticles.

References

[1] Oberdörster G *et al* 2009 Nanoparticles and the Brain: Cause for Concern? *Journal of nanoscience and nanotechnology* **9** pp4996-5007

[2] Savolainen K *et al* 2010 Risk assessment of engineered nanomaterials and nanotechnologies *Toxicology* **269** pp 92-104

[3] Schrurs F and Lison D 2012 Focusing the research effort *Nature nanotechnology* **7** pp 546-548

[4] Pietroiusti A 2012 Health implications of engineered nanomaterials *Nanoscale* **4** pp1231-1247

[5] Lin W *et al* 2006 In vitro toxicity of silica nanoparticles in human lung cancer cells *Toxicology and applied pharmacology* **217** pp 252-259

[6] Fubini B *et al* 2010 Physico-chemical features of engineered nanoparticles relevant to their toxicity *Nanotoxicology* **4** pp 347-363

[7] Schulze C *et al* 2011 Interaction of metal oxide nanoparticles with lung surfactant protein A *European Journal of pharmaceutics and biopharmaceutics* **77** pp 376-383

[8] Lundqvist M *et al* 2008 Nanoparticle size and surface properties determine the protein corona with possible implications for biological impacts *Pnas* **105** pp 14265-14270

[9] Walczyk D *et al* 2010 What the Cell "Sees" in Bionanoscience *Journal of the American Chemical Society* **132** pp 5761-5768

[10] Lynch I *et al* 2007 The nanoparticle - protein complex as a biological entity; a complex fluids and surface science challenge for the 21st century *Advances in Colloid and Interface Science* **134-35** pp 167-174.
[11] Kamali P P and Simberg D 2011 Interaction of nanoparticles with plasma proteins: implication on clearance and toxicity of drug delivery systems *Expert opinion drug delivery* **8** pp 343-357
[12] Sugio S *et al* 1999 Crystal structure of human serum albumin at 2.5 Å resolution *Protein engineering* **12** pp 439-446
[13] Jung SY *et al* 2003 The Vroman effect:a molecular level description of fibrinogen displacement *Journal american chemical society* **125** pp 12782-12786
[14] Jackson S P 2007 The growing complexity of platelet aggregation *Blood* **109** pp 5087-5094
[15] Thevenot P *et al* 2008 Surface chemistry influences implant biocompatibility *Current Topics in Medicinal Chemistry* **8** pp 270-280
[16] Deng ZJ *et al* 2011 Nanoparticle-induced unfolding of fibrinogen promotes Mac-1 receptor activation and inflammation *Nature nanotechnology* **6** pp 39-44
[17] Nel A *et al* 2009 Understanding biophysicochemical interactions at the nano-biointerface *Nature Materials* **8** pp 543-557
[18] Nel A *et al* 2006 Toxic potential of materials at the nanolevel *Science* **311** pp 622-627
[19] Rivera Gil P *et al* 2010 Correlating physico-chemical with toxicological properties of nanoparticles: the present and the future *Acs Nano* **4** pp 5527-5531
[20] Fenoglio I *et al* 2011 Multiple aspects of the interaction of biomacromolecules with inorganic surfaces *Advanced Drug Delivery Reviews* **63** pp 1186-1209
[21] Liu H *et al* 2007 Nanomedicine for implants: A review of studies and necessary experimental tools *Biomaterials* **28** pp 354-369
[22] Bolis V *et al* 2012 Hydrophilic/hydrophobic features of TiO_2 nanoparticles as a function of crystal phase, surface area and coating, in relation to their potential toxicity in peripheral nervous system *Journal of colloid and interface science* **369** pp 28-39
[23] Suttiponparnit K *et al* 2011 Role of surface area primary particle size and crystal phase on titanium dioxide nanoparticles dispersion properties *Nanoscale research letters* **6**
[24] Turci F *et al* 2010 An integrated approach to the study of the interaction between proteins and nanoparticles *Langmuir* **26** pp 8336-8346
[25] Strehle M A *et al* 2004 a raman spectroscopy study of the adsorption of fibronectin and fibrinogen on titanium dioxide nanoparticles *Phys. Chem. Chem. Phys.* **6** pp 5232-5236
[26] Fenoglio *et al* 2009 Non-Uv-induced radical reactions at the surface of TiO_2 nanoparticles that may trigger toxic responses *chem. Eur. J.* **15** pp 4614-4621
[27] Vroman L 1962 Effect of absorbed proteins on the wettability of hydrophilic and hydrophobic solids *Nature* **196** pp 476-7

Surface ligand dependent toxicity of zinc oxide nanoparticles in HepG2 cell model

D Bartczak[1], M-O Baradez,[1]S Merson[1], H Goenaga-Infante[1] and D Marshall[1]

[1]LGC, Queens Road, Teddington, Middlesex TW11 0LY, UK

E-mail: dorota.bartczak@lgcgroup.com

Abstract. Physicochemical properties of nanoparticles (NP) strongly affect their influence on cell behaviour, but can be significantly distorted by interactions with the proteins present in biological solutions. In this study we show how different surface functionalities of zinc oxide (ZnO) NP lead to changes in the size distribution and dissolution of the NP in serum containing cell culture media and how this impacts on NP toxicity. NPs capped with weakly bound large proteins undergo substantial transformations due to the exchange of the original surface ligands to the components of the cell culture media. Conversely, NP capped with a tight monolayer of small organic molecules or with covalently conjugated proteins show significantly higher stability. These differences in ligand exchange also affect the toxicity of the NP to the HepG2 liver cell model, with the NP capped with small organic molecules being more toxic than those capped with large proteins. This study highlights the importance of characterising NPs in biological media and the effect the media has during in-vitro analysis.

1. Introduction

To date there are over 1300 commercial products which incorporate NP, in a global market that is expected to be worth €30 billion by 2015[1]. These products range from cosmetics and food additives through to paints or clothing and impact on almost all areas of human life. Inevitably this raises safety concerns due to the potential toxicity of the NP released from the products and absorbed, ingested or inhaled into the body[2]. This can be a particular problem for organs such as the liver where bioaccumulation can rapidly occur following exposure to NP[3], increasing and localising any toxic effect. Measurement of potential NP toxicity is usually performed using cell based screens. However, while this approach can be high throughput it is often challenging to study due to the interactions between the NP and serum proteins in the cell culture media[4]. These interactions lead to structural and functional changes of the NP, affecting their physicochemical properties and bioactivity. Therefore, it is important to understand these exchange reaction and how they affect the NP behaviour, cellular uptake and corresponding toxicity.

The ligand exchange reaction is directly related to the NP chemical composition, size, shape, net charge and most importantly, the initial surface functionality[5]. In this study we show how changes to the surface functionality of ZnO NP affect the physicochemical properties of the material over 24 hours, with size distribution profiles and ligand exchange reactions dependent upon the initial surface chemistry. We also show how different surface functionalities then impact the kinetics of cellular toxicity using impedance spectroscopy in a cellular model of the liver.

2. Experimental

2.1. Preparation of plain, APTMS, BSA and APTMS-BSA capped ZnO NP

Zinc oxide nanopowder (10-30nm, US Research Nanomaterials) was dispersed in water or capped with a selection of ligands using the following protocol. Weighted 5mg of powder was sonicated for 15min at room temperature (plain, APTMS and APTMS-BSA) or at 4°C (BSA) in an ultrasonic bath along with 50µl of isopropanol. Next, 5ml of ultrapure water (plain), 5ml of 2% 3-aminopropyltrimethoxysilane (APTMS) in isopropanol (APTMS and APTMS-BSA) or 5ml of 0.05% bovine serum albumin fraction V in ultrapure water (BSA) was introduced and the suspension was sonicated for 1h at room temperature (plain, APTMS and APTMS-BSA) or with a gradual temperature increase to room temperature (BSA). NP were then reacted for additional 24h at room temperature (plain and BSA) or at 60°C (APTMS and APTMS-BSA) with shaking, prior to purification from excess ligands and/or organic solvent residues by triple centrifugation/decantation (13,000rpm, 15min, room temperature). Capped NP were then redispersed by sonication in ultrapure water at 5mg/ml. BSA was covalently conjugated to APTMS capped NP (APTMS-BSA) using EDC/sulfo-NHS coupling agents. Briefly, 0.34mg of APTMS capped NP was redispersed by sonication in 5ml of 0.01M borate buffer pH9. Then, 100µl of BSA fraction V stock (1mM in ultrapure water), 50µl of EDC (0.198M in ultrapure water) and 100µl of sulfo-NHS (0.198M in ultrapure water) were introduced. After, 48h stirring at room temperature, the conjugates were purified by triple centrifugation/decantation (13,000rpm, 15min, room temperature) and redispersed by sonication in ultrapure water at 5mg/ml.

2.2. ICP-MS analysis of the dissolution rate of plain and capped ZnO NP

The supernatants remaining after purification of plain, APTMS, BSA and APTMS-BSA ZnO NP (triple centrifugation/decantation at 13,000rpm, 15min, room temperature) were collected. NP were then stored for 24h at room temperature in the ultrapure water and centrifuged once more at 13,000rpm, 15min, room temperature to precipitate the NP. The supernatants containing dissolved ions were combined with the supernatants collected during the NP purification (also here the supernatants contained the dissolved ions only, since the NP were precipitated by centrifugation). The total content of ^{66}Zn and ^{68}Zn in the collected supernatants (diluted 1/50 by weight with HNO_3 just before analysis) were measured using an Agilent 7700x spectrometer operating in the He mode.

2.3. TEM imaging of plain and capped ZnO NP

10µl of NP suspension (5mg/ml) was deposited on a TEM grid (carbon film 400 mesh Cu 25, Agar Scientific) and air dried. TEM grids were imaged with Hitachi H7000 transmission electron microscope operating at 75kV voltage with 100,000x magnification (Biomedical Imaging Unit, Southampton, UK).

2.4. UV-visible and fluorescent spectroscopy of plain and capped ZnO NP

UV-visible and fluorescent spectra of ZnO NP suspensions in ultrapure water (0.1-1mg/ml) were recorder with Tecan Infinite M200 plate reader using high quality flat bottom black microplates (Sigma-Aldrich), over 350-400nm and 320-480nm range with 250nm excitation wavelength, respectively.

2.5. NTA sizing of plain and capped ZnO NP

Suspensions of NP (30-70µg/ml) were prepared in ultrapure water or cell culture media (EMEM 10% FBS) by sonication. The movement of NP in suspensions under Brownian motion was recorded with NanoSight NS500 over 60-90s at room temperature, then analyzed with NTA2.2 software using detection threshold value 9 and a minimum expected particle size of 30nm. The experiments were performed in triplicates with different batches of NP, processed with Microsoft Excel Office 2003 and are presented as mean ± standard deviation.

2.6. RT-CES testing of plain and capped ZnO NP toxicity in HepG2 cell model
HepG2 cells (LGC Standards), were seeded directly onto the gold microelectrode of the E-Plate (Acea Bioscience), at a density of 20,000 cells/well, as determined by an automated cell counter (ViCell, Beckman). The plate was then loaded into an impedance station housed at 37°C in 5% CO_2 environment and the impedance measurements were taken every hour over the duration (48h) of the experiment. Following the initial 24h after seeding the cells were exposed to the varied concentration of NP (30-70μg/ml) suspended in cell culture media (EMEM 10%FBS) and returned to the impedance station for the additional 24h. Untreated HepG2, as well as cells treated with APTMS (25μg/ml), BSA (50μg/ml) or 1% H_2O_2 served as controls. The experiments were performed in triplicates and are expressed as mean ± standard deviation.

3. Results and Discussion

3.1. Characterisation of ZnO NP suspensions in water.
ZnO NP with a primary crystal size of 10-30nm (as specified by the manufacturer) were functionalised with BSA, APTMS or a combination of both (APTMS-BSA). Using this approach, APTMS serves as a silinisation agent[6], forming a tight monolayer of organics encapsulating the inorganic core and increasing protection against dissolution in aqueous media. Conversely, NP stabilized with loosely bound large protein (BSA) were less protected from ions and small organics in the surrounding environment and could readily undergo structural and functional changes, including destabilization and replacement of the original ligands[5,7]. Proteins, however, offered additional functionality, acting as targets that can be recognized by cells and in some cases modify their toxic response profiles[8].

Table1. Physicochemical characteristics of ZnO NP.

Surface ligand	Dissolution	Morphology	
		Size (nm)*	Shape
plain	6.35%	20-70	near-spherical
APTMS	0.68%	18-60	near-spherical
BSA	3.90%	16-65	near-spherical
APTMS-BSA	2.95%	25-75	near-spherical

* as determined by TEM

To measure the effect of the capping process on the dissolution of zinc oxide, the NP samples were incubated at room temperature for 24 hours in ultrapure water and analysed with ICP-MS (Table1). In all cases the capping process provided protection to the NP against dissolution. Control NP with no functionalisation (plain) have a dissolution rate of 6.35% after 24 hours, compared to 0.68% after silinisation with APTMS and 3.9% following capping with BSA. This demonstrates the protective effect of the capping process. Surprisingly, the NP capped with a combination of APTMS and BSA (APTMS-BSA) had a higher dissolution rate (2.95%) than NP capped with APTMS alone. This is likely due to the BSA coupling step, which involved 48h incubation in a higher ionic strength environment with organics present which can destabilise ZnO NP and promoted dissolution[9]. Examination of NP using TEM (summary of TEM data displayed in the morphology section of Table1; images not shown) indicated that the capping process did not affect their morphology. All types of NP were near-spherical in shape with rough edges present and polydispersed in size with the diameter of primary crystals in a range between 16 and 75nm and no significant differences observed between the samples. It is important to note that the NP agglomerated/aggregated on the surface of the TEM grids regardless of the NP surface functionality, however the size of individual NP within the agglomerates/aggregates could still be measured manually (size data in Table1). The overall size of agglomerates/aggregates is not given, since as it was not possible to distinguish between the agglomerates/aggregates formed in solution (if any) prior to deposition on the TEM grids and those formed during the grids preparation (Note: NP agglomerates/aggregates present in solution, as well as the primary NP and the relative proportion of the two, were measured with NTA, see 'Ctrl' in Figure 2).

Further characterisation of the capping process outcome was performed using fluorescence and UV-visible spectroscopy (Figure 1). These analysis showed that the fluorescent spectra of all types of ZnO NP had similar characteristics with a typical band around 400nm, meaning that the semiconductor properties were not lost or changed during capping[10]. Extinction characteristics of capped NP were altered only slightly in comparison to plain NP due to differences in the refractiveness of the surrounding media[11]. Clear sharpening of the UV-visible spectra and a slight blue shift of the extinction maxima indicate narrower size distribution of the overall population of NP (primary crystals and aggregates combined) and/or larger proportion of the primary fraction present.

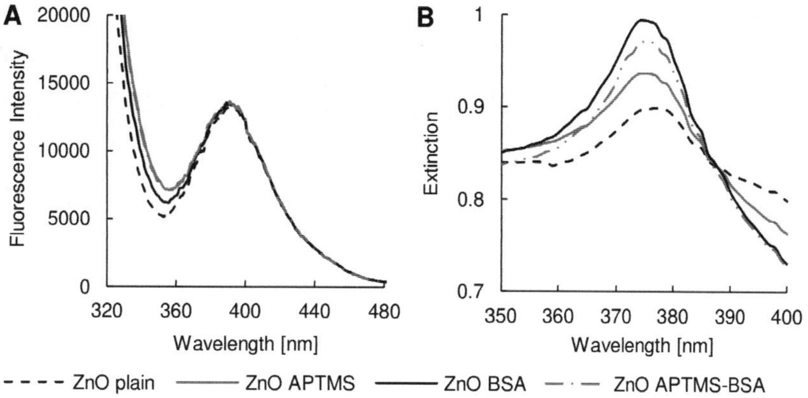

Figure 1. Fluorescence (A) and UV-visible (B) spectra of plain, APTMS, BSA and APTMS-BSA ZnO NP suspensions in water.

3.2. Characterisation of ZnO NP suspensions in cell culture media.

Measurement of NP toxicity usually involves incubating the cells with the NP dispersed in serum containing cell culture media for a period of 24 hours. To measure how ZnO NP change during this time, capped and plain NP suspensions were examined for changes in the proportion of primary and aggregated fractions over 24 hours incubation at 37°C (Figure 2). Measurements of the area, height and distribution were performed using an image based nanoparticle tracking analysis (NTA) system to allow sizing of polydispersed samples (Figure 2). Dotted bars represent the characteristics of ZnO NP in water (Ctrl), solid bars show changes in NP size profiles during 24 hour incubation. In all cases the aggregated fraction accounts for a substantial proportion of the overall population of NP, with significantly lower number of plain compared to APTMS and BSA capped NP. When dispersed in the cell culture media APTMS and APTMS-BSA capped NP retained their original (Ctrl) size distribution profiles, while plain and BSA capped NP underwent substantial changes. At first the proportion of NP in the primary size fraction significantly dropped, which can be attributed to the initial stage of capping (plain) or ligand exchange (BSA) reactions with cell culture media components, happening directly on the NP surface and leading to destabilization and agglomeration/aggregation. After 4h in the culture media, the proportion of plain and BSA NP increased as the capping/ligand exchange propagated and the new organic corona was formed in equilibrium with the surrounding environment. Since no significant differences in the size distribution profiles of APTMS and APTMS-BSA NP over time was observed, it can be assumed that silane protected NP are generally less prone to the ligand exchange reactions with the cell culture media components.

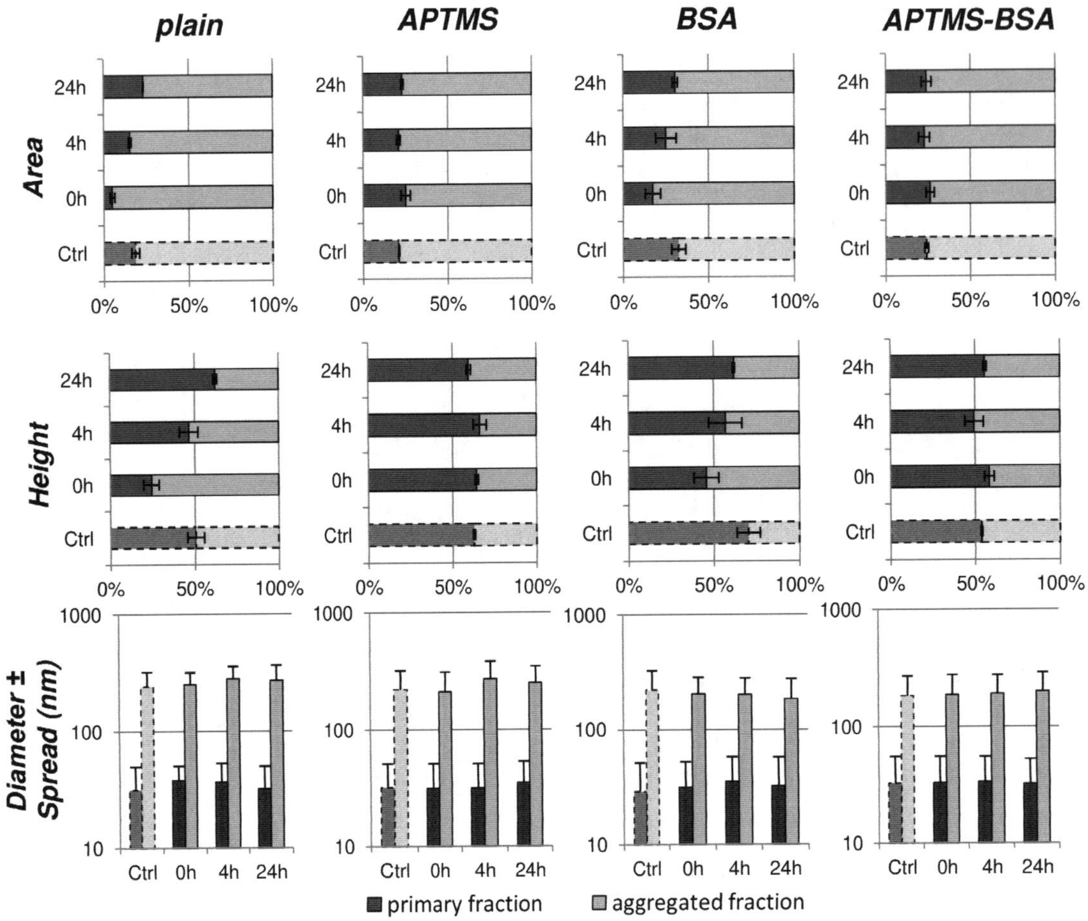

Figure 2. Size changes of plain, APTMS, BSA and APTMS-BSA ZnO NP over 24h incubation in cell culture media, determined by the relative proportion of primary (0-100nm) and aggregated (>100nm) fractions. Area (top row) represents the sum of trapezoids under the size distribution curve, while height (middle) is the peak maximum. Position of the maximum peak determines the fraction diameter, while spread derived from the peak width (bottom). Ctrl are NP water suspensions.

3.3. Characterisation of ZnO NP toxicity to liver cells.

The effect of capping on the toxicity of ZnO NP was measured in the HepG2 liver cell model using impedance spectroscopy (Figure 3), a label free method shown to be comparable to standard reagent based toxicity assays[12]. Data for 48 hours is displayed, with the first 24 hours showing the growth of cells prior to exposure, indicated by an increase in impedance as the cells divide and increase in number. After 24 hours, the different capped NP suspensions in the cell culture media were introduced and elicit dose dependant toxic responses from the cells. Plain (uncapped) NP at 30µg/ml (lowest dose) did not have a direct toxic effect on the cells but changed their behaviour compared to control (untreated) samples, as indicated by a initial increase in impedance followed by a plateau through to the 48 hour time point. Plain NP at 40µg/ml were toxic to the cells and caused a rapid decrease in impedance over the first 8 hours of exposure as approximately 50% of the cells were killed by the NP. Concentrations of plain ZnO NP above 40µg/ml were extremely toxic killing the HepG2 cells within 6 hours. In comparison ZnO NP capped with either BSA or APTMS were more toxic than plain NP with even the lowest dose (30μg/ml) eliciting a mild toxic response to the cells and increased toxicity at 40μg/ml and above. Capping the ZnO NP with both APTMS and BSA reduced their toxicity with NP concentrations up to 50μg/ml having no toxicity, NP at a concentration of 60μg/ml having mild toxicity and the highest concentration (70μg/ml) having moderate toxicity. The BSA and APTMS ligands alone were not toxic to the cells, while Zn^{+} ions have been shown previously to be non-toxic[12].

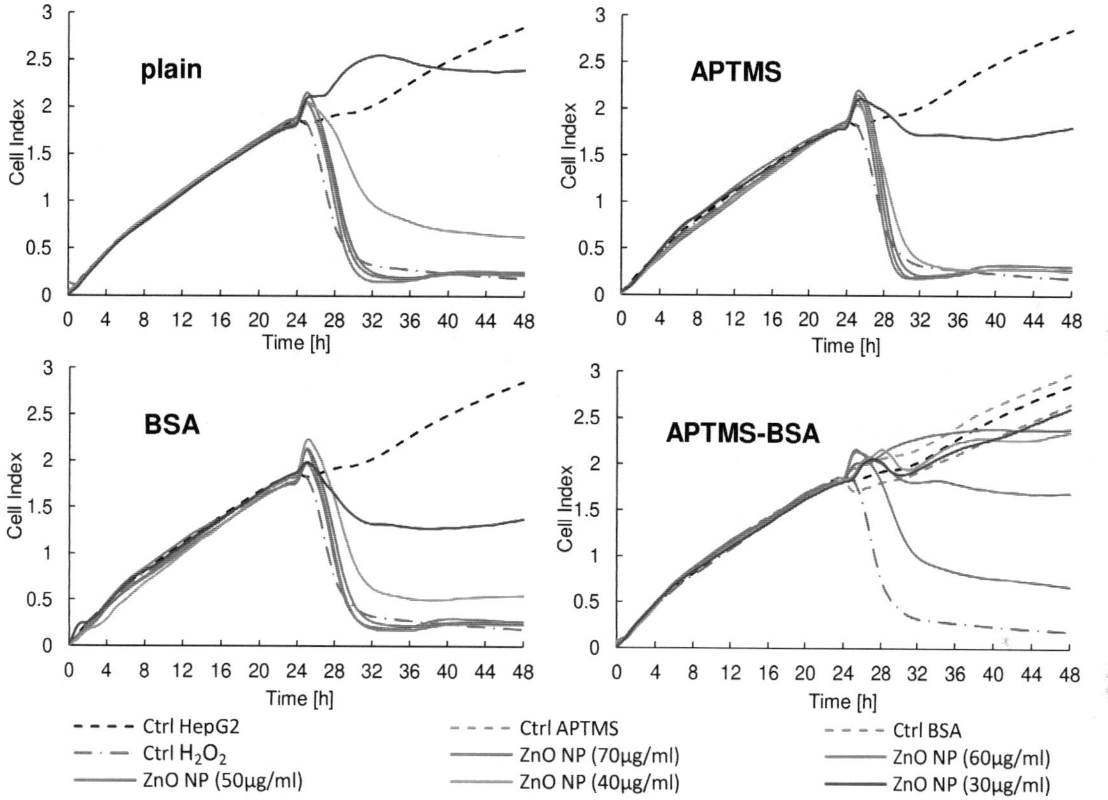

Figure 3. Dynamic monitoring of changes in RT-CES impedance over 24h following exposure of sub-confluent HepG2 cells (after 24h of initial incubation) to 30-70 μg/ml suspensions of plain, APTMS, BSA and APTMS-BSA ZnO NP. The figure represents normalized cell index values with standard deviations less than 5% (not shown).

Changes in impedance provide valuable information about how the cells are responding to the presence of the NP. For all of the NP samples tested, there is an initial increase in impedance immediately after the NP have been added to the cells (after 24 hours). In previous reports this has been shown to occur due to biochemical changes and increased active transporter activity which modifies the cells impedance properties as they try to remove the internalised NP[12]. These defence mechanisms are energy dependant and eventually exhaust the resources, so that after approximately 3 hours the toxic affect of the NP can be observed. At low concentrations of NP (30 and 40μg/ml) BSA and APTMS capped NP appear more toxic to cells than plain NP (cause bigger drop in the cell index value). This is most likely related to the higher content of NP in the primary size fraction within the first 4h, when the cell response takes place. In general, smaller (primary sized) NP are taken up by the cells more readily than larger aggregates and as more NP are internalised the effect on cells becomes stronger[13]. When IC_{50} values are taken into consideration (NP concentration at which cell impedance drops to 50% of the value recorded for untreated control), APTMS and BSA capped ZnO NP show the highest toxicity with IC_{50} of 30µg/ml and 31µg/ml, respectively. Plain NP are slightly less toxic (IC_{50} of 34 µg/ml) while APTMS-BSA are the least toxic to HepG2 cells, with and IC_{50} of 58µg/ml. Significantly lower toxicity of APTMS-BSA NP compared to other types of NP, cannot be explained by the proportion of NP in the primary size range (higher than plain) or their diameter (not significantly different from other types of NP), thus it is most likely related to other characteristics of NP such as net charge and the effect of the functionality on the entry route of the NP into the cells and their overall uptake rate.

4. Conclusions

Changes in the surface functionality of NP achieved with different capping approaches affect the physicochemical behaviour of the NP in cell culture media and their cellular toxicity. Within the first 4h, which was the most important in generating the toxic response (as shown by impedance maximum peaks), relatively low proportion of BSA capped and plain ZnO NP was found in primary size fraction. This, combined with a high dissolution rate (ions which can be easily transported out of the cell) could be one reason why they showed slightly lower IC_{50} values than APTMS capped NP. Significantly lower toxicity of APTMS-BSA NP than other types of NP, is most likely related to other than size characteristics of NP, e.g. differences in the net charge and the type of ligands present during the first 4h of the capping/exchange reaction in the cell culture media. Organic corona can have an effect on the NP entry route and the overall NP internalization rate, while appropriate surface functionality may even reduce the toxic responses[7].

References

[1] G A McWilliams A. *BCC Market Research Reports* **2010**, NAN031D, 1-276.
[2] S Sharif, S Behzadi, S Laurent, M L Forrest, P Stroeve and M Mahmoudi *Chemical Society Reviews* **2012**, 41, 2323-43.
[3] C M Lee, H J Jeong, K N Yun, D W Kim, M H Sohn, J K Lee, J Joeng and S T Lim *International Journal of Nanomedicine* **2012**, 7, 3203-09.
[4] D A Giljohann, D S Seferos, P C Patel, J E Millstone, N L Rosi and C A Mirkin *Nano Letters* **2007**, 7, 3818-21.
[5] C You, A Chompoosor and V M Rotello *Nanotoday* **2007**, 2, 34-43.
[6] (**A**) F Grasset, N Saito, D Li, D Park, I Sakaguchi, N Ohashi, H Haneda, T Roisnel, S Mornet and E Duguet *Journal of Alloys and Compounds* **2003**, 360, 298–311; (**B**) Y L Wu, A I Y Tok, F Y C Boey, X T Zeng and X H Zhang *Applied Surface Science* **2007**, 253, 5473-79.
[7] R A Sperling, W J Parak *Philosophical Translations of the Royal Society* A **2010,** 368, 1333-83.
[8] (**A**) T Xia, M Kovochich, M Liong, L Madler, B Gilbert, H Shi, J I Yeh, J I Zink and A E Nel *ACS Nano* **2008**, 2, 2121-34; (**B**) A M Derfus, W C W Chan, S N Bhatia *Nano Letters* **2004**, 4, 11-18.
[9] (**A**) S-W Bian, I A Mudunkotuwa, T Rupasinghe and V H Grassian *Langmuir*, **2011**, 27, 6059-68; (**B**) A Khorsand Zak, R Razali, W H Abd and Majid Darroudi *International Journal of Nanomedicine* **2011**, 6, 1399-1403.
[10] S Tachikawa, A Noguchi, T Tsuge, M Hara, O Odawara and H Wada *Materials* **2011**, *4*, 1132-43.
[11] K L Kelly, E Coronado, L L Zhao, G C Schatz *the Journal of Phyical Chemistry B* **2002**, *107*, 668-77.
[12] J M Seiffert, M-O Baradez, V Nischwitz, T Lekishvili, H Goenaga-Infante, D Marshall *Chemical Research in Toxicology* **2012**, 25, 140-152.
[13] A Lankoff, W J Sandberg, A Wegierek-Ciuk, H Lisowska, M Refsnes, B Sartowska, P E Schwarze, S Meczynska-Wielgosz, M Wojewodzka, M Kruszewski *Toxicology Letters* **2012**, 208, 197-213.

Biocompatibility and cytotoxicity study of nanophotonic rigid gas permeable contact lens material

M Tomić[1], J Munćan[1], D Stamenković[2], M Jokanović[3] and L Matija[1]
[1]NanoLab, Biomedical Engineering, Faculty of Mechanical Engineering, Kraljice Marije 16, Belgrade, Serbia
[2]Optix, Oračka 13, Zemun, Serbia
[3]Academy of Sciences and Arts of the Republic of Srpska, Bana Lazarevića 1, Banja Luka, Republic of Srpska

E-mail: mmtomic@mas.bg.ac.rs

Abstract. Since materials on nanoscale have different characteristics from materials on macro scale their biocompatibility should be precisely and specifically investigated. Fullerenes, the third carbon allotrope, are one of the most used nanomaterials. The least stable and the most common is fullerene C_{60}. One of the main disadvantages of fullerene is its low solubility in water. In order to make it soluble, it must be functionalized with polar groups such as –OH and –COOH. From all the water soluble fullerenes the most important ones are those with –OH groups attached named fullerols. We have developed new materials for contact lenses by adding fullerene (C_{60}) and fullerol ($C_{60}(OH)_{24}$) into PMMA. The aim of our investigation was to compare the influences of those materials on aqueous solutions similar to tear film. For the analysis of the solutions we used opto-magnetic imaging and IR spectroscopy. The acquired spectrums were commented and compared with the standard contact lens material, which was analyzed by the same methods. The ISO 10993 cytotoxicity test on extract of nanophotonic material with incorporated C_{60} was done as well. This research contributes to better understanding of the biocompatibility of new rigid gas permeable contact lens materials.

1. Introduction

Due to more often application of nanomaterials, the studies of their interactions with tissue have become very important. Clinical applications of biomaterials shouldn't make any unwanted reactions in the organism and shouldn't compromise the life of a patient therefore every material must be biocompatible. Biocompatibility of nanomaterials should be specifically tested since they have different characteristics from the bulk materials and due to the size of the particles their penetration into tissues is easier.

When designing a contact lens few things should be considered, but probably the most important one is biocompatibility. A biocompatible contact lens is the one that doesn't damage surrounding ocular tissue during contact. The contact lens should permit oxygen from air to the cornea because unlike other tissues the cornea doesn't get any oxygen from the blood vessels. If the cornea doesn't get enough oxygen, it would swell and become infected. Another important property is wettability because the lens is in contact with tear film which consists of proteins, lipids, enzymes and other molecules that could be deposited on the contact lens surface.

In this paper we present the investigation of influence of three sorts of contact lenses on different aqueous solutions which are similar to tear film. Opto-magnetic imaging spectroscopy method and IR spectroscopy were used for characterizing the samples. Also the cytotoxicity test on extract of nanophotonic contact lenses with incorporated C_{60} was conducted to evaluate the potential for cytotoxicity.

Journal of Physics: Conference Series **429** (2013) 012016 doi:10.1088/1742-6596/429/1/012016

2. Materials

Fullerenes are a big family of super-atomic three-dimensional molecules. They were discovered by H. W. Kroto, R. F. Curl and R. E. Smalley in 1985, but the gram quantities became available in 1990 when Krätschmer and Huffman set the procedure for their production. Fullerenes consist of sp^2 hybridized carbon atoms distributed in hexagons and pentagons. Geodesic dome in Montreal made by Richard Buckminster Fuller was an inspiration for the name of these molecules, because this dome has the same structure as fullerene cage. The most popular fullerene is C_{60} also known as the *Buckyball* which is shown on figure 1 [1].

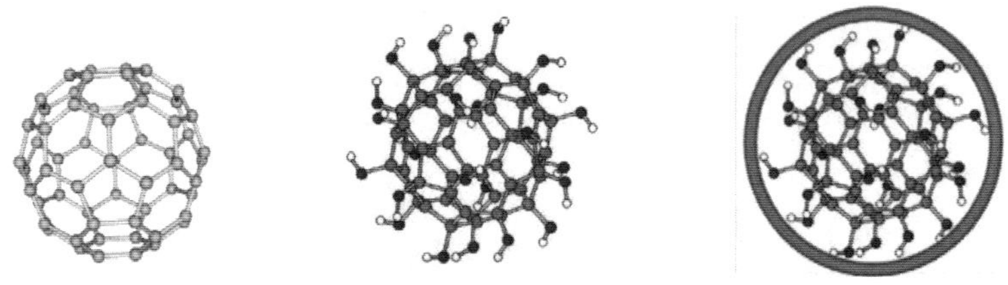

Figure 1. Fullerene C_{60} (left), fullerol $C_{60}(OH)_{24}$ (middle) and NHS-fullerol: $C_{60}(OH)_{24}$ @ $X(H_2O)$ or water shell fullerol (right) [3,4].

Fullerenes show strong affinity to electrons and act like ''radical sponges", because they easily enter addition reactions with nucleophiles. Spectroscopic characteristics of fullerenes are in strong relation with their symmetry. Structural information can be acquired from the number of bonds, for example, in IR spectra. Fullerene C_{60} is a good optical limiter but has low solubility and poor transparency. Fullerene materials are a group of new optical filters with such remarkable attributes as easy fabrication, predictable wavelength tuning, and excellent performance stability [2].

One of the main disadvantages in fullerene applications is its low solubility in water. In order to make them soluble, they must be functionalized with polar groups such as –OH and –COOH. From all the water soluble fullerenes the most important ones are those with –OH groups attached, named fullerols or fullerenols. When stable water layers (water shell) are surrounding $C_{60}(OH)_{24}$, in notation $C_{60}(OH)_{24}$ @ $X(H_2O)$, it is named nano harmonized supstance (NHS) (figure 1) [3]. They are free radicals scavengers and have anti-oxidative properties. Modified fullerenes are water soluble because they interact with water through hydrogen bonds. These fullerenes are unstable and can be degraded in contact with chemical agents from the environment. Modified fullerenes can be stabilized in the process of harmonization [3].

The properties of C_{60} can also be modified by its incorporation into polymers making it a material that can be easily handled and produced (figure 2). Therefore, combination of both systems has led to a wide variety of new materials showing appealing features based on the possibility of tuning their properties by modifying the chemical nature of the components or the chemical linkage between them [1].

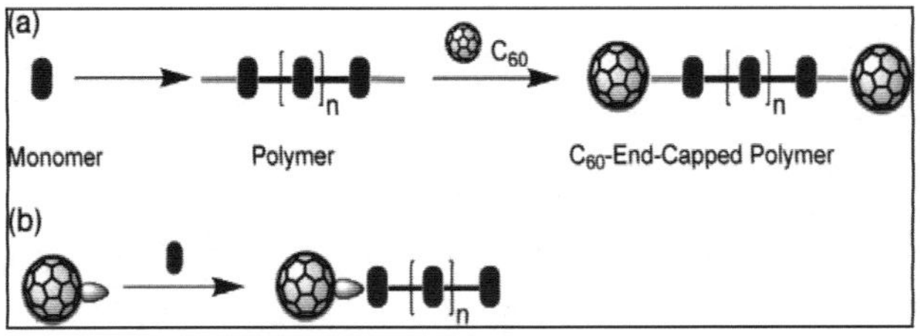

Figure 2. Synthetic strategies for the synthesis of C_{60}-end-capped polymers [2].

In our investigation of new contact lens materials we used two different sorts of nanomaterials: fullerene C_{60} and fullerol $C_{60}(OH)_{24}$. One gram of each nanomaterial was added during the polymerization process to the standard polymethylmethacrylate material, SolekoSP40™. The percentage of nanoparticles was 0.33%, but they were not completely dissolved in methylmethacrylate. Polymerization was homogenous in all samples, so the new nano-photonic material for rigid gas permeable lenses was made. The third polymerization was done without nanomaterials, so the standard Soleko SP40™ RGP material was made and was used as a referent sample.

For biocompatibility investigation we used three sorts of solutions: aqua purificata, saline and drops for dry eyes. Aqua purificata is a drinking water which is purified by double reverse osmosis. This water is completely demineralized. The saline is Natrii Chloridi infundibile 0.9% (Sodium chloride solution). The artificial drops were Refresh Contacts drops that consist of sodium-carboxymethylcellulose of 0.5%, as well as PURITE®. This is the unique preservative which breaks down into natural tear components – sodium and chloride ions, oxygen and water. Since these components are also in natural tears, the risk of irritating the eye or damaging cornea is minimized. Carboxymethylcellulose is a polymer which is ideal cytoprotective lubricant for the ocular surface. It provides long-lasting relief and protection from the symptoms of dry eye. This polymer is a mucoadhesive that shares similar protective qualities against harmful agents as natural ocular mucin.

3. Methods

For the characterization of different aqueous solutions we used two different types of spectroscopy methods: opto-magnetic imaging spectroscopy (OMIS) and IR spectroscopy based on 12 vibration modes, called Aquaphotomics [4].

3.1. Opto-magnetic imaging spectroscopy method (OMIS)

Each type of matter has special, different, angle value of light polarization. When the incident white light is diffuse, the reflected white light is composed of electrical and magnetic components, whereas diffuse incident light that is inclined under certain angle will produce reflected light which contains only electrical component of light (figure 3). This angle is called Brewster's angle and it represents the magnitude of the angle of incidence under which the sample polarizes the incident light. Taking the difference between white light (electromagnetic) and reflected polarized light (electrical) yields magnetic properties of matter based on light-matter interaction. Because such measurement can identify the conformational state and change in tissue on molecular level we named this method the opto-magnetic imaging spectroscopy [5].

The equipment for recording was a Canon digital camera, model IXUS 105, 12.1 MP. The light solution was accomplished by diffuse white diode and a lighting composition at Brewster's angle (three LED set at angle of 53° in regard to vertical axes). The recording region is circular with diameter of about 25 mm. In addition to customized camera, a software solution is used for analyzing the obtained images, yielding a characteristic diagram – diagnostic result – showing the intensities of light in correspondence with wavelength difference. Since this light is polarized by the sample that means that the character of polarization describes the character of the material. In this way, by characterizing the reflected light we can actually characterize the properties of the sample. This method is very sensitive since it detects magnetic properties on the basis of the response to visible light excitation which is relatively low in energy [5].

In this part of the research we used three samples of three types of contact lenses: standard SP40, SP40 with C_{60} incorporated and SP40 with $C_{60}(OH)_{24}$ incorporated. The solutions, aqua purificata, saline and eye drops, were poured into cups of 20 ml, and the contact lenses were put into those solutions. The solutions with contact lenses were left for 72 hours. After 72 hours, the contact lenses were taken out of the solutions, and the solutions were poured in other cups for taking pictures. Every sample was shot 10 times, with white and reflected polarized light. The pictures of solutions, without influence of contact lenses, were taken as well, for comparison.

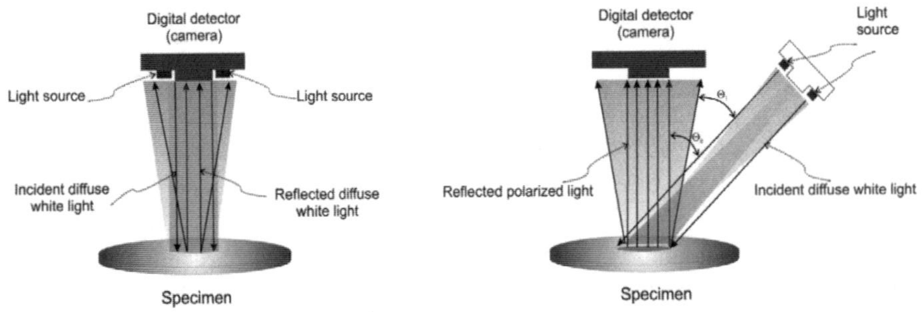

Figure 3. Operational principle of opto-magnetic imaging spectroscopy method [5].

3.2. Set of 12 infrared spectroscopic modes - Aquaphotomics

Aquaphotomics is a term, recently introduced to describe the concept in which water as multi-element system could be well described by its multi-dimensional spectra. For dynamic, non-invasive studies, vis-NIR spectroscopy has proved to be a powerful tool and source of information and it facilitates the establishment of Aquaphotomics. It discovers new water hydrogen bonds in biological systems under various perturbations and relates water absorbance patterns to respective biofunctionalities. To visualize the changes of water absorbance pattern a star chart named "Aquagram" is used. Aquaphotomics has been successfully applied in various fields from water characterization, food quality control to early diagnostics in medicine [5].

For this part of the research we used two types of rigid gas permeable contact lenses with incorporated nanomaterials and a standard SP40 contact lens. For investigating the behaviour of nanophotonic materials in aqueous solution we used saline (Natrii Chloridi infundibile 0.9%) and as a reference sample we used purified water (aqua purificata). First of all the spectrums of pure aqua purificata and pure saline were acquired by infrared spectrometer. Then all three sorts of contact lenses were left in saline. The spectrums were acquired after 1h and 40 minutes, after two days and finally after five days.

4. Results and discussion

4.1. Opto-magnetic imaging spectroscopy method results

The diagrams acquired by opto-magnetic imaging spectroscopy method show the differences and similarities for the same class of samples as well as for different classes. From the average spectrum diagram for aqua purificata and for aqua purificata under influence of SP40 contact lens we can see there are four peaks, two positive and two negative ones.

For aqua purificata average wavelengths and amplitudes for the first and the second positive peak are 111.71/100.728 and 120.71/72.91, respectively, and for the first and the second negative peak they are 110.7/-121.95 and 121.91/-68.49. On the other hand, those peaks under the influence of standard contact lens are at 113.57/85.05 and 119.22/52.91, and negative ones at 112.28/-103.05 and 120.68/-64.69, which is shown on figure 4.

For aqua purificata under influence of contact lens with incorporated C_{60} and $C_{60}(OH)_{24}$, there are three peaks, two negative and one positive. The wavelengths are slightly different (approximately 0.6 nm of difference) however the intensities (amplitudes) of the peaks are very different (the differences are up to 77 units). For aqua purificata under influence of contact lens with incorporated C_{60} the wavelength and amplitude of the positive peak are 115.9/164.2, and for the first and the second negative peak are 114.99/-63.54 and 116.73/-131.25, respectively. The same peak for the case with incorporated $C_{60}(OH)_{24}$ is at 115.29/86.38, and for the negative ones at 114.32/-38.11 and 116.26/-111.81.

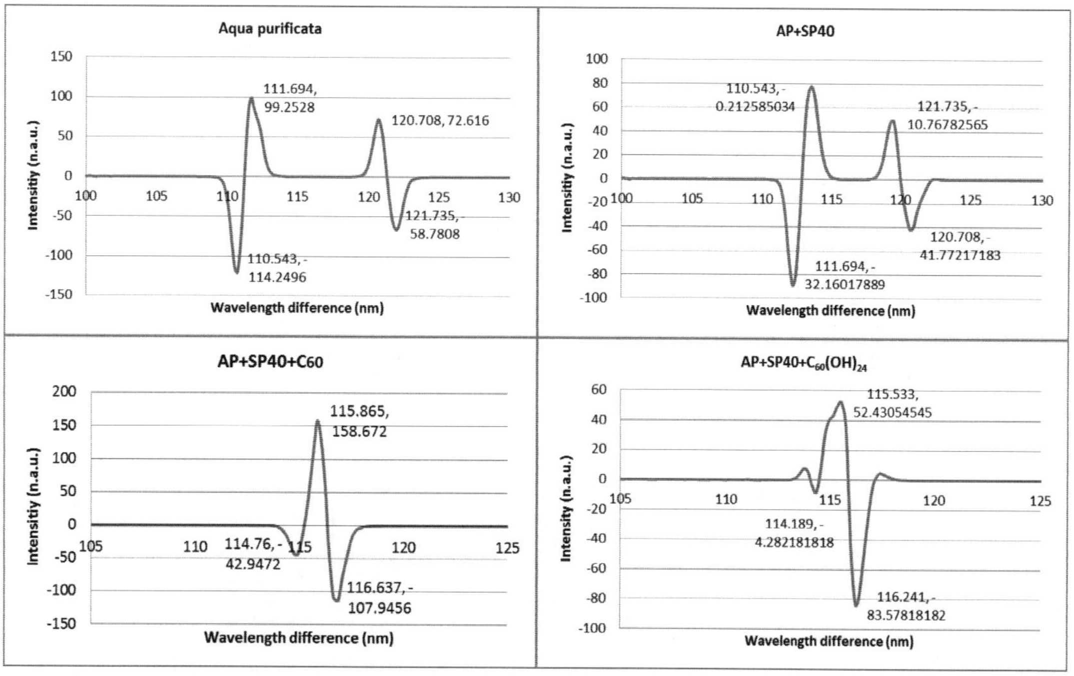

Figure 4. Comparison of wavelength difference (nm)-Intensity (n.a.u.) diagrams for aqua purificata, aqua purificata with SP40 contact lens (AP+SP40), aqua purificata with contact lens with C_{60} (AP+SP40+C_{60}) and aqua purificata with contact lens with $C_{60}(OH)_{24}$(AP+SP40+$C_{60}(OH)_{24}$).

For the class of saline the diagrams are slightly different, there are four peaks, two positive and two negative. Average wavelengths and intensities of the first and the second peak are 109.14/66.87 and 121.609/58.16, respectively, and the negative ones are 111.196/-37.53 and 119.51/-137.48. However, when this liquid was under the influence of the contact lenses the diagrams show two or three peaks. For the case of a standard contact lens SP40 and the one with incorporated C_{60} there are two peaks, one negative and one positive, at almost the same average wavelengths, and at slightly different amplitudes.

The positive peak for the saline under influence of SP40 contact lens is at 115.51/78.03, and the negative one is 116.63/-72.89. When saline was under influence of C_{60} the peaks were at 115.73/86.62 and 116.92/-73.73, respectively. The differences in wavelengths are approximately 0.2 nm, and for intensities approximately 0.8 units for the negative peak and 8 units for the positive one. In the case of influence of the contact lens with incorporated $C_{60}(OH)_{24}$ on the saline there are three peaks, two positive and one negative (figure 5). The wavelengths of the peaks are approximately the same as in two previous cases. The first positive peak is at 114.91/99.74, and the negative one is at 116.19/-113.29.

In the case of eye drops the diagrams are similar (figure 6). For the eye drops only there are four peaks, two positive and two negative, as well as for the case of aqua purificata and saline. The first and the second positive peaks are at 111.94/91.855 and 121.28/69.611, respectively. The negative peaks are at 110.37/-99.73 and 122.34/-81.26. However, in this class we have four peaks in the case of eye drops under influence of contact lens with incorporated C_{60} as well.

Here the positive peaks are at 114.87/84.76 and 118.12/50.08, and the negative ones are at 115.372/-74.86 and 118.48/-104.73. When eye drops were under influence of contact lens SP40 and the one with incorporated $C_{60}(OH)_{24}$ there are three peaks. For the SP40 contact lens the peaks are at 114.39/105.89, 115.84/-112.02 and 118.25/22.18, and for the one with $C_{60}(OH)_{24}$the peaks are at 114.91/79.3, 116.09/-89.03 and 117.08/34.02.

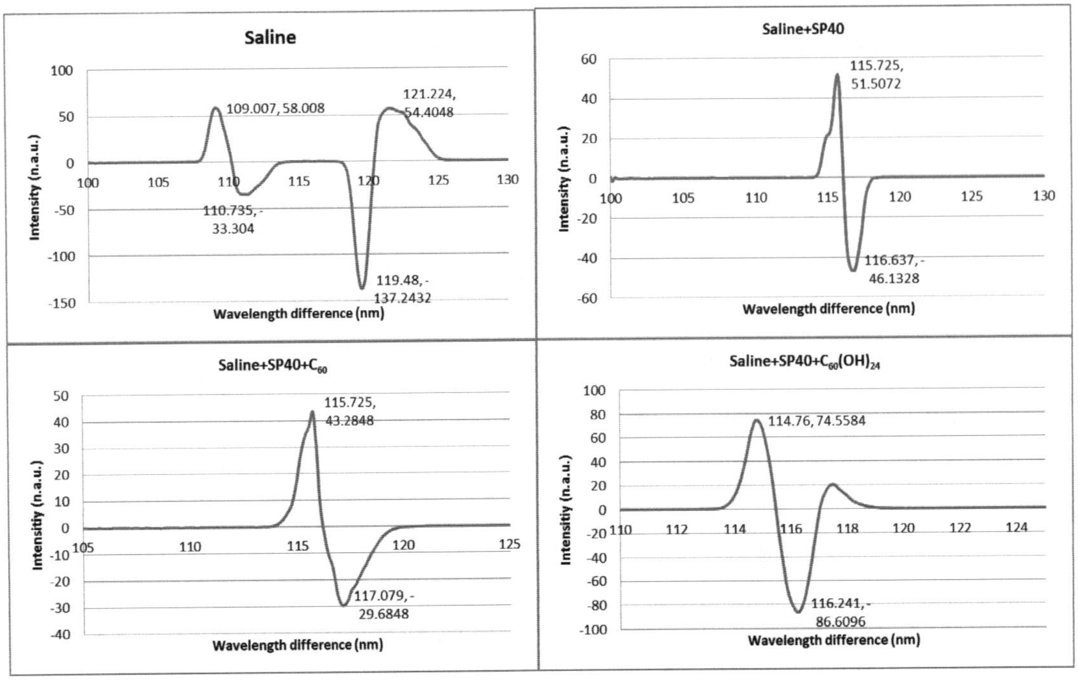

Figure 5. Comparison of wavelength difference (nm)-Intensity (n.a.u.) diagrams for saline, saline with SP40 contact lens (Saline+SP40), saline with contact lens with C_{60} (Saline+SP40+C_{60}) and saline with contact lens with $C_{60}(OH)_{24}$ (Saline+SP40+$C_{60}(OH)_{24}$).

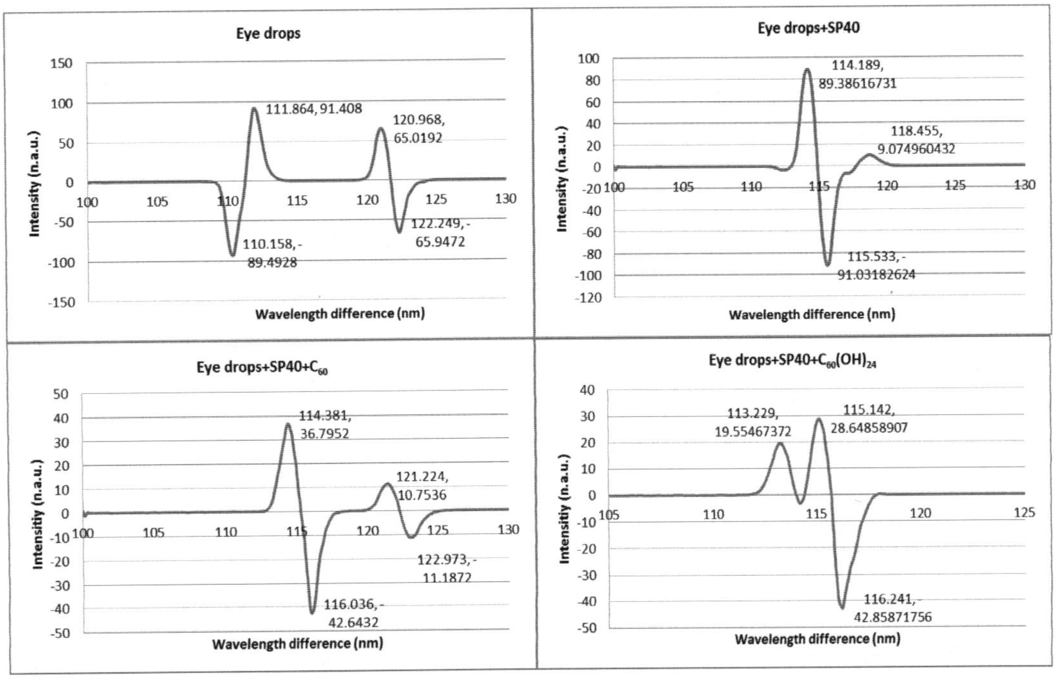

Figure 6. Comparison of wavelength difference (nm)-Intensity (n.a.u.) diagrams for eye drops, eye drops with SP40 contact lens (Eye drops+SP40) eye drops with contact lens with C_{60} (Eye drops+SP40+C_{60}) and eye drops with contact lens with $C_{60}(OH)_{24}$ (Eye drops+SP40+$C_{60}(OH)_{24}$).

4.2. Aquaphotomics results

The first results were acquired after 1h and 40 minutes (figure 7). Comparing to the pure water, the aquagram shows the decrease in strongly bounded water (the left part of the aquagram, 1493-1517 nm), an increase in free water molecules (S0) as well as the existence of water molecules in fist water shell (1360 nm). The presence of SP40 contact lens in saline induced a slight increase of water molecules in the first water shell, free water molecules and water molecules with free OH group which is different from the case of pure saline. In the presence of doped contact lenses there is an increase on free water molecules and molecules with free OH group which implies on breaking strongly bounded water. The highest absorption could be seen in the case of a material doped with fullerol ($SP40+C_{60}(OH)_{24}$).

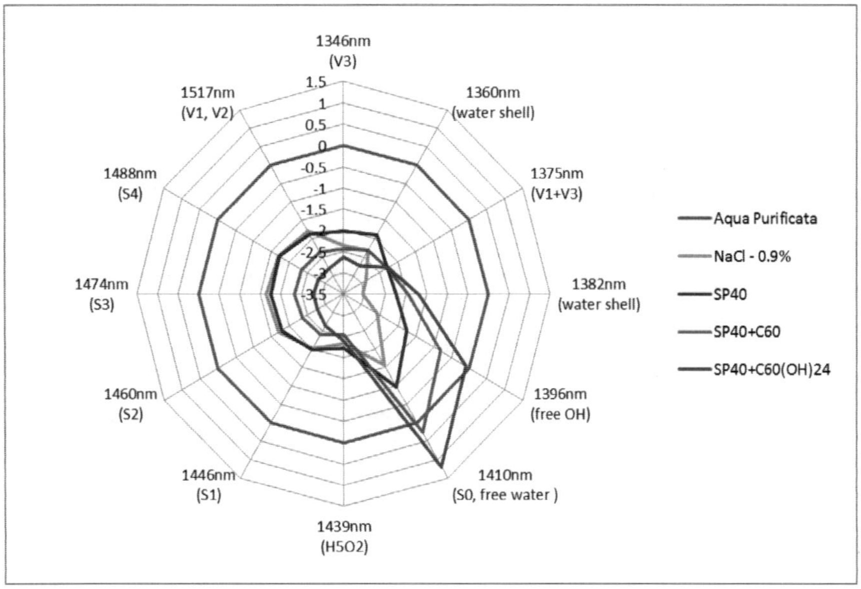

Figure 7. The aquagram: nanophotonic contact lenses in saline after 1h 40 min.

The second spectrum acquisition (figure 8) was made after two days and the aquagram shows the reorganization of water in NaCl 0.9%. This is most likely the consequence of evaporation i.e. the breaking of hydrogen bonds between the water molecules. Therefore there is an increase of free water molecules and molecules with free OH groups (1360 nm, 1382 nm – hydration shells).

Comparing to the basic material (SP40) the presence of the material doped with fullerene ($SP40+C_{60}$) induces more water molecules responsible for hydration and even more for the material doped with fullerol ($SP40+C_{60}(OH)_{24}$). The reason for this is that free water molecules and molecules with free OH groups form stable structures and take part in hydration of NaCl molecules and of nanoparticles that are probably released from the material. The proof for this is the highest absorption in the presence of $SP40+C_{60}(OH)_{24}$ because it has the most OH groups, it is very hydrophilic and it easily forms bonds with available water molecules.

After five days (figure 9) NaCl 0.9% showed reorganization of water with slightly decreased number of free water molecules and molecules with free OH groups, but there is a significant amount of hydration shells at 1360 nm and 1382 nm. There is also strongly bounded water (left part of the aquagram at 1460 nm and 1488 nm), which is not the case for saline and for the solution in presence of all types of contact lenses.

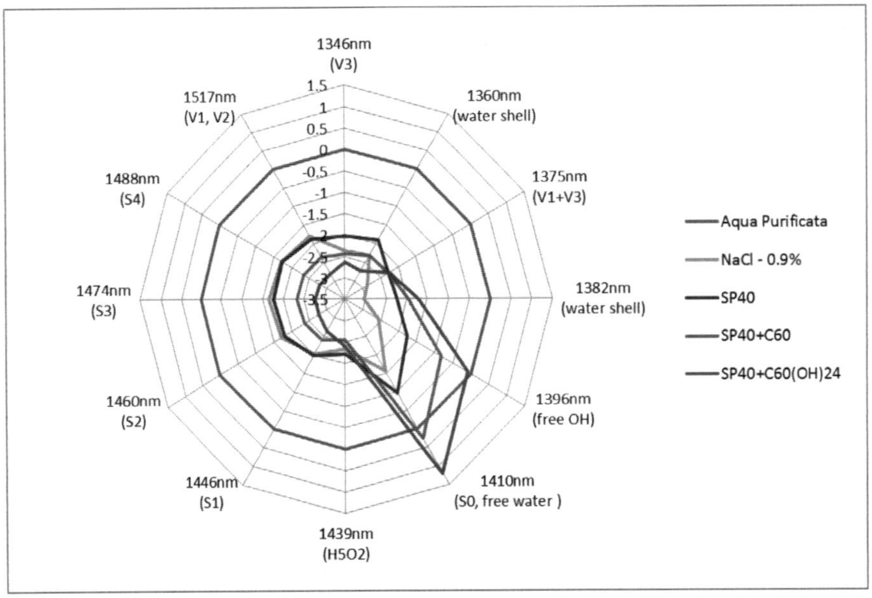

Figure 8. The aquagram: nanophotonic contact lenses in saline after 2 days.

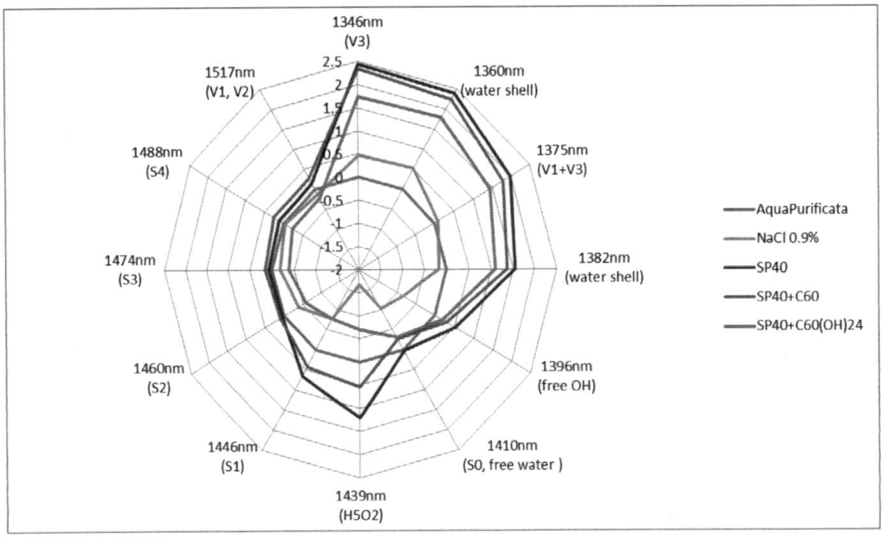

Figure 9. The aquagram: nanophotonic contact lenses in saline after five days.

In the presence of the basic material and the material with fullerene C_{60}, the organization of water is almost the same, in both cases there is large amount of hydration shells and there is absorption at 1439 nm (H5O2). In the presence of the material with $C_{60}(OH)_{24}$ the difference is significant due to the presence of strongly bounded water in a left part of the aquagram (1446 nm, 1460nm, 1474 nm and 1488 nm) which implies the presence of water molecules forming 1, 2, 3 and 4 hydrogen bonds.

4.3. Cytotoxicity study

For the cytotoxicity study we have chosen PMMA with C_{60} incorporated, as the "worst case". A cytotoxicity test on extract of nanophotonic contact lenses with incorporated C_{60} was conducted to evaluate the potential for cytotoxicity. This study was conducted according to the requirements of the United States Pharmacopeia and ISO 10993 standard: Biological Evaluation of Medical Devices, Part 5 (2009): Tests for in vitro cytotoxicity. Under the conditions of this study, the extract of the test article showed no evidence of causing cell lysis or toxicity greater than a grade 2 (mild reactivity), grade 0 for the test article (table 1). The extract of the test article met the requirements of the USP and ISO 10993 standard.

Table 1. Cytotoxicity test results.

	GRADE	REACTIVITY
	0	None
Test article	0	None
	0	None
	4	Severe
Positive control	4	Severe
	4	Severe
	0	None
Negative control	0	None
	0	None

5. Conclusion

Biocompatibility of fullerene materials is still insufficiently investigated and according to the existing results precaution is necessary. According to the study of cytotoxicity, contact lenses of standard material with incorporated C_{60} have no toxic effect on mouse fibroblast cells, therefore it can be concluded that other types of materials should not have cytotoxic effect since they contain functionalized fullerenes.

By the method of opto-magnetic imaging spectroscopy we have detected the differences in influences, but the meaning of those differences is not known until comparative toxicological investigations are done. The acquired Wavelength difference (nm) - Intensity (n.a.u.) diagrams show that all three sorts of contact lenses have almost the same influence on all three used liquids. Therefore, for aqua purificata before adding the contact lenses there are four peaks, two positive and two negative. The diagrams are the same for saline and eye drops as well. Also, it is obvious that the SP40 contact lens didn't make significant changes for aqua purificata, and for saline and eye drops there is a change, there are three peaks. For the case of contact lenses with incorporated C_{60} and $C_{60}(OH)_{24}$ in aqua purificata the peaks are at approximately the same wavelengths, only different intensities, which can imply to the same influence of both sorts of contact lenses.

In the case of saline under influence of contact lenses with incorporated C_{60} and $C_{60}(OH)_{24}$ the diagrams are slightly different. It is obvious that those sorts of contact lenses have different influences, because there are two and three peaks, at different wavelengths. It is the same for the eye drops that were under influence of contact lenses with incorporated C_{60} and $C_{60}(OH)_{24}$, but here we have four and three peaks. Also it is marked that the peaks appear at wavelengths in range of 108 nm to 122 nm, while intensities have bigger variations.

According to the aquagrams it can be concluded that there is a certain influence of the materials on saline. The water is organized around some molecules, but it is not certain whether these molecules are nanoparticles from the material or some other molecules. Infrared spectroscopy (aquaphotomics) method showed specific influences of nanophotonic materials on saline solution.

After acquiring first aquagrams it could be concluded that doped materials induce restructuring of saline meaning the number of strongly bounded water molecules decreases and the number of free water molecules and molecules with free OH groups increases. After two days the water restructured even more and the absorbance was higher for the saline with doped materials, especially in the case of a material doped with $C_{60}(OH)_{24}$. After five days it is obvious that the material doped with fullerol (SP40+$C_{60}(OH)_{24}$ has the lowest influence.

The future investigation should be focused on eye irritation and sensitivity testing, acute systemic and sub chronic toxicity and on *in vitro* genotoxicity testing according to the appropriate ISO standards.

Acknowledgments

This research has been partially funded by Ministry of Science and Technological Development of Republic of Serbia, through Project III 45009.

References

[1] Giacalone F and Martín N 2006 Fullerene polymers: synthesis and properties *Chem. Rev.* **106** 5136-5190

[2] Giacalone F and Martín N 2009 *Fullerene Polymers: Synthesis, Properties and Applications* Wiley-VCH VerlagGmbH & Co. Weinheim

[3] Koruga Dj 2011 Composition of Matter Containing Harmonized Hydroxyl Modified Fullerene Substance, US Patent 8,058,483 B2, November 15, 2011.

[4] Tsenkova R 2009 Introduction to aquaphotomics: dynamic spectroscopy of aqueous and biological systems describes peculiarities of water *J. Near Infrared Spectrosc* **17** 303-314

[5] Koruga Dj and Tomić A 2009 System and Method for Analysis of Light-matter Interaction Based on Spectral Convolution *US Patent* Pub. No.: 2009/0245603

[6] Tomić M 2011 Investigation of Classical and Fullerene Contact Lenses Influence on Biological Liquids by Optomagnetic Spectroscopy Method *MSc Thesis* Faculty of Mechanical Engineering University of Belgrade

ECSIN's methodological approach for hazard evaluation of engineered nanomaterials

Lisa Bregoli, Federico Benetti, Marco Venturini, Enrico Sabbioni

ECSIN-European Center for the Sustainable Impact of Nanotechnology, Veneto Nanotech S.C.p.A., Viale Porta Adige 45, I-45100 Rovigo, ITALY

lisa.bregoli@venetonanotech.it

Abstract. The increasing production volumes and commercialization of engineered nanomaterials (ENM), together with data on their higher biological reactivity when compared to bulk counterpart and ability to cross biological barriers, have caused concerns about their potential impacts on the health and safety of both humans and the environment. A multidisciplinary component of the scientific community has been called to evaluate the real risks associated with the use of products containing ENM, and is today in the process of developing specific definitions and testing strategies for nanomaterials. At ECSIN we are developing an integrated multidisciplinary methodological approach for the evaluation of the biological effects of ENM on the environment and human health. While our testing strategy agrees with the most widely advanced line of work at the European level, the choice of methods and optimization of protocols is made with an extended treatment of details. Our attention to the methodological and technical details is based on the acknowledgment that the innovative characteristics of matter at the nano-size range may influence the existing testing methods in a partially unpredictable manner, an aspect which is frequently recognized at the discussion level but oftentimes disregarded at the laboratory bench level. This work outlines the most important steps of our testing approach. In particular, each step will be briefly discussed in terms of potential technical and methodological pitfalls that we have encountered, and which are often ignored in nanotoxicology research. The final aim is to draw attention to the need of preliminary studies in developing reliable tests, a crucial aspect to confirm the suitability of the chosen analytical and toxicological methods to be used for the specific tested nanoparticle, and to express the idea that in nanotoxicology,"devil is in the detail".

1. Introduction

ECSIN – European Centre for the Sustainable Impact of Nanotechnology - is a research center part of Veneto Nanotech, located in the city of Rovigo, Veneto Region of Italy. ECSIN's most relevant activities are in the fields of nanotoxicology and risk assessment. In particular, the main objective of the facility is to apply the most recent and innovative scientific findings to the study of the impact of nanomaterials on the environment and human health as well as to investigate ethical and social aspects of this emerging technology. Final aim is to support companies in the production of non-toxic products, and to provide support for survey systems and measurement standards required by companies, public authorities and investors.

As standard protocols for nanotoxicity assessment are still lacking, we have optimized some of the most frequently used toxicology protocols, adapting them for the use with nanoparticles (NPs).
In this report we review the most important issues that should be addressed in the optimization of nanotoxicity testing. We highlight some of the most important technical considerations that should drive the development of standardized protocols for nanotoxicology, suggesting control experiments and protocol modifications to avoid artifacts.

2. Development and optimization of nanotoxicity *in vitro* tests

In vitro testing represents a first common step for assessing ENM related health hazards. The most commonly used vitality and functional assays (e.g. MTT, WST-1, XTT, MTS, LDH, $DCFH_2$ and caspase 3/7) rely on absorbance or fluorescence measurements for quantification. These *in vitro* assays have been established for small molecules of drugs and chemicals. Their application to Engineered Nanomaterials (ENM) opens new questions regarding interferences with the expected response of the assays because of the unique physico-chemical properties. Potential interferences are summarized in Table 1.

Table 1

	Artifact
Optical interferences	NPs absorbance adds to the assay absorbance
	NPs absorbance decreases the excitation light and the emission spectrum of fluorescent probes
Physical interferences: adsorption on NPs surface	Reduction of end products and/or enzymes decreases absorbance or fluorescence and prevents enzymatic reactions
Physico-chemical interferences: NPs reactivity	NPs or their ions catalyze or inhibit *in vitro* assays
Interferences by systems used	Culture medium and serum may catalyze oxidation of several substrates used in the assays

2.1. Optical interferences

Most metal NPs, carbonaceous and some metal oxides NPs absorb and scatter light over the spectral regions used in the assay systems [1–3]. Therefore, *in vitro* assays could be affected by the presence of NPs interacting with visible light. Data generated using colorimetric *in vitro* assays may derive by a linear combination of chromophore and NPs contribution, resulting in false positives. NPs may also affect data obtained with fluorescence-based assays, reducing the incident light absorbed by fluorophores and generating false negatives.

In vitro assays used for evaluating metabolic activity are based on tetrazolium salts reduction and purple formazan formation. Kroll et al (2011) investigated the optical interferences of twenty-three ENM – CeO_2, TiO_2, Ti-Zr, ZrO_2, $SrCO_3$, carbon black, AlOOH and Al-Ti-Zr – on three standardized in vitro assays: MTT, LDH and $DCFH_2$ [4]. They found interferences of the nanomaterials with all the *in vitro* toxicity test systems. These interferences were abolished removing NPs from formazan solution, at least at low NPs concentrations used.

Optical interferences may also affect quantification of lactate dehydrogenase (LDH) release. In the assay, levels of extracellular LDH released from damaged cells is measured as an indicator of cytotoxicity. The assay is based on the ability of LDH to convert lactate into pyruvate with the subsequent reduction of NAD to NADH. Although quantification of LDH can be measured recording absorbance induced by NADH at 340 nm, several commercial kits have been developed. These kits are based on the reduction of iodotetrazolium chloride to formazan with an absorption peak between 500 nm and 600 nm. Wilhelmi and colleagues reported alterations in LDH quantification due to the NPs used [2]. In presence of fine and ultrafine TiO_2, zinc oxide and SiO_2 NPs, an increase in absorption and then an overestimation of LDH release occurred. Also in this case, interferences were abolished removing NPs from formazan containing solution.

Caspase 3/7 assay represents the most commonly used test to investigate the molecular mechanisms involved in cell death induced by NPs, as it indicates apoptosis induction. It is based on the cleavage of non-fluorescent caspase substrate by caspase 3/7 to create a fluorescent molecule. Since excitation wavelengths of cleaved substrate overlap with NPs absorption, the resulting emission spectrum decreases in the presence of NPs. An example is reported in Figure 1 where fluorescence of cleaved substrate decreases in the presence of silver NPs, while it returns to the positive control values after NPs removal.

In conclusion, artifacts due to optical interferences can be solved removing NPs from the chromophore/fluorophore containing solutions. In the case of colorimetric assays, NPs contribution can be conveniently subtracted by sample absorbance.

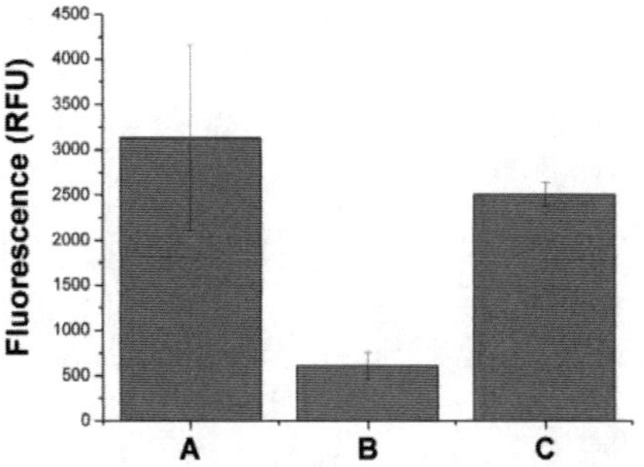

Figure 1. Caspase 3/7 assay on A549 cells treated with staurosporine 0.5 μM (Positive control) (A), after addition of silver NPs (B) and after their removal (C).

2.2. Physical interferences: adsorption on NPs surface

Some NPs may adsorb formazan or other end products as well as LDH or other enzymes reducing the assay absorbance or preventing enzymes from being measured, respectively [5]. Moreover, they can carry pyrogenic contaminations, difficult to detect but biologically strongly active [6]. If this interference occurs, alternative methods to test NPs cytotoxicity need to be sought.

2.3. Physico-chemical interferences: NPs reactivity

Metal NPs dissolution may interfere with *in vitro* assays either directly or indirectly. Oxidative dissolution of metal NPs can generate formazan increasing the assay absorbance. Holder et al (2012) analyzed the potential interference of several carbonaceous and one TiO_2 NPs on formazan formation [5]. Activated carbon did not react with MTT at any concentration, while soot, oxidized soot and TiO_2 produce formazan only at the highest concentration used (50 μg/mL). Diesel particles generated formazan at all concentration tested. Reduction of tetrazolium salts by TiO_2 was also demonstrated by Wang et al (2011) [7]. Authors reported that in comparison with trypan blue exclusion assay, MTT and XTT predict inaccurately cell toxicity or overestimate cell viability, respectively. The superoxide produced by TiO_2 [8,9] resulted in the conversion of MTT and XTT to formazan causing misleading results. Ion released from oxidative dissolution of metal NPs may inactivate enzymes involved in formazan production denaturing them or blocking their active site [10]. Han et al (2011) investigated the effects of copper NPs, silver NPs and TiO_2 NPs on LDH activity [11] as a function of their dissolution rates. All the NPs inhibited LDH activity, although in a different way. Because of the high dissolution rate, copper NPs could interfere with the LDH assay by inactivating the enzyme. Conversely, TiO_2 NPs were found to adsorb LDH molecules. LDH inhibition by silver NPs was attributed to the carbon matrix used to cage the particles during the synthesis.

2.4. Ion release: methods and limitations

As described above, ions released from NPs may inhibit enzymatic assays. Therefore, it is important to establish the dissolution rate of NPs in working conditions. This data is also important to disclose molecular mechanisms underlying NPs cytotoxicity. Several methods are currently used, though they provide different types of information: ultracentrifugation [12] or ultrafiltration [13] coupled with ICP-MS, fluorescent probes [14,15] and ion-selective electrode potentiometry [16]. Although it has been calculated that ultracentrifugation at 100000g for 60 minutes is sufficient to move particles with diameter larger than 4 nm at least 2.5 cm, NPs can be present in the supernatant causing an overestimation of the total ion released. On the other side, ultrafiltration permits to measure only ions bound to small molecules, while those bound to large molecules or proteins such as, for instance, serum are retained. In the presence of complex media, ultrafiltration does not provide the total amount of released ions but only free-ions or ions chelated with small molecules. The information obtained using molecular probes depends on their affinity constants. Molecular probes with low affinity constants detect only high concentrations of free-ions, while they are not able to chelated ions bound to other molecules. The use of ion-selective electrode potentiometry can be invalidate by interferences with complex matrix such as culture media. In this case, it is very important to perform titration curves in the complex matrix in order to avoid interferences.

2.5. Culture medium and serum interferences

Evaluation of NPs-generated oxidative stress is based on oxidation of dichlorofluorescein ($DCFH_2$) or its derivatives. Several critical issues may occur: $DCFH_2$ oxidation by culture medium during loading into cells or leakage from cells; $DCFH_2$ adsorption on NPs surface; $DCFH_2$ oxidation by NPs and optical interferences. $DCFH_2$ can be deacetylated and subsequently oxidized by culture medium as reported in Figure 2. Therefore, increased fluorescence intensity may depend on aspecific oxidation by medium or NPs occurring during $DCFH_2$ loading or its leakage. To avoid aspecific $DCFH_2$ oxidation, it is important to evaluate culture medium and serum quality and, eventually, load it into cells in a very short-time using phosphate buffer and before NPs addition. NPs ability to oxidize $DCFH_2$ has to be tested. Contribution of oxidized $DCFH_2$ as a function of leakage can be avoided by rinsing cells with phosphate buffer before the assay.

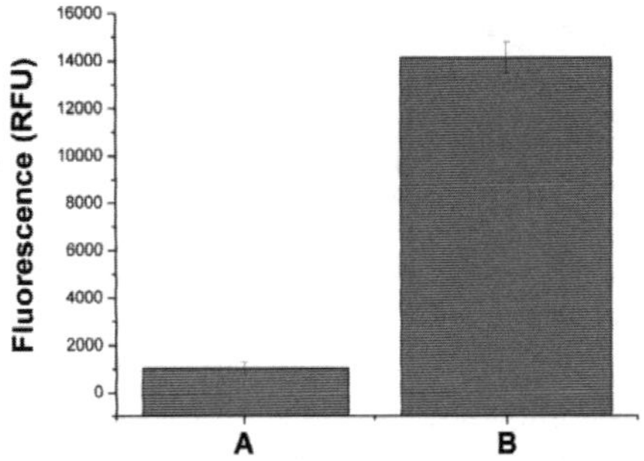

Figure 2. Aspecific oxidation of $DCFH_2$ in phosphate buffer (A) and cell culture medium F12 (B).

3. Cell models and optimization of culturing procedures

3.1. Primary cells versus cell lines

Primary cells are considered to be the closest *in vitro* cellular model for human organs, and the use of cell lines as surrogates of primary cells has been questioned by different authors. Cell lines often derive from transformed or transfected cells which have lost their basic proliferation control mechanisms, and show a great number of genetic and chromosomal aberrations which potentially increase at each passage. Their phenotypical difference compared to the primary cells results for example in differences in the basal levels of enzymes, as in the case of hepatocytes [17]. Also the response to cytotoxic exogenous stimuli can be different between primary cells and cell lines. In the case of hematopoietic cells, we have shown that the level of cytotoxicity induced by antimony trioxide nanoparticles exposure can be very different between primary cultures of bone marrow hematopoietic progenitors and seven human cell lines of hematopoietic origin: K562, HL-60, CEM, CEM-R, Thp-1, Jurkat, and Molt-4 [18].

Nevertheless, nano-cytotoxicity assays need to be standardized and cells need to be as stable as possible for use in different laboratories all over the globe. For this purpose, the use of primary cells present strong limitations such as limited life span, phenotypical instability, large variability between different human donors, limited availability and high costs. These limitations have brought to extensive use of cell lines for nanotoxicity screenings, mainly due to the fact that cell lines are easy to maintain and can be sub-cultured for many passages, batch-to-batch variability is supposed to be low and their cost is affordable. In a recent report, Kermanizadeh et al (2012) have compared primary human hepatocytes with the C3A cell line in respect to their response to a panel of nanomaterials [19]. The cell line was very comparable with the primary hepatocytes with regards to their cytotoxic response to ZnO, MWCNTs, Ag and positively functionalized TiO_2 NPs. This confirms that some cell lines can be used as a model in place of primary cells for cytotoxicity screenings. Taken together, these studies bring to the consideration that confirmation analysis should ideally be done for each cell line and relative to each cell assay, keeping in mind that it is not possible to predict what the interaction of NPs with cells in living organisms will be, solely by extrapolating from nanotoxicology studies on cell models.

3.2. Choice of cell type

The choice of the cell model is based on the evaluation of the potential exposure route to individual nanoparticle, depending on the specific application. Table 2 shows the cell lines we have chosen as representative of cellular models in reference to different nanoparticles applications and routes of exposure.

Table 2

Examples of NP applications	Exposure Route	Primary target organs	Cell lines used at ECSIN
Paintings and any type of surface coatings, air treatment and purification, textiles, worker exposure to dry nanoparticles	Inhalation	Respiratory tracts	A549 (lung) NR8383 (rat alveolar macrophages)
Food industry and processing, food contact material and kitchen utensils, dental care products	Ingestion	Digestive tract	Caco2 (colon)
Cosmetics, textiles, bandages and wound care, sunscreen lotions, surface coatings	Dermal	Skin	L929 (fibroblasts from subcutaneous connective tissue) HaCaT (keratinocytes)
Nanomedicine, secondary organ exposure following uptake and biodistribution	Blood Circulation	Blood cells, endothelial cells, internal organs	HepG2 (liver) HEK293 (kidney) MDCK (dog kidney) MG63 (osteoblasts) Thp1(monocytes)

3.3. Proliferating versus differentiating cells

One issue that we would like to bring to light here regards the choice of cell cycle state at which cells are exposed to NPs for cytotoxicity assays. In a previous study we have treated primary hematopoietic progenitor cells during proliferation and during differentiation state with Sb_2O_3 NPs at 5 ppm. When NPs were added during the proliferation phase, either at start of culture or at later point, cells looked morphologically altered within 6 hours, were mostly dead at 24 hours and completely dead within 96 hours after start of treatment. When, instead, NPs were added during the differentiation phase we observed no effect on differentiation as confirmed with flow cytometry analysis and real-time PCR analysis of differentiation markers. We then performed the exposure experiment on Thp-1 cell line. Sb_2O_3 NPs had no significant toxic effect on proliferating Thp-1 but inhibited the PMA-induced differentiation of Thp-1 cells into macrophages, driving them to death. These results demonstrate that the choice of cell proliferating state may greatly influence the cytotoxicity response, as the cytotoxic effect can be very different if cells are in highly proliferating versus differentiating states.

This is not surprising, as it is well known that cells in different states of their life express different biomolecules, and may respond in different manners to exogenous stimuli. Nevertheless, if we consider some of the recently published studies where A549 cells are used for nanotoxicity assays, we see that the same cell line is not always used at the same cell cycle state [4, 20-24]. Cells of the A549 human lung adenocarcinoma cell line have an epithelial morphology and are widely used as an *in vitro* model for NPs cytotoxicity assay due to inhalation exposure. This cell line is hypotriploid with a modal chromosome number of 66 which occurs in 24% of cells, with higher ploidies occurring at an infrequent rate (0.4%). This brings to the possibility to select clones if sub-culturing is not performed when cells are below confluency. When these cells are kept growing to confluency, they form a monolayers with Type II characteristic morphology and tannic acid staining for typical lamellar bodies [25], indicating differentiation.

For nanotoxicity screening we noticed heterogeneity of the experimental procedures in regard to the number of cells/cm^2 at time of plating. Some researchers plate these cells at low concentration, so that cells are not confluent at time of NPs exposure [20–22]. In other words, cells are in their proliferating phase during exposure to NPs. Other authors, instead, indicate plating of cells at high concentration in order to reach confluency at time of exposure [4,23,24], so that cells have formed monolayers and are starting to differentiate. These two different plating protocols for nanotoxicity screenings on the same cell line may allow to different results due to the different physiological state of cells at time of treatment.

3.4. Optimization of assays with cells

In our laboratory, we have chosen to run the cytotoxicity assays MTS and LDH on A549, MDCK and HepG2 cells after 6, 24 and 48 hours of exposure to NPs, according to the ASTM standard protocol E2526–08 (Evaluation of Cytotoxicity of Nanoparticulate Materials in Porcine Kidney Cells and Human Hepatocarcinoma Cells) with modifications.

Our first step of the optimization phase of the procedure has been focused on the selection of the best cell plating concentrations for each of the 3 cell lines, so that we can perform both MTS and LDH assays on the same cultures. Exposure is preceded by a 24 hours incubation to allow cell attachment and start of proliferation. In order to select a plating concentration that would allow us to expose NPs to cells that are proliferating during the whole incubation period, we have run preliminary experiments. Figure 3 shows an MTS assay performed with A549 cells that were plated at 3,000, 5,000 and 8,000 cells/well in 96 well plates (surface area of approximately 0,35 cm^2 each well), in a total volume of 200 μl. After 24 hours, medium was changed to mimic start of exposure, and cells were incubated for the additional indicated time (6, 24 and 48 hours after medium change). At the end of incubation, supernatant was removed and assayed for LDH activity (Biovision kit). MTS assay was performed following providers instructions (Promega Cell Titer Kit).

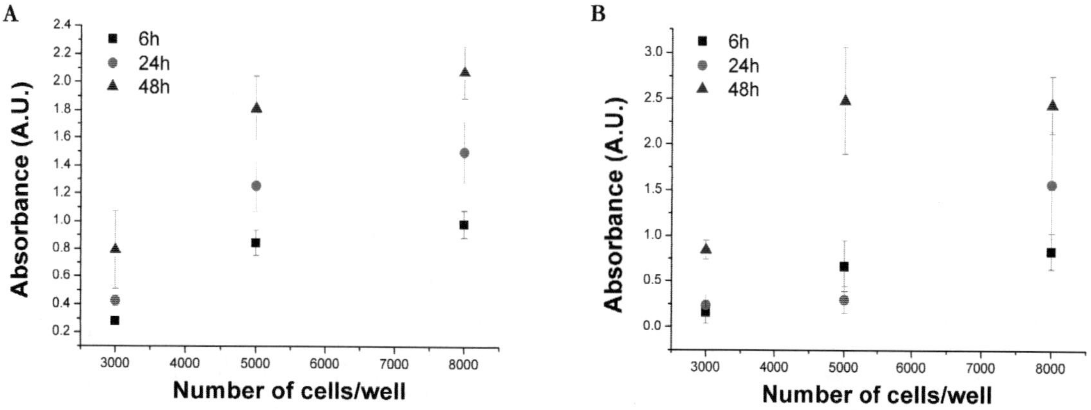

Figure 3. MTS (A) and LDH (B) assays on A549 cells at different incubation times after medium change: 6 hours (black squares), 24 hours (red dots) and 48 hours (blue triangles). X axis: number of cells. Y axis: absorbance values at 510 nm (MTS assay) or 500 nm (LDH assay), after subtraction of blank (cell free well). Values represent the mean of 3 replicates. Error bars represent standard deviation.

From results shown in Figure 3, we have selected 5,000 cells/well as the best starting concentration of cells for these assays, as at this concentration we can see a linearity both in terms of absorbance reading and in terms of incubation time. A density of 3,000 cells/well was not chosen because absorbance values are too low to allow a detectable difference in case of toxic stimulus, especially for samples at 6 hours incubation. On the other hand, 8,000 cells/well is too high because cells are already starting to enter the stationary phase, where alterations of metabolic activities would not be detectable properly.

Following the same procedure, we have performed the optimization test to select the best cell concentration on dog renal MDCK and human hepatocellular carcinoma HepG2 cell lines. Figures 4 and 5 show their results.

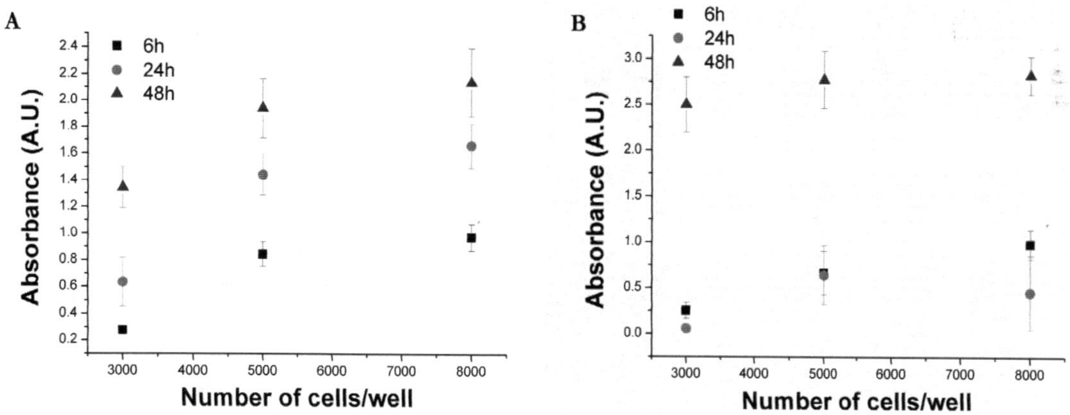

Figure 4. MTS (A) and LDH (B) assays on MDCK cells at different incubation times after medium change: 6 hours (black squares), 24 hours (red dots) and 48 hours (blue triangles). X axis: number of cells. Y axis: absorbance values at 510 nm (MTS assay) or 500 nm (LDH assay), after subtraction of blank (cell free well). Values represent the mean of 3 replicates. Error bars represent standard deviation.

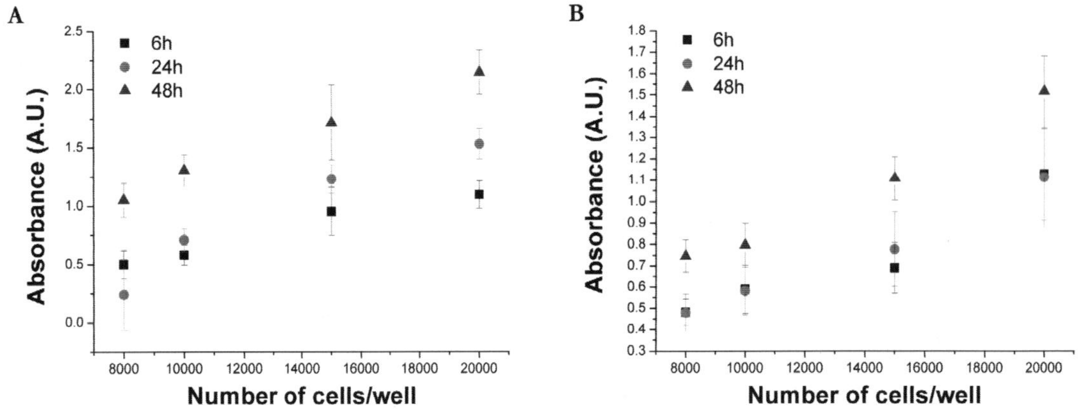

Figure 5. MTS (A) assay and LDH (B) on HepG2 cells at different incubation times after medium change: 6 hours (black squares), 24 hours (red dots) and 48 hours (blue triangles). X axis: number of cells. Y axis: absorbance values at 510 nm (MTS assay) or 500 nm (LDH assay), after subtraction of blank (cell free well). Values represent the mean of 3 replicates. Error bars represent standard deviation.

For MDCK cells, instead, we have selected 3,000 cells/well as these cells have a higher metabolic activity than A549 cells. For HepG2 cells we have chosen 15,000 cells/well as starting cell concentration for plating, in fact these cells are smaller and have a lower metabolic activity compared to A549 cells.

This first step of optimization is performed only once for each cell type and assay type. On the contrary, the second step of the optimization procedure is repeated for each NP type, as it is focused on the interaction of NPs to the assay. For this purpose, we add the NP being tested into the cell-free medium in the presence or absence of a positive control and run the assay. If the assay results show that the positive control is not influenced by the presence of the NPs at different concentrations, then we conclude that the NPs do not interfere with the assay. If, on the contrary, we observe an interference, then we consider all the possible interference mechanisms that have been described above. For the optimization of assays that require the reading of a fluorescent dye which is trapped inside live cells (such as microplate $DCFH_2$ assay for ROS determination), we apply a few additional technical considerations. In particular, as proliferating cells do not cover the whole well substrate surface, it is necessary that the reading occurs at different points of the well, to avoid reading only at one site where cells are not representative of the whole well. To solve this problem, it is necessary to use a reader that allows to perform an area scan.

4. Conclusions

In line with the actual multi-disciplinary efforts to develop and validate standard nanotoxicity assay protocols, our work at ECSIN has been focused on optimization of a few frequently used protocols. These assays have been borrowed from classical toxicity testing but require a careful consideration of all components and steps, in order to be adapted to the use with NPs. This is due to the peculiar physico-chemical characteristics of NPs, which often cause problems at different steps of the tests, such as optical interferences with the readout, adsorption of assay components onto NPs surface, NPs reactivity with assay molecules. Control experiments have to be performed for each type of NP being tested, in order to determine whether any of these interferences could be induced. Some of these issues can be overcome by removing NPs from the readout solutions, subtracting appropriate blanks or reducing the concentrations of NPs. When these modifications are not sufficient, for example in case of interference due to the released ions, then the assay should be abandoned and the same molecular endpoint should be sought using different methods. The choice of the cell model and cell density has to be carefully evaluated. The former is based on the evaluation of the potential exposure route to individual nanoparticle, while cell density at time of plating is important to compare results obtained from different laboratories.

References

[1] Amendola V and Meneghetti M 2009 Size Evaluation of Gold Nanoparticles by UV–vis Spectroscopy. *The Journal of Physical Chemistry C* 113: 4277–4285.

[2] Wilhelmi V *et al.* 2012 Evaluation of apoptosis induced by nanoparticles and fine particles in RAW 264.7 macrophages: facts and artefacts. *Toxicology in vitro : an international journal published in association with BIBRA* 26: 323–334.

[3] Zook JM *et al.* 2011 Measuring silver nanoparticle dissolution in complex biological and environmental matrices using UV-visible absorbance. *Analytical and bioanalytical chemistry* 401: 1993–2002.

[4] Kroll A *et al.* 2011 Cytotoxicity screening of 23 engineered nanomaterials using a test matrix of ten cell lines and three different assays. *Particle and fibre toxicology* 8: 9.

[5] Holder AL *et al.* 2012 Particle-Induced Artifacts in the MTT and LDH Viability Assays. *Chemical research in toxicology* 25: 1885–1892.

[6] Hartung T and Sabbioni E 2011 Alternative in vitro assays in nanomaterial toxicology. *Wiley interdisciplinary reviews Nanomedicine and nanobiotechnology.*

[7] Wang S *et al.* 2011 Limitation of the MTT and XTT assays for measuring cell viability due to superoxide formation induced by nano-scale TiO_2. *Toxicology in vitro : an international journal published in association with BIBRA* 25: 2147–2151.

[8] Gurr J-R *et al.* 2005 Ultrafine titanium dioxide particles in the absence of photoactivation can induce oxidative damage to human bronchial epithelial cells. *Toxicology* 213: 66–73.

[9] Singh S *et al.* 2007 Endocytosis, oxidative stress and IL-8 expression in human lung epithelial cells upon treatment with fine and ultrafine TiO_2: role of the specific surface area and of surface methylation of the particles. *Toxicology and applied pharmacology* 222: 141–151.

[10] Schmitt J and Ritter H 1975 Lactate dehydrogenase: inhibition of subunit A by the sulfhydryl reagent $AgNO_3$. *Humangenetik* 28: 239–243.

[11] Han X *et al.* 2011 Validation of an LDH assay for assessing nanoparticle toxicity. *Toxicology* 287: 99–104.

[12] Kennedy AJ *et al.* 2010 Fractionating nanosilver: importance for determining toxicity to aquatic test organisms. *Environmental science & technology* 44: 9571–9577.

[13] Lok C-N *et al.* 2007 Silver nanoparticles: partial oxidation and antibacterial activities. *Journal of biological inorganic chemistry : JBIC : a publication of the Society of Biological Inorganic Chemistry* 12: 527–534.

[14] Chatterjee A *et al.* 2009 Selective fluorogenic and chromogenic probe for detection of silver ions and silver nanoparticles in aqueous media. *Journal of the American Chemical Society* 131: 2040–2041.

[15] Kang J *et al.* 2002 New fluorescent chemosensors for silver ion. *The Journal of organic chemistry* 67: 4384–4386.

[16] Benn TM and Westerhoff P 2008 Nanoparticle silver released into water from commercially available sock fabrics. *Environmental science & technology* 42: 4133–4139

[17] Hansson PK *et al.* 2004 Glucose and fatty acid metabolism in McA-RH7777 hepatoma cells vs. rat primary hepatocytes: responsiveness to nutrient availability. *Biochimica et biophysica acta* 1684: 54–62.

[18] Bregoli L *et al.* 2009 Toxicity of antimony trioxide nanoparticles on human hematopoietic progenitor cells and comparison to cell lines. *Toxicology* 262: 121–129.

[19] Kermanizadeh A *et al.* 2012 Primary human hepatocytes versus hepatic cell line: assessing their suitability for in vitro nanotoxicology. *Nanotoxicology*: 1–17.

[20] Bhattacharya K *et al.* 2012 Reactive oxygen species mediated DNA damage in human lung alveolar epithelial (A549) cells from exposure to non-cytotoxic MFI-type zeolite nanoparticles. *Toxicology letters*: 1–10.

[21] Michael Berg J *et al.* 2012 Comparative cytological responses of lung epithelial and pleural mesothelial cells following in vitro exposure to nanoscale SiO(2). *Toxicology in vitro : an international journal published in association with BIBRA*.

[22] Jugan M-L *et al.* 2012 Titanium dioxide nanoparticles exhibit genotoxicity and impair DNA repair activity in A549 cells. *Nanotoxicology* 6: 501–513.

[23] Kroll A *et al.* 2012 Interference of engineered nanoparticles with in vitro toxicity assays. *Archives of toxicology* 86: 1123–1136.

[24] Lanone S *et al.* 2009 Comparative toxicity of 24 manufactured nanoparticles in human alveolar epithelial and macrophage cell lines. *Particle and fibre toxicology* 6: 14.

[25] Foster K *et al.* 1998 Characterization of the A549 cell line as a type II pulmonary epithelial cell model for drug metabolism. *Experimental cell research* 243: 359–366.

Aerosol size distribution estimation and associated uncertainty for measurement with a Scanning Mobility Particle Sizer (SMPS)

L. Coquelin[1,2,3,4], N. Fischer[1], C.Motzkus[2], T. Mace[2], F. Gensdarmes[3], L. Le Brusquet[4] and G. Fleury[4]

[1] Département de mathématiques et statistiques, Laboratoire National de Métrologie et d'Essais, 29 avenue Roger Hennequin, 78197 Trappes, France
[2] Département de qualité de l'air et débitmétrie gazeuse, Laboratoire National de Métrologie et d'Essais, 1 rue Gaston Boissier, 75724 Paris, France
[3] Institut de Radioprotection et de Sûreté Nucléaire, BP 68, 91192 Gif-sur-Yvette. France
[4] Supélec Sciences des Systèmes - EA4454 (E3S), 3 rue Joliot Curie, 91192 Gif-sur-Yvette, France

E-mail: `Loic.Coquelin@lne.fr`

Abstract. Scanning Mobility Particle Sizer (SMPS) is a high resolution nanoparticle sizing system that has long been hailed as the researcher's choice for airborne nanoparticle size characterization for nano applications including nanotechnology research and development. SMPS is widely used as the standard method to measure airborne particle size distributions below 1 μm. It is composed of two devices: a Differential Mobility Analyzer (DMA) selects particle sizes thanks to their electrical mobility and a Condensation Particle Counter (CPC) enlarges particles to make them detectable by common optical counters. System raw data represent the number of particles counted over several classes of mobility diameters. Then, common inversion procedures lead to the estimation of the aerosol size distribution. In this paper, we develop a methodology to compute the uncertainties associated with the estimation of the size distribution when several experiences have been carried out. The requirement to repeat the measure ensures a realistic variability on the simulated data to be generated. The work we present consists in considering both the uncertainties coming from the experimental dispersion and the uncertainties induced by the lack of knowledge on physical phenomena. Experimental dispersion is quantified with the experimental data while the lack of knowledge is modelled via the existing physical theories and the judgements of experts in the field of aerosol science. Thus, running Monte-Carlo simulations give an estimation of the size distribution and its corresponding confidence region.

1. Introduction

Among devices using electrical mobility methods to measure aerosol particle size distribution, the Scanning Mobility Particle Sizer (SMPS) is the most widely used to characterize particles smaller than $1\mu m$. Since it can be applied in fields like: indoors air quality measurement, vehicle exhaust pipes, atmospheric studies, toxicology testing, etc., it has become of interest to assess the uncertainty associated with its outputs. So far, although some works have been carried on the adequacy of particle size measurements made by this device, such a dimensional feature uncertainty cannot be provided. SMPS is composed of a Differential Mobility Analyser (DMA) and a Condensation Particle Counter (CPC). First one selects particles mobility diameters while the other is a counter used for the detection. First, aerosol passes through an inertial impactor to avoid largest particles to enter the DMA column, then the aerosol enters the charge neutraliser to be conditioned. Indeed, aerosol particles are usually charged. In this case, a bipolar charge distribution is created by exposing the aerosol to a radioactive source, so, particles that carry several charges lose their charge excess. Once aerosol is well conditioned particles are selected using electrical classification inside DMA column: an electric field is created, and the airborne particles drift in the DMA according to their electrical mobility. Once selected, particles reach the CPC to be counted. In this paper, we focus on providing the users with a methodology to compute the uncertainties associated to the size distribution estimated by an inversion procedure. The article is organized as follows: section 2 defines the technical protocol to produce the data, section 3 describes the physical model to relate the experimental data to the size distribution and the inversion procedure to estimate the size distribution. Improvements are mainly brought in section 4 where the sources of uncertainty arising from both the measure and the estimation of the size distribution are decomposed and propagated. Thus, the experimental dispersion is represented by errors on the raw data. Section 4.1 explains how the model of these errors is deduced from some observed signals. Then, section 4.2 describes how the estimation of the size distribution is realized via an inversion model whose variability will refer to the lack of knowledge on the parameters and other physical laws of the system. Preliminary results obtained when this methodology is carried out on SiO_2 airborne nanoparticles are presented in section 5 and it is followed by a discussion about the limits of the methodology and the possible improvements.

2. Technical protocol

Measurements using a SMPS can be performed via several kind of devices. The measurement process can be split into phases: the generation of the aerosol, the neutralization of the charges, the selection of the particles and their detection. In this case, the measurement is assured by a pneumatic atomizer 3076, a charge neutraliser using Krypton source ($^{85}K_r$), a Long-DMA 3081, and a butanol based CPC 3022. Each equipment comes from TSI company. As an example, an other choice could have been an electrospray, a corona charger, a nano-DMA and a water based CPC. The results that will be shown later in this paper refer to the chosen devices.
The protocol that is used is defined as follows:

- Instruments are warmed up for at least 30 minutes
- TSI 3076 operates in the following conditions: air flow rate: 3 L.min^{-1} with inlet pressure at 2.5 kg.cm^{-2}, liquid flow rate: 20 cm^3.min^{-1} at 2.5 kg.cm^{-2} which are nominal settings for the device
- Nanoparticles generated by the atomizer pass through the charge neutralizer 3077, and then, are introduced into the DMA
- DMA settings only consist in defining the following equalities for the flow rates: $Q_{sh} = Q_{exc}$ and $Q_a = Q_{cpc}$ so that the flows are symmetric inside the classifier
- CPC low flow mode is activated such that the inlet flow is 5 cm^3.s^{-1}

Figure 1 presents a schematic diagram of the experimental setup.

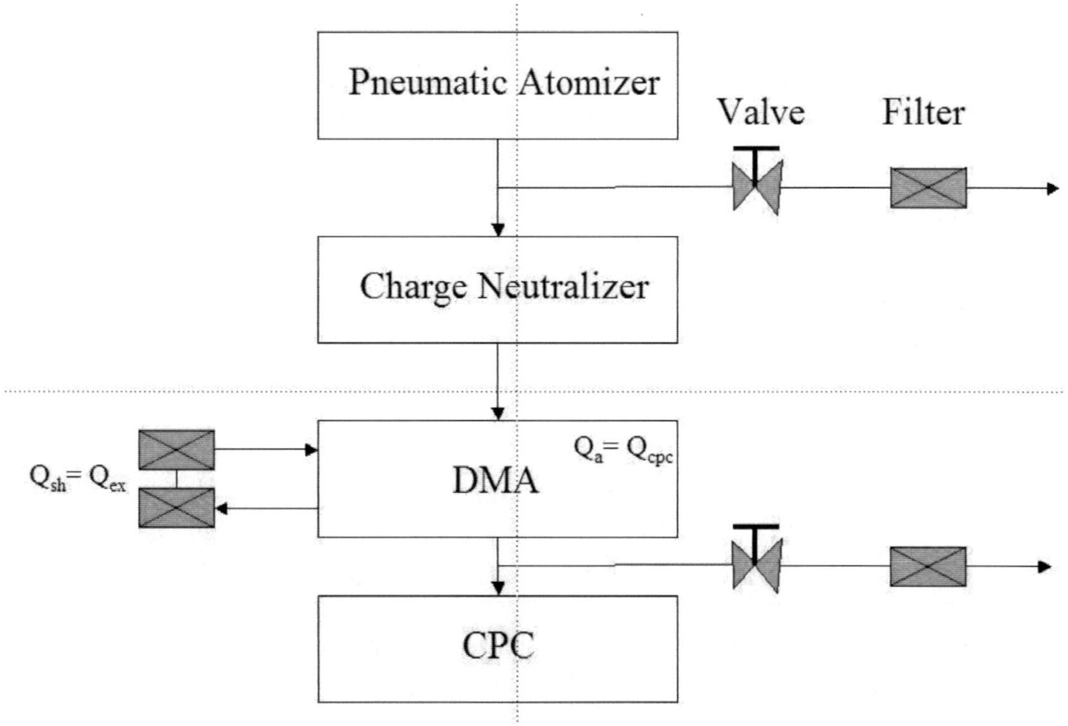

Figure 1. Schematic diagram of the experimental setup

3. Model definition and Inversion procedure

The recovery of size distributions, which is referred as data inversion, requires the solution of the set of Fredholm integral equations. Let y_i be the number of particles counted by the CPC in the ith channel, k_i the nonnegative kernel function for the time range corresponding to the channel i, d_p the mobility diameter of the particle, u the logarithm function, ϵ_i the measurement error in the ith channel and finally f the aerosol size distribution, then:

$$y_i = \int_0^{+\infty} k_i(d_p)f(u(d_p))d(u(d_p)) + \epsilon_i, \qquad i = 1, ..., m \tag{1}$$

where each kernel function is expressed as:

$$k_i(d_p) = Q_a t_c^i \sum_{p=1}^{+\infty} \Phi(p, d_p)W(d_p)\bar{\Omega}_i(d_p, p). \tag{2}$$

Components Φ, W and $\bar{\Omega}_i$ of the kernel function respectively represent the bipolar charge distribution function, the CPC detection efficiency and the mean transfer function for the time range corresponding to the channel i, p being the number of charges carried by the particle, Q_a the aerosol flow rate entering the DMA and t_c^i the counting time for the ith channel.
The principle of selection is based on the relation between the voltage applied at the center rod of the column of the classifier and the mobility diameter. This relationship is expressed via the electrical mobility. The experimental electrical mobility, denoted as Z_v, can be computed according to the voltage V, the flow rates (Q_{sh} the sheath flow rate entering the DMA and

Q_{ex} the excess flow rate exiting the DMA) and the geometry of the DMA (L the length of the column, r_1 and r_2 respectively the inner and outer radii of the electrodes). The theoretical electrical mobility, noted Z_d, relates Z_v to a specific mobility diameter d_p. Finally considering the equality of both quantities, one can get the mobility diameter associated to the applied voltage.

$$Z_v = \frac{(Q_{sh} + Q_{ex}) log(r_2/r_1)}{4\pi LV}, \quad Z_d = \frac{peC_c(d_p)}{3\pi\mu_g d_p}, \tag{3}$$

where e is the charge of the electron, μ_g the dynamic gas viscosity and C_c the slip correction factor. The expression of the slip correction factor is based on the triplet (α, β, γ) and its expression is as follows:

$$C_c(d_p) = 1 + K_n(d_p)\left[\alpha + \beta \exp(-\gamma/K_n(d_p))\right], \tag{4}$$

$$K_n(d_p) = (2\lambda_m)/d_p, \tag{5}$$

where K_n is called the Knudsen number and λ_m is the mean free path of a particle.

Since the DMA voltage is being continuously scanned, the times of the system become crucial. Indeed, particles are selected by the DMA and after transport through the tubing (transport time is denoted as t_d), the counter detects them. An error in the time specification corresponding to the time for particles to pass from the DMA exit slit to the CPC counting chamber lead to an error in the estimation of f on the diameter space. The same kind of statement can be made in what concerns the time for particles to pass through the DMA column (denoted as t_f). This time is computed as the time for the aerosol to pass through the DMA.
The discrete form of equation 1 after applying a numerical quadrature to approach the integral numerically can be written as:

$$y_i = \sum_{j=1}^{j=n} w_{ij} k_{ij} f_j + \epsilon_i, \quad i = 1, ..., m, \tag{6}$$

where $k_{ij} = k_i(d_{p,j})$, $f_j = f(d_{p,j})$, $(d_{p,j})_{j=1,...,n}$ being the vector of points used for the discretization of the diameter space. Once the relationship between the inputs and the outputs of the measurement system is well-defined, next step is to estimate each component of the vector f. Several algorithms lead to comparable estimations of the size distribution. Most commonly used algorithms in the field of aerosol are those developed by Seinfeld [1], Twomey [2],[3] and more recently Collins *et al.* [4] or Talukdar [5]. Following Seinfeld's method, inversion with regularization techniques has been chosen here. It leads to a fast and accurate estimation of f and it also takes into account the judgements of the experts through the prior. Let H the matrix such that $H = (w_{ij}k_{ij})_{i=1,...,m}^{j=1,...,n} \in R^{m \times n}$, $y = (y_i)_{i=1,...,m} \in R^m$ and $f = (f_j)_{j=1,...,n} \in R^n$, then the inversion procedure becomes the minimization problem: find $f^* \in R^n$ such that

$$f^* = \arg\min_{f \geq 0} \left\{ \|Hf - y\|_2^2 + \lambda \|D_2 f\|_2^2 \right\}, \tag{7}$$

$$= \arg\min_{f \geq 0} \left\{ G(H, f, y, \lambda) \right\}, \tag{8}$$

where D_2 is the second-order difference matrix and $f \geq 0$ means $f_j \geq 0$, $\forall j$.

Here the prior penalizes solutions that are not smooth. This type of penalty seems justified in this application because aerosol processes often tend to smooth rough distributions. The parameter λ will be a trade off between the prior we associate to the solution to be estimated and the solution that minimizes the quadratic error term. The L-curve method is chosen here to determine the regularization parameter and it is computed via the pruning algorithm developed in [6]. To ensure robustness, λ is calibrated through the all set of data. The solution of the minimization is computed via the fast non negative least-square algorithm (*fnnls* [7]) algorithm that is an optimized version of Lawson and Hanson non negative least-square algorithm (*nnls* [8]). Several references [1],[2],[3],[4],[5] define how to estimate the size distribution, this article does not intend to bring improvements in the inversion procedure but mainly in the following section dedicated to the uncertainty analysis.

4. Uncertainty analysis

In this part, the aim is to quantify and to propagate the uncertainties coming from both the measurement process and the inversion. The experimental dispersion is difficult to model since each component of uncertainty that it involves is not accurately modelled yet. Thus, choice has been made to quantify it with experimental data. It can be seen as a behavioural model and it has the advantage that it warranties a realistic quantification of this source. On the other hand, the uncertainty associated with the inversion procedure is modelled. Unlike for the experimental dispersion, the uncertainty it induces cannot be track down into the data because the error it produces is systematic. The major source is the lack of knowledge on the physical models to be used to define the matrix H. Indeed, several theories compete to define the factors of Cunningham, the bipolar charging law or the diffusive transfer function. Moreover, elements of this group are different in terms of nature: scalars and functions. Some scalars can be modelled with corresponding uncertainty because they can be measured (L, r_1, r_2). Some cannot be measured but the existing theories can be used to bound them (α, β, γ) or concerning the functions ϕ and Ω, comparisons between data and models have been performed under controlled conditions. In this paper, we will only consider the ignorance on parameters, so Fuch's theory [9] is chosen to define ϕ and Stolzenburg's work [10] is used to model the diffusional transfer function. To sum up, the experimental dispersion, denoted as the first group of uncertainty represents 3 major components:

- the instability of the aerosol generation,
- the partial variability of the physical parameters (flow rates, environmental conditions, etc.),
- the noise coming from the CPC counting process,

while the inversion uncertainty that is considered as the second group is made of:

- the lack of knowledge associated with the choice of the physical parameters,
- the variability induced by the choice of the regularization parameter,
- the bias due to the method of regularization.

The evaluation of the bias requires an aerosol of reference from a metrological point of view. Since it is not available due to the instabilities of the generation process, it is not yet possible to account for it in the uncertainty budget.

4.1. The experimental dispersion

In practice, a set of measurements have been made and the user's got observations $(y_{il})_{i=1,\dots,m}^{l=1,\dots,n_l}$, i being the index for the class of diameter and l the index that corresponds to the number of experiences. We consider that the experiences have been repeated n_l times, so that enough information can be extracted from the data. The idea we propose is to infer a model based on this information by considering simulated data modelled as a correlated Gaussian process noted $\tilde{y}$. Experiences tend to justify the Gaussian assumption and authors usually abound

in that sense. Let C be the empirical correlation matrix, the autocorrelated Gaussian kernel is computed by performing a SVD (Singular Value Decomposition) on C such that it can be decomposed as $C = U\Sigma V^*$.
$\tilde{y}$ is then created as follows:

$$\tilde{y} = \left(U\sqrt{\Sigma}z\right)\sigma + \mu, \tag{9}$$

where $z = (z_1, z_2, ..., z_m)$ with $z_i \sim N(0,1)$, μ and σ are respectively the mean and standard deviation of the real observations: $\mu = (\mu_1, \mu_2, ..., \mu_m) \in R^m$ such that $\mu_i = \frac{1}{n_l}\sum_{l=1}^{l=n_l} y_{il}$, $\sigma = (\sigma_1, \sigma_2, ..., \sigma_m) \in R^m$ such that $\sigma_i = \frac{1}{n_l-1}\sum_{l=1}^{l=n_l}\sqrt{(y_{il}-\mu_i)^2}$. To propagate the uncertainty arising from the experimental dispersion, the methodology is to generate a set of simulated data via Monte-Carlo simulations and to estimate the vector f via the minimization described in (eq 7). Let $\tilde{y}^{[q]} \in R^m$ be the simulated data generated for the qth simulation, n_q the total number of simulations and $\bar{\lambda}^{Lcurve}$ the regularization parameter calibrated thanks to the set of real data y via the L-curve algorithm, then the estimation of the solution vector f for the qth simulation denoted as $\hat{f}^{[q]} \in R^n$ is computed as:

$$\hat{f}^{[q]} = \arg\min_{f\geq 0}\left\{G\left(H, f, \tilde{y}^{[q]}, \bar{\lambda}^{Lcurve}\right)\right\}, \quad q = 1, ..., n_q. \tag{10}$$

4.2. The inversion uncertainty

A parametric model is developed in order to take into account the uncertainty associated with the lack of knowledge on parameters to be used for the definition of the matrix H. These parameters are selected such that they are not well-known or because several theories compete to define them and especially because they have a significant effect on the final estimation. In a previous paper [11], sensitivity analysis has been carried out to determine the most significant parameters, significant means here that a small error in their definition produces a large variability on the reconstruction of the solution vector. The chosen parameters are: L, r_1, r_2, α, β and γ. Nominal values for α, β and γ are taken from experiments made by Kim *et al.*. [12]. Experiences realized by Hutchins *et al.*. [13] and Allen and Raabe [14] are chosen as bounds around the nominal values. Subscripts $_k$, $_h$ and $_{ar}$ refer to each author. Nominal values for the geometrical parameters of the DMA are given by the constructor and their associated uncertainty is taken from the ISO standard (reference ISO 15900:2009(E)). To account for the uncertainty the selected parameters produce, they are modelled as random variables and propagate through the inversion via Monte-Carlo simulations as for the experimental dispersion. Table 1 summarizes the choices made for the definition of the random variables. Let $\tilde{H}^{[q]}$ be the matrix drawn by the random variables for the qth simulation, for a chosen experimental measurement $y = (y_{il})_{i=1,...,m}^{l=l'}$, $\hat{f}^{[q]}$ is now computed as:

$$\hat{f}^{[q]} = \arg\min_{f\geq 0}\left\{G\left(\tilde{H}^{[q]}, f, y, \bar{\lambda}^{Lcurve}\right)\right\}, \quad q = 1, ..., n_q. \tag{11}$$

Table 1. Definition of the parameters that model the lack of knowledge. U and N respectively refer to Uniform and Normal distributions. Subscript $_0$ refers to the nominal value given by the constructor for the corresponding parameter.

Name	Law	Mean	Standard deviation	Min	Max
r_1	U	$r_{1,0}$	-	$r_{1,0} - 0.001r_{1,0}$	$r_{1,0} + 0.001r_{1,0}$
r_2	U	$r_{2,0}$	-	$r_{2,0} - 0.0006r_{2,0}$	$r_{2,0} + 0.0006r_{2,0}$
L	U	L_0	-	$L_0 - 0.005L_0$	$L_0 + 0.005L_0$
α	N	α_k	$\frac{\max(\|\alpha_h-\alpha_k\|,\|\alpha_{ar}-\alpha_k\|)}{\alpha_k}$	-	-
β	N	β_k	$\frac{\max(\|\beta_h-\beta_k\|,\|\beta_{ar}-\beta_k\|)}{\beta_k}$	-	-
γ	N	γ_k	$\frac{\max(\|\gamma_h-\gamma_k\|,\|\gamma_{ar}-\gamma_k\|)}{\gamma_k}$	-	-

The decomposition of the sources of uncertainty as presented above is required when the purpose is to characterize their respective effect on the reconstruction of the solution vector f. Nevertheless, the final result is the estimation of f with the associated uncertainty and it has to be computed when both sources are simultaneously considered. Combining the sources is straightforward since the procedure is the same as for each source taken individually. Thus, when both sources are combined, the estimation of the solution vector f for the qth simulation is obtained as:

$$\hat{f}^{[q]} = \arg\min_{f\geq 0} \left\{ G\left(\tilde{H}^{[q]}, f, \tilde{y}^{[q]}, \lambda^{[q],Lcurve}\right)\right\}, \quad q = 1, ..., n_q, \tag{12}$$

where $\lambda^{[q],Lcurve}$ is the regularization parameter computed for the qth simulation.

5. Results

Starting from real observations, the aim of this part is to show the uncertainty associated with the mean estimate, denoted $\bar{\hat{f}}$, that is computed with the methodology presented above for each source taken individually and combined. As an example, SiO_2 airborne nanoparticles are presented. The aerosol presents 2 populations in terms of size: the first one is around 30 nm while the second population stands around 40 nm. Dispersions around the peaks are similar. These observations will prove the robustness of the inversion procedure, indeed, both peaks are really close and a bad estimation of the regularization parameter shall lead to over-smooth the solution. Figure 2 shows the real observations y ($l = 40$ in this case) and the corresponding simulated data generated via Monte-Carlo simulations. Following the developments proposed in section 4, experimental dispersion and its effects on the estimation of the vector f as well as the lack of knowledge and its impact on the reconstruction are illustrated in figure 3. Combined uncertainties of both sources are also presented in this figure. The uncertainties are shown through a confidence region with a confidence level of 95 %. The results suggest that the experimental dispersion affects the estimation of f on the y axis (concentration of particles) with the same amplitude through the all diameters while the lack of knowledge on the physical phenomena does not significantly affect the concentration of particles but it induces a translation

Journal of Physics: Conference Series 429 (2013) 012018 doi:10.1088/1742-6596/429/1/012018

of the reconstruction. Other experiences are actually being realized on different aerosols in order to confirm these results.

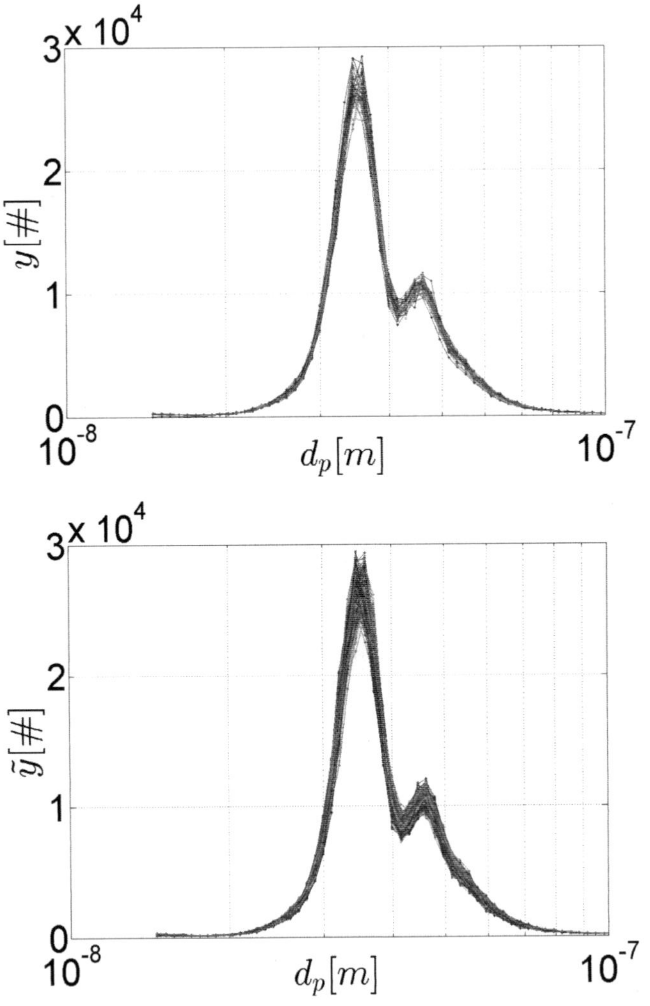

Figure 2. Real data (top) versus simulated data (bottom) for SiO_2 airborne nanoparticles

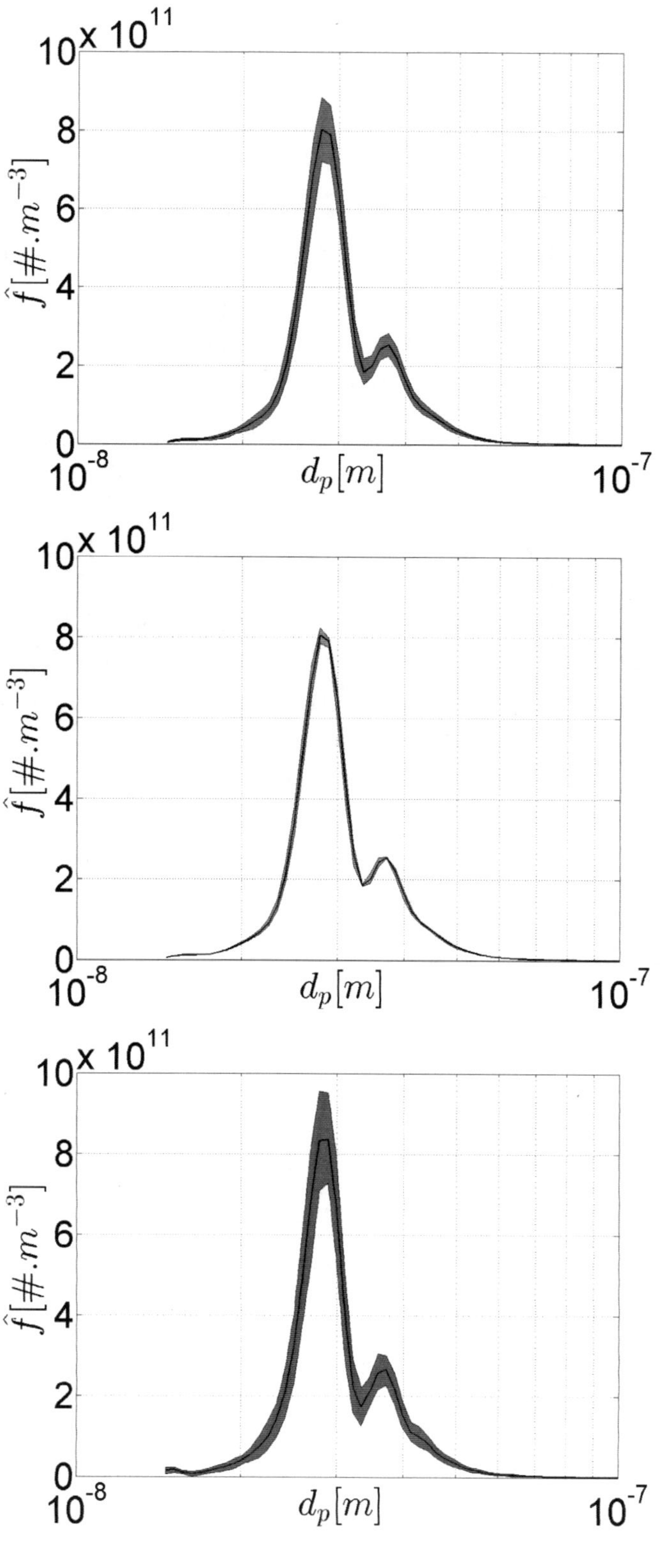

Figure 3. Illustrations of the 95 % confidence region computed thanks to Monte-Carlo simulations for each source of uncertainty. The figures show respectively the uncertainty associated with the experimental dispersion (top), the lack of knowledge (middle) and both sources combined (bottom)

6. Conclusion

A methodology has been presented to compute the confidence region on the estimated size distribution when several experiences have been carried out to characterize an aerosol. The statistical model that accounts for the experimental dispersion has been inferred from real observations because spatial correlations are difficult to model precisely. The statistical inversion procedure takes into account the uncertainties coming from the lack of knowledge on parameters that are used to relate the data to the size distribution through the matrix of kernel functions. An example of aerosol composed of SiO_2 airborne nanoparticles is treated using this methodology. Results reveal that the uncertainty arising from each source affects respectively the reconstruction on the concentration of particles and on the diameters. Some sources still need to be added especially the impact of the prior in the minimization problem and the bias that it introduces on the estimation of f. Next step will be to consider analytical size distributions to estimate the bias induced by the prior since an aerosol of reference is not yet available. An other improvement would be to precisely model and propagate the uncertainties when only one experience is available.

References

[1] Seinfeld J H and Wolfenbarger J K 1990 *Journal of Aerosol Science,* **21** 227–247

[2] Twomey S 1963 *Journal of the Association for Computing Machinery* **10** 97–101

[3] Twomey S 1975 *Journal of Computational Physics* **18** 188–200

[4] Collins D R, Flagan R and Seinfeld J H 2002 *Aerosol Science and Technology* **36** 1–9

[5] Talukdar S S and Swihart M T 2003 *Aerosol Science and Technology* **37** 145–161

[6] Hansen P C and O'Leary D P 1993 *SIAM Journal on Scientific Computing* **14** 1487–1503

[7] Bro R 1997 *Journal of Chemometrics* **11** 393–401

[8] Lawson C L and Hanson R J 1995 *Society for Industrial and Applied Mathematics* **15** 345–349

[9] Fuchs N A 1963 *Geofis. Pura Appl* **56** 185–193

[10] Stolzenburg M R 1988 *An Ultrafine Aerosol Size Distribution Measuring System* Ph.D. thesis University of Minnesota, Minneapolis

[11] Coquelin L, Fischer N, Le Brusquet L, Fleury G, Motzkus C and Gensdarmes F 2012 *Instrumentation and Measurement Technology Conference (I2MTC)* (2012 IEEE International) pp 1718–1721

[12] Kim J H, Mulholland G W, Kukuck S R and Pui D Y H 2005 *Journal of Research of National Institute of Standards and Technology* **110** 31–54

[13] Hutchins D K, Harper M H and Felder R L 1995 *Aerosol Science and Technology* **22** 202–218

[14] Allen M D and Raabe O G 1985 *Journal of Aerosol Science* **4** 269–285

Using a PBPK model to study the influence of different characteristics of nanoparticles on their biodistribution

D Li[1], C Emond[2], G Johanson[3], and O Jolliet[1]

[1] University of Michigan, Ann Arbor, USA
[2] BioSimulation Consulting Inc., Newark, USA
[3] Karolinska Institutet, Stockholm, Sweden

dingsli@umich.edu

Abstract. The studies on potential health risks possessed by engineered nanoparticles (NPs) have been growing rapidly. However, detailed and systemic knowledge on the uptake and biodistribution of NPs in body is still limited. Moreover, there is a need to characterize the relation between the characteristics of NPs (size, surface modifications, etc.) and their behaviours in the body. The aim of this study is to explore how these characteristics will influence the NPs uptake and biodistribution. We have successfully developed a Physiologically Based Pharmacokinetic (PBPK) model for the biodistribution of polyethylene glycol-coated polyacrylamide NPs in rats, modelling the capture and removal of NPs by phagocytizing cells. Based on this PBPK model, the behaviours of other nanoparticles (polymeric, quantum dot, silver, titanium oxide and cerium oxide NPs) are investigated, based on data from several experiments published in the literature. Size is one of the important properties to consider. Our model parameterization suggests that the uptake rate by phagocytizing cells will decrease as the size of nanoparticles increases when the removal rates for these nanoparticles are similar. This could indicate that the phagocytizing cells are saturated by the number of NPs rather than absolute mass. Nevertheless, surface modification, such as polyethylene glycol coating, may reduce the uptake rate by phagocytizing cells. With phagocytizing cells serving as a deposit of NPs, these influences of different characteristics of NPs to the behavior of phagocytizing cells could affect the fate of NPs in the body not only during the initial uptake within the first hour but also in long-term at the kinetic and dynamic levels.

1. Introduction

Wide production and use of nanoparticles (NPs) increase the likelihood of unintentional exposure at workplaces as well as in the general environment [1-4]. *In vivo* and *in vitro* studies have shown that NPs have the potential to induce oxidative stress, leading to inflammatory reactions [5, 6]. The degree of inflammatory response to NPs strongly depends on the biodistribution of NPs in the body [6, 7]. A few detailed investigations have revealed that at least some types of inhaled or intravenously injected NPs rapidly migrate to, and deposit in different organs via the systemic circulation [8-12]. However, the determinants of NP behavior in the body still need to be explored in a systematic way. It is therefore necessary to determine what extent and by which mechanisms nanomaterials translocate and bioaccumulate.

Recently a few studies have employed physiologically based pharmacokinetic (PBPK) models to describe the biodistribution of NPs. However, these models have not specifically addressed the role of phagocytizing cells [13, 14], although it has been observed that phagocytosis is a major element of NP biodistribution [7, 15]. The interaction between phagocytizing cells and NPs has been modeled based on *in vitro* experiments [16, 17], but such *in vitro* knowledge has hitherto not been incorporated in the attempts to describe the overall biodistribution in the body.

To further develop our knowledge of the biodistribution of NPs, several issues need to be addressed in the PBPK models, such as: a) how to account for nano-specific phenomena, such as phagocytosis; b) what are the most influential determinants affecting the biodistribution; and c) how would different characteristics (size, surface modification, etc.) impact these determinants.

In this study we build a PBPK model with the emphasis on the importance of phagocytosis using experimental data from rats given single doses of polyethylene glycol-coated polyacrylamide (PAA-peg) by intravenous injection. Then we adapted the model to other types of NPs without major structural changes in order to explore the influences of different characteristics of NPs on their biodistribution.

2. Method

2.1. Experimental data

The model was developed based on anatomical and physiological considerations and on an experimental biodistribution study published by Wenger *et al* [12]. In brief, five groups of three rats per group received a single intravenous dose of 7,000 μg ^{14}C-labeled PAA-peg NPs via the tail vein. Urine, feces, cage residues, and blood samples were collected at designated times and tissue samples of the liver, spleen, kidney, heart, lungs, brain, lymph nodes (mesenteric, inguinal, and popliteal), and bone marrow (both femurs) were collected following euthanasia by CO_2. Radioactivity levels were determined by liquid scintillation counting with correction for background chemiluminescence.

To adapt the model to other types of NPs, experimental data were taken from various studies: Wenger *et al* [12] for polyacrylamide (PAA) NP, Panagi *et al* [11] for poly(Lactide-co-glycolide)-monomethoxypoly(ethyleneglycol) (PLGA-mPEG) NP, Liu *et al* [18] for coated breviscapine-loaded poly(D, L-lactic acid) (BVP-PLA) NP, and Lankveld *et al* [14] for three different sizes of silver (Ag) NP.

2.2. PBPK model

The PBPK model consists of eight compartments: blood, lungs, heart, brain, kidneys, liver, immune organs (spleen, lymph nodes, and bone marrow), and the rest of the body. Within each compartment, there are three sub-compartments representing capillary blood, tissue and phagocytizing cells (PCs) and remaining tissue. The conceptual framework of the model is illustrated in Fig. 1.

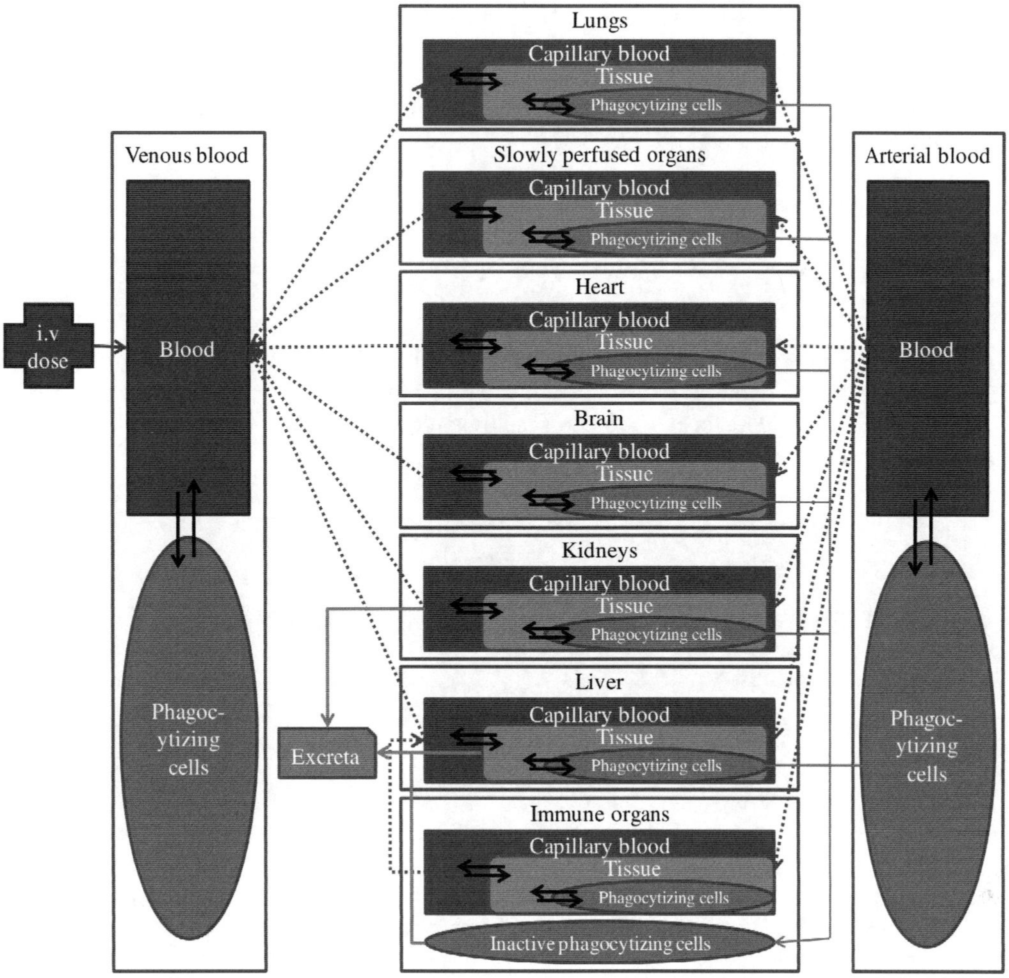

Fig. 1 Framework of the PBPK model

The exchange of nanoparticle between blood and tissue in each organ is described as a flow- and diffusion-limited process and the latter is controlled by permeability parameters, which limits the effective blood flow [19]. The permeability coefficient for the brain compartment is set to zero, under the assumption of a highly efficient blood-brain barrier. The possibility of uneven distribution between blood and tissue at steady-state is taken care of by a blood-tissue partition coefficient.

Some of the nanoparticles that enter the tissue are captured by the tissue PCs. The uptake rate is a function of the efficacy and saturation level of the PCs and decreases as the PAA-peg nanoparticles in the PCs become saturated. For now, we assume that all PCs have the same efficacy and saturation level, independent of their location. However, the numbers of PCs differ between organs, causing different saturation levels. The saturated PCs become inactive and migrate to the immune organs. A removal rate of NPs, which includes excretion from the body and possibly biodegradation for some NPs, is adopted in the model.

The PBPK model was implemented in Berkely-Madonna^{TM} (version 8.3.18). The model was optimised by fitting the unknown model parameters against the PAA-peg experimental data first. The following parameters were fitted: phagocytosis saturation level (one for each compartment), blood: tissue partition coefficient, permeability coefficients, uptake rate and desorption rate by PCs, migration rate of inactive PCs, and clearance rate from kidneys and liver. All other parameter values were taken from the scientific literature. Then the model was adapted for other types of NPs. For this adaption, the phagocytosis saturation levels among all organs remained a constant ratio but were scaled either up or down to the same degree, expressed in the uptake capacity (UC), when fitting different datasets.

2.3. Model evaluation and sensitivity analysis

For model evaluation we determine the deviation from the 1:1 line between the $\log_{10}$ of measured and predicted values, calculating the standard deviation and the squared geometric standard deviation as well as the corresponding R^2.

Model sensitivity for each model parameter can be evaluated with the area under the curve (AUC) of blood and liver for the entire duration of each study. To test the sensitivity, each parameter's value is increased by 10% and the AUC recalculated. A sensitivity coefficient can then be calculated using the following equation

$$\text{Sensitivity coefficient} = \frac{dAUC_i \;/\; dp_j}{AUC_i \;/\; p_j} \tag{1}$$

Where AUC_i represents area under the curve (AUC) for compartment i, and p_j represents the parameter j. In this study, we chose the PCs uptake capacity (UC), uptake rate by PCs (k_{ab0}), and the removal rate of NPs ($k_{removal}$) for the sensitivity analysis.

3. Results

3.1. Time course of the PAA-peg NP model

The time course for the PAA-peg NP is shown in Fig. 2 below. According to the experimental data, the fast perfused organs (including the lungs, liver, kidneys, heart, immune organs, and brain) exhibits a pattern of saturation of NPs and the model captures this phenomenon with the sub-compartment of PCs. For other types of NPs in this study, this saturation phenomenon is also observed, especially in the liver which has a high level of PCs.

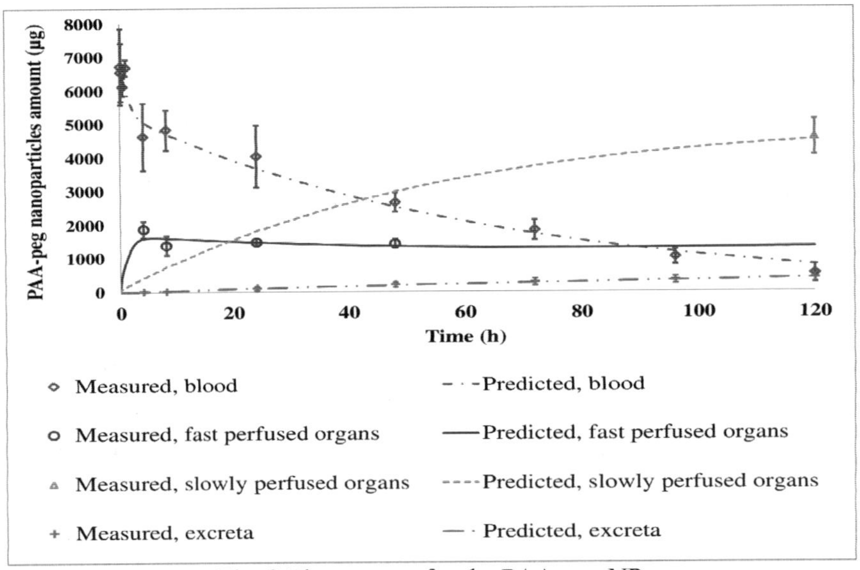

Fig. 2 Time course for the PAA-peg NP

3.2. Evaluation of the model

The predictions from the model are plotted against the measured data from the studies as shown in Fig. 3. The results for PAA NP and Ag NPs in sizes of 80 nm and 110 nm in diameters are not shown but they are very similar to the results for PAA-peg NP and Ag NP 20 nm. The R^2 range from 0.707 to 0.994, this shows the model's predictions fit the experiment measurements relatively well. The two outliers for Ag NP 20 nm are in the blood. According to the measured data, towards the end of the study [14], no NPs were detected in the blood. In order to make these two data points available on the $\log_{10}$ scale, their values were set to 1 ng (the smallest value found in blood was 359 ng).

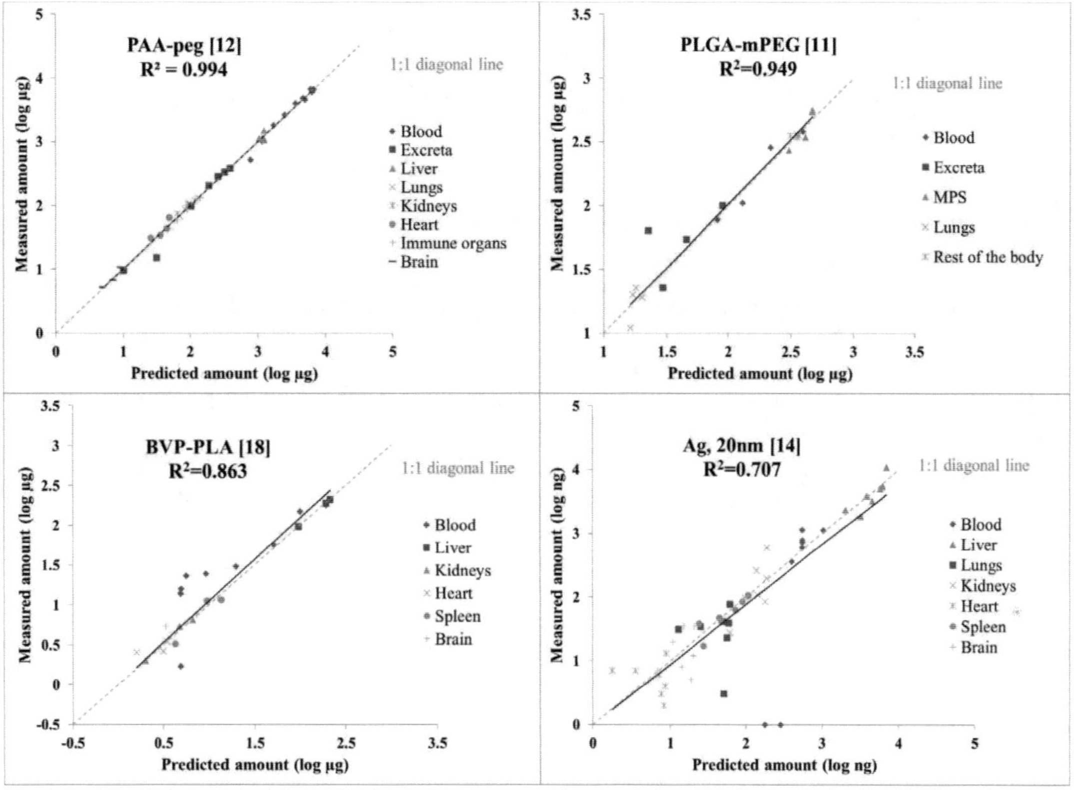

Fig. 3 Model predictions versus experimental data for different types of NP

3.1. Parameterization of the model

For the different types of NPs in this study, Table 1 summarizes their characteristics (sizes in diameter, whether have surface modification), injected doses, and the values of three fitted model parameters – UC, k_{ab0}, and $k_{removal}$. These NPs fall into two groups, one with low $k_{removal}$ (PAA-peg, PAA, and PLGA-mPEG) and one with high $k_{removal}$ (BVP-PLA and Ag). When comparing PAA-peg and PLGA-mPEG NPs (both have surface modification and low $k_{removal}$), the UC increased by 1.4 times and the k_{ab0} decreased by 30 times when the diameter increased by 2.7 times; while the UC decreased slightly while k_{ab0} increased for PAA NP when compared to PAA-peg NP (only difference is PAA-peg has surface modification). For BVP-PLA NPs and the Ag NPs, there seems to be a trend of increasing UC with increasing diameter but there is no clear pattern for k_{ab0}.

Table 1. Different nanoparticles and fitted parameters [1]

Nanoparticle	Diameter (nm)	Surface modification	Dose (ug)	UC (-) [2]	k_{ab0} (per h) [3]	$k_{removal}$ (per h) [4]
PAA-peg	31	yes	7000	1	17	0.0028
PAA	31	no	11300	0.72	27	0.0051
PLGA-mPEG	114	yes	1050	2.4	0.54	0.025
BVP-PLA	319	yes	2200	5.8	1.7	40.7
Ag	20	no	119	0.019	15.2	25.1
Ag	80	no	132	0.055	30.7	26.3
Ag	110	no	138	0.031	57.6	17.0

1: All phagocytizing cells are assumed to behave the same; the number of phagocytizing cells per organs is independent of nanoparticles types, but parameters governing its dynamic may change.

2: UC="uptake capacity". Representing the relative uptake capacities of PCs to different NPs.

3: k_{ab0} is the maximum uptake rate by PCs

4: $k_{removal}$ is removal rate from the body which may be caused by degradation or excretion.

3.2. Sensitivity of the model

The sensitivity coefficients calculated according to equation (1) for all types of NPs in this study are listed in Table 2. For all types of NPs except for BVP-PLA, the model is most sensitive to the parameter UC. k_{ab0} only becomes a sensitive parameter to the model for BVP-PLA. For the NPs with high $k_{removal}$ (BVP-PLA and Ag), the model is also sensitive to this parameter. The reason for the model for BVP-PLA is not sensitive to UC but to k_{ab0} can be explained by the short duration of the BVP-PLA NP study, which lasted only for 3 hours while other experiments for the other NPs were at least 24 hours. It is possible that the PCs cannot be saturated within 3 hours and therefore making the UC an insensitive parameter but the k_{ab0} a sensitive one.

Table 2. Sensitivity coefficients for some parameters [1]

Nanoparticle	UC		k_{ab0}		$k_{removal}$	
	Blood	Liver	Blood	Liver	Blood	Liver
PAA-peg	-0.20	0.59	-0.03	-0.07	-0.03	-0.01
PAA	-0.07	0.62	-0.0004	0.0004	-0.04	-0.02
PLGA-mPEG [2]	-0.15	0.13	-0.03	0.07	-0.05	-0.04
BVP-PLA	0.22	0.004	-0.06	0.48	-0.90	-1.78
Ag(20)	0.97	0.45	0.0001	0.33	-0.03	-0.48
Ag(80)	0.98	0.30	0.01	0.18	-0.02	-0.52
Ag(110)	0.97	0.73	-0.0003	0.07	-0.02	-0.18

1: The sensitivity coefficients were calculated by the percentage change divided by the percentage change in the parameter (set at 10%).

2: Also includes spleen in the columns for "liver" due to the nature of the experiment for this NP [11].

4. Conclusions and outlook

Introducing the phagocytizing cells enables the model to explain saturation phenomena in several compartments and model the dynamic well. With the increase in nanoparticle sizes, the phagocytizing cells uptake capacity (expressed as mass) increases in general when removal rates are similar. This could indicate that the phagocytizing cells are saturated by the number of nanoparticles rather than absolute mass. Compared to the other nanoparticles in this study, the BVP-PLA and silver nanoparticles are removed from the body 3 to 4 orders of magnitude faster than the other nanoparticles in this study.

For further study, it is worth designing more detailed experiments to test the hypothesis of increase in nanoparticle size would increase the uptake capacity of phagocytizing cells. It would also be of interest to pay more attention to the excretion within 1 hour of dosing since for the silver nanoparticles, a very fast removal of the nanoparticles from the body was observed but little is known about their fate.

Acknowledgement

This study was supported by U.S. EPA grant EPA-G2010-STAR-N1, the Swedish Council for Working Life and Social Research, and BioSimulation Consulting Inc., DE, USA.

References

[1] Warkad A *et al* 2008 *J. Ind. Ecol.* **12** 435-448
[2] Abbott L C and Maynard A D 2010 *Risk Anal.* **30** 1634-1644
[3] Weinberg H *et al* 2011 *TRAC-Trend Anal. Chem.* **30** 72-83
[4] Nel A *et al* 2006 *Science* **311** 622-627
[5] Zhang H G *et al* 2011 *Environ. Sci. Technol.* **45** 3725-3730
[6] Warheit D B *et al* 2007 *Toxicology* **230** 90-104
[7] Cho M *et al* 2009 *Toxicol. Lett.* **189** 177-183
[8] Nemmer A *et al* 2002 *Circulation* **105** 411-414
[9] Semmler M *et al* 2004 *Inhal. Toxicol.* **16** 453-459
[10] Oberdörster G *et al* 2002 *J. Toxicol. Environ. Healh, Part A* **65** 1531-1543
[11] Panagi Z *et al* 2001 *Int. J. Pharm.* **221** 143-152
[12] Wenger Y *et al* 2011 *Toxicol. Appl. Pharm.* **251** 181-190
[13] Lin P *et al* 2008 *Environ. Sci. Technol.* **42** 6264-6270
[14] Lankveld D P K *et al* 2010 *Biomaterials* **31** 8350-8361
[15] Moghimi S M *et al* 2001 *Pharmacol. Rev.* **53** 283-318
[16] Wilhelm C *et al* 2002 *Langmuir* **18** 8148-8155
[17] Luciani N *et al* 2009 *J. Matter. Chem.* **19** 6373-6380
[18] Liu M *et al* 2007 *Arch Pharm Res* **31** 547-554
[19] Krishnan K 2007 *Physiologically based pharmacokinetic and toxicokinetic models Principles and Methods of Toxicology* ed A W Hayes (Taylor & Francis) chapter 5 pp 231-292

Journal of Physics: Conference Series **429** (2013) 012020 doi:10.1088/1742-6596/429/1/012020

Cytotoxicity and genotoxicity of biogenic silver nanoparticles

R Lima [1,2], L O Feitosa [2], D Ballottin[3], P D. Marcato [4], L Tasic [3] and N Durán [3,5]

[1] Sorocaba University, Sorocaba, SP., Brazil,
[2] Universidade Federal de São Carlos, Sorocaba, SP., Brazil,
[3] Chemistry Institute, Universidade Estadual de Campinas, Campinas, SP, Brazil,
[4] Faculty of Pharmaceutical Sciences of Riberão Preto, Universidade de São Paulo, SP, Brazil,
[5] Center of Natural and Human Sciences, Universidade Federal do ABC, Santo André, SP, Brazil.

E-mail: renata.lima@prof.uniso.br

Abstract. Biogenic silver nanoparticles with 40.3 ± 3.5 nm size and negative surface charge (- 40 mV) were prepared with *Fusarium oxysporum*. The cytotoxicity of 3T3 cell and human lymphocyte were studied by a TaliTM image-based cytometer and the genotoxicity through *Allium cepa* and comet assay. The results of BioAg-w (washed) and BioAg-nw (unwashed) biogenic silver nanoparticles showed cytotoxicity exceeding 50 μg/mL with no significant differences of response in 5 and 10 μg/mL regarding viability. Results of genotoxicity at concentrations 5.0 and 10.0 ug/mL show some response, but at concentrations 0.5 and 1.0 μg/mL the washed and unwashed silver nanoparticles did not present any effect. This in an important result since in tests with different bacteria species and strains, including resistant, MIC (minimal inhibitory concentration) had good answers at concentrations less than 1.9 μg/mL. This work concludes that biogenic silver nanoparticles may be a promising option for antimicrobial use in the range where no cyto or genotoxic effect were observed. Furthermore, human cells were found to have a greater resistance to the toxic effects of silver nanoparticles in comparison with other cells.

1. Introduction

Metal nanoparticles (NP) have been increasingly recognized for their important industrial applications, especially in the fields of medicine, food, and agriculture. Most metals can be used to produce metal nanoparticles; however, the most common are nickel, copper, palladium, platinum, gold and silver [1-10].

Silver nanoparticles (silver NP) are employed in textile engineering and water treatment, as well as in medicine, where their small size and high surface area enhances their antimicrobial activity. The particles can be synthesized by different techniques, including biological methods using fungi, bacteria, and plants, as well as by chemical methods involving the reduction of organic or inorganic salts, and by physical and photochemical procedures [11-17].

In addition to their bactericidal action, silver NP can also be used to combat viruses and fungi [18-19]. Its high antimicrobial activity can be attributed to their affinity for chemical groups containing the elements phosphorus and sulfur present in cell membrane, causing damage and interrupting cellular respiration, and interacting with DNA and its interference in cell division [20].

In recent years, there has been increased interest in silver NP and their applications, while at the same time there are concerns regarding the possible impacts of using NP on a large scale, including problems of toxicity and risks to the environment and human health. In response, there have been a number of published *in vivo* studies concerning potential inflammatory and other effects of silver NP [22-27]. However, little is known concerning the genotoxicity of silver nanoparticles to different organisms and cell types [28]. This study is very relevant due to broad application of silver NP and to high industrial interest in these particles. The importance of these silver NP in the area of health and materials is due to their antimicrobial activity. More than 400 products are already on the market. This large investment reflects the efficacy of silver in combating resistant microorganisms [29], in enhancing the effects of antibiotics such as gentamicin [30], and in producing antimicrobial fabrics [2]. Furthermore, this great investment and the number of products indicate the importance of study of genotoxic potential of these particles [28].

2. Material And Methods

2.1 Silver nanoparticle production

Fungal inoculates were prepared at 28°C in Petri dishes with a medium containing malt extract (2%) and yeast extract (0.5%). Silver reduction was carried out by placing approximately 10 g of *F. oxysporum* biomass in a conical flask containing 100 mL of distilled water, and leaving for it 72 h at 28°C. The aqueous solution components were then separated by filtration, which was followed by addition of $AgNO_3$ (10^{-2} mol/L) to the solution and leaving for several hours at 28°C [31].

2.2 Characterization of the biogenic silver nanoparticles

2.2.1 Transmission electron microscopy (TEM)

The AgNP were characterized using a Carl Zeiss Libra transmission electron microscope, operated at 120 keV. One drop of the particle dispersion was deposited onto carbon-coated Parlodion films supported on 300 mesh copper grids (Ted Pella).

2.2.2 Particle size and zeta potential

The mean diameter (z-average diameter) and size distribution of the nanoparticles were measured by photon correlation spectroscopy (PCS) with a Zetasizer Nano ZS (Malvern Instruments, UK), at 25 °C, using polystyrene cuvettes with a path length of 10 mm. The zeta potential was measured with the same instrument using capillary cells with a path length of 10 mm.

2.3 Cell culture and *Allium cepa* test conditions

Analyses were performed using human lymphocytes and 3T3 cell line fibroblasts. The lymphocyte tests employed RPMI medium supplemented with 10% fetal bovine serum, and the 3T3 cell analyses used DMEN medium supplemented with 15% fetal bovine serum. Both media were fortified with antibiotics, and the cultures were kept at 37 °C in a 5% CO_2 humidified atmosphere. Contact with the nanoparticles was maintained for 1 h in both the cytotoxicity analyses (evaluated by Tali scan) and the comet tests. For the *Allium cepa* analyses, the seeds were germinated in ultrapure water at room temperature. When the roots reached a length of 2 cm, the seedlings were removed from the germination box and placed into contact with the test materials. The analyses performed are summarized in Figure 1.

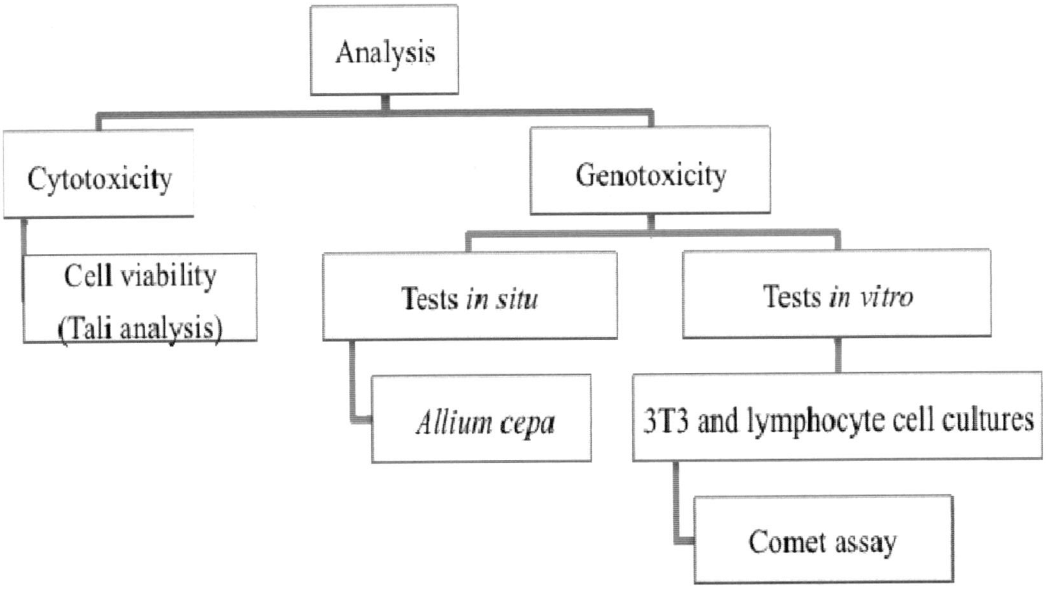

Figure 1 – Scheme of the experimental protocol adopted.

2.4 Tali analysis

Cell viability was measured using a Tali™ Apoptosis Kit consisting of Annexin V Alexa Fluor® 488 and Propidium Iodide (Invitrogen), and a Tali™ image-based cytometer, which enabled the numbers of viable, apoptotic, and dead cells to be counted. The cells that had been treated with nanoparticles were centrifuged and concentrated to 1 x 10^6 cells per mL. 100 μL aliquots of sample were prepared according to the specifications of the kit, and the tests were performed in triplicate.

2.5 *Allium cepa* assays

In the *Allium cepa* assays, ultrapure water was used as a negative control and trifluoraline (0.075 g/L) as the positive control. After 24 hours of contact with the silver nanoparticles, the roots were fixed with Carnoy's reagent and subjected to acid hydrolysis. They were then washed in distilled water and submitted to a Schiff-base reaction. The meristematic region of the roots was cut, placed on a slide together with a drop of 2% acetocarmine, and covered with a cover slip, which was used to gently press and spread the cells. The slides were examined under an optical microscope. The results of the *Allium cepa* tests were used to calculate the MI, RMI (treatment mitotic index / negative mitotic index), CAS (number of damaged cells / number of cells in division), and RCAS (treatment chromosomal aberration index / negative treatment chromosomal aberration index), as described by Parida et al. [32].

2.6 Comet assays

The lymphocytes and 3T3 cells were exposed to the silver nanoparticles for a period of 1 hour. Negative and positive controls employed phosphate buffered saline (PBS) and H_2O_2 (200 μmol/L), respectively. The comet assay was performed as described by Azqueta et al. [33]. Each treatment used 10 μL of lymphocytes in 110 μL of low melting point agarose (0.8%), and the mixture was placed onto microscope slides that had been pre-coated with normal melting point agarose (1.5%). After polymerization, the cover slips were removed and the slides were treated with an ice-cold lysis solution for 20-30 minutes. All treatments were then incubated in an electrophoresis buffer for 20 minutes, followed by electrophoresis for 20 minutes at 1.3 V/cm.

After electrophoresis, the slides were stained and analyzed by optical microscopy. Scores of between 0 and 4 were assigned according to the quantity of DNA in the tail and the length of the tail: Score 0 corresponded to intact cells, with no damage caused by the exposure; Score 1 corresponded to cells with minimal damage; Score 2 to average damage; Score 3 to severe damage; and Score 4 to cells with maximum damage (Figure 2). In this visual method, the number of cells found for each score was multiplied by the value of the score and the values were summed at the end of the analysis of each slide. Since the score depended on the number of cells observed, an index of tail damage (TD) was created by dividing the score given for the slide by the number of cells analyzed on the slide [33-36].

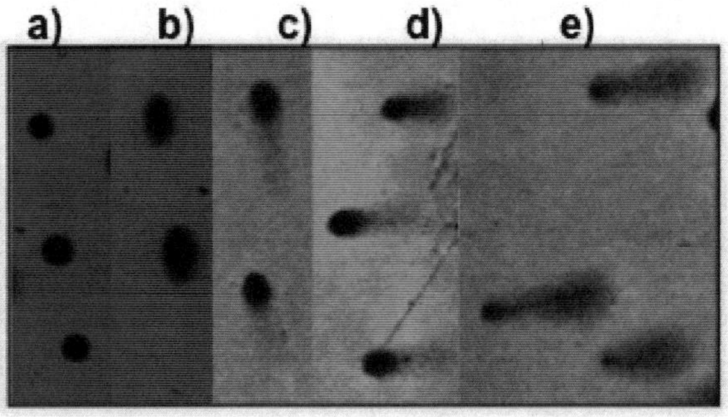

Figure 2 - Representative images (at x40 magnification) illustrating the tail intensity for the fibroblast comet test scores: a) score = 0 (negative control); b) score = 1; c) score = 2; d) score = 3; e) score = 4.

3. Results and Discussion

3.1. Biogenic silver nanoparticle characterizations

The spherical biogenic silver NP prepared using *Fusarium oxysporum* was obtained with average size of 40.3 ± 3.5 nm as shown Figure 3.

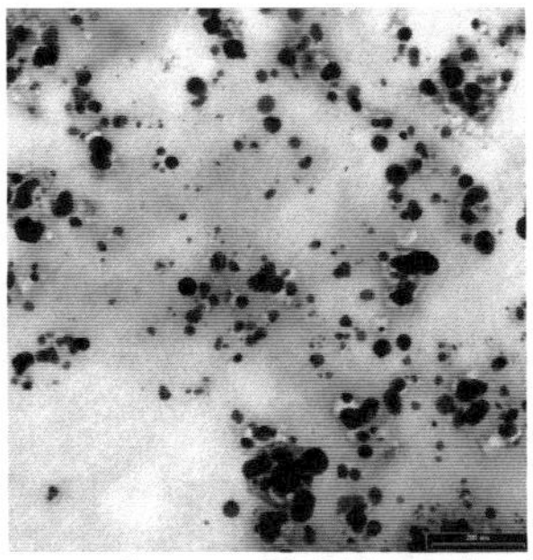

Figure 3 - TEM image of unwashed biogenic silver nanoparticles.

The particles dispersion was stable for several months (6 months) due to protein (from fungi) capping around the particles as described by Durán et al. [31]. To separate the protein bound with silver NP from protein present in the dispersion, the particles were washed (3 times) and centrifuged at 5000 rpm for 30 minutes. The particle size, polidispersity index (PdI) and zeta potential of these washed particles (BioAg-w) was analyzed and compared with unwashed silver NP (BioAg-uw) as shown in Table 1. The BioAg-w exhibited larger sizes than BioAg-uw, indicating particle agglomeration in the centrifugation process. The PdI for both particles was around 0.2, showing low polydispersity. In relation to the zeta potential, an increase in modulus was observed when the particles were washed. This difference can be attributed the protein lost (i.e. not bound to the NP) from the dispersion in the washing processes.

Table 1 – Size, polydispersivity, and zeta potential of washed and unwashed silver nanoparticles.

Particle type	**Size (nm)**	**Polydispersivity**	**Zeta potential (mV)**
BioAg-uw (unwashed)	106.2 ± 13.0	0.231	-37.1 ± 2.6
BioAg-w (washed)	145.1 ± 4.5	0.206	-47.8 ± 1.1

4. Cytotoxicity and genotoxicity

4.1. Cytotoxicity

The cytotoxicity analyses using 3T3 cells and lymphocytes were performed by the Tali method for concentrations of 5, 10, and 50 µg/mL of the washed (BioAg-w) and unwashed (BioAg-uw) biogenic silver nanoparticles. For both nanoparticles, the results revealed cytotoxicity at concentrations exceeding 50 µg/mL (not shown), while for concentrations of 5 and 10 µg/mL neither of the cell cultures showed any significant differences in response in terms of viability (Figure 4).

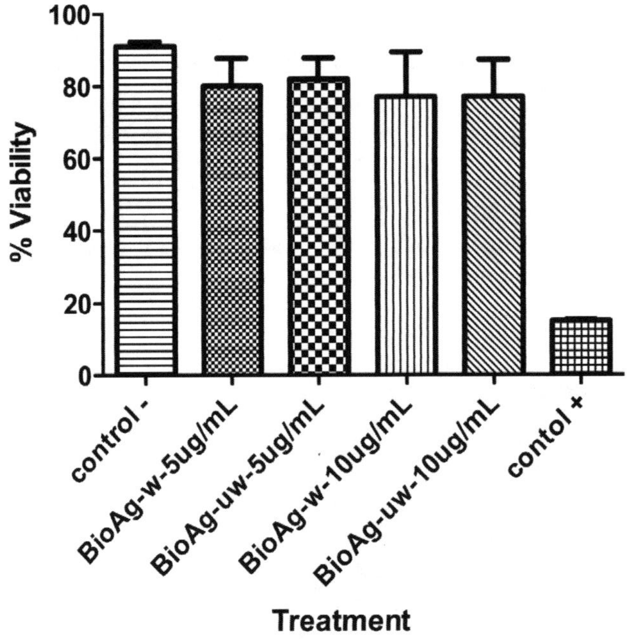

Figure 4 – Results of the Tali analyses using lymphocyte cells.

4.2 Genotoxicity

4.2.1 Allium cepa assays

The results of the *Allium cepa* cytogenetic tests revealed no significant differences in mitotic indices between the treated groups and the negative control (Figure 5). However, in the case of the alteration indices, significantly higher values were obtained for the treatments with BioAg-uw, compared to the negative control and BioAg-w (Figure 6). This effect can be related to nanoparticle aggregation in the washing processes, as verified by PCS analysis. Silver nanoparticles with 20 nm were more toxic in all endpoints evaluated than larger silver nanoparticles (80 and 113 nm). Park et al. (2011) [37] verified a correlation between silver nanoparticle sizes and cyto and genotoxicities.

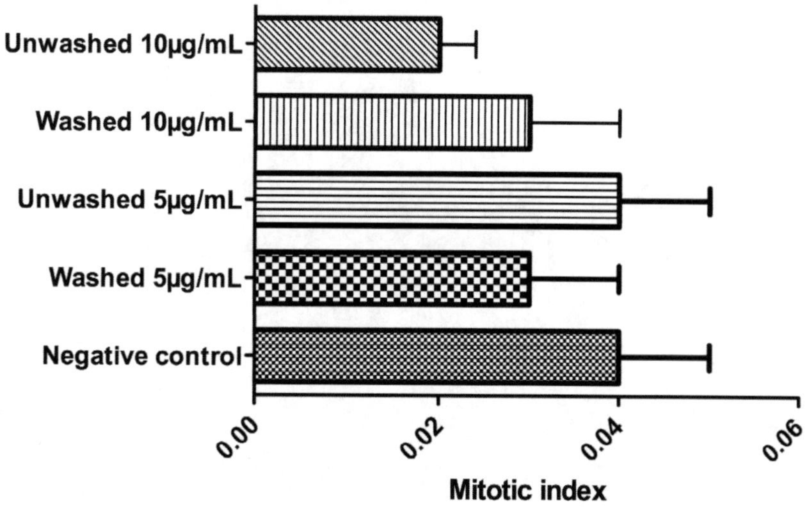

Figure 5 - Results of the mitotic index analyses.

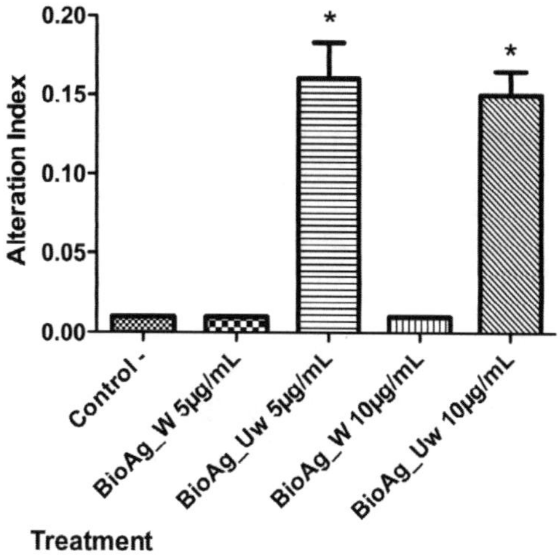

Figure 6 – Results of the alteration analyses using the *Allium cepa* test.

4.2.2 Comet assays

The comet assays identified a genotoxic response at nanoparticle concentrations of 5 and 10 µg/mL. However, at concentrations of 0.5 and 1 µg/mL, neither washed (BioAg-w) nor unwashed silver nanoparticles (BioAg-uw) exhibited any genotoxicity (Figure 7). This is an important finding, since tests with different bacteria species and strains, including resistant strains, have found MIC (minimum inhibitory concentration) values of less than 1.9 µg/mL (data not shown).

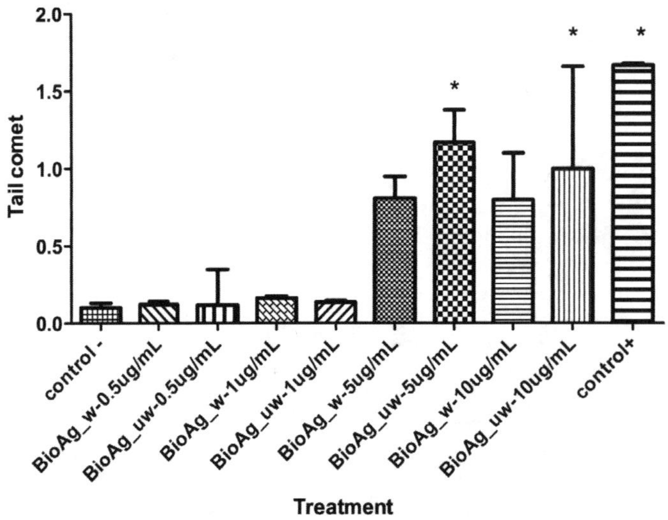

Figure 7 – Results of the comet assays.

The results suggest that a concentration of 5 µg/mL could be a critical level for genotoxicity for biogenic silver NP. As verified in the *Allium cepa* assays (alteration indices), BioAg-w showed lower genotoxic effects than BioAg-uw at concentrations of 5 and 10 µg/mL. Further studies will be needed to investigate the relation between nanoparticle aggregation and alteration in genotoxicity.

5. Conclusions

Silver biogenic nanoparticles showed similar response to cytotoxicity tests in which both (washed and unwashed) were not cytotoxic up to the concentration of 10ug/mL. In relation to genotoxicity the washed and unwashed silver nanoparticles did not show significant differences related to the mitotic index at the *Allium cepa* test; but at the damage analysis unwashed nanoparticles showed significant damage index. The same was observed for the comet assay in which unwashed nanoparticles were genotoxic while the washed silver nanoparticles were not. Deeper studies are necessary in relation to biogenic nanoparticles, mainly regarding the substances that cap the nanoparticles.Despite nanoparticles show genotoxic in concentration of 5µg/mL we remind that their MIC is between 0.9-3.8 µg/mL.

Acknowledgments: The authors would like to thank FAPESP, CAPES and CNPq (Brazil) for financial support and INOMAT (MCT/CNPq) and Brazilian Network in Nanotoxicology (MCT/CNPq), Brazil.

6. References

[1] Baroli B, Ennas M G, Loffredo F, Isola M, Pinna R and López-Quintela M A 2007 *J. Invest. Dermatol.* **127** 1701
[2] Durán N, Marcato P D, De Souza G I H, Alves O L and Esposito E 2007 *J. Biomed. Nanotechno.* **3** 203
[3] Pavon L F and Akamoto O K 2007 *Einstein* **5** 74
[4] Pradeep T and Anshup 2009 *Thin Solid Films* **517** 6441
[5] Uehara N 2010 *Anal. Sci.* **26** 1219
[6] Kumar H, Ghosh S, Avasthi D K, Kabiraj D, Mücklich A, Zhou S, Schmidt H and Stoquert J P 2011 *Nanoscale Res. Lett.* **6** 155
[7] Kurek A, Grudniak A M, Kraczkiewicz-Dowjat A and Wolska K I 2011 *Polish J. Microbiol.* **60** 3
[8] Nomura M, Yoshimura Y, Kikuiri T, Hasegawa T, Taniguchi Y, Deyama Y, Koshiro K, Sano H, Suzuki K and Inoue N 2011 *J. Pharmacol. Sci.* **117** 243

[9] Silva A C, Oliveira T R, Mamani J B, Malheiros S M, Malavolta L, Pavon L F, Sibov T T, Amaro E Jr, Tannús A, Vidoto E L, Martins M J, Santos R S and Gamarra L F 2011 *Inter. J. Nanomed.* **6** 591
[10] Conde J, Doria G and Baptista P 2012 *J. Drug Delivery* **2012** 1
[11] Furno F, Morley K S, Wong B, Sharp B L, Arnold P L, Howdle S M, Bayston R, Brown P D, Winship P D and Reid H J 2004 *J. Antimicrob. Chemother.* **54** 1019
[12] Nair S L and Laurencin 2007 *J. Biomed. Nanotechnol.* **3** 301
[13] Khaydarov R R, Khaydarov R A, Estrin Y, Evgrafova S, Scheper T, Endres C and Cho S Y 2009 *Silver Nanoparticles: Environmental and Human Health Impacts* (Nanomaterials: Risks and Benefits) ed Linkov I and Steevens J (Springer Science) pp 287-297
[14] Durán N, Marcato P , Conti R, Alves O L, Costa F T M and Brocchi M 2010 *J. Braz. Chem. Soc* **21** 949
[15] Ahmad M Sharma S, Singh V N, Shamsi S F, Fatma, A and Mehta B R 2011 *Biotechnol. Res. Inter.* **2011** 1
[16] Srivastava A A, Kulkarni A P, Harpale P M and Zunjarrao R S 2011 *Inter. J. Eng. Sci. Technol.* **3** 8342
[17] Li G He D, Qian Y, Guan B, Gao S, Cui Y, Yokoyama K and Li Wang 2012 *Int. J. Mol. Sci.* **13** 466
[18] Elechiguerra J L, Burt J L, Morones J R, Bragado A C, Gao X, Lara H H and Miguel J Yacama 2005 *J. Nanobiotechnol.* **3** 2005
[19] Galdieiro S, Falanga A, Vitiello M, Cantisani M, Marra V and Galdiero M 2011 *Molecules* **16** 8894
[20] Neto E A B, Ribeiro C and Zucolotto V 2008 *Technical Communication - Embrapa Agricultural Instrumentation* **99** 1
[21] Seth D, Choudhury S R, Pradhan S, Gupta S, Palit D, Das S, Debnath N and Goswami A 2011 *Curr. Microbiol.* **62** 715
[22] Wise Sr J P, Goodale B C, Wise S S, Craig G A, Pongan A F, Walter R B, Thompson W D, Ng A K, Aboueissa A M, Mitani H, Spalding M J and Mason MD 2010 *Aquatic Toxicol.* **97** 34
[23] Griffitt R J, Hyndman K, Denslow N D and Barber D S 2009 *Toxicol. Sci.* **107** 404
[24] Gaiser B K, Fernandes T F, Jepson M A, Lead J R, Tyler C R, Baalousha M, Biswas A, Britton G J, Cole P A, Johnston B D, Ju-Nam Y, Rosenkranz P, Scown T M and Stone V 2011 *Environ. Toxicol. Chem.* **31** 144
[25] Park H, Yu M, Kang S K, Yang S I and Kim Y 2010 *Mol. Cell. Toxicol.* **7** 67
[26] Nair P M, Park S Y, Lee S W and Choi J 2011 *Aquatic Toxicol.* **101** 31
[27] Kim J S, Sung J H, Ji J H, Song K S, Lee J H, Kang C S, and Yu I J 2011 *Safe Health Work* **8** 34
[28] Lima R, Seabra A B, Durán N 2012 *J. Appl. Toxicol.* **32** 867
[29] Sundaramoorthi C, Kalaivani M, Mathews D M, Palanisamy S, Kalaiselvan V and Rajasekaran A 2009 *Int. J. Pharm. Tech. Res.* **1** 1523
[30] Gade A K, Bonde P, Ingle A P, Marcato P D, Durán N and Rai M K 2008 *J. Biobased Mat. Bioener.* **2** 243
[31] Durán N, Marcato P D, Alves O L, De Souza G I H and Esposito E 2005 *J. Nanobiotechnol* **3** 1
[32] Parida U K, Bindhani B K and Nayak P 2011 *World J. Nano Sci. Eng.* **1** 93
[33] Azqueta A, Gutzkow K B, Brunborg G and Collins A R 2011 *Mutat. Res.* **724** 41
[34] Dusinská M, Collins A, Kazimírová A, Barancoková M, Harrington V, Volkovová K, Staruchová M, Horská A, Wsólová L, Kocan A, Petrík J, Machata M, Ratcliffe B and Kyrtopoulos S 2004 *Mutat. Res.* **553** 91
[35] Dusinska M and Collins A R 2008 *Mutagenesis* **23** 191
[36] Collins A R, Oscoz A A, Brunborg G, Gaivão I, Giovannelli L, Kruszewski M, Smith C C and Stetina R 2008 *Mutagenesis* **23** 143
[37] Park M V D Z, Neigh A M, Vermeulen J P, de la Fonteyne L J J, Verharen H W, Briedé J J, van Loveren H and de Jong W H 2011 *Biomaterials* **32** 9810

Iron oxide nanoparticles show no toxicity in the comet assay in lymphocytes: A promising vehicle as a nitric oxide releasing nanocarrier in biomedical applications

R de Lima[1], J L Oliveira[1], P S K Murakami[2], M A M Molina[3], R Itri[3], P Haddad[2] and A B Seabra[2*]

[1]Departamento de Biotecnologia, Universidade de Sorocaba, Rodovia Raposo Tavares Km 92 , CEP 18023-000, Sorocaba, SP, Brazil
[2]Departamento de Ciências Exatas e da Terra, Universidade Federal de São Paulo, Rua São Nicolau 210, CEP 09913-030, Diadema, SP, Brazil
[3]Instituto de Física, Universidade de São Paulo, CP 66318, CEP 05314-970, SP, Brazil

*E-mail: amedea.seabra@unifesp.br

Abstract. This work reports the synthesis and toxicological evaluation of surface modified magnetic iron oxide nanoparticles as vehicles to carry and deliver nitric oxide (NO). The surface of the magnetic nanoparticles (MNPs) was coated with two thiol-containing hydrophilic ligands: mercaptosuccinic acid (MSA) or dimercaptosuccinic acid (DMSA), leading to thiolated MNPs. Free thiols groups on the surface of MSA- or DMSA-MNPs were nitrosated leading to NO-releasing MNPs. The genotoxicity of thiolated-coated MNPs was evaluated towards human lymphocyte cells by the comet assay. No genotoxicity was observed due to exposure of human lymphocytes to MSA- or DMSA-MNPs, indicating that these nanovectors can be used as inert vehicles in drug delivery, in biomedical applications. On the other hand, NO-releasing MPNs showed genotoxicity and apoptotic activities towards human lymphocyte cell cultures. These results indicate that NO-releasing MNPs may result in important biomedical applications, such as the treatment of tumors, in which MNPs can be guided to the target site through the application of an external magnetic field, and release NO directly to the desired site of action.

1. Introduction

Nitric oxide (NO) is a key molecule involved in several important physiological processes, such as the dilation of blood vessels, inhibition of platelet adhesion and aggregation, cell communication, wound healing, immune responses, apoptosis and anti-cancer activities, among others [1]. NO is a gaseous free-radical and one of the smallest endogenous molecules with the ability to function as a chemical messenger, particularly in cells of the vascular endothelium and immune system. Medical and scientific interest in NO has grown exponentially since 1992, when it was nominated "Molecule of the Year" [2]. Its documented physiological impacts are ever expanding, and NO is now known as one of the most important mediators of intra- and extracellular processes and is one of the major targets of the pharmaceutical industry [3]. Administration of high doses of NO (micro and milli concentrations) are known to play a key role in the immunological system and thus to initiate tumor regression [4].

Although NO is used in many biomedical applications, its utility is limited by its short half-life and instability during storage. As a free radical, in the biological medium, NO can ready react with biomolecules, such as hemoglobin, leading to its inactivation. Therefore, there is great interest in the development of NO-releasing drugs and matrices which are capable of stabilizing and releasing NO locally, directly into different tissues and organs [5]. Among the nanostructured materials aimed at drug delivery, magnetic nanoparticles (MNPs), in particular iron oxide nanoparticles, such as magnetite (Fe_3O_4), are considered a promising material for biomedical applications [6]. Indeed, MNPs have been used for targeted drug delivery to specific cells, tissues or organs, for enhancing magnetic resonance imaging contrast and for hyperthermia treatments [7]. In drug delivery, iron oxide MNPs can be easily transported *in vivo* to the desired site of action by application of an external electrical-magnetic field. Magnetization disappears upon removal of the magnetic field, leading the iron oxide MNPs to remain at the target site for a certain period [8].

However, in order to propose the use of MNPs in biomedical applications, such as vehicles for drug delivery, it is mandatory to investigate the toxicological aspects of these nanomaterials. In this context, nanomaterial safely, in particular nanogenotoxicology, is an increasing issue in the field of nanotechnology [9,10]. Recent papers highlighted that many nanomaterials induce cytotoxicity and oxidative stress. Conflicting data have been reported [9,10]. Therefore, there is great interest in the characterization of possible toxic effects of nanomaterials, in particular MNPs, aimed for pharmacological applications. In this context, this work reports the synthesis and evaluation of genotoxicity of surface-modified MNPs, as vehicles to carry and deliver NO in biomedical applications.

2. Methods

2.1. Synthesis of magnetic nanoparticles (MNPs)

A co-precipitation method was used to synthesize MNPs [11]. Briefly, 4.0 mL of $FeCl_3 \cdot 6H_2O$ solution and 1.0 mL of $FeCl_2 \cdot 4H_2O$ solution (molar ratio 2:1) in 1.0 mol/L HCl were mixed and stirred, while a volume of 50 mL of NH_4OH (0.7 mol/L) was added as precipitator. The solution was centrifuged and the precipitate was decanted, followed by the addition of 6.0 mL of oleic acid. This mixture was stirred for 20 minutes. The solution was centrifuged several times and the new precipitate was washed several times with ethanol and acetone, leading to MNPs covered with oleic acid.

Oleic acid coated-MNPs (~10.0 mg) were suspended in 1.0 mL of toluene while mercaptosuccinic acid (MSA) or 2,3 dimercaptosuccinic acid (DMSA) (molar ratio Fe_3O_4:MSA or DMSA = 1:40) was dissolved in dimethyl sulfoxide (DMSO) in a second tube. The solutions were mixed and vigorously stirred for 14 h producing a black powder that was isolated by centrifugation. This procedure led to ligand exchange and, hence, to the formation of water stable thiol-containing MNPs (SH-NPs).

2.2. Nitrosation of thiolated magnetic nanoparticles

The thiol groups (SH) on the surface of MSA- or DMSA-coated MNPs were nitrosated by sodium nitrite ($NaNO_2$) leading to the formation of S-nitroso-MNPs, which are spontaneous NO donors. In this step, filtered MSA- or DMSA-MNPs were suspended in deionised water. A volume of 200 μL of aqueous sodium nitrite (60 mmol/L) was added to the MNPs. After 15 min of incubation at room temperature, the nanoparticle suspension was filtered by centrifugal ultrafiltration by using a Microcon centrifugal filter device containing ultrafiltration membranes (MWCO 10-kDa molar mass cut-off filter, Millipore, Billerica, MA, USA) and washed with deionised water to remove excess unreacted nitrite.

2.3. Cell description
For the analyses involving lymphocytes, the cells were separated from whole blood using Ficoll-Paque™ PLUS medium (GE Healthcare, Little Chalfont, UK). The blood was provided by donors aged between 18 and 24 years (who freely signed terms of agreement forms) and the project was approved by the Ethics Committee of the University of Sorocaba (protocol #008/08). Blood samples were collected using disposable materials throughout the procedure. The lymphocytes were placed in Roswell Park Memorial Institute (RPMI) 1600 culture medium (Cultilab) containing 300 µg/mL of L-glutamine and 200 µg/mL of $NaHC_3$, supplemented with 5% bovine fetal serum, 50 µg/mL of gentamicin sulfate (antibiotic) and 2 µg/mL of amphotericin B (antifungal). The culture was kept at 37°C, under a humidified atmosphere containing 5% CO_2.

2.4. Incubation of human lymphocytes with MNPs
Human lymphocytes were incubated with MSA- or DMSA-MNPs (thiol-containing MNPs, as control groups) and nitrosated-MSA (SNO-MSA) or nitrosated-DMSA (SNO-DMSA)- MNPs at the following nanoparticle concentrations: 0.01; 0.1 and 0.5 mg/mL, for 1 and 24 h. Negative and positive controls employed phosphate buffered saline (PBS) and H_2O_2 (200 µmolol/L), respectively.

2.5. Comet assay
Each treatment group involved 10 µL of lymphocytes in 110 µL of low melting point agarose (0.6%) and the mixtures were placed onto microscope slides that had been pre-coated with normal melting point agarose (1.5%). Coverslips were positioned over the materials, and the slides were placed in a refrigerator for polymerization. After polymerization, the coverslips were removed, and the slides were treated for 90 minutes with an ice-cold (4°C) lysis solution (2.5 mol/L NaCl, 0.1 mol/L ethylenediaminetetraacetic acid (EDTA), 10 mmol/L tris(hydroxymethyl)aminomethane (Tris), 1% Triton X-100TM, pH 10). All groups were incubated in an electrophoresis buffer (0.3 mol/L NaOH, 1 m mol/L EDTA, pH 13) for 20 minutes, followed by electrophoresis for 20 minutes at 1.3 V/cm. After electrophoresis, the slides were covered with a neutralizing solution (0.4 mol/L Tris, pH 7.5) for 5 min, washed three times with distilled water, and allowed to rest overnight at room temperature. Prior to staining, the dry slides were left in a fixing solution (15% w/v trichloroacetic acid, 5% (w/v) zinc sulfate, and 5% glycerol) for 10 minutes, and were further washed three times with distilled water. After these procedures, the slides were allowed to rest at room temperature for 1.5 hours. They were then rehydrated with distilled water and stained for approximately 15 minutes with silver staining solution consisting of 34 mL of solution A (0.2% w/v ammonium nitrate, 0.2% w/v silver nitrate, 0.5% w/v tungstosilicic acid, 0.15% v/v formaldehyde, and 5% w/v sodium carbonate) and 66 mL of solution B (5% sodium carbonate), followed by a bath in distilled water and a bath in stop solution. Lastly, the slides were further washed with distilled water and allowed to dry at room temperature. Staining using silver is analogous to fluorescence, where the positive charge of the silver enables it to bind with DNA and DNA fragments, producing the characteristic color. Throughout the procedures involving cellular material, both natural light and light from fluorescent lamps were avoided to prevent possible influence on the results.

Analyses were performed using the Zeiss Axovert optical microscope and at least 100 cells were counted on each slide, with 3 slides for each test (around 300 cells). The experiment was performed in triplicate, giving a total of 900 cells analysed for each sample tested. The comet assay analyses were performed by assigning a score of 0 to 4, according to the quantity of DNA in the tail, and the length of the tail: Class 0 corresponded to intact cells, Class 1 corresponded to cells with minimal damage; Class 2 to average damage; Class 3 to severe damage; and Class 4 to cells with maximum damage [12]. For this visual method, the number of cells found for each score was multiplied by the value of the score and the values were summed at the end of the analysis of each slide. Since the score depended on the number of cells observed, an index of tail damage (TD) was created by dividing the score given to the slide by the number of cells analysed on the slide.

3. Results and Discussion

3.1. Synthesis of MSA- or DMSA-coated MNPs

Iron oxide MNPs comprised by Fe_3O_4 were coated with two thiol-containing hydrophilic ligands: MSA or DMSA (Figure 1).

Figure 1. Chemical representation of MSA (i) and DMSA (ii) used as coating for MNPs.

Figure 2 represents schematically the nanoparticle preparation. The presence of thiol-containing ligands (MSA or DMSA) on the nanoparticle surface is responsible to increase water-solubility of the nanoparticles, to provide thermal stabilization, and to avoid nanoparticle aggregation, in addition, it represents an important site for bioconjugation. Indeed, in this work, free thiols groups (SH) on the nanoparticle surfaces were used to load NO to the nanoparticle.

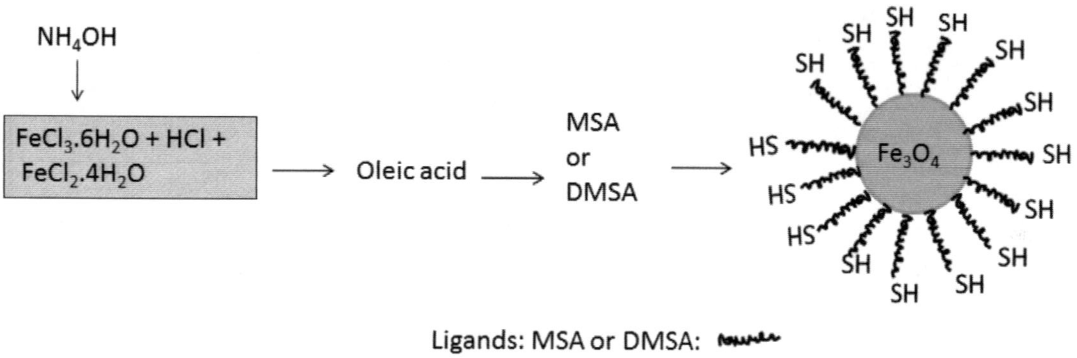

Figure 2. Schematic representation of the synthesis of MSA- or DMSA-coated MNPs, leading to the formation of thiol-containing nanoparticles.

3.2. Nitrosation of thiols groups on MSA- or DMSA-coated MNPs

Thiol groups on MSA- or DMSA-coated nanoparticles were nitrosated by the addition of sodium nitrite, leading to the formation of S-nitroso-MNPs (SNO-MNPs) (Figure 3). In acidified aqueous solution, nitrite (NO_2^-) is in equilibrium with nitrous acid (HNO_2), which is considered the nitrosating agent of -SH groups leading to the formation of -SNO [13]. SNO moieties act as spontaneous NO donors, due to homolitic S-N bond cleavage, with free NO release [13], as represented in Figure 3.

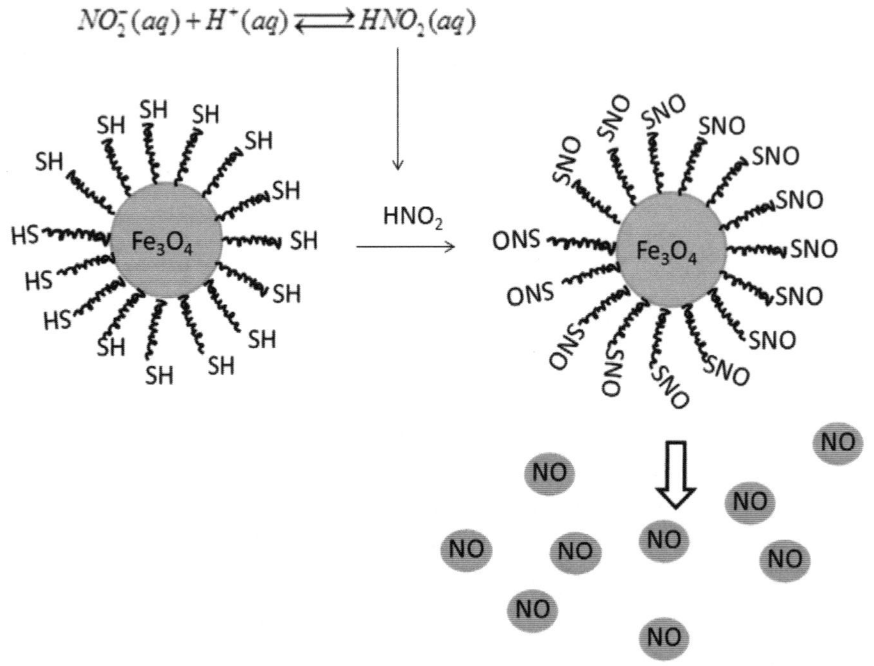

Figure 3. Schematic representation of nitrosation of thiols groups of MSA- or DMSA-MNPs upon incubation with nitrite, leading to the formation of SNO-MNPs, which are spontaneous NO donors.

3.3. Genotoxicity of MNPs (Comet Assay)

The genotoxicities of thiolated and nitrosated MNPs toward human lymphocytes were evaluated by comet assay that detects DNA single strand breaks, alkaline labile sites, and excision processes in individual cells. Figure 4 shows the DNA damage (as expressed in % tail DNA content) due to exposure of lymphocytes to MSA-MNPs (thiolated NPs) and SNO-MSA-MNPs (S-nitroso-MSA-MNPs) for 1 and 24 h, at different nanoparticle concentrations (0.5; 0.1 and 0.01 mg/mL), as indicated in the Figure. As can be observed, MSA-MNPs were found to be not genotoxic at all tested concentrations, after 1 and 24 h of exposure to human lymphocytes (Figure 4). The values of tail DNA were shown to be similar to the values obtained for the negative control, indicating the absence of genotoxicity of these MNPs.

However, NO-releasing MNPs (SNO-MSA-MNPs) were found to be genotoxic, under the same experimental conditions (Figure 4). The genotoxicity was higher after 24 h of cell exposition, in comparison to 1 h of exposure, indicating a time-dependence toxicity of NO-releasing MNPs. No concentration-dependence was observed for SNO-MSA-MNPs, since the values of DNA damage were similar for all tested MNPs concentrations.

Figure 5 shows the DNA damage due exposure of lymphocytes to DMSA-MNPs (thiolated NPs) and SNO-DMSA-MNPs (S-nitroso-DMSA-MNPs) for 1 and 24 h, at different nanoparticle concentrations, as indicated in the Figure. MSA-MNPs exhibited low genotoxicity. Although, the intensity of DNA damage was found to be discrete, a slight toxicity was observed for DMSA-MNPs at all concentrations studied, after 1 and 24 h of exposition. By comparing the genotoxicity of MSA-MNPs (Figure 4) with DMSA-MNPs (Figure 5) it can be noted that MSA-coated MNPs are less genotoxic in comparison with DMSA-MNPs, although both nanpparticles can be considered not toxic to cells.

On the other hand, the genotoxicity of SNO-DMSA-MNPs was higher related to DMSA-NPs. As expected, the presence of NO on NP surface led to toxic effects. Moreover, the DNA damage depended on the time of exposure, since 24 h of cell incubation with SNO-DMSA-MNPs resulted in a more intensive genotoxicity, compared with exposure of only 1 h. SNO-DMSA-MNPs at different concentrations caused similar DNA damage (Figure 5).

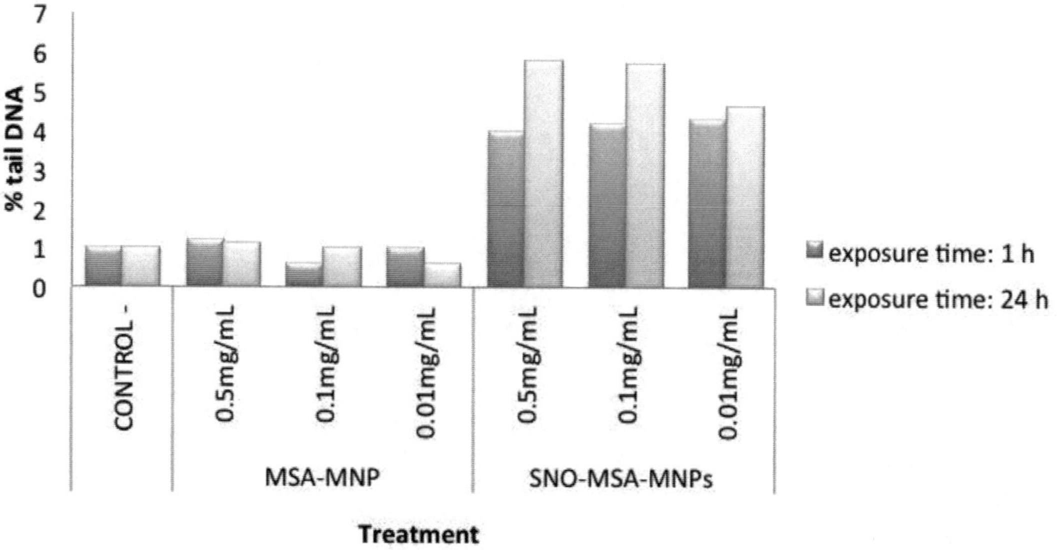

Figure 4. Results of DNA tail damage obtained for human lymphocyte cells after treatments with: MSA-MNP (MSA-coated magnetic nanoparticles) and SNO-MSA-MNP (S-nitrosated MSA-coated magnetic nanopartilces) at concentrations 0.5, 0.1 and 0.01 mg/mL. Exposure time: 1 and 24 h, as indicated in the Figure.

Uncoated iron oxide MNPs were found to have toxic effects towards cells [14]. However, MNPs toxicity can be greatly reduced by coating their surface [14]. In this context, this report shows that MSA- and DMSA-coated iron oxide MNPs are not genotoxic towards human lymphocytes. This result can be understood by assuming that both MSA and DMSA (the later to a lesser extend) are responsible to create a barrier for direct contact between the MNPs and the exposed cell, inhibiting possible toxic effects. Therefore, MSA- or DMSA-coated MNPs offer a high potential in biomedical applications, in particular, in drug delivery.

Both SNO-MSA-MNPs and SNO-DMSA-MNPs exhibited genotoxic effects on lymphocytes, under the experimental conditions employed in this study. Indeed, NO can both promote and inhibit cell damage, apoptosis, acting as anti- or pro- oxidant agents [1,4]. The biological roles of NO are greatly influenced by its source of production, concentration and flux. In fact, large concentrations of NO (i.e., micromolar) produce reactive nitrogen species, which along with reactive oxygen species result in oxidative and nitrosative stress, leading to DNA base deamination, nitrosylation/nitrosation of enzymes, impaired cellular function, enhanced inflammatory reactions, inhibited mitochondrial respiration and cell apoptosis [1,4]. On the other hand, lower NO concentrations (i.e., nano to picomolar) present anti-apoptotic effects, promote angiogenesis, wound healing and inhibition of platelet adhesion [1,4]. The concentrations used in this present work (0.01 to 0.5 mg of SNO-MNPs/mL) were high enough to promote genotoxic effects towards cells.

Based on the results reported here, NO-releasing MNPs might find important application against tumor cell lines. Indeed, in cancer science, NO therapies have been tested against numerous healthy and tumor cell lines, revealing that susceptibility depends on concentration, cell type, and the oxygen environment surrounding the site of NO delivery [4]. Due to the ability to have genotoxic effects against cells, SNO-MNPs can be used to carry and to delivery NO directly to the target site, since the MNPs can be guided to tumors cells, upon application of an external magnetic field. These nanoparticles would be able to release NO direct to cancer cells, where NO has its toxic effects. At the present moment, toxicological evaluations of NO-releasing MNPs towards tumors cell lines are under progress.

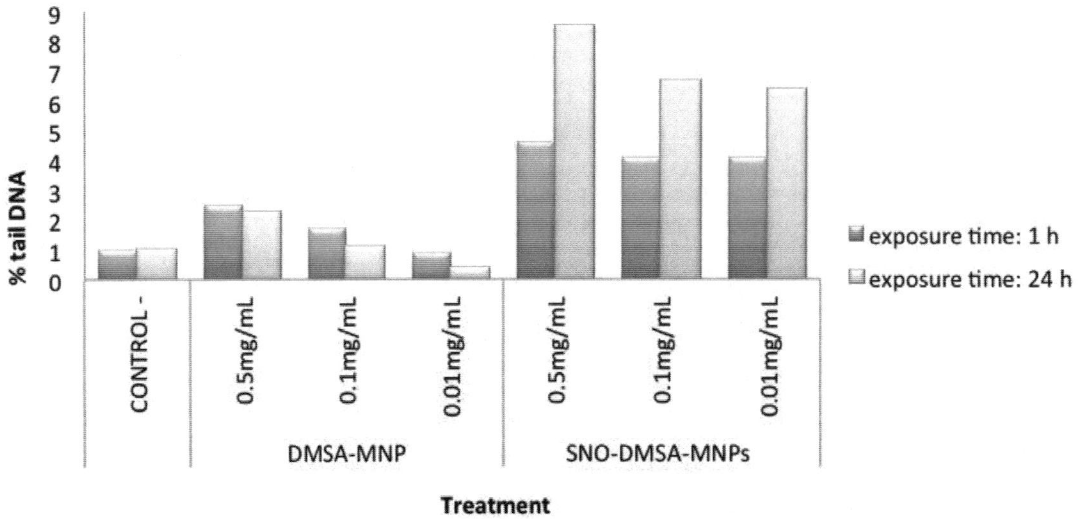

Figure 5. Results of DNA tail damage obtained for human lymphocyte cells after treatments with: DMSA-MNP (DMSA-coated magnetic nanoparticles) and SNO-DMSA-MNP (S-nitrosated DMSA-coated magnetic nanopartilces) at concentrations 0.5, 0.1 and 0.01 mg/mL. Exposure time: 1 and 24 h, as indicated in the Figure.

4. Conclusions

This work describes the genotoxicity of thiolated and nitrosated iron oxide MNPs. Thiol-coated MNPs showed no genotoxicity towards human lymphocyte cells, while nitroso-MNPs were found to lead to toxic effects on the cells, under the same experimental conditions. These results indicate the promising use of MSA or DMSA for coating MNPs aimed for biomedical applications, such as drug delivery. Moreover, due to the genotoxicity of SNO-MNPs, these nanoparticles might find important applications in cancer treatment, since these nanomaterials could be employed to deliver high amounts of NO directly to the target site of application, where NO can have its cytotoxic effects.

5. Acknowledgements

FAPESP, CNPq, CAPES, INOMAT (MCT/CNPq) and Brazilian Network in Nanotoxicology (MCT/CNPq). The authors would like to thanks Prof. Mauricio S. Baptista (IQ-USP) for the use of laboratory facilities and Prof. Carol H.Collins for revising the manuscript.

6. References

[1] Carpenter A W and Schoenfisch M H 2012 *Chem. Soc. Rev.* **41** 3742
[2] Cullota E and Koshland Jr. 1992 *Science* **258** 1861
[3] Serafim R A M, Primi M C, Trossini G H G and Ferreira E I 2012 *Curr. Med. Chem.* **19** 386
[4] Choudhari S K, Sridharan G, Gadbail A and Poornima V 2012 *Oral Oncol.* **48** 475
[5] Seabra A B and Duran N 2010 *J. Mat. Chem.* **20** 1624
[6] Majewski P and Thierry B 2007 *Crit. Rev. Solid State* **32** 203

[7] Mahmoudi M, Sant S, Wang B, Laurent S, Sen T, Lord O T, Walter M J, Dasgupta R, Walker D and Clark S M 2011 *Adv. Drug Delivery Rev.* **63** 24
[8] Klostergaard J and Seeney CE 2012 *Nanomed. Nanotechnol. Biol. Med.* **8** S37
[9] Li Y-F and Chen C 2011 *Small* **7** 2965
[10] Mahmoudi M, Hofmann H, Rothen-Rutishauser R and Petri-Fink A 2012 *Chem. Rev.* **112** 2323
[11] Cha Y J, Kim M J, Choa Y H, Kim J, Nam B, Lee J, Kim D H and Kim K 2010 *HIEEE Trans. Magn.* **46** 443
[12] Lima R ,Feitosa L, Maruyama C R Barga M A, Yamawaki P C, Vieira I J Teixeira E M, Corrêa A C, Mattoso L H C and Fraceto L F 2012 *Int. J. Nanomed.* **7** 3555
[13] Seabra A B, Martins D M, da Silva R, Simões M M S G, Brocchi M and de Oliveira MG 2010 *Artif. Organs* **34** E204
[14] Auffan M, Decome L, Rose J, Orsiere T, De Meo M, Briois V, Chaneac C, Olivi L, Le Franc J L B, Botta A, Wiesner M and Bottero J Y 2006 *Environ. Sci. Technol.* **40** 4367

Translocation of SiO2-NPs across in vitro human bronchial epithelial monolayer

I George[1], S Vranic[1], S Boland[1], M C Borot[1], F Marano[1], A Baeza-Squiban[1]

[1]*Laboratoire des Réponses Moléculaires et Cellulaires aux Xénobiotiques*, Unité Biologie Fonctionnelle et Adaptative, EAC CNRS 4413, Université Paris Diderot, Sorbonne Paris Cité, 5 rue Thomas Mann 75205 Paris cedex 13.

E-mail: isabelle.george.rmcx@gmail.com

Abstract. Safe development and application of nanotechnologies in many fields require better knowledge about their potential adverse effects on human health. Evidence of abilities of nanoparticles (NPs) to cross epithelial barriers and reach secondary organs via the bloodstream led us to investigate the translocation of SiO_2 NPs of 50 nm (50 nm-SiO_2-NPs) across human bronchial epithelial cells that are primary targets after exposure to inhaled NPs. We quantified the translocation of fluorescently labelled SiO_2 NPs at non-cytotoxic concentrations (5 and 10 µg/cm^2) across Calu-3 epithelial monolayer. After 14 days in culture Calu-3 cells seeded onto 3 µm-polycarbonate Transwell membranes formed an efficient bronchial barrier assessed by measurement of the transepithelial electric resistance and quantification of the permeability of the monolayer. After 24 hours of exposure, we observed a significant translocation of NPs that was more important when the initial NP concentration decreased. Confocal microscopy observations revealed NP uptake by cells and an important NP retention inside the porous membrane. In conclusion, 50 nm-SiO_2-NPs can cross the human bronchial epithelial barrier without affecting the integrity of the epithelial cell monolayer.

1. Introduction

The production of nanomaterials and their use in many fields has considerably increased in the last decade. Nanomaterials are used in a large variety of applications such as electronics, cosmetics, drug delivery etc. Their widespread use present a potential risk for human health due to unintentional environmental and/or occupational exposure [5].

The respiratory tract is one of the primary targets of NP exposure. It represents the major route for unwanted exposure with an important surface area of gas exchange that promotes NP penetration. According to their size, inhaled NPs could be deposited in three regions of the respiratory tract: nasopharyngeal, tracheobronchial and alveolar regions [10].

After inhalation, NPs can be eliminated by two types of clearance: i) mucociliary clearance for NPs trapped within the mucus in the airways and ii) alveolar-macrophage-clearance in the alveoli. However the uptake efficiency by alveolar macrophages is suspected to be lower for NPs than for larger particles [8]. Despite these defence systems, the smallest particles can persist throughout the respiratory system [5].

A key question in nanotoxicology concerns the NP abilities to cross epithelial barriers at the site of exposure [11]. Several biodistribution studies have shown that inhaled NPs can translocate through the respiratory barrier reaching the blood circulation and secondary organs. Although the amount of translocation is generally low (1% or less), chronic exposures could lead to an accumulation of biopersistent NPs in secondary organs [12, 13]. Translocation depends on NP size, composition, or surface charge [9]. For instance, a study of a panel of near infrared fluorescent NPs instilled in rats, have shown that NPs with a hydrodynamic size less than 34 nm rapidly moved from the lung to lymph nodes and that for particles below 34 nm, the presence of negative charges prevent their translocation [4].

With the increasing synthesis of a huge number of NPs differing in size, chemical composition or surface properties, a thorough assessment of their abilities to cross the different biological barriers has to be performed as well as the determination of the physicochemical characteristics favouring their translocation.

In this context, our aim was to develop an *in vitro* model of a human bronchial epithelial monolayer to study NP translocation. For this purpose we used the Calu-3 cell line grown in Transwell membranes and characterized for its ability to form an efficient barrier. Calu-3 monolayers were exposed for 24 hours to increasing concentrations of 50 nm-fluorescently labelled SiO_2 NPs used as a model of NPs. SiO_2 NP uptake by epithelial cells was investigated by confocal microscopy and their translocation through the epithelial monolayer was quantified measuring the fluorescence in the apical and basolateral compartments.

2. Materials and Methods

2.1. Nanoparticles

The fluorescein-doped 50 nm-SiO_2-NPs were synthesized as already described by *Vranic et al., 2012* [14]. Briefly, fluorescein isothiocyanate (FITC) was covalently attached to (3-aminopropyl)-trimethoxysilane (APS) by reaction of the amino group with the isothiocyanate group. 5 mg of FITC was dissolved in 5 mL of 42.7 mM of APS in ethanol. After 12 hours of stirring, the fluorescent silane was added to 250 mL ethanol, 5 mL TEOS (Tetraethyl orthosilicate), 7.6 mL ammonium hydroxide (28 %) and 10.9 mL water. The reaction was performed at 50°C for 12 hours in the dark under magnetic stirring. The so-prepared particles have an average diameter of about 30 nm. Further silica growth was performed to obtain the expected diameter of 50 nm and to make the NP more spherical with a pure silica surface. Dynamic light scattering and zeta potential values were measured by Zetasizer (nano ZS, Malvern Instruments, USA) providing hydrodynamic diameter distribution curves showing small agglomerates (242.7 nm) and a negative surface charge in culture medium (zeta-potential of -9.8 mV).

2.2. Cell culture

2.2.1. Calu-3 cells

Human lung adenocarcinoma (Calu-3) cells were purchased from the American Type Culture Collection (ATCC, Sigma-Aldrich, Saint Quentin Fallavier, France) and grown in DMEM/F12 (Dulbecco's Modified Eagle Medium) medium with phenol red (Life Technologies, Saint Aubin, France), containing 10% fetal calf serum (FCS, Life Technologies) and 1% glutaMAX (Life Technologies), subsequently referred to as complete cell culture medium. All experiments were performed with cells from passages 21 to 33. Cells were grown in T75-flasks (Costar, Sigma-Aldrich) for cellular expansion and onto 3 µm-polycarbonate Transwell membranes (12 mm in diameter) in two compartment chambers for translocation experiments. Cells were seeded at 500,000 cells/cm^2 on 3 µm polycarbonate Transwell membranes inside 12-wells plate (Costar, Sigma-Aldrich). 500 µL of complete culture medium were added in the apical chamber and 1,500 µL in the basolateral chamber. Cell cultures were maintained in a humidified atmosphere of 5% CO_2 and 95% air at 37°C.

2.2.2. Assessment of cell confluence

To assess Calu-3 confluence, the TransEpithelial Electrical Resistance (TEER) was measured and the monolayer permeability to Lucifer Yellow (LY) (Sigma-Aldrich, Saint Quentin, France) was assessed three times per week, during 14 days.

Before TEER measurements, cells were rinsed by fresh complete cell culture medium. After 1 hour of incubation at 37°C in fresh culture medium, the TEER values were determined using the electronic circuit of the Epithelial VoltOhmMeter (EVOM) and the STX2 electrode (World Precision Instruments, Hertfordshire, United Kingdom). Mean TEER value of ten wells was calculated. The mean resistance of a cell-free Transwell insert was subtracted from the resistance measured across each cellular monolayer to yield the TEER value of the cell monolayer.

For LY measurements, cells were rinsed by HBSS (with calcium and with magnesium) and incubated for 1 hour at 37°C with 500 µL of LY (0.2 µg/mL final in HBSS). FluoSart Galaxy (BMG, Ortenberg, Germany) was used to determine fluorescence (485 nm / 520 nm) of apical and basolateral media.

2.2.3. Viability assay

Viability of cell cultures was determined by the mitochondrial metabolic assay, WST-1 according to the method described by *Hussain et al., 2010* [7].

2.3. Translocation studies

14-days confluent cultures were rinsed with HBSS (Sigma-Aldrich) to eliminate trace amounts of FCS. 50 nm-SiO_2-NP stock (23.2 mg/mL in water) was vortexed shortly before making final dilutions (5 or 10 µg/cm^2 corresponding to 10 or 20 µg/mL) in complete cell culture medium without phenol red and FCS.

After 24 hours of exposure, 100 µL of apical and basolateral media were deposited inside a white 96-well plate with clear bottom (Greiner, Courtaboeuf, France) as well as serial dilutions of 50 nm-SiO_2-

NPs (in a concentration range of 0 to 10 µg/cm^2). Fluorescence was quantified with Infinite 200 Pro (TECAN, Lyon, France). 50 nm-SiO_2-NPs were excited at 488 nm and detected at 521 nm.
Blank fluorescence value (medium without NPs) was removed from each fluorescence value obtained. Fluorescence values were converted in concentration values or in percentage values through the use of a standard curve. Results for the "filter" or "cells/filter" chambers were determined by subtracting values of apical and basal media from the initial NP concentration applied. Membranes were removed and mounted for imaging. For confocal microscopy experiments without cells, the membrane was observed through a reflection mode.

2.4. Immunofluorescence and Microscopy

After 24 hours of exposure, cells were fixed with paraformaldehyde 4% in PBS (Santa-Cruz, Heidelberg, Germany) for 20 minutes at room temperature and incubated for 10 minutes with NH_4Cl (50 mM, Sigma-Aldrich), permeabilized in 0.05% PBS-Tween 20 (Sigma-Aldrich) for 5 minutes. Fixed cells were incubated over night at 4°C with either MUC5AC or ZO-1 primary antibodies (Life Technologies) diluted to 1:500 followed by secondary antibodies stained by Alexa 488 or 568 fluorochromes (1:500, Life Technologies) for 2 hours at room temperature. For actin staining, fixed cells were incubated for 30 minutes with TRITC-phalloidin (0.9 nM in PBS, Life Technologies). Cell nuclei were stained with DAPI (4',6-Diamidino-2-Phenylindole, Dihydrochloride, 0.25 µg/mL in PBS, Life Technologies) for 1 minute. After mounting in polyvinyl alcohol mounting medium with DABCO (Sigma-Aldrich), cells were examined under a Zeiss LSM 710 confocal microscope using objectives 40X and 63X. Image treatment was done with Image J software (Image J 1.42 NIH, USA).

2.5. Transmission Electron Microscopy

Cell cultures were fixed and analyzed by transmission electron microscopy according to the method described by *Hussain et al., 2010* [7].

2.6. Statistical analysis

Every experiment was repeated at least twice with triplicates of each condition. Data are represented as means ± SD and were analyzed on commercially available software SigmaStat (Version 3.0, Systat software Inc, San Jose, California, USA) analysis of variance (one-way ANOVA) followed by Dunnett's test for multiple comparisons with $p<0.05$ (two tailed) considered as significant.

3. Results

3.1 50 nm-SiO_2-NP distribution across 3 µm-polycarbonate Transwell membranes in absence of cells

Translocation could be studied using an epithelial monolayer seeded onto a porous membrane in an insert delimiting two chambers allowing to get access to the lower (basolateral) chamber where NPs translocate after their application in the upper (apical) chamber. Membranes of different porosities are commercially available but the choice of the porosity is dictated by their abilities to allow NP translocation and cell development. Low porosity (0.4 µm) is better for cellular growth but proved to prevent an efficient Transwell crossing of the 50 nm-SiO_2-NPs (data not shown). We evaluate the suitability of 3 µm-polycarbonate Transwell membranes by quantifying NP concentrations in the apical and the basolateral chambers and by studying the NP behaviour inside the porous membrane by confocal microscopy after 24 hours of incubation.

As shown in Figure 1, 50 nm-SiO_2-NPs can cross the 3 µm-polycarbonate Transwell membranes after 24 hours of treatment. The NP concentration in the basolateral chamber increased as the NP concentration initially applied in the apical chamber increased (Fig. 1A). By contrast when results were expressed as percentage of translocation, we noted that the higher initial apical concentration led to a lower percentage of translocation. For concentrations of 5 or 10 µg/cm², we determined a translocation of 4.04% and 3.33% respectively (Fig. 1B) and an important trapping inside the filter of 91.08% and 92.47% respectively (Fig. 1B).

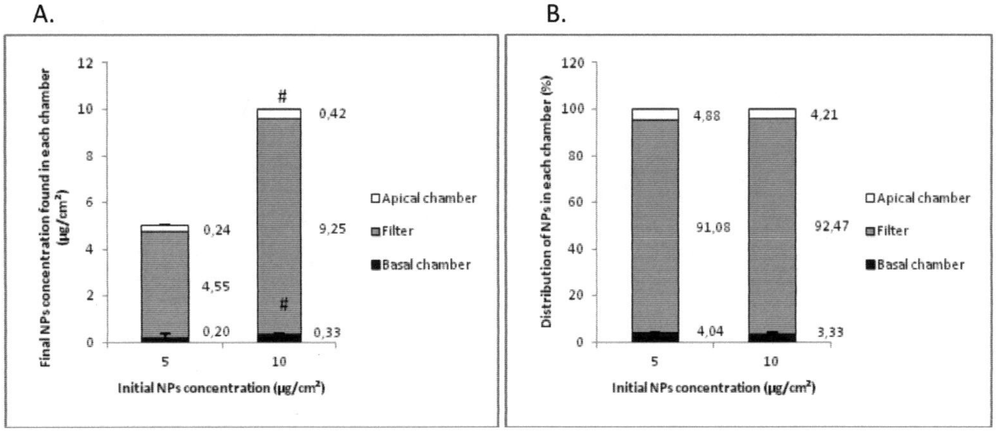

Figure 1: **50 nm-SiO_2-NP translocation study across 3 µm-polycarbonate Transwell membranes without cells, after 24 hours of incubation**. (A) 50 nm-SiO_2-NP concentrations found in each chamber and in the porous membrane (filter) (in µg/cm²); (B) 50 nm-SiO_2-NP distribution in each chamber (in percentage). *n=3 for each condition; # indicates that the condition 10 µg/cm² is statistically different from the condition 5 µg/cm² ($p<0.05$)*

NP retention inside 3 µm-polycarbonate Transwell membranes after 24 hours of incubation was confirmed by confocal microscopy observations. Figure 2 shows a large amount of NPs on the top of the membrane and to a lower amount inside and at the bottom of the membrane.

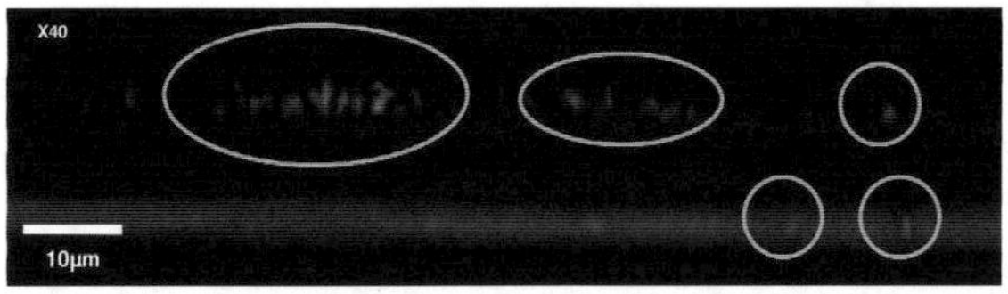

Figure 2: **50 nm-SiO_2-NPs retention on the top and on the bottom of a 3 µm-polycarbonate Transwell membrane** after an incubation of 24 hours at the concentration of 10 µg/cm². SiO_2-NPs labelled in green and filter in red.
Scale bar: 10 µm; X40.

3.2 Calu-3 epithelial monolayer

In order to evaluate the capacity of Calu-3 cells seeded onto 3 µm-polycarbonate Transwell membranes to form an efficient barrier, a time course study was performed over 14 days after seeding measuring the TEER and the epithelial permeability (Fig. 3A). The 3rd day TEER-value was around 100 Ω/cm² and progressively increased to reach 1,200 Ω/cm² at the 14th day. The passage of LY in the basolateral chamber progressively decreased from 800 AU at the 3rd day to 0 at the 14th day demonstrating an absence of paracellular transport across the cellular monolayer (Fig. 3A). The TEER increase associated to the decrease of LY flux over time shows the establishment of an efficient Calu-3 barrier. These functional studies were confirmed by ultrastructural observations of Calu-3 cells which exhibited cellular junctions between cells (Fig. 3B).

A. B.

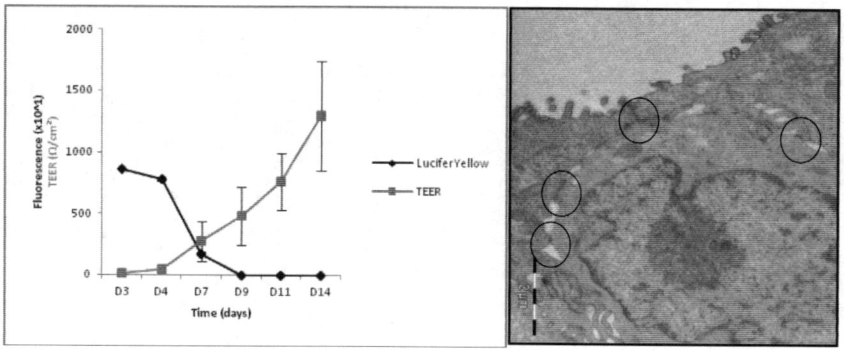

Figure 3: (A) **TEER and Lucifer Yellow measurements.** Calu-3 cells grown onto 3 µm-polycarbonate Transwell membranes for 3 to 14 days ; (B) **Observation by transmission electron microscopy of the ultrastructure of Calu-3 cells** seeded onto 3 µm-polycarbonate Transwell membranes after 14 days of culture. The circles show different types of junctions.

Particularly, tight junctions are present as shown by confocal microscopy observations of the specific immunostaining with ZO-1, a polypeptide exclusively associated with tight junctions (Fig. 4A), in a confluent cellular monolayer shown by actin filament staining (Fig. 4B). Moreover, the culture of Calu-3 onto 3 µm-polycarbonate Transwell membranes allowed their differentiation as they express the mucin MUC5AC (Fig. 4C). For further studies, Calu-3 cells were only used after 14 days of culture.

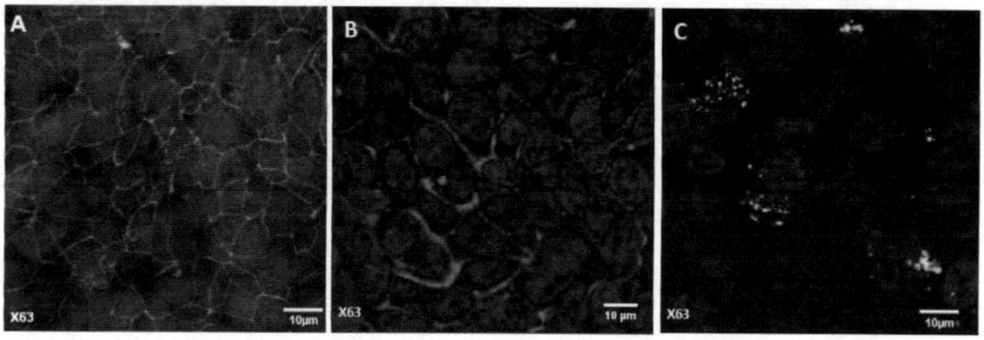

Figure 4: **Confocal microscopy observation of Calu-3 cells** seeded onto 3 µm-polycarbonate Transwell membranes after 14 days of culture. (A) Tight junctions labelled in green by ZO-1 specific antibody (B) Actin filaments in red with TRITC-phalloidin and (C) Mucus in green by MUC5AC specific antibody. Cell nuclei labelled in blue with DAPI. *Scale bar: 10 µm*

3.3 50 nm-SiO_2-FITC-NP distribution across Calu-3 epithelium

A viability assay was performed showing that NPs were not cytotoxic for Calu-3 cells (data not shown). Furthermore observations by confocal microscopy revealed that the integrity of the monolayer was conserved although NPs were present inside the cells. This is confirmed by the staining of actin filaments which preserve a proper structure (Fig. 5B & 5D). Moreover NP uptake seemed to increase with the exposure-concentration (Fig. 5A & 5C). Orthogonal views of Z-stack merge showed the presence of NPs around cell nucleus and close to the basolateral level of the cell (Fig. 5B & 5D).

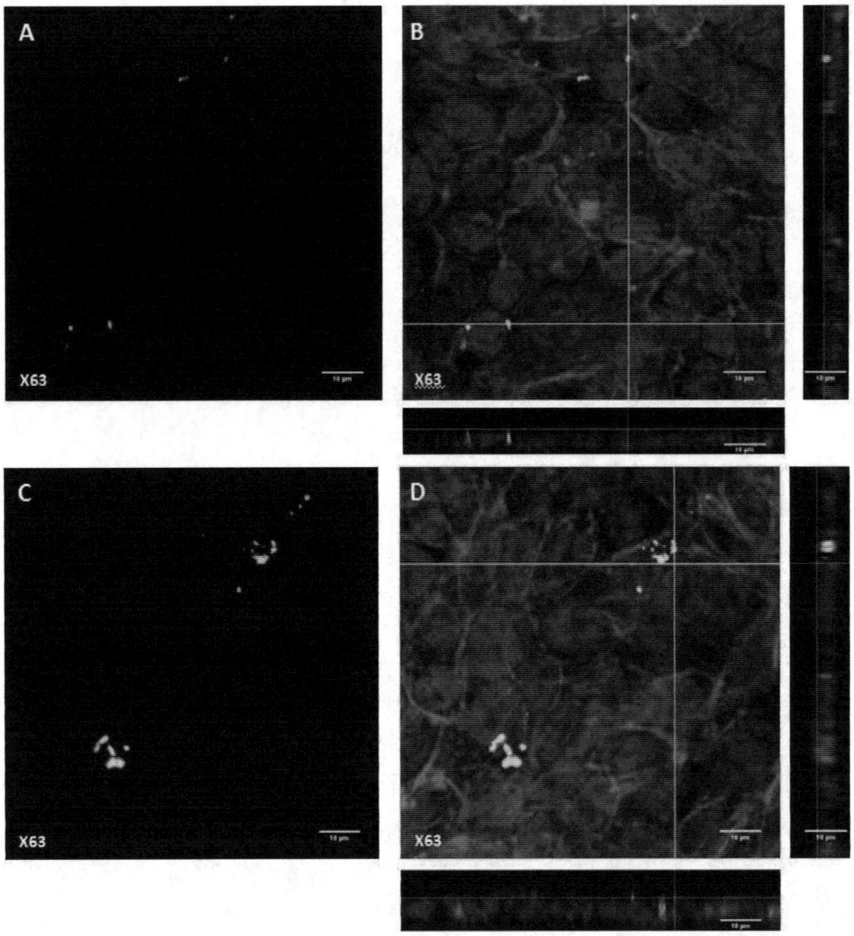

Figure 5: **50 nm-SiO_2-NP detection inside Calu-3 cells seeded onto 3 µm-polycarbonate Transwell membranes** after an exposition of 24 hours at 5 µg/cm² (**A**: z-stack merge of green channel & **B**: z-stack merge of blue, red and green channels + Orthogonal Views) and 10 µg/cm² (**C**: z-stack merge of green channel & **D**: z-stack merge of blue, red and green channels + Orthogonal Views). NPs labelled in green, cell nuclei in blue, and actin filaments in red. *Scale bar: 10 µm; X63.*

After studying NP internalization by Calu-3 cells, NP translocation studies were carried out. A NP crossing through the Calu-3 monolayer model was observed. Basolateral NP concentrations were similar whatever the initial NP concentration was, around 0.2 µg/cm² (or 0.4 µg/mL) (Fig. 6A), and

NPs were mainly trapped within the "cells/filter" compartment (Fig. 6B). The percentage of NP translocation was 1.8 times more important when the cells were treated at 5 µg/cm² compared to a treatment at 10 µg/cm² (Fig. 6B).

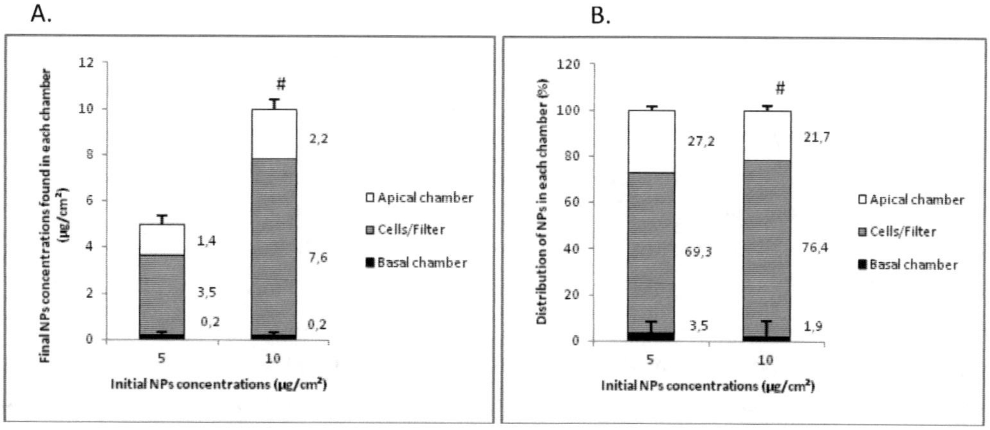

Figure 6: **50 nm-SiO$_2$-NP translocation study across 3 µm-polycarbonate Transwell membranes with confluent cultures after 24 hours of treatment.** (A) 50 nm-SiO$_2$-NP concentrations found in each compartment (in µg/cm²); (B) 50 nm-SiO$_2$-NP distribution in each compartment (in percentage). *n=3 for each condition; # indicates that the condition 10 µg/cm² is statistically different from the condition 5 µg/cm² (p<0.05)*

4. Discussion

In this study we provided evidence that Calu-3 cell line grown on 3 µm-polycarbonate Transwell membranes is a valuable model to investigate the NP translocation through the bronchial epithelial monolayer. We have shown that 50 nm-SiO$_2$-NPs were internalized and translocated to a low but significant amount after 24 hours of exposure.

To easily quantify NP translocation without time-consuming and cost-efficient techniques, we have chosen to use fluorescently labelled NPs. By comparison with other quantitative techniques such as ICP-OES (Inductively Coupled Plasma Optical Emission Spectrometry) based on elemental analysis, the use of fluorescently labelled NPs allows a combination of experiments such as measurements of NP flux and morphological studies of NP internalization. 50 nm-SiO$_2$-NPs were chosen as a relevant NP model as they can be easily labelled with fluorochromes. Moreover they can be considered as relevant for human exposure as they are widely used in different industrial applications as well as in nanomedicine [2].

To characterize the NP abilities to cross physiological barriers after different routes of exposure, *in vitro* models have to be developed. To mimic the lung barrier, different cell lines were available such as Calu-3, NCI-H292, 16HBE14o$^-$ or A549. We have chosen the Calu-3 cell line as the other cell lines did not exhibit effective tight junctions when grown on 3 µm Transwell membranes [6] (*Isabelle George personal communication*). By contrast Calu-3 cells have the ability to form an efficient barrier developing tight junctions as we demonstrated by morphological (transmission electron microscopy and ZO-1 immunolabelling) and functional studies (TEER and LY flux). Moreover the Calu-3 cell line is known to express xenobiotic metabolizing enzymes such as CYP1A1 or CYP2B and to produce mucus (and more specifically the MUC5AC mucin) [1] as we confirmed in our cultures.

The use of 3 µm-polycarbonate Transwell membranes was a prerequisite as 0.4 µm membranes prevented 50 nm-SiO_2-NPs Transwell crossing (data not shown). A lower NP translocation through 0.4 µm-polycarbonate Transwell membranes was also observed in studies using Fluoresbrite® polystyrene latex NPs [3] or amine-modified or carboxyl-modified polystyrene beads [6]. Such membrane porosity seemed to be a good compromise as it allows combining a proper development of the Calu-3 cells with a detectable NP translocation. Nevertheless a large amount of NPs (around 91%) was retained inside the membrane, leading to an underestimation of the real NP flux to the basolateral chamber. The amount of NP translocation through the 3 µm-polycarbonate Transwell membranes was also shown to be low for 37 nm Fluoresbrite® polystyrene latex NPs [3] but to be more important for 46 nm amine-modified or carboxyl-modified polystyrene beads suspended in the presence of 1% bovine serum albumin (BSA) [6]. These data suggest that the NP composition, their size as well as conditions of exposure influence NP translocation. Moreover, we noticed that increasing NP concentrations reduced NP translocation, suggesting that at high concentrations NP agglomeration could be favoured preventing the efficient crossing of the membrane by NPs. This also shows that an increase of NP flux in the basolateral compartment cannot be achieved by increasing initial NP concentrations. Modifying the suspension medium adding BSA as *Geys et al., 2006* or serum could increase NP translocation [6].

Using not cytotoxic concentrations, confocal microscopy experiments have revealed that 50 nm-SiO_2-NPs are internalized by the Calu-3 monolayer without inducing structural modifications of actin filaments after 24 hours of exposure. This is in agreement with another study using the same 50 nm-SiO_2-NPs that were shown to be dose-dependently internalized by NCIH-292 cells grown as islets on plastic (*Sandra Vranic personal communication*).

Our results demonstrated that exposing Calu-3 monolayers to relevant NP concentrations induced a low but a significant NP translocation (0.2 µg/cm^2 corresponding to 0.4 µg/mL). Considering the important NP retention inside the porous membrane that we noticed with experiments of NP flux in the absence of cells, the real translocation would likely be higher. However the respective NP distribution inside cells and membranes is unknown as the amount of NPs trapped in cells remains to be measured. The 1.9 to 3.5% of translocation that we observed through Calu-3 monolayer and membrane with 50 nm-SiO_2-NPs is in the same range (6%) as the one observed with the same cell line for 46 nm amine-modified or carboxyl-modified polystyrene beads [6].

To conclude, we have shown that the Calu-3 cell line proved to be a relevant *in vitro* model for the study of NP translocation across an *in vitro* human bronchial epithelial monolayer. Experiments are in progress to investigate the role of NP physicochemical characteristics and NP environment in NP translocation.

5. Acknowledgement

This work was supported by grants from Anses project N° PNR-EST-2010/2/79 (TiSiTrans) and National Grant "Nanotrans" (N°EST-2010/2/079 *and by Région Ile de France / DIM SEnT for I. George PhD funding and FP7 project ENPRA* (n°228789) *for S. Vranic PhD funding. We acknowledge S. Mornet for the NP synthesis and the ImagoSeine BioImaging Core Facility for confocal and transmission electron microscopy at the Institute Jacques Monod, University Paris Diderot, France.*

6. References

[1] Borchard G 2002 Calu-3, a valid model for the airway epithelium ? (INIST-CNRS: Editions de santé, Paris, FRANCE (1991-2003); (Revue)) pp pp. 205-11

[2] BéruBé K, Gibson C, Job C and Prytherch Z 2011 Human lung tissue engineering: a critical tool for safer medicines. *Cell Tissue Bank* **12** 11-3

[3] Cartwright L, Poulsen M S, Nielsen H M, Pojana G, Knudsen L E, Saunders M and Rytting E 2012 In vitro placental model optimization for nanoparticle transport studies *Int J Nanomedicine* **7** 497-510

[4] Choi H S, Ashitate Y, Lee J H, Kim S H, Matsui A, Insin N, Bawendi M G, Semmler-Behnke M, Frangioni J V and Tsuda A 2010 Rapid translocation of nanoparticles from the lung airspaces to the body. *Nat Biotechnol* **28** 1300-3

[5] Geiser M and Kreyling W G 2010 Deposition and biokinetics of inhaled nanoparticles. *Part Fibre Toxicol* **7** 2

[6] Geys J, Coenegrachts L, Vercammen J, Engelborghs Y, Nemmar A, Nemery B and Hoet P H 2006 In vitro study of the pulmonary translocation of nanoparticles: a preliminary study. *Toxicol Lett* **160** 218-26

[7] Hussain S, Thomassen L C, Ferecatu I, Borot M C, Andreau K, Martens J A, Fleury J, Baeza-Squiban A, Marano F and Boland S 2010 Carbon black and titanium dioxide nanoparticles elicit distinct apoptotic pathways in bronchial epithelial cells *Part Fibre Toxicol* **7** 10

[8] Kreyling W G, Semmler-Behnke M, Takenaka S and Möller W 2012 Differences in the Biokinetics of Inhaled Nano- versus Micrometer-Sized Particles *Acc Chem Res*

[9] Lipka J, Semmler-Behnke M, Sperling R A, Wenk A, Takenaka S, Schleh C, Kissel T, Parak W J and Kreyling W G 2010 Biodistribution of PEG-modified gold nanoparticles following intratracheal instillation and intravenous injection. *Biomaterials* **31** 6574-81

[10] Oberdörster G, Oberdörster E and Oberdörster J 2005 Nanotoxicology: an emerging discipline evolving from studies of ultrafine particles *Environ Health Perspect* **113** 823-39

[11] Pietroiusti A, Campagnolo L and Fadeel B 2012 Interactions of Engineered Nanoparticles with Organs Protected by Internal Biological Barriers *Small*

[12] Semmler M, Seitz J, Erbe F, Mayer P, Heyder J, Oberdörster G and Kreyling W G 2004 Long-term clearance kinetics of inhaled ultrafine insoluble iridium particles from the rat lung, including transient translocation into secondary organs *Inhal Toxicol* **16** 453-9

[13] Semmler-Behnke M, Takenaka S, Fertsch S, Wenk A, Seitz J, Mayer P, Oberdörster G and Kreyling W G 2007 Efficient elimination of inhaled nanoparticles from the alveolar region: evidence for interstitial uptake and subsequent reentrainment onto airways epithelium. *Environ Health Perspect* **115** 728-33

[14] Vranic S, Garcia-Verdugo I, Darnis C, Sallenave J M, Boggetto N, Marano F, Boland S and Baeza-Squiban A 2013 Internalization of SiO(2) nanoparticles by alveolar macrophages and lung epithelial cells and its modulation by the lung surfactant substitute Curosurf(®) *Environ Sci Pollut Res Int*

Nanosafe 2012: International Conferences on Safe Production and Use of Nanomaterials IOP Publishing
Journal of Physics: Conference Series **429** (2013) 012023 doi:10.1088/1742-6596/429/1/012023

Development of a dose-controlled multiculture cell exposure chamber for efficient delivery of airborne and engineered nanoparticles

Akrivi Asimakopoulou[1], Emmanouil Daskalos[1], Nastassja Lewinski[2], Michael Riediker[2], Eleni Papaioannou[1], Athanasios G. Konstandopoulos[1,3]

[1]Aerosol and Particle Technology Laboratory, CPERI/CERTH, P.O. Box 60361, 57001, Thessaloniki, Greece, [2]Institut universitaire romand de Santé au Travail, IST, Rue du Bugnon 21, 1011, Lausanne, Switzerland, [3]Department of Chemical Engineering, Aristotle University, P.O. Box 1517, 54006, Thessaloniki, Greece

[*]e-mail: asimak@cperi.certh.gr

Abstract. In order to study the various health influencing parameters related to engineered nanoparticles as well as to soot emitted by Diesel engines, there is an urgent need for appropriate sampling devices and methods for cell exposure studies that simulate the respiratory system and facilitate associated biological and toxicological tests. The objective of the present work was the further advancement of a Multiculture Exposure Chamber (MEC) into a dose-controlled system for efficient delivery of nanoparticles to cells. It was validated with various types of nanoparticles (Diesel engine soot aggregates, engineered nanoparticles for various applications) and with state-of-the-art nanoparticle measurement instrumentation to assess the local deposition of nanoparticles on the cell cultures. The dose of nanoparticles to which cell cultures are being exposed was evaluated in the normal operation of the *in vitro* cell culture exposure chamber based on measurements of the size specific nanoparticle collection efficiency of a cell free device. The average efficiency in delivering nanoparticles in the MEC was approximately 82%. The nanoparticle deposition was demonstrated by Transmission Electron Microscopy (TEM). Analysis and design of the MEC employs Computational Fluid Dynamics (CFD) and true to geometry representations of nanoparticles with the aim to assess the uniformity of nanoparticle deposition among the culture wells. Final testing of the dose-controlled cell exposure system was performed by exposing A549 lung cell cultures to fluorescently labeled nanoparticles. Delivery of aerosolized nanoparticles was demonstrated by visualization of the nanoparticle fluorescence in the cell cultures following exposure. Also monitored was the potential of the aerosolized nanoparticles to generate reactive oxygen species (ROS) (e.g. free radicals and peroxides generation), thus expressing the oxidative stress of the cells which can cause extensive cellular damage or damage on DNA.

1. Introduction

Research on airborne nanoparticles has already proven that they can have adverse impacts on health, affecting primarily the respiratory and the cardiovascular systems [1-3]. This body of research naturally has led to increased concerns on the potential of Engineered Nanoparticles (ENP) to also cause adverse health effects and occupational safety impacts, since increasing use of ENP is caused by the growing nanotechnology market. As any adverse public or ethical reaction to nanotechnology-based products is thus bound to create a very large economic impact, the scientific assessment of the potential health risks associated with nanotechnology-based materials and products is of utmost importance. Exposure scenarios to ENP and risk assessment at the workplace by EU organizations [4] and/or leading chemical industries [5] indicate that the potential health risk of ENP is mostly concerned with inhalation exposure of ENP dusts or aerosols (if e.g. sprayed), while dermal exposure of industrially relevant ENP formulations was found to pose essentially no risk [6]. Several studies [7, 8] point to the high importance of redox interactions between nanoparticles and cells and the relevance of oxidative stress as a very appealing paradigm for discriminating the adverse effects of different nanoparticles at the cellular and molecular level [9]. ENP made from mixed/doped oxides with varying redox activities find numerous applications in catalytic engineering, energy production, composite materials, electronic and optical devices and are largely unstudied in the context of health impacts compared with commodity ENP such as titania, silica and carbon black. Exposure to such redox ENP may eventually result in systemic inflammation and potentially promote the progression of atherosclerosis and precipitate acute cardiovascular responses as in the case of airborne incidental nanoparticles [10].

Since main contact with nanoparticles occurs during inhalation at the lung epithelium, there is an urgent need for appropriate sampling devices and methods for *in vitro* cell exposure studies. Traditionally, these *in vitro* experiments have been performed under liquid cover conditions, where nanoparticles are added to the culture medium [11]. Despite the simple and fast toxicity screening procedure that these experiments may offer, liquid cover conditions do not resemble the real exposure of the lung to airborne nanoparticles which occurs at air-liquid interface (ALI) *in vivo*. Moreover, the dose of nanoparticles which finally reach the cell culture remains undefined as the nanoparticles are not directly applied to the cells [12]. On the other hand, direct exposure of the cell cultures to aerosolized nanoparticles at the air-liquid interface not only represents more realistic conditions of lung exposure, but it also overcomes the uncertainty of the dose determination. The nanoparticles remain unchanged before deposition and directly contact the layer of cells which are grown on a porous membrane and are fed nutrients from below. This configuration better resembles the *in vivo* condition of exposure of airway epithelial cells to nanoparticles.

A variety of in-house and commercially available ALI cell culture exposure systems have been described and tested in the literature and summarized in recent papers (e.g. [13, 14]). In most of these techniques, the design is relying on diffusion and/or gravitational settling deposition mechanisms of the nanoparticles, as the aerosol flows directly to the cell culture surface by an inlet tube ending above it (basic configuration of commercial CULTEX® exposure system) using a stagnation point flow system [15]. Despite the advantageous operation compared with the liquid cover exposure, in this configuration the deposition efficiency of nanoparticles is typically low over a broad size range of 50 to 500 nm in (mobility) diameter; deposition efficiencies from 0.7% [16] to 2% [17] (for 200nm diameter particles) have been reported in the literature for systems with the aforementioned deposition mechanisms.

In MAAPHRI project [9], a specific concept for a cell culture exposure chamber had been introduced to allow the uniform exposure of cell cultures to Diesel aerosols (Figure 1a). The concept of this design is based on the stagnation point flow, but only to direct the aerosol flow through an inlet slit to be parallel to the cell culture surface, resembling in a more realistic way the *in vivo* lung exposure [18]. This design has been improved in an upgraded Multiculture Exposure Chamber (MEC) for *in vitro* continuous flow cell exposure tests (compared to the previous MAAPHRI version) regarding the geometrical arrangement of exposed cell cultures in order to increase the throughput screening possibility by 50%. Thus, the newest version of MEC can accommodate testing of up to 36 cell cultures in comparison with the 24 cell cultures in MAAPHRI concept (Figure 1b) (considering 6-well plates). The objective of the present work is the further advancement of the aforementioned MEC into a dose-controlled system for efficient delivery of nanoparticles to cells and its validation with various types of nanoparticles (Diesel engine soot aggregates, engineered nanoparticles for various applications) with state-of-the-art nanoparticle measurement instrumentation to assess the local deposition of nanoparticles on the cell cultures.

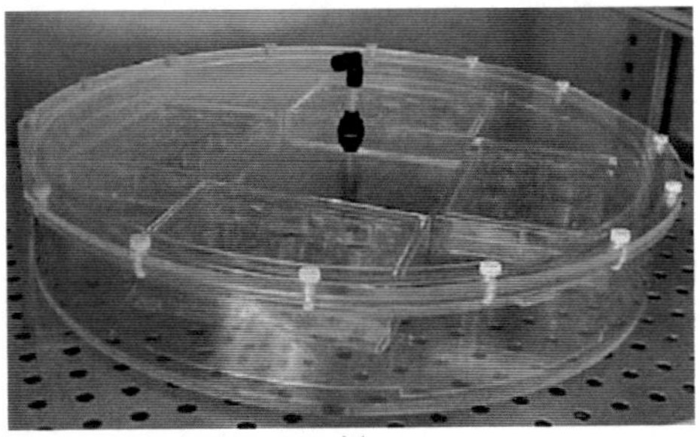

(a)

(b)

Figure 1. Multiculture Exposure Chamber (MEC) for *in vitro* continuous flow cell exposure tests with a) 24 cell cultures (MAAPHRI version), b) 36 cell cultures (considering the case of 6-well plates in both versions)

2. Materials and methods

2.1. Exposure system with Diesel engine exhaust

In order to characterize and evaluate the Multiculture Exposure Chamber, exposure experiments of a free of cells device to Diesel engine exhaust have been performed considering the experimental setup shown below (Figure 2).

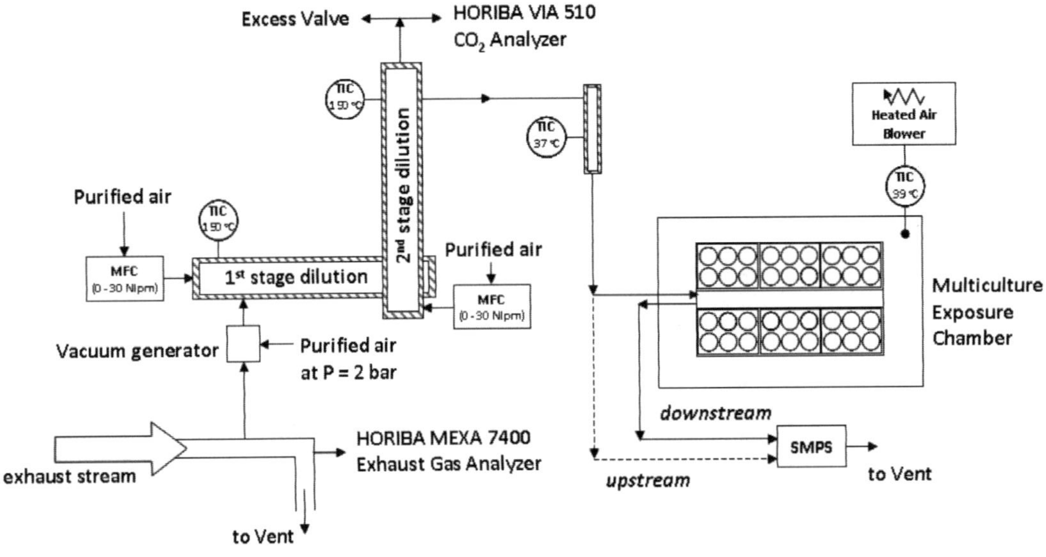

Figure 2. Schematic view of the exposure system with Diesel engine exhaust

A sample from the Diesel engine exhaust (at a steady engine operation point: speed and torque, 2250 rpm and 101 Nm, respectively) is being diluted (dilution ratio 1:250) and thermally pretreated before it enters the exposure chamber. The dilution of the sample is considered as a crucial experimental factor and the final dilution ratio is selected in order to adequately simulate the "real" exposure of human lung cells to the environmental particulate pollution [18]. The temperature of the exposure chamber is constantly kept at 37±1°C with the aid of a specially constructed oven. Soot particle concentration and size distribution in the aerosol phase from Diesel engine exhaust are being recorded "upstream" and "downstream" of the exposure chamber with a Scanning Mobility Particle Sizer (SMPS, TSI model 3936). The total particle concentration is being recorded with a Condensation Particle Counter (CPC, TSI model 3022). The diluted upstream total soot particle concentration is $2.5{\cdot}10^5 \#/cm^3$, in average. The exhaust flowrate is controlled at 2.2 L/m and the duration of the exposure experiment is 3 hours.

2.2. Flow visualization experiments

In order to assess the uniformity of the flow distribution as well as soot the particle deposition, the performance of the exposure chamber was studied using Computational Fluid Dynamics (CFD) and particle transport simulation within the chamber volume in a previous study [19]. In the present study, flow visualization experiments have been performed in order to facilitate our visual perception of the aerosol velocity uniformity in the MEC by adding a tracer. More specifically, a gaseous stream of colored (blue) smoke is driven in the MEC at the flowrate 2.2 L/m. The flowrate is controlled by means of an ejector type vacuum generator in the exit of MEC (Figure 3). The distribution of the flow inside the MEC is being recorded by a digital camera through the transparent top ports of the MEC.

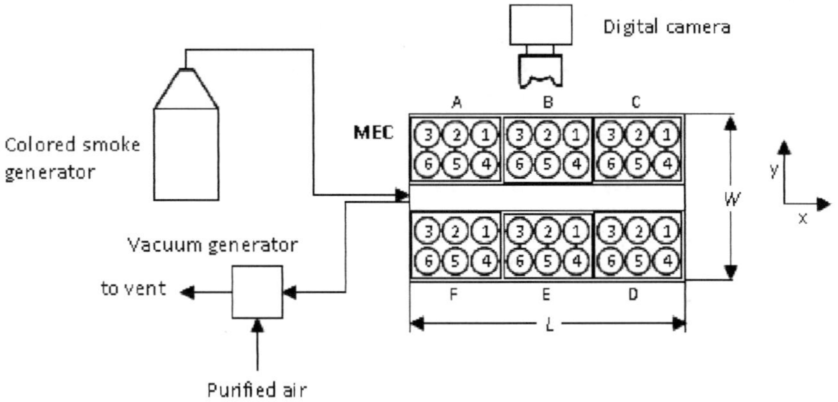

Figure 3. Schematic view of the flow visualization experiment setup

In these experiments only aerosol flow distribution and velocity are being visualized and not the particle deposition in the MEC, since, firstly, the smoke particles are much larger than the nanoparticles under investigation (Diesel engine soot nanoparticles and/or ENP) and, then, the concentration is much higher than the one used in the exposure experiments.

2.3. Control of particle deposition

During the acellular exposure experiments described in section 2.1, samples of Diesel engine soot nanoparticles are collected to be analyzed by Transmission Electron Microscopy (TEM, Jeol JEM 2010). Holey Carbon Copper TEM grids (EMS®) were placed in the 6-well plate inserts either in empty wells, or in wells containing double distilled water. In both cases, the exact position of the TEM grids resembles the level of the cell surface; proper sample holders are placed in the wells to "elevate" the TEM grids in the desirable height. The purpose of this study is to evaluate the deposition of nanoparticles on the cell surface.

2.4. Exposure system with nebulized nanoparticles

Characterization of the MEC for exposure experiments on cultured lung cells to aerosolized ENPs have been performed using the experimental setup shown below (Figure 4).

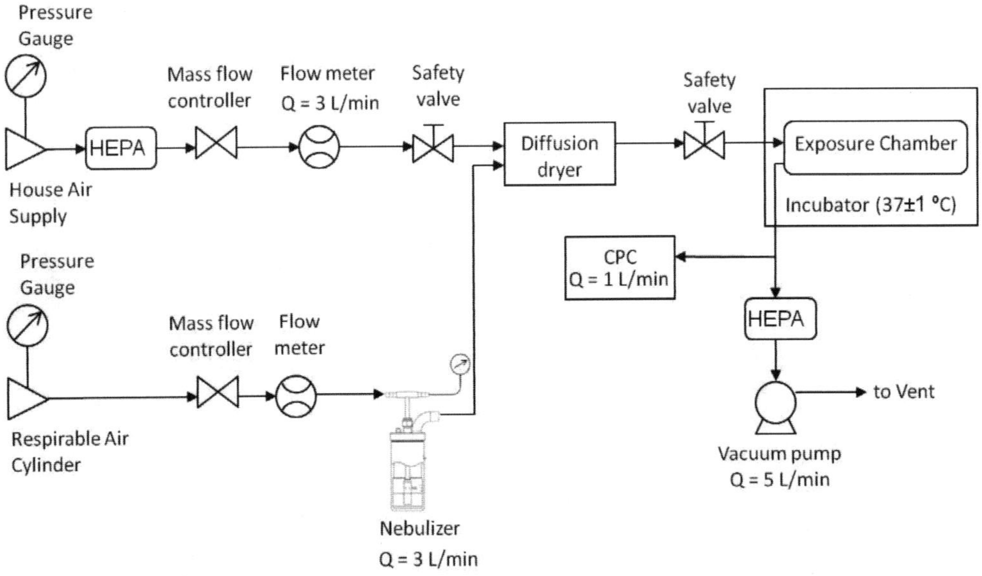

Figure 4. Schematic view of the experiment setup for cell exposures to aerosolized nanoparticles

A 100 mL solution of $20 \pm 2 \cdot 10^8$ #/mL Fluoprobe532 (Interchim) labeled silica nanoparticles (Kisker Biotech) was loaded into a 1-jet Collison nebulizer (BGI Inc.) to generate the ENP aerosol. The ENP hydrodynamic diameter of 89±7 nm was determined by nanoparticle tracking analysis (NanoSight Ltd).

2.5. Cell exposure experiments

A549 cells (American Type Culture Collection, ATCC) were grown at 37°C in a 5% CO_2 atmosphere using Dulbecco's Modified Eagle Medium supplemented with 1% penicillin / streptomycin and 10% fetal bovine serum (FBS). The cell line was maintained in 25 cm^2 culture flasks until confluent, rinsed once with 1×PBS, and incubated with trypsin-EDTA for 5 min at 37°C in a 5% CO_2 atmosphere. The trypsin-EDTA was then neutralized with culture medium and the cells were subsequently counted using a hemocytometer. A cell suspension with an estimated concentration of $5 \cdot 10^5$ cells/mL was prepared and transwells designed to fit 6 well plates were seeded with 1 mL of cell suspension. Cells were cultured on transwells under liquid cover for the first 24 hours. Afterwards, the culture media on the apical side was removed and the cells were incubated for an additional 24 hours. One hour prior to loading in the MEC, the cells were incubated with 10 μM 2,7-dichlorodihydrofluorescein diacetate (DCFH-DA), a commonly used fluorescent probe for reactive oxygen species detection, in 1×Hank's buffered salt solution (HBSS, Gibco). The dye solution was then removed, the well plates reconstituted with HBSS on the basolateral side of the transwells, and the well plates loaded into the MEC. The cells were exposed for 1 hour and then removed for fluorescence readings (DCF excitation/emission @ 485/530 nm and Fluoprobe532 excitation/emission @ 530/555 nm) using a microplate reader (Tecan Infinite M200). After fluorescence readings, the cells were fixed with 4% formaldehyde and processed for imaging with fluorescence microscopy (Zeiss AxioSkop with Filter sets 49/DAPI, 38HE/FITC, 43HE/Rhodamine). Images were acquired with a monochrome camera (Zeiss Axiocam) and overlayed using ImageJ.

3. Results and Discussion

3.1. Collection efficiency of MEC

Diluted exhaust flow from the Diesel engine is entering into the chamber (upstream) with total soot particle concentration at $2.5 \cdot 10^5 \#/cm^3$. In Figure 5 the number concentration dN/dlogDp upstream and downstream of the exposure chamber are shown as a function of the soot particle diameter, Dp.

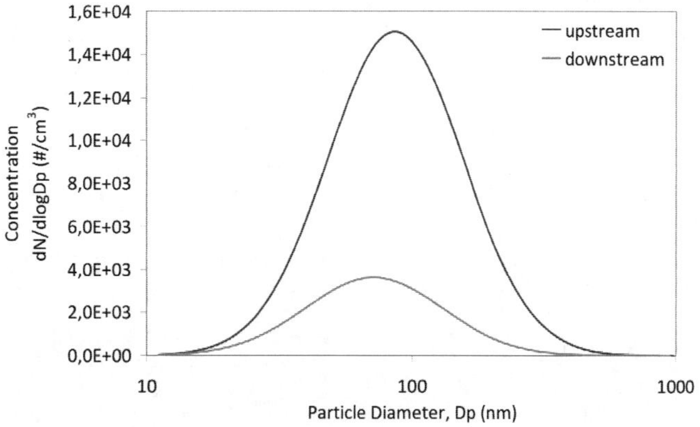

Figure 5. Number concentration of soot particles upstream and downstream of the exposure chamber versus soot particle diameter

The average collection efficiency of the exposure chamber is defined as the ratio between the number of particles deposited in the chamber and those that are being recorded in the exit, i.e. the relative difference between upstream and downstream particle number concentrations. As a result of all repetition tests (see Table 1) the average efficiency in delivering nanoparticles in the MEC is approximately 82%.

Table 1. Total collection efficiency of the MEC

	Total collection efficiency of the MEC
Test No 1	84%
Test No 2	79%
Test No 3	86%
Test No 4	80%
Average	**82%**

This high value of MEC collection efficiency is referring to the particle deposition on the whole inner surface of the chamber, at both cell growth area and "dead" surface (the stagnation plate and the margins of the insert cases[1]). However, considering firstly that the growth area per insert is $5.71 \cdot 10^{-3}$ m² for the case of 6-well plate inserts, then that the total growth area which can be accommodated in MEC is sixfold this surface area, and lastly that the total surface area of the exposure chamber is $8.21 \cdot 10^{-2}$ m², then we can easily conclude that the ratio between the total growth area of MEC and the total surface area of MEC is really high (approximately 42%). Consequently, the cell-specific deposition efficiency is roughly half of the inner deposition efficiency, i.e. 35%, which is much greater than the collection efficiencies reported in the literature [16, 17]. Comparing MEC with the previous

[1] We assume that the effective design of the MEC (position of inlet ports, distances between inserts and the inlet slit etc) has minimized the particle losses in other possible "dead" surface areas.

in-house constructed exposure chamber (MAAPHRI version), the cell-specific collection efficiency of the current version has significantly increased by 120%, since in MAAPHRI version the inner deposition efficiency was 76%, but the ratio of growth area per the total surface of the chamber was only 21% (cell-specific collection efficiency 16%).

3.2. Flow and velocity uniformity in MEC

Several snapshots from the video recording during of the flow visualization test are shown in Figure 6. As t = 0 sec is considered the time that the flow of blue smoke/tracer is entering the exposure chamber. We notice that the flow is being distributed along the two thirds of the central inlet tube of the chamber until t = 4 sec and then the flow starts to be distributed along the vertical axis towards the inserts on both sides of the central tube, from t = 4 sec to t = 9 sec. At t = 9 sec, the flow has reached the end of the central inlet tube and starts to distribute along the most "remote" inserts from the inlet. After 34 sec, the flow has been distributed in the whole exposure chamber surface. For a typical cell exposure experiment that lasts from 1 to 24 hours, depending on the end point, this short time needed for the flow to contact each well in the chamber is not considered to cause significant effects in the uniformity and the efficiency of the exposure results.

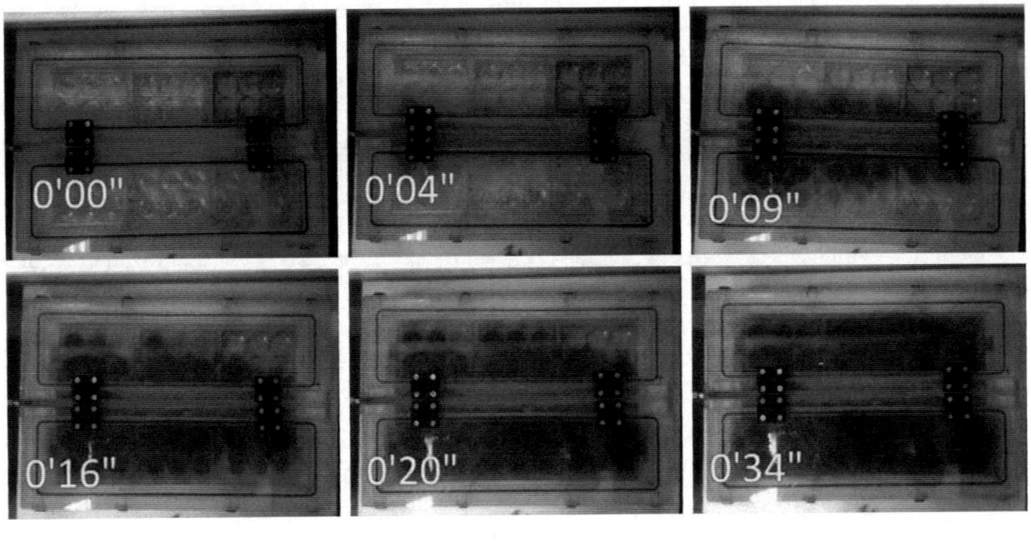

Figure 6. Snapshots from the video recording during flow visualization test at several intervals

Nevertheless, we notice from the snapshot at the times t = 4 sec that a flow front starts to develop on both sides of the central tube, which is attributed to a slight non-uniformity in the flow velocity, u, along the central inlet tube. One way to quantify the non-uniformity is to calculate a uniformity metric ξ as the ratio of the velocities between position I (at x = 0) and position II (at x = L, where L is the length of the chamber in the detail in Figure 3), or to the corresponding ratio of the times needed for the blue smoke/tracer to reach the peripheral regions of the chamber (time needed from y = 0 to y = ½W, where W is the width of the chamber, in the detail in Figure 3):

$$\xi = \frac{u_I}{u_{II}} = \frac{\frac{\frac{1}{2}W}{T_I}}{\frac{\frac{1}{2}W}{T_{II}}} = \frac{\frac{1}{T_I}}{\frac{1}{T_{II}}} = \frac{\frac{1}{(20\text{sec} - 0\text{sec})}}{\frac{1}{(34\text{sec} - 9\text{sec})}}$$

The estimated uniformity metric ξ is 1.25 for the specific geometry of the exposure chamber, revealing a slight difference in velocities along the central inlet tube which can be 25% maximum. It is believed that, in future work, this difference can be easily reduced further by manipulation of the flow field inlet through minor modification in the chamber geometry.

3.3. Soot particle deposition observation with TEM

TEM grids were placed in the 6-well plate inserts either in empty wells (Figure 3 positions B4 and E1), or in wells containing double distilled water (Figure 3 positions E3 and F2). In Figure 7, a selection of TEM pictures of the grids located both in water and in air, is shown.

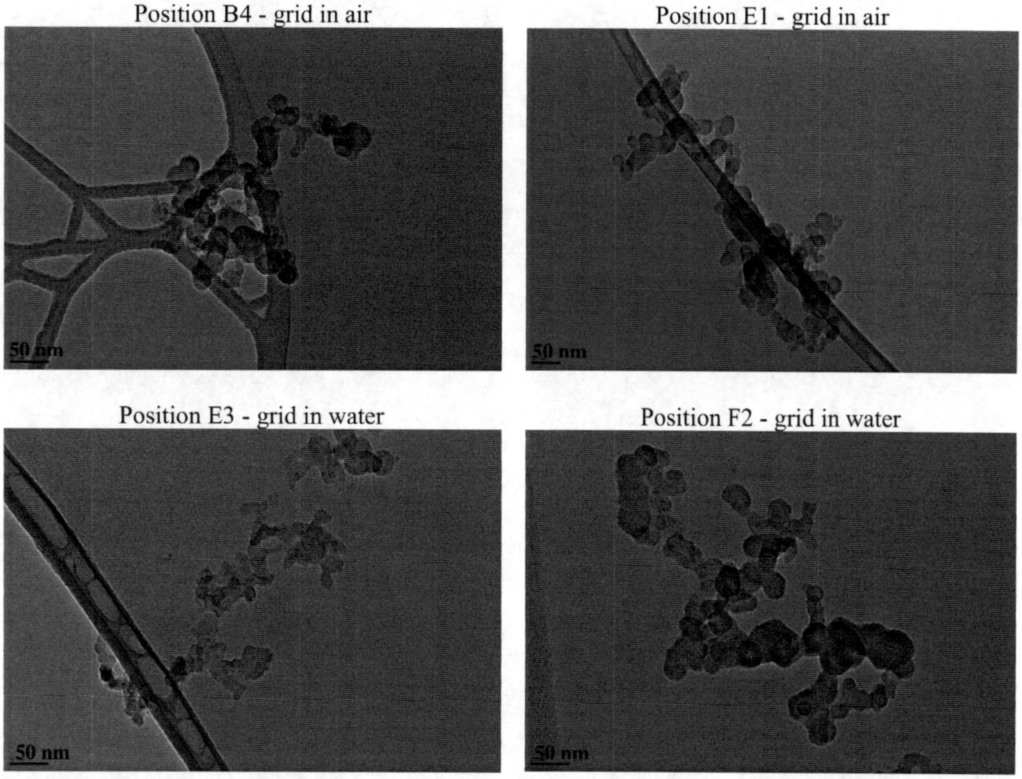

Figure 7. A selection of TEM pictures from soot particles (aggregates and agglomerates) both in air and in water

The overall picture of the TEM grids used in our experiments reveal a systematic deposition of soot particles regardless the exact position of the grid in the chamber, both in water and in the air. However, the agglomeration differs between the two cases; in the case of the empty wells, we assume that agglomerates have been formed in the wells in a greater degree due to the presence of electrostatic forces, while in water the agglomeration was attributed to the presence of van de Waals forces. In parallel, the aqueous environment has somewhat affected the external surface of the soot particles, a phenomenon which, however, is not of interest in the present study.

3.4. Cell exposure experiments

ROS production was monitored by DCFH-DA activation in A549 cells exposed to aerosolized nanoparticles for 1 hour. Comparing the fluorescence before and after exposure, elevated ROS signals were measured and visually confirmed using fluorescence microscopy (Figure 8). However, this 39%

increase in fluorescence intensity was attributed to the expected increase in DCFH-DA activation that occurs in unexposed cells loaded with the dye alone for 1 hour.

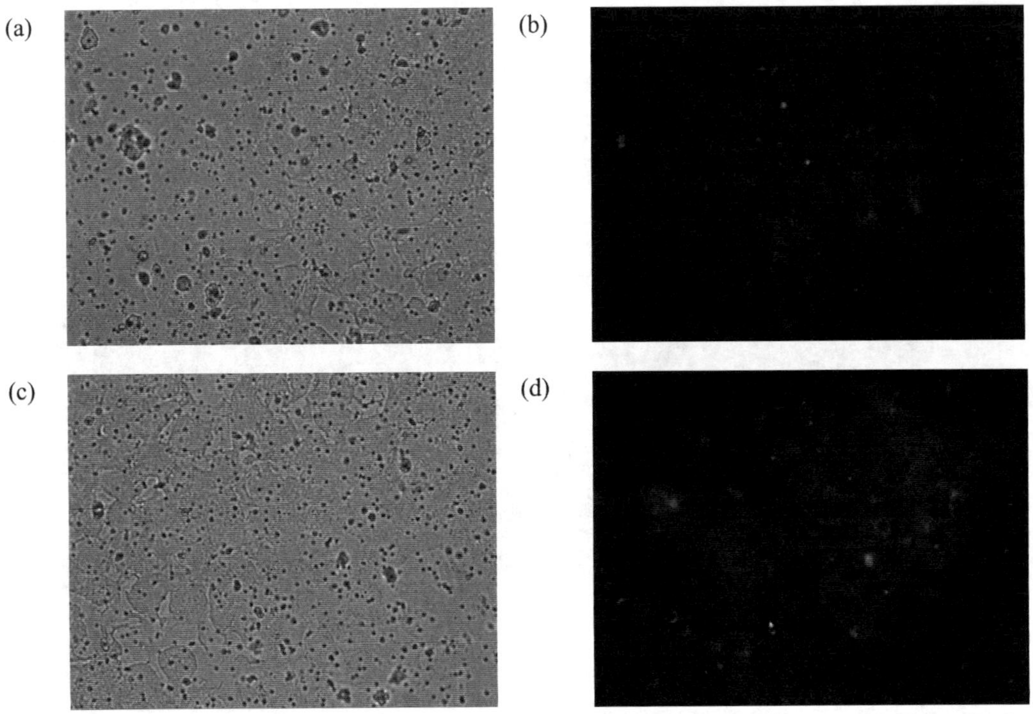

Figure 8. Microscope images of A549 cells loaded with DCFH-DA before (a & b) and after (c & d) exposure to Fluoprobe532 labeled SiO_2 nanoparticles in brightfield (left) and fluorescence (right). Magnification = 400x, Green = ROS (DCF).

Fluorescence corresponding to the Fluoprobe532 labeled silica nanoparticles appears distributed throughout the center of the transwell with aggregates visible as shown in Figure 9. Similar pictures have been taken from four more transwells in different positions in the chamber (positions A1, B5, C1, D4, E5, F3 in the detail of Figure 3).

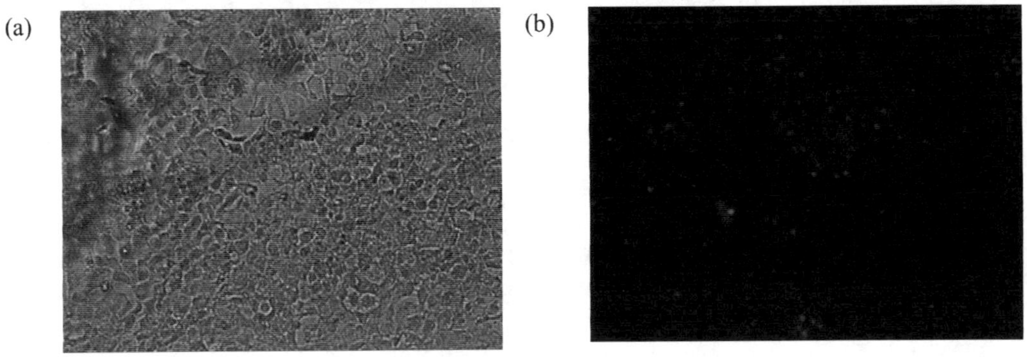

(c) (d)

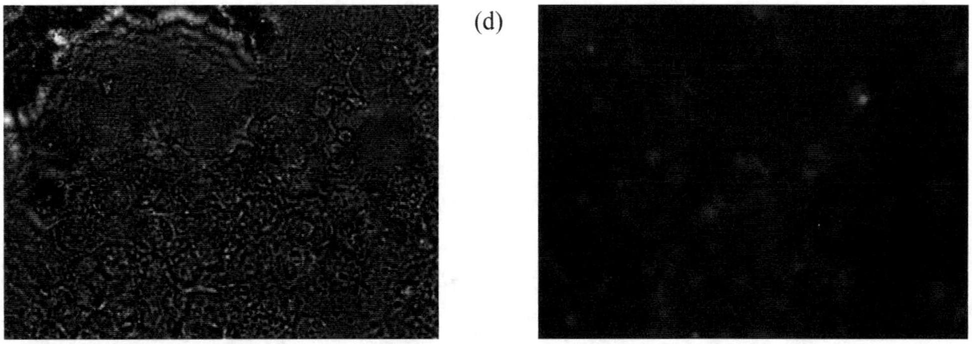

Figure 9. Microscope images of A549 cells at positions A1 (a & b) and B5 (c & d) exposed to Fluoprobe532 labeled SiO_2 nanoparticles. Magnification = 200x (a & b) 400x (c & d), Blue = nuclei (DAPI), Green = ROS (DCF), Red = nanoparticles (Fluoprobe532).

4. Conclusions

The advancement of an upgraded Multiculture Exposure Chamber (MEC) into a dose-controlled system for efficient delivery of nanoparticles to cells and its validation with various types of nanoparticles (Diesel engine soot aggregates, SiO_2 nanoparticles) is presented. The average efficiency in delivering nanoparticles in the MEC is approximately 82%, while the cell-specific deposition efficiency is roughly half of the inner deposition efficiency, i.e. 35%, which is much greater than the collection efficiencies reported in the literature. A high degree of flow homogeneity has been observed with flow visualization tests, while a slight non-uniformity in the velocity (uniformity metric = 1.25) was attributed to defects in the inlet channel, which, however is easily eliminated by design modification in future work. Nanoparticle deposition was demonstrated by TEM and final validation of the dose-controlled cell exposure system was performed with exposure tests of A549 cell cultures to SiO_2 nanoparticles. The efficient delivery of nanoparticles as well as the homogeneity in particle deposition among the 6-well plate inserts was demonstrated by fluorescence microscopy.

References

[1] Donaldson K and Borm P 2007 *Particle Toxicology* (CRC Press Boca Raton FL)
[2] Seaton A *et al* 2010 *J. R. Soc. Interface* **7** (Suppl_1)
[3] Wei Yang W *et al* 2008 *Int. J. Pharm* **356** 239
[4] Organization for Economic Cooperation and Development (OECD) Paris, France 2010 **Report 01.12.10**
[5] Nasterlack M *et al* 2006 *Proc. Int. Conf. on Nanotech., Occup. & Env. Health & Safety* (Cincinnati)
[6] Gamer A O *et al* 2006 *Toxinvit.* **20(3)** 301
[7] Nel A *et al* 2006 *Science* **311** 622
[8] Stone V *et al* 2006 *Nature* (Nanotech.) **1** 23
[9] MAAPHRI, Quality of Life, CONTRACT No. QLRT2001-02357
[10] Delfino R J *et al* 2005 *Environ. Health Perspect.* **113** 934
[11] Lenz A G *et al* 2009 *Part. Fibre Toxicol.* **6(32)**
[12] Paur H-R *et al* 2008 *J.Verbr. Lebensm.* **3** 319
[13] Muller L *et al* 2010 *Environ. Sci. Technol.* **44** 2632
[14] Paur H-R *et al* 2011 *J. Aerosol Sci.* **42** 668
[15] Aufderheide M and Mohr U 2000 *Exp. Toxicol. Pathol.* **52(3)** 265
[16] Desantes J M *et al* 2006 *J. Aerosol Sci.* **37** 1750
[17] Bitterle E *et al* 2006 *Chemospere* **65(10)** 1784
[18] Papaioannou E *et al* 2006 *SAE Tech. Paper* **2006-01-1075** (SP-2024) 389
[19] Asimakopoulou A *et al* 2011 *J. Phys.: Conf. Series* **304(1)**, 012005

Predictive tests to evaluate oxidative potential of engineered nanomaterials

Mara Ghiazza[1], Emanuele Carella[1], Simonetta Oliaro-Bosso[2], Ingrid Corazzari[1], Franca Viola[2], Ivana Fenoglio[1]

[1] University of Torino, Dipartimento di Chimica, "G. Scansetti" Interdepartmental Center for Studies on Asbestos and other Toxic Particulates and Interdepartmental Center for Nanostructured Interfaces and Surfaces, , via P. Giuria 7, 10125 Torino, Italy

[2] University of Torino, Dipartimento di Scienza e Tecnologia del Farmaco, , V. P. Giuria 9, 10125 Torino, Italy

e-mail: ivana.fenoglio@unito.it

Abstract. Oxidative stress constitutes one of the principal injury mechanisms through which particulate toxicants (asbestos, crystalline silica, hard metals) and engineered nanomaterials can induce adverse health effects. ROS may be generated indirectly by activated cells and/or directly at the surface of the material. The occurrence of these processes depends upon the type of material. Many authors have recently demonstrated that metal oxides and carbon-based nanoparticles may influence (increasing or decreasing) the generation of oxygen radicals in a cell environment. Metal oxide, such as iron oxides, crystalline silica, and titanium dioxide are able to generate free radicals via different mechanisms causing an imbalance within oxidant species. The increase of ROS species may lead to inflammatory responses and in some cases to the development of cancer. On the other hand carbon-based nanomaterials, such as fullerene, carbon nanotubes, carbon black as well as cerium dioxide are able to scavenge the free radicals generated acting as antioxidant. The high numbers of new-engineered nanomaterials, which are introduced in the market, are exponentially increasing. Therefore the definition of toxicological strategies is urgently needed. The development of acellular screening tests will make possible the reduction of the number of in vitro and in vivo tests to be performed. An integrated protocol that may be used to predict the oxidant/antioxidant potential of engineered nanoparticles will be here presented.

1. Introduction

Nanotechnology rapidly developed in recent years and a large number of new nanomaterials for a wide range of applications have been introduced in the market [1,2]. Concerns exist about possible adverse health effects following human exposure to those nanomaterials, which may release nanoparticles [3,4]. Because of a growing number of new NPs is expected to be produced in the future is urgent to define strategies to predict their possible impact on health.

Oxidative stress constitutes one of the principal injury mechanisms through which engineered NPs can induce adverse effects. Zhang and coworkers [5] recently reported the development of a quantitative structure-activity relationship (QSAR) model to predict the acute pulmonary inflammation potential of various metal oxide NPs based on the values of the conduction band energy levels. In fact, when the biological and material energetic states are similar, the permissive electron transfer could lead to the formation of oxidizing or reducing molecules that influence the level of antioxidants and/or increase the production of ROS. However, the proposed predictive tool may be not applicable to materials other than oxides. On the other hand processes different from electron transfer between surface and cells (e.g. physical interaction with biomolecules or mechanical damage to cells) likely participate to the overall mechanism of toxicity.

The aim of this study is to get insight on the capability of three different nanomaterials to induce oxidative damage by generating ROS or by directly damaging biomolecules. Two types of metal oxides, i.e. titanium oxide and amorphous silica nanoparticles, were compared with a carbon soot sample. These samples were chosen since they are expected to be very different in surface reactivity. TiO_2 is known to be a potent photo-catalyst while amorphous silica is an inert covalent material. Carbon exhibits a reactivity that depends upon its allotropic form. An integrated protocol to evaluate the oxidative potential of the materials has been used. The capability of NPs to interfere with the ROS was evaluated by means of EPR spectroscopy/spin-trapping or probing technique while the ability to cause oxidative damage to lipids, proteins and DNA were evaluated by means of UV-Vis spectroscopy, SDS-PAGE electrophoresis and agarose gel electrophoresis, respectively.

2. Materials and methods

2.1. Samples

The rutile–anatase TiO2 (Aeroxide P25), hereafter named TiO_2, was purchased by Evonik, (Essen, Germany). Carbon soot, hereafter named C, was purchased from Sigma-Aldrich s.r.l (St. Louis, MO; lot number 390127-25G) was obtained by resistive heating of graphite. The amorphous silica (Aerosil OX 50), hereafter named SiO_2, was purchased by Degussa Frankfurt A.M., Germany.

2.2. Generation of free radicals

Free radical release was monitored by means of EPR spectroscopy (Miniscope 100 EPR spectrometer, Magnettech) using the spin trapping technique (5,5-dimethyl-pirroline-N-oxide, DMPO, as spin trapping molecule) with a procedure described in a previous study [6]. Briefly, hydroxyl radicals were detected by using DMPO (5,5-dimethyl-1-pyrroline-N-oxide, Alexis Biochemicals, San Diego, CA) in the presence of hydrogen peroxide, as target molecule.

2.3. *Irradiation conditions.*

Irradiation experiments were performed with a 500 W mercury/xenon lamp (Oriel instruments) equipped with an IR water filter to avoid the overheating of the suspensions. Simulated solar light was obtained by applying a 400nm cut-off filter that let to pass about 5% of UV light in the UV A region.

2.4. *Photo-generation of Reactive Oxygen Species.*

The photo-generation of radical species by TiO_2 powder has been evaluated following a procedure already reported by some of us [6-8]. Briefly, hydroxyl and carboxylate radicals were detected by using DMPO (5,5-dimethyl-1-pyrroline-N-oxide, Alexis Biochemicals, San Diego, CA) and superoxide anion radicals by using PBN (N-tert-butyl-α-phenylnitrone, Sigma-Aldrich, St.Louis, MO) as spin trapping agents. The generation of singlet oxygen was monitored by employing 4-oxo-TMP (2,2,6,6-tetramethyl-4-piperidone, Sigma-Aldrich) as spin probing agent. All tests were performed with an amount of dust corresponding to a surface area of 1.4 m^2.

2.5. *Scavenging activity.*

Scavenging activity has been evaluated following the procedure already reported [9,10]. Briefly, hydroxyl radicals were generated by Fenton reaction or by irradiating with a UV lamp (ThermoOriel UV lamp) a solution of hydrogen peroxide directly in the EPR spectrometer cavity. The reaction was repeated in the presence of carbon soot. All the experiments were repeated at least twice.

2.6. *Oxidative damage to plasmid DNA.*

The reactivity of powders toward double stranded supercoiled plasmid pYES2 (Invitrogen) were been investigated in order to evaluate the potential of samples to cause direct oxidative damage [11]. All experiments were performed with ~ 0.2 mg of powder suspended in 30 μl of milliQ water and then vortexed. To this suspensions 3 μL of DNA solution (concentration 100-150ng/μL) were added and then exposed to a simulated solar light (UV-Vis lamp using a filter having a cut-off of 400nm) for 20 minutes. As a control, DNA was also irradiated in the absence of any powders in order to exclude a direct damage to this molecule. After irradiation time the suspension was centrifuged and the supernatant used for agarose gel electrophoresis analysis. The samples were loaded on a 1% agarose gel (Promega) and, after electrophoresis; DNA bands were stained and visualized with ethidium bromide (Promega).

2.7. Oxidative damage to linoleic acid

The reactivity of powders toward linoleic acid has been investigated in order to evaluate the potential of sample to cause oxidative damage directly to lipids following a procedure previously reported [7]. Briefly, powders and linoleic acid were continuously stirred under the ambient light at 37 °C for 72 h and then the formation of MDA was evaluated. The assay is based on the reactivity of MDA, a colorless end product of degradation, with tiobarbituric acid (TBA) to produce a pink adduct that absorbs at 535 nm.

2.8. Oxidative damage to proteins

Bovine serum albumin (BSA) (Sigma-Aldrich, Germany) has been chosen as model protein. All experiments were performed with ~ 5 mg of powder suspended in 50 μl of phosphate buffer 5 mM pH 7.4 and then sonicated for 5 minutes. To this suspensions 50 μL of BSA solution (1 mg/ml in phosphate buffer 5 mM pH 7.4) were added and then exposed to a simulated solar light (UV-Vis lamp using a filter having a cut-off of 400nm) for 1 hour. After irradiation time 10 μL of SDS 10% were added in order to eliminate adsorbed proteins at the powder surface and then centrifuged (10 minutes at 10000 rpm). The supernatant, boiled at 100°C in the presence of LAEMMLI solution, was used for SDS-PAGE electrophoresis analysis.

3. Results and discussion

Particles may generate reactive oxygen species by a direct mechanism (surface-derived ROS), or by an indirect one relying on the alteration of mitochondrial functions or the activation of cells of the immune systems (cell-derived ROS) [12-14]. Oxidative stress may also derive by the release of redox-active ions from particles to biofluids [15] or following the depletion of endogenous antioxidants by adsorption or reaction with the particle surface [16,17]. Finally, damage may follow the direct reaction of biomolecules with particles.

Both generation of ROS and direct damage to biomolecules are related to the existence of surface sites accessible to the fluid and able to undergo redox cycling. The chemical nature of these reactive sites depends upon the type of solid. One of the most important reactions occurring at the surface of an inorganic material leading to the generation of ROS is the Fenton reaction, i.e. the generation of hydroxyl radicals through the reaction of hydrogen peroxide with metal ions at a low oxidative state:

$$M^{(n)} + H_2O_2 \longrightarrow M^{(n-1)} + HO^{\bullet} + HO^{-}$$

This reaction was reported to occur in the case of iron containing minerals [18] or particles deriving by grinding covalent solid e.g. quartz [12].

3.1. Fenton-like reactivity

The ability to generate hydroxyl radicals in the presence of hydrogen peroxide has been evaluated by using EPR spectroscopy/spin trapping technique. Figure 1 reports the EPR spectra obtained by incubating titanium dioxide, amorphous silica or carbon nanoparticles in the presence of the spin trap DMPO and hydrogen peroxide (Figure 1). $FeSO_4$ was used as positive control: in this case the typical four lines EPR spectra of DMPO/ $HO^{\bullet}$ was observed.

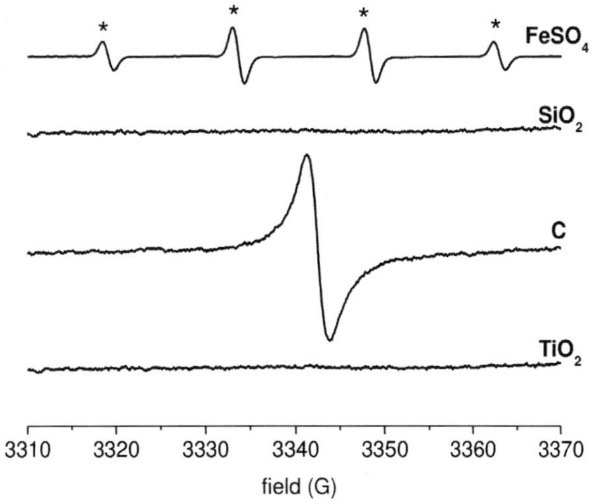

Figure 1: Generation of hydroxyl radical from hydrogen peroxide in the presence of $FeSO_4$, SiO_2, carbon particle (C) and TiO_2. * corresponds to the typical signal of DMPO-HO adduct.

Conversely, no signal was observed in the presence of the powders suggesting that all these materials are unable to reduce hydrogen peroxide. Note that the isometric sharp signal at field 3330-3340 G observed for carbon soot is due to intrinsic carbon centred free radicals, in the bulk of the material [19].

The reaction was performed in the dark since TiO_2 reacts with hydrogen peroxide if activated by UV light generating hydroxyl and superoxide radicals thorough a different mechanisms [7]. Titanium dioxide is in fact known to be a potent photo-catalyst. It generates high amount of reactive oxygen species (ROS) also in the absence of hydrogen peroxide when exposed to UV light in wet conditions. Under UV irradiation charge separation occurs in the bulk of the oxide leading to the promotion of an electron in the conduction band and to the formation of a hole in the valence band. When the charge carriers reach the surface of the solid reduction and oxidation reactions may occur following the interaction with the surrounding medium. The redox potential of charge carriers allow forming highly reactive radical species such as superoxide anion radicals ($O_2^{\bullet -}$), through electron transfer to O_2, and hydroxyl ($HO^{\bullet}$) radicals through hole interaction with water and singlet oxygen ($^1O_2$). The photo-generated holes may also oxidize organic molecules generating carbon-centered free radicals. A comprehensive evaluation of the ROS generated by TiO_2 may be obtained by EPR spectroscopy. The EPR spectra obtained by incubating the TiO_2 sample in solutions containing different spin trap or probe molecules under simulated solar light are reported in figure 2 and compared with the spectra obtained in the dark. Three types of reactions have been considered.

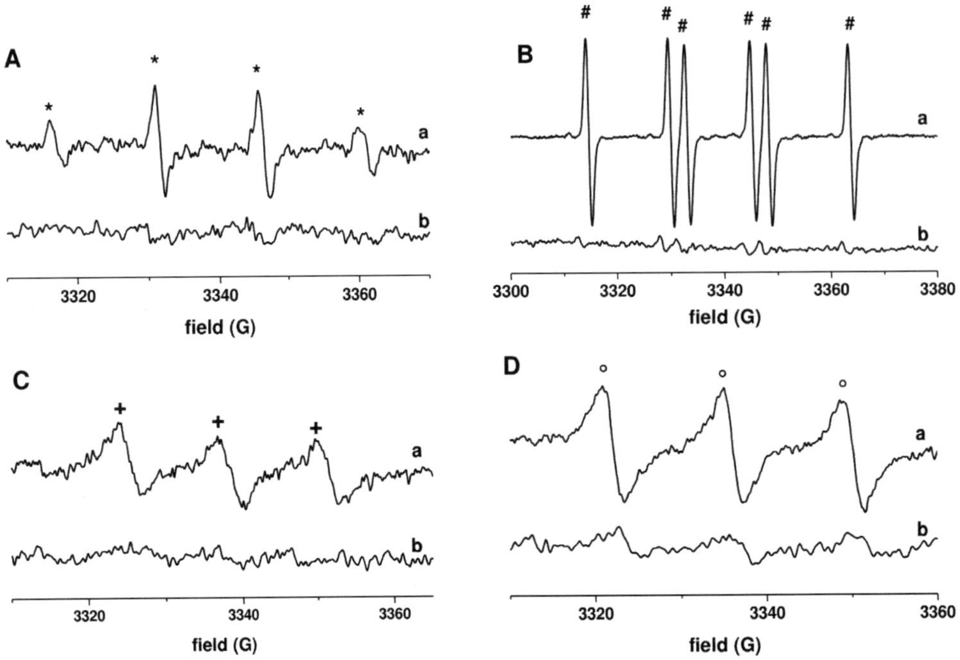

Figure 2: Generation of free radicals by TiO_2 under simulated solar light irradiation (spectra a) or in the dark (spectra b). A) hydroxyl radicals ($HO^•$) B) carboxylate radicals ($CO_2^{•-}$) and C) superoxide anion radicals ($O_2^{•-}$) and D) generation of the reactive specie singlet oxygen ($^1O_2$). *, #, + and ° correspond to the typical signal of hydroxyl, carboxylate, superoxide and singlet oxygen signal respectively.

1. The generation of hydroxyl radicals, among ROS the most aggressive species, through oxidation of water by photo-generated holes (h^+):

$$h^+ + H_2O \rightarrow HO^• + H^+$$

In agreement with what was previously reported [6-8,20] under irradiation TiO_2 generates large amount of hydroxyl radicals while no radical generation occurs in the dark (panel A).

2. The generation carbon centered radicals by organic molecules (here sodium formiate as model molecule) reacting with photo-generated holes (panel B):

$$h^+ + HCO_2^- \rightarrow CO_2^{•-} + H^+$$

A remarkably intense six lines EPR spectrum correspondent to the trapped $CO_2^{•-}$ radicals was observed under UV irradiation. Albeit much less intense a signal was observed also in the dark.

3. The reduction of oxygen to superoxide radicals:

$$Ti^{4+} + e^- + O_2 \rightarrow O_2^{•-}$$

The typical signal of $O_2^{•-}$ was observed under UV irradiation (Panel C) while no signal was detected in the dark.

4. The generation of singlet oxygen. Being diamagnetic this species is not detected by EPR. However, it may react with the spin probe 4-oxo-TMP to give a nitroxide radical as previously reported by other authors [8]. TiO_2 generates large amount of singlet oxygen when irradiated by UV light (Panel D). A residual photocatalytic activity was also observed in the dark.
Oppositely to titanium dioxide carbon soot and amorphous silica, albeit irradiated with simulated solar light, did not show any activity in generating free radicals.

3.2. Scavenging of free radicals

Beside several materials exhibit at the surface redox active sites able to induce oxidative stress, there are some other that are intrinsically unable to generate free radicals. Amorphous silica and carbon are among them. However, there are numerous evidences that some carbon-based materials (e.g. fullerenes, carbon nanotubes and carbon black) may act as free radical acceptors [9,21,22]. The susceptibility of carbon-based materials to radical attack is well known and has been exploited to introduce functionalities at their surface [23-25]. Some of us previously reported that MWCNTs was able to scavenge oxygenated free radicals and in particular hydroxyl radicals, the most reactive among ROS [9,10]. This property makes these materials promising in all applications where radical reaction need to be controlled such as stabilizing additive for composite [26,27] and in medicine to prevent free-radicals mediated diseases (tumor, cardiovascular diseases and neurodegenerative disorder) [28].

Carbon soot, used in this study as model for carbon particles, derive from incomplete combustion processes or pyrolysis of carbon-containing materials, such as waste or fuel oils, diesel fuel, coal, wood, paper plastic and rubber. It is mainly composed by elemental carbon partially organized in graphenic/graphitic structures. Like other carbon based nanomaterials it scavenges $HO^{\bullet}$ free radicals. Hydroxyl radicals may be generated by irradiating with a UV lamp a solution of hydrogen peroxide or by using the Fenton reaction (H_2O_2, $FeSO_4$). If the reaction is performed in the presence of a spin trap molecule an intense EPR signal is obtained (Figure 3 spectra a). When the reaction is repeated in the presence of carbon soot the signal disappear (Figure 3 spectra b) confirming the ability of carbon based material to scavenge free radicals.

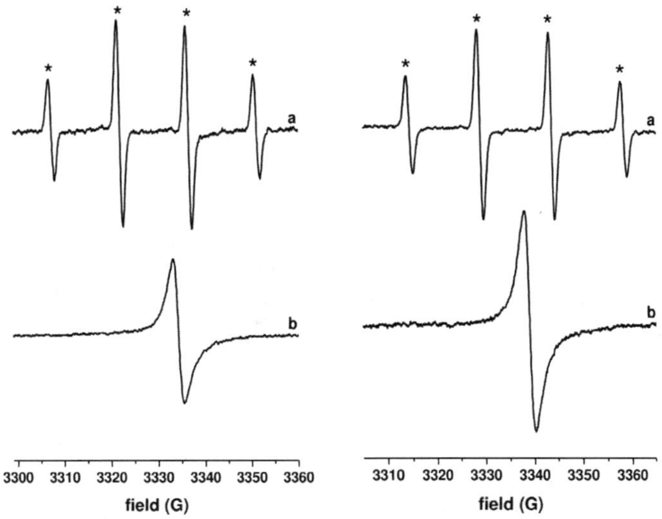

Figure 3: Scavenging activity of soot toward hydroxyl radicals generated by photolysis of H_2O_2 (A) and Fenton reaction (B). a) EPR spectra recorded in the absence of carbon soot; b) EPR spectra recorded in the presence of carbon soot. * corresponds to the typical signal of DMPO-HO adduct.

As reported elsewhere the mechanism of scavenging likely occurs through the addition of radicals to the particles [18]. Amorphous silica did not shown any free radicals scavenging activity in agreement with what previously reported [9].

3.3. Damage to biomolecules

Reactive surface sites also react with organic molecules; the reaction with sodium formiate reported above is an example. Biomolecules may also undergo degradative reactions initiated by surface reactive sites. Therefore the ability to generate ROS is not the only parameter to be considered in the evaluation of the oxidative potential. The capability to directly damage lipids, protein and DNA may be evaluated by cell-free experiments.

Lipids peroxidation was here evaluated by measuring the amount of malonyldialdehyde (MDA) generated after the reaction of linoleic acid with titanium oxide and amorphous silica nanoparticles after 72 h of incubation under ambient light irradiation (Figure 4A). The oxidative potential damage of proteins by titanium oxide and amorphous silica nanoparticles was evaluated under simulated solar light irradiation for 1 h, by using bovine serum albumin (BSA) as model protein (Figure 4B).

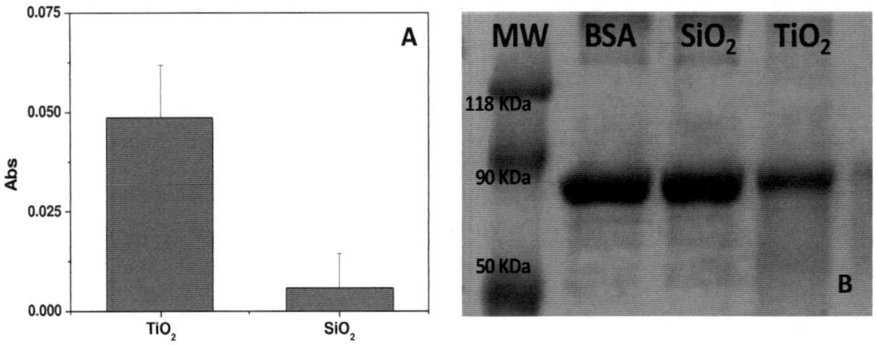

Figure 4: Oxidative degradation of (A) linoleic acid and (B) BSA protein.

As expected under irradiation TiO_2 oxidized both linoleic acid and BSA while amorphous silica did not.

The oxidative potential on DNA by titanium dioxide, amorphous silica and carbon nanoparticles was evaluated under simulated solar light irradiation. Results are reported in figure 5.

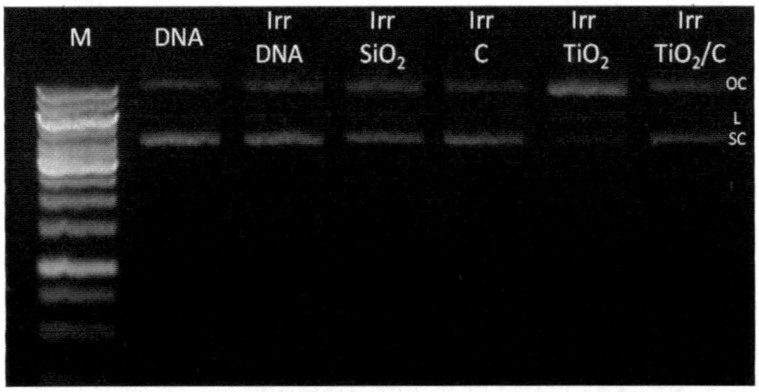

Figure 5: Photo induced damage to double strand supercoiled plasmid DNA. M correspond to DNA weight marker. SC corresponds to supercoiled DNA form, L to linear DNA form and OC to open circular DNA form.

DNA damage (breaks) was performed by irradiating DNA in the presence of powders for 20 minutes and analyzing the supernatant by agarose electrophoresis. Irradiated DNA (Figure 5, column 2) mainly remained in the supercoiled circular (SC) form as well as non-irradiated DNA (column 1) suggesting that direct irradiation did not damage DNA. Moreover, in both cases a band corresponding to open circular form (OC) was present, indicating the presence of DNA already damaged, probably during the plasmid DNA preparation. Addition of both amorphous silica (column 3) and carbon soot (column 4) did not modify the supercoiled/open circular DNA pattern suggesting that no damage occur in the presence of both powders. On the contrary, addition of TiO_2 powder (column 5) caused an increase of open circular form DNA and a partial conversion into linear (L) form of DNA (very low intensity band) suggesting damage to DNA caused by this powder. The contemporary addition of TiO_2 and carbon soot (column 6) partially reversed the damaging effects of TiO_2 powder on DNA. In fact, the bands corresponding to SC and OC forms of DNA show equal intensity indicating that a damage still occurred, but the absence of L form indicates that the damage is lower than that caused by TiO_2 alone suggesting a protective effect of carbon soot. The present data suggest that, oppositely to silica, carbon nanoparticles are not inert but are able to actively interact with the cellular ROS homeostasis by acting as active antioxidants.

4. Conclusions

The growing number of engineered nanoparticles which will potentially enter in the market makes urgent the need of screening protocols able to predict their toxic potential. The availability of validated high throughput screening tests will accelerate both the definition of SARs for nanomaterials and the assessment of the risk related to their exposure. However, when performing screening tests the peculiar chemical properties of each nanomaterial need to be considered. Furthermore, a detailed knowledge of the chemical processes, which may occurs at the nanoparticle/bio-systems interfaces, may help in understanding their behavior in vivo. Integrated chemical screening tests like the one proposed here might be promising tools for the understanding of such processes at a molecular level.

Ackwnoledgements

This research has been carried out with the financial support of the University of Torino (Progetti di Ricerca finanziati ex 60%- 2012) project title: “Integrated chemical tests for the evaluation of the oxidative potential of nanopowders”

M.G. was recipient of a post-doc fellowship from Progetti di Ricerca di Ateneo-Compagnia di San Paolo-2011-Linea 1A, progetto ORTO11RRT5

References

[1] Vashist S K, Zheng D, Pastorin G, Al-Rubeaan K, Luong J H T and Sheu F S 2011 *Carbon* 49(13) 4077-97.
[2] http://www.observatorynano.eu/project (from 2008-04-01 to 2012-03-31)
[3] Sharifi S, Behzadi S, Laurent S, Forrest M L, Stroeve P and Mahmoudi M 2012 *Chem. Soc. Rev.* 41, 2323–43
[4] Pietroiusti A 2012 *Nanoscale* 4, 1231-47
[5] Zhang 2012 *ACS Nano*
[6] Fenoglio I, Greco G, Livraghi S and Fubini B 2009 *Chem.–Eur. J.* 15 4614-21.
[7] Livraghi S, Corazzari I, Paganini M C, Ceccone G, Giamello E, Fubini B and Fenoglio I 2010 *Chem.Commun*. 46 8478-80
[8] Corazzari I, Livraghi S, Ferrero S, Giamello E, Fubini B and Fenoglio I 2012 *J. Mater. Chem.* 22, 19105-12.
[9] Fenoglio I, Tomatis M, Lison D, Muller J, Fonseca A, Nagy J B, and Fubini B 2006 *Free Rad Biol Med*, 40(7), 1227-33
[10] Fenoglio I, Greco G, Tomatis M, Muller J, Rayrnundo-Pinero E, Beguin F and Fubini B 2008

Chem. Res. Toxicol., 21(9), 1690-97
[11] Wang Y, Zhang X, Zhang Q and Yang Z 2010 *Biometals* 23, 265-73
[12] Fubini B and Hubbard A 2003 *Free Rad Biol Med* 34 (12), 1507-16
[13] Nel A, Xia T, Madler L and Li N 2006 *Science* 311 622–27
[14] Fenoglio I, Fubini B, Ghibaudi E and Turci F 2011 *Adv. Drug Deliver. Rev.* 63, 1186-1209
[15] Welch K D, Davis T Z, Van Eden M E and Aust S D 2002 *Free Rad Biol Med* 32, 577–83
[16] Moller P, Jacobsen N R, Folkmann J K, Danielsen P H, Mikkelsen L, Hemmingsen J G, Vesterdal L K, Forchhammer L, Wallin H and Loft S 2010 *Free Rad Res* 44 1–46
[17] Fenoglio I, Fonsato S and Fubini B 2003 *Free Radic. Biol. Med.* 35, 752-62
[18] Ayres J G, Borm P, Cassee F R, Castranova V, Donaldson K, Ghio A, et al., 2008 *Inhal Toxicol* 20, 75-99
[19] Carella E, Ghiazza M, Alfè M, Gargiulo V, Ciajolo A, Fubini B, Fenoglio I 2012 *submitted to Bio-Nano Interface*
[20] Brezova V, Gabcova S, Dvoranova D and Stasko A 2005 *J. Photochem. Photobiol.* B 79(2) 121-34
[21] Dugan L L, Turetsky D M, Du C, Lobner D, Wheeler M, Almili CR, Shen CKF, Luh TY, Choi DW and Lin TS 1997 *Proc. Natl. Acad. Sci.* 94:9434–39
[22] Karakoti A, Singh S, Dowding J M, Seal S, Self W T 2010 *Chem. Soc. Rev.* 39 4422–32
[23] Ying Y M, Saini R K, Liang F, Sadana A K and Billups W E 2003 *Organic Letters* 5(9)
[24] Banerjee S, Hemraj-Benny T and Wong S S 2005 *Adv Materials*, 17(1), 17-29.
[25] Tasis D, Tagmatarchis N, Bianco A and Prato M 2006 *Chem Rev* 106(3) 1105-36
[26] Mwila J, Miraftab M and Horrocks A R 1994 *Polymer Degradation and Stability* 44 351-6
[27] Watts P C P, Fearon P K, Hsu W K, Billingham N C, Kroto H W and Walton D R M 2003 *J Mat Chem* 13(3) 491-5
[28] Firuzi O, Miri R, Tavakkoli M and Saso L 2011 *Curr Med Chem* 18(25) 3871-88

Comparison of toxicity of uncoated and coated silver nanoparticles

K C Nguyen[1], V L Seligy[1], A Massarsky[2], T W Moon[2], P Rippstein[3], J Tan[3], and A F Tayabali[1]

[1] Biotech Lab, Mechanistic Studies Division, Environmental Health Science and Research Bureau, Healthy Environments and Consumer Safety Branch, Health Canada, 50 Colombine Driveway, Ottawa, Ontario, Canada, K1A 0K9

[2] University of Ottawa, Department of Biology and Centre for Advanced Research in Environmental Genomics, 30 Marie-Curie, Ottawa, Ontario, Canada K1N 6N5

[3] University of Ottawa, Heart Institute, 40 Ruskin Street, Ottawa, Ontario, Canada K1Y 4W7

Email: Azam.Tayabali@hc-sc.gc.ca; Kathy.Nguyen@hc-sc.gc.ca

Abstract: This study compares toxic effects of uncoated (20, 40, 60 and 80 nm) and OECD (Organization for Economic Co-operation and Development) standard citrate- and polyvinylpyrrolidone (PVP)-coated (10, 50, and 75 nm) silver nanoparticles (Ag-NPs) in J774A.1 macrophage and HT29 epithelial cells. The cells were exposed to different concentrations (silver content) of Ag-NPs for 24 h. Analysis showed that uncoated Ag-NPs, at a concentration of 1 µg/ml, decreased cell viability by 20-40% and that 20 and 40 nm particles were 10% more cytotoxic than the 60 and 80 nm particles. In exposures to coated Ag-NPs, cell viability dropped at 25 µg/ml or higher concentrations, and the effects were also size-dependent. PVP-coated particles induced greater cytotoxicity than citrate-coated particles. Changes in sub-cellular architecture were observed in J774A.1 cells upon exposure to test Ag-NPs. Furthermore, uncoated Ag-NPs (1 µg/mL) decreased the expression of selected cytokines including TNF-α, IL-1β, and IL-12 (p70) in J774A.1 and IL-8 in HT29 cells. In contrast, both citrate- and PVP-coated Ag-NPs increased the expression of these cytokines at higher concentrations (25 µg/ml), and PVP-coated particles elevated cytokine levels the most. Moreover, while uncoated Ag-NPs resulted in decreased glutathione (GSH) content and increased superoxide dismutase (SOD) activity in test cells in a size-dependent manner at 1 µg/ml, coated Ag-NPs caused non-significant changes in GSH and SOD, even at the highest test concentrations. Lastly, uncoated (20 and 40 nm) at 1 µg/ml and coated Ag-NPs (10 nm PVP) at 50 µg/ml slightly increased the production of reactive oxygen species (ROS). Our data showed that uncoated Ag-NPs are more toxic than coated Ag-NPs. While uncoated Ag-NPs appear to suppress inflammatory responses and enhance oxidative stress in the test cells, coated Ag-NPs induce toxic effects through up-regulation of cytokines. Our findings support the toxicity of Ag-NPs as being size- and coating- dependent while providing additional insight on the health impact of Ag-NPs.

1. Introduction

Silver-containing nanoparticles (Ag-NPs), like other nano-sized particles, exhibit special properties relative to their bulk counterparts partly due to their very small size and greater surface area. In addition, Ag-NPs possess unique optical activities that have been exploited in many applications such as catalysis, electronics, and biosensors [1, 2]. Further, due to their antimicrobial properties, these NPs have been used in a variety of consumer and medical products, leading to increased human and ecosystem exposures [3, 4]. Examples of Ag-NP-containing products include wound dressing, surgical instruments, disinfectants, clothing, cosmetics, paints, home appliances, food containers, toys, and computer accessories [5]. As a consequence of production and distribution of a large number of consumer products containing different types of Ag-NPs and new design applications under development, rigorous knowledge is needed about the potential impacts of these NPs on human health and the environment.

Over the past decade, research on biological effects of various Ag-NP types has been performed on microorganisms, various cell lines, aquatic organisms and animal models [6]. The antimicrobial activity of Ag-NPs is fairly well documented in a broad spectrum of bacteria and fungi [7]. *In vitro* toxicity studies have indicated effects of Ag-NPs in immune, skin, lung, liver, and neuron cells [8, 9, 13, 10, 11, 12]. Exposures of Ag-NPs *in vivo* have shown toxic effects in various non-mammalian models, including zebrafish (*Danio rerio*) [13], rainbow trout (*Oncorhynchus mykiss)* [14], nematode (*Caenorhabditis elegans)* [15], and crustacean *(Daphnia magna)* [16]. Moreover, studies in rats and mice have demonstrated toxicity of Ag-NPs in a variety of target organs including lung, liver and brain [17, 18, 19]. From these *in vitro* and *in vivo* studies, the proposed mechanisms of Ag-NP-induced toxicity are oxidative stress, damage to cellular components including mitochondria and DNA, and cytokine induction [20, 21, 22, 23]. Even though there is growing evidence that Ag-NPs are toxic in various test models, the exact mechanisms of Ag-NP toxicity remain unclear.

Toxicological studies of NPs have demonstrated that physicochemical properties of NPs including size, shape, surface coating, surface charge, solubility, and chemical composition could dramatically affect NP behaviour in biological systems and thus influence the toxicity of NPs [24, 25]. A number of studies reported that Ag-NP toxicity was dependent on their size and surface coating [20, 26, 27, 28]. Similarly to other NPs, the size of Ag-NPs has been linked to cellular uptake processes [27]. The surface coating has been shown to affect the affinity of Ag-NPs for the cell surface and the dissolution or the release of Ag^+ ions from Ag-NPs, which was also proposed as a toxicity mechanism for Ag-NPs [28, 29]. The investigation of these physicochemical characteristics and their impact on the biological effects of Ag-NPs is important for their safe design, development, and applications.

In this study, we investigated and compared the toxicity of uncoated and coated Ag-NPs of different sizes in J774A.1 macrophage and HT29 epithelial cells. The comparative toxicity was assessed by metabolic activity, cell morphology, sub-cellular structures, inflammatory responses, ROS generation, and oxidative stress induction upon exposure to these Ag-NPs. The study provides scientific insight relating physical/chemical characteristics of Ag-NPs to cell damage and possible toxicity mechanisms of these NPs.

2. Materials and Methods

2.1. Materials:

Cell lines modelling murine macrophage (J774A.1) and human colonic epithelial (HT29) were obtained from the American Type Culture Collection (Manassas, VA). Uncoated Ag-NPs (20, 40, 60, and 80 nm) were purchased from Ted Pella, Inc. (Redding, CA). Ag-NPs marketed as OECD (Organization for Economic Co-operation and Development) standard material with either a citrate- or a PVP-coating (10, 50, and 75 nm) were purchased from NanoComposix (San Diego, CA). Silver nitrate ($AgNO_3$), (3-(4,5-dimethylthiazol-2-yl) 2,5-diphenyl tetrazolium bromide (MTT), menadione, DMSO, glutaraldehyde, paraformaldehyde, osmium tetroxide, and Triton-X 100 were obtained from Sigma-Aldrich (St. Louis, MO). Rhodamine-Phalloidin, Sytox-Red, dihydroethidium (DHE), and Prolong antifade were purchased from Molecular Probes-Invitrogen (Carlsbad, CA). Bio-Plex cytokine kits and reagents were purchased from Bio-Rad (Hercules, CA).

2.2. Ag-NP size characterization

Test Ag-NPs were characterized using transmission electron microscopy (TEM) and dynamic light scattering (DLS). For TEM, Ag-NP stocks were diluted 10-fold with ddH_2O, deposited on carbon-coated grids, which were air-dried overnight at RT and then examined with a JOEL JEM 1230 operating at 60 kV. DLS analysis was performed on a Zetasizer Nano (Malvern Instruments Ltd). Ag-NP stocks were diluted 10-fold with ddH_2O or Dulbecco's modified Eagle's medium (DMEM). One millilitre was added to a clean 12 mm polystyrene cuvette (DTS0012) and inserted into the instrument to obtain the readings. Each sample was measured at least 3 times.

2.3. Cell cultures and Ag-NP treatments

J774A.1 and HT29 cells were cultured in DMEM, supplemented with 10% fetal bovine serum (FBS), 1 mg/ml glutamine, and 100 μg/ml gentamicin at 37°C in a humidified atmosphere with 5% CO_2. Cells were treated with different concentrations (based on silver content) of Ag-NPs for 24 h. For assays involving measurement of MTT bioreduction, glutathione, and cytokines, cells were seeded into 96-well plates at a density of $5x10^4$ cells/well. For confocal microscopy, cells were seeded on cover-slips in 12-well plates at a concentration of $1x10^5$ cells/ml/well. Cells were pre-cultured for 24 h to 80% confluency and media were replaced before exposure regimes. $AgNO_3$ treatments of with equivalent silver content that used for Ag-NPs, were done in some assays for comparison purposes. The silver content of Ag-NPs was provided by the manufacturers. The silver content of $AgNO_3$ was calculated as the total silver mass, which is 63.5% of the total $AgNO_3$ molecular mass. Menadione (25 μM) was used as a positive control in ROS detection assays. Working solutions of Ag-NPs and $AgNO_3$ were prepared by diluting stock solutions in cell culture medium.

2.4. Cell viability – MTT assay

After exposure to Ag-NPs, the culture media was removed and replaced with fresh media (100 μl/well) and 10 μl/well of MTT stock (10 mg/ml), followed by incubation for 1 h at 37°C. Media was removed, cells were rinsed with PBS (100 μl/well), and the formed formazan was solubilized with DMSO (100 μl/well) and quantified spectrophotometrically (Molecular Devices) at 505 nm. All measurements were done in duplicate in three independent experiments.

2.5. Cell morphology by confocal microscopy

Cells, after being exposed to Ag-NPs, were fixed with 4% paraformaldehyde in PBS containing 1% Triton X-100 for 15 min and were washed twice with PBS for 5 min. Cells were stained for actin by treatment with rhodamine-phalloidin (1:40) for 1 h. After two washes with PBS, cells were stained for nuclei by treatment with Sytox-red (1:1000) for 15 min and were washed again with PBS. Each coverslip was inverted onto a drop of Prolong™ antifade (Molecular Probes) placed on a glass slide and was then cured overnight in the dark before observation with a Nikon TE2000 microscope attached to a C1 confocal unit.

2.6. Cell morphology by TEM

After treatments, cells were fixed with 2.5% glutaraldehyde in 66.7 mM cacodylate buffer (pH 7.4) for 1 h, washed twice with 100 mM cacodylate buffer for 10 min, and subsequently post fixed with 1% osmium tetroxide in 100 mM cacodylate, pH 7.4. Following dehydration in a series of ascending alcohols the cell samples were infiltrated and processed in Spurr's epoxy resin and embedded onto resin-filled Beem capsule molds. Specimen blocks were then ultrathin-sectioned on a Leica EM UC6 ultramicrotome and the resulting sections stained with uranyl acetate and lead citrate. Sections were analysed with a Jeol 1230 TEM equipped with AMT imaging software

2.7. ROS detection
Cells were grown to 80% confluency on glass cover-slips inside 12-well cell culture plates. After treatment, cells were rinsed with PBS and incubated with dihydroethidium (DHE) at a concentration of 30 μM for 30 min. Cells were again washed with PBS and the cover-slip containing the monolayer of cells was mounted on a slide and viewed immediately with a Nikon TE2000 microscope attached to a C1 confocal unit (Nikon Canada Inc. Mississauga ON). Fluorescence areas from confocal micrographs were analysed using Nikon Imaging Software (NIS) element (Nikon Canada Inc. Mississauga ON). The average area of fluorescence of three micrographs was plotted to quantify the level of ROS production in the control and treatments.

2.8. Total GSH measurement
A Total GSH Colorimetric Assay kit (Oxford Biomedical Research) was used to estimate the total glutathione content in cell lysates. Briefly, after treatment, cells were washed with PBS pH 7.2. To each well, 100 μl of 5% metaphosphoric acid (MPA) was added, and the plate was frozen at −80°C and subsequently thawed at 37°C. After two freeze/thaw cycles, 50 μl of each cell lysate was transferred to new microtiter wells, and 50 μl each of 5,5′-dithiobis (2-nitrobenzoic acid) (DTNB, 0.5 mg/ml) and glutathione oxidoreductase solution was added to each well. The microtiter plate was incubated for 10 min at RT, and 50 μl of the reduced form of β-nicotinamide adenine dinucleotide phosphate (β-NADPH, 0.6 mg/ml) was added to each well. The absorbance was measured at 405 nm after a 3 min incubation at room temperature.

2.9. SOD activity measurement
After treatment, cells were collected, homogenized in cold 20 mM HEPES buffer (pH 7.2) containing 1 mM EGTA, 210 mM mannitol, and 70 mM sucrose, and centrifuged at 1,500 x g for 5 min at 4°C. The supernatants were collected and centrifuged at 10,000 x g for 15 min at 4°C to yield cytosolic SOD samples. The pellets were homogenized in cold HEPES buffer to yield mitochondrial SOD samples. SOD samples were assayed using a SOD colorimetric enzyme assay kit from Cayman Chemical (Ann Arbor, MI) according to the manufacturer's protocol.

2.10. Cytokine measurement
Levels of cytokines/chemokines produced by the two cell lines were estimated by using multiplex bead assays based on mouse 23-plex cytokine kits and human 8-plex cytokine kits. The cytokine/chemokine levels were measured using the Luminex-based Bio-Plex array system (Bio-Rad, Hercules, CA) according to the manufacturer's instructions. Briefly, anti-cytokine/chemokine conjugated beads were added to individual wells of a 96-well filter plate. Beads were then sedimented using vacuum filtration and washed briefly with kit wash buffer. After washing, 50 μl of pre-diluted standards or cell culture supernatants were added and incubated for 30 min at RT with gentle shaking. The filter plate wells were then washed before adding 25 μl/well of pre-diluted detection antibody and incubating for 30 min at RT. After a further washing, 50 μl/well of pre-diluted streptavidin-conjugated phycoerythrin was added and the plate was shaken for 10 min. The wells were again washed and 125 μl/well of assay buffer was added. The plate was shaken for one minute and analyzed with the Bio-Plex 100 Array System.

2.11. Statistical analysis
Results were compared by one-way analysis of variance (ANOVA) followed by Dunnett's multiple comparison test for comparison to the controls. All data were expressed as mean ± standard deviation. A value of $p<0.05$ was considered statistically significant.

3. Results

3.1. Ag-NP size distribution characterization
Ag-NPs from both sources were characterized for size distribution and shape by TEM image analysis (Figure 1). Uncoated Ag-NPs appeared as individual spherical particles. The size was obtained by measuring the minimum and maximum dimensions of each NP. The average particulate size in either dimension for uncoated Ag-NPs was slightly greater than the size indicated by the supplier (Table 1). Coated Ag-NPs also appeared as spherical particles and many particles existed as aggregates. The average particulate sizes for coated Ag-NPs are also shown in Table 1. TEM images showed forms of agglomerated NPs for all coated Ag-NPs (Figure 1).

A

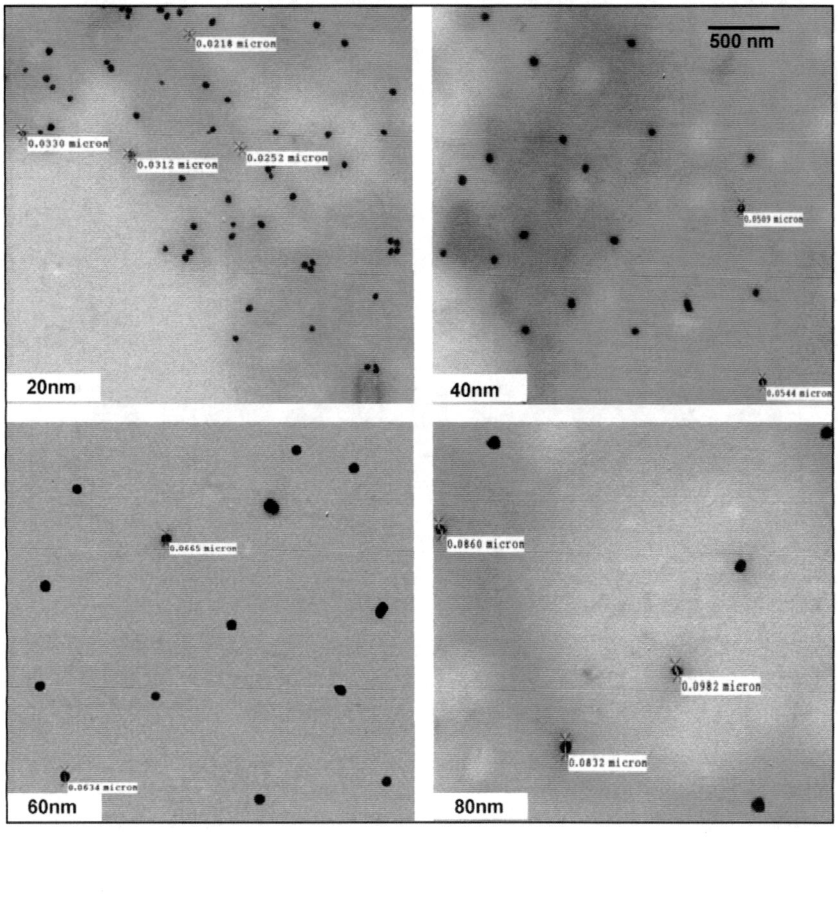

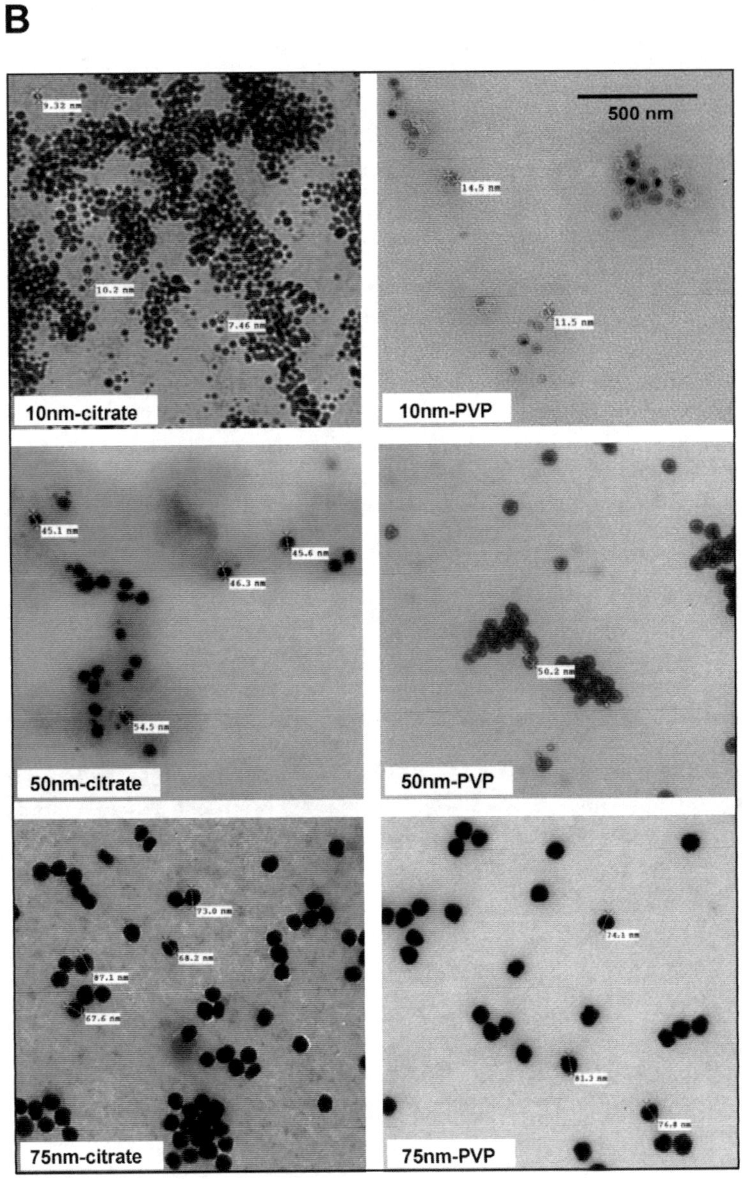

Figure 1: TEM micrographs of test Ag-NPs. (A) uncoated Ag-NPs (20, 40, 60, and 80 nm). (B) coated Ag-NPs (10 nm-citrate, 10 nm-PVP, 50 nm-citrate, 50 nm-PVP, 75 nm-citrate, and 75 nm-PVP).

DLS results showed bell-curved size distributions for all Ag-NPs (data not shown). The mean sizes for the uncoated and coated Ag-NPs in ddH_2O were similar to claims made by respective manufacturers, except for the 10 nm coated Ag-NPs whose mean sizes were generally larger than the sizes provided by the manufacturer. For most test particles, DLS particle sizes in cell medium appeared larger than those measured in ddH_2O indicating agglomeration (Table 1).

Table 1: Average Ag-NP sizes determined by TEM and DLS

Ag-NP	**TEM (nm)**	**DLS in water (nm)**	**DLS in cell medium (nm)**
20 nm	27.81 ± 5.23	19.49 ± 2.13	20.71 ± 3.60
40 nm	52.63 ± 2.41	41.68 ± 5.12	52.75 ± 7.56
60 nm	71.60 ± 9.82	58.07 ± 7.82	83.89 ± 9.23
80 nm	89.10 ± 7.92	69.26 ± 3.95	84.23 ± 6.12
10 nm-citrate	9.86 ± 1.22	16.57 ± 6.12	14.99 ± 7.34
10 nm-PVP	11.95 ± 1.55	15.90 ± 3.11	10.06 ± 3.21
50 nm-citrate	50.58 ± 4.97	45.85 ± 6.74	50.30 ± 4.12
50 nm-PVP	52.60 ± 3.23	48.31 ± 5.53	53.75 ± 1.22
75 nm-citrate	78.60 ± 8.66	72.32 ± 6.43	74.17 ± 1.68
75 nm-PVP	78.38 ± 3.02	74.19 ± 3.96	82.87 ± 7.67

3.2. Cell viability

Cell viability was assessed by measuring the cellular capacity to reduce MTT to its formazan form. Decreased viability in J774A.1 and HT29 cells in response to Ag-NPs exposure was dose-dependent (Figures 2A, B, C and D). Uncoated Ag-NPs resulted in cell viability changes at levels as low as 1 μg/ml of silver content (Figures 2A and C), while coated Ag-NPs resulted in changes starting at 25 μg/ml (Figures 2B and D). With both uncoated and coated Ag-NPs,

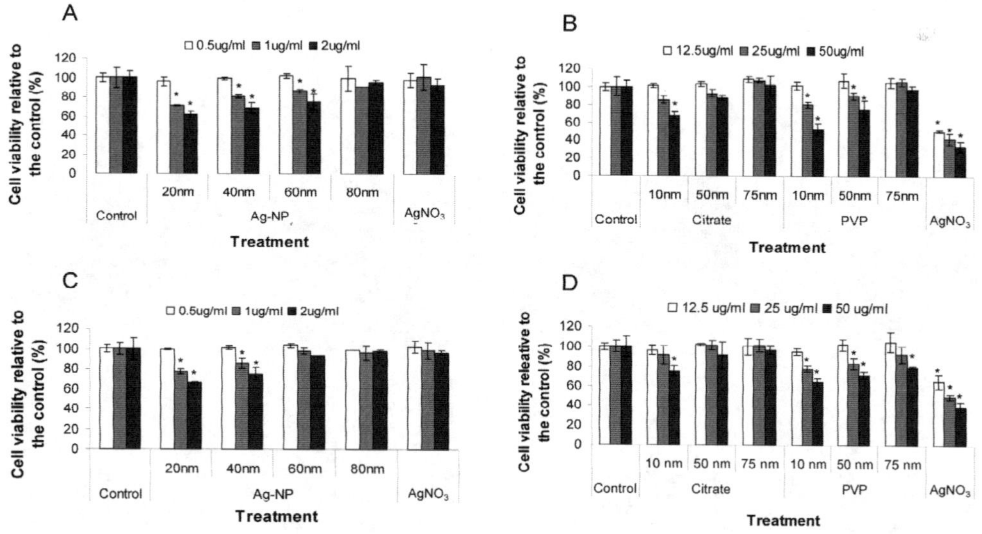

Figure 2: Cell viability of Ag-NPs treated cultures relative to the controls. J774A.1 macrophages (A, B) and HT29 epithelial cells (C, D) were exposed to uncoated Ag-NPs (A, C) or coated Ag-NPs (B, D) for 24 h. The asterisks (*) indicate statistically significant differences compared to the controls ($p<0.05$).

smaller sized particles induced a greater loss in viability, compared to their larger counterparts. For coated Ag-NPs, PVP-coated NPs were 10-20% ($p<0.05$) more cytotoxic compared to citrate-coated NPs of the same size (Figures 2B and D). Moreover, at the equivalent concentrations of silver that were used for uncoated Ag-NPs, $AgNO_3$ caused no observable changes in cell viability indicating that test uncoated Ag-NPs were more cytotoxic than their bulk material. By contrast, at the same silver contents of test coated Ag-NPs, $AgNO_3$ treatments resulted in greater cytotoxic effects (Figure 2).

3.3. Cell morphology

The effects of test Ag-NPs on J774A.1 cell morphology were examined using confocal microscopy and TEM. At 1 µg/ml of silver content, 20 and 40 nm uncoated Ag-NPs caused changes in cell shape and structure (Figures 3A and B). Cell damage including shrinkage, deformation, and enlargement of mitochondria were observed in response to NP treatments. Both the 10 and 50 nm citrate- and PVP-coated Ag-NPs also induced changes in cell morphology at 50 µg/ml (Figures 3A and B). Cells treated with these coated NPs appeared elongated and enlarged. Similar to uncoated particles, these coated versions also resulted in enlarged mitochondria. Both uncoated and coated Ag-NPs were visible inside the cells on the TEM images (arrows in Figures 3B).

3.5. ROS generation

The effect of Ag-NPs on the production of ROS was studied by quantifying the fluorescence from confocal micrographs of oxidised DHE in J774A.1. Treatments with uncoated 20 and 40 nm Ag-NPs at 1 µg/ml resulted in slightly increased intensity and area (1.2-1.3 fold, $p<0.05$) of fluorescence from DHE oxidation compared to the controls, indicating slight ROS generation (Figures 4A and B). Similar observation was made for treatments with 10 nm citrate-coated Ag-NPs (50 ug/ml). PVP-coated 10 and 50 nm Ag-NPs also induced slight increases in ROS generation, but the effects were not statistically significant (Figures 4A, and C). Treatment with menadione was used as the positive control for the assay (Figure 4A).

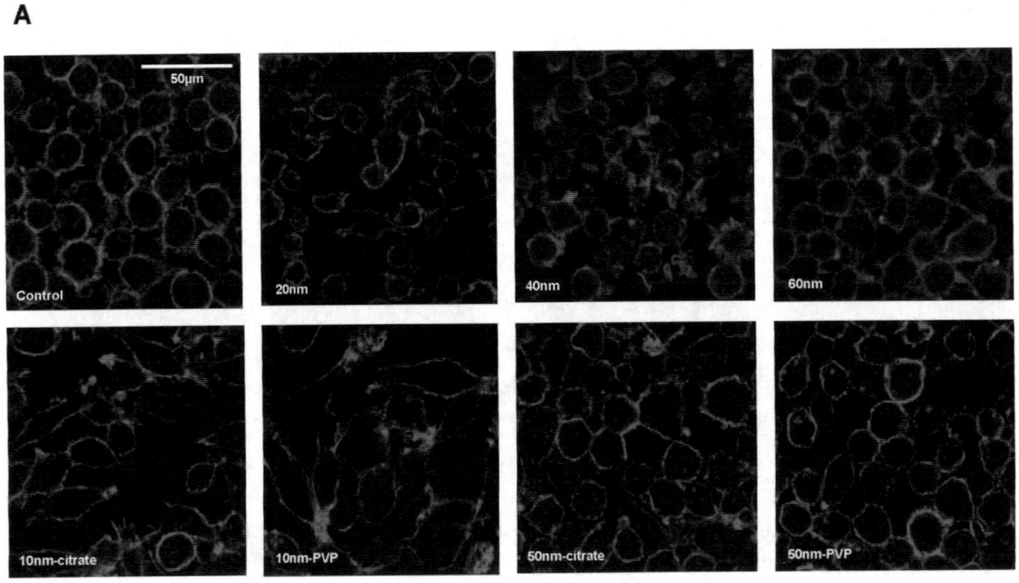

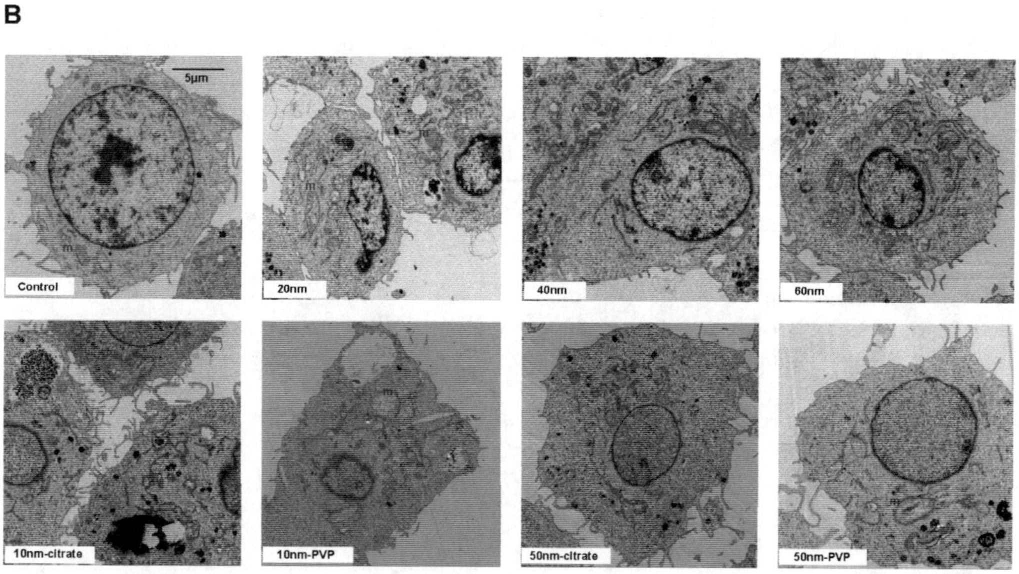

Figure 3: Change in cell morphology and sub-cellular structure induced by Ag-NPs. (A) Confocal micrographs of J774A.1 cells treated with uncoated and coated Ag-NPs for 24 h. The cells were stained for F-actin (pink); nucleus is stained (blue). (B) TEM micrographs of J774A.1 cells treated with uncoated and coated Ag-NPs for 24 h. "m" stands for mitochondrion. Arrows point to the presence of intracellular Ag-NPs.

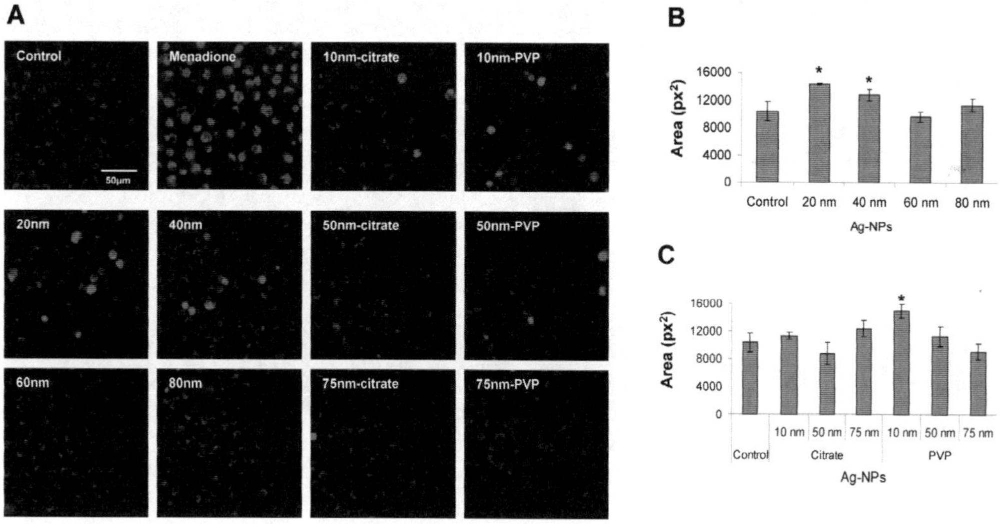

Figure 4: ROS production in J774A.1 cells. (A) Representative micrographs of control cells, cells treated with menadione, and cells treated with test Ag-NPs. The presence of ROS is indicated by red fluorescence. (B, C) The total area of red fluorescence (pixels squared: px^2) was analysed with image analysis software (NIS Elements) using three micrographs for each sample. Each data point represents the mean $\pm$ standard deviation. The asterisks (*) indicate statistically significant differences compared to the control.

3.6. Oxidative stress induction
The total GSH levels and SOD activities were measured to examine oxidative stress induction in cells exposed to Ag-NPs. At 1 μg/ml, uncoated Ag-NPs significantly reduced GSH levels (1.2-1.7 fold, $p<0.05$) in both J774A.1 and HT29 cells. At the same dose, uncoated Ag-NPs also significantly increased SOD activities (1.6-1.8 fold, $p<0.05$) in J774A.1, but not in HT29 cells. These effects appeared to be size-dependent (Figures 5A and C). In contrast, citrate- and PVP-coated Ag-NPs did not affect GSH levels or SOD activities even at the highest test concentrations (Figures 5B and D).

3.4. Cytokine production
The effects of Ag-NPs on selected cytokine levels in test cells were investigated. Treatments of cells with 20 and 40 nm uncoated Ag-NPs at 1 μg/ml resulted in decreased TNF-α levels (1.2-1.4 fold, $p<0.05$) in J774A.1 cells and in IL-8 levels (1.3-1.4 fold, $p<0.001$) in HT29 cells, compared to the control (Figures 6A and C). In contrast, treatments of cells with coated NPs at 25 μg/ml led to elevated (1.2-1.5 fold, $p<0.05$) levels of these cytokines (Figures 6B and D). Similar effects were observed for IL-1β, and IL-12 (p70) levels in J774A.1 (Data not shown).

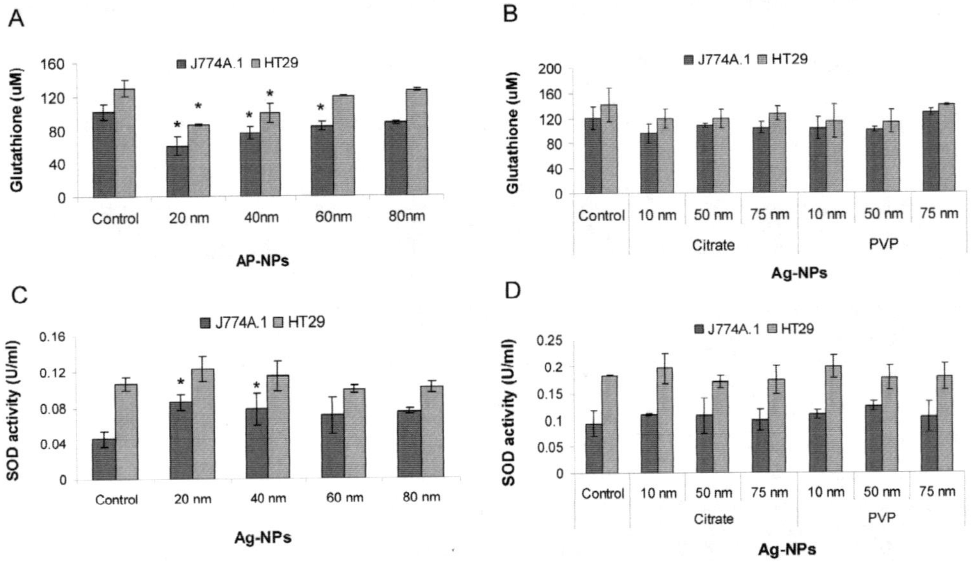

Figure 5: Effects of Ag-NPs on biomarkers of oxidative stress. (A) GSH levels in J774A.1 and HT29 cells after treatment with uncoated Ag-NPs for 24 h. (B) SOD activities in J774A.1 cells after treatment with uncoated Ag-NPs for 24 h. (C) GSH levels in J774A.1 cells after treatment with coated Ag-NPs for 24 h. (D) SOD activities in J774A.1 cells after treatment with coated Ag-NPs for 24 h. Data points represent the means of three independent experiments done in duplicate ± standard deviations. The asterisks (*) indicate statistically significant differences compared to the control at $p<0.05$.

4. Discussion and Conclusions

This study aimed at investigating the impact of physicochemical properties, such as surface coating and size, on the toxicity of Ag-NPs in two different mammalian cell lines, namely

J774A.1 macrophage and HT29 epithelial cells. These cell lines are well-characterised cell models and have been previously used to assess the pathogenic effects of bacteria [30] and the toxicity of other NPs such as quantum dots [31].

Initial work showed that reduction in cell viability by uncoated Ag-NPs occurred at lower concentrations than for coated Ag-NPs, revealing that uncoated NPs are more cytotoxic than their coated counterparts. For coated Ag-NPs of the same size, PVP-coated NPs resulted in greater loss in cell viability compared to citrate-coated NPs, indicating surface coating-dependent toxicity. Our findings agree with a previous study, which compared the toxic effects of citrate-, PVP- and gum arabic (GA)-coated Ag-NPs on the nematode *Caenorhabditis elegans*. The authors reported that PVP-coated Ag-NPs were more toxic than citrate-coated Ag-NPs and demonstrated a linear correlation between Ag-NP toxicity and dissolved Ag^+ ions. The study also suggested that citrate from the coating was protective by forming chelation complexes with free Ag^+ ions and thus reducing availability of dissolved Ag^+ [29].

The greater cytotoxicity of uncoated Ag-NPs compared to coated Ag-NPs observed in the current study is contrary to a recent study which reported that uncoated Ag-NPs were less toxic

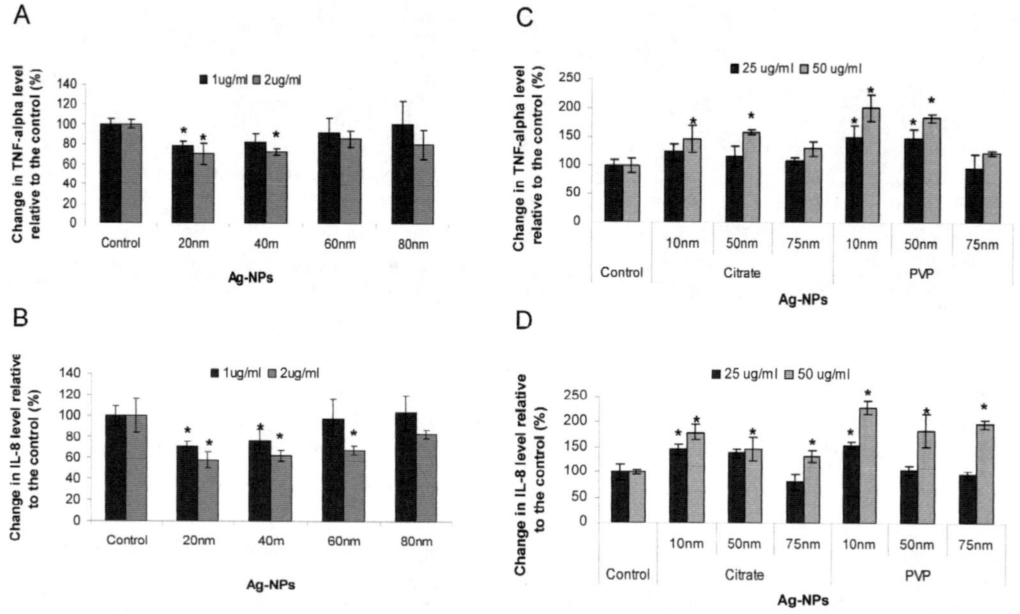

Figure 6: Effects of a 24 h exposure to Ag-NPs on the production of cytokines in test cells. (A) Levels of TNF-α in J774A.1 cells after treatment with uncoated Ag-NPs. (B) Levels of IL-8 in HT29 cells after treatment with uncoated Ag-NPs. (C) Levels of TNF-α in J774A.1 cells after treatment with coated Ag-NPs. (D) Levels of IL-8 in HT29 cells after treatment with coated Ag-NPs. The asterisks (*) indicate statistically significant differences compared to the control ($p<0.05$).

than coated Ag-NPs in similar cell lines such as macrophage RAW264.7 and lung epithelial C-10 [28]. This previous study compared poly(diallyldimethylammonium) chloride, bionic, oleate-coated and uncoated colloidal Ag-NPs sized from 4.0 to 9.0 nm and found that uncoated Ag-NPs were least toxic compared to coated counterparts. The reasons for the discrepancy between our findings and this report could be the difference in the composition of the coating materials and the size of test NPs examined. A number of studies have proposed that the toxicity of Ag-NPs is associated with the release of Ag^+ ions from Ag-NPs [32, 33, 34, 35]. While the dissolution of Ag-NPs was not measured in this study, $AgNO_3$ was used as the Ag^+ ion source for comparative purposes. Our study showed that at equivalent silver concentrations, uncoated Ag-NPs showed greater toxicity than $AgNO_3$. This suggests that silver dissolution cannot be the only factor contributing to uncoated Ag-NP-induced effects in the test cells. Additionally, coated Ag-NPs were less toxic than $AgNO_3$ at equivalent silver concentrations, suggesting that coatings might have protective effects on the release of Ag^+ ions or the mechanisms of toxicity of the Ag-NPs are different from the bulk material.

Coated and uncoated Ag-NPs induced loss in cell viability in a dose- and particle size-dependent manner. Our observations support claims from previous reports on dose- and size-dependent cytotoxicity of Ag-NPs [20, 26, 27]. A study in alveolar macrophages reported that exposure to 15 nm hydrocarbon-coated Ag-NPs led to a significant decrease in metabolic activity, as assessed by the MTT redox assay, compared to a marginal or non-significant effect from exposures to 30 and 55 nm particle sizes [20]. Another study examined the toxicity of Ag-NPs of different sizes (5, 20, and 50 nm) on human cell models including human lung adenocarcinoma A459, gastric carcinoma SGC-790, liver hepatocellular HepG2, and breast adenocarcinoma MCF-7 and reported that 5 nm Ag-NPs exposure resulted in highest cell viability loss (also assessed by MTT assay) and greater changes in cell morphology compared to its larger sized counterparts [26]. Similar findings were reported in a recent study that examined cellular toxicity of differently sized Ag-NPs on mouse osteoblastic MC3T3-E1 and rat adrenal pheochromocytoma PC12 cell lines [27].

Changes in gross cell morphology as examined by confocal microscopy corroborated our observations on the loss in cell viability. However, while uncoated Ag-NPs resulted in cell shrinkage, coated Ag-NPs induced cell elongation and enlargement, suggesting different mechanisms of cell damage. Examination of sub-cellular structures using TEM revealed mitochondrial enlargement after exposures to both uncoated and coated Ag-NPs at high test concentrations. Mitochondria are known to play a role in NP toxicity. For example, following NP exposures, mitochondrial structural damage, loss of mitochondrial membrane integrity, opening permeability transition pores, ROS production, and cell death may occur [36, 37]. Our observations support previous studies that linked Ag-NP toxicity to mitochondrial damage [10, 11, 21, 38, 39]. Further studies are needed to determine whether mitochondrial damage is the primary effect of Ag-NPs or is secondary to cell death stimulation from other cellular targets of the NPs.

Oxidative stress is one of the proposed mechanisms of Ag-NP toxicity [20, 40, 41]. Oxidative stress can be the consequence of the generation of intracellular ROS [38, 20, 10]. The depletion of GSH and a reduction of anti-oxidant enzyme activity might also be causes of oxidative stress in cells and animals [41, 42]. Ag-NPs have been reported to increase ROS generation from Ag^+ ions released from the NPs or from the Ag^+ on the NP surface [11, 20, 40, 35]. Alternatively, there is evidence that Ag-NPs are able to bind to thiol groups with high affinity which are found in thiol proteins/peptides such as GSH restricting the availability of these molecules for ROS neutralization and resulting in an oxidant-mediated response to ROS [2, 43]. A recent study using Chang liver cells also reported that Ag-NPs decreased GSH levels through the inhibition of GSH synthesizing enzymes [42]. In the present study, uncoated Ag-NPs at high test concentrations reduced GSH content and slightly increased ROS generation. This may be explained by the possible direct interaction of these Ag-NPs with GSH or by the inhibition of GSH-synthesizing enzymes by the Ag-NPs leading to an accumulation of ROS. The increase in SOD activities observed with uncoated Ag-NPs could be a result of excess ROS and oxidative stress in cells. In contrast, coated Ag-NPs exposure lead to non-significant changes in GSH and SOD suggesting oxidative stress might not be the primary toxic mechanism of these Ag-NPs.

Ag-NPs have been shown to have both stimulatory and suppressive effects on the production of cytokines [9, 20, 44, 45, 46]. In the present study, uncoated Ag-NPs decreased pro-inflammatory cytokines such as TNF-α, IL-1β, IL-12 (p70) and IL-8, while cells exposed to coated Ag-NPs had elevated levels of these cytokines. Inhibition of cytokine production by Ag-NPs has been reported [45, 46]. These reports found that at doses similar to ours, uncoated Ag-NPs strongly inhibited cytokine production of IL-1β, IL-6, and TNF-α in J774A.1 cells [45]. The inhibitory effects of Ag-NPs on the production of IL-8 and IL-11 were also reported in human mesenchymal stem cells (hMSCs) and on the production of IL-5, IFN-γ, and TNF-α in the peripheral blood mononuclear (PBMC) cell line [44, 46]. The inhibitory effects of Ag-NPs on cytokines have been suggested to result from the interaction of Ag-NPs with toll-like receptors (TLRs) located on the cell surface or in the endocytic compartments [47]. However, Ag-NPs were also found to increase the production of TNF-α, MIP-2 and IL-1β in alveolar macrophages [20]. Similarly, exposures of Ag-NPs to human epidermal cells led to an increase in IL-1β, IL-6, IL-8 and TNF-α [9]. These differential effects on cytokine production have been suggested to be dependent on dose and cell type [48]. In our study, for the same cell type, the threshold doses used for uncoated Ag-NPs were much lower than those used for coated particles. This difference in dose could explain the different effects observed on cytokine production. These effects need to be further investigated.

In summary, the present study demonstrates the effects of physicochemical properties on the toxicity of Ag-NPs. Our data suggest that uncoated Ag-NPs are more toxic than coated Ag-NPs and that the size and coat-type are also important contributing factors to NP toxicity. Cells treated with uncoated Ag-NPs appeared to undergo oxidative stress while coated-Ag-NPs seemed to induce cell damage through inflammatory pathways by up-regulating selected cytokines. This study provides toxicological data which might be helpful not only for assessing the potential health and environmental risks of Ag-NPs, but also for the safe design and use of these NPs.

Acknowledgements

The authors would like to thank Dr. Hongyan Dong and Dr. Guillaume Pelletier for reviewing the manuscript. This work was supported by Canadian Regulatory System for Biotechnology and intramural Chemical Management Plan (CMP) Monitoring and Surveillance Funding and the Natural Sciences and Engineering Research Council (NSERC) of Canada.

References

[1] Murphy C J, Sau T K, Gole A M, Orendorff C J, Gao J, Gou L, Hunyadi S E and Li T 2005 *J Phys Chem B* **109** 13857–13870.
[2] Lewis L N 1993 *Chem Rev* **93** 2693– 2730.
[3] Marambio-Jones C and Hoek E MV 2010 *J Nanopart Res* **12** 1531–1551.
[4] Chen X and Schluesener H J 2008 *Toxicol Lett* **176** 1–12.
[5] Maynard AD 2006 Washington DC: WoodrowWilson International Center for Scholars.
[6] Ahamed M, Alsalhi M and Siddiqui M 2010 *Clinica Chimica Acta* **411** 1841–1848
[7] Lara H H, Garza-Trevino E N, Ixtepan-Turrent L and Singh D K 2011 J. *Nanobiotechnology*. **9** 30.
[8] Eom H J and Choi J 2010 *Environ Sci Technol* **44** 8337–8342.
[9] Samberg M E, Oldenburg S J and Monteiro-Riviere N A. *Environ Health Perspect* **118** 407–413.
[10] Asharani P V, Mun G L K, Hande M P and Valiyaveettil S 2009 *ACS Nano* **3** 279–290.

[11] Hussain S M, Hess K L, Gearhart J M, Geiss K T and Schlager J J 2005 *Toxicol In Vitro* **19** 975–983.
[12] Yang Z, Liu Z W, Allaker R P, Reip P, Oxford J, Ahmad Z and Ren G 2010 *J R Soc Interface* **7** 411–422.
[13] Asharani P V, Wu Y L, Gong Z and Valiyaveettil S 2008 *Nanotechnology* **19** 255102.
[14] Scown T M, Santos E M, Johnston B D, Gaiser B, Baalousha M, Mitov S, Lead J R, Stone V, Fernandes T F, Jepson M, van Aerle R and Tyler C R 2010 *Toxicol Sci* **115** 521–534.
[15] Roh J Y, Sim S J, Yi J, Park K, Chung K H, Ryu D Y and Choi J 2009 *Environ Sci Technol* **43** 3933–3940.
[16] Asghan S, Johan S A, Lee J H, Kim Y S, Jeon Y B, Choi H J, Moon M C and Yu I J 2012 *J. Nanobiotechnolpgy* **10** 14.
[17] Hyun J S, Lee B S, Ryu H Y, Sung J H, Chung K H and Yu I J 2008 *Toxicol Lett* **182** 24–28.
[18] Sung J H, Ji J H, Yoon J U, Kim D S, Song M Y, Jeong J, Han B S, Han J H, Chung Y H, Kim J, Kim T S, Chang H K, Lee E J, Lee J H and Yu I J 2008 *Inhal Toxicol* **20** 567–574.
[19] Rahman M F, Wang J, Patterson T A, Saini U T, Robinson B L, Newport G D, Murdock R C, Schlager J J, Hussain S M, Ali S F 2009 *Toxicol Lett* **187** 15–21.
[20] Carlson C, Hussain S M, Schrand A M, Braydich-Stolle L K, Hess K L, Jones R L and Schlager J J 2008 *J Phys Chem B* **112** 13608–13619.
[21] Hsin Y H, Chen C F, Huang S, Shih T S, Lai P S and Chueh P J 2008 *Toxicol Lett* 179 130–139.
[22] Ahamed M, Karns M, Goodson M, Rowe J, Hussain S M, Schlager J J and Hong Y 2008 *Toxicol Appl Pharmacol* **233** 404–410.
[23] Samberg M E, Oldenburg S J and Monteiro-Riviere N A 2010 *Environ Health Perspect* **118** 407–413.
[24] Oberdörster G, Oberdörster E and Oberdörster J 2005 *Environ Health Perspect* **113** 823–839.
[25] Maynard A D, Warheit D B and Philbert M A 2011 *Toxicol Sci* **120** S109–S129.
[26] Liu W, Wu Y, Wang C, Li H, Wang T, Liao C Y, Cui L, Zhou Q F, Yan B and J G B 2010 *Nanotoxicology* **4** 319–330.
[27] Kim T-H, Kim M, Park H-S, Shin U S, Gong M-S and Kim H-W 2012 *J Biomed Mater Res Part A* **100A** 1033–1043.
[28] Suresh A K, Pelletier D, Wang W, Morrell-Falvey J L, Gu B, and Doktycz M J 2012 *Langmuir* **28** 2727–2735.
[29] Yang X Y, Gondikas A P, Marinakos S M, Auffan M, Liu J, Hsu-Kim H and Meyer J N 2012 *Environ. Sci. Technol.* **46** 1119–1127.
[30] Tayabali A F and Seligy V L 2000 *Environ Health Perspect* **108** 919-930.
[31] Nguyen K C, Seligy VL and Tayabali AF 2012 *Nanotoxicology* **Early Online** 1-10.
[32] Park E J, Yi J, Kim Y, Choi K and Park K 2010 *Toxicol In Vitro* **24** 872-878.
[33] Miura N and Shinohara Y 2009 *Biochem Biophys Res Commun* **390** 733–737.
[34] Meyer J N, Lord C A, Yang X Y, Turner E A, Badireddy A R, Marinakos S M, Chilkoti A, Wiesner M R and Auffan M 2010 *Aquat Toxicol* **100** 140–150.
[35] Foldbjerg R, Olesen P, Hougaard M, Dang D A, Hoffmann H J and Autrup H 2009 *Toxicol Lett* **190** 156–162.
[36] Li N, Sioutas C, Cho A, Schmitz D, Misra C, Sempf J, Wang M, Oberley T, Froines J and Nel A 2003 *Environ Health Perspect* **111** 455–460.
[37] Xia T, Kovochich M and Nel A E 2007 *Front Biosci* **12** 1238–1246.
[38] Foldbjerg R, Dang D A, and Autrup H 2010 *Arch Toxicol* **85** 743–750.
[39] Teodoro J S, Simões A M, Duarte F V, Rolo A P, Murdoch R C, Hussain S M and Palmeira

C M 2011 *Toxicol. In Vitro* **25** 664–670.
[40] Kim S, Choi J E, Choi J, Chung H, Park K, Yi J and Ryu D Y 2009 *Toxicol In Vitro* **23** 1076–1084.
[41] Arora S, Jain J, Rajwade J M, and Paknikar K M 2008 *Toxicol Lett* **179** 93–100.
[42] Piao M J, Kang K A, Lee I K, Kim H S, Kim S, Choi J Y, Choi J and Hyun J W 2011 *Toxicol. Lett.* **201** 92–100.
[43] Navarro E, Piccapietra F and Wagner B, Marconi F, Kaegi R, Odzak N, Sigg L and Behra R 2008 *Environ Sci Technol* **42** 8959–8964.
[44] Yen H J, Hsu S H and Tsai C L 2009 *Small* **5** 1553–1561.
[45] Greulich C, Kittler S, Epple M, Muhr G and Koller M 2009 *Langenbecks Arch Surg* **394** 495–502.
[46] Shin S H, Ye M K Kim H S and Kang H S 2007 *Int Immunopharmacol* **7** 1813–1818.
[47] Castillo P M, Herrera J L, Fernandez-Montesinos R, Caro C, Zaderenko A P, Mejias J A and Pozo D 2008 *Nanomed* **3** 627–635.
[48] Stensberg M C, Wei Q, McLamore E S, Porterfield D M, Wei A and Sepúlveda M S 2011 *Nanomedicine* **6** 879–898.

Journal of Physics: Conference Series **429** (2013) 012026 doi:10.1088/1742-6596/429/1/012026

An impedance-based high-throughput method for evaluating the cytotoxicity of nanoparticles

M R Cimpan[1], T Mordal[1], J Schölermann[2], Z E Allouni[1], U Pliquett[3] and E Cimpan[4]

[1] Biomaterials – Department of Clinical Dentistry, Faculty of Medicine and Dentistry, University of Bergen, Årstadveien 19/21, N-5009 Bergen, Norway

[2] Department of Biomedicine, Faculty of Medicine and Dentistry, University of Bergen, Jonas Lies vei 91, N-5009 Bergen, Norway

[3] Institute for Bioprocessing and Analytical Measurement Techniques, D-37308 Heilbad Heiligenstadt, Germany

[4] Department of Electrical Engineering, Bergen University College, Nygårdsgaten 112, N-5020 Bergen, Norway

E-mail: mihaela.cimpan@iko.uib.no

Abstract. Impedance-based assays can constitute a reliable alternative to the conventional methods used in nanotoxicology due to the important advantages of being label-free and monitoring the cells in real-time. In this study, the suitability of impedance-monitoring for the screening of nanoparticle (NP)-induced cytotoxicity was assessed. The effect of titanium dioxide (TiO_2)-NPs on cellular proliferation, viability, spreading, and detachment from substrate was evaluated by continuous impedance-based measurements made with an xCELLigence system. Fibroblasts seeded in microelectrode-embedded E-plates were exposed to spherical anatase nano-TiO_2 (5, 10, and 40 nm in diameter) for up to 120 h. An alternative excitation signal (20 mV control voltage amplitude) was applied at 10, 25, and 50 kHz to the microelectrodes in the E-plates. Cells attached to the electrode surfaces act as insulators and lead to an increase in impedance. For validating the impedance-method, Trypan Blue exclusion and ultrahigh resolution imaging (URI) were employed. The general trend observed was a decrease in impedance following exposure to TiO_2-NPs. Impedance-based results were in most instances in accordance with those from the Trypan Blue exclusion and URI assays indicating that the impedance-based approach has merit. Further studies are needed to validate it as a high-throughput method for evaluating NPs' cytotoxicity.

1. Introduction

A reliable and efficient evaluation of the biological effects of nanomaterials requires the development of appropriate high-throughput screening methods [1, 2]. In general, the same methods that have been used for in vitro toxicity testing of soluble and macro-scale substances are also employed for evaluating the toxicity of NPs [3, 4]. A considerable number of nanotoxicity studies have generated confounding or conflicting data, indicating that due to their physicochemical properties, NPs interfere

[1] To whom any correspondence should be addressed.

with assay components and/or detection systems [4]. Label-free techniques would eliminate the interferences between NPs, substances used for labeling and the detection systems [4]. Another common problem shared by traditional methods is that they assess the effects only at a specific point in time and do not give any information on the biokinetic behavior and the interaction between living cells and toxicants over time [5]. As a result, the need for label-free real-time detection methods for investigating the toxicity of NPs on living systems has emerged.

Assays that measure the electrical properties of cells, such as impedance-based assays, are label-free and can represent a reliable alternative to the traditional methods [6]. These methods analyze matter either by their active or passive electrical properties [7]. A method that takes advantage of the passive properties is impedance-based monitoring which measures how much an object impedes a flow of electrical current. In impedance-based monitoring, an alternating current is used to generate information about the properties of the cell as a function of signal frequency. A variety of cellular properties such as proliferation, growth, morphology, and adhesion can be measured by applying an electric source to an electrode covered by cells, and measuring how the cells impede the current [7].

Titanium dioxide (TiO_2) NPs are widely used in paints, pharmaceuticals, food colorants, and cosmetics. Titanium and its alloys are commonly used as implant materials [8] and TiO_2 NPs can be released from implants and cause toxic effects [9, 10, 11]. In this study, the suitability of impedance-monitoring for the screening of NPs -induced cytotoxicity was assessed. Impedance-based monitoring of L929 fibroblasts exposed to TiO_2 NPs was performed by using an xCELLigence system, which is able to monitor cells "in situ" without the use of markers. The system monitors electrical impedance by sending an alternative current through microelectrodes at the bottom of the wells. When the cells cover the microelectrodes, the impedance goes up. The increase in impedance leads to an increase in voltage drop across the electrodes. Based on the measured impedance, it can be inferred whether the amount of cells is increasing. The number, the development, morphology, strength and the degree of adhesion of the cells can also be deduced by the instrument to a varying degree. To validate the method, cellular proliferation and viability were evaluated by manual and automated cell counting (the Trypan Blue assay) and by ultrahigh resolution microscopy (URI).

2. Material and Methods

2.1. TiO_2 nanoparticles

Three spherical anatase TiO_2-NPs (Nanostructured & Amorphous Mat., USA) with diameters of 5 (Ti1), 10 (Ti2) and 40 nm (Ti3), were characterized by X-ray diffraction, transmission electron microscopy (TEM), N_2-BET adsorption, dynamic light scattering, and electophoretic mobility [11] (table 1). NP suspensions were prepared as described by Allouni et al. [11, 12].The concentrations used, i.e., 0.05, 0.5, and 5 mg/L were in the range of those that have been reported in serum, blood, and in dry tissue surrounding Ti-based implants [13, 14, 15].

2.2. Cell culture

L929 mouse fibroblasts (CCL-1; ATCC, Manassas, VA, USA) were cultured in 25 cm^2 flasks (Costar, Switzerland) in RPMI 1640 medium (BioWhittaker, Lonza, Belgium) supplemented with 10% heat inactivated (30 min, 56°C) fetal calf serum (FCS) (Millipore, MA, USA), 2% Penicillin/Streptomycin/Amphotericin, and 1% L-glutamine (BioWhittaker, Lonza, Belgium). The cells were kept at 37°C in humidified air with 5% CO_2. Only cells from cultures with a viability >90% (tested by exclusion of 0.2% Trypan Blue) were used in the experiments. All subsequent experiments were run in duplicate and repeated three times.

2.2. Impedance-based monitoring of cells

The system used in this study was the xCELLigence Real Time Cell Analyzer Dual Plate instrument (RTCA DP) (Roche Diagnostics GmbH, Mannheim, Germany). The system measures electrical impedance across interdigitated microelectrodes situated at the bottom of culture wells. The RTCA-DP Analyzer has three stations for electrode-plates (E-plates). The measurements are done by applying an alternative excitation signal (20 mV control voltage amplitude) at three different frequencies (10, 25 and 50 kHz) through the microelectrodes in the E-plates while monitoring the voltage drop across the electrodes where the quotient voltage/current yields the impedance. It was previously demonstrated that the excitation signal used does not affect the physiological state of the cells [16]. The E-plate 16 is a single-use plate with 16 wells (5 mm diameter) with microelectrodes which cover about 80% of the bottom of the wells. The RTCA DP Analyzer was located inside an incubator at 37°C in humidified air with 5% CO_2. The RTCA DP Analyzer is connected to a control unit consisting of a computer with software installed for controlling the unit and for receiving, analyzing and displaying the data obtained from the measurements. The software displays the measured impedance, which is normalized by a frequency factor Zn, as a Cell Index (CI) for each point in time. The CI is a measure of the electrical impedance both in the presence and absence of cells (equation 1):

$$CI = \frac{R(f_n,t)-(f_n,t_0)}{Z_n} \qquad (1)$$

where f_n is the frequency at which the impedance measurement is carried out, $R(f_n,t)$ is the measured impedance at frequency f_n at time-point t, $R(f_n,t_0)$ is the measured impedance at frequency f_n at t_0 (time of background measurement), and Z_n is the corresponding frequency factor of f_n. The xCELLigence system measures the impedance at three discrete frequencies: f_1=10 kHz, f_2=25 kHz and f_3=50 kHz. The corresponding frequency factors are Z_1= 15 Ω, Z_2= 12 Ω, and Z_3= 10 Ω respectively. The CI is a dimensionless number since the measured impedance is divided by the frequency factor. A change in CI can be indicative of changes in cell number, cell adhesion, spreading or cell morphology.

When cells are not present the impedance measured from the microelectrodes in the wells is dependent on the electrode geometry, the ion environment and the electrical properties of the material in the well. When cells are attached to the sensor surfaces of the microelectrodes they act as insulators, and lead to an increase in impedance.

Table 1. Physiochemical characteristics of TiO_2 NPs used in the study

Sample	S_{BET}[a] (m^2/g)	Crystal Structure	ζ-potential ± SD [mV]	IEP[b]	Hydrodynamic[c] diameter (PDI) (nm) (after 24 h)
Ti1	131	86.6% **A.** 13.4% **R.**	-13.2 ± 0.6	3.52	423.7 ± 9.7 (408.0 ± 10.0)
Ti2	105	100% **A.**	-11.8 ± 1.1	6.88	463.1 ± 12.1 (445.5 ± 10.9)
Ti3	20	81% **A.** 18.9% **R.**	not given	3.45	446.2 ± 6.1 (400.0 ± 9.2)

[a] Specific surface area. [b] Isoelectric point from titration curve: zeta potential vs. pH in aqueous solution of 0.14 M NaCl. [c] DLS measurements in RPMI cell culture medium with supplemented 10% FBS, immediately after dispersing the particles, and after 24 hours (from Allouni et al. [11]).

2.2.1. Determination of optimal cell density. A titration experiment was performed in order to determine the cell concentration that gave optimum cell viability and proliferation. The following cell densities were tested: 3125, 6250, 12500, 25000, 50000, and 100000 cells/cm^2.

2.2.2. Impedance-based monitoring of cells exposed to TiO_2 nanoparticles. L929 cells (100,000 cells/cm^2) needed 48 h to reach the log phase, which was the time-point when they were exposed to 0.05, 0.5, and 5 mg/L suspensions of TiO_2 NPs in cell culture medium. Impedance measurements were carried out at 15 min intervals for a total period of 7 days, including the initial attachment phase.

2.2.3. Trypan Blue assay
The viability of the cells was evaluated at the beginning and end of each experiment by Trypan Blue (0.2%). The cells were seeded in 24-well plates (TPP, Switzerland) and exposed to TiO_2 NPs under the same conditions as in the impedance assay. The viable and total cell number was counted manually with a Neubauer chamber and with the Countess Automated Cell Counter (Invitrogen, CA, USA). The cells on the E16 plates were trypsinised at the end of each assay and their viability evaluated by manual counting.

2.3. Ultra-high resolution microscopy
For investigating morphological changes after interactions with NPs, an Ultrahigh Resolution Imaging (<100 nm) (URI) system (CytoViva™, Auburn, USA) was used [17]. The cells were fixed and then visualized with a 100x oil immersion objective as described by Allouni et al. [11].

2.3.1. Data analysis
The Mann-Whitney U test for paired comparison of independent samples ($p<0.05$) and the SPSS statistical software were used. The CI-values of both the control and treated cells were compensated for the impedance of the medium and for the impedance of the NPs plus medium, respectively by performing the following subtractions at each time-point:

$$\mathrm{CI(control\ cells)} = \mathrm{CI(control\ cells + medium)} - \mathrm{CI(medium)}$$

$$\mathrm{CI(treated\ cells)} = \mathrm{CI(treated\ cells + medium + NPs)} - \mathrm{CI(medium + NPs)}$$

For cells exposed to TiO_2 NPs, the Mean Cell Index (MCI) values for all experiments were calculated at each time-point by using equation (2):

$$\mathrm{MCI} = \frac{1}{N}\sum_{I=1}^{N} CI_i \qquad (2)$$

where N was the number of measurements and CI_i were the individual measurements of the CI at a given time. The MCI-values were normalized by choosing the MCI at 7 h after exposure as the basis for normalization. This was done to correct for the differences which occurred before the normalization time. The reason for choosing this time-point for normalization was that variations due to changes in the temperature had settled down at this time. The Normalized Cell Index (NCI) was calculated as follows:

$$NCI = MCI\ (specific\ time)\ /\ MCI\ (normalization\ time)$$

A Mean Normalized Cell Index (MNCI) for the different control cells and treated cells was subsequently calculated for all time-points t up to t_{final}, as follows:

$$\mathrm{MNCI} = \frac{\frac{1}{N}\sum_{I=1}^{N} NCI_i}{\Delta t} \qquad (3)$$

where N was the number of measurements, NCI_i were the NCIs for each time-point t, and Δt was the time between the time-point of normalization and the time-point where the cells started to plateau t_{final}. The calculated MNCI was used to determine the impact of TiO_2 NPs cell proliferation.

3. Results

3.1. Determination of optimal cell density

The MCI reached a peak 2-3 h after cell seeding and then declined and reached a minimum after about 20 h (figure 1). The MCI then started to rise as a result of cells attaching to the surface and dividing. No growth could be seen for 3,125 and 6,250 cells/cm², while a minimal growth was noted at 12,500 cells/cm². Tested densities leading to the highest rate of cell growth were 50,000 and 100,000 cells/cm².

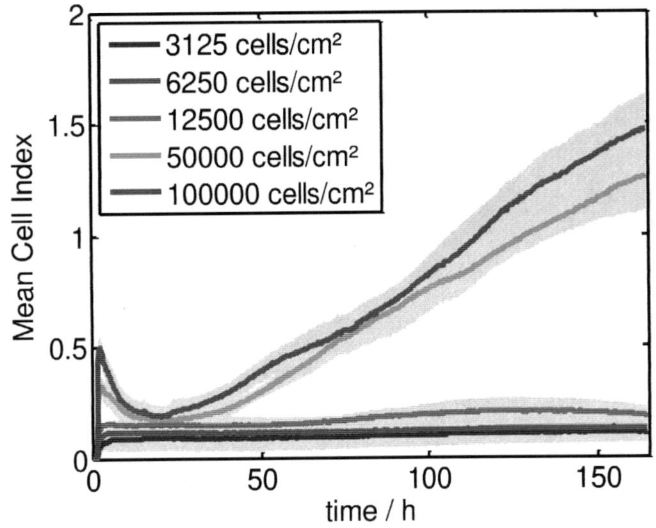

Figure 1. Impedance-based monitoring of L929 cells seeded at densities between 3,125 and 100,000 cells/cm². The highest MCI is seen for the highest seeding densities, indicating that the cells grew best at these densities. Each curve represents an average of six measurements (gray area represents the standard deviation).

3.2. Impedance-based monitoring of cells exposed to TiO_2 nanoparticles.

Plots of MCI vs. time and NCI vs. time are shown for cells exposed to Ti3 in figures 2 and 3. In the first plot, the MCI vs. time is shown for the different treatments for the duration of the experiment (figure 2). The second plot shows the NCI starting at 7 h after the exposure to TiO_2 NPs (figure 3). The MNCI was calculated starting at 7 h after exposure and until the cells reached a plateau.

At 12 h, cells started to spread and divide and differences between the curves became visible, which are possibly due to small differences in seeding densities. TiO_2 NPs were added at 48 h when cells were in the log phase. When NPs were added, a double peak was observed (B). The time at 55 h was selected as the time point for normalizing the MCI-curves from which the relative growth of the cells can be observed (figure3). As mentioned earlier, this specific time-point was chosen for normalization because peaks due to changes in the temperature had settled down at this time. By normalization, it was possible to correct for the different MCI of the cells that were observed before the exposure. The relative growth of the NCI for each of the curves was plotted from the normalization point at 55 h onward up to t_{final} (figure 3). The highest NCI at the end was observed for the control, while the rest distributed according to the concentration of TiO_2 NPs added. It could be seen that the increased NPs concentration induced a decrease in NCI.

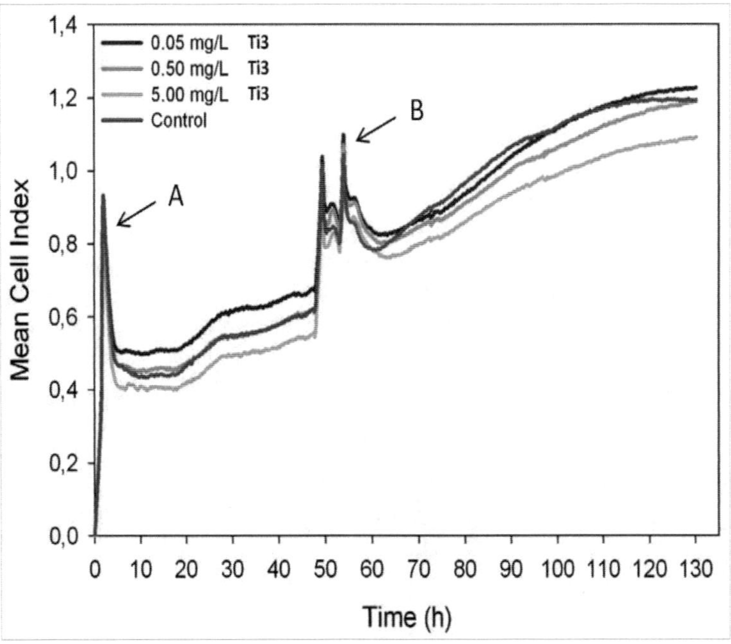

Figure 2. MCIs for L929 cells exposed to Ti3 48 h after being seeded at a density of 100,000 cells/cm^2. A) The spike at time 0 is due to cells settling onto the electrodes and starting to attach to the surface. B) The spike at the time of exposure (48h) is due to temperature changes following the E-plate being taken out of the RTCA analyzer to be exposed to TiO_2 NPs. The curves represent the average of six measurements.

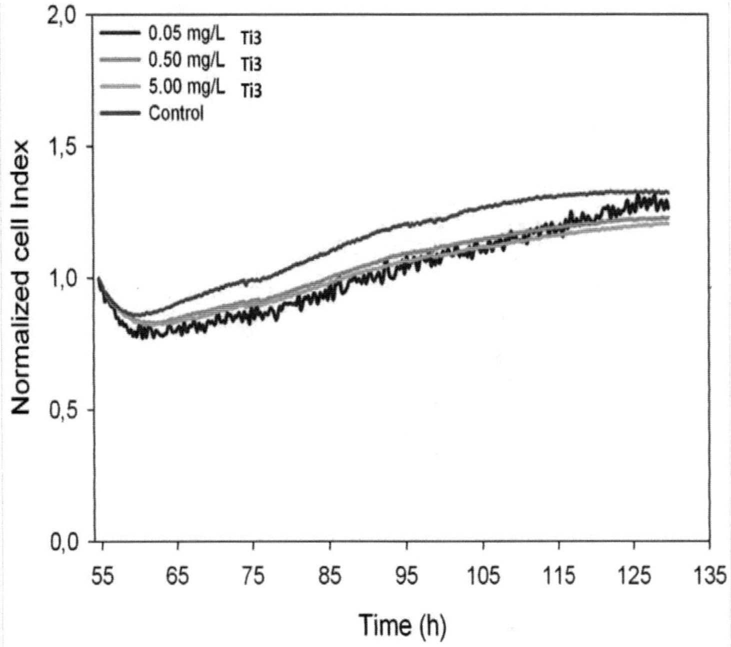

Figure 3. NCIs for L929 cells exposed to 0.05, 0.5, and 5 mg/L of Ti3. The cells were exposed to the particles 48 h after being seeded at a density of 100,000 cells/cm^2. The cell indexes were normalized based on the MCI at 55 h.

3.3 Effects of TiO2 NPs on the Mean Normalized Cell Index.

For comparing the effects of the NPs, the MNCI was calculated for the interval of 55-130 h. The endpoint of the interval was considered to be where cell proliferation reached a plateau. All exposures induced a decreased MNCI compared to controls (figure 4). For Ti1, the MNCI increased with an increase in the concentration of particles. For the remaining particles, the only trend that could be observed was that the treatments had lower MNCIs than the controls.

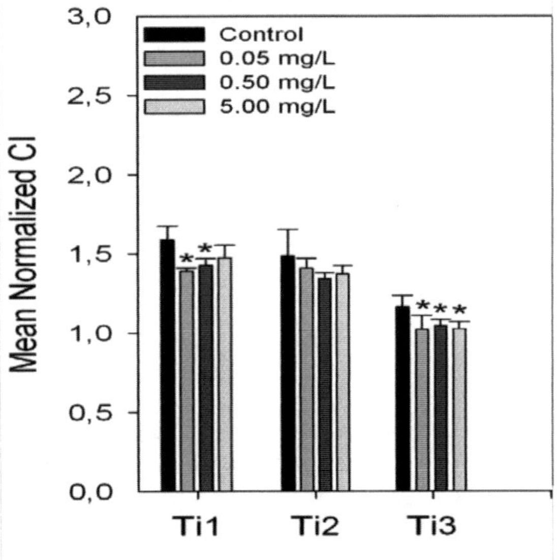

Figure 4. MNCIs of L929 cells exposed to TiO_2 NPs 48 h after being seeded at a density of 100,000 cells/cm^2. The average cell indexes were calculated starting 7 h after exposure until the cells plateaud at 75 h postexposure to NPs. The cell indexes were normalized based on the MCI 7 h after exposure to NPs. The results are expressed as mean ± SD of the mean. (*) shows statistical significance ($p<0.05$) compared to the control.

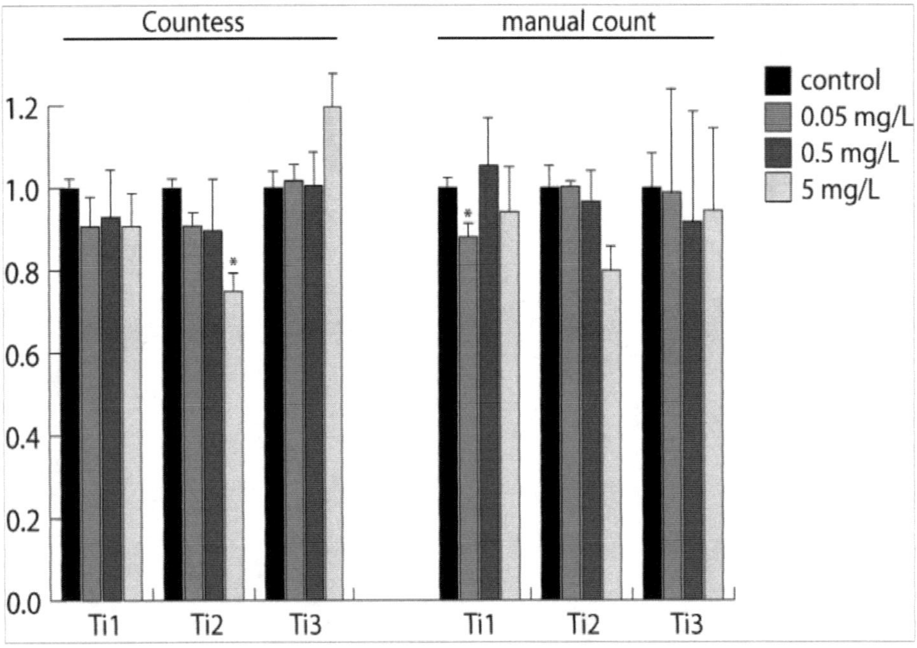

Figure 5. Total number of viable L929 fibroblasts after 120 hrs of incubation with Ti1, 2 and 3. Cells were seeded at an initial density of 100,000 cells/cm^2, allowed to attach during 48 hrs before NPs were added for 120 hrs. At the end of incubation, cells were trypsinized, stained with Trypan Blue and the number of viable cells was counted using the Countess automated cell counter as well as by manual counting using a Neubauer counting chamber. Total cell count was normalized to control for each NP type. (*) shows statistical significance ($p<0.05$) compared to the control.

The most common trend appears to be that the MNCI (impedance measurements) and the number of viable cells (Trypan Blue) decrease after exposure to particles. Statistically significant differences could be noted for Ti1 at 0.5 and 5 mg/L and for Ti3 at all concentrations in the impedance assay. In the Trypan Blue assay, significant differences were noted only for Ti2 5 mg/L and Ti1 0.05 mg/L. Even for the same assay, i.e., Trypan Blue, differences could be seen between the results obtained by manual and automated counting. The impedance measurements were more in accordance with the ones obtained by manual counting.

3.4. URI microscopy.

The uptake of TiO_2 NPs by L929 fibroblasts and putative changes in cellular morphology were explored by URI-microscopy. Agglomerates of TiO_2 NPs could easily be observed due to their intense light scattering and were associated with cells at all three concentrations [11]. The particles were situated on the membrane and inside the cells. The cells exposed to 0.5 mg/L and 5 mg/L were rounder and smaller than the control cells, which could be indicative of cellular death by apoptosis (figure 6).

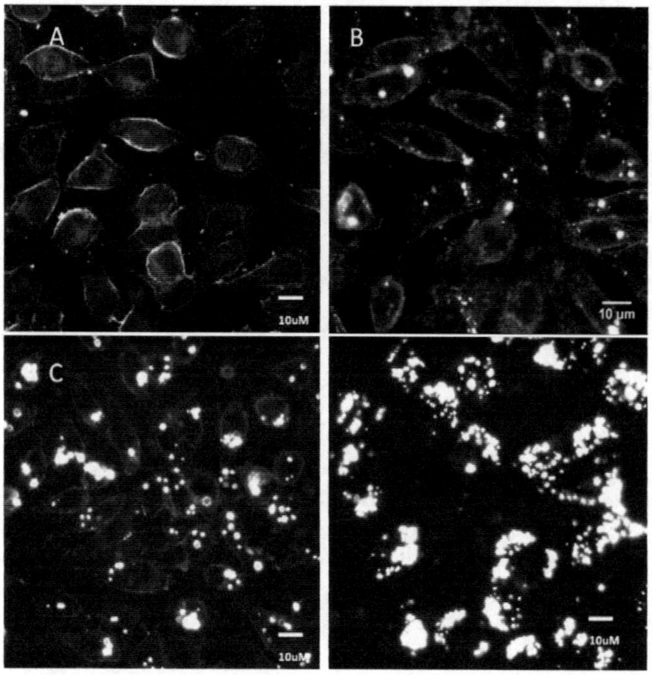

Figure 6. URI images of L929 cells at 100x magnification after 24 h exposure to Ti1: A) unexposed cells. B) 0.05 mg/L. C) 0.5 mg/L. D) 5 mg/L. The bright spots are TiO_2 NPs, either attached to the cell membrane or internalised by the cells.

4. Discussion

The objectives of this study were to evaluate the impedance-based sensing approach for nanotoxicity assays and to investigate the effect of TiO_2 NPs on cellular viability and proliferation.

The xCELLigence system monitored the proliferation and adhesion of cells positioned on microelectrodes by measuring the cell-sensor impedance. Several studies have indicated that the xCELLigence system is advantageous for testing the cytotoxicity of compounds compared to end-point assays [16, 18, 19]. Nevertheless, the need for validating the xCELLigence results with results obtained from traditional assays has also been pointed out [18]. For evaluating the impedance-based approach in this study, cell proliferation and viability measured by the Trypan Blue assay were used for comparison. To control for possible interferences of NPs, both manual and automated (Countess, Invitrogen) counting were employed. URI was used as a qualitative assay for visualizing the putative modifications in cellular morphology after exposure to TiO_2 NPs.

The variations of the MCI represented in the xCELLigence plots as curves, could be explained as follows: after being seeded on the E-plates, the cells sedimented to the bottom of the wells and started to attach to the surface of the sensors. This was observed as a sharp increase in the MCI around 2-3 h after cell seeding. This was followed by a lag phase with cells preparing to enter cell division. This was seen as an early plateau in the MCI plots. Once the cells found their place, they attached better to the surface and started to stretch and spread, and then entered the log phase where they proliferated. The proliferation was reflected by a steep MCI increase. Then, the curves reached a final plateau, also known as the stationary phase, where the cells stop growing due to constraints like too little space or cell culture medium. During this phase, some cells will die and MCI consequently declines. Nanoparticles can affect the adhesion and spreading of cells as well as cell growth and viability. Changes in the cell surface/electrode sensor contact will lead to a change in the measured impedance [20]. When cells die by apoptosis, they round up and detach from the substrate and the measured impedance decreases. If cells die by necrosis, their membranes become permeable and the measured impedance decreases as in the case of apoptosis.

To correct for differences in CI before exposure, the MCI was normalized. The MNCI was used as a measure for the overall proliferation of the cells. The lowered MNCI is mainly attributed to cellular shrinking and cell detachment, which corroborated with the reduced proliferation and viability seen in the Trypan Blue assay suggesting that the cause for the lower impedance was cell death. Morphological modifications were observed in the cells exposed to TiO_2 NPs by URI microscopy. It was observed that, while the control cells were well spread and stretched, the treated cells appeared to be smaller and rounder, signs indicative of apoptosis [21, 11]. Observations made by URI, as well as those for viability, indicate that cell death, both by apoptosis and necrosis, occurred. These results are in accordance with results from previous studies [10, 11, 22, 23, 24]. In the absence of a clear dose-response effect and given the fact that a relatively high standard deviation was seen in the manual counting, especially for Ti3, a definitive conclusion could not be drawn with regard to the validity of the impedance-based assay for nanotoxicity testing. However, the fact that all the methods showed the same low level of toxicity induced by the chosen nano-TiO_2 indicates that the impedance-based approach deserves to be taken into consideration.

5. Conclusions

Impedance-based monitoring provided valuable information, in real-time, about the effects of TiO_2 NPs on fibroblasts' proliferation and viability. The results from impedance measurements were in most instances in accordance with those given by the Trypan Blue assay, proving that the impedance-based approach for nanotoxicity has merit. Further studies are needed in order to validate it as a high-throughput method for evaluating the cytotoxicity of NPs.

Acknowledgments

Richard Davies for impedance measurements and for financial support: the Department of Clinical Dentistry, Faculty of Medicine and Dentistry, the NanoUiB platform, University of Bergen, and the Michelsen Centre for Industrial Measurement Science and Technology, Bergen, Norway.

References

[1] Dusinska M 2009 *Nanomedicine* **4** 605-7

[2] Arora S, Rajwade J M and Paknikar K M 2012 *Toxicol. and Applied Pharmacol.* **258** 151-165

[3] Oberdörster G, Maynard A, Donaldson K, Castranova V, Fitzpatrick J, Ausman K, Carter J, Karn B, Kreyling W, Lai D, Olin S, Monteiro-Riviere N, Warheit D and Yang H 2005 *Part Fibre Toxicol,* **2** 8

[4] Kroll A, Pillukat M H, Hahn D and Schankenburger J 2009 *Eur. J. Pharm. and Biopharm.* **72** 370-77

[5] Ceriotti L, Ponti J, Colpo P, Sabbioni E and Rossi F *Biosensors & Bioelectronics* **22** 3057-63

[6] Cheran L E, Chung S, Wang X, Thompson M 2008 *Electrochemica Acta* **53** 6690-97

[7] Pliquett U 2010 *Food Eng. Rev.* **2** 74-94
[8] Rack H J and Quazi J I 2006 *Materials Sci. and Eng.* C **26** 1269-77
[9] Urban R M, Tomlinson M J, Hall D J and Jacobs J J 2004 *J. of Arthroplasty* **19** 94–101
[10] Vamanu C I, Cimpan M R, Hol P J, Sornes S, Lie S A and Gjerdet N R 2008 *Toxicol. in vitro* **22** 1689-96
[11] Allouni Z E, Hol P J, Cauqui M A, Gjerdet N R and Cimpan M R 2012 *Toxicol. in vitro* **26** 469-79
[12] Allouni Z E, Cimpan M R, Hol P J, Skodvin T and Gjerdet N R 2009 *Colloids and Surfaces* B: *Biointerfaces* **68** 83-87
[13] Brien W W, Salvati E A, Betts F, Bullough P, Wright T, Rimnac C, Buly R and Garvin K 1992 *Clin. Orthop. and Related Res.* **276** 66–74
[14] Buly R L, Huo M H, Salvati E, Brien W and Bansal M 1992 *J. of Arthropl.* **7** 315–23
[15] Jacobs J J, Skipor AK, Campbell P A, Hallab N J, Urban R M and Amstutz H C 2004 *J. of Arthroplasty* **19** 59-65
[16] Urcan E, Haertel U, Styllou M, Hickel R, Schertan H and Reichl F X 2010 *Dental Materials* **26** 51-58
[17] Vainrub A, Pustovyy O and Vodyanoy V 2006 *Optics Letters* **31** 2855-57
[18] Moodley K, Angel C E, Glass M and Graham E S 2011 *J. Neurosc. Meth.* **200** 173-80
[19] Quereda J J, Martinez-Alarcon L, Mendoca L, Majado M J, Herrero-Medrano J M, Pallares, F J, Rios A, Ramirez P, Munoz A and Ramis G 2010 *Transplantation Proceeding* **42** 3239-43
[20] Xing J Z, Zhu L, Gabos S and Xie L 2006 *Toxicol. in vitro* **20** 995-04
[21] Cimpan M R, Matre R, Cressey L I, Tysnes B, Lie S A, Gjertsen B T and Skaug N 2000 *Acta Odontol. Scand.* **58** 217–28
[22] Long T C, Saleh N, Tilton R D, Lowry G V and Veronesi B 2006 *Environ. Sci. and Technol.* **40** 4346-52
[23] Sayes C M, Wahi R, Kurian P A, Liu Y, West J L, Ausman K D, Warheit D B and Colvin V L 2006 *Toxicol. Sci.* **92** 174-85
[24] Yoo K C, Yoon C H, Kwon D W, Hyun K H, Woo S J, Kim R K, Lim E J, Suh Y J, Kim M J, Yoon T H and Lee S J 2012 *Int. J. Nanomed.* **7** 1203–14

Comparative study of neurologic effects of nano-TiO_2 versus SiO_2 after direct intracerebral exposure in mice

A Balvay[1], N Thieriet[2], L Lakhdar[3] and A Bencsik[1]

[1] ANSES Lyon Unité MND 31 avenue Tony Garnier 69364 France
[2] ANSES Maisons-Alfort, 27-31 avenue du général Leclerc, 94701, France
[3] ANSES Lyon PFEA, 31 avenue Tony Garnier 69364 France
French Agency for Food, Environmental and Occupational Health Safety (ANSES)

E-mail: anna.bencsik@anses.fr

Abstract. Titanium and silicon dioxide nanoparticles (TiO_2 and SiO_2 NPs) are now in daily use in many commercial products of which food, sunscreens, toothpastes or cosmetics. However, their effects on human body, especially on the central nervous system, are still unclear. The aim of this study was to determine whether direct exposition of the brain to TiO_2 and SiO_2 NPs results in alternations in nervous system function. C57Bl6 mice were exposed to 5 and 10 µg doses of TiO_2 and SiO_2 NPs through intracerebroventricular administration using a stereotaxic approach. Then the neurologic effects were investigated using motor performance parameters, measured on a rotarod at 20 rpm or at an accelerating rod (from 4 to 40 rpm). Before and after injection, motor activity is registered individually for each mouse exposed, once a week, for 8 weeks. Besides, a group of 3 mice is culled at 1, 2, 3, 4 and 8 weeks after exposure in order to study the time dependant effect on the histopathology of the brain (gliosis, inflammatory process...). Both rotarod tests (accelerating and at 20 rpm) showed that TiO_2 and SiO_2 NPs exposure could significantly impair the motor performances, even several weeks after initial acute exposure. The first examination of the brain histopathology revealed microglial activation. As it appeared to grow throughout the brain in a time dependant manner this suggests the induction of a long lasting neuroinflammation. These primary findings indicated that exposure to TiO_2 and SiO_2 NPs could possibly impair the locomotor ability and this deficit may be possibly attributed at least to an inflammatory process maintained till 8 weeks after exposure in the mouse brain. To fully investigate the neurotoxicological consequences of TiO_2 and SiO_2 NPs exposure, brain contents in these NPs will be also investigated as well as other alterations like neurotransmitter levels. These preliminary data already underline the necessity of more *in vivo* studies to better characterize TiO_2 and SiO_2 NPs exposure effects especially on human brain for long-term and low-dose treatment.

1. Introduction

The expanding development and production of engineered nanomaterials (ENMs) cover varied and extensive potential benefits in consumer products, food, drugs… and thus increase consumer's exposure to ENMs. The unique properties of ENMs have also raised concerns about the potential non intended consequences on human health and the environment. A potential risk for neurotoxicity arises since ENMs are able to reach the brain and to accumulate there. NPs inhaled through the nose can translocate to the brain via olfactory neurons [1, 2], similarly NPs from the lung are able to reach the brain after lung-blood barrier crossing [3]. Recently NPs were demonstrated as being able to cross the brain blood barrier [4] thus whatever the way of exposure, if NPs enter the systemic circulation they can reach the brain. Considering the susceptibility of the brain towards different kind of injuries and because of the limited regenerative capability of the brain, NPs presence in the brain should be considered cautiously. After a certain time of exposure, NPs might as well participate to the appearance of neurodegenerative diseases like Parkinson's or Alzheimer's diseases. However, neurotoxicity remains poorly documented *in vivo*. Thus, to consider whether any brain exposure may trigger a neurotoxic effect, a first approach based on direct injection of nanoparticles (NP) into the brain of mice using a stereotaxic apparatus is proposed. Even if this is not reflecting any natural way of exposure, this way, as other classical experimental routes of administration like intra peritoneal, allows assessing quickly neurotoxicity. Ventricles were chosen as the site of NPs delivery, because through the cerebro-spinal-fluid, it brings the advantage to ease the dissemination of NPs to the entire brain in a short time. In this preliminary study the doses of NP tested were 5 to 100% lower than those detected in the brain of mice exposed by intragastric route for consecutive 60 days [5], in an attempt to to avoid acute toxic effect. Two different ENMs were selected, titanium dioxide (TiO_2) a metal oxide and silicon dioxide (SiO_2) a non-metal oxide as they are already highly used in several manufactured products as diverse as additives in cosmetics, food, inks, textile...In addition they offer to study different properties of these NPs such as their metal or non-metal nature, their opposite charges in biological media (pH 7-8). The objective of this paper was to set up these studies on *in vivo* evaluation of potential neurotoxicity of ENMs in C57Bl6 mouse line through the analysis of neurological tests. Neurologic effects of NP brain exposure were studied quantifying motor performance using a Rotarod system. Before and after injection motor activity is registered individually for each mouse exposed, once a week, for 8 weeks. Besides, a group of 3 mice is culled at 1, 2, 3, 4 and 8 weeks after exposure in order to study the time dependant effect on the histopathology of the brain like on gliosis, inflammatory process.

2. Material and Methods

2.1. Nanoparticles

TiO_2 (99% purity, aeroxide TiO_2 P-25, spherical 22 nm, 85% anatase, 15% rutile) and SiO_2 (96.5% purity, primary particle of 5-35 nm, specific surface 220 m^2/g), are both provided by the EC-JRC in the framework of the OECD sponsorship program as representative nanomaterial for research purposes. The dispersion protocol developed by Nanogenotox Joint Action was used in the present study. This standard operation procedure was determined in order to harmonize and standardize the dispersion of ENMs. This generic Nanogenotox dispersion protocol produces ENM exposure media suitable for *in vitro* and *in vivo* toxicity testing (www.nanogenotox.eu). Briefly, NPs were suspended and highly dispersed in sterile-filtered 0.05% w/v BSA-water at a fixed concentration using a high energy probe sonication at 4°C. The protocol ensures stable dispersion for 0.5 to 1 hour. The pH value was 7 for each suspended NPs solution.

2.2. Mice exposure

All procedures were carried out in complete compliance with the guidelines laid down by the French (Decree 87-848) Ethical Committee and European Community Directive 86/609/EEC. The intracerebral exposure protocol was approved (11-0042) by the ComEth (National Committee on the Ethics of Animal Experiments). Animal experiments were performed in the ANSES animal facilities (PFEA) which have the relevant approval to carry out animal work (A 69 387 0801) by licensed people working in the animal experiment unit (license numbers AB: 69 387 531, LL: 69 387 191).

Groups of 15 female C57Bl6 mice, 6-week-old (average weight 19 g), were deeply anesthetized before unique acute administration of NPs. A volume of2 µL of 2.5 or 5 µg/µL NPs solution freshly sonicated was stereotaxically injected into the right side of lateral ventricle over a period of 10 min under the control of a motorized microinjection pump. A sham experiment was performed with the same volume 2 µL of solvent (sterile-filtered 0.05% w/v BSA-water) into the right side of lateral ventricle of a control group of C57Bl6 mice. None of the mice used in this study presented an overt neurological impairment at the time of NPs injection. At 1, 2, 3, 4, and 8 weeks after initial exposure, 3 mice were euthanized by lethal injection of pentobarbital. Brains were removed, fixed in 10% buffered formalin solution and either processed for histochemical analysis, or kept for programmed study of elemental content analysis (Ti brain mapping using μ-X-ray fluorescence).

2.3. Histopathology
Fixed brains were routinely embedded in paraffin blocks. Brain slices of 5 μm thickness were placed onto glass slides. Once dewaxed, brain slices were rehydrated, stained using hematoxylin-eosin (HE) or immunostained using markers for glial or microglial cells following procedures described elsewhere [6]. Slides were observed under a light microscope coupled to an image analysis station.

2.4. Nanoparticules impact on motor performances
To assess the effect of NPs direct exposure of the brain on motor function, mice were trained to remain on a rotarod (Bioseb, *in vivo* Research Instruments, Spain), through a 3-day training program on a 3-cm diameter rotarod (mice-sized). During the training period, each mouse was placed on a horizontal rod rotating at a gradually increasing speed from 4 to 40 rotations per minute (rpm) for a maximum of 10 min by which time a steady baseline level of performance was attained. This is the acceleration test. A second test consisting in measurement of endurance at a fixed speed of 20 rpm was similarly taught to mice. The day after training, the motor performances (coordination and endurance) were recorded for each animal during three trials. The latency to fall off the rotarod was recorded and the time limit was fixed at 3 min. One day prior surgery, baseline values were registered. After NPs or sham exposure, motor performances were registered individually for each mouse, once a week, for 8 weeks following the same procedure.

Statistical analyses were done using Software R. Rotarod results were compared with reference to baseline data before injection using ANOVA, and subsequently Barlett test, and for the last data registered 8 weeks post surgery, the number of animals being <5, a non parametric Kruskal-Wallis test was applied. Bonferroni method was also used when needed (test the means two by two). A p-value<0.05 was considered as significant.

3. Preliminary results and discussion

Figure 1 shows the results on motor performances registered in the control experiment. Performances on accelerated rotarod (ACC) were slightly progressing reflecting a possible positive effect of the regular week training of the mice. On the rotarod test fixed at 20 rpm there was no effect at all on the motor performance before and after surgery. This shows that the surgery by itself does not induce loss of motor performances. Compared with sham controls, NPs exposed mice had neurological deficits as revealed by the results of rotarod tests.

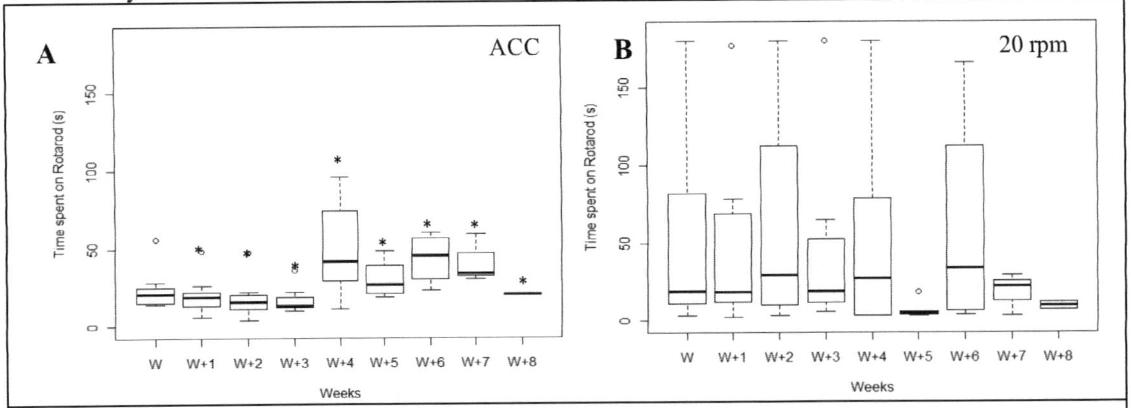

Figure1. Direct brain exposure: control experiment using water + 0.05% BSA. Motor activity

histograms represent the time that mouse stays on the bar of rotarod one week before surgery and then every week till 8 weeks after surgery. (A) ACC test reveals no reduction on motor performances, on the contrary a slight positive progression was observable especially from week 4 post surgery. (B) 20 rpm test induces no changes on motor performances. Values are the mean ± SEM. Data from exposed mice (*n*=9) were compared with control baseline (before surgery) using ANOVA, Barlett test, and a non parametric Kruskal-Wallis test. * $P<0.05$

3.1. Impact of SiO_2-NPs on motor performances

Performances recorded on an accelerated rotarod (ACC) were early significantly reduced at 1, 2, 3 and 4 weeks after surgery compared to control mice (before injection) (*p<0.05) (Figure 2). Performances were even more reduced until 8 weeks after initial exposure whatever the dose injected (5 and 10 µg). On the rotarod fixed at 20 rpm, whereas control mice run at least 129 s +/- 105 (means +/- SEM), nano-sized SiO_2 exposition induced a deficit in motor performance as soon as 1 week after surgery (29 s +/- 52 *p<0.05). Then the significant effect is observed solely for the 10 µg dose that lead to progressive and regular loss of motor performances until week 4 after surgery, which became null thereafter.

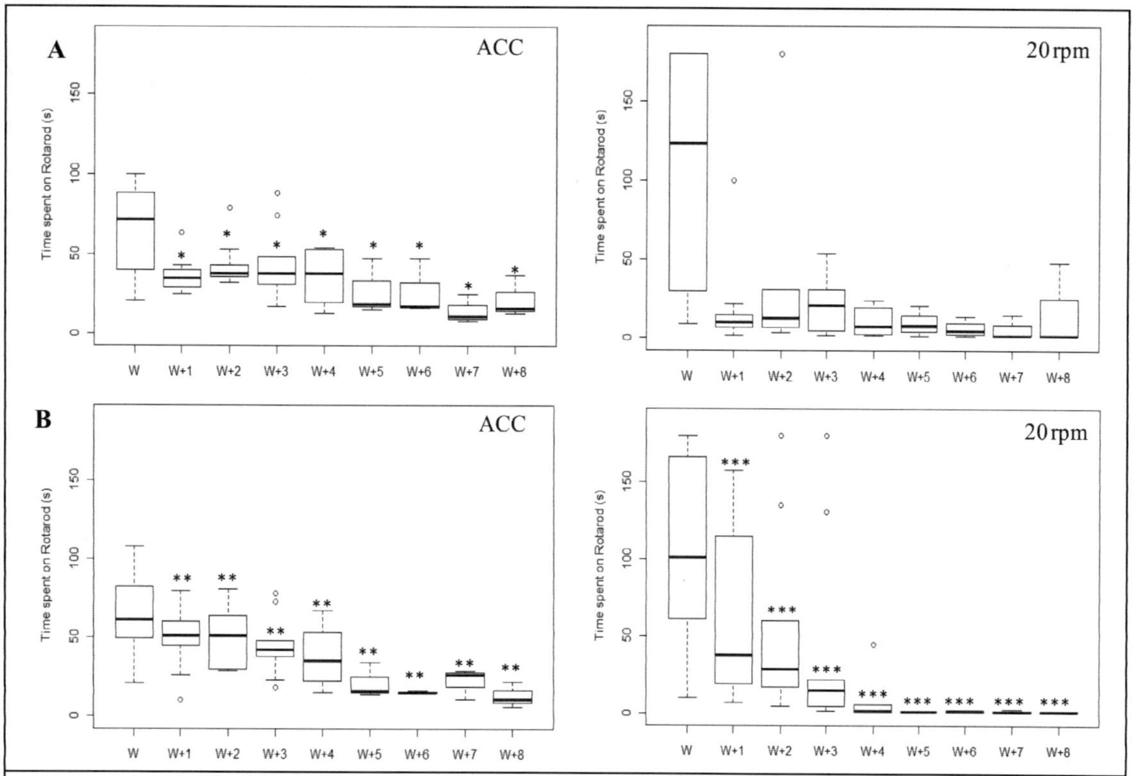

Figure 2. SiO_2 direct brain exposure reduced motor performances measured using a rotarod equipment. Motor activity histograms represent the time that mouse stays on the bar of rotarod one week before surgery and then every week till 8 weeks after surgery. (A) 5 µg and (B) 10 µg of freshly suspended SiO_2 NPs. Values are the mean ±SEM. Data from NP exposed mice (*n*=15) were compared with control baseline (before surgery) using ANOVA, Barlett test, and for the last data registered 8 weeks post surgery, the number of animals being <5, a non parametric Kruskal-Wallis test. * $P<0.05$, ** $P<0.01$, *** $P<0.001$

3.2. Impact of TiO_2-NPs on motor performances

Nano-TiO_2 induced a more progressive deterioration of each motor performance (*p=0.0001) (Figure 3). At 20 rpm the initial performance of 104 s +/-86 decline slowly to 84 +/- 100 at 1 week and became null after 4 weeks. The only test that reveals significant difference depending on the dose injected is the endurance test at 20 rpm as shown in figure 3.

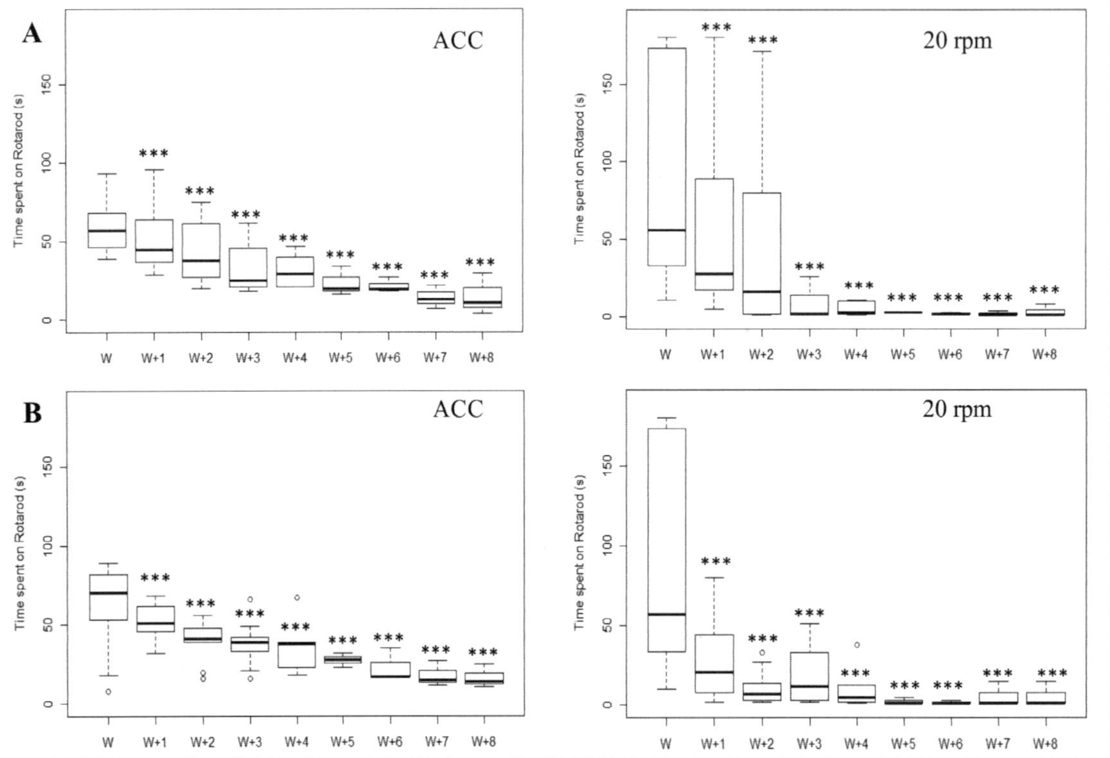

Figure 3. TiO_2 direct brain exposure reduced motor performances measured using a rotarod equipment. Motor activity histograms represent the time that mouse stays on the bar of rotarod one week before surgery and then every week till 8 weeks after surgery. (A) 5 µg and (B) 10 µg of freshly suspended SiO_2 NPs. Values are the mean ±SEM. Data from NP exposed mice (*n*=15) were compared with control baseline (before surgery) using ANOVA, Barlett test, and for the last data registered 8 weeks post surgery, the number of animals being <5, a non parametric Kruskal-Wallis test. * *P*<0.05, ** *P*<0.01, *** *P*<0.001

These first results indicate that both NPs studied were able to induce neurological effects after a direct intracerebral exposure, for several weeks after exposure. Both appear to induce a decline in motor performance but the effects appears to be different in the time-dependant response. Compared to SiO_2, TiO_2 NPs seem to induce a more progressive motor deterioration. Nano-TiO_2 and nano-SiO_2 were selected as a large body of literature already exists in terms of toxicity. However, neurotoxicity remains poorly documented and in particular if some neurotoxic effects were reported for TiO_2 [7, 8], less is known about SiO_2 potential neurotoxicity. In addition, nano-TiO_2 and SiO_2 were also selected because they are probe particles with positive (TiO_2) and negative charges (SiO_2) in biological media (pH 7-8). Surface charges are important to probe since they will be strongly related to the transfer of NPs [9]. With both nano-TiO_2 and SiO_2, two opposite possible trends were tested; the motor abilities were affected in both cases but in a different manner that may be linked to one or several of these specificities that will be studied later.

Hematoxylin-eosin (H&E) staining was performed to examine brain histological changes at 1, 2, 3, 4 and 8 weeks after exposure. At the moment, only one/three mouse per each time point was analyzed, thus no quantitative and definitive data are yet available. Still, several preliminary histopathological data could be collected. As an example, it was possible to visualize the site of injection, to follow the cicatrisation of the mechanical lesion induced by the needle during injection. At 1 week after surgery, the needle scar was already replaced by a fibroglial tissue as shown in figure 4. Immunohistochemical studies showed the activation of astrocytes and microglial cells thus demonstrating a local gliosis with an inflammation at the initial point of injury.

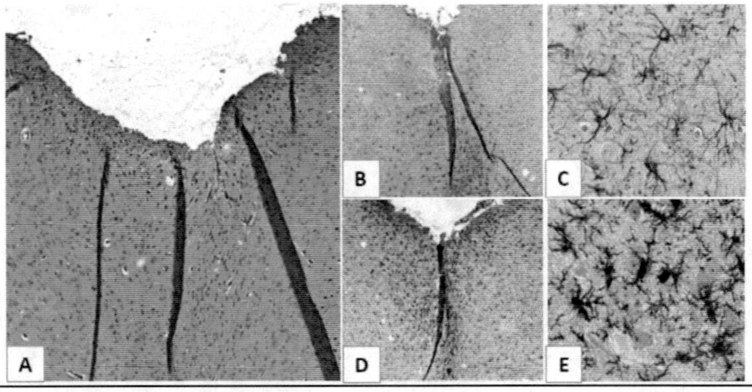

Figure 4. A. Site of injection, cerebral cortex one week after surgery (H&E). B. & C. Local glial scar: astrocytes (in brown) increase their synthesis of glial fibrillary acidic protein (GFAP) and extend their processes. D. & E. Activated microglial cells are the second most prominent cell type present within the glial scar.

This was not linked to the NPs themselves as this feature was also observed in brain of sham injected mice. Obviously the early loss of motor performance in NPs exposed mice, cannot be attributed to this local histopathological change as sham operated mice did not present this reduction of motor function. Interestingly, in the weeks after, whereas gliosis progressively vanished, microglial activation seems to grow throughout the brain suggesting an induction of a long-lasting neuroinflammation (figure 5).

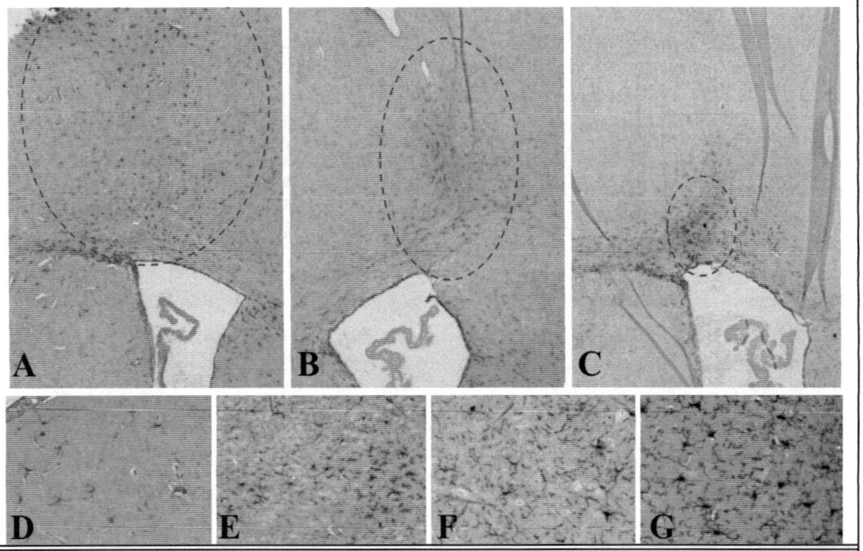

Figure 5. A.-C. Time-dependant loss of glial activation at the injection site at 1, 3 and 8 weeks post-injection. D.-G. microglial activation at 1, 2, 3 and 4 weeks after SiO_2 injection, in other brain areas than the injection site.

This may most probably be linked to the NP, but this needs to be checked on all the other brains available. As well, the results of NPs brain distribution to be studied by µ-X-ray fluorescence will allow easing the possible link between NP presence and induction of neuroinflammation. To take into consideration the loss of motor performances observed in the present study, which could be related to a neuroinflammatory process it will be also relevant to study disturbances in other different neurotransmitter systems.

4. Conclusion and future work

This preliminary study indicates already that nano-sized SiO_2 and TiO_2 are both able to induce some neurological effects after direct exposure into the brain as shown by loss of motor performance of exposed mice. Comparison with sham experiment indicates that NP exposure is most probably at the origin of these and not the initial needle injury itself. Microglial activation may contribute to these behavioral deficits. These results indicate that rotarod tests are useful in assessing brain response to NP presence within cerebral tissue and therefore could be commonly used in rodent model of nanotoxicology. Still, to fully investigate the neurotoxicological consequences of TiO_2 and SiO_2 NPs exposure, brain contents in these NPs will be also investigated. Other alterations like neurotransmitter levels will be studied as they may be more directly involved in motor deficits. The necessity of more *in vivo* studies to better characterize TiO_2 and SiO_2 NPs exposure effects especially on human brain for long-term and low-dose treatment are required and other experiments using peripheral exposure are already going on.

5. Acknowledgements

Aurélie Balvay was supported by a grant from ANSES. We thank Émilie Antier, Coralie Pulido, Rabah Belkheir, Damien Gaillard, members of the plateforme expérimentation animale (PFEA) of l'ANSES - Lyon, for their respective contributions to this work. We thank Jean-Michel Bridon for technical assistance in preparation of NP solutions as well as Marion Gay for histotechnical assistance.

References

[1] Elder A, Gelein R, Silva V, Feikert T, Opanashuk L, Carter J, Potter R, Maynard A, Ito Y, Finkelstein J and Oberdorster G 2006 *Environ Health Perspect* **114** 1172

[2] Oberdorster G, Sharp Z, Atudorei V, Elder A, Gelein R, Kreyling W and Cox C 2004 *Inhal Toxicol* **16** 437

[3] Shimada A, Kawamura N, Okajima M, Kaewamatawong T, Inoue H and Morita T 2006 *Toxicol Pathol* **34** 949

[4] Brun E, Carriere M and Mabondzo A 2012 *Biomaterials* **33** 886

[5] Hu R, Gong X, Duan Y, Li N, Che Y, Cui Y, Zhou M, Liu C, Wang H and Hong F 2010 *Biomaterials* **31** 8043

[6] Relano-Gines A, Lehmann S, Bencsik A, Herva M E, Torres J M and Crozet C A 2011 *J Infect Dis* **204** 1038

[7] Wang J, Chen C, Liu Y, Jiao F, Li W, Lao F, Li Y, Li B, Ge C, Zhou G, Gao Y, Zhao Y and Chai Z 2008 *Toxicol Lett* **183** 72

[8] Zhang L, Bai R, Li B, Ge C, Du J, Liu Y, Le Guyader L, Zhao Y, Wu Y, He S, Ma Y and Chen C 2011 *Toxicol Lett* **207** 73

[9] Lockman P R, Koziara J M, Mumper R J and Allen D D 2004 *J Drug Target* **12** 635

Application of in vitro BBB model to measure permeability of nanoparticles

S Hanada[1], K Fujoka[2], Y Inoue[3], F Kanaya[1], Y Manome[2] and K Yamamoto[1]

[1] Research Institute, National Centre for Global Health and Medicine, Shinjuku-ku, Tokyo 162-8655, Japan
[2] Department of Molecular Cell Biology, Jikei University School of Medicine, Minato-ku, Tokyo 105-8461, Japan
[3] Department of Anatomy, School of Medicine, Faculty of Medicine, Toho University Tokyo, 143-8540, Japan

E-mail: hanada@ri.ncgm.go.jp

Abstract. In both pharmaceutical and toxicological fields, one of major issues has been the possibility of nanoparticle uptake to central nerve system. For the safe use of nanoparticles, it is integral to evaluate the permeability of nanoparticles through BBB. In our collaborative research group reported that a few nanoparticles accumulated in brain in animal experiment, as an in vitro model, we applied commercially available cell-based BBB model for establishing evaluation method, which is quick, quantitative and equivalent to in vivo assay. We assayed 30-1500 nm silica and surface charge dependent Qdots. Our results showed the size-dependency and the surface modification dependency. We compared our assay to several animal experiments. There are both equivalence and discrepancy with animal experiments. Our BBB model can be useful tools for evaluating size-dependent permeability, but not for surface modification-dependent permeability. Our BBB assay is non-serum assay and we have not adequately reflected the serum-related interaction between nanoparticles and cell surfaces. To clear up the discrepancy of our BBB model, serum-based assay and low-concentration detection will be needed.

1. Introduction

Nanoparticles (NPs) have great functionalities and potentials in many industrial fields. In also a bio-medical field, these are benefits of NPs for contributing to various novel medicines. For example, NPs are potential carriers of drug delivery systems for treatment [1-2] and some fluorescent, magnetic or radiolabelled NPs are useful as imaging materials for diagnosis [3-5]. In contrast, researchers are concerned about NPs give any stress to biological systems. Not a few research groups reported the risk. For instance, NPs in sun cream can stress brain cells [6], and NPs can induce brain edema formation by influencing BBB breakdown in vivo [7].

There are many parameters to evaluate the risk of NPs. NPs have many characteristics like size, materials (metal, oxide, semiconductor and organic), surfaces (charge and specific interaction) and shapes (round and sharp), and there are many exposure routes to the body, like oral, dermal, via pulmonary and via systemic circulation. The most important thing is that there are also many evaluation indexes such as inflammation, brain activities, NPs' accumulation in brain and BBB permeability. To evaluate all the possible risks, we need an evaluation method, which is quick, quantitative and equivalent to in vivo experiment (We would like to call "3Q" evaluation method.).

Our collaborative research group, previously, reported that, in animal experiment, surface-modified Qdots, which we know as one of the fluorescent NPs, are able to cross the BBB and remain in the brain parenchyma at 6 hour after intra-peritoneal (i.p.) administration [8]. In the research, 0.1% of NPs accumulated in brain by the i.p. injection in mice at 6hr while 8.8% of NPs stayed in blood circulation. In the histological analysis, red fluorescent Qdots accumulated in many areas of brain parenchyma, such as olfactory bulb, cerebral cortex, hippocampus, thalamus and brainstem. From this result, we know that, though the amount is few and small, NPs can be certainly in brain. We do not know so far where NPs transport into brain from exactly, but blood-brain barrier (BBB) will be one of solid pathways into brain.

Recently, we applied new cell-based in vitro BBB model to evaluate NPs permeability into brain, which is a commercially available culture system (PharmaCo-cell company, Japan), and is culturing both rat brain micro-vascular endothelial cells and pericytes separated by millicell membrane for mimicking the actual BBB [9]. The model has equivalence to in vivo experiment in pharmaceutical drug screening test. This assay procedure is as follows: (1) Reconstruction of the BBB model under astrocyte pre-cultivation, (2) Doing BBB permeability assay and measuring drug concentrations and (3) Calculating permeability coefficient (Papp). Papp represents apparent permeability of a total BBB membrane (both cell layers and a plastic membrane).

We can assay BBB permeability of NPs by using this BBB model. We evaluated 30-1500 nm silica and surface charge dependent Qdots by Papp. The Papp had size-threshold between 30- and 100 nm. Cationic surface-modified NPs can pass through the BBB model.

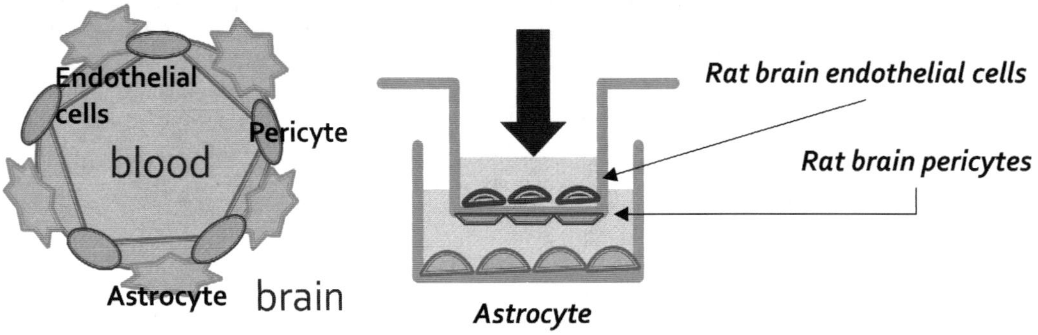

Figure 1. The left image is a picture of actual BBB model. Endothelial cells form vascular walls and pericyte supported cell-cell contact of the endothelium. Astrocyte is also important for maintaining tight-junction of vascular cells. The right image shows the cell-based BBB model. Endothelial cells are at the top of the membrane, pericytes are at the bottom of the membrane, and before doing the assay, this model pre-cultures with astrocytes.

2. Aim of this study

We compared between our cell-based BBB assay and animal experiments performed by other research group to confirm the equivalence of our in vitro assay to in vivo experiment.

3. Results-validation of the BBB model as 3Q evaluation method

This assay is a quick (within 30 min) and quantitative (usage of Papp) evaluation method. We validated the equivalence to animal experiment. Concerning the size-dependent permeability, our assay has size-threshold for BBB permeability between 30 and 100-nm; while, in animal experiment, a research group reported that 70-nm silica accumulate in the brain [10]. We should consider BBB permeability of nanoparticles with tens of nano-meter in diameter. Our assay can be qualitatively equivalent to in vivo experiment.

Concerning the quantitative relevance between our assay and animal experiment, we briefly calculated brain-blood amount ratio form the amount (weight/mole) of nanoparticles in the blood and the brain sides. The value in our in vitro assay is 0.018 in 30 min., while the value in previous animal experiment is 0.011 in 6 hr. Though the administration concentration in our research is much higher than that of the realistic exposure, these values are comparable to each other.

We have to report the limitation of the model. There is a discrepancy between our results and those of a previous in vivo study. In our study, the cationic nanoparticles passed through the BBB model more easily than the anionic and neutral nanoparticles. The reason for this result is that the cationic NPs strongly interacted with the negative charged cell surface.

So far, there are some shortcomings in our BBB model for evaluation of NPs transportation into brain as an alternative to animal experiments. For example, in animal experiment, anionic nanoparticles accumulated in brain more easily, in contrast to the result in our BBB assay. In our assay, we did not use serum in the assay buffer by following conventional protocols, but in the human body, NPs are thought to absorb serum proteins, which enhance cell-NPs interaction, like a "protein corona" in the Figure 2 [11]. Furthermore, in the human body, we have to consider the effect of NPs that accumulate in the blood vessels or in the endothelial cells. Further experiments will be needed to fill gaps between in vitro and in vivo experiments. In future, we will try serum-based assay or long time evaluation, and improve our model like a brain slice culture and a low-concentration detection method.

Table 1. Comparison between in vitro assay and in vivo experiment

	In vitro	**In vivo**
Experimental results		
Size specificity	Between 30- and 100-nm	70 nm
Surface modification	Cationic > Anionic	Anionic > Cationic
Evaluation methods		
Quickness	30 min	6 hr
Quantitativity	Papp	Disposition in organs
Equivalence	To animal experiment	Unknown to human

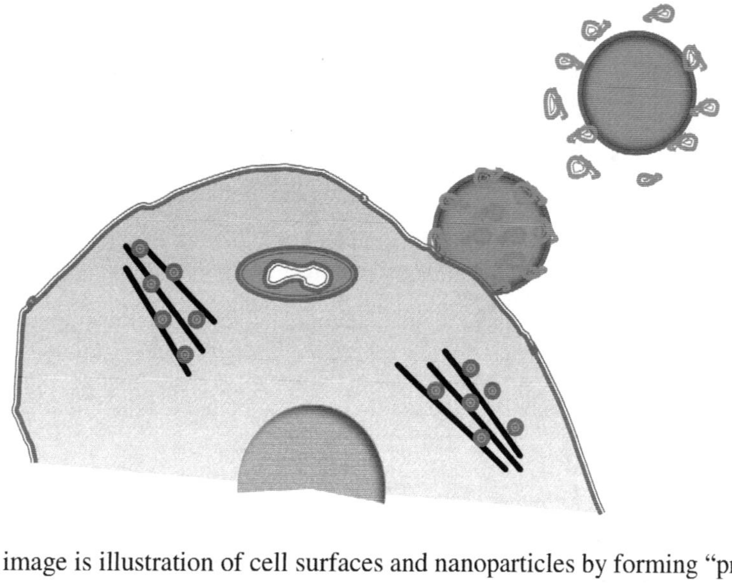

Figure 2. The image is illustration of cell surfaces and nanoparticles by forming "protein corona"

4. Conclusion

The in vitro BBB model can be "3Q" evaluation method of NPs' permeability through BBB; we observed size-threshold for passable NPs into brain.
Further improvement will be needed, like serum-based assays and low-concentration detection methods.

Acknowledgement

This work was supported by Health and Labour Sciences Research Grants from the Ministry of Health, Labour and Welfare of Japan (H22-chemical-young-009). Dr. Fujioka is a principal investigator in the grant. We thank Mr. Kazuyuki Itou for daily support in our laboratory.

References

[1] Manabe N, Hoshino A, Liang Y, Goto T, Kato N and Yamamoto K 2006 *IEEE Trans. Nanobiosci.* **5** 263-7
[2] DDS
[3] Yamamoto S, Manabe N, Fujioka K, Hoshino A and Yamamoto K 2007 *IEEE Trans. Nanobiosci.* **6** 94-8
[4] Mornet S, Vasseur S, Grasset F and Duguet E 2004 *J Mater Chem.* **14** 2161–75
[5] Banerjee T, Singh A K, Sharma R K, Maitra A N 2005 Int J Pharmaceu **289** 189–95
[6] Long T C, Tajuba J, Sama P, Saleh N, Swartz C, Parker J, Hester S, Lowry G V, and Bellina Veronesi B 2007 *Env Health Perspect.* **115** 1631–37
[7] Sharma HS, Hussain S, Schlager J, Ali SF, Sharma A 2010 Brain Edema XIV **106** 359–364
[8] Kato S, Itoh K, Yaoi T, Tozawa T, Yoshikawa Y, Yasui H, Kanamura N, Hoshino A, Manabe N, Yamamoto K, Fushiki S 2010 *Nanotechnology* **21** 335103
[9] Nakagawa S., Deli M A, Kawaguchi H, Shimizudani T, Shimono T, Kittel A, Tanaka K, Niwa M 2009 *Neurochem Int.* **54** 253–63
[10] Nabeshi H, Yoshikawa T, Matsuyama K, Nakazato Y, Matsuo K, Arimori A, Isobe M, Tochigi S, Kondoh S, Hirai T, Akase T, Yamashita T, Yamashita K, Yoshida T, Nagano K, Abe Y, Yoshioka Y, Kamada H, Imazawa T, Itoh N, Nakagawa S, Mayumi T, Tsunoda S I, Tsutsumi Y 2011 *Biomaterials* **32** 2713–24
[11] Praetner M, Rehberg M, Bihari P, Lerchenberger M, Uhl B, Holzer M, Eichhorn M E, Fürst R, Perisic T, Reichel C A, Welsch U, Krombach F 2010 *Biomaterials* **31** 6692–700

ROS evaluation for a series of CNTs and their derivatives using an ESR method with DMPO

S Tsuruoka[1], K Takeuchi[2], K Koyama[2], T Noguchi[1], M Endo[1], F Tristan[1], M Terrones[1], H Matsumoto[3], N Saito[4], Y Usui[2], D W Porter[5] and V Castranova[5]

[1]Shinshu University, Research Center for Exotic Nanocarbons, 4-17-1 Wakasato, Nagano, Japan
[2]Shinshu University, Faculty of Engineering, *ditto*
[3]Department of Organic and Polymeric Materials Tokyo Institute of Technology, 2-12-1 Ookayama, Meguro-ku, Tokyo, JAPAN
[4]Department of Applied Physical Therapy, Shinshu University, School of Health Sciences, 3-1-1 Asahi, Matsumoto, Nagano, Japan
[5]National Institute for Occupational Safety and Health, Morgantown, WV, USA

s_tsuruoka@shinshu-u.ac.jp

Abstract. Carbon nanotubes (CNTs) are important materials in advanced industries. It is a concern that pulmonary exposure to CNTs may induce carcinogenic responses. It has been recently reported that CNTs scavenge ROS though non-carbon fibers generate ROS. A comprehensive evaluation of ROS scavenging using various kinds of CNTs has not been demonstrated well. The present work specifically investigates ROS scavenging capabilities with a series of CNTs and their derivatives that were physically treated, and with the number of commercially available CNTs. CNT concentrations were controlled at 0.2 through 0.6 wt%. The ROS scavenging rate was measured by ESR with DMPO. Interestingly, the ROS scavenging rate was not only influenced by physical treatments, but was also dependent on individual manufacturing methods. Ratio of CNTs to DMPO/ hydrogen peroxide is a key parameter to obtain appropriate ROS quenching results for comparison of CNTs. The present results suggest that dangling bonds are not a sole factor for scavenging, and electron transfer on the CNT surface is not clearly determined to be the sole mechanism to explain ROS scavenging.

1. Introduction

Carbon nanotubes (CNTs) are expected to have broad potential to materialize new and prominent technologies over the next decades [1-3], and CNT hybrid materials with the other carbon groups and/or doped elements are also under development [4]. For instance, a specific rubber composite of CNTs is being successfully commercialized, which significantly increases recovery from oil wells [5] and may significantly impact the global energy market. For another instance, particular types of CNTs exhibit super conductivity without addition of other elements [6]. These results will accelerate the product development with CNTs further.

It is critically important to evaluate the material safety and develop appropriate in-use regulations for CNTs and nano carbon materials prior to the major industrialization. There are many reports on the evaluation of CNT safety which discuss the similarity to asbestos in shape and suggest that CNTs may exhibit asbestos-like carcinogenicity [7]. It has been proposed that conventional approaches be employed using physicochemical characteristics and in vitro or in vivo assays to evaluate toxicity of nanoparticles [8]. Maynard et al. [9] summarized the early stage studies on nanotechnology and occupational health following intensive discussions of possible adverse effects of CNTs [10-15]. The first systematic evaluation on CNT toxicity was reported by Shvedova et al. [16] followed by an inhalation [17] and bolus instillation [18, 19, 20] studies. CNTs were demonstrated to induce fibrosis and granulomatous lesions at relatively low exposure doses. In addition, CNT-induced mesothelioma was reported at higher doses after installation into the abdomen or scrotum of rodents [21, 22]. Results suggest that inflammation of the mesothelial lining is dependent on the length of CNTs [23].

It is essentially required to utilize physicochemical properties of materials to reinforce the conventional toxicological evaluations. However, it has been recently argued whether those conventional methods are applicable for fibril materials such as CNTs because bolus instillation and inhalation evaluation results do not correspond with *in vitro* tests (as discussed at the annual conference of Society of Toxicology Mar. 2012 in San Francisco [24]). Results of reactive oxygen species (ROS) measurement appear to correspond with CNT *in vitro* results. Shvedova et al. [25] conducted examination of free radical generation after exposure of HaCaT cells to Single-Walled CNTs (SWCNTs) using electron spin resonance (ESR) spin trapping. Fenoglio et al. [26] discussed the relationship between ROS and biological evaluations. The current spin trapping method is pursued using indirect measurement with a spin adduct that forms the stable amine-oxide radical and relatively-long-lived free radicals. The adduct gives a measurable hyperfine coupling constant of spin, and the concentration of free radical is obtained analyzing area of the ESR signal spectrum. Measurement of HO and HO_2 free radicals with nitrones (the N-oxide of an imine) was proposed by Harbor et al. [27]. Among various spin trapping compounds, 5, 5-dimethyl-1-pyrroline-N-oxide (DMPO) is often used for biological radical spin trapping because it dissolves into aqueous solution well with selectivity of radical molecules captured [28]. The author, nevertheless, addressed the limitations of DMPO. Shvedova et al. [25] likewise discussed the issue that ferrous irons associated with SWCNTs can cause the decomposition of hydrogen peroxide or alkoxyl radicals from lipid peroxide, since the unpurified SWCNTs used in that study contained up to 30% iron residue. They concluded that raw SWCNTs have a relatively high risk potential to induce inflammatory lesion at their manufacturing places. Murray et al. [29] reported thereafter that the OH radical generation was mostly caused by the iron residue via the Fenton reaction. Furthermore, Porter et al. [18] demonstrated that purified Multi-Walled CNTs (MWCNTs) scavenge OH radicals generated by the Fenton reaction and do not generate ROS in an acellular system containing hydrogen peroxide. Fenoglio et al. [26, 30, 31] conducted ROS-ESR measurement of CNTs by several methods and concluded that purified MWCNTs quench OH radicals.

The present study aims to evaluate ROS generation by number of CNTs and verifies the assumption that surface morphology of CNTs affects ROS scavenging potential. In addition, we evaluated the importance of the CNT to DMPO/hydrogen peroxide ratio to obtain appropriate scavenging rates.

2 Materials and Methods

2.1. Materials

Seventeen kinds of CNTs were prepared and titanium dioxide was used as a reference. As shown in Table 1, Group A consists of a cup-stack type of MWCNTs (cs-CNTs) and their derivatives. Particular CNTs identified as CS1 of Group A were the pivotal ones in the present study and were prepared and characterized intensively. CNTs identified as CS2 through CS5 were prepared to investigate effects of surface morphological changes. CS2 was prepared by annealing at 2800°C in inert gas atmosphere.

CS3 and CS4 were chopped by machines to shorten the length. CS5 was rubbed to de-bulk the agglomerated CS1 mechanically so that the outer surfaces were damaged significantly. CS4-PT and CS5-PT were deposited with platinum (20 wt%) after annealing at 2800 °C. Figure 1 shows a transmission electron micrograph of CS1 and CS5-PT. As CS1 has a lot of dangling bonds on the surface in comparison with usual CNTs, it was assumed that physical treatments might give quantitative differences by significantly causing morphological changes as suggested by Fubuni et al. [32]. Group B consists of commercially available MWCNTs. This group was designed to compare the dependence on manufacturing processes of CNTs. CT1 and CT1-N were synthesized to evaluate nitrogen-doping effects only.

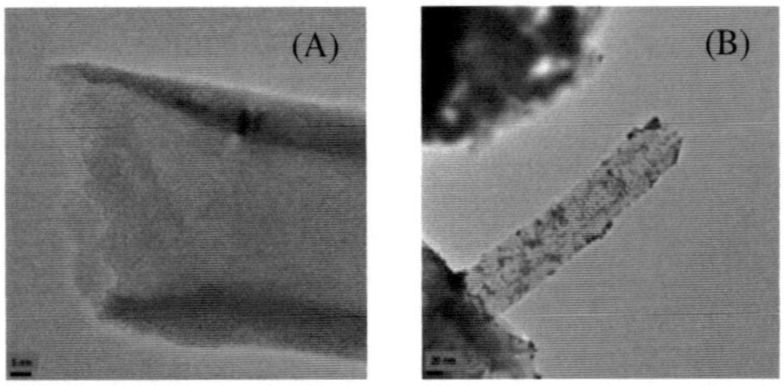

Figure 1. Transmission electron micrograph: (A) cs-MWCNT (CS1), (B) platinum deposited cs-MWCNT (CS5-PT); dots are Pt particles deposited on the CNT surface.

2.2 Measuring method
First of all, dispersibility of the prepared CNTs was tested to avoid agglomeration during the ESR measurements. All CNTs were reasonably dispersed in a DMPO and hydrogen peroxide solution with surfactant after sonication, with no clumping by CNT agglomeration in the pipet or the measuring cell of ESR equipment observed.

The generation and scavenging of radical oxygen species by CNTs were measured at room temperature by ESR (JES-FA200, JEOL) using DMPO as the spin trapping agent. ESR settings were: frequency 9415.404 MHz, power 0.998 mW, field center 335 mT, sweep time 2 min., width +/- 5 mT, and modulation frequency 100 kHz. Fresh frozen DMPO (Dojindo Laboratories, Kumamoto Japan) was thawed at room temperature and diluted to 100 mM with ultrapure water. Hydrogen peroxide (hydrogen peroxide 30%, Wako Pure Chemical Industries, Ltd. Japan) was diluted to 1 mM. Ferrous chloride (Iron (II) Chloride Tetrahydrate, Wako Pure Chemical Industries, Ltd. Japan) was dissolved into ultrapure water at 0.1 mM. Titanium dioxide (Evonik Degussa P25) and CNTs listed in Tables 1-A and 1-B were dispersed into aqueous solution made of 16 mg surfactant (Sodium Dodecyl Benzensulfonate, Kanto Chemical Co., Inc. Japan) and 10 ml ultrapure water. ROS generation by CNTs was measured with the mixed solution (Mixture A) consisting of 0.1 ml of each solution of DMPO, surfactant, hydrogen peroxide and CNTs and the addition of ultrapure water for a total volume of 5 ml. Scavenging ROS generated by Fenton reaction was measured with the mixed solution (Mixture B) consisting of 0.1 ml of each solution of DMPO, ferrous chloride, hydrogen peroxide and CNTs, and ultrapure water for a total volume of 5 ml. CNT concentration was determined by weight in the total volume of 5ml. The blank solution was made with the ultrapure water instead the CNT solution. Each suspension was dispersed by sonication. All of measuring samples were prepared just before ESR measurement to avoid any change over time. Each sample was measured five or more times repeatedly and averaged after elimination of the highest and lowest values.

3. Results and Discussion

The effect of surfactant on the ESR signal is shown in Figure 2. The results show OH radical quenching was proportional to the concentration of surfactant present. Thus, the surfactant concentration was kept constant throughout further experiments. Figure 3 shows a typical ESR spectrum of ROS generation by the Fenton reaction and quenching by CNTs. CNTs and TiO_2 in the absence of ferrous chloride did not produce ROS (Figure 3(A)). Figure 3(B) shows the spectra that indicate radical quenching by CNTs. OH radicals generated by the Fenton reaction between $FeCl_2$ and hydrogen peroxide were apparently quenched by CNTs in a concentration dependent manner.

Table 1-A. CNT properties used for ROS evaluations (Group A)

CNTs	Description	SSA[a] (m^2/g)
CS1	Cup-Stack type MWCNT, diameters: outer = 80 ~100 nm, inner = 50 ~ 70 nm, length = 5 μm ave., purity > 95wt%	49.0
CS2	Graphitized CS1 annealed at 2800 °C	49.5
CS3	Mechanically chopped CS1, average length = 1.5 μm	52.5
CS4	Mechanically chopped CS1, average length = 1.0 μm	74.2
CS5	Mechanically de-bulked CS1, slightly shorter length than CS1	45.8
CS4-PT	Graphitized at 2800 °C of CS4 and deposited with Pt (20wt%)	57.1
CS5-PT	Graphitized at 2800 °C of CS5 and deposited with Pt (20wt%)	39.4

[a] SSA stands for specific surface area measured by BET isotherm adsorption method.

Table 1-B. CNT properties used for ROS evaluations (Group B)

CNTs	Properties[a]			Making methods
	Ave. outer diameter (nm)	Ave. Length (μm)	Purity (wt%)	
DA1	< 1.5	N/A	> 95	CVD[b]
CA1	13 ~ 16	>1	> 95	CVD[b]
CA2	13 ~ 16	>1	> 95	CVD[b]
CA3	~ 13	1.5	> 90	CVD[b]
CA4	10 ~ 15	0.1 ~ 10	> 90	CVD[b]
CA5	0.7 ~ 70	10	> 95	CVD[b]
CA6	15	3	> 95	CVD[b]
CB1	40 ~ 90	4.9[d]	99	FCM[c]
CB1-T	40 ~ 90	4.9[d]	99	FCM[c]
CB2	80	N/A	> 99	FCM[c]

[a] Properties of manufactured CNTs are from manufacturers' specifications.
[b] Catalytic chemical vapor deposition
[c] Floating catalyst method
[d] The measured length by scanning electron micrograph

Figure 4(A) and 4(B) show the scavenging rates for various CNTs at concentrations of 0.2, 0.4 and 0.6 wt%, and those normalized figures by CNT weight concentrations, respectively. The scavenging rate was calculated in comparison to the blank. Figure 4(A) clearly shows that radical quench ability depends on the particular CNT tested. CA1 and CA2 were obtained from the same manufacturer but exhibited different dispersibilities in media. They did not quench OH radicals, but enhanced OH radical generation when compared to the blank. CA3 behaves as an OH radical scavenger at low concentrations, but generates radicals at the highest CNT concentration. As CA4 through CA6 were produced from the similar process but did not generate OH radicals, it suggests that CA1 through CA3 included something reacting with $FeCl_2$. CB1-T, according to the manufacturer, was a reduced bulk volume product of CB1 using glue. However, as CB1-T behaved differently, the glue or the gluing process lowered OH radical quenching at the low concentration. CT1-N was the same as CT1 but nitrogen-doped. Nitrogen doping reduced quenching. These results suggest that different CNTs exhibit different surface activity. Thus, each product must be evaluated individually in terms of occupational health and environmental safety. Figure 4(B) specifically tells us that balance between CNTs and

DMPO/hydrogen peroxide has to be considered carefully when conducting ROS evaluations. Scavenging rate may reach almost 100% at the lowest concentration of CNTs but quenching per weight decreases at the higher CNT concentration

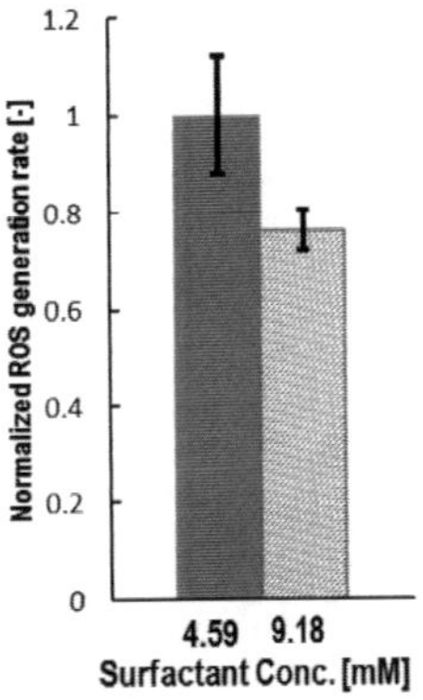

Figure 2. Interaction of surfactant to OH radical quench using Mixture B without CNTs; The vertical axis indicates that the normalized average intensity of the ESR signal was set at 1.00. The deviation bars indicates standard deviations.

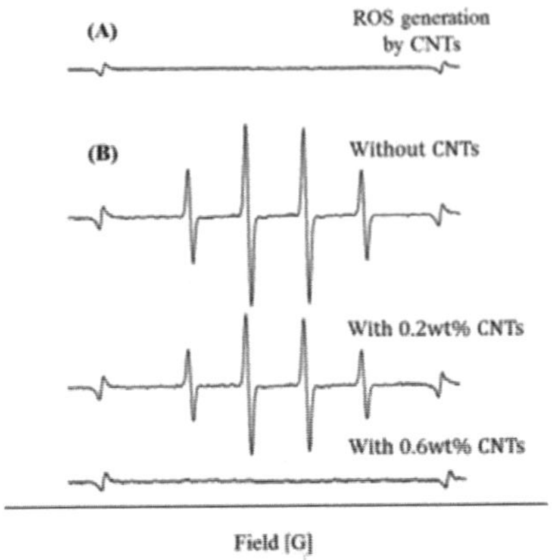

Figure 3. Typical ESR spectrum intensity of free radical adducts with DMPO using CS1; (A) OH radicals generated by CNTs without $FeCl_2$. (B) OH radicals generated by the Fenton reaction with $FeCl_2$. The more radicals were scavenged with the higher CNTs concentration.

The relationship between CNT surface morphology and OH radical quenching was evaluated using CS1 and its derivatives. Figure 5 shows results normalized by the individual specific surface area. It demonstrates that surface morphology has a small effect on OH radical quenching as does addition of Pt, i.e., augmentation of quenching. Since platinum catalyzes hydrogen peroxide, synergetic decomposition of OH radicals may occur. Figure 5 suggests that the shortest CNT (CS4) exhibited the lowest scavenging rate, and that radical quenching corresponds to length in addition to the surface properties. Meanwhile the graphitization that increases crystallinity and decreases the number of dangling bonds apparently affected quenching. DA1, double-walled CNTs, were also strong ROS quenchers (Figure 4). Further investigation is necessary to determine the role of electron charge transfer on the CNT surface with and without dangling bonds.

The present study demonstrates two major points. First, commercial CNTs have different characteristics, depending on processing, which affect ROS quenching. Second, CNT scavenging rate depends on the surface morphology, length and particles deposited on the surface. In addition, since CNTs have high OH radical quench ability, concentration ratio between CNTs and DMPO/ hydrogen peroxide has to be considered carefully in order to obtain appropriate ROS results. The present ROS measuring method by ESR is simple and reproducible but chemical reactions of CNTs with DMPO have to be reviewed.

4. Conclusion

Seventeen CNTs at three concentrations were evaluated to determine their ROS generation or scavenging rate. The majority of CNTs tested quenched OH radicals. The scavenging rate depended on individual characteristics determined by the CNT synthesis process and surface morphology. CNTs from Group A demonstrate a dependence on surface morphology, which supports the assumption discussed by Fenoglio et al. [30] and Fubini et al. [32]. On the other hand, the OH radical degeneration reaction is far more complicated than predicted. Parameters to describe CNT scavenging characteristics have to include not only metal impurities but also surface morphology, such as the number of dangling bonds, length, and deposition of particles. The results suggest necessity of the further investigation of the role of impurities and CNT surface features in hydroxyl radical generation or scavenging.

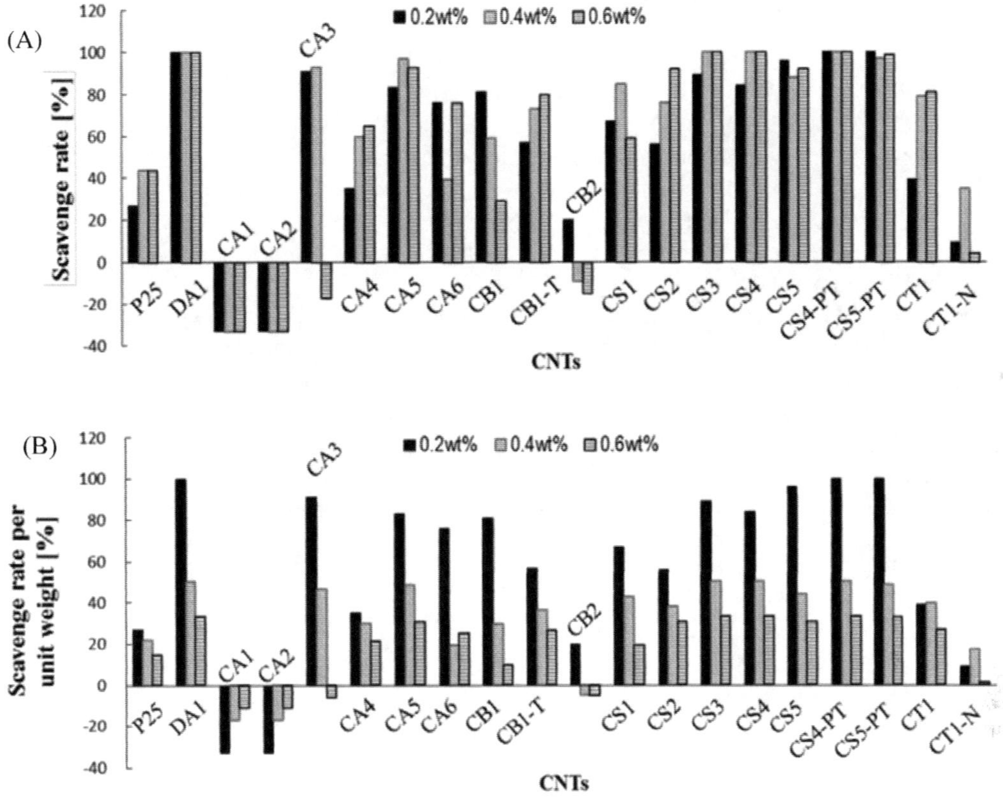

Figure 4. Scavenge rate of OH radicals by various CNTs with concentration change: (A) Scavenge rate at 0.2, 0.4 and 0.6 wt% of CNTs. (B) Scavenge rate per unit CNT weight (0.2 wt%) at 0.2, 0.4 and 0.6 wt % of CNTs; OH radicals were generated by the Fenton reaction.

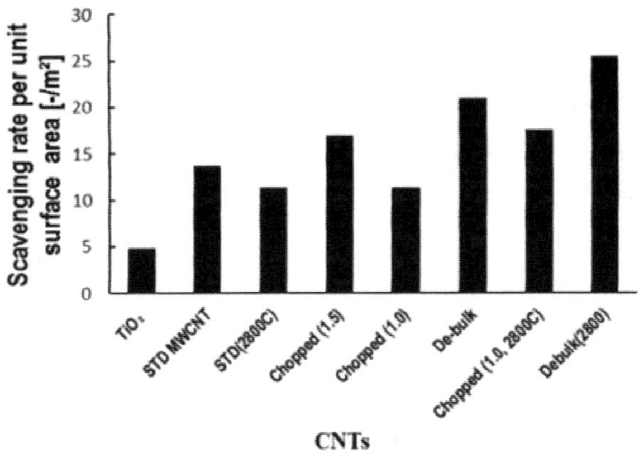

Figure 5. Scavenging rate of cs-CNT derivatives; scavenging rates at 0.2 wt% of CNT were normalized by specific surface area measured by BET isotherm adsorption.

References

[1] Endo M, Strano M S, Ajayan P M 2008 Potential applications of carbon nanotubes *Carbon Nanotubes: Advanced Topics in the Synthesys, Structure, Properties and Applications, Topics in Applied Physics* **111**, ed Jorio A, Dresselhaus M S and Dresselhaus G (Springer-Verlag, Berlin Heidelberg) pp 13–61

[2] Saito Y 2011 *Handbook of Carbon Nanotubes and Graphene*, ed The Fullerene, Nanotubes and Graphene Research Society (Corona Publishing Co. Ltd., Tokyo)

[3] Van Noorden R 2011 *Nature* **469** 14

[4] Mayne M, Grobert N, Terrones M, Kamalakaran R, Rühle M, Kroto H W and Walton D R M 2001 *Chemical Physics Letters* **338**, Issues 2–3 101

[5] Deng F, Ito M, Noguchi T, Wang L, Ueki H, Niihara K, Kim Y A, Endo M and Zheng Q S, 2011 *ACS Nano* **5** 3858

[6] Shi W, Wang Z, Zhang Q, Zheng Y, Ieong C, He M, Lortz R, Cai Y, Wang N, Zhang T, Zhang H, Tang Z, Sheng P, Muramatsu H, Kim Y A, Endo M, Araujo P T and Dresselhaus M S 2012 *Scientific Reports* **2:625** DOI: 10.1038/srep00625

[7] The Royal Sciety 2004 *Nanoscience and nanotechnologies: opportunities and uncertainties* RS Policy document 19/04

[8] Oberdörster G, Maynard A, Donaldson K, Castranova V, Fitzpatrick J, Ausman K, Carter J, Karn B, Kreyling W, Lai D, Olin S, Monteiro-Riviere N, Warheit D, Yang H and A report from the ILSI Research Foundation/Risk Science Institute Nanomaterial Toxicity Screening Working Group 2005 *Particle and Fibre Toxicology* **2:8** doi:10.1186/1743-8977-2-8

[9] Maynard A D and Pui D Y H 2007 *Nanoparticles and Occupational Health* (Springer Netherlands)

[10] Donaldson K, Aitken R, Tran L, Stone V, Duffin R, Forrest G and Alexander A 2006 *Toxicol Sci* **92** 5

[11] Warheit D B 2006 *Carbon* **44** 1064

[12] Lison D and Muller J 2008 *Toxicol Sci* **101** 179

[13] Shvedova A A, Kisin E R, Porter D, Schulte P, Kagan V E, Fadeel B and Castranova V 2009 *Pharmacol Ther* **121** 192

[14] Shvedova A A and Kagan V E 2010 *J Intern Med* **267** 106

[15] Pacurari M, Castranova V and Vallyathan V 2010 *J Toxicol Environ Health A* **73** 378

[16] Shvedova A A, Kisin E R, Mercer R, Murray A R, Johnson V J, Potapovich A I, Tyurina Y Y, Gorelick O, Arepalli S, Schwegler-Berry D, Hubbs A F, Antonini J, Evans D E, Ku B K, Ramsey D, Maynard A, Kagan V E, Castranova V and Baron P 2005 *Am J Physiol Lung Cell Mol Physiol* **289** L698

[17] Ma-Hock L, Treumann S, Strauss V, Brill S, Luizi F, Mertler M, Wiench K, Gamer A O, van Ravenzwaay B and Landsiedel R 2009 *Toxicol Sci* **112** 468

[18] Porter D W, Hubbs A F, Mercer R R, Wu N, Wolfarth M G, Sriram K, Leonard S, Battelli L, Schwegler-Berry D, Friend S, Andrew M, Chen B T, Tsuruoka S, Endo M and Castranova V 2010 *Toxicology* **269** 136

[19] Mercer R R, Hubbes A F, Scabilloni J F, Wang L, Battelli L A, Schwegler-Berry D, Castranova V and Porter D W 2010 *Particle and Fibre Toxicol* **7:28** doi:10.1186/1743-8977-7-28

[20] Muller J, Delos M, Panin N, Rabolli V, Huaux F and Lison D 2009 *Toxicol Sci* 110 442

[21] Takagi A, Hirose A, Nishimura T, Fukumori N, Ogata A, Ohashi N, Kitajima S and Kanno J 2008 *J Toxicol Sci* **33** 105

[22] Sakamoto Y, Nakae D, Fukumori N, Tayama K, Maekawa A, Imai K, Hirose A, Nishimura T, Ohashi N and Ogata A 2009 *J Toxicol Sci* 34 65

[23] Poland C A, Duffin R, Kinloch I, Maynard A, Wallace W A H, Seaton A, Stone V, Brown S, MacNee W and Donaldson K 2008 *Nature Nanotechnology* 3 423

[24] Workshop session 2012 Chaired Elder A *State of the science and the future for predictive power of in vitro and in vivo methods for nanomaterials toxicity testing* 51st Annial Meeting,

Society of Toxicology (March 11-15, 2012 San Francisco, CA)
[25] Shvedova A A, Castranova V, Kisin E R, Schwegler-Berry D, Murray A R, Gandelsman V Z, Maynard A and Baron P 2003 *J Toxicol Environ Health A* **66** 1909
[26] Fenoglio I, Aldieri E, Gazzano E, Cesano F, Colonna M, Scarano D, Mazzucco G, Attanasio A, Yakoub Y, Lison D and Fubini B 2012 *Chem Res Toxicol* 25 74
[27] Harbour J R. Chow V and Bolton J R 1974 *Can J Chem* **52** 3549
[28] Kohno M 2006 *Dojin News* **117** www.dojindo.co.jp/letterj/117/index.html
[29] Murray A R, Kisin E, Leonard S S, Young S H, Kommineni C, Kagan V E, Castranova V and Shvedova A A 2009 *Toxicology* **257** 161
[30] Fenoglio I, Tomatis M, Lison D, Muller J, Fonseca A, Nagy J B and Fubini B 2006 *Free Radic Biol Med* **40** 1227
[31] Fenoglio I, Greco G, Tomatis M, Muller J, Raymundo-Piñero E, Béguin F, Fonseca A, Nagy J B, Lison D and Fubini B 2008 *Chem Res Toxicol* **21** 1690
[32] Fubini B, Fenoglio I, Tomatis M and Turci F 2011 *Nanomedicine* **6** 899

Acknowledgement
We thank GSI Creos Corporation to provide the specific CNT samples and evaluate their properties. This work was supported by a Grant-in-Aid for Specially Promoted Research from the Ministry of Education, Culture, Sports, Science and Technology of Japan (MEXT) (No. 19002007) (ME), the Exotic Nanocarbon Project, Japan Regional Innovation Strategy Program by the Excellence, JST (Japan Science and Technology Agency) (ME, ST, KT).

Disclaimer
The findings and conclusions in this article are those of the authors and do not necessarily represent the view of the National Institute for Occupational Safety and Health.

Journal of Physics: Conference Series **429** (2013) 012030 doi:10.1088/1742-6596/429/1/012030

Violacein/poly(ε-caprolactone)/chitosan nanoparticles against bovine mastistis: Antibacterial and ecotoxicity evaluation

E Berni[1], P D Marcato[1], G Nakazato[2], R K T Kobayashi[2], F I Vacchi[3], G A Umbuzeiro[3], N Durán[1,4]

[1]Chemistry Institute, Campinas State University (UNICAMP), Campinas, Brazil
[2] Department of Microbiology, Londrina State University (UEL), Londrina, Brazil
[3] Laboratory of Ecotoxicology and Environmental Microbiology, Technology Faculty, UNICAMP, Limeira, SP, Brazil
[4]Center of Natural and Human Sciences, Universidade Federal do ABC, Santo André, Brazil

e-mail: eliasberni@gmail.com

Abstract: The nanocarrier was synthesized by nanoprecipitation, using poly(ε-caprolactone) (PCL) as polymer, Tween 80 as surfactant and the biopolymer chitosan (CS) as a charge modification agent. Charge, size and morphology were analyzed by zeta potential, photo correlation spectroscopy (PCS), scanning electron microscopy (SEM) and differential scanning calorimetry (DSC). Bactericidal assays were carried out using a resistant strain of *Staphylococcus aureus*, and the acute ecotoxicity tests were performed with *Daphnia similis*. The nanoparticle without CS (PCLnp) exhibited an average size of 200 nm and zeta potential of -4.28 mV, while the nanoparticle with 0.04% (w/v) of CS (CS_PCLnp) had 250 nm and +21.3 mV. Both were stables for at least 30 days. 200 μg mL^{-1} violacein was encapsulated in CS_PCLnp, which was dissolved in the polymer matrix, a shown by DSC analysis. The minimal inhibitory concentration against *S. aureus* of CS_PCLnp-vio was 25 μmol L^{-1}, while for free violacein it was > 25 μmol L^{-1}. Nanoparticles exhibited an EC50 between 0.3 – 1.1 μmol L^{-1} with *Daphnia*, while free violacein was around 3.3 – 5.0 μmol L^{-1}. Thus, it was possible to control the charge of the nanoparticles, without extreme changes in size and that it is possible also to encapsulate a powerful antibactericidal compound such as violacein in nanoparticle.

1. Introduction

Mastitis occurs when the udders become inflamed due a microbial infection that invades the teat canal, multiplies and produces toxins that are harmful to the mammalian gland. This disease can be classified in two ways: subclinical, without visible signs, and clinical, with visible signs such as flakes or clots in the milk, slight swelling, secretion, fever, loss of appetite, dehydration and depression. Bovine mastitis is a complex problem because there are several causative agents. Over 100 microorganisms can cause mastitis, the most common being *Staphylococcus aureus* [1, 2]. Bovine mastitis is the main economic loss in milk production, estimated at $ 185/cow [2] and reach $ 1.8 billion/year in the United States. In addition, this kind of infection of the mammary glands also affects humans. Therefore new treatment approaches should be attempted. Among several possibilities, drug delivery has been promising approach against several diseases in recent years.

The extensive research in drug delivery is due to several advantages that these systems present, compared with conventional forms of drug administration. Drug delivery presents such interesting features as the possibility to introduce poorly soluble drugs in formulations, increase diffusion through the skin [3], and prolong residence time of the drug and its action in specific cells [4], besides allowing decreases in toxicity. The carriers employed in drug delivery can be in micro or nano size ranges, including liposomes [5], emulsions [6], and lipid nanoparticles [7], dendrimers [8], oxides [9] and polymer nanoparticles [10]. Concerning polymeric nanoparticles, poly(ε-caprolactone) (PCL) is widely used, being compatible with a wide range of drugs [11]. Its slow degradation rates allow a long releasing period, up to several months [11] and, as PCL is lipophilic, it allows a good interaction with lipophilic compounds. Usually the lipophilic drugs are uniformly distributed in the matrix while hydrophobic drugs generally go to the polymer-solvent interface [11]. Regarding a potential candidate to encapsulate in PCL nanoparticles, violacein is a lipophilic compound and acts against bovine mastitis, since one of its characteristic is effectiveness against *S. aureus* [12].

Violacein is a violet pigment produced by *Chromobacterium violaceum*, which exhibits antiviral, antiparasitic, antibiotic and antibactericidal activities [13-15]. However, violacein shows low bioavailability due its poor solubility. Therefore, violacein usually needs an organic solvent in order to be dissolved; in many cases toxic ones such as dimethyl-sulfoxide (DMSO) [14].

Since, the drug delivery method is an interesting option to increase the bioavailability of violacein, the aim of this work was the nanoencapsulation of violacein in a PCL nanoparticle and in order to control the charge of the nanoparticles the mucoadhesive chitosan (CS) [4] was also used.

2. Methods

2.1 *Preparation of polymeric nanoparticles*

The PCL nanoparticles coated with chitosan were prepared by the nanoprecipitation technique – solvent displacement method, modified from Govender *et al.* [16]. In a typical synthesis of unloaded nanoparticles, PCL was dissolved in acetone. The organic phase was poured into an aqueous phase containing 1% acetic acid (v/v), 1 mg mL^{-1} Tween 80 and 0 to 0.08% of chitosan (w/v). After mixing, the acetone was eliminated by evaporation under reduced pressure. To produce nanoparticles containing violacein, 8 mg of the active compound was dissolved in the organic phase and the same synthesis procedure used for unloaded nanoparticles was followed. The unloaded nanoparticles were called CS_PCLnp and the loaded nanoparticles as CS_PCLnp_vio.

2.2 *Physical characterization*

Particle size distribution (z-average), polydispersity index and zeta potential were studied by dynamic light scattering, using a Zetasizer Nano instrument, ZEN3600 (Malvern Instruments). All samples were diluted in 1 mmol L^{-1} solution of KCl, ratio 1:19 of sample:KCl solution, then placed in capillary polystyrene cuvettes. Measurements were performed at 25°C.

The morphology of the nanosystem was evaluated using scanning electron microscopy, with a JSM-6360LV (Jeol) instrument. A drop of sample diluted in MilliQ® water (1:19, for sample:water) was placed on the carbon ribbon.

2.3 *Stability of nanoparticles*

The stability of the system was analyzed with respect to size distribution and zeta potential, varying the temperature and pH. The pH was modified by an external module, MPT – 2 (Malvern Instrument). The modification of size and zeta potential were monitored using DLS, using a Zetasizer Nano instrument, ZEN3600 (Malvern Instrument). To correct the pH, acidic (HCl 0.25 mol L^{-1}) and basic (KOH, 0.5 mol L^{-1} and 0.25 mol L^{-1}) solutions were used.

2.4 *Differential scanning calorimetry*

The interaction between compounds in the nanostructure and their physical state were studied by a differential scanning calorimetry DSC-Q10 (TA Instruments). The samples were lyophilized, placed in aluminum pans and sealed. Samples were heated from 30°C up to 180°C, at 10°C min^{-1} and held isothermally for 1 min, in a nitrogen atmosphere (50 mL min^{-1}).

2.5 *Loaded efficiency*

The entrapment efficiency and total amount of violacein were analyzed by UV-vis spectrometry, Hitachi U-2000, using Beer´s law (ε_{575} in ethanol 3.13 x 10^{-2} mL μg^{-1} cm^{-1}) [14]. To obtain the total amount of encapsulated violacein, 100 µL of suspension was dissolved in ethanol (1:15) and centrifuged at 10,000 rcf for 5 min. The supernatant was recovered and the UV-vis spectra measured. To determine the percentage of loaded violacein, 1000 µL from the sample was centrifuged at 10,000 rcf for 5 min, the supernatant was recovered and analyzed in UV-vis spectrophotometer.

2.6 *Antibacterial assay*

The isolates of *S. aureus* (MBSA) were obtained from farms located in the central region of the State of São Paulo, Brazil, with histories of chronic subclinical and clinical Bovine mastitis problems [17]. The diagnosis of clinical mastitis was performed using the strip cup test [18]. Subclinical mastitis was diagnosed based on the California Mastitis Test-CMT [19]. The *S. aureus* used were: MBSA 4 if it was subclinical mastitis and resistant to penicillin and MBSA19 for subclinical mastitis and resistant to penicillin/erythromycin. The determination of MIC was performed using the micro-dilution assay in 96-well plates, according to CLSI [20]. Single colonies were diluted in saline solution and adjusted to 0.5 index ont the MacFarland scale (2 x 10^{8} cfu ml^{-1}).

2.7 *Ecotoxicicity assay*

Free violacein, PCL and violacein nanoparticles (CS_PCLnp_vio and PCLnp_vio) were tested for acute toxicity with *Daphnia similis* according to OECD 202 guidelines [21]. Twenty neonates (<24 h old) were placed in each concentration of the test solutions. Experiments were performed in four replicates at 21 ± 0.3°C in the dark. DMSO was used to dissolve free violacein and negative control was performed accordingly. After 48 h, the number of immobile organisms was recorded. The results were statistically analyzed using the Trimmed Spearman–Karber method for estimating the effective concentration EC50 [22].

3 Results and discussion

Before encapsulating the violacein, the nanoparticle system produced by the nanoprecipitation method was optimized by varying the amount of CS, in order to obtain a positively charged surface, with a satisfactory nanoparticle size. The effect of CS addition on mean diameter size, polydispersity and zeta potential of the nanoparticles were evaluated using Zetasizer Nano. Studies carried out by Quemeneur *et al.* [23] showed that CS with molar masses between 50,000 and 500,000 g mol^{-1} present the same adsorption mechanism onto polymer, thus the same zeta potential behavior. In this way, we used only one molar mass of CS, changing just the amount of CS in the synthesis.

Table 1. Effect of chitosan on particle sizes and zeta potential in the nanoparticles

Chitosan concentration (% w/v)	Zeta potential (mV)	Particle sizes (nm)	PdI
0.00	- 5 ± 2	210 ± 10	0.24 ±0.08
0.04	+ 40 ± 3	280 ± 30	0.25 ±0.03
0.08	+ 44 ± 5	320 ± 50	0.21 ±0.06

In this first part of the study, each system containing a specific amount of CS (from 0.00 to 0.04 and 0.08% w/v) was synthesized at pH 3.5. Table 1 shows the effect of chitosan on particle size, polydispersity and zeta potential of the systems. As expected, the addition of chitosan increased the mean diameter and zeta potential of the particles, due to the increase of molecule adsorption on the nanoparticle surfaces.

The sizes and zeta potentials of the nanoparticles were evaluated by varying the pH since, due to the pKa of CS (pKa=6.5), it was expected that variation of pH could cause modification in the mean diameter size and zeta potential. Figure 1 (a) shows that higher amounts of CS cause agglomeration above pH 8. However, the zeta potential behavior was the same for different amounts of CS, Figure 1 (b). The nanoparticles without CS presented a negative zeta potential, which increased with basic pH. Regarding the DLVO (Derjaguin, Verwey, Landau and Overbeek) theory [24] the systems containing chitosan (0.04% and 0.08%) were stable up to pH 6, while the systems containing only Tween 80 was stable above pH 7.5, Figure 1 (b). Comparing the systems with 0.04% and 0.08% of CS, it is possible to observe a more stable behavior of the system containing 0.04% of CS.

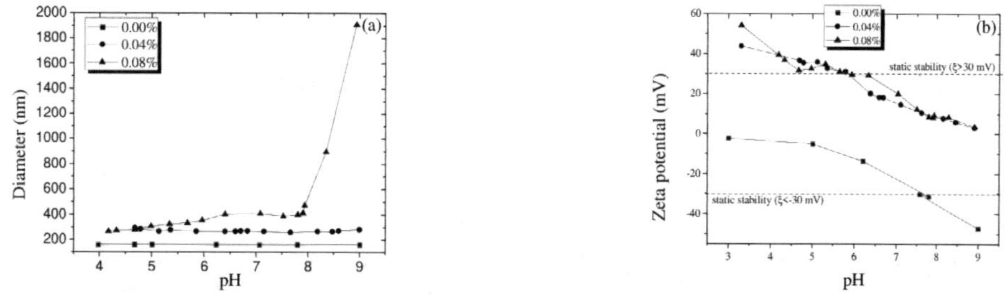

Figure 1: Mean diameter size (a) and zeta potential (b) changes with pH, from 3 up to 9 of three systems with different amounts of chitosan.

After these results, the system containing 0.04% of chitosan was selected for further studies. The system having only nanoparticle was named CS_PCLnp and the system containing violacein was named CS_PCLnp-vio. In this latter system 180 μg mL^{-1} of violacein were encapsulated with a yield of 91 ± 1%, having violacein at 11% in mass of PCL (w/w). The same study of mean diameter size and zeta potential was carried out. Figure 2 shows the effect of loading violacein on the size and zeta potential with pH variation. Probably the observed changes are due the presence of violacein in the nanoparticle surfaces. SEM images of CS_PCLnp_vio showed spherical shape without agglomeration, Figure 2(c). Moreover, the presence of violacein did not affect the mean diameter in relation to the temperature, Figure 3.

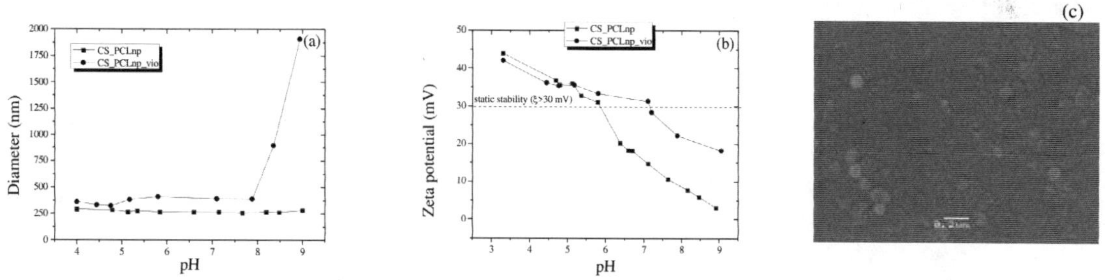

Figure 2: Mean diameter (a) and zeta potential (b) of unloaded and loaded nanoparticles with violacein in relation to the pH, (c) SEM image of CS_PCLnp_vio.

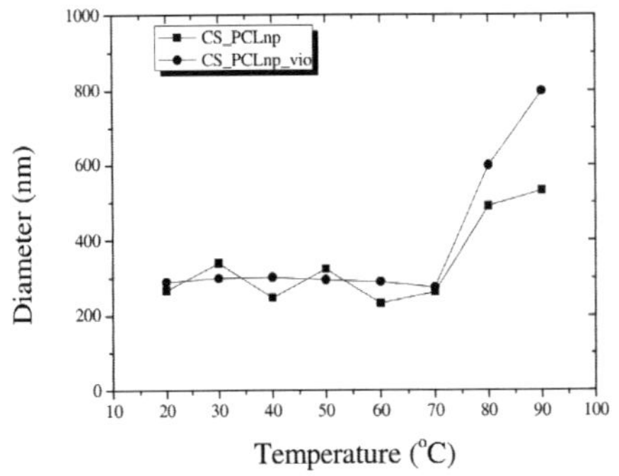

Figure 3: Mean diameter in relation to the temperature variation of unloaded and loaded nanoparticles with violacein.

A differential scanning calorimetric measurement was carried out to determine the state of violacein in polymer nanoparticle matrix and the results are presented in Figure 4. Violacein exists in one polymorphic form (melting point of 135°C), which was absent in the CS_PCLnp-vio systems. The absence of an observable melting point of violacein in the nanostructure can be attributed to complete drug dissolution in the polymer matrix [25].

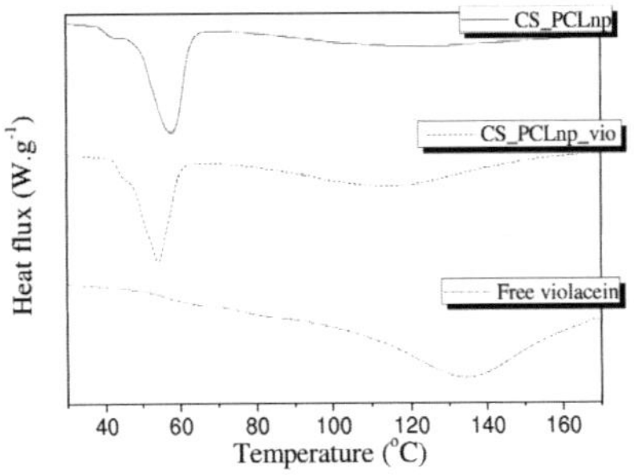

Figure 4: DSC curves for CS_PCLnp, CS_PCLnp_vio and free violacein, obtained at 10°C min^{-1} in a nitrogen atmosphere.

The free violacein and the violacein nanosystems, presented antibacterial activity. The CS_PCLnp_vio system presented a minimal inhibitory concentration of 25 μmol L^{-1} against two resistant *S. aureus* strains, MBSA4 and MBSA19, while free violacein was less effective (minimal inhibitory concentration greater than 25 μmol L^{-1}).

Toxicity tests with Daphnia revealed that CS_PCLnp did not present acute toxicity to Daphnia, Figure 5 (a). In order to facilitate the dissolution of free violacein, 1% of DMSO was used. Controls performed only with 1% of DMSO did not show any toxicity to the test organism. EC50 for CS_PCLnp_vio was in the range of 0.3 – 1.1 μmol L^{-1}; and EC50 for free violacein was 4 μmol L^{-1}, with a confidence interval of 3.3 – 5.0 μmol L^{-1}, Figure 5(a). The higher toxicity of CS_PCLnp_vio in relation to the free violacein could be explained by the mucoadhesive property of CS [4] that would enhance the retention time of the nanoparticles in the organism's gut. A purple color was observed in the gut of the organisms exposed to the nanosystem, Figure 5 (b), in contrast to no color for the free violacein exposed organisms. Both systems, free violacein and CS_PCLnp-vio, were more toxic to Daphnia in comparison with *S. aureus* strains, but at the same order of magnitude (μmol L^{-1}).

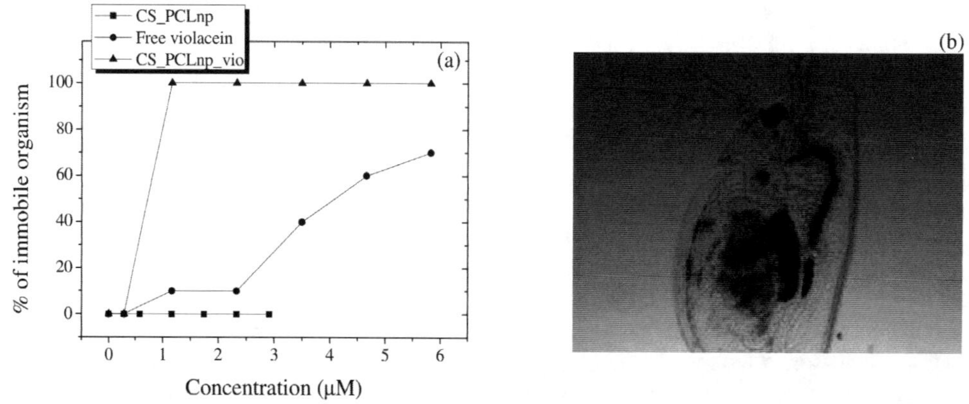

Figure 5: Results of the acute toxicity test *Daphnia similis* of CS_PCLnp, CS_PCLnp_vio and free violacein (a) and a photograph of the test organism exposed to CS_PCLnp_vio (b).

4. Conclusions

The control of charge was successfully achieved using chitosan, changing the negative zeta potential to positive. The nanoparticle was increased due chitosan addition. The violacein was loaded with a high efficiency in the CS_PCLnp_vio nanosystem and showing a better result against Bovine mastitis infection with *S. aureus*, compared with free violacein. Related to the ecotoxicity test, CS_PCLnp did not present acute toxicity to Daphnia. Both free violacein and CS_PCLnp_vio were toxic to Daphnia being the violacein nanosystem the most toxic one. Therefore, attention should be given when environmental exposures to those systems are expected.

Acknowledgement: Support from the Brazilian Network of Nanotoxicology (MCT/CNPq), INOMAT (CNPq/CNPq), and FAPESP is acknowledged

5. References

[1] Schoroeder J W 2010 *Extension Service* NDSU (**www.ag.ndsu.edu**) AS-1129.
[2] Pereira U P, Oliveira D G S, Mesquita L R, Costa G M and Pereira L J 2011 *Rev. Microbiol.* **148** 117.
[3] Teeranachaideekul V, Souto E B, Junyaprasert V B and Müller R H 2007 *Intern. Symp. Control. Rel. Bioact. Mater.* **34** 1956.
[4] Mazzarino L, Travelet C, Ortega-Murillo S, Otsuka I, Pignot-Paintrand I, Lemos-Senna E and Borsali R 2012 *J. Colloid. Interface Sci.* **370** 58.
[5] Torchilin V P 2007 *Pharmaceut. Res.* **24** 1.
[6] He W, Tan Y, Tian Z, Chen L, Hu F, and Wu W 2011 *Int. J. Nanomedicine* **6** 521.
[7] Müller R H, Peterson R D, Hommoss A and Pardeike J 2007 *Adv. Drug Deliver. Rev.* **59** 522.
[8] Huynh L, Neale C, Pomès R and Allen C 2012 *Nanomed. Nanotech. Biol. Med.* **8** 20.
[9] Jiang H T, Wang T, Wang L, Sun C, Jiang T, Cheng G and Wang S 2012 *Microp. Mesopor. Mat.* **153** 124.
[10] Rao J P and Geckeler K E 2011 *Prog. Polym. Sci.* **36** 887.
[11] Dash T K and Konkimalla V B 2012 *J. Control. Release* **158** 15.
[12] Cazoto L L, Martins D, Ribeiro M G, Durán N and Nagazato G 2011 *J. Antibiot.* **64** 395.
[13] Durán M, Faljoni-Alario A and Durán N 2010 *Folia Microbiol.* **55** 535
[14] Martins D, Costa F T M, Brocchi M and Durán N 2011 *J. Nanopart. Res.* **13** 355.
[15] Durán M, Ponezi A N, Faljoni-Alario A, Teixeira M F S, Justo G Z and Durán N 2012 *Med. Chem. Res.* **21** 1524.
[16] Govender T, Stolnik S, Garnett M C, Illum L and Davis S S 1999 *J. Control. Release* **57** 171.

[17] Cabral K G, Lammler C, Zschock M, Langoni H, De Sa Me, Victoria C and Da Silva A 2004 *Can. J. Microbiol.* **50** 901.

[18] Radostits O M, Gay C C and Hinchcliff K W 2007 *Veterinary Medicine: a textbook of the diseases at cattle, horses, sheep, pigs and goats. 10th ed.* (Saunders, Philadelphia).

[19] Schalm O W and Noolander D O 1957 *J. Am. Vet. Med. Assoc.* **139** 199.

[20] CLSI - Clinical Laboratory Standards Institute 2011 *Methods for dilution antimicrobial susceptibility tests for bacteria that grow aerobically 8th ed* (Wayne, PA, USA).

[21] OECD 2004 *Daphnia sp. Acute Immobilisation Test. Organisation for Economic Co-operation and Development - Guideline for testing of Chemicals*, **202** 12.

[22] Hamilton M A, Russo R C and Thurfton R B 1977 *Sci. Technol.* **11** 714.

[23] Quemeneur F, Rinaudo M and Pepin-Donat B 2008 *Biomacromolecules* **9** 396.

[24] Zetasizer Nano series technical note *Zeta potential an introduction in 30 minutes*, **MRK654-01** available from www.malvern.co.uk.

[25] Mundargi R C, Srirangarajan S, Agnihotri S A, Patil S A, Ravindra S, Setty S B and Aminabhavi T M 2007 *J. Control. Release* **119** 59.

In vivo toxicity of enoxaparin encapsulated in mucoadhesive nanoparticles: Topical application in a wound healing model

S C Huber[1*], P D Marcato[2, 3], R M Barbosa [3], N Duran [2,5], J M Annichino-Bizzacchi[1]

[1] Hemocentro, Universidade Estadual de Campinas, Campinas, SP, Brazil
[2] Chemistry Institute, Universidade Estadual de Campinas, Campinas, SP, Brazil
[3] Faculty of Pharmaceutical Sciences of Ribeirão Preto, Universidade de São Paulo, SP, Brazil
[4] Biology Institute, Universidade Estadual de Campinas, Campinas, SP, Brazil
[5] Center of Natural and Human Sciences, Universidade Federal do Abc, Santo André, SP, Brazil

* E-mail: stephany_huber@yahoo.com.br

Abstract: Wound healing comprises four distinct phases and involves many cell events and biologic markers. The use of nanoparticles for topical application has gaining attention due to its deeper penetration in the skin and the retention capacity of the drug in the site of application. In this study the effect and toxicity of mucoadhesive polymeric nanoparticles loaded with enoxaparin was evaluated in *in vivo* model of skin ulcer. Our results showed an interesting formulation based on mucoadhesive nanoparticles with enoxaparin that improved wound healing without cytotoxicity *in vitro* in all endpoint evaluated. Then, this semi-solid formulation is a promising option for skin ulcer treatment.

1. Introduction

Wound healing is a dynamic and complex process that results in anatomical and functional restoration of the injured area [1]. It comprises four distinct phases: hemostasis, inflammation, proliferation and remodeling. There are some pathological conditions which this efficient process is altered, resulting in not healed ulcers, like in diabetes [2-4]. An interesting drug for treatment of these ulcers is low molecular weight heparin (enoxaparin) due to its anti-inflammatory property, analgesia, neoangiogenesis and epithelization [5, 6]. An area that has been widely investigated is the topical application of nanoparticles, due to its deeper penetration in the skin and the retention capacity of the drug in the site of application [7]. The encapsulation of enoxaparin in polymeric nanoparticles for topical use can help in the healing process, increasing the drug stability and skin penetration and decreasing its toxicity. To increase their adhesiveness and to helps in the wound healing, chitosan is being used. Chitosan is a biocompatible and biodegradable polymer, nontoxic and low cost, and also presents interesting biological properties like antimicrobial activity, bioadhesiveness, potent topical analgesic action and acceleration of the healing process [8,9].

In this study polymeric nanoparticles recovered with chitosan were prepared in one step, with the aim to improve the economical aspect of the process and the properties of these nanostructures (increase of the skin adhesion and antimicrobial effect) with enoxaparin, to promote an efficient wound healing ulcers.

Actually, the interest for the safe use of nanometric materials became valuable. The risk/benefit about the use of nanoparticles must be measured by medical and technological methods [10]. One way to test nanoparticles *in vitro* is through cytotoxicity evaluation. Fibroblasts from mouse BALB/c 3T3 could be consider a cell model to estimate the toxicity of nanoparticles. The evaluation of toxicity *in vivo* was made through hepatic and renal enzymes, hemogram and oxidative stress. Wound healing is affected according the quantity of oxygen reactives species (ROS), since the ROS suppression will result in infection and their elevation could result in destruction of healthy stromal tissue [11].

In this study the *in vivo* effect in wound healing and *in vitro* toxicity of mucoadhesive nanoparticles loaded with enoxaparin was evaluated. The results showed an interesting nanocarrier system for enoxaparin inproving the wound healing and did not exhibit *in vitro* toxicity.

2. Material and Methods

2.1. Preparation of polymeric nanoparticles recovered with chitosan in the encapsulation of enoxaparin

The nanoparticles were prepared with the biodegradable polymer poly(ε-carpolactona) (PCL) (Aldrich) recovered with the natural polymer chitosan (105 g/mol ~81% acetylation, Polymar). The preparation of nanoparticles loaded enoxaparin was held by the double emulsion modified technique and evaporation of the solvent as describe by Meng et al [12].

2.2. Characterization of the nanoparticles

2.2.1. *Size, distribution and Zeta potential*

The dispersion of nanoparticles was diluted with water and the average diameter (Z-average), distribution and superficial charge was measured in Zeta Sizer Manlvern®.

2.2.2. Encapsulation efficiency

The encapsulated enoxaparin in nanoparticles was quantified through the modified Azure II colorimetric method as describe by Lam et al [13].

2.3. *In vivo* evaluation

2.3.1. Experimental animals

This study was performed using 50 male Wistar rats with age of 6-8 weeks. All the experiments were carried out with the standards established by Brazilian law for the animal experiments approved by Ethics Committee on the use of Animals (CEUA – Unicamp) certificate number 2290-1.

After intramuscular anesthesia with ketamine (75 mg/kg) and xylazine (10 mg/kg), a 1 cm^2 quad format dermatological lesion was performed with scissor and tweezer. Posteriorly, the animals were divided in five groups of daily treatment, with 10 animals each: G1- free ENOXA gel (2mg/g gel), G2- Sham, G3- pure gel, G4- empty nanoparticles gel (2 mg/g gel), G5- ENOXA nanopartcles gel (2m/g gel).

Macroscopically, the evaluation of the wound healing was made by measurement of wound contraction in established days (4th day, 7th day and 10th day) and analyzed through Image J software. After 10 days of the lesion, the animals were anaesthetized and sacrificed. The blood was collected by cardiac puncture.

2.3.2. Enzymatic evaluation, hemogram and oxidative stress

The hemogram was accomplished in the automation system Cell Dyn 1700 ABBOTT.

The blood was washed with the cold buffer to catalse analysis, as describe by Aebi [14]. The plasm was utilized to evaluate lipid peroxidation (TBARS) according Yagi methodology [15] and nitrite dosage was realized through Griess reaction [16]. The plasm was also analyzed to verify the presence of enoxaparin in the bloodstream with the evaluation of the anti-Xa activity (Kit IEL-equipament Top 500).

The serum was used to analyze aspartate aminotransferase (AST) and alanine aminotransferase (ALT) both using the kinetic method (Kit Laborlab). Urea was measured by enzymatic colorimetric method (Kit Laborclin) and creatinine was measured by kinetic method (Kit Laborlab), as indicators of hepatic and renal function.

2.4. Free-enoxaparin cytotoxicity, empty nanoparticles and encapsulated enoxaparin

The cytotoxicity test of free-enoxaparin, empty nanoparticles and encapsulated enoxaparin was made in mouse embryo fibroblasts (3T3) from BALB/c. The cells were plated in the density of 2 x 10^4/mL in 96 wells plates and incubated for 48 hours until reach the semi-confluency phase. Then, the cultures were exposed to free-enoxaparin, empty and encapsulated nanoparticles, in different concentrations (0-100 μM for free and encapsulated enoxaparin and 0-400 μM for empty nanoparticles). The cells were incubated for 2 hours with the samples. Each concentration was tested in 6 replicates. In the end of incubation, MTT was used to evaluate the cell viability.

3. Results and discussion

3.1. Preparation and characterization of mucoadhesive enoxaparin nanoparticles

The preparation of PCL/chitosan nanoparticles had a relation 6:1, with or without enoxaparin, and presented a average particles size of 496 e 491 nm, respectively, and positive surface charges with average values of +25 mV (without enoxaparin) e +20 mV (with enoxaparin) (Table 1).The encapsulation efficiency was 98% (Table 1) and the particles were stable during 90 days.

Table 1. Particles sizes, surface charge (Zeta potential), and encapsulation efficiency of mucoadhesive nanoparticles with and without enoxparin

PCL/Chitosan ratios	Enoxaparin (mg)	Zeta Potential (mV)	Particles sizes (nm)	PdI	Encapsulation efficiency (%)
6:1	0	+25	491	0.357	---
6:1	200	+20	496	0.369	98

3.2. Macroscopically analysis, ezymatic evaluation, hemogram and oxidative stress

In the macroscopically analysis the values are expressed in percentage of size lesion through mean ± SD in the different groups according the analyzed days (d4- 4th day, d7- 7th day, d10- 10th day) (Table 2). The animals treated with mucoadhesive nanoparticles with ENOXA showed in the 10th day a minor wound (0.034 cm^2) compared with the group treated with free ENOXA (0.091 cm^2), being statistically significant ($p=0.0238$) according the Mann-Whitney test.

Table 2: Macroscopically evaluation of the wound size in the animals of different groups

Groups	D4	D7	D10
Free-ENOXA gel	64.6% ± 6.7	34.9% ± 8.8	9.2% ± 5.9
Sham	70.3% ± 10.3	46.2% ± 12.4	18.5% ± 2.5
Pure gel	80.6% ± 13.9	35.9% ± 10.9	9.1% ± 6.1
Empty nanoparticles gel	58.4% ±12.6	26.0% ± 5.8	3.5% ± 2.7
ENOXA-nanoparticles gel	69.8% ± 13.9	28.4% ± 12.8	4.7% ± 2.7

The animals hemogram presents all the values in the normal range according Melo et al [17] (Table 3), evidencing that no collateral effects was present due to the use of enoxaparin in the animals (normal number of platelets). The enzymes also presents concentrations in the reference range as describe by Melo et al [17]. The mean of the results ± SD is evidenced in the table below (Table 3).

Table 3: Results of the hemogram and enzymatic evaluation of the animals

	WBC (k/µL)	RBC (M/µL)	Hb (g/dL)	HCT (%)	Plt (k/µL)	ALT (U/L)	AST (U/L)	Uréia (mg/dL)	Creatinina (U/L)
Média	5.23	7.02	13.70	39.10	555.03	45.5	135.9	51.5	0.45
DP	1.87	0.62	1.06	3.34	63.80	3.5	15.3	3.6	0.03

It was not found significant differences in catalase activity, the groups presents values close having a mean of 0.01031 (k/gHb/min) ± 0.0007542.

The TBARS and nitrite analysis revealed a higher concentration of both in the groups treated with enoxaparin, suggesting a increase in the pro-oxidants concentrations by the use of enoxaparin, being statically significant according ANOVA test ($p= 0.0022$ for TBARS e $p< 0.0001$ for nitrite), as show the Table 4.

Table 4: Evaluation of stress oxidative markers TBARS e NO_2 in plasm

Groups	µmol NO_2^-/µg de ptn	MDA (nM/mg ptn)
Gel free-ENOXA	32.1± 11.7	0.000268 ± 0.000076
Sham	24.1± 9.8	0.000109 ± 0.000075
Pure gel	25.4 ± 11.8	0.000202 ± 0.000076
Empty nanoparticles gel	12.4 ± 6.1	0.000055 ± 0.000027
ENOXA-nanoparticles gel	27.0 ± 6.3	0.000494 ± 0.000285

The anti-Xa activity was not detected, showing enoxaparin did not reach the bloodstream, having only a local effect in the lesion area, offering no adverse risk in the concentration tested for topical application (2mg/g gel).

3.3 Cytotoxicity of free enoxaparin, empty and encapsulated nanoparticles

Free enoxaparin showed toxicity in fibroblasts (3T3) evaluated through MTT (42.5% of toxicity in the concentration of 100 µM) and this is observed in the figue 1A. However, when the enoxaparin is encapsulated in nanoparticles its toxicity decreases a lot (5% of toxicity in the concentration of 100 µM), presenting a significative value through Mann-Whitney test (p=0.0095) evidencing the protector effect of nanoparticles with the drug. The figure 1B presents the non toxicity of the polymers used in the preparation of the nanoparticles showing no toxicity until 200 µM of PCL, demonstrating a interesting drug delivery system.

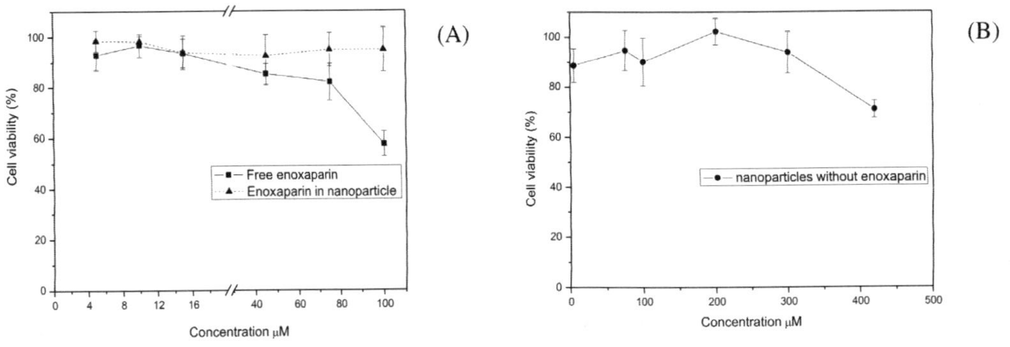

Figure 1: Viability graph in 3T3 cells with the cellular target MTT of the free enoxaparin, enoxaparin in nanoparticle (A) and nanoparticles without enoxaparin (B).

4. Conclusions

Our results showed no toxicity *in vitro*. There was an increase in the concentration of the oxidative stress markers in the groups treated with ENOXA. This study demonstrated an interesting formulation based on mucoadhesive nanoparticles with enoxaparin that improved wound healing. Furthermore, the formulation showed a good effect on wound healing in diabetic animals, indicating a promising treatment for ulcers.

5. Acknowledgments

Support from FAPESP, CNPq, INCTs, INOMAT (MCT/CNPq) and Brazilian Network of Nanotechnology (MCT/CNPq) are acknowledged.

References

[1] Lazarus G S *et al* 1994 *Arch. Dermatol.* **130** 489
[2] Diegelmann R F and Evans M 2004 *Frontiers Biosci.* **9** 283
[3] MacKay D and Miller A L 2003 *Alternative Med. Rev.***8** 359
[4] Mishra M *et al* 2008 *Dig. J. Nanomater. Bios.* **3** 49
[5] Barreto M G P *et al* 2010 *Rev. Assoc. Méd. Bras.* **56** 51
[6] Ravikumar T *et al* 2006 *Ann. Burn Fire Disasters.* **19** 123
[7] Guterres S S *et al* 2007 *Drug Target Insights* **2** 147
[8] Silva H S R *et al* 2006 *Quím. Nova.* **29**, 776
[9] Okamoto Y *et al* 1992 *Elsevier* 70
[10] Eidi H *et al* 2010 *Inter. J. Pharm.* **396,** 156
[11] Bryan N *et al* 2012 *Eur.Cells Mat.* **24**, 249
[12] Meng F T *et al* 2003 *J.Control. Release.* **91**, 407
[13] Lam L H *et al* 1976 *Biochem Biophys Res Commun.* **69,** 570
[14] Aebi H 1984 *Methods Enzymol.* **105**, 121
[15] Yagi K 1987 *Chem. Phys Lip.* **45**, 337
[16] Green L C *et al* 1982 *Analytical Biochem*estry **126**, 131
[17] Melo M G D *et al* 2012 *Scientia Plena.* **8**. 1

Immunological Assays as an Opportunity of Assessment of Health Risks of Airborne Particle Mixture Including Nanoparticles

Táňa Brzicová[1], Ivo Lochman[2], Pavel Danihelka[1], Alexandra Lochmanová[3], Karel Lach[2], Vladimír Mička[1,2]

[1] VŠB – Technical University of Ostrava, Faculty of Safety Engineering, Lumírova 13, 700 30 Ostrava-Výškovice, Czech Republic

[2] Institute of Public Health Ostrava, Partyzánské náměstí 7, 702 00 Ostrava, Czech Republic

[3] University of Ostrava, Faculty of Medicine, Syllabova 19, 703 00 Ostrava-Zábřeh, Czech Republic

E-mail: tana.brzicova@vsb.cz

Abstract. The aim of this pilot study was to evaluate perspectives of the assessment of non-specific biological effects of airborne particulate matter including nanoparticles using appropriate immunological assays. We have selected various *in vitro* immunological assays to establish an array allowing us to monitor activation of the cell-mediated and humoral response of both the innate and adaptive immunity. To assess comprehensive interactions and effects, the assays were performed in whole blood cultures from healthy volunteers and we used an original airborne particle mixture from high pollution period in Ostrava region representing areas with one of the most polluted air in Europe. Even if certain effects were observed, the results of the immunological assays did not prove significant effects of airborne particles on immune cells' functions of healthy persons. However, obtained data do not exclude health risks of long-term exposure to airborne particles, especially in case of individuals with genetic predisposition to certain diseases or already existing disease. This study emphasizes the *in vitro* assessment of complex effects of airborne particles in conditions similar to actual ones in an organism exposed to particle mixture present in the polluted air.

1. Introduction

Nanoparticles have always been present in airborne solid matter, but contemporary development of engineered nanoparticles, together with an increasing number of unintentionally produced ultrafine particles and nanoparticles, draw attention to corresponding adverse health effects. The same properties that make nanomaterials unique and useful may be on the other side associated with unforeseen adverse health effects [1]. Since the nanomaterials are widely used and actually ubiquitous in the environment, research on the suitable methods of their potential toxicity evaluation represents a global challenge. The experience with studies on adverse health effects of airborne particles implies the importance of the influence of immune system functions, especially through the reactive oxygen species generation and pro-inflammatory action [2].

The essential role of the immune system is the recognition of foreign potential dangerous agents and their subsequent elimination from the organism. However, these processes can be accompanied by various harmful actions on the cells and tissues and can be the cause of the adverse health effects relating to the nanoparticle exposure both in workplaces and the environment. The health risk assessment is complicated by the high number of nanomaterial characteristics which can influence their effects on the immune system and the human health. The recognition versus non-recognition of nanomaterials by immune-competent cells can determine their distribution as well as their toxicological potential [3]. Particles can either stimulate or suppress immune responses or act indirectly on components of the immune system, for instance through modulation of secretion of cytokine and other immune system regulatory molecules [3].

With regard to the marginal role of the immune system on the homeostasis in an organism, there is a presumption that an array of appropriate immunological assays which monitor activations of various immune system components and functions should represent a suitable screening tool for the assessment of harmful nanomaterial action on the human health.

Nanoparticle compatibility with the immune system and thus the ability of the immune system to react on the presence of nanoparticles in an organism is largely determined by the particle surface chemistry. Small size of nanoparticles provides large surface area, where more other pollutants can be adsorbed and in this form transported into the human body. Moreover, the smaller the particles are, the deeper they can penetrate into the respiratory tract and there are evidence that some nanoscale particles cross the biological membranes [4].

2. Methods and Materials

2.1. Particulate matter sample preparation

Owing to the restricted amount of available nanoparticles, an integral sample containing fine airborne particles less than 2.5 micrometers in diameter ($PM_{2.5}$) was used in this pilot study. In case of positive assay results, the ultrafine fraction of airborne particles ($PM_{0.1}$) would be collected and tested. Particulate matter was obtained from industrial measuring station located in Ostrava-Radvanice (the district of the Ostrava City in the Moravian-Silesian region, the Czech Republic), representing area with one of the most polluted air in Europe. Recent measurements have shown significant number of nanoparticles present in the polluted air in Ostrava region [5]. Data from epidemiologic studies suggest adverse health impact on local population associated with heavy airborne particle pollution [6,7]. The average $PM_{2.5}$ concentration during the sampling period was 175 $\mu g/m^3$. The main sources of the air pollution in this region are industry (especially metallurgy, fuel processing and chemical industry), local heating, traffic and emission transport from industrial areas in Poland. The situation is even worsened by geomorphological condition supporting concentration of the air pollution [8]. With regard to the presence of heavy industry in this region, the metals bound on the airborne particle can be presumed. This presumption was confirmed by chemical analysis performed on the particle sample suspension. Significant amount of lead, zinc, cadmium, arsenic and antimony was detected. These elements are a common part of the fly ash from combustion processes [9].

$PM_{2.5}$ fraction was collected on mixed cellulose ester (MCE) membrane filters (Millipore Co., Ltd., USA, 47 mm diameter, 0.8 μm pore size). The sampling flow rate was 16.7 l/min. Particulate matter was collected on November 13–14, 2011 (19.11 hours, total air volume 19.185 m^3).

The filter exempt from sampler head was kept in Petri dish covered with paper box to restrain light exposition. After stabilizing at constant temperature (22–24 °C) and humidity (40–50 %), the filter was weighed on an electronic balance (Sartorius BP211D).

To obtain particulate matter suspension, the filter was sonicated in beaker filled with 250 ml of distilled water using Ultrasonic compact cleaner 4L PS04000A. Sonication time was one and one minute consequently, with the pause to prevent overwarming of the water suspension. Obtained suspension was retained in the refrigerator. Just before the use, the suspension was resonicated and diluted. A part of the suspension was chemically analyzed by Inductively Coupled Plasma –

Mass Spectrometry (ICP-MS) using model THERMO X-Series2. Chemical analysis was aimed to detection of selected metals.

2.2. Biological material withdrawing and preparation

The immunological assays were performed on the whole blood culture of healthy volunteers. Such an approach is suitable in case of screening studies due to comprehensive evaluation of interactions and effects of the particle sample, as well as easy taking of the biological material. When positive assay results occure, it is possible to continue in testing using isolated blood cell populations.

Blood samples were collected from three healthy volunteers (two men, one woman, aged from 25 to 55 years) into Vacutainer Tubes containing heparin as an anticoagulant. Immunological assays were performed within three hours after blood taking. Whole blood analysis (total red blood cell count, total and differential white blood cell counts) was determined using Sysmex XS-800i analyser.

2.3. Chemiluminescence assay

The assay was performed according to SOP 25.47 CKL of Institute of Public Health Ostrava [10]. Briefly, the sample was diluted to obtain four tested concentrations (30 pg/ml, 300 pg/ml, 3,000 pg/ml, 134 μg/ml) of original dust in Hank´s Buffered Salt Solution (HBSS). 2 ml of HBSS and 10 μl of luminol (used to enhance detection of the light) were pippeted into all scintillation viols. The viol representing negative control was left without application of any stimulant. Into the viol representing positive control, 50 μl of opsonized zymosan were added. The tested particle sample in four various concentrations was added into the other viols. 20 μl of blood were pippeted into the viols in 40-second intervals. After vortexing, all viols were placed into Tri-Carb Scintillation Counter and left to temperate for 15 minutes at 37 °C.

The measured results, expressed in the terms of cpm (counts per minute) were read 30, 60 and 90 minutes after the adding of the blood. The test results were statistically analyzed using pared Student´s t-test and one-way ANOVA.

2.4. Basophil degranulation assay

Commerce kit BasoFlowEx® was used to evaluate basophil degranulation. The assay was performed according to the kit instructions [11]. Briefly, the sample was diluted to obtain four tested concentrations (30 pg/ml, 300 pg/ml, 3,000 pg/ml, 134 μg/ml) of original dust in Physiological Buffered Solution (PBS).

The tube representing negative control was left without application of any allergen. Into positive control tube, 10 μl of Stimulation Control were added. The tested airborne particle sample in four varying concentrations was added into the others tubes. Into all tubes, 100 μl of whole blood were added and stimulated by adding of 100 μl of stimulation buffer for 15 minutes at 37 °C. Then, 300 μl of Staining Reagent were added and incubated for 20 minutes at 6 °C. Thereafter 300 μl of Lyzing Solution were added, mixed and incubated for 5 minutes at laboratory temperature. To red blood cell lysis, 3 ml of demineralized water were added into each tube and incubated for 5 minutes at laboratory temperature. Then the tubes were centrifuged for 5 minutes at 300 g, the supernatant was removed and pellet was resuspended in 0.2 ml of PBS. Cell surface expression of CD63 was measured by flow cytometry within 2 hours after staining. Obtained data were visualized on the side-scatter (SSC) versus fluorescence intensity in PE channel (FL2) dot-plot. The gate for basophil population (CD203c positive, SSC low) was set and the gated basophils were brought to histogram.

2.5. Lymphocyte proliferation assay

The assay was performed according to SOP 25.47 CKL of Institute of Public Health Ostrava [12]. Briefly, the sample was diluted to obtain four tested concentrations (30 pg/ml, 300 pg/ml, 3,000 pg/ml, 134 μg/ml) of original dust in Minimum Essential Medium (MEM). The assay was performed in microplate wells in triplicates. Into three wells representing negative controls, only 200 μl of MEM were added. Into positive control wells, 200 μl of cultivated media with phytohemaglutinine (PHA) as a standard mitogen were added. 200 μl of the tested sample in four varying concentrations were pippeted into the other wells. Then 10 μl of blood were added into each well, vortexed and incubated in atmosphere with 5 % CO_2 for 72 hours. After incubation, 10 μl of [^{3}H]thymidine were added into each well and the microplate was returned into the CO_2 incubator for 24 hours. Finally, the content of each well was recovered with a cell harvestor. A stripe of filter paper was put onto the sturdy paper and let drying. Drought rounds of the filter paper were taken out with tweezers and placed into particular flasks fulfilled with 1.5 ml of scintillation liquid. Then the flasks were transported into Tri-Carb scintillation counter to measure the incorporated radioactivity. The assay results were counted as an average of three values for a individual well, and expressed in the terms of cpm (counts per minute). The test result was statistically analyzed using pared Student´s t-test and one-way ANOVA.

2.6. *Cytokine release evaluation*

Commerce kit Human Cytokine 25-Plex Panel was used to evaluate the release of 25 selected cytokines. The assay was performed according to the assay kit´s instructions [13]. Briefly, the airborne particle sample was diluted to obtain the tested concentration of 3,000 pg/ml in Roswell Park Memorial Institute Medium (RPMI). The filter plate was pre-wetted with 200 μl of working Washing Solution. The Washing Solution was aspirated using a vacuum manifold. 25 μl of the beads with defined spectral properties conjugated to protein-specific capture antibodies were pipetted into each well and incubated for 2 hours. Thereafter, protein-specific biotinylated detector antibodies were added and after 1-hour incubation the exceeding antibodies were washed and the wells drought again. Streptavidin conjugated to the fluorescent protein, R-Phycoerythrin (Streptavidin-RPE), were added and washed after 30-minute incubation. Finally, the beads were analyzed using Luminex detection system which enables monitoring of the spectral properties of the beads and the amount of associated R-Phycoerythrin (RPE) fluorescence, therefore the concentration of proteins can be determined.

Polynomial regression was used for the calibration curves. Positivity index (values of $PM_{2.5}$ stimulated cultures in relation to negative control values) was used to reciprocal comparison of individual cytokine release.

3. Results

3.1. *White blood cell counts*

Total red blood cell and total white blood cell counts and differential white blood cell counts of all three blood donors correspond with reference values [14]. Therefore, the results of immunological assays should not be influenced by abnormalities of haematological parameters of the donors.

3.2. *Chemiluminescence assay*

Airborne particle sample was cultivated in four concentrations in the range from 30 pg/ml to 134 μg/ml. Stimulation effect was compared with the negative control (sample without any stimulant). Positive control with application of standard stimulant opsonized zymosan was used to confirm the assay functionality. The assay results are expressed in terms of cpm (counts per minute). The values were read 30, 60 and 90 minutes after the incubation had begun.

As seen on Figure 1, the presence of airborne particle sample did not significantly enhance cpm values compared with the values of negative controls. Despite the increased values observed in some measurements, the no effect hypothesis was statistically confirmed. Pared Student's t-test was used to compare results measured for single concentrations with negative control values. Only in one case (concentration of 3,000 pg/ml, incubation time 30 minutes) significant difference between negative control and the particle sample was recorded. With regard to low number of measurements and the other negative results the response cannot be attributed to the action of airborne particles. The results of one-way ANOVA confirmed that the sample did not have stimulant effect.

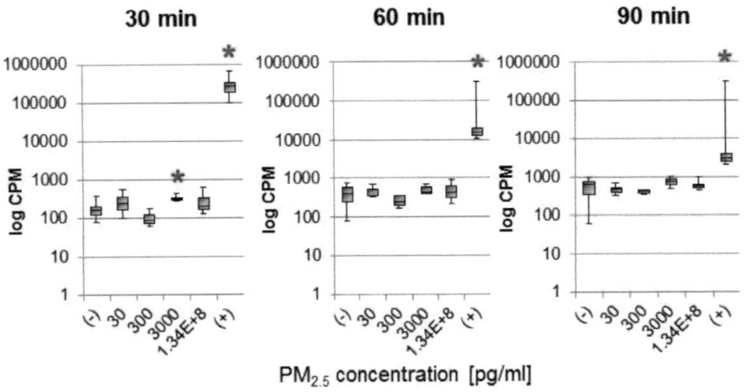

Figure 1. *In vitro* generation of reactive oxygen species by neutrophils from whole blood after stimulation with $PM_{2.5}$ sample. Boxplots are constructed for particular measurement times from values of three healthy volunteers.
∗ p-value < 0.05

3.3. Basophil degranulation assay

Airborne particle sample was cultivated in four concentrations in the range from 30 pg/ml to 134 µg/ml. Stimulation effect was compared with the negative control (sample without any stimulant). Positive control with application of standard stimulant was used to confirm the assay functionality. The assay results are expressed in percentage of activated basophils.

For a right interpretation of the basophil degranulation assay results, it is necessary to achieve convenient values of both negative and positive control. According to BasoFlowEx® Kit user manual, negative control result should be lower than 10 % of activated basophils and contrarily positive control should be higher than 10 % of activated basophils. These conditions were fulfilled.

The cut-off limit between negative and positive response against inhalation allergens when allergies are assessed was established on 15 % of activated basophils. As seen on Figure 2, percentage values of activated basophils did not exceed this limit in any of tested airborne particle concentrations. Therefore, the tested sample did not stimulate significant basophil degranulation.

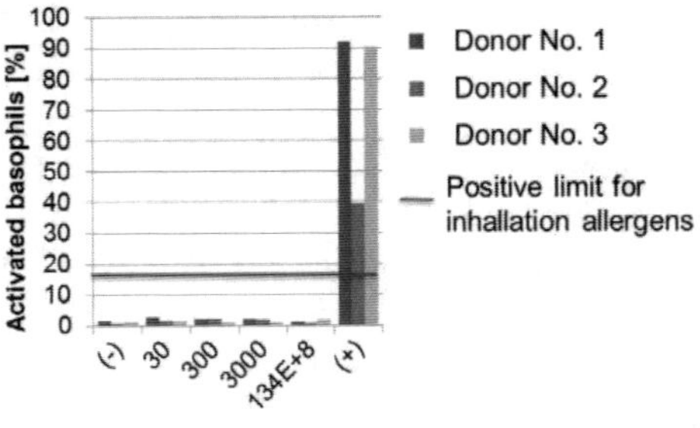

Figure 2. *In vitro* basophil activation with $PM_{2.5}$ sample. Each column represents percentage of activated basophils for one donor and one $PM_{2.5}$ sample concentration.

3.4. Lymphocyte proliferation assay

Airborne particle sample was cultivated in four concentrations in the range from 30 pg/ml to 134 µg/ml. Stimulation effect was compared with the negative control (sample without any stimulant). Positive control with application of standard stimulant PHA was used to confirm the assay functionality. The assay results are expressed in terms of cpm (counts per minute) as an average from triplicate values.

Slight growth in cpm compared with negative control was recorded with particle concentrations of 300 pg/ml and 134 µg/ml, as seen on Figure 3. Pared Student´s t-test as well as one- way ANOVA confirmed significant stimulant effects on p-value < 0,05 of airborne particle sample in concentrations of 300 pg/ml and 134 µg/ml on lymphocyte activation.

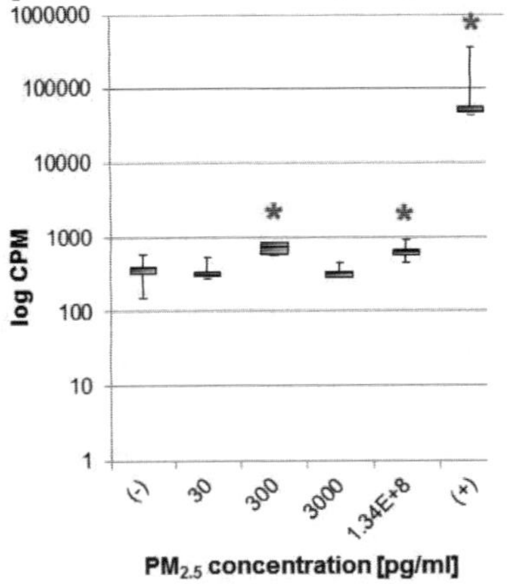

Figure 3. *In vitro* activation of lymphocytes in whole blood after stimulation with $PM_{2.5}$ sample. Boxplots are constructed for particular measurement times from values of three healthy volunteers.
∗ p-value < 0.05

Journal of Physics: Conference Series **429** (2013) 012032 doi:10.1088/1742-6596/429/1/012032

3.5. Cytokine release evaluation

We used commerce kit Human Cytokine 25-Plex Panel enabling to evaluate production of 25 cytokines divided according to their biological function in four panels:

- Inflammatory Panel: GM-SCF, IL-1β, IL-1RA, IL-6, IL-8, TNF-α
- Th1/Th2 Panel: INF-γ, IL-2, IL-2R, IL-4, IL-5, IL-10
- Cytokine II Panel: INF-α, IL-7, IL-12, IL-13, IL-15, IL-17
- Chemokine Panel: Eotaxin, IP-10, MIG, MCP-1, MIP-1α, MIP-1β, RANTES

Airborne particle sample was cultivated in a concentration of 3,000 pg/ml. Stimulation effects were compared with the negative control (sample without any stimulant). Positive control with application of standard stimulant PHA was used to confirm the assay functionality. The assay results were expressed as positivity index, which allow us the comparison of even very different concentrations of individual cytokines. The assay was performed with three incubation periods: 6, 24 and 72 hours. The assay results are shown on Figure 4.

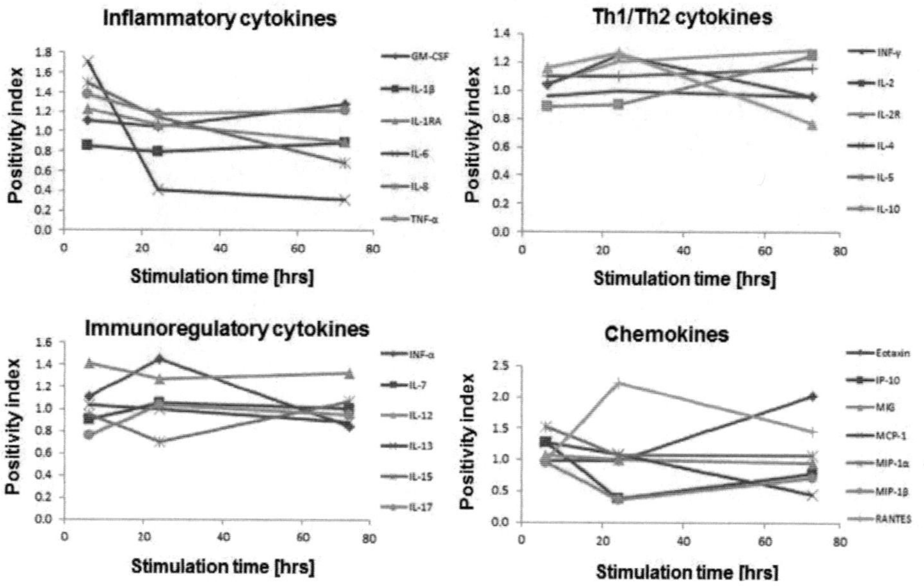

Figure 4.

In vitro cytokine release evaluation in cultures of whole blood after stimulation with $PM_{2.5}$ sample. The results are expresses as a positivity index (with and without $PM_{2.5}$ stimulation concentration ratio). Plots are constructed for particular stimulation times from values of three healthy volunteers.

4. Discussion

Nanoparticle toxicity assessment arises from studies concerning the toxicity effect of airborne particulate matter. Epidemiological and toxicological studies indicate that airborne particulate matter can considerably contribute to morbidity and mortality, especially in regions heavy laden by industry and traffic [6, 15, 16]. Suspended airborne particles form a multicomponent mixture with particles varying in their size and chemical composition. Particle diversity together with their possible synergy effects complicates the health risk assessment. This is why adverse health effects of airborne particulate matter cannot be derived only from mass concentrations in the air. Therefore, there is a need to search for suitable methods able to assess negative particulate matter action on human health.

Presented study deals with the short-term effects of real airborne particulate matter on the human immune cells' functions. With regard to the integrating and protective role of the immune system, there is a presumption, that influence of the immune system functions can be the cause of adverse health effects relating to the airborne particle exposure.

To assess immunomodulatory effects, four *in vitro* immunological assays allowing us to monitor both the humoral a cell-mediated components of the innate and acquired immune responses were used.

The basic idea was that if significant immunological response was observed for integral fine and ultrafine particle sample, the next step would be the separation of nanoparticles from PM in sufficient amount for the performance of the immunological assays.

4.1. Chemiluminescence assay

Stimulation of phagocytes involves generation of reactive oxygen species (ROS) called respiratory burst. The return of the excited chemical groups to the basic state is accompanied by foton emission - chemiluminescence. The chemiluminescence assay was used to measure of light production by phagocyte cells in whole blood cultures representing the release of reactive oxidant species [10].

This assay enables us to detect the impact of the tested particle sample on the oxidative burst of phagocytic cells in whole blood represented mainly by polymorphonuclears. Donaldson et al. [17] suppose that the increase of phagocyte activity induced by the presence of airborne particles in the respiratory tract can support particle clearance. However, nonspecific action of ROS can cause inflammation, cell and tissue damage. Epidemiological as well as experimental data show that reactive molecules play an important role in progress of some respiratory disease [15, 17]. Therefore, there is a need for a method able to measure the rate of oxidative molecule generation, as a screening tool for possible airborne particle associated oxidative damage evaluation.

In our study, the particle sample only in concentration of 3,000 pg/ml after 30-minute incubation significantly increased phagocytosis associated metabolic activity of leukocytes in whole blood cultures compared to the values of negative control. However, one way ANOVA did not confirm the stimulation effect of the particle sample. With regard to the low number of measurements and the negative results with the other particle sample concentrations and incubation times, the difference cannot be contributed to the action of airborne particles.

The negative assay results could be caused by absence of ligands on the particle surface able to bind on the phagocyte receptors. Both *in vivo* and *in vitro* studies have shown that particularly the presence of a biological component on the airborne particles is associated with phagocyte activation [18]. However the presence of the biological components was not investigated in our study.

Antonini et al. [19] consider luminol-dependent chemiluminescence to be a suitable method to measure the earliest events in the inflammatory process, based on their experiments with inflammatory effects of silica nanoparticles. Nevertheless in other study [20], pathogenic respirable mineral dusts did not induce higher production of ROS than non-pathogenic ones. Donaldson et al. [21] relate the pathogenicity of various nanomaterials rather to their biopersistence, than *in vitro* ROS production.

Negative chemiluminescence assay results obtained in presented study indeed do not exclude possible adverse effect of the airborne particle matter on the ROS generation. The situation in a real organism is more complex, the influence of other important factors, such as retention, lysosomal enzyme release, opsonisation, etc., needs to be taken into account, particularly in case of people with already existing health problems accompanied by the increase of ROS levels in the organism (COPD, asthma). Epidemiological studies confirm that those people are generally more threatened with the exposure of particle polluted air [16, 22].

4.2. Basotest

Recent increase in incidence and prevalence of allergic disease is probably caused by environmental factors, particularly by polluted air [23]. Particulate matter can serve as a vector carrying allergens (pollen grains, mites, moulds) into the respiratory tract, where they can be deposited [24]. Futhermore, studies have shown that particulate matter enhance the susceptibility to common environmental allergens and trigger sensibilization to neoallergens [25, 26].

Basotest, used in this study to evaluate the allergenic properties of airborne particle sample, allows us to monitor the basophil activation after the allergen binding to IgE present on the surface of the basophils. Basophil activation led to release of inflammatory mediators and consequent manifestation of allergic inflammation [27]. Basophil degranulation assay is based on the measurement of exposition of CD63 transmembrane protein on the surface of activated basophils [11].

The cut-off limit between negative and positive response against inhalation allergens was established on 15% of activated basophils [11]. Our results show that the airborne particle sample did not stimulate significant basophil degranulation since none of the tested airborne particle concentrations exceeded this limit.

Schober et al. [28] investigated possible synergy between organic extracts of urban aerosol and major allergen of birch pollen grains (rBet v 1) on basophil activation in birch pollen-allergic donors and healthy controls. Synergy was only observed in blood samples from allergic individuals. Samples from healthy controls did not showed upregulation of CD63 by rBet v 1. Affirmatively other studies indicates that airborne particulate matter, or its organic extract, enhances allergic inflammatory mediator realize from basophils in the presence of antigen [27]. The negative results of our study can be caused by the absence of antigens in the tested particle sample. Chemical composition, especially content of organic matter and heavy metals, plays an important role in the adjuvant effect of airborne particulate matter on the allergic response. On the other side, there are experimental works proving that even inert particles (e.g. polystyrene nanoparticles) influenced the allergy reaction [29].

Despite many studies, the accurate mechanism by which airborne particles worsen the allergic reaction is still unknown. Moreover, gas pollutants present in the polluted air can enhanced the effect of the airborne particles [30]. Recent research has revealed new mechanism by which TiO_2 nanoparticles cause mast cell degranulation of histamine even without previous IgE antibody-based sensitization through the activation of Ca^{2+} channels [31].

4.3. Lymphocyte proliferation assay

Lymphocyte proliferation is caused by sensible lymphocyte exposition to antigens or allergens. To detect lymphocyte proliferation, the method of [^{3}H]thymidine incorporation into newly synthetized lymphocyte DNA was used [12]. Enhanced lymphocyte proliferation reflects more effective immune response and thus possible more intensive removing of particles from the organism. On the other side, increased lymphocyte proliferation can be related to inflammatory and autoimmune disease.

Slight increase in cpm compared with negative control values was recorded using the particle concentrations of 300 pg/ml and 134 µg/ml. However, regarding to the negative results in concentration of 3,000 pg/ml, another evaluation is needed to make a decision on the possible stimulation properties of the airborne particles on the lymphocyte proliferation.

Longer incubation time can enhance assay efficiency in case of less potent stimulants. Incubation period influenced the assay results in study of Miller et Zarkower [32]. T-lymphocytes in the spleen were found to be depressed after 7 days of pre-exposure to carbon dust. However, after 14, 21 and 28 days of pre-exposure to carbon dust 30 to 40 % of enhancement over control group responses were recorded. Worsened T-lymphocyte activity can be caused by oxidative stress related to enhanced apoptosis of T-lymphocytes.

4.4. Cytokine release evaluation

Multiplex methods, especially ALBIA, enable to analyze the cell functionality and reactivity by evaluation of the release of biologically active molecules into the cultivation media. For this pilot study we used commerce kit Human Cytokine 25-Plex Panel enabling to evaluate production of 25 cytokines divided according to their biological function in four panels: Inflammatory Panel, Th1/Th2 Panel, Cytokine II Panel and Chemokine Panel. In case of positive assay results, the set of suitable cytokines would be established to monitor the effects of particulate matter on the immune cells. Effects of the $PM_{2.5}$ sample on the release of 25 various cytokines were assessed in the particle concentration of 3,000 pg/ml. Particular immune mechanisms are involved gradually, based on the published data, we evaluated the cytokine release after 6, 24 and 72 hours after incubation [33].

With regard to the uncertainty of the assay expected on 20 %, the slight stimulation effect was found at some inflammatory cytokines (IL-6, IL-8, TNF-α) and chemokines (eotaxin, RANTES,

MIP-1α). These effects were recorded only after certain incubation periods. However, the ultimate values were very low compared to the stimulation effect of PHA used as a positive control and in real organism probably would not have any impact.

5. Conclusion

Even if certain effects of the particle sample were observed compared to the results of negative controls, changes in evaluated immune response parameters were not high enough to take effect in a real organism. Thus, the results of the immunological assays did not prove significant effects of airborne particles on immune cells´ functions of healthy persons.

Results of our experiments are related to short-term action of the real airborne particle mixture on the whole blood cultures of healthy people. Obtained data indeed do not exclude health risks of long-term exposure to airborne particles, especially for individuals with genetic predisposition to certain diseases or already existing disease. Age, co- pollutants and existing disease of respiratory tract can modify the immune response to particulate matter exposure.

Suitability of selected immunological assays should be also taken into account. The main challenge of immunological studies of nanomaterials as well as airborne particles is to select an experimental approach that is free of false positive or false negative results. Modification of available assays to be applicable for nanomaterial immunomodulatory effect assessing can be an objective of our future research.

Acknowledgements

Authors thank to project INEF CZ.1.05/2.1.00/01.0036 for the support.

References

[1] Fadeel B 2012 Clear and present danger? Engineered nanoparticles and the immune system *Swiss. Med. Wkly.* **26** 142–w13609

[2] Gilmour I M, Stevens T and Saxena R K 2007 Effect of particles on the immune system In Donaldson K and Borm P (Eds.) 2007 *Particle Toxicology* CRC Press, NY 245–58

[3] Krug H F 2011 *Quality handbook standard procedures for nanoparticle testing* [S.l.], Nanommune

[4] Geiser M, Rothen-Rutishauser B, Kapp N, Schurch S, Kreyling W, Schulz H, Semmler M, Im Hof V, Heyder J and Gehr P 2005 Ultrafine particles cross cellular membranes by nonphagocytic mechanisms in lungs and in cultured cells *Environ. Health Perspect.* **113** 1555–60

[5] Kaličáková Z 2012 *Air contamination of Moravia-Silesia region by nanoparticles* Master's thesis VŠB – Technical University of Ostrava, Faculty of Mining and Geology

[6] Tomášková H, Tomášek I, Šlachtová H and Šebáková H 2011 Impact assessment of PM_{10} concentrations on mortality and morbidity in Ostrava during smog episodes *Hygiena* **1** 5–10

[7] Dostál M and Šrám R J 2010 The study on the health status of children from Ostrava-Radvanice and Ostrava-Bartovice *Ochrana Ovzduší* **5–6** 35–8

[8] Jančík P, Pavlíková I and Bitta J 2009 Analýza kvality ovzduší na území města Ostravy *Ochrana ovzduší* **4** 24–7

[9] Meij R, Janssen L H J M and van der Kooij J 1986 Air pollutant emissions from coal fired power stations *KEMA Sci. Tech. Rep.* **6** 651–69

[10] Lochmanová A 2005 *Polymorphonuclear Metabolic Activity Evaluation Using Chemiluminescence Assay* SOP 25.47 CKL ZÚ Ostrava

[11] EXBIO *BasoFlowEx® Kit Cat. No: ED7036*

[12] Lochmanová A 2005 *Lymphocyte proliferation Evaluation using Lymphocyte proliferation Assay with ^{3}H thymidin incorporation* SOP 25.46 CKL ZÚ Ostrava

[13] Invitrogen Corporation 2010 *Human Cytokine 25-Plex Pane. Cat. No: lhc0009 User manual*

[14] Klener P (Ed.) 2006 *Vnitřní lékařství* Praha Galén

[15] Pope III C A and Dockery D W 2006 Health effects of fine particulate air pollution: lines that connect *J. Air Waste Manag. Assoc.* 6 709–42

[16] Katsouyanni K, Touloumi G, Samoli E, Gryparis A, Le Tertre A, et al. 2001 Confounding and effect modification in the short–term effects of ambient particles on total mortality: results from 29 European cities within the APHEA2 project *Epidemiology* **5** 521–31

[17] Donaldson K, Slight J and Bolton J E 1998 The effect of products from bronchoalveolar–derived neutrophils on oxidant production and phagocytic activity of alveolar macrophages *Clin. Exp. Immunol.* **74** 477–82

[18] Antonini J M, Dyke K V, Ye Z, Dimatteo M and Reasor M J 1994 Introduction of luminol–dependent chemiluminescence as a method to study silica inflammation in the issue and phagocytic cells of rat lung *Environ. Health. Perspect.* Suppl. **10** 37–42

[19] Antonini J M, Roberts, J R, Yang H M, Barger M W, Ramsey D, Castranova V and Ma J Y 2000 Effect of silica inhalation on the pulmonary clearance of a bacterial pathogen in Fischer 344 rats *Lung.* **6** 341–50

[20] Gormley P, Kowolik M J and Cullen R T 1985 The chemiluminescent response of human phagocytic cells to mineral dusts *Br. J. Exp. Pathol.* **66** 409–16

[21] Donaldson K, Borm P J A, Castranova V and V Gulumian M 2009 The limits of testing particle–mediated oxidative stress in vitro in predicting diverse pathologies, relevance for testing nanoparticles *Part. Fibre Toxicol.* **27** 6–13

[22] Kim C S and Kang C W 1997 Comparative measurement of lung deposition of inhaled fine particles in normals and patients with obstructive airway disease *Am. J. Respir. Crit. Care Med.* **155** 899–905

[23] Gutová V 2007 Alergie – systémové onemocnění s lokálními projevy *Dermatologie pro praxi* **1** 27–30

[24] Jin C, Shelburne C P, Li G, Potts E N, Riebe K J, Sempowski G D and Foster M A 2011 Particulate allergens potentiate allergic asthma in mice through sustained IgE–mediated mast cell activation *J. Clin. Invest.* **3** 941–55

[25] Ishizaki T, Koizumi K, Ikemori R, Ishiyama Y and Kushibiki E 1987 Studies of prevalence of Japanese cedar pollinosis among the residents in a densely cultivated area *Ann. Allergy.* **58** 265–70

[26] Diaz–Sanchez D, Tsien A, Fleming J and Saxon A 1997 Combined diesel exhaust particulate and ragweed allergen challenge markedly enhances human in vivo nasal ragweed–specific IgE and skews cytokine production to a T helper cell 2–type pattern *J. Immunol.* **5** 2406–13

[27] Kepley C L, Lauer F T, Oliver J M and Burchiel S W 2003 Environmental polycyclic aromatic hydrocarbons, benzo(a)pyrene (BaP) and BaP-quinones, enhance IgE-mediated histamine release and IL-4 production in human basophils *Clin. Immunol.* **107** 10–9

[28] Schober W, Lubitz S, Belloni B, Gebauer G, Lintelmann J, Matuschek G, Weichenmeier I, Eberlein–Konig B, Buters J and Behrendt H 2007 Environmental polycyclic aromatic hydrocarbons (PAHs) enhance allergic inammation by acting on human basophils *Inhal. Toxicol.* Suppl. **1** 151–6

[29] Granum B and Løvik M 2002 The effect of particles on allergic immune responses *Toxicol. Sci.* **65** 7–17

[30] Diaz–Sanchez D and Riedl M 2005 Diesel effects on human health: a question of stress? *Am. J. Physiol. Lung Cell Mol. Physiol.* **289** L722–3

[31] Chen J–C and Schwartz J 2008 Metabolic Syndrome and Inflammatory Responses to Long–Term Particulate Air Pollutants *Environ. Health. Perspect.* **5** 612–7

[32] Miller S D and Zarkower A 1974 Effects of Carbon Dust Inhalation on the Cell–Mediated Immune Response in Mice *Infect. Immun.* **3** 534–9

[33] Largelius M, Jones P, Franck K and Gaines H 2006 Cytokine detection by multiplex technology useful for assessing antigen specific cytokine profiles and kinetics in whole blood cultured up to seven days *Cytokine* **33** 156–65

Morphological and cytohistochemical evaluation of renal effects of cadmium-doped silica nanoparticles given intratracheally to rat.

T Coccini[1], E Roda[1], S Barni[2] and L Manzo[1]

[1]Toxicology Division, Salvatore Maugeri Foundation IRCCS and University of Pavia, 27100 Pavia, Italy,
[2]Department of Biology and Biotechnology, University of Pavia, Pavia, Italy.

Abstract. Renal morphological parameters were determined in rats intratracheally instilled with model cadmium-containing silica nanoparticles (Cd-SiNPs, 1mg/rat), also exploring whether their potential modifications would be associated with toxicogenomic changes. Cd-SiNP effects, evaluated 7 and 30 days post-exposure, were assessed by (i) histopathology (Haematoxylin/Eosin Staining), (ii) characterization of apoptotic features by TUNEL staining. Data were compared with those obtained by $CdCl_2$ (400µg/rat), SiNPs (600µg/rat), 0.1 ml saline. Area-specific cell apoptosis was observed in all treatment groups: cortex and inner medulla were the most affected regions. Apoptotic changes were apparent at 7 days post-exposure in both areas, and were still observable in inner medulla 30 days after treatment. Increase in apoptotic frequency was more pronounced in Cd-SiNP-treated animals compared to either $CdCl_2$ or SiNPs. Histological findings showed comparable alterations in the renal glomerular (cortex) architecture occurring in all treatment groups at both time-points considered. The glomeruli appeared often collapsed, showing condensed, packed mesangial and endothelial cells. Oedematous haemorrhagic glomeruli were also observed in Cd-SiNPs-treated animals. Bare SiNPs caused morphological and apoptotic changes without modifying the renal gene expression profile. These findings support the concept that multiple assays and an integrated testing strategy should be recommended to characterize toxicological responses to nanoparticles in mammalian systems.

1. Introduction

A variety of nanomaterials have been studied and utilized in widespread biological applications [1]. Among them, silica/cadmium containing nanomaterials have attracted much attention in the latest years for their applications in medicine and industrial manufacturing, synthesis and engineering [2, 3, 4].

Information on toxicological and health impact of these nanomaterials is still limited although occupational respiratory exposure to cadmium (Cd) nanoparticles may occur in many sectors, including manufacturing, research laboratories and work sites in industry.

Most of the studies on the biological effects of nanoparticles are often focused on pulmonary toxicity being the lungs the first port of entry by inhalation and the tissue in which the nanoparticles can easily disperse due to their very low density. Following the process of translocation across biological barriers, nano-SiO_2 (silica) can be transported into the blood and deposit in target organs where the nanoparticles can exert potential toxic effects [5]. Kidney could be a target because of its role in elimination of xenobiotics. Specifically to nano-SiO_2, it has been reported that these nanoparticle exhibit prominent distribution in several tissues including kidney after intravenous injection in mice [6]. In rats, Cho et al. [7] demonstrated that, after absorption, silica nanoparticles are efficiently internalized in renal cells and rapidly excreted in urine. Recently, cadmium nanoparticles have been shown to deposit in rats kidney after intraperitoneal injection [8] and both free cadmium and nanoparticles accumulated and increased with doses in this tissue [8].

Cd has well-established pneumotoxicity and nephrotoxicity as documented by epidemiological studies and reports in workers chronically exposed by inhalation [9, 10]. Experiments using animal models indicated that translocation of pulmonary cadmium to the kidney occurs readily for inhalation of cadmium chloride [see review 11].

Our recent studies on model cadmium-containing silica nanoparticles (Cd-SiNPs) indicated long-lasting gene expression changes occurring in kidney after intratracheal instillation (i.t.) of these nanoparticles in rats [12]. Compared with $CdCl_2$ the effect of Cd-SiNPs was more selective. Delayed gene expression alterations linked to apoptosis and regulatory processes were observed 30 days post-treatment in rats given Cd-SiNPs but not in those receiving $CdCl_2$.

In this work, cyto-histochemical methods were used to examine renal responses in rats after i.t. application of Cd-SiNPs as well as to determine whether morphological and biochemical parameters were modified by these nanoparticles in association with toxicogenomic changes. The effect of Cd-SiNPs was evaluated at 7 and 30 days post-exposure in comparison with that caused by administration of equivalent amount of $CdCl_2$ or SiNPs.

2. Materials and Methods

2.1. Nanoparticles

The method used to produce Cd-SiNPs was previously described [13]. The particles presented amorphous and crystalline structure (confirmed by X-ray diffraction analysis), spherical form, primary particle size range of 20-80 nm and specific surface area of about 200 m^2/g.

Dynamic light scattering (DLS) determination of the Cd-SiNPs size distribution showed tendency to form aggregates and agglomerates of about 350 nm, and a zeta potential of – 23 mV (in deionized water). Flame-atomic absorption analysis was used to determine metal impurities and the release of cadmium from nanoparticles dispersed in physiological solution. Maximum cadmium release was 15% after 16 hr. Further metal release was negligible in the subsequent 10-day period. Main impurities were Ca (0.3%), Na (0.2%), K (0.2%), Fe (0.04%) and Mn (0.001%). Other metals were present in quantities less than 1 ‰. Cadmium and silica contents in the nanoparticles were 32.5% and 24.1%, respectively. $CdCl_2$ was purchased from Sigma Aldrich (Milan, Italy) and SiNPs from Degussa GmbH (Germany).

2.2. In vivo experiment

Adult male Sprague–Dawley rats were treated with a single intratracheal (i.t.) instillation of Cd-SiNPs (1 mg/rat, corresponding to about 250 µg Cd/rat). Separate groups of animals received i.t. an equivalent cadmium dose as $CdCl_2$ (400 µg/rat), SiNPs (600 µg/rat), or 0.1 ml saline/rat (control). In previous studies [14, 15], i.t. instillation of a single dose of 400 µg $CdCl_2$/rat was shown to induce in rats moderate lung injury evolving into chronic inflammation and fibrosis. Cd-SiNP suspension was vortexed on ice just before the exposure to force nanoparticle dispersion and avoid formation of agglomerates. No surfactants or solvents were used. Seven and 30 days after treatment, rats were anesthetized by i.p. injection of 35% chloral hydrate (100 µl/100 g b.w.) and kidney was carefully removed, dissected and processed for (i) histopathology (Haematoxylin/Eosin Staining) and (ii) TUNEL assay.

All experimental procedures were performed in compliance with the European Council Directive 86/609/EEC on the care and use of laboratory animals. Rats were anesthetized with pentobarbital sodium before i.t. instillation with the test materials.

2.3. Rat renal preparation for morphohistochemical evaluations

2.3.1. Histopathology. At necropsy the kidneys from control and treated animals were removed and hemi-dissected. Both half of each kidney were washed in NaCl 0.9% and post-fixed by immersion for 7 hr in 4% paraformaldehyde in 0.1 M phosphate buffer (pH 7.4), dehydrated through a graded series of ethanol and finally embedded in Paraplast. Seven micrometer thick sections of the samples were cut in the coronal plane and collected on silane-coated slides.

Kidney sections were then stained with Haematoxylin/Eosin (H&E) to evaluate overall structural changes.

2.3.2. *TUNEL staining: characterization of apoptotic features.* In addition to morphological criteria, apoptotic cell death was assayed by *in situ* detection of DNA fragmentation using the terminal deoxynucleotidyl-transferase (TUNEL) assay (Oncogene Res. Prod., Boston, MA, USA). The kidney sections were incubated for 5 min with 20 μg ml^{-1}proteinase-K solution at room temperature, followed by treatment with 3% H_2O_2 to quench endogenous peroxidase activity. After incubation with the TUNEL solution (90 min with TdT/biotinylated dNTP and 30 min with HRP-conjugate streptavidin) in a humidified chamber at 37 °C, the reaction was developed using 0.05% 3-amino-9-ethylcarbazole (AEC) in 0.1 M TRIS buffer (pH 7.6) with 0.2% H_2O_2; in some specimens the reaction was developed using a 0.1% DAB solution. The specimens were lightly counterstained with Haematoxylin. As a negative control, in some sections the TdT incubation was omitted; no staining was observed in these conditions.

The quantification of apoptotic cells was calculated as the percentage (TUNEL Labelling Index) of a total number (about 500) of TUNEL-cytochemically positive nuclei of nephron-epithelial cells (from the three kidney portion namely cortex, inner and outer medulla) for each animal and experimental condition, in a minimum of 10 randomly selected high-power microscopic fields. Statistical analyses among the different biological situations was performed by two-way analysis of variance (ANOVA) followed by the Bonferroni test.

The slides were observed and scored with a bright-field Zeiss Axioscop Plus microscope. The images were recorded with an Olympus Camedia C-2000 Z digital camera and stored on a PC running Olympus software.

3. Results and Discussion

Region-specific renal cell apoptosis was observed in all treatment groups, the cortex and inner medulla being the most affected regions (Figure 1). The apoptotic phenomena were apparent already at 7 days post-exposure in both affected areas. Notably, these effects were still observable in the inner medulla 30 days after all types of treatment, while in the cortex merely Cd-SiNPs produced a long lasting apoptotic effect. Additionally, the increase in apoptotic frequency was more pronounced in Cd-SiNP-treated animals compared to either $CdCl_2$ or SiNPs groups. The outer medulla was not significantly affected by any treatment.

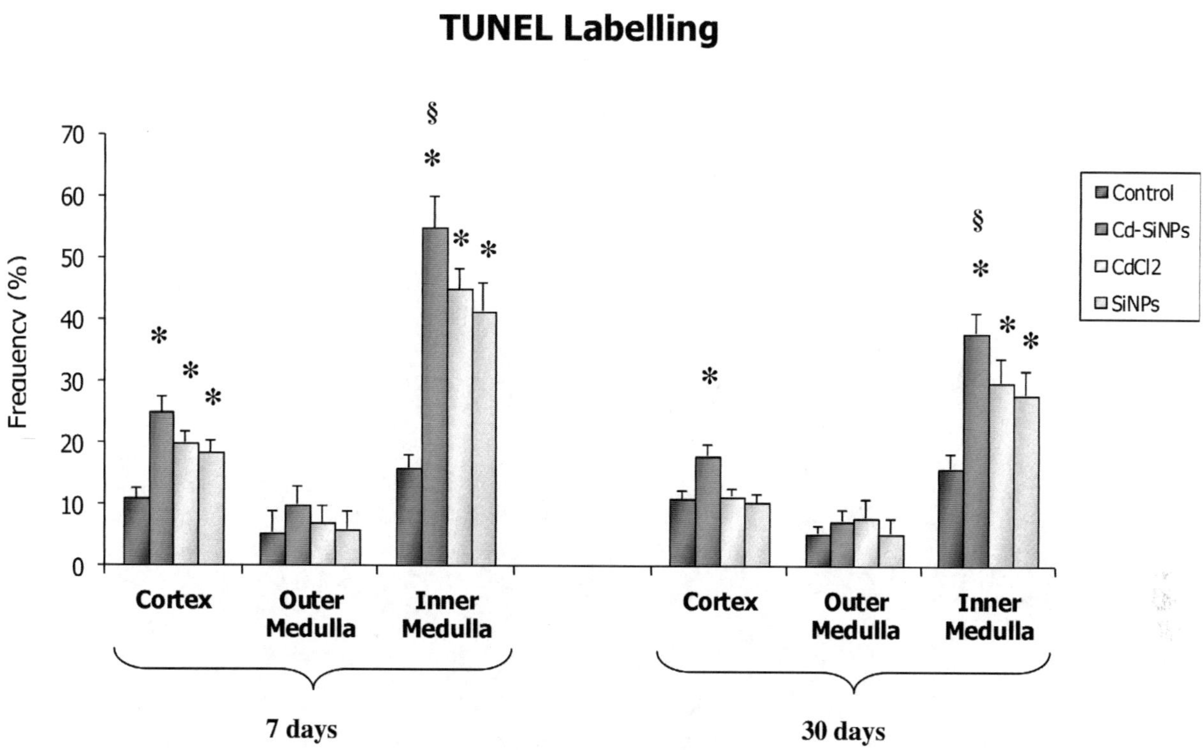

Figure 1. Histograms show changes in percentage of TUNEL Labelling Index of renal cortical, outer medullar, and inner medullar cells as a consequence of the i.t. exposure to the different test materials (Cd-SiNPs, SiNPs, $CdCl_2$). Data are expressed as mean ± S.D. *P<0.05 compared to control; § P<0.05 compared to $CdCl_2$ or SiNPs.

Coherently with the above reported quantitative data, Figure 2 shows representative light microscopy photomicrographs depicting significant enhancement of the apoptotic phenomenon, characterized by nuclear pyknosis, and TUNEL-positivity mainly in the cortex, both in proximal, distal tubule and glomerular region, and in the inner medulla of rats 7 days post-exposure. No evidence of necrotic tissue degeneration was detected.

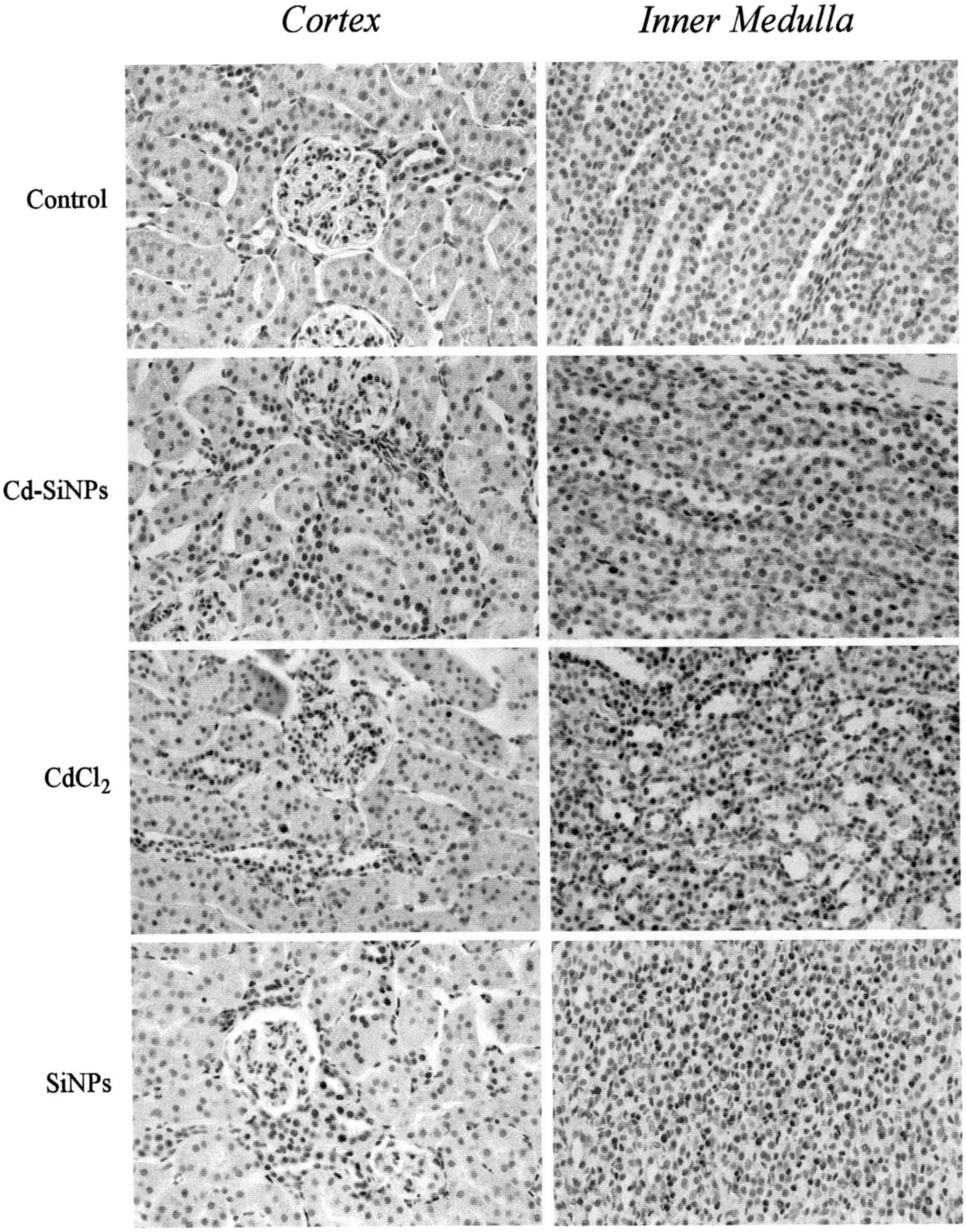

Figure 2. Cell apoptosis detected by TUNEL staining in the renal cortex and inner medulla, 7 days post-exposure to Cd-SiNPs, $CdCl_2$, SiNPs compared to control. *Objective magnification: 40x.*

Histological findings demonstrated comparable morphological alterations of the renal glomerular architecture (cortex) occurring after all treatments at both time-points considered. Thirty days after both $CdCl_2$- or Cd-SiNPs-i.t. exposure, the glomeruli appeared deformed, often collapsed showing packed mesangial and endothelial cells. Oedematous, haemorrhagic glomeruli were also found showing the alterations of the vascular component of the renal mesangium, with the capillary loops sometimes distorted. Hypertophic glomeruli were observed after exposure to SiNPs (Figure 3).

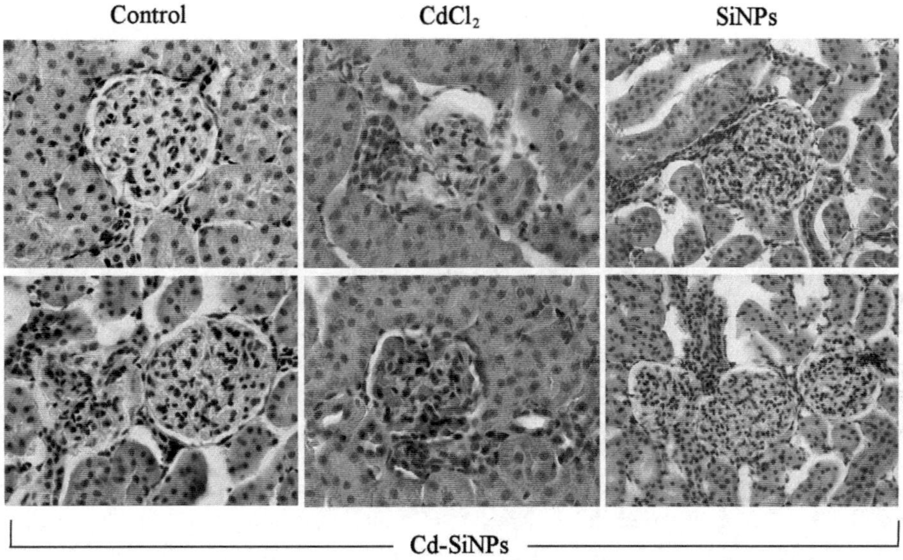

Figure 3. Representative Haematoxylin and Eosin-stained cortical renal specimens from control animals and rats exposed to Cd-SiNPs, SiNPs, $CdCl_2$ (30 days post exposure). *Objective magnification: 40x.*

Altogether, these cytohistochemical results indicate the occurrence of apoptotic phenomena and morphological changes in the renal cortex of animals i.t. instilled by all types of the test materials (Cd-SiNPs, $CdCl_2$, SiNPs). These effects appeared early (at 7 days) and lasted until 30 days post-exposure. Notably, the kidney insult was more pronounced after Cd-SiNPs.

Our recent investigations on whole genome expression evaluated in the renal tissues after i.t. instillation of these nanoparticles, indicated that Cd-SiNPs but not $CdCl_2$, induced changes in the expression of genes linked to apoptosis and regulatory processes at 30 days. The action of Cd-SiNPs was also evidenced at the earlier time point (7 days post-exposure) by modifying genes involved in immune function, stress response and inflammation processes. Similar effects were also observed 1 week post-exposure to $CdCl_2$. These overall data indicate that Cd-SiNPs induced in kidney (i) long-lasting gene expression changes, (ii) a more selective effect compared to $CdCl_2$, (iii) delayed gene expression alterations linked to apoptosis and regulatory processes differently from $CdCl_2$ that did not modulate genes related to these endpoints. The approach to compare the gene expression results with biochemical and histopathological evaluation at the renal tissue may provide a direct linkage between a change in gene expression and the type and outcome of the toxic response. Both methodological approaches confirmed long-lasting renal effects of Cd-SiNPs on apoptosis and regulatory processes.

With respect to i.t. administration of bare SiNPs, we observed morphological and apoptotic changes without modification of the gene expression profile in the kidney [12]. Extra-pulmonary organ toxicity (e.g., nephrotoxicity) caused by inhaled nanoparticles have been documented in recent studies with SiNPs [5, 16] and in experiments using carbon nanotubes [17].

The early effects caused by both Cd-SiNPs and $CdCl_2$ on immune, inflammation and oxidative stress processes characterized by trascriptomics still need to be specifically examined. Further investigations are in progress in our laboratories to assess renal responses to these nanomaterials applying an extended battery of assays to test more comprehensive endpoints (e.g. IL-6, TGF-beta1) comparable with those suggested by trascriptomics.

These findings support the concept that multiple assays and an integrated testing strategy should be recommended to characterize toxicological responses to nanoparticles in mammalian systems.

Translocation is an important component in the toxicity of Cd which has been shown to accumulate in kidney and affect renal function after chronic inhalation exposure other than produce toxic effects on the primary target organ, i.e. lung, being the site of entry. Indeed, we observed renal gene expression changes and morphological alterations (collapsed glomeruli, showing condensed, packed mesangial and endothelial cells) not only by Cd-SiNPs but even after $CdCl_2$. The total amount of Cd given to rats in this study (about 250 μg Cd/rat) was likely below the dose capable to generate critical organ concentration (100-200 μg Cd/g tissue) correlated with onset of cadmium-induced nephrotoxicity [18, 19]. In a more recent study a renal Cd level of about 8 μg/g tissue, obtained after 30 day-exposure to Cd in rat, was associated with an evident morphological changes of the renal cortex including alterations in both proximal and distal tubular cells and glomerular cells as well as in extracellular matrix [20]. Although renal Cd content in rat was not measure in the present study, we can assume that about 12 μg Cd/kidney (0.6 g) could have been reached in renal tissue of the $CdCl_2$-treated animals, taking into consideration that approximately 5% of the administered Cd could be detected in rat kidney 30 days after a single i.t. instillation of $^{109}CdCl_2$ [11].

In summary, the present results contribute to point out the involvement of a secondary organ (i.e. the kidney, a known target organ for cadmium toxicity) in the action of Cd-SiNPs given by pulmonary instillation. The renal response may be due to migration of the nanoparticles from lung to the systemic circulation or to secondary organ changes caused by circulating inflammatory factors (e.g., IL-6, IP-10 and TGF-beta1) released from lung following local insult [13]. A direct renal action of cadmium ions released from the absorbed Cd-SiNPs cannot also be excluded although a good stability of the Cd-doped material used in this study was observed with no more than 15% of cadmium released from the nanoparticles dispersed in saline. Furthermore, Dynamic Light Scattering experiments with the nanoparticles dispersed in physiological solution showed agglomeration and aggregation of Cd-SiNPs to extent greater than that found with SiNPs. The more pronounced tendency of Cd-SiNPs to form aggregates and agglomerates may offer an additional mechanistic explanation of our experimental findings.

Acknowledgements

This work was supported by the Italian Ministries of Health, Research and Education and grant from Cariplo Rif. 2011-2096. The authors wish to acknowledge Mr. Davide Acerbi for his excellent technical assistance.

References

[1] Chen N, He Y, Su Y, Li X, Huang Q, Wang H, Zhang X, Tai R, Fan C 2012 The cytotoxicity of cadmium-based quantum dots. *Biomaterials* **33**, 1238-1244.

[2] Rzigalinski BA, Strobl JS 2010 Cadmium-containing nanoparticles: perspectives on pharmacology and toxicology of quantum dots. *Toxicol. Appl. Pharmacol.* **238,** 380-388.

[3] Vivero-Escoto JL, Slowing II, Trewyn BG, Lin VS 2010 Mesoporous silica nanoparticles for intracellular controlled drug delivery. *Small* **6**, 1952-1967.

[4] Mamaeva V, Sahlgren C, Lindén M 2012 Mesoporous silica nanoparticles in medicine – Recent advances. *Adv. Drug Deliv. Rev.* http://dx.doi.org/10.1016/j.addr.2012.07.018.

[5] Fruijtier-Polloth C 2012 The toxicological mode of action and the safety of synthetic amorphous silica - a nanostructured material. *Toxicology* **294**, 61-79.

[6] He X, Nie H, Wang K, Tan W, Wu X, Zhang P 2008 In vivo study of biodistribution and urinary excretion of surface-modified silica nanoparticles. *Anal. Chem.* **80**, 9597-9603.

[7] Cho M, Cho WS, Choi M, Kim SJ, Han BS, Kim SH, Kim HO, Sheen YY, Jeong J 2009 The impact of size on tissue distribution and elimination by single intravenous injection of silica nanoparticles. *Toxicol. Lett.* **189**, 177-183.

[8] Arslan Z, Ates M, McDuffy W, Agachan MS, Farah IO, Yu WW, Bednar AJ 2011 Probing metabolic stability of CdSe nanoparticles: alkaline extraction of free cadmium from liver and kidney samples of rats exposed to CdSe nanoparticles. *J. Hazard. Mater.* **192**, 192-199.

[9] Nogué S, Sanz-Gallén P, Torras A, Boluda F 2004 Chronic overexposure to cadmium fumes associated with IgA mesangial glomerulonephritis. *Occup. Med.* **54**. 265-267.

[10] Nordberg GF, Onawa K, Nordberg M, Friberg LT 2007 Cadmium. In: Nordberg GF, Fowler BA, Nordberg M, Friberg LT (Eds), Handbook of Toxicology of Metals, 3rd ed. Elsevier Publishers, Amsterdam.

[11] Oberdörster G 1992 Pulmonary deposition, clearance and effects of inhaled soluble and insoluble cadmium compounds. In: Nordberg GF, Herber RFM, Alessio L. (eds.) Cadmium in the Human Environment: Toxicity and Carcinogenicity. Lyon: International Agency for Research on Cancer, pp 189-204.

[12] Coccini T, Roda E, Fabbri M, Sacco MG, Gribaldo L, Manzo L 2012 Gene expression profiling in rat kidney after intra-tracheal exposure to cadmium-doped nanoparticles. *J. Nanopart. Res.* **14**, 925, doi: 10.1007/s11051-012-0925-2.

[13] Coccini T, Barni S, Vaccarone R, Mustarelli P, Manzo L, Roda E 2012 Pulmonary toxicity of instilled cadmium-doped silica nanoparticles during acute and subacute stages in Rats. *Histol. Histopathol., in press.*

[14] Damiano VV, Cherian P V, Frankel F R, Steeger J R, Sohn M, Oppenheim D, Weinbaum G 1990 Intraluminal fibrosis induced unilaterally by lobar instillation of CdCl2 into the rat lung. *Am. J. Pathol.* **137**, 883-894.

[15] Driscoll K E, Maurer J K, Poynter J, et al. 1992 Stimulation of rat alveolar macrophage fibronectin release in a cadmium chloride model of lung injury and fibrosis. *Toxicol. Appl. Pharmacol.* **116**, 30-37.

[16] Passagne I, Morille M, Rousset M, Pujalté I, L'Azou B 2012 Implication of oxidative stress in size-dependent toxicity of silica nanoparticles in kidney cells. *Toxicology* **299**, 112-124.

[17] Reddy ARN, Krishna DR, Reddy YN, Himabindu V 2010 Translocation and extra pulmonary toxicities of multi wall carbon nanotubes in rats. *Toxicol. Mech. Method.* **20**, 267-272.

[18] Goyer RA, Cherian MG, Delaquerriere-Richardson L 1984 Correlation of parameters of cadmium exposure with onset of cadmium-induced nephropathy in rats. *J. Environ. Pathol. Toxicol. Oncol.* **5**, 89-100.

[19] Nogue´ S, Sanz-Galle´n P, Torras A, Boluda F 2004 Chronic overexposure to cadmium fumes associated with IgA mesangial glomerulonephritis. *Occup. Med.* **54**, 265-267.

[20] Asar M, Kayisli ÜA, Izgüt-Uysa VM, Akkoyunlu AG 2004 Immunohistochemical and ultrastructural changes in the renal cortex of cadmium-treated rats. *Biol. Trace Elem. Res.* **97**, 249-263.

Nitric oxide releasing iron oxide magnetic nanoparticles for biomedical applications: cell viability, apoptosis and cell death evaluations

R de Lima[1], JL de Oliveira[1], A Ludescher[2], MM Molina[3], R Itri[3], AB Seabra[2] and PS Haddad[2]*

[1] Departamento de Biotecnologia, Universidade de Sorocaba, CEP 18023-000, Sorocaba, S.P., Brazil

[2] Departamento de Ciências Exatas e da Terra, Universidade Federal de São Paulo, Diadema, SP, Brazil

[3]Instituto de Física, Universidade de São Paulo, CEP 05314-970, São Paulo, SP, Brazil

*E-mail: haddadps@gmail.com

Abstract. Nitric oxide (NO) is involved in several physiological and pathophysiological processes, such as control of vascular tone and immune responses against microbes. Thus, there is great interest in the development of NO-releasing materials to carry and deliver NO for biomedical applications. Magnetic iron oxide nanoparticles have been used in important pharmacological applications, including drug-delivery. In this work, magnetic iron oxide nanoparticles were coated with thiol-containing hydrophilic ligands: mercaptosuccinic acid (MSA) and dimercaptosuccinic acid (DMSA). Free thiol groups on the surface of MSA- or DMSA- coated nanoparticles were nitrosated, leading to the formation of NO-releasing iron oxide nanoparticles. The cytotoxicity of MSA- or DMSA-coated magnetic nanoparticles (MNP) (thiolated nanoparticles) and nitrosated MSA- or nitrosated DMSA- coated MNPs (NO-releasing nanoparticles) were evaluated towards human lymphocytes. The results showed that MNP-MSA and MNP-DMSA have low cytotoxicity effects. On the other hand, NO-releasing MNPs were found to increase apoptosis and cell death compared to free NO-nanoparticles. Therefore, the cytotoxicity effects observed for NO-releasing MNPs may result in important biomedical applications, such as the treatment of tumors cells.

1. Introduction

The increasing exploration of nanotechnology in biological and medical applications has led to significant improvements in diagnosis, prevention, and treatment of diseases. Iron oxide magnetic nanoparticles (MNPs) have been explored in biomedical applications such as drug delivery [1-3] and therapy [4-6], since they are considered less toxic than their metallic counterparts. Moreover, MNPs offer further advantages such as high saturation magnetization and superparamagnetic behaviour. Among iron oxide MNPs, magnetite is a very promising choice due to its already proven biocompatibility.

Co-precipitation of Fe^{2+} and Fe^{3+} ions under alkaline conditions has been routinely used for the synthesis of magnetite nanoparticles (Fe_3O_4) [7], although more well-designed procedures involving high temperature decomposition of iron(0) or iron(III) precursors stabilized by surfactant molecules have been reported [8]. The use of Fe_2O_3 and Fe_3O_4 nanoscale particles also includes drug targeting of cancer cells, tracking target cells using labeling, and imaging techniques like magnetic resonance tomography. *In vivo* studies with Fe_3O_4 nanoscale particles have demonstrated severe inflammatory and toxicity responses in rats exposed to nanoscale particles through inhalation [9-10]. Other studies have demonstrated that MNPs (Fe_3O_4) cause oxidative trauma to human bronchoalveolar epithelial and murine neuronal cells leading to loss of cell viability, and causing a change in electrical activity [11-12]. In contrast to these findings, other studies performed with microscale and nanoparticulate Fe_3O_4 described them both to be nontoxic under *in vitro* test conditions in human small airway epithelial and mouse fibroblast cells [13]. Therefore, it is of interest to observe the difference in the bio-nano interaction between Fe_3O_4 nano- and microscale particles in human lung cells [14] *in vitro* and to correlate the effect with their physicochemical properties.

Most studies under *in vitro* conditions consider the basic physicochemical characteristics of the nanoscale particles, such as shape, size, and surface coating. However, recent studies have shown that the surface of the nanoscale particles changes after interaction with the surrounding environment through solvation, protein corona formation, and agglomeration [15-16]. The biomedical applicability of MNPs is greatly improved by coating their surfaces with biocompatible ligands [17]. MNPs having suitable surface characteristics have a great potential for *in vitro* and *in vivo* applications. The effects of MNPs coated with biocompatible ligands containing functional groups such as COOH and SH as well as different sizes of MNPs are critical determinants in toxicological studies [18]. In this context, a good candidate to interact on the surface of these nanoparticles covered with biocompatible ligands is the nitric oxide (NO) molecule.

NO is involved in several physiological and pathophysiological processes, such as the control of vascular tone, the inhibition of platelet aggregation, and the immune response against microbes [19]. Thus, there is a great interest in the development of NO-releasing vehicles that are able to stabilize and release NO locally direct to the target site, in diverse biomedical applications.

This work reports the synthesis of surface-modified MNPs as vehicles to carry and deliver NO molecules and the evaluation of the cytotoxicity of these nanoparticles. MNPs, synthesised through the co-precipitation technique [7], were functionalized with NO. To this end, MNPs were coated with two thiol-containing ligands: mercaptosuccinic acid (MSA) and 2,3-dimercaptosuccinic acid (DMSA), leading to the formation of stable aqueous dispersions of thiolated nanoparticles (SH-MNPs). Free thiol (SH) groups were used as sites to covalently bind NO to the surface of MNPs. Cytotoxicity of SH-MNPs and NO-releasing MNPs were evaluated towards human lymphocytes. The results show that SH-MNPs were not cytotoxic. However, the presence of NO on the MNPs surface was responsible to increase cell death, apoptosis and cytotoxic effects, in a concentration-dependence manner. These results indicate that NO-releasing nanoparticles may result in important biomedical applications, such as the treatment of tumors.

2. Methods

2.1 Synthesis of MNPs

MNPs were synthesized by using the co-precipitation method, as previously reported [20, 21]. In brief, 4.0 mL of $FeCl_3 \cdot 6H_2O$ and 1.0 mL of $FeCl_2 \cdot 4H_2O$ (molar ratio 2:1), prepared in 1.0 mol/L HCl, were mixed and stirred while a volume of 50 mL of NH_4OH (0.7 mol/L) was added as precipitator. At this stage, the solution was centrifuged and the precipitate was decanted, followed by the addition of 6.0 mL of oleic acid. This mixture was then stirred for 20 minutes. The solution was centrifuged several times and the new precipitate was washed several times with ethanol and acetone, leading to MNP covered with oleic acid.

2.2 Adsorption of thiolated ligands

Oleic acid coated-MNPs (~10.0 mg) were then dissolved in 1.0 mL of toluene while MSA or DMSA (molar ratio Fe_3O_4:MSA or DMSA = 1:40) was dissolved in dimethyl sulfoxide (DMSO). The solutions were mixed and vigorously stirred for 14 h producing a black powder that was isolated by centrifugation. This procedure led to ligand exchange and, hence, to the formation of water stable thiol-containing MNPs (SH-MNPs) [21].

2.3 Nitrosation of thiolated magnetic nanoparticles

Thiolated nanoparticles (SH-MNPs) were nitrosated leading to the formation of SNO-MNPs by adding excess of sodium nitrite solution [21]. Excess of unreacted nitrite was removed from MNP suspension by centrifugal ultrafiltration.

2.4 Cell description

For the analyses involving lymphocytes, the cells were separated from whole blood using Ficoll-Paque™ PLUS medium (GE Healthcare, Little Chalfont, UK). The blood was provided by donors aged between 18 and 24 years (who freely signed terms of agreement forms) and the project was approved by the Ethics Committee of the University of Sorocaba (protocol #008/08). Blood samples were collected at a suitable location by a qualified professional, using disposable materials throughout the procedure. The lymphocytes were placed in RPMI 1600 culture medium (Cultilab) containing 300 µg/mL of L-glutamine and 200 µg/mL of $NaHC_3$, supplemented with 5% bovine fetal serum, 50 µg/mL of gentamicin sulfate (antibiotic), and 2 µg/mL of amphotericin B (antifungal). The culture was kept at 37°C, under a humidified atmosphere containing 5% CO_2.

2.5 Incubation of cells with MNPs

Human lymphocytes cells were incubated with: SH-MNPs (both MSA- and DMSA-coated MNPs) and SNO-MNPs at nanoparticle concentrations (0.01; 0.05 and 0.1 mg/mL) for 1 h. In addition, the cells were also incubated with the nanoparticle ligands, (free MSA or free DMSA) at the same concentrations for 1 h. Negative and positive controls employed phosphate buffered saline (PBS) and H_2O_2 (200 µmolol/L), respectively.

2.6 Tali analysis

Cell viability was measured using a Tali™ Apoptosis Kit consisting of Annexin V Alexa Fluor® 488 and Propidium Iodide (Invitrogen), and a Tali™ image-based cytometer, which enabled the numbers of viable, apoptotic, and dead cells to be counted. The cells that had been treated with SH-MNPs (both MSA- or DMSA-coated MNPs) and SNO-MNPs at nanoparticle concentrations (0.01; 0.05 and 0.1 mg/mL), nanoparticles were centrifuged and concentrated to 1 x 10^6 cells per mL. 100 µL aliquots of sample were prepared according to the specifications of the kit, and the tests were performed in triplicate.

3. Results and Discussion

3.1 Nitrosation of thiolated MNPs leading to nitrosated MNPs (SNO-MNPs)

Thiol groups on the surface of nanoparticles (SH-MNPs) were nitrosated by the addition of sodium nitrite, leading to the formation of SNO-MNPs. Indeed, nitrous acid is considered the nitrosating agent of SH groups leading to the formation of SNO groups [22]. Figure 1 shows schematically a representation of the coating of MNPs with MSA or DMSA (a), and subsequent functionalization of the nanoparticle surface with NO (b).

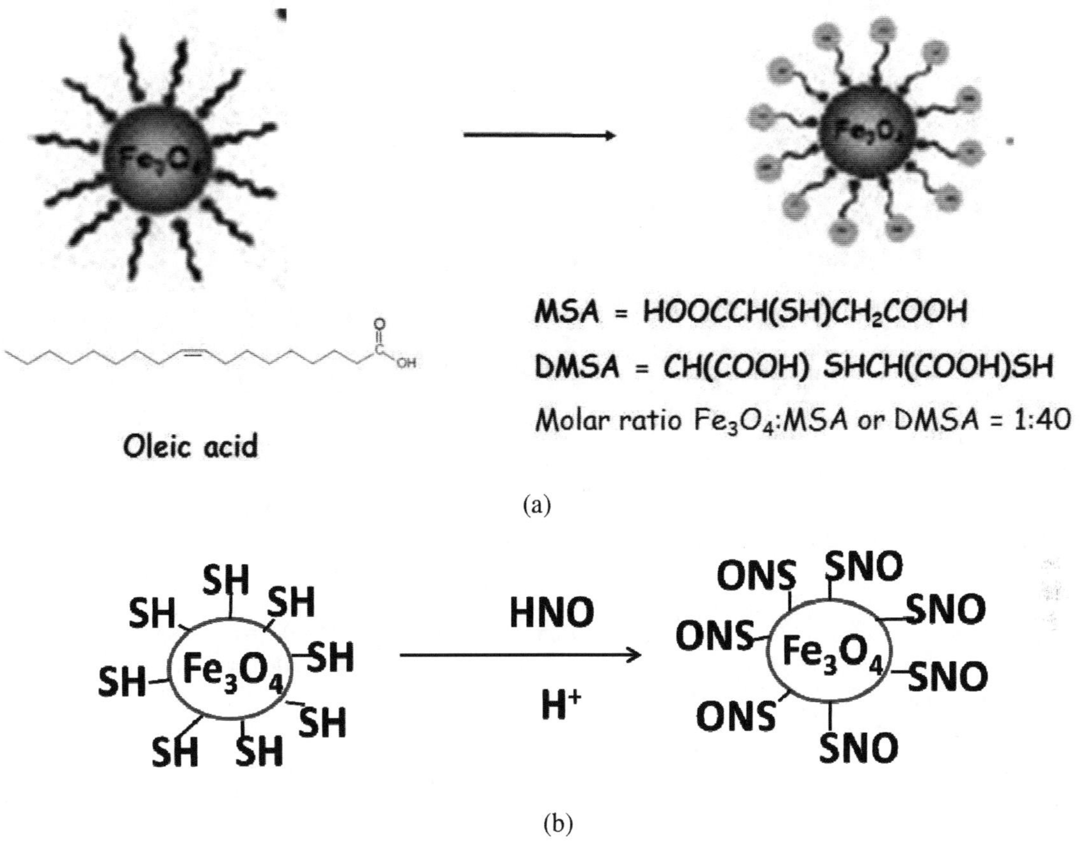

Figure 1. a) Ligand exchange of MNPs with oleic acid leading to the formation of SH-MNPs (coated either with MSA or DMSA; b) Functionalization of SH-MNPs with NO molecules [21].

3.2 Tali analysis – Cell viability

Overall, cell viability studies show that SH-MNPs were not toxic towards human fibroblasts (Figure 2). A detailed observation of the results indicates that MSA-MNPs can be considered to have slightly higher cell viability, compared to DMSA-MNPs. These results are consistent with published work, in which little cytotoxicity was observed upon incubation of fibroblasts with DMSA-MNPs [23]. In addition, exposure of lymphocytes to free MSA or DMSA (used as MNPs ligands) showed no toxic effects on cells. Although preliminary, these promising results indicate that thiolated MNPs (coated with either MSA or DMSA) can be used as an inert vehicle in drug delivery aimed at biomedical applications.

On the other hand, nitrosated nanoparticles (SNO-MNPs) decrease cell viability in comparison with the thiolated MNPs, for all tested concentrations (Figure 2). Thus, NO-releasing MNPs can be used to promote cytotoxicity.

3.3 Apoptosis and cell death

Figure 3 shows apoptosis and cell death upon incubation of human lymphocytes with thiolated MNPs (MSA- or DMSA-MNPs) and nitrosated MNPs (NO-releasing MNPs). It can be observed that NO-releasing MNPs caused higher apoptosis indices and cell death, compared with thiolated nanoparticles. Moreover, for NO-releasing MNPs, apoptotic cells indices were found to be higher in comparison with cell death. This preliminary result suggests that NO-releasing MNPs may have genotoxic effects on cell, under the experimental conditions.

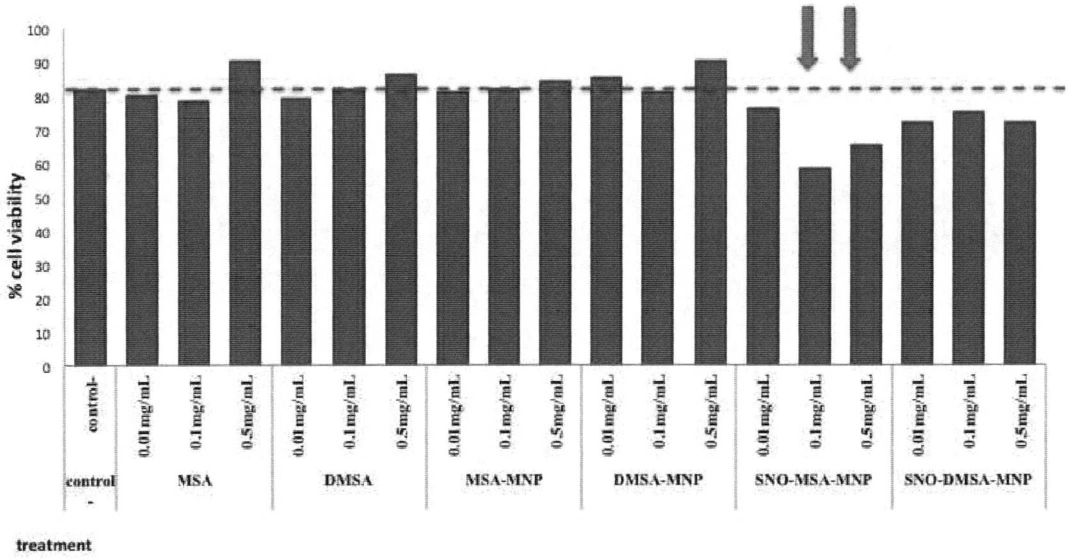

Figure 2. Cell viability of human lymphocytes treated with thiolated MNPs (MSA-MNPs or DMSA-MNPs), and NO-releasing MNPs (SNO-MSA-MNPs or SNO-DMSA-MNPs), at different concentrations, as indicated in the Figure. Incubation time: 1 h. MNPs were also incubated with only the ligands: MSA or DMSA, under the same experimental conditions.

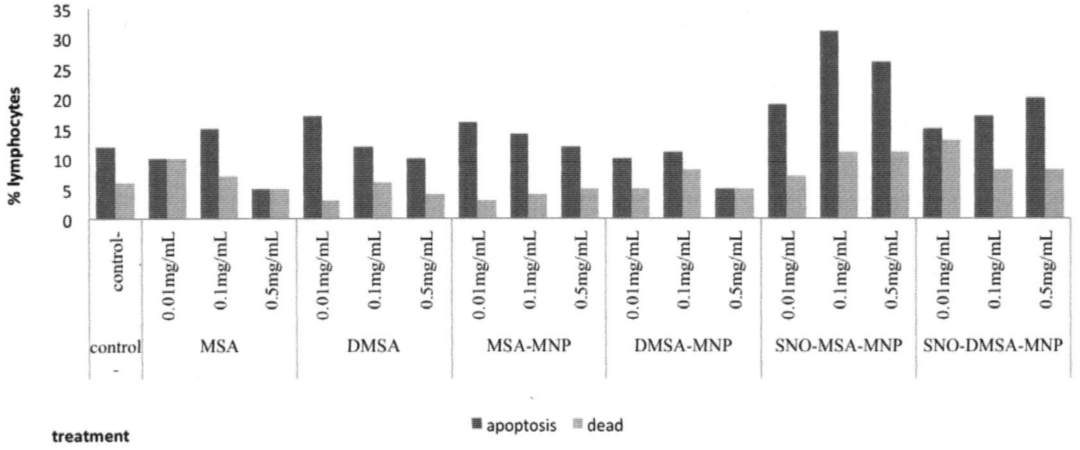

Figure 3. Percentage of apoptosis and cell death of human lymphocytes incubated with: thiolated MNPs coated with MSA (MSA-MNP) or DMSA (DMSA-MNP), nitrosated MNPs (SNO-MSA-MNP and SNO-DMSA-MNP), at different concentrations, as indicated in the Figure. Incubation time: 1 h. Cells were also incubated only with the ligands: MSA or DMSA, under the same experimental conditions.

Taken together, the results show a low value of apoptosis and cell death upon exposure of lymphocytes to MNPs. This result is in accordance with published work that reported a lower susceptibility of lymphocytes in comparison with other cell types [16].

On the other hand, these initial results show that NO-releasing MNPs increased cell apoptosis and death.

4. Conclusions

This work describes the preparation of MNPs as a new NO-delivery platform. Firstly, NPs were prepared through the co-precipitation method, and coated with thiol groups, leading to the formation of a stable aqueous dispersion of thiolated nanoparticles (SH-MNPs). Free thiols groups on the MNPs surfaces were used as sites to covalently bind NO. Toxicological evaluations of thiolated and NO-releasing MNPs were investigated using human lymphocytes. The results showed that thiolated MNPs (MSA- or DMSA-coated MNPs) were not cytotoxic, and did not cause apoptosis and cell death, indicating that these nanoparticles can be safely used as drug delivery vehicles. On the other hand, NO-releasing MNPs were found to cause toxic effects on cells, and to increase apoptosis and cell death. In this regard, NO-releasing MNPs might be considered a promising approach to combat tumors, since the nanoparticles can be guided to the target site through the application of an external magnetic field, and release NO directly to the desired tissue/cell, where NO can have its cytotoxic effect.

5. Acknowledgements

Supported from FAPESP, CNPq, CAPES, Brazilian Network in Nanotoxicology (MCT-CNPq). The authors would like to thank Prof. Mauricio S. Baptista (IQ-USP) for the use of laboratory facilities and Prof. Carol H. Collins for revising the manuscript.

6. References

[1] Yokoyama M 2011 J. *Artif. Organs*. **8**, 238.
[2] Uhrich K E, Cannizzaro S M, Langer R S and Shakesheff K M 2011 *Chem. Rev.*, **99**, 3181.
[3] Nguyen T K, Thanha L and Greena, A W, *Nano Today* 2002, **5**, 213.
[4] Fang B, Gon S, Park M, Kumar K N, Rotello V M, Nusslein K and Santore M M 2011 *Colloids and Surfaces B: Biointerfaces*, **87**, 109.
[5] Rosen J E and Gu F X 2011 *Langmuir*. **27**, 10507.
[6] Pankhurst, Q A, Connolly J, Jones S K and Dobson J. 2004 J. *Phys. D:Appl. Phys*. 2003, **36**, R167.
[7] Haddad P S, Duarte E L, Baptista M S, Goya G F, Leite, C A P and Itri R 2004 *Progr. Colloid Polym.Sci.*,**128**, 232.
[8] Wang L, Ding W and Zhang F 2010 *J. Nanosci. Nanotechnol.* **10**, 8617.
[9] Zhu M T, Feng W Y, Wang B, Wang T C, Gu Y Q, Wang M, Wang Y, Ouyang H, Zhao Y L and Chai Z F 2008 *Toxicology*. **247**, 102.
[10] Bhattacharya K, Davoren M, Boertz J, Schins R P, Hoffmann E and Dopp E 2009 P*art. Fibre Toxicol.* **6**, 17.
[12]Gramowski A, Flossdorf J, Bhattacharya K, Jonas L, Lantow M, Rahman Q, Schiffmann D, Weiss D G and Dopp E 2010 *Environ. Health Perspect.* **118**, 1363.
[13] Mahmoudi M, Simchi A, Milani A S and Stroeve P 2009 *J. Colloid Interface Sci.* **336**, 510.
[14] Bhattacharya K, Hoffmann E, Schins R F P, Boertz J, Prantl E M, Alink G M, Byrne H J, Kulbusch T A J, Rahman Q, Wiggers H, Schultz C and Dopp E 2012 *Toxic. Sci.* **126**,173.
[15] Lynch I, Cedervall T, Lundqvist M, Cabaleiro-Lago C, Linse S and Dawson K A 2007 *Adv. Colloid Interface Sci.*, **134**, 167.
[16] Nel A E, Madler L, Velegol D, Xia T, Hoek E M, Somasundaran P, Klaessig F, Castranova V and Thompson M 2009 *Nat. Mater.* **8**, 543.

[17] Murray C B, Norris D J and Bawendi M G 1993 *J. Am. Chem. Soc.* **115**, 8706.
[18] Hong S C, Lee J H, Lee J, Kim H Y, Park J Y, Cho J, Lee J, Han D H 2011 *Int. J. Nanomed.* 6, 3219.
[19] Seabra A B and Duran N 2010 *J. Mat. Chem.* **20**, 1624.
[20] Cha Y J, Kim M J, Choa Y H, Kim J, Nam B, Lee J, Kim D H and Kim H 2010 *IEEE Trans. Magn.*,**46**, 443.
[21] Molina M M, Seabra A B, de Oliveira MG, Itri R, Haddad PS. 2013 *Mat. Sci. Engin. C.* **33**, 746.
[22] Kemp M, Go Y M and Jones D P 2008 *Free Radic. Biol. Med.* **44** 921.
[23] Aufann M, Decome L, Rose J, Orsiere T, De Meo M, Briois V, Chaneac C, Olivi L, Berge J L, Lefranc B, Botta A, Wiesner M R and Bottero J Y 2006 *Environ. Sci. Technol.* **40**, 4367.

Cytotoxicity of solid lipid nanoparticles and nanostructured lipid carriers containing the local anesthetic dibucaine designed for topical application

R M Barbosa[1], C M G da Silva[1], T S Bella[1], D R de Araújo[4], P D. Marcato[2], N Durán[3, 4] and E de Paula[1]

[1]Biochemistry Department, Institute of Biology, State University of Campinas, (UNICAMP) Campinas, SP, Brazil
[2]Faculty of Pharmaceutical Sciences of Ribeirão Preto, University of São Paulo, Ribeirão Preto, SP, Brazil
[3]Biological Chemistry Laboratory, Institute of Chemistry, UNICAMP, Campinas, SP, Brazil
[4]Center of Natural and Human Sciences, Universidade Federal do ABC, Santo André, SP, Brazil

E-mail: m.g.barbosafernandes@gmail.com

Abstract. Dibucaine (DBC) is powerful long-lasting local anesthetic, but it is also considered fairly toxic to the CNS. Solid lipid nanoparticles (SLN) and nanostructured lipid carriers (NLC) have attracted attention as carriers for drug delivery. The aim of this study was to develop and to evaluate the cytotoxic activity of DBC-loaded SLN and NLC against 3T3 fibroblast and HaCat keratinocyte cells. The SLN and NLC had myristyl myristate and Liponate®GC as their lipid matrices, respectively, plus a surfactant. SLN and NLC were characterized in terms in their diameter, size distribution, surface charge and DBC encapsulation efficiency. The particle size of SLN and NLC were around 234.33 and 166.62 nm, respectively. The polydispersity index was kept below 0.2 for both nanomaterials. Negative surface charges were observed for both nanoparticles, which decreased in the presence of the anesthetic. Encapsulation efficiency reached 76% and 90%, respectively, in SLN and NLC. DBC alone was found to be toxic to 3T3 and HaCat cells in culture. However, NLC and SLN loaded DBC decreased its intrinsic cytotoxic effect against 3T3 and HaCat cells. In conclusion, encapsulation of DBC in SLN and NLC decreased the in vitro toxicity of the local anesthetic, indicating the potential of these nanocarriers for clinical applications.

1. Introduction

Pain management continues to be one of the greatest challenges in medicine. Local anesthetics (LA) are widely used to reversibly block neural transmission [1, 2] and they differ from general anesthetics that cause loss of consciousness and inhibition of sensory and autonomic reflexes [3]. LA are amphiphilic medicines for which potency and toxicity are highly correlated with the molecule's hydrophobicity [4, 5].

Dibucaine (2-butoxy-N-(2-diethylaminoethyl) quinoline-4-carboxamide) belongs to the amino-amide class of local anesthetic. The DBC chemical structure (Figure 1) differs from those of the LA commonly used in clinical practice, due to its large quinolinic ring, to which a butyl ether group is attached [6]; this rigid ring imposes restrictions on the molecule, and modulates its interaction with model membranes [6-8].

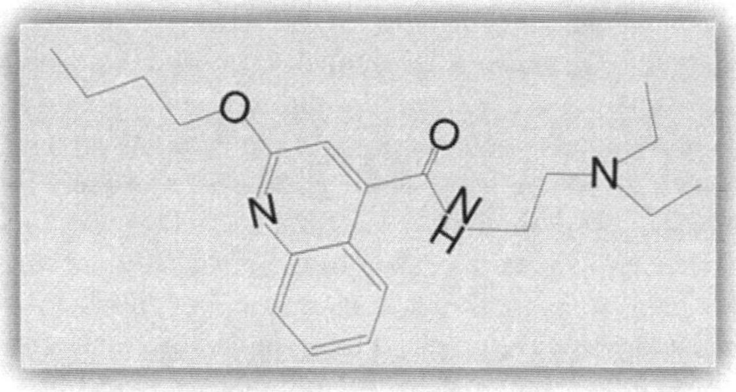

Figure 1: Chemical structure of dibucaine (deprotonated form).

Dibucaine is one of the most powerful long-lasting local anesthetics, being 15 times more potent than procaine [9]. However, it is also considered to be fairly toxic to the CNS and its use is limited to surface anesthesia, mainly in the form of creams applied to the skin or mucous membranes. The symptoms of DBC overdoses can include convulsions, hypoxia, acidosis, bradycardia, arrhythmia, and cardiac arrest [10-12]. These problems have led us to investigate suitable delivery systems for the encapsulation of dibucaine.

Drug encapsulation in nanocarriers has been extensively studied for the transport of bioactive agents of low bioavailability and/or high toxicity [13-19]. Regarding local anesthetics, macrocyclic carriers (e.g. cyclodextrins), polymers (nanospheres, nanocapsules), and lipid-based carriers such as liposomes [19] and solid lipid nanoparticles [20] have provided effective drug delivery systems with prolonged anesthesia and reduced toxicity.

Solid lipid nanoparticles (SLN) have been used for the encapsulation of different active agents [20-27]. This type of carrier offers the same benefits as traditional ones such as liposomes and polymeric nanoparticles, without their disadvantages (such as physical instability, scale up problems, leakage of the active agent, and, in some cases, cytotoxicity). SLN are attractive to the pharmaceutical and cosmetic industries because they are stable, versatile, use safe excipients, and are readily scalable [21, 28-30].

Nanostructured lipid carriers (NLC) are considered to be the second generation of lipid-based particles (SLN being the first generation). They present at least one liquid lipid at room temperature to form their inner core. Recently, Gallarate & Battaglia [31] classified these structures as nanocapsules for having this liquid core and an outer envelope. Although both types of lipid nanoparticles, SLN and NLC, are able to encapsulate high amounts of hydrophobic drugs, NLC have not shown the limitations of SLN, regarding drug expulsion during storage caused by the crystalline rearrangements of the solid lipid [32, 33].

Encapsulation of dibucaine in SLN and NLC should render the drug available in higher doses at the site of administration, prolonging anesthesia and reducing its systemic toxicity due to controlled sustained release. In order to better understand the potential of such formulations, we prepared SLN and NLC containing dibucaine and characterized them in terms of their encapsulation efficiencies, particle sizes, zeta potentials and cytotoxicities against 3T3 fibroblasts and HaCat keratinocytes, in culture.

2. Material and methods

2.1. Materials

Dibucaine (DBC) and poloxamer 188 (Pluronic F68) were supplied by Sigma (USA), myristyl myristate (MM) was provided by Croda (Brazil) and Liponate® GC was from Lipo do Brasil Ltda (Brazil). Acetonitrile (J.T. Baker, USA), triethylamine (Vetec, Brazil), and orthophosphoric acid (Ecibra, Brazil) were of HPLC grade. Deionized water (18 MΩ) was obtained from an Elga USF Maxima Ultra Pure Water equipment. The solutions used in HPLC were filtered through 0.22 μm nylon membranes (Millipore, Belford, USA). All other chemicals and solvents used were of analytical grade.

2.2. Preparation of the solid lipid nanoparticles (SLN)

SLN were prepared by the hot emulsion technique. The solid lipid, myristyl myristate, was heated 10°C above its melting point (39°C, according to the manufacturer's MSDS). To obtain the anesthetic-loaded formulations, dibucaine was solubilized into this oily phase after complete fusion. After that, this phase was carefully added (3 min.) to a hot aqueous solution of Pluronic F68 (0.02 mmol/L) under high speed agitation (10,000 rpm) using a Turrax blender (IKA WerkeStaufen, Germany) up to a final myristyl miristate concentration of 47 mmol/L. This pre-emulsion was then homogenized at 600 bar - 3 cycles - in a Panda homogenizer (NiroSoavi, Italy). The Panda homogenizer and the sample's milieu were insulated to keep the temperature above the melting point of the lipid. Finally, the SLN were cooled to room temperature, and conditioned in glass flasks at 4°C [34], as shown in Figure 2.

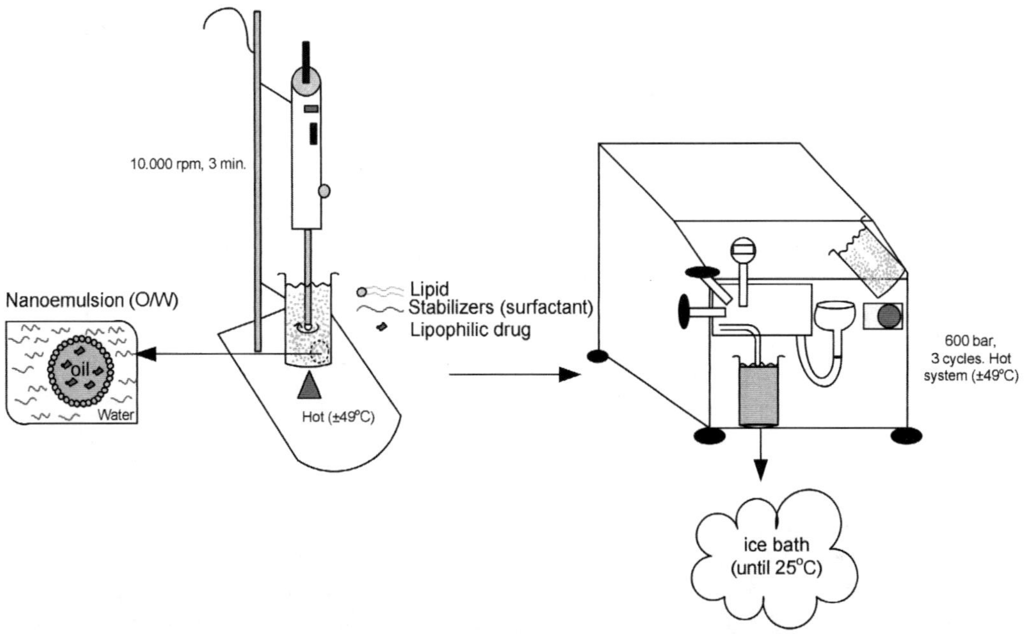

Figure 2: Schematic representation of the methodology used in the preparation of SLN , using high pressure homogenization.

2.2. Preparation of the nanostrutured lipid carriers (NLC)
NLC were prepared using the same methodology described for preparing SLN (Figure 2) with modifications in the lipid components: in addition to myristyl myristate, Liponate® GC (a mixture of triglycerides of caprylic and capric acids that is liquid at room temperature) was used to form the oily phase. This oily phase was added to 0.06 mmol/L Pluronic F68 at 49°C. The final concentration of myristyl myristate and Liponate® GC in the formulation were 47 mmol/L and 0.5% w/w (referring to the total myristyl myristate lipid), respectively.

2.3. Particle size, polydispersity and zeta potential
The size distribution and average particle size (z-average) of the SLN and NLC particles were measured by photon correlation spectroscopy (PCS) in a Zetasizer Nano ZS (Malvern Instruments, UK), at 25°C with 10 mm path length polystyrene cuvettes. The samples were diluted in Milli-Q water prior to the analysis (sample/ Milli-Q water ratio of 1:100, v/v). The zeta potential was measured in a solution of 0.1 mmol/L of sodium chloride. The size, polydispersity, and zeta potential of the SLN and NLC samples (with and without DBC) were observed one day after preparation.

2.4. Transmission electron microscopy (TEM)
The nanoparticle morphology was analyzed using a Zeiss LEO-906, 60 kV transmission electron microscope. A drop of each nanoparticle sample was placed in a 200 mesh copper grid and a drop of a 2% aqueous uranyl solution added; excess volumes were removed with filter paper. Samples were incubated for 4 hours to dry at room temperature prior to TEM analysis.

2.5 High performance liquid chromatography (HPLC)
Dibucaine was quantified at 247 nm, using a Varian ProStar HPLC fitted with a PS 325 UV-Vis detector, a PS 210 solvent delivery module, and an automatic injector. Galaxy Workstation software was used for data collection and processing. The chromatographic conditions were optimized using a Merck reversed phase C18 column (LiChroCART 100 RP-18, 5 µm, 110 Å, 125 mm x 4 mm). The mobile phase consisted of a mixture of acetonitrile and 0.04 mol/L triethylamine phosphate buffer (55:45 v/v), pH 2.3. The flow rate was 1.0 mL/min, the detection and the injection volumes were 30 µL. Peak areas were measured. The analyses were

conducted at 35°C [35]. Quantification of free DBC employed a calibration curve (Y=–0.52868+ 58.52676 X, r=0.999) obtained in the 1.5-30.0 µg/mL concentration range.

2.6 Encapsulation efficiency

The encapsulation efficiency (%EE) of dibucaine into the nanoparticles (SLN, NLC) was determined according to equation 1 [29]. The samples were diluted 2:100 v:v in Milli-Q water and transferred to a 10 kDa cellulose filtration unit (Millex, Millipore) and centrifuged (4000 *g*, 20 min). Free dibucaine in the filtered solution was quantified by HPLC (section 2.5).

$$\%EE = {DBC_{bound}}/{DBC_{total}} \; x100 \qquad \text{(Eq 1)}$$

Where, DBC_{bound} is the amount of DBC entrapped in the SLN or NLCs and DBC_{total} is the DBC added to the formulation, prior to phase separation. $DBC_{bound} = DBC_{total}$ - [free DBC], detected by HPLC.

2.7 In vitro cytotoxicity

The cytotoxicity tests were performed using mouse 3T3 fibroblasts and human keratinocytes (HaCaT) cultured cells. MTT dye (3- (4, 5-dimethylthiazol-2-yl)-2,5-diphenyltetrazolium bromide) was used to assess cell viability, according to the capacity of the cells to reduce MTT to its formazan derivative. This reduction is catalyzed mainly by mitochondrial and cytoplasmic dehydrogenases, so that any alteration in mitochondrial function can be detected in the ability of the cell to reduce MTT to the purple formazan dye. The mouse embryo BALB/c 3T3 fibroblast cell line was obtained from the National Institutes of Health (Baltimore, USA) and the immortalized human keratinocytes, HaCaT, from the Academic Medical Center, Amsterdam University.

The cells were cultivated in DMEM medium supplemented with 10% fetal bovine serum and 1% penicillin/streptomycin antibiotic. The cells were plated out onto 96-well plates, at a density of $1x10^4$ cells/mL, and incubated at 37°C under a humid atmosphere with 3.5% CO_2. The medium was modified using different concentrations of dibucaine (0.02-0.30mM, free of encapsulated) or non-DBC loaded nanoparticles (SLN or NLC, 0.53-4.20mM).

In order to improve the solubility of free DBC 0.1% v:v of dimethyl sulfoxide was added to its water solution. The treatment medium was removed after 2 hours of incubation, and replaced by a medium without serum but containing 0.5 mg/mL MTT . After incubation for 3 hours at 37°C, the medium was carefully removed, and 100 µL of ethanol was added to solubilize the formazan derivative. The plates were agitated for 10 min. and the absorbance corresponding to each well was measured at 570 nm, using a BIO-TEK Elx 800 spectrophotometer. The values were expressed as the percent MTT reduction, relatively to a sample of control cells, not exposed to the tested agents [36]. Averages (± standard deviations) were calculated from six separate measurements.

3. Results and discussion

3.1. Production and characterization of nanoparticles

SLN and NLC were prepared with myristyl myristate (MM) and myristyl myristate/Liponate® GC, respectively, and Pluronic F-68. High pressure homogenization was the technique employed to prepare the nanoparticles. The mean diameters (Z-averages), polydispersity indices (PI), and zeta potentials of the SLN and NLC obtained are presented in Table 1. The diameters of the SLN and NLC were 188.02 ± 7.07 and 167 ± 14 nm, respectively. Although, in both cases, incorporation of dibucaine slightly increased the nanoparticle size, statistical analysis (unpaired t-test) revealed no significant differences ($p > 0.05$) between the formulations, with and without DBC. Moreover, as shown in Table 1, the type of nanoparticle had no significant effect on the particle features such as size, PI and zeta potential ($p>0.05$). In all formulation the polydispersity index was smaller than 0.25, which is within the range expected for homogeneously dispersed colloids [23]. The measured zeta potentials were negative (ca. 30 mV) and, upon addition of dibucaine, became more positive (~15-18 mV), for both SLN and NLC formulations. This result was taken as an indication that that at least a fraction of the DBC molecules resides at the nanoparticle surfaces, affecting their surface charge distribution. All formulations exhibited high encapsulation efficiencies, as expected from the high lipophilicity of DBC, which favors its interaction with lipids [6]. The encapsulation efficiency of NLC was significantly higher than that observed with SLN (Table 1).

Table 1. Size (Z-Averages), polydispersity indices (PIs) and zeta potentials of SLN and NLC, prepared by high pressure homogenization, with and without DBC. The encapsulation efficiency (EE%) of such nanoparticles for the anesthetic is also given (pH 8.2).

Sample	Z-Averages (nm)	PI	zeta potential (mV)	EE%
SLN	188.02± 7.07a^{ns}	0.15± 0.02a^{ns}	-26.91± 7.72a^{ns}	
SLN DBC	234.33± 42.87b^{ns}	0.23± 0.06b^{ns}	-18.47± 2.55b^{ns}	76.58 ± 7.88c^{**}
NLC	166.62± 14.96	0.12± 0.01	-28.83± 4.41	
NLC DBC	177.39± 15.87	0.20± 0.08	-14.89± 2.78	90.54± 0.95

(Mean ± SD, n=3) Z-Averages, PI and zeta potential: a. SLN *vs* NLC, b. SLN DBC *vs* NLC DBC. EE%: c. SLN DBC *vs* NLC DBC, (unpaired t test, **$p < 0.01$ and ns (not significant), $p > 0.05$).

TEM images of SLN and NLC containing DBC (Figure 3 A, B) revealed spherical structures, with delineated surfaces and homogenous size distribution. The nanoparticle sizes in TEM micrographies are in good agreement with those determined by photon correlation spectroscopy.

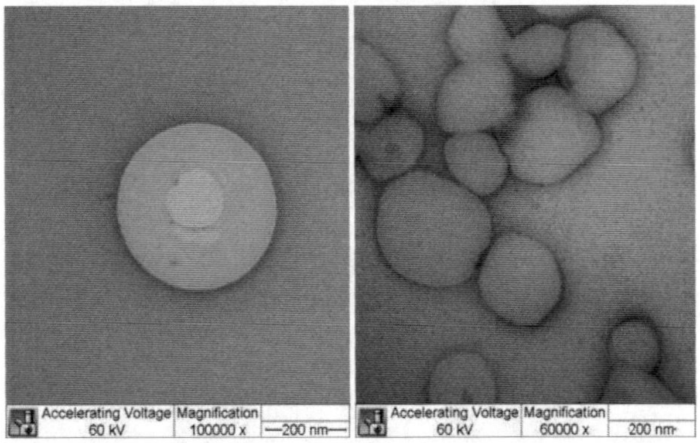

(A)

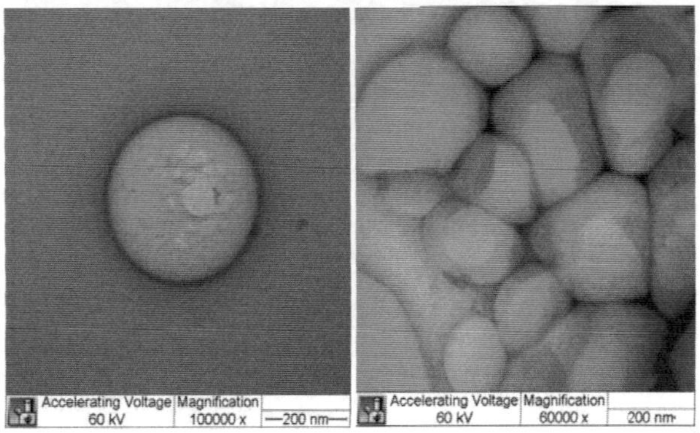

(B)

Figure 3.Morphology of the nanoparticles containing dibucaine. (A) SLN DBC, (B) NLC DBC. The images in the right (60,000 x magnification) and the images in the left (100,000 x magnification) give details of one chosen particle.

3.2 Cytotoxicity tests

The cytotoxicity of the nanoparticles and their DBC formulations on two distinct cell lines (3T3 fibroblasts and HaCaT) were tested (Figure 4). Under the experimental conditions (0.5 - 4.2 mmol/L lipid concentration), SLN and NLC can be considered safe, since they did not affect the survival of either cell lines (~100% cell viability) (Figure 4 A).

Free dibucaine (Figure 4B) decreased the viability of 3T3 and HaCat cells: the concentration required to reduce cell viability by 50 %, IC50 = 0.25 mmol/L and 0.24 mmol/L for 3T3 fibroblasts and HaCaT, respectively. The cytotoxic effect of DBC has also been observed with other anesthetic agents [37] and seem to be related to the partitioning of the amphiphile molecule inside the biomembrane [38] leading to membrane destabilization [7].

Encapsulation of DBC into the nanoparticles significantly decreased its intrinsic cytotoxic effect ($p > 0.001$, ANOVA and Turkey-Kramer test), as shown in Figure 4B. The protection was even more evident when NLC were used in comparison to SLN ($p < 0.05$, ANOVA).

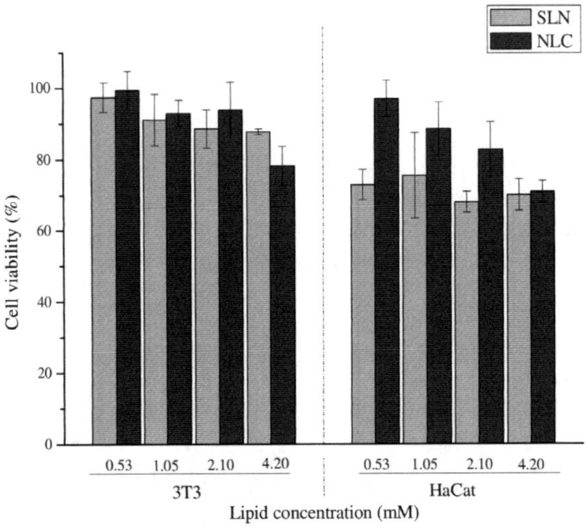

(A)

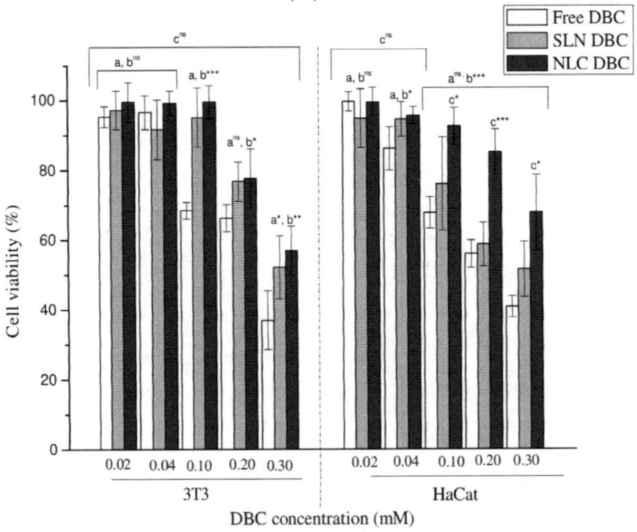

(B)

Figure 4.Cytotoxicity tests: cell viability of 3T3 fibroblasts (left) and HaCat cells (right) (mean ± SD, n = 6) treated with: (A) SLN and NLC; (B) free DBC and DBC-loaded SLN or NLC. a. free DBC *vs* SLN DBC, b. free DBC *vs* NLC DBC and c. SLN DBC *vs* NLC DBC, ANOVA and Turkey-Kramer test (ns: no significant, $p > 0.05$, * $p < 0.05$,** $p < 0.01$, *** $p < 0.001$.

4. Conclusions and perspectives

Solid lipid nanoparticles and nanostructured lipid carriers containing dibucaine, were prepared by the high pressure homogenization technique. The high pressure process is attractive to the pharmaceutical industry, since it avoids scalability problems, and the equipment required can be easily acquired at low cost. A high DBC encapsulation efficiency (90.54± 0.95%) was achieved with NLC. The nanoparticles presented high encapsulation efficiency and reduced the intrinsic toxic effect of the anesthetic against 3T3 fibroblasts and HaCat keratinocyte cells. Overall, these results show that encapsulation of dibucaine in NLC and SLN promotes sustained release of the anesthetic, while the *in vitro* cytotoxicity decrease supports the potential of this type of drug delivery system for future clinical (dermal anesthesia) applications. In vivo studies are being carried in our laboratory to test the antinociceptive effects of both drug delivery systems.

Acknowledgments

The authors would like to thank Fapesp (# 06/00121-9), CAPES and CNPq (Brazil) for financial support and INOMAT (MCT/CNPq) and Brazilian Network in Nanotoxicology (MCT/CNPq), Brazil.

5. References

[1] de Jong R H 1994 *Local anesthetics* (St. Louis: Mosby-Year Book, Inc)
[2] Malamed S F 2004 *Handbook of Local Anaesthesia*(5th ed St. Louis: Mosby-Year Book, Inc)
[3] Katzung B G, Masters S and Trevor A 2009 *Basic & Clinical Pharmacology* (11th ed. Local Anesthetics) ed. L.B.S. Series(McGraw Hill Lange) pp1232
[4] Butterworth J F T and Strichartz G R 1990 *Anesthesiology* **72** 711
[5] Gupta S P 1991 *Chem Rev* **91** 1109
[6] de Paula E and Schreier S 1995 *Biochim.Biophys.Acta* **1240** 25
[7] Lorite G S, Nobre T M, Zaniquelli M E, de Paula E and Cotta M A 2009 *Biophys. Chem.* **139** 75
[8] Kuroda Y, Ogawa M, Nasu H, Terashima M, Kasahara M, Kiyama Y, Wakita M, Fujiwara Y, Fujii N and Nakagawa T 1996 *Biophys J.* **71** 1191
[9] Abdel-Ghani N T, Youssef A F and Awady M A 2005 *Farmaco* **60** 419
[10] Dayan P S, Litovitz T L, Crouch B I, Scalzo A J and Klein B L 1996 *Ann. Emerg. Med.* **28** 442
[11] Curtis L A, Dolan T S and Seibert H E 2009 *J. Emerg. Med.* **37** 32
[12] Hanzlicek A S and Van der Merwe D 2010 *J.Med. Toxicol.* **6,** 44
[13] Mitri K, Shegokar R, Gohla S, Anselmi C and Mueller, R H 2011 *Int. J. Pharm.* **414** 267
[14] Silva A C, Gonzalez-Mira E, Garcia M L, Egea M A, Fonseca J, Silva R, Santos D, Souto E B, Ferreira D 2011 *Colloids Surf. B Biointerf.* **86** 158
[15] Gonzalez-Mira E, Nikolic S, Garcia M L, Egea MA, Souto E B and Calpena AC 2011 *J. Pharm. Sci.* **100** 242
[16] Ridolfi D M Marcato P D, Justo G Z, Cordi L, Machado D and Durán N 2012 *Colloids Surf. B Biointerf.* **93** 36
[17] Korting H C and Schafer-Korting M 2010 *Handb. Exp. Pharmaco.l* **197** 435
[18] Neubert R H 2011 *Eur. J. Pharm. Biopharm.* **77** 1
[19] de Paula E, Cereda C M, Tofoli G R, Franz-Montan M, Fraceto L F and de Araujo, D R 2010 *Recent Patents Drug Deliv. Formulation* **4** 23
[20] zur Muhlen A, Schwarz C and Mehnert W 1998 *Eur. J. Pharm. Biopharm.* **45** 149
[21] Müller R H, Mader K and Gohla S 2000 *Eur. J. Pharm. Biopharm.* **50** 161
[22] Mehnert W and Mader K 2001 *Adv. Drug Deliv. Rev.* **47,** 165
[23] Müller R H, Radtke M and Wissing S A 2002 Adv. Drug Deliv. Rev. **54, Suppl 1** S131
[24] Schubert M A and Muller-Goymann C C 2005 *Eur. J. Pharm. Biopharm.* **61** 77
[25] Pardeike J, Hommoss A and Mueller R H 2009 *Int. J. Pharm.* **366,** 170
[26] Aji Alex M R, Chacko A J, Jose S and Souto E B 2011 *Eur. J. Pharm. Sci.* **42** 11
[27] Marcato P D, Caverzan J, Rossi-Bergmann B, Pinto E F, Machado D, Silva R A, Justo G Z, Ferreira C V and Duran N 2011 *J. Nanosci.Nanotechnol.* **11** 1880
[28] Maia C S, Mehnert W and Schafer-Korting M 2000 *Int. J. Pharm.* **196** 165
[29] Müller R H and Keck C M 2004 *J. Biotechnol.* **113** 151
[30] Puglia C, Blasi P, Rizza L, Schoubben A, Bonina F, Rossi C and Ricci M, 2008 *Int. J. Pharm.* **357** 295
[31] Battaglia L and Gallarate M 2012 *Expert Opin. Drug Deliv.* **9** 497
[32] Wissing S A, Saupe A and Müller R 2003 *Eurocosmetics* **5** 14
[33] Wissing S A, Kayser O and Müller R H 2004 *Adv. Drug Deliv. Rev.* **56** 1257
[34] Müller R H, Schwarz, C, zur Mühlen, A and Mehnert, W 1994 *Proc.Intern. Symp. Control. Rel. Bioactive Mat.* **21** 146
[35] Elsayed MM 2007 *Biomed. Chromatogr.* **21** 491

[36] Mosmann T 1983 *J. Immunol. Methods* **65** 55
[37] Cereda C M, Tofoli G R, Maturana L G, Pierucci A, Nunes L A, Franz-Montan M, de Oliveira A L, Arana S, de Araujo D R and de Paula E 2012 *Anesth. Analg.* **115** 1234
[38] Malheiros S V, Pinto L M, Gottardo L, Yokaichiya D K, Fraceto L F, Meirelles N C and de Paula E 2004 *Biophys.Chem.* **110** 213

Journal of Physics: Conference Series **429** (2013) 012036 doi:10.1088/1742-6596/429/1/012036

Nanoparticles in paints: A new strategy to protect façades and surfaces?

J-P Kaiser, L Diener and P Wick
Empa - Swiss Federal Laboratories for Materials Science and Technology, CH - 9014 St. Gallen, Switzerland

E-mail: jean-pierre.kaiser@empa.ch

Abstract. The paint and lacquer industries consider the use of nanosilver, photocatalytic active nanotitanium dioxide or nanosilica dioxide as additives for the protection of surfaces, against microbial, physical and chemical deterioration, as alternative to conventional organic based additives. Nowadays it is not clear, if nanoparticles in paints will achieve the proposed effects, since there are no long time studies available. Another fact is that the potential risks of nanoparticles for the environment and the human health is still controversial discussed.
The most sensitive entry port for nanomaterials is the lung. However other human organs/systems may also be affected by nanoparticles. Therefore the aim of the study was to assess the potential hazard effects of the three most interesting particles for paints on the gastro-intestinal tract and the immune system in vitro.
In our study we could show that:
i) Nanosilver (TEM size 25 nm) was far less toxic than silver ions of comparable concentrations tested with cells representing the gastro-intestinal tract (CaCo-2) and immune cells (Jurkat, T-lymphocytes). A significant amount of necrotic cells could be observed after exposure of CaCo-2 cells to 27 µg/ml nanosilver for 48 h.
ii) Nanotitanium dioxide can adsorb UV-light and in the presence of water hydroxyl radicals are generated photocatalytically. The exposure of CaCo-2 cells and Jurkat cells to photocatalytically active nanotitanium dioxide (Hombikat UV 100, TEM-size 15 nm) under dark conditions, didn't affected the cells significantly. However, the cells were able to incorporate nanotitanium dioxide, especially when cells were exposed to higher concentrations.
iii) Nanosilica dioxide improves the properties of the paints by increasing the water repellence. When cells were exposed to 243 µg/ml nanosilica dioxide (TEM-size 19 nm) for up to 48 h no cytotoxic effect could be observed.

1. Introduction

Microbial growth on façades is an aesthetical problem. Microorganisms are among others actively involved in the weathering and deterioration of façades and paints. In order to save energy more and more buildings have a thermal insulation. This increases the duration of wet periods due to dew condensation. This makes the system particularly vulnerable to biological attack, necessitating the use of biocides [1].
Since the ecotoxic biocides with a wide-protection range are forbidden by law (Biocidal Products Directive 98/8/EC), the paint industry is forced to use non-ecotoxic, non-persistent biocidal substances, e.g. organic based biocides [2]. Therefore the paint industry also considers using nanoparticles to increase the paint properties and to create new paint formulations, which fulfil the Biocidal Products Directive 98/8/EC regulations.
Paints containing biodegradable biocides might lose their protective impact, because the biocides could get washed out or could get inactivated by physical/chemical sources. Therefore the paint industry is focussing on different strategies to overcome this problem. The replacement of organic based, degradable biocides by nanomaterials is one of the different strategies, which the paint industry has taken in consideration. The paint industry considers therefore to use nanomaterials ,such as nanosilver, photo-catalytically active nanotitanium dioxide or nanosilica dioxide to improve paint

properties, such as water repellence, scratch resistance, increase of the hydrophobicity and antimicrobial properties [3].
Silver nanoparticles incorporated into the paints act as depot of silver ions. The release of the ions is tuned by particle size and surface coating. How silver ions interact with microorganisms is not fully understood. However, silver ions are able to penetrate into bacteria and once in the cell they will interact with thiol groups of vital enzymes and inactivate them, which leads to dysfunction and cell death [4]. The biocidal potency of a silver additive is therefore directly related to the potential for releasing silver ions. In addition silver ions are interacting with DNA and can inhibit respiratory enzymes, as well as induce oxidative stress upon generation of reactive oxygen species (ROS) [4, 5]. Ionic silver is also able to bind to sulphur and phosphor containing molecules involved in the cell antioxidant defence [6].
The lacquer and paint industry uses photo-catalytically active nanotitanium dioxide as biocide. Photo-catalytically active nanotitanium dioxide absorbs UV-light and produces in the presence of water hydroxyl radicals. These hydroxyl radicals are strong oxidizers and can induce oxidative stress and inflammation leading to oxidative attacks of cellular constituents and by that protecting the painted surfaces from microbial growth [7, 8]. Cytotoxicity of photo-catalytically active nanotitanium dioxide is controversially discussed in literature. Size, chemistry, coating and functionalization determine the cytotoxicity of titanium dioxide. Especially the modification of the nanoparticles can change toxicological properties completely compared to the parent compound.
Nanosized silica dioxide is widely used in wood preservation (hydrophobation) and coating applications. Nanosilica dioxide in paints improves scratch resistance, increases the water repellence, protects the paint against corrosion and provides the products a high gloss [9]. Nanosilica dioxide does not agglomerate and is therefore very persistent in the water-phase. Exposure to high silica dioxide dust should be avoided. Dry blasting hazardous paints can still lead to an elevated release of silica dioxide and other metals. Exposure to high concentrations of respirable silica dioxide can lead to silicosis and cancer [10, 11]. Nanosilica dioxide can in addition induce necrosis at high concentration, when nanosilica dioxide doped paints are coming in contact with human tissue [12].
The most sensitive entry port for nanomaterials is the lung. The pulmonary uptake is the most probable scenario of exposure and therefore intensively studied [13]. However, other organs/systems may also be affected by nanoparticles. Nanoparticles might be released from the façade and transferred by run-off water into the surface water, and accidently consumed by animals and/or children. Therefore nanoparticles might come in contact with cells of the gastro intestinal tract and might affect also our immune system.
The aim of this study was therefore to focus on possible effects of nanomaterials present in surface water to cells of the gastro intestinal tract (CaCo-2 cells), as well as on our immune system (Jurkat cells). Reports about possible effects that can occur, when CaCo-2 cells or Jurkat cells are exposed to nanoparticles is sparse.
Metallic nanomaterials, which are releasing metallic ions, seemed to be the most toxic nanomaterials among all synthetic nanomaterials. The bioavailability and toxicity of these metallic nanoparticles (e.g. copper) is highly dependent on the amount of adsorbing particles in the water column [14]. Nanoparticles bound to other compounds forming larger complexes are less bioavailable and therefore less toxic. Nanoparticles, which are released and transferred into the surface water, can cause toxicological problems, as long as they are not adsorbed to other particles [15].
Fortunately the concentrations of nanometals (nanosilver, nanotitanium dioxide, nanozinc oxide) found in surface waters are in the range of ng/l, which is still low [16]. A goal of this study was to estimate the cytotoxic concentrations of different nanoparticles (nanosilver, nanotitanium dioxide and nanosilica dioxide), which might be used by the paint industry in the near future to improve the properties of the paints. The nanoparticles might be released from the paints and might be transferred into the surface water. Therefore CaCo-2 cells and Jurkat cells were exposed to different concentrations of nanoparticles and the cytotoxic concentrations of these particles were estimated.

2. Materials and Methods

2.1. Cell cultures

In this study the effects of three nanoparticles was evaluated on two different cell types (CaCo-2 cells and Jurkat cells). CaCo-2 cells were derived from the gastrointestinal tract and were used to investigate the cytotoxic effect of ingested nanoparticles via drinking water. CaCo-2 cells (order number 86010202) were obtained from Health Protection Agency Culture Collections (HPACC), Salisbury, UK. Jurkat cells are human T-lymphocytes and were used to investigate the immunotoxic effect of nanoparticles and paints. This cell line was provided by Dr. S. J. Korsmeyer, Harvard Medical School, Boston, USA.

2.2. Nanoparticles

Nanosilver, TEM-size 25 nm.
Nanotitanium dioxide, TEM-size 15 nm.
Nanosilica dioxide, TEM-size 19 nm.

Table 1. Nanoparticle characterization

Material	Nanosilver	Nanotitanium dioxide	Nanosilica dioxide
Function	Biocide	Biocide	Scratch resistance
Endotoxin contaminated	No	No	No
Primary particle size and morphology (TEM)	25 nm (spherical) to 80 – 90 nm (rods)	15 nm	19 nm
Hydrodynamic diameter (nm)	91 nm	386 nm	195 nm
Zeta potential (mV)	- 42 mV	- 25 mV	- 14 mV
Shape	Some spherical, others rods	Spherical	Spherical

The nanoparticles were examined by different physical-chemical methods, such as dynamic light scattering (DLS), zeta potential analysis, transmission electron microscope analysis (TEM), according to Smulders et al. [17].

2.3. Endotoxin measurement

The investigated nanoparticles were tested for potential endotoxin contaminations prior starting the cytotoxicity studies. In this study a kinetic chromogenic LAL assay (Limulus amebocyte lysate endosafe endochrome-K; US License 1197; Charles River Laboratories, Inc., Charleston, SC, USA) was used for testing the endotoxin content in the different nanoparticle-samples. The test procedure was performed according to the protocol of Charles River Laboratories, USA.

2.4. Apoptosis/necrosis

Cell viability was quantitatively examined by flow cytometry analysis (FACS) based on the binding of annexin V fluorescein and propidium iodide incorporation to distinguish between apoptotic and late apoptotic/necrotic cells. The staining procedure was performed according to the user protocol (PF032) of Calbiochem, D. The procedure was described in detail by Kaiser et al. [18].

2.5. Reactive oxygen species (ROS)
The release of reactive oxygen species (ROS) was investigated with the 2, 7-dichlorodihydro-fluorescein reduction assay. The fluorescence intensity, which is a function of liberated reactive oxygen species was analysed with a fluorescence plate reader at a wave length 485/528 nm. The procedure was described in detail by Limbach et al. [19].

2.6. Release of cytokines
Nanoparticles may also activate stress-related kinases, which can induce the release of proinflammatory cytokines, such as tumour necrosis factor α (TNF-α) and interleukin 8 (IL-8). Interleukin 8 was shown to increase significantly in response to a variety of chemical irritants, following the activation of primary proinflammatory cytokines, such as TNF-α. The release of proinflammatory cytokines (TNF-α and IL-8) was analysed with an enzyme linked immunosorbent assay (Elisa), (eBioscience, Inc. San Diego, CA, USA).

2.7. Cell morphology
Cellular uptake, respectively cell morphology was analysed by bright field microscopy; phase contrast method at a magnification of 100 x (Nikon-Diaphot, Egg, CH). Cell morphology was in addition investigated by fluorescence microscopy after staining the actin cytoskeleton with phalloidin-Alexa Fluor 488 (Invitrogen, Lucerne, CH) and the nucleus with Drag5 (Biostatus, Leicestershire, UK).

2.8. Genotoxicity
Genotoxicity was measured by the single cell gel electrophoresis (comet) assay. The electrophoresis takes place under alkaline conditions. The DNA fragments start to move in the agarose gel and were forming, depending on the damage of the genome, a comet. The DNA is thereafter stained with ethidium bromide and comets were analysed by fluorescence microscope. Quantification of the head and tail intensities was done using the "Comet Assay IV" software. The procedure was described in detail by Hirsch et al. [20].

3. Results and Discussion

3.1. Endotoxin measurement
Since the production and the storage of nanomaterials are not occurring under sterile conditions, it cannot be excluded that nanomaterials got contaminated with endotoxins. Therefore, prior starting our cytotoxicity studies, the nanomaterials (nanosilver, nanotitanium dioxide and nanosilica dioxide) were analysed. The three tested nanomaterials were endotoxin free or the endotoxin level was not detectable.

3.2. Apoptosis/necrosis
After treatment with the three well characterized nanoparticles (nanosilver, nanotitanium dioxide and nanosilica dioxide) cell viability on epithelial cells (CaCo-2) and T-lymphocytes (Jurkat) was analysed by life/dead measurements (apoptosis/necrosis). After the cells were exposed to nanotitanium dioxide or nanosilica dioxide in different concentrations (1, 3, 9, 27, 81 and 243 µg/ml) for 24 h and 48 h no significant effect was observed. However, a significant amount of necrotic CaCo-2 cells were observed after exposure to 27 µg/ml nanosilver for 48 h (Figure 1). Jurkat cells were more robust against nanosilver than CaCo-2 cells. The observed amount of dead Jurkat cells was not significant, when these cells were exposed to the same nanosilver concentration for 48 h. However, after exposure of Jurkat cells to a nanosilver concentration of 81 µg/ml for 48 h a significant amount of dead cells could be observed.

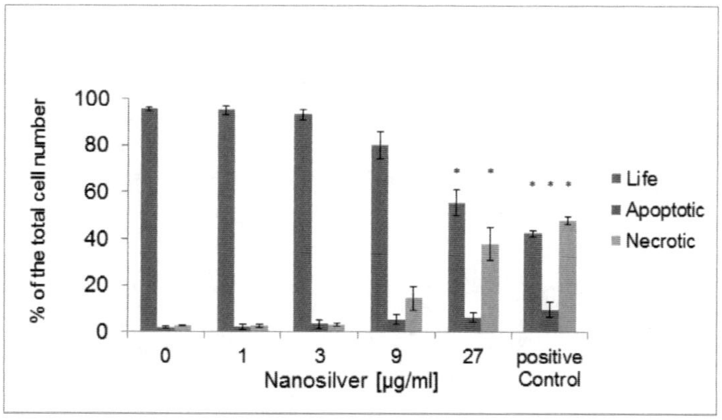

Figure 1. Apoptosis/necrosis has been analysed quantitatively after CaCo-2 cells were exposed to nanosilver in different concentrations (1, 3, 9, 27, 81 and 243 µg/ml) for 48 h. $CdSO_4$ (25 µM) was used as positive control and the cells were exposed for 24 h. Data were represented as mean ± SEM, n = 3.

3.3. Release of reactive oxygen species (ROS)

CaCo-2-cells exposed to nanosilver and nanosilica dioxide in a concentration range of 1 to 243 µg/ml for 1 h induced only a low amount of reactive oxygen species (ROS). Jurkat cells exposed to nanosilver and nanosilica dioxide reacted in a similar way (data not shown). However, Jurkat cells showed a higher ROS release, when exposed to nanotitanium dioxide. An elevated ROS generation was already noticed, when cells were exposed to 81 µg/ml nanotitanium dioxide (Figure 2).

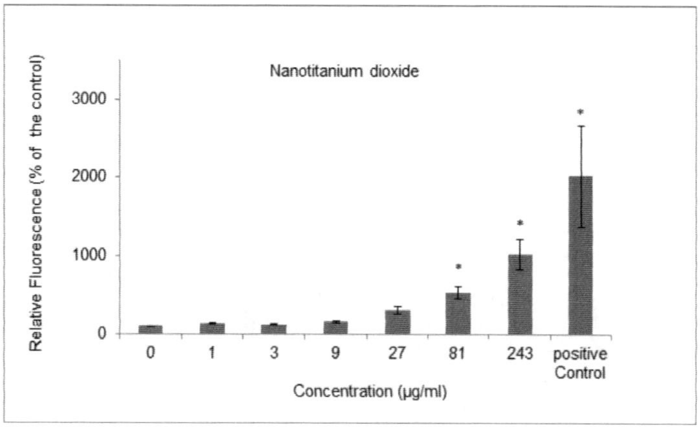

Figure 2. Jurkat cells exposed to nanotitanium dioxide incorporated a high amount of the nanotitanium dioxide, which resulted in a significant release of ROS. 3-morpholinosydnonimine hydrochloride (5 µM) was used as positive control. Data represented as mean ± SEM, n = 3.

3.4. Release of proinflammatory cytokines, e.g. interleukins

When CaCo-2 cells were exposed to nanotitanium dioxide and nanosilica dioxide no significant amount of IL-8 was released at all three investigated exposure times. However, it was different when cells were exposed to nanosilver. Higher concentrations of released IL-8 were found after cells had been exposed to 27 µg/ml nanosilver for 24 h and 48 h (Figure 3). This was the nanosilver concentration, where a significant amount of necrotic cells could be observed. Jurkat cells again behaved different. Jurkat cells exposed to nanosilver, nanotitanium dioxide or nanosilica dioxide released only minor amounts of IL-2.

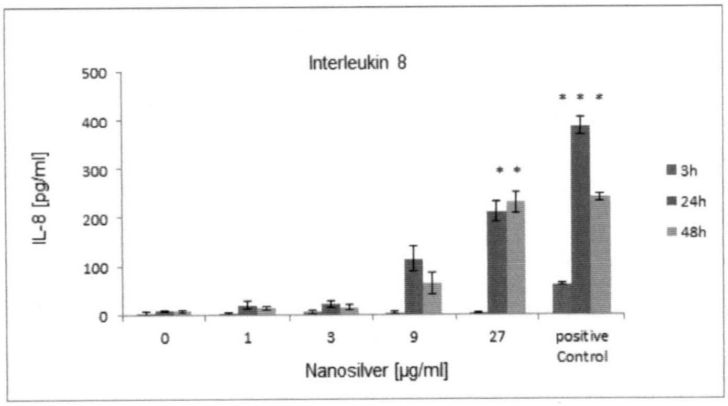

Figure 3. Release of interleukin 8 during exposure of CaCo-2 cells to different concentrations nanosilver. Tumor necrosis factor α (TNF-α, 50 ng/ml) was used as positive control; the cells were exposed to TNF-α for 24 h. Data represented as mean ± SEM, n = 3.

3.5. Genotoxicity

Genotoxicity was investigated with the comet assay. When CaCo-2 cells were exposed to 80 µg/ml nanosilica dioxide no significant effect on cell viability could be observed. When cell cultures were exposed to the same concentration (80 µg/ml) nanotitanium dioxide for 48 h some DNA strand brakes could be observed, however the effect was not significant. When CaCo-2 cells were exposed to 5 µg/ml nanosilver for 48 h more DNA strand brakes occurred. The effect was however not significant. But cells exposed to 10 µg/ml nanosilver for 48 h showed a significant genotoxic effect. The tail intensities of the cells in the comet assay reached similar values as the positive control. This indicates that nanosilver at these concentrations leads to high DNA strand brakes (Figure 4).

When CaCo-2 cells were exposed to 9 µg/ml nanosilver for 48 h, less than 20 % of the cells died and cell death was not significant. Thus the results obtained in the genotoxicity test were in accordance with the results obtained with the life/dead measurements (apoptosis/necrosis). When CaCo-2 cells were exposed to lower nanosilver concentrations no significant cell death as well as no significant genotoxicity could be observed.

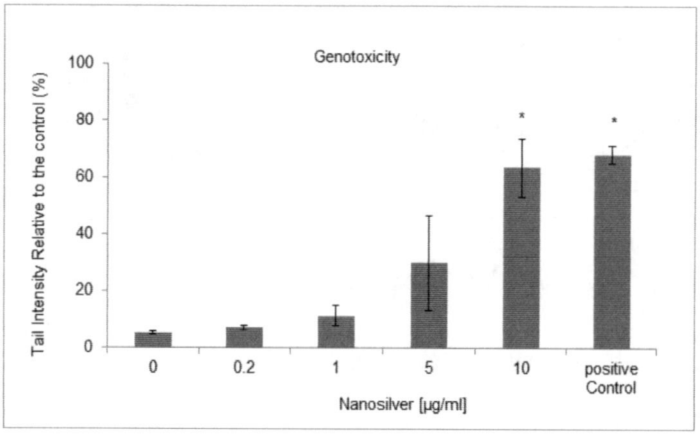

Figure 4. DNA strand brakes after exposure of CaCo-2 cells to nanosilver for 48 h. Ethyl methyl sulfate (EMS, 15 mM) was used as positive control; the cells were exposed to EMS for 30 min. Data represented as mean ± SEM, n = 3.

3.6. Cell morphology

Cell morphology was analyzed after exposure of CaCo-2 cells and Jurkat cells to these three nanoparticles (nanosilver, nanotitanium dioxide and nanosilica dioxide) at concentrations up to 243 µg/ml for 48 h. Nanosilica dioxide (up to 243 µg/ml) had no visible effect on cell morphology. This was also true for nanotitanium dioxide. However, bright field microscopic pictures showed that the nanotitanium dioxide got either adsorbed to the extracellular matrix of the CaCo-2 cells and Jurkat cells and/or got incorporated into the cells. Cell morphology was not notably affected by the high incorporation of nanotitanium dioxide (Figure 5A, 5B). These observations were not unusual. Thus it has been reported by other scientists that nanotitanium dioxide was taken up by various cell types without causing cytotoxic effects during the observation period [21]. CaCo-2 cells and Jurkat cells exposed to nanosilver at a concentration of 27 µg/ml for 48 h didn't show any visible effects on cell morphology. At higher nanosilver concentrations cell growth decreased and/or the adherent cells started to detach from the surface.

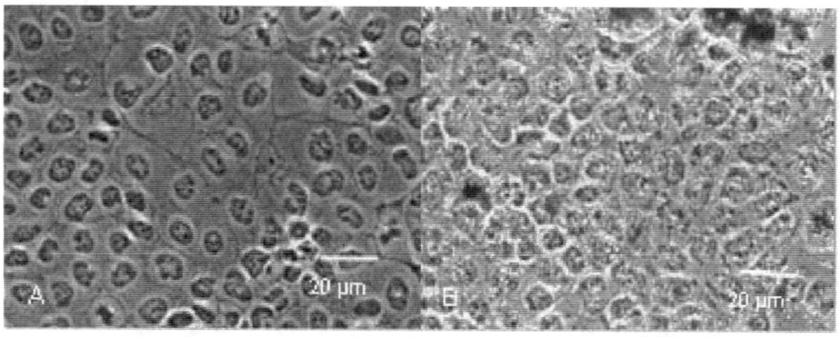

Figure 5A. Bright field microscopic pictures. A: CaCo-2 cells grown in absence of nanotitanium dioxide, B: CaCo-2 cells exposed to 243 µg/ml nanotitanium dioxide for 48 h.

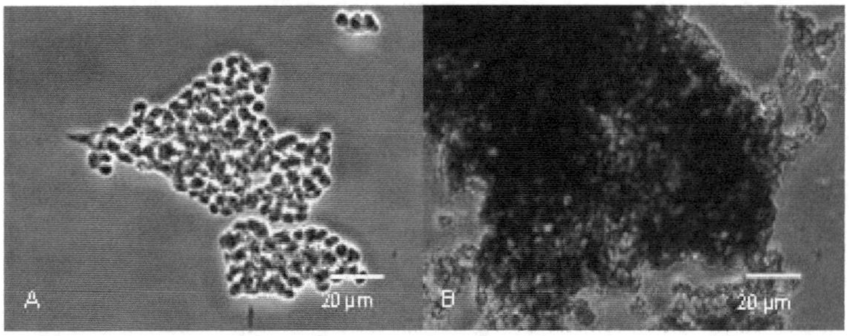

Figure 5B. Bright field microscopic pictures. A: Jurkat cells grown in absence of nanotitanium dioxide, B: Jurkat cells exposed to 243 µg/ml nanotitanium dioxide for 48 h.

Fluorescence microscopic pictures show that CaCo-2 cells exposed to 27 µg/ml nanosilver for 48 h started to detach from the underlying surface (Figure 6A). CaCo-2 cells, which were still alive after these 48 h of exposure, were not spread anymore. The actin cytoskeleton was disrupted and the remaining actin fibres were located in the near vicinity of the nucleus.

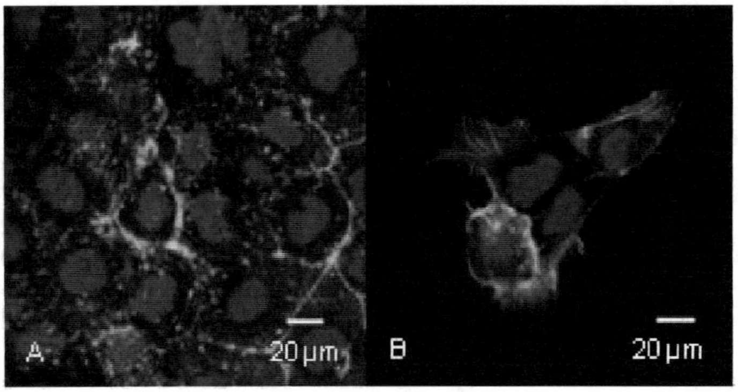

Figure 6A. Fluorescent microscopic pictures. A: CaCo-2 cells grown in absence of nanosilver, B: CaCo-2 cells exposed to 27 µg/ml nanosilver for 48 h. Staining: nucleus (blue), actin filaments (green).

During the exposure of Jurkat cells to 81 µg/ml nanosilver for 48 h the development of the actin cytoskeleton (green) of these cells were severely affected (Figure 6B).

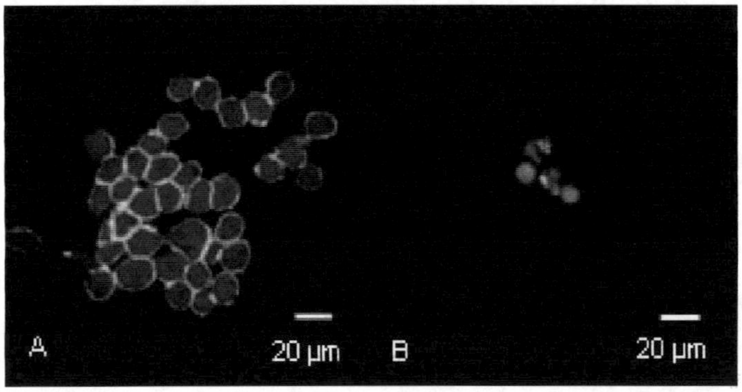

Figure 6B. Fluorescent microscopic pictures. A: Jurkat cells grown in absence of nanosilver, B: Jurkat cells exposed to 27 µg/ml nanosilver for 48 h. Staining: nucleus (blue), actin filaments (green).

4. Summary / Conclusion

When CaCo-2 cells and Jurkat cells were exposed to nanotitanium dioxide or nanosilica dioxide (up to 243 µg/ml for 48 h) no severe effects on cell death could be observed. However, bright field microscopic pictures showed that the nanotitanium dioxide got either adsorbed to the extracellular matrix of the cells and/or incorporated into the cells. This might cause dysfunction after a prolonged exposure period. Severe effects were found, when CaCo-2 cells and Jurkat cells were exposed to nanosilver. When CaCo-2 cells were exposed to 27 µg/ml nanosilver for 48 h different cell parameters, such as cell death, cytokine release and cell morphology were affected. The same was true for Jurkat cells. When Jurkat cells were exposed to nanosilver (81 µg/ml) for 48 h a significant amount of necrotic cells was observed and cells showed a change in cell morphology.
Thus, at the time it is not clear, if the paint industry will replace the biodegradable, organic based biocides by nanoparticles. The efficiency of antimicrobial active nanoparticles in paints is still questionable. Nevertheless first preliminary studies show that the mechanical abrasion of nanoparticle-containing paints should not result in an increase of free nanoparticles in the environment, compared to the sanding of conventional paints without nanoparticles. The small amounts of nanoparticles, which may still be released during surface treatment do not induce any acute toxic response. Therefore we consider a rather low risk, however to support our findings further investigations on chronic or long term effects were needed.

Acknowledgments

The work was financially supported by the Seventh Framework Program of the European Commission (NanoHouse project; Grant Agreement No. 207816).

References

[1] Wangler TP, Zuleeg S, Vonbank R, Bester K, Boller M, Carmeliet J et al 2012 Laboratory scale studies of biocide leaching from facade coatings *Building Environ.* **54** 168-73
[2] Schoknecht U, Gruycheva J, Mathies H, Bergmann H and Burkhardt M 2009 Leaching of biocides used in facade coatings under laboratory test conditions *Environ. Sci. Technol.* **43** 9321-8
[3] Koponen IK, Jensen KA and Schneider T 2011 Comparison of dust released from sanding conventional and nanoparticle-doped wall and wood coatings *J. Expo. Sci. Environ. Epidemiol.* **21** 408-18
[4] Morones JR, Elechiguerra JL, Camacho A, Holt K, Kouri JB, Ramirez JT et al 2005 The bactericidal effect of silver nanoparticles *Nanotech.* **16** 2346-53
[5] Matsumura Y, Yoshikata K, Kunisaki S and Tetsuaki T 2003 Mode of bactericidal action of silver zeolite and its comparison with that of silver nitrate *Appl. Environ. Microbiol.* **69** 4278-81
[6] Pappa A, Franco R, Schoneveld O, Galanis A, Sandaltzopoulos R and Panayiotidis MI 2007 Sulfur-containing compounds in protecting against oxidant-mediated lung diseases *Curr. Med. Chem.* **14** 2590-6
[7] Lai JC, Lai MB, Jandhyam S, Dukhande VV, Bhushan A, Daniels CK et al 2008 Exposure to titanium dioxide and other metallic oxide nanoparticles induces cytotoxicity on human neural cells and fibroblasts *Int. J. Nanomed.* **3** 533-45
[8] Shukla RK, Sharma V, Pandey AK, Singh S, Sultana S and Dhawan A 2011 ROS-mediated genotoxicity induced by titanium dioxide nanoparticles in human epidermal cells *Toxicol. Vitro* **25** 231-41
[9] Saber AT, Jensen KA, Jacobsen NR, Birkedal R, Mikkelsen L, Moller P et al 2012 Inflammatory and genotoxic effects of nanoparticles designed for inclusion in paints and lacquers *Nanotox.* **6** 453-71
[10] Kaegi R, Ulrich A, Sinnet B, Vonbank R, Wichser A, Zuleeg S et al 2008 Synthetic TiO2 nanoparticle emission from exterior facades into the aquatic environment *Environ. Pollut.* **156** 233-9

[11] Locker J and Carstens A 2010 High pressure injection of silica-based paint into a finger *N. Z. Med. J.* **123** 104-6

[12] Kaiser JP, Zuin S and Wick P 2013b Is nanotechnology revolutionizing the paint and lacquer industry? A critical opinion *Sci. Total Environ.* **442** 282-9

[13] Krug HF and Wick P 2011 Nanotoxicology: An Interdisciplinary Challenge *Angew. Chem. Int. Ed. Engl.* **50** 2-21

[14] Kiaune L and Singhasemanon N 2011 Pesticidal copper (I) oxide: environmental fate and aquatic toxicity *Rev. Environ. Contam. Toxicol.* **213** 1-26

[15] Wijnhoven SWP, Peijnenburg WJ, Herberts CA, Hagens WI, Oomen AG, W. Heugens EH et al 2009 Nano-silver: A review of available data and knowledge *Nanotox.* **3** 109-138

[16] Gottschalk F, Sonderer T, Scholz RW and Nowack B 2009 Modeled environmental concentrations of engineered nanomaterials (TiO(2), ZnO, Ag, CNT, Fullerenes) for different regions *Environ. Sci. Technol.* **43** 9216-22

[17] Smulders S, Kaiser JP, Zuin S, Van Landuyt KL, Golanski L, Vanoirbeek J et al 2012 Contamination of nanoparticles by endotoxin: evaluation of different test methods *Part. Fibre Toxicol.* **9** 41

[18] Kaiser JP, Buerki-Thurnherr T and Wick P 2013a Influence of single walled carbon nanotubes at subtoxical concentrations on cell adhesion and other cell parameters of human epithelial cells *J KSU Science* **25** 15-27

[19] Limbach LK, Wick P, Manser P, Grass RN, Bruinink A and Stark WJ 2007 Exposure of engineered nanoparticles to human lung epithelial cells: influence of chemical composition and catalytic activity on oxidative stress *Environ. Sci. Technol.* **41** 4158-63

[20] Hirsch C, Kaiser J-P, Wessling F, Fischer K, Roesslein M and Wick P et al 2011 A novel comprehensive evaluation platform to assess nanoparticle toxicity in vitro *J. Phys. Conf. Series* **304** 1-6

[21] Pujalte I, Passagne I, Brouillaud B, Treguer M, Durand E, Ohayon-Courtes C et al 2011 Cytotoxicity and oxidative stress induced by different metallic nanoparticles on human kidney cells *Part. Fibre Toxicol.* **8** 10

Toxicokinetics of zinc oxide nanoparticles in rats

HE Chung[1], J Yu[1], M Baek[1], JA Lee[1], MS Kim[2], SH Kim[2], EH Maeng[2], JK Lee[3], J Jeong[3] and SJ Choi[1]

[1] Department of Food Science and Technology, Seoul Women's University, 621 Hwarang-ro, Nowon-gu, Seoul 139-774, Republic of Korea
[2] Korea Testing & Research Institute, Health Care Research Lab. 7-6 Gomak-ri, Wolgot-myeon, Gimpo-si, Gyeonggi-do, 415-817, Republic of Korea
[3]Toxiclogical Research Division, National Institute of Food and Drug Safety Evaluation, KFDA, Osong-eup, Chungchungbuk-do 363-700, Republic of Korea

E-mail: sjchoi@swu.ac.kr

Abstract. Zinc oxide (ZnO) nanoparticle have been extensively applied to diverse industrial fields because they possess UV light absorption, catalytic, semi-conducting, and magnetic characteristics as well as antimicrobial property. However, up to date, toxicological effects of ZnO nanoparticles in animal models have not been completely determined. Moreover, little information is available about kinetic behaviors of ZnO nanoparticles in vivo, which will be crucial to predict their potential chronic toxicity after long-term exposure. The aim of this study was, therefore, to evaluate the pharmacokinetics and toxicokinetics of ZnO nanoparticles after single-dose and repeated dose 90-day oral administration in male and female rats, respectively. The blood samples were collected following administration of three different doses (125, 250, and 500 mg/kg) and ZnO concentration was assesed by measuring zinc level with inductively coupled plasma-atomic emission spectroscopy (ICP-AES). The result showed that the plasma ZnO concentration significantly increased in a dose-dependent manner, but decreased within 24 h after single-dose oral administration up to 500 mg/kg, without any significant difference between gender. However, when repeated dose 90-day oral toxicity study was performed, the elevated plasma concentrations did not return to normal control levels in all the cases, indicating their toxicity potential. These findings suggest that repeated oral exposure to ZnO nanoparticles up to the dose of 125 mg/kg could accumulate in the systemic circulation, thereby implying that the NOAEL values could be less than 125 mg/kg via oral intake.

1. Introduction

Zinc is an essential trace element in human body and ZnO has been used as a source of zinc in food industry. Since nano-sized particles have high reactivity related to large surface area compared to bulk-sized ones, ZnO nanoparticles have been extensively applied to many industrial domains such as alloys, ceramics, paints, rubber as well as biological fields including medicine, personal care products, sunscreens, and food additives [1-4]. The wide range application of ZnO nanoparticles is attributed to their unique properties such as UV light absorption, antimicrobial, catalytic, semi-conducting and magnetic characteristics [5-7]. Along with rapid development of nanotechnology in human-related products, the potential toxicity of nanomaterials in human health has been raised concern. Some studies were conducted to evaluate the toxicity of ZnO nanoparticles in cell lines [8-10] and in animal

[1] To whom any correspondence should be addressed.

models [11-12]. However, there is little information about their chronic or sub-chronic toxicity after the long-term exposure. In this study, we investigated the pharmacokinetics and toxicokinetics of ZnO nanoparticles after single-dose and repeated dose oral administration for 90 days to male and female rats, respectively, which will provide critical information about their potential toxicological effects after long-time exposure.

2. Experimental methods

2.1. Preparation of nanoparticle suspension

20 nm-sized ZnO nanoparticles were purchased from Sumitomo Osaka (Tokyo, Japan). For oral administration of ZnO nanoparticles, the nanoparticles were suspended in pH 7 20 mM HEPES buffer containing 1 % sodium citrate, and then dispersed by vortexing for one minute. The final pH of the suspension was adjusted to 7.3, and 20 % of the ZnO nanoparticle suspension were used as a stock solution. The suspension was vortexed for 10 seconds just before administration, and then diluted with distilled water.

2.2. Characterization of nanoparticles

Particle size was determined by transmission electron microscope (TEM: JEM-1010, JEOL, Tokyo, Japan). The surface charge (zeta potential) of thus prepared ZnO nanoparticles was determined using a zeta potentiometer (Zetasizer Nano ZS system, Malvern Instruments, Worcestershire, UK)

2.3. Animals

Five-week-old male and female Sprague-Dawley rats weighing 120-140 g were purchased from G-bio (Seoul, Republic of Korea). Animals were acclimatized as follow for 7 days before receiving experimental treatment; they were housed in plastic laboratory animal cages in a ventilated room. The room was maintained at 20°C±2℃ and 60%±10% relative humidity on a 12-hour light/dark cycle. Water and commercial laboratory complete food for rats were available ad libirum. All animal experiments were performed in compliance with the animal and ethics review committee of the Korea Testing & Research Institute.

2.4. Dosing and sample collection

Three groups of male and female rats (n=9 per group) were daily administered via oral gavage for 90 days with a dose of 125, 250, 500 mg/kg of ZnO nanoparticles. And an additional group of 3 rats received an equivalent volume of citrate/HEPES buffer as controls for all the experiments. Changes in body weight and behaviours including food and water intake, as well as possible appearance of symptoms in the rats were carefully recorded every day after administration of the ZnO nanoparticles. Blood samples were collected via the tail vein at several time points (0, 0.5, 1, 2, 4, 6, 10, 24 hours) on 0, 28, and 90 days after oral administration to evaluate the plasma ZnO nanoparticle concentration. The blood samples were centrifuged at 12,000 rpm for 3 minutes at 4°C to obtain the plasma. Several toxicokinetic parameters such as maximum concentration (C_{max}), time to maximum concentration (T_{max}), area under the plasma concentration-time curve (AUC), half-life (T1/2), and mean residence time (MRT), were estimated using the Kinetica program (version 4.4, Thermo Electron Corpporation, Waltham, MA).

2.5. Sample preparation for ICP-AES analysis

The plasma samples were pre-digested in 5 ml of ultrapure nitric acid, heated at about 180°C, and then, 0.5 mL of H_2O_2 was added. Each sample mixture was heated until completely digested. The remaining solution was then removed by heating until the solutions were colorless and clear.

2.6. ICP-AES analysis

Quantitative analysis was carried out using external five-point calibration with internal standard correction spiking experiments. Stock solutions were diluted with 20% ultrapure nitric acid to designated concentrations. Analysis of standards and biological samples was undertaken using inductively coupled plasma-atomic emission spectroscopy (ICP-AES: Jobin Y von Horiba, JY 2000 Ultrace). The instrument operating conditions were as follows: RF power and nebulizer gas flow were set at 1000 W and 0.02 mL per minute of argon, respectively.

2.7. Statistical analysis
Statistical analysis were performed using SAS software (Tukey's test, version 11.0, Cary, NC) for unpaired data and p values of less than 0.05 were considered significant.

3. Results and discussion

3.1. Characterization of ZnO nanoparticles
Particle size and surface charge of ZnO nanoparticles were measured by TEM and zeta potentiometer, respectively, and presented in Table 1. The ZnO nanoparticles were capped with citrate molecules, widely used as capping agents for inorganic nanoparticles, giving rise to negatively charged surface property. Citrate was also used as a dispersing agent for oral administration. The result showed that the average size of ZnO nanoparticles was determined to be 20±9 nm with surface charge of -28.1 mV.

Table 1. Particle size and surface charge of ZnO nanoparticles

	Measured particle size (nm)	**Zeta potential (mV)**
ZnO dispersed in citrate/HEPES	20 ± 9	-28.1

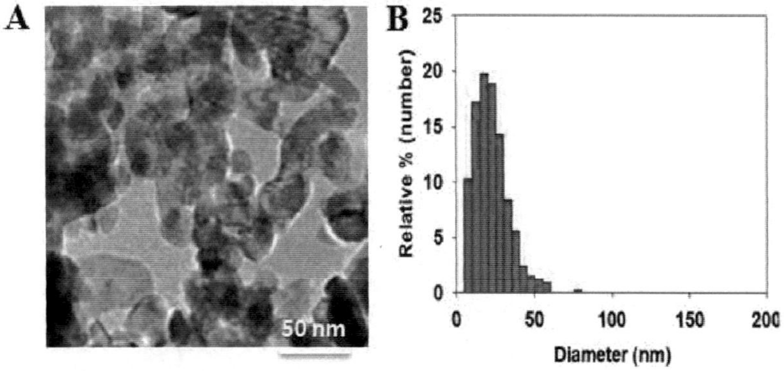

Figure 1. TEM image (A) and size distribution (B) of ZnO nanoparticles dispersed in citrate/HEPES buffer. Average particle size and size distribution were measured by randomly selecting 500 particles from transmission electron micrographs

3.2. Effects of ZnO nanoparticles on body weight changes and symptoms
During oral administration of the ZnO nanoparticles for 90 days, changes in body weight, mortality, behaviors, and appearance of symptoms were observed and recorded daily. No mortality was observed in all the ZnO nanoparticles-treated groups up to the dose of 500 mg/kg. Any statistically significant difference in body weight was not found between the controls and the groups administered ZnO nanoparticles as shown in Fig. 2. However, piloerection and alopecia was observed after 8-day and 13-day oral administration, respectively, both in male and female rats, and these symptoms were continued in all the experimental groups during all the periods of treatment. Salivation was also

observed after 14-day repeated administration in male and female rats when they were treated with 500 mg/kg, and this was found in all treated groups by the end of the experiment. Abnormal symptoms such as soft stool, wound, diarrhea, staining around mouth, spoiled fur, and opacity of eyeball were observed in some groups. It seems that the present nanoparticles up to the dose of 500 mg/kg did not cause acute oral toxicity, but can induce some toxicological effects.

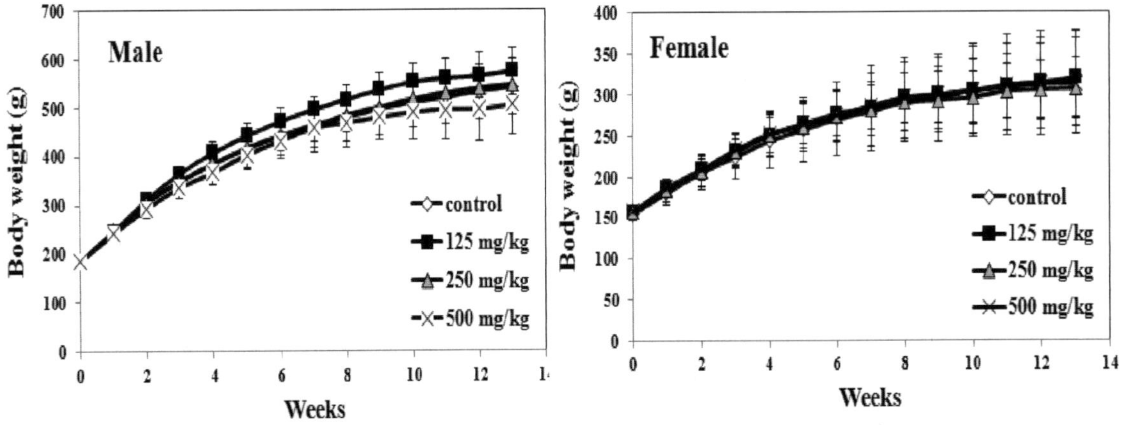

Figure 2. Body weight gain in male and female rats treated with ZnO nanoparticles. The data show no significant difference from the control group ($p < 0.05$).

3.3. Food and water intake in rats

Fig. 3 showed the mean amount of food and water intake in rats administered ZnO nanoparticles. Overall food intake gradually increased both in male and female rats during all the periods of treatment, in comparison with control group. On the other hand, water intake amount significantly increased during the first 1 to 5 weeks of treatment, especially when 500 mg/kg were repeatedly administered. However, increase in food and water intake in a dose-dependent manner was not found. Together with the result on body weight gain (Fig. 2), it is likely that the changes in food and water intake were not resulted from the toxicological effects of ZnO nanoparticles.

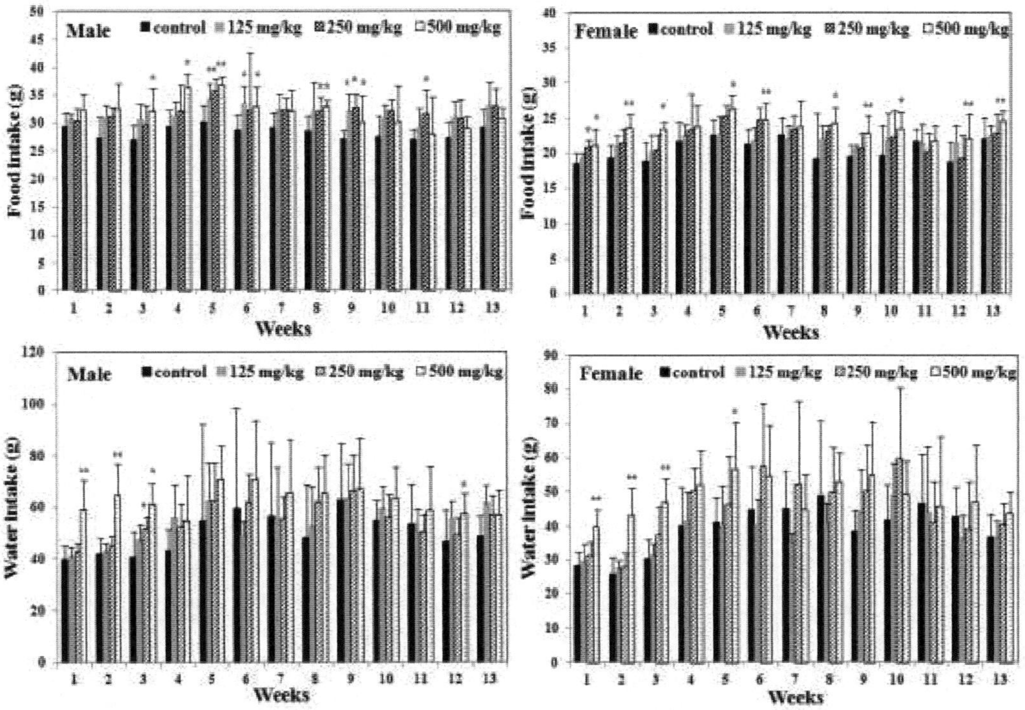

Figure 3. Food and water intake in rats during the administration of ZnO nanoparticles. *denote significant differences from the control group (*$p < 0.05$, **$p < 0.01$).

3.4. Plasma concentration–time curve

ZnO concentration in plasma after oral administration to male or female rats was evaluated by measuring the zinc level using ICP-AES. The plasma concentration-time curve was highly dependent on the dose administered (Fig. 4). After single-dose administration, elevated zinc levels returned to normal levels with 24 h. However, when 125 mg/kg were administered, there was slight increase in zinc level at 90 day compared to 0 day in male and female rats (Fig. 4A). Meanwhile, both 250 and 500 mg/kg doses resulted in statistically increased plasma zinc level after 28 and 90 day-repeated oral administration (Fig. 4B and 4C). This result suggests that the present nanoparticles could be accumulated in the blood circulation after repeated exposure.

Based on this result, the toxicokinetic parameters of ZnO nanoparticles were presented in Table 2. All parameters increased as the doses administered increased. AUC values, representing total amount of the nanoparticles absorbed into the blood circulation seemed to be slightly higher in female rats than those in male rats, in case of 250 and 500 mg/kg. But different significance was not found between genders. It is worthy to note here that $T_{1/2}$ and MRT values, indicators for systemic half-life and mean residence time in the body, respectively, increased gradually as ZnO nanoparticles were repeatedly exposed to rats in all the cases. This is an indication for systemic accumulation of the present nanoparticles. It is strongly probable that 90-day repeated oral administration up to the dose of 125 mg/kg could cause toxicological effects associated with gradually increased ZnO concentration at the systemic level. This may explain some abnormal behaviors or symptoms as observed in section 3.2.

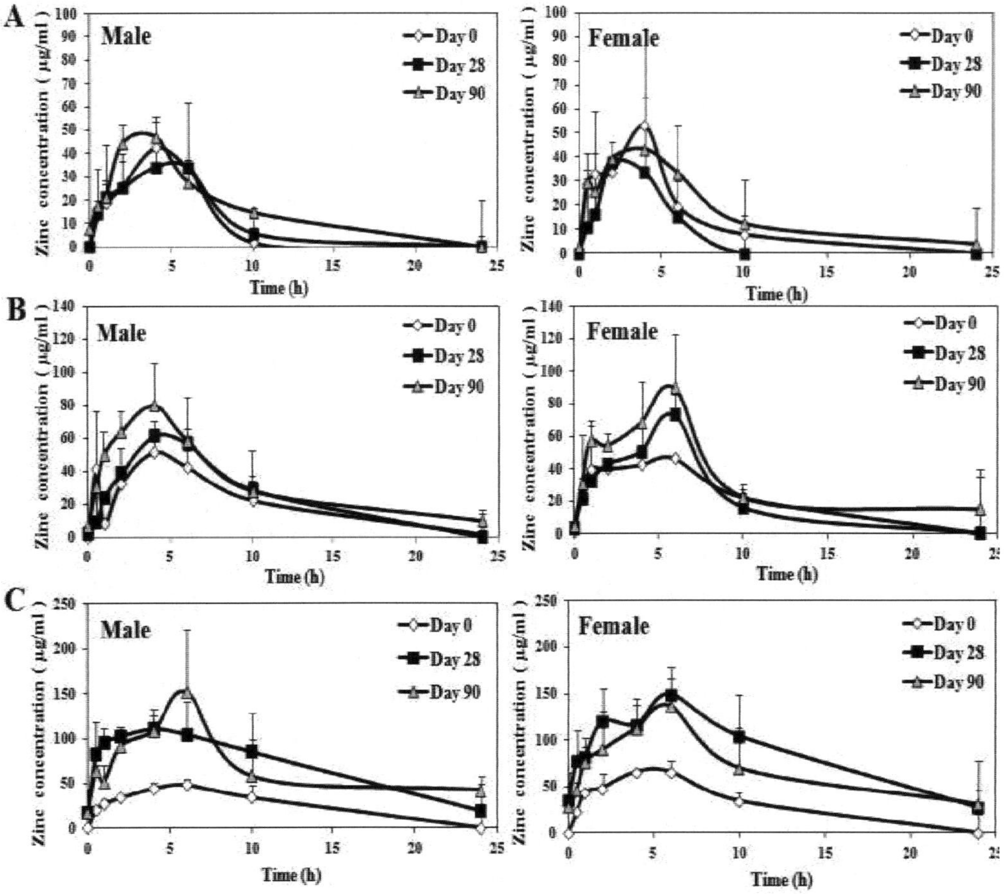

Figure 4. Zinc concentration in plasma after 90 day-repeated oral administration to rats. A, 125 mg/kg; B, 250 mg/kg; C, 500 mg/kg. The data are presented as increased zinc levels after subtraction of the basal zinc level in the control group.

Table 2. Toxicokinetic parameters of 20 nm ZnO nanoparticles in rats.

	125 mg/kg						**250 mg/kg**						**500 mg/kg**					
	Male			**Female**			**Male**			**Female**			**Male**			**Female**		
	0	**28**	**90**	**0**	**28**	**90**	**0**	**28**	**90**	**0**	**28**	**90**	**0**	**28**	**90**	**0**	**28**	**90**
C_{max}	42.53	33.89	46.88	51.10	37.89	43.34	51.10	60.95	79.66	46.18	72.54	89.79	48.40	110.45	151.78	65.82	149.84	137.17
T_{max}	4	4	4	4	2	4	4	4	4	6	6	6	6	4	6	6	6	6
AUC	249.66	296.73	528.58	469.60	216.04	438.66	480.82	647.92	913.23	693.33	789.13	1106.14	788.58	1945.24	2395.69	823.57	2409.64	2205.75
$T_{1/2}$	3.69	5.06	8.81	4.63	5.01	7.12	6.25	5.23	9.45	9.00	9.72	12.51	8.16	10.21	12.74	6.04	9.53	11.61
MRT	5.52	7.65	12.59	8.11	6.13	9.83	8.77	9.54	12.86	13.57	13.81	17.22	13.82	14.39	19.80	10.34	14.43	17.33

Notes: C_{max}, maximum peak concentration (μg/mL); T_{max}, time to peak maximum concentration (h); AUC, area under the plasma concentration-time curve (h×μg/mL); MRT, mean residence time (h); $T_1/2$, half-life time.

4. Conclusion

The present study demonstrated the toxicokinetics of ZnO nanoparticles after single-dose and repeated 90-day oral administration to male and female rats, respectively. The results showed that the present nanoparticles up to the highest dose tested, 500 mg/kg, did not cause mortality and body weight change after 90-day oral exposure. But, some abnormal symptoms such as piloerection and alopecia were observed in the experimental groups. Pharmacokinetic study of ZnO nanoparticles after single-dose oral administration demonstrated that their plasma concentration increased in a dose-dependent manner, but returned to normal levels with 24 h. However, when repeated dose 90-day administration was performed, slightly increased levels of the nanoparticles were found in the groups treated with 125 mg/kg, and moreover, significantly elevated systemic accumulation was evident in rats administed 250 and 500 mg/kg. Significant difference in toxicokinetics between genders was not found. These results suggested that ZnO nanoaprticles could be accumulated up to the dose of 125 mg/kg in the systemic circulation in rats. It is, therefore, concluded that the NOAVEL (no observed adverse effect level) values may be less than 125 mg/kg via oral ingestion.

5. Acknowledgement

This research was supported by a grant (10182KFDA991) from Korea Food & Drug Administration in 2011.

6. References

[1] Djurisic AB and Leung YH 2006 *small* **2** 944-961
[2] Fan Z and Lu JG 2005 *J.Nanosci. Nanotechnol.* **5** 1561
[3] Serpone N, Dondi D and Albini A 2006 *Inorg. Chim. Acta* **360** 794-802
[4] Chen W and Zhang J 2006 *J. Nanosci. Nanotechnol.* **6** 1159-1166
[5] Qian JL 2011 *Appl. Surf Sci.* **180** 308-314
[6] Su YK, Peng SM, Ji LW, Wu CZ, Cheng WB and Liu CH 2009 *Langmuir.* **26** 603-606
[7] Kumari L and Li WZ 2010 *Cryst. Res. Technol.* **45** 311-315
[8] Yang H, Liu C, Yang D, Zhang H and Xi Z 2009 *J. Appl. Toxicol.* **29** 69-78
[9] Xia T, Kovechich M, Liong M, Madler L, Gilbert B, Shi H, Yeh J, Zink J and Nel A 2008 *ACS Nano* **2** 2121-2134
[10] Baek M, Kim MK, Cho HJ, Lee JA, Yu J, Chung HE and Choi SJ 2011 *J. Phys. Conf. Ser.* **304** 012044
[11] Wang B, Feng W, Wang M, Wang T, Gu Y, Zhu M, Oryang H, Shi J, Zhang F and Zhao Y 2008 *J. Nanopart. Res.* **10** 263-276
[12] Sayes CM, Reed KL and Warheit DB 2007 *Toxicol. Sci.* **97** 163-180

Nanosafe 2012: International Conferences on Safe Production and Use of Nanomaterials IOP Publishing
Journal of Physics: Conference Series **429** (2013) 012038 doi:10.1088/1742-6596/429/1/012038

Enhanced cytotoxicity of anticancer drug delivered by novel nanoscale polymeric carrier

R Stoika[1], N Boiko[2], Y Senkiv[3], Y Shlyakhtina[4], R Panchuk[5], N Finiuk[6], Y Filyak[7], R Bilyy[8], Y Kit[9], N Skorohyd[10], O Klyuchivska[11], A Zaichenko[12], N Mitina[13], A Ryabceva[14]

[1] Head of Department, Institute of Cell Biology, Lviv, Ukraine
[2] Scientist, Institute of Cell Biology, Lviv, Ukraine
[3] PhD Student, Ivan Franko Lviv National University, Lviv, Ukraine
[4] MS, Ivan Franko Lviv National University, Lviv, Ukraine
[5] Scientist, Institute of Cell Biology, Lviv, Ukraine
[6] PhD Student, Ivan Franko Lviv National University, Lviv, Ukraine
[7] Scientist, Institute of Cell Biology, Lviv, Ukraine
[8] Senior Scientist, Institute of Cell Biology, Lviv, Ukraine
[9] Head of Laboratory, Institute of Cell Biology, Lviv, Ukraine
[10] Researcher, Institute of Cell Biology, Lviv, Ukraine
[11] Researcher, Institute of Cell Biology, Lviv, Ukraine
[12] Associate Professor, Lviv National Polytechnic University, Lviv, Ukraine
[13] Senior Scientist, Lviv National Polytechnic University, Lviv, Ukraine
[14] PhD Student, Lviv National Polytechnic University, Lviv, Ukraine

E-mail: stoika@cellbiol.lviv.ua

Abstract. We compared *in vitro* action of highly toxic anticancer drug doxorubicin under its delivery to the mammalian tumor cells in free form and after encapsulation in novel bio-functionalized nanoscale polymeric carrier. Such encapsulation was found to enhance significantly drug uptake by the targeted cells, as well as its cytotoxic action. 10 times higher cytotoxicity of the carrier-immobilized doxorubicin comparing to its free form was demonstrated by direct cell counting, and 5 times higher cytotoxicity of encapsulated doxorubicin was shown by FACS analysis. The polymeric carrier itself did not possess significant toxicity *in vitro* or *in vivo* (laboratory mice). The carrier protected against negative side effects of doxorubicin in mice with experimental NK/Ly lymphoma. The life duration of tumor-bearing animals treated with doxorubicin-carrier complex was significantly longer than life duration in animals treated with free doxorubicin. Besides, the effective treatment dose of the carrier-delivered doxorubicin in tumor-bearing mice was 10 times lower than such dose of free doxorubicin. Thus, novel nanoscale polymers possess high potential as drug carrier.

1. Introduction

Rapidly developing production and growing scales of application of nanocomposites in different fields of industry, agriculture, medicine, and common human life also predict potential dangers of these materials for human and environment. In order to improve the biological characteristics of novel nanocomposites, they are frequently covered with a polymeric shell. Such coating was found to improve the bio-compatibility of the nanoparticles containing very toxic anticancer drugs, as well as to enhance the antineoplastic activity of those drugs [1]. There are also other reasons for using the nanocomposites with specific functionalized polymer coating, such as: 1) extremely high toxicity of many drugs for selected normal tissues and organs; 2) low water solubility of many drugs that often requires applying higher doses of these drugs in order to achieve positive treatment effects; 3) natural and acquired drug resistance in many tumor cells; 4) un-capability of many drugs to cross the existing biological barriers in the treated organism, for example blood-brain barrier. Since most anticancer agents cannot discriminate between tumor and normal cells, they can accumulate non-selectively in healthy tissues resulting in severe clinical toxicities. Doxorubicin (Dox) is one of the most effective and widely used anticancer agents whose clinical application is of potential danger because of frequent development of acute or chronic cardiotoxity. Doxil is a PEGylated liposomic nano-container of Dox that is known to lack most of the cardiotoxicity of that drug [2].

In this study, we compared *in vitro* action of anticancer drug Dox when delivered in free form or by the developed nanoscale polymer coated with bio-functionalized shell. The results of bio-evaluation of novel polymeric nanoscale carrier VEP-GMA that was additionally functionalized with PEG and phospholipid in Dox delivery *in vitro* into the mammalian tumor cells are presented. Significant enhancement of the cytotoxic action of Dox *in vitro* and *in vivo* was demonstrated when this anticancer drug was delivered by the developed carrier, while the carrier itself did not possess significant toxicity towards tumor cells *in vitro* or toxic action in the laboratory mice. Similar nanoscale polymer BG-2 was also shown to be non-mutagenic in Ames test.

2. Materials and methods

Synthesis of the nanoscale polymers was performed by radical copolymerization of vinyl pyrrolidone (NVP), 5-(*tert*butylperoxy)-5-methyl-1-hexene-3-yne (VEP) and dimethyl aminoethyl methacrylate (DMAEM) (molar ration = 1:1:1) in the dimethyl formamide solution, as described [3, 4]. A schematic structure of the obtained polymeric compound VEP-GMA is presented in Fig. 1 A.

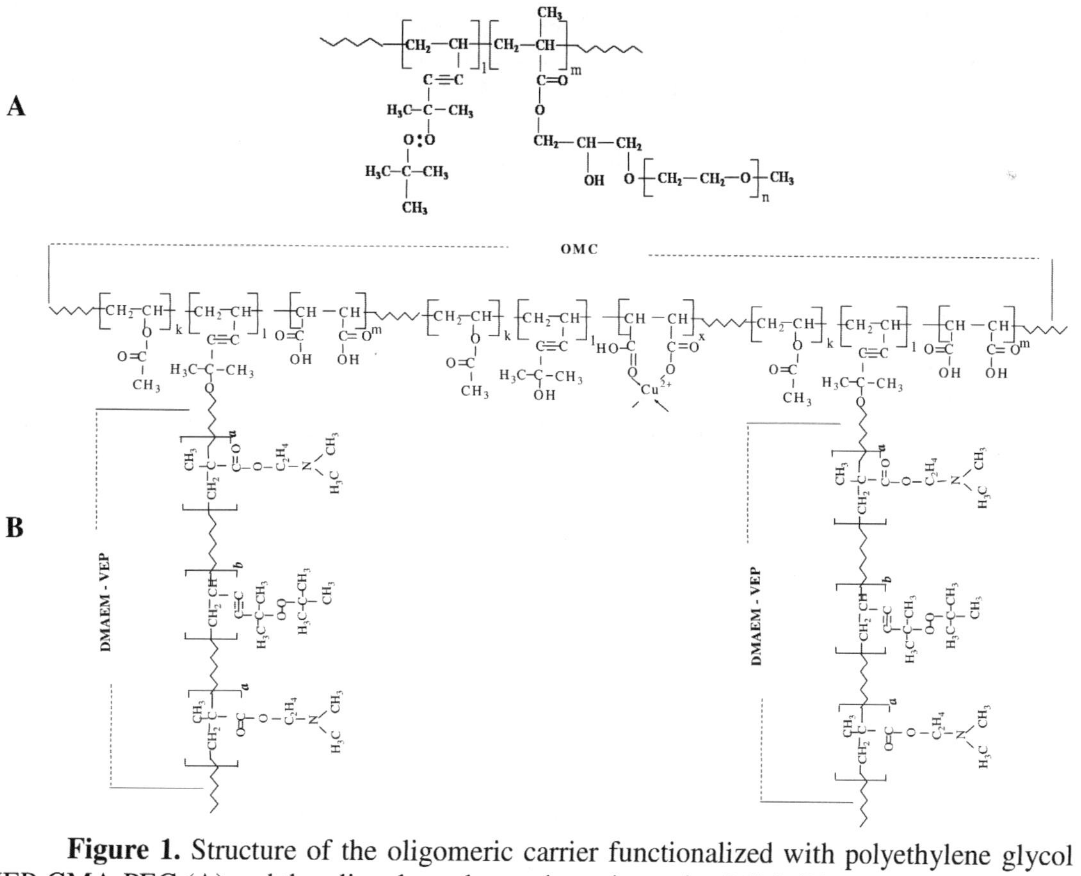

Figure 1. Structure of the oligomeric carrier functionalized with polyethylene glycol VEP-GMA-PEG (A) and the oligoelectrolyte polymeric carrier BG-2 (B).

The size of phospholipid-containing nanoscale polymers VEP-GMA-PEG and BG-2 (see below) was measured by a dynamic light scattering on a Zetasizer Nano ZS (Malvern Instruments GmbH, Stuttgart, Germany) and photon correlation spectra using the NIBS (Non-Invasive Back Scatter) technology at 25°C. Their diameter (dynamic light scattering) was shown to be 58 nm and 19 nm, correspondingly.

For preparing of the nanoscaled PEG- and phosphatidyl choline-containing Dox delivery system (PC-Dox), 100 mg of PC were dissolved in 1.0 ml DMSO. 50 mg of L-α-phosphatidylcholine (PL) from egg yolk (Fluka, Germany) were dissolved in 1.0 ml chloroform. Then chloroform was evaporated to form a thin film of phosphatidylcholine. The polymer solution was added to the phosphatidylcholine film and incubated for 30 min. Notably, only 3.0±0.3 mg of phosphatidylcholine was incorporated into the total amount of PC. 3.0 mg of Dox (Arterium, Ukraine), were dissolved in 0.5 ml DMSO. The resulting solutions of Dox and polymer with phosphatidylcholine were mixed and added drop by drop to 8.5 ml water containing 0.9 % NaCl. Subsequently, the solution was sonicated for 20 seconds. Average diameter of the nanocomposites with immobilized doxorubicin was 28 nm (dynamic light scattering). A hypothetic schematic structure of phospholipid (PL)-containing PC-Dox is presented in Figure 1 A. PC-Dox was analyzed by UV-VIS and fluorescence spectroscopy, as well as surface tension measurements indicating the formation of Dox-containing polymeric system. In addition, drug release measurements showed that the nanoparticles were stable for at least one month without significant drug release.

The novel carrier BG-2 is a copolymer of a comb-like structure that combines an oligoelectrolyte chain of the anionic type as a backbone with 1 to 3 of grafted side chains of the cationic type (Figure 1 B). This oligoelectrolyte polymer was synthesized by controlled radical polymerization initiated by the oligoperoxide metal complex (OMC) in a polar organic media. Synthesis of the comb-like oligoelectrolyte polymer was carried out as follows: a monomer mixture of dimethyl aminoethyl methacrylate (DMAEM) (from 17.13 to 28.18 g, with 28.18 g being optimal) and 5-(tertbutylperoxy)-5-methyl-1-hexen-3-yne (VEP) (from 3.19 to 17.13 g, with 3.19 g being optimal) was injected into round bottom glass reactor equipped with impeller mixer and backflow condenser. Then a solution of OMC (1.66 g) in ethanol or dimethyl formamide (67 g) was added with stirring. The temperature of incubation was increased to 333 K, and the reactor was actively maintained at this temperature for 8 hrs under Argon flow. The monomer conversion was controlled by using a gravimetric technique; after achievement of the desired conversion efficiency (i.e., 60 %), the solvent was evaporated under vacuum. The synthesized comb-like polyelectrolyte was purified by multiple precipitations from acetone solution into hexane, and dried till constant weight under vacuum. The diameter of its particles equaled 19 nm (dynamic light scattering).

Cell culture. The murine leukemia cell model L1210 was purchased from the American Type Culture Collection (Manassas, VA). The cell line was cultured in RPMI-1640 medium (Sigma-Aldrich, Austria). Culture media was supplemented with 10% fetal bovine serum (Sigma-Aldrich, Austria).

In Vivo Studies. Animal experiments were conducted in accordance with the requirements of the European Convention on protection of vertebrate animals (November 13, 1987) and The Law of Ukraine "On Protection of Animals From Cruel Behavior" (March 28, 2006). BALB/c mice were obtained from R.E. Kavetsky Institute of Experimental Pathology, Oncology and Radiobiology, NAS of Ukraine (Kyiv, Ukraine). Animals were kept in a controlled environment in the conventional cages. NK/Ly lymphoma in BALB/c mice (8-10-

week old female, 20–22 g) was used as an experimental tumor model. Animals were injected intraperitoneally with 20-25 mln NK/Ly lymphoma cells, and chemotherapy treatment was performed by the intraperitoneal injection of doxorubicin or its complex with the polymeric carrier (total volume did not exceed 0.2 ml after drug dissolving in saline). Injections were conducted every day for 8 days, starting from the 2nd day after tumor cell inoculation. Treatment doses are noted in Figure 3. In control, mice were injected with saline using the same treatment scheme. Tumor growth was monitored by animal weighting.

Cytotoxicity Assay. Mammalian cells used in the study were plated (2x10^3 cells in 100 µl/well) in 96-well plates (IWAKI) and allowed to recover for 24 h. Drugs were added in another 100 µl growth medium and cells exposed for the indicated time periods. After 24 h or 48 h h drug treatment, the proportion of viable cells was determined by MTT assay following the manufacturer's recommendations (EZ4U, Biomedica, Vienna, Austria).

Ames testing for mutagenicity was conducted for the BG-2 nanopolymer. TA 98 and TA 100 strains of *Salmonella typhimurium* bacteria were tested in the presence of this polymer with or without the microsomal activation by S9 rat liver homogenate. The ratio of numbers of the revertants produced per plate in the presence of various concentrations of BG-2 to control was measured. Positive controls were benzidine (100 µg/plate), nitrosoguanidine (1 mg/plate) in the presence of the metabolic activation and sodium azide (1.5 µg/plate) in the absence of S9 [5].

Western blot analysis. After 24 h exposure of cell to studied substances, cellular proteins were isolated, resolved by electrophoresis in SDS-PAAG, and transferred onto a polyvinylidenedifluoride membrane for Western blotting, as described [6]. The following antibodies were used in this study at a 1:1,000 dilution: anti-cleaved PARP and anti-cleaved caspase 3 (both polyclonal rabbit and from Cell Signaling Technology, Beverly, MA). Equal loading of each lane was evaluated by the immunoblotting of the same membrane with anti-β-actin monoclonal mouse AC-15 (Sigma). Secondary peroxidase-labelled antibodies were from Pierce and used at working dilution of 1:10,000.

Statistical analysis. The results were expressed as means ± standard deviation and significant difference at $p<0.05$.

3. Results and Discussion

In this study, two versions (VEP-GMA-PEG and BG-2) of novel polymeric nanoscale carrier were studied, and VEP-GMA-PEG was used for the delivery of anticancer drug doxorubicin (Dox). It was found that such encapsulation allows achieving cytotoxic effect (24 hrs) of this drug towards the mammalian tumor cells *in vitro* in 10 times less concentration than for achieving such effect at using free drug - 0.1 µg/ml of the encapsulated Dox versus 1 µg/ml of free Dox (Fig. 2 A). In 48 hrs, the inhibitory effect of such drug complexes towards L1210 leukemia cells was even more pronounced (Fig. 2 B). The encapsulated drug was much faster up-taken by the targeted cells, accumulated in their nucleus, and its cytotoxic effects were accompanied by typical pro-apoptotic changes. There were no significant effects on cell survival and proliferation caused by the developed polymeric drug-free carriers *in vitro*. No visible negative side effects were found in BALB/c mice injected with the drug-free nanopolymers (data not presented).

The results of the bacterial reversion assay (Ames test) with using two concentrations (0.1% and 0.01%) of BG-2 oligoelectrolyte did not demonstrate its mutagenicity. According to the EPA and GenPharmTox guidelines, the mutagenic potential of Ames test item is considered to be significant when the mutant frequency (experiment versus control) equals or is above 2.0 [5]. It is probable if the quotient ranges from 1.7 to 1.9, while the lack of mutagenicity is assumed if the quotients range from 1.0 (and lower) to 1.6. It was found that almost all quotients were below 1.0, comparing to positive controls: benzidine = 2.54, nitrozoguanidine = 2.69, and sodium azide = 2.14 - all exceeding the critical value of 2.0. Therefore, no mutagenic activity was observed in any of BG-2 type oligoelectrolyte samples. Although an increase in induction factor can be seen in presence of S9 homogenate (+S9), there was no statistical significance in genotoxic potentials of any of the samples comparing to the control value.

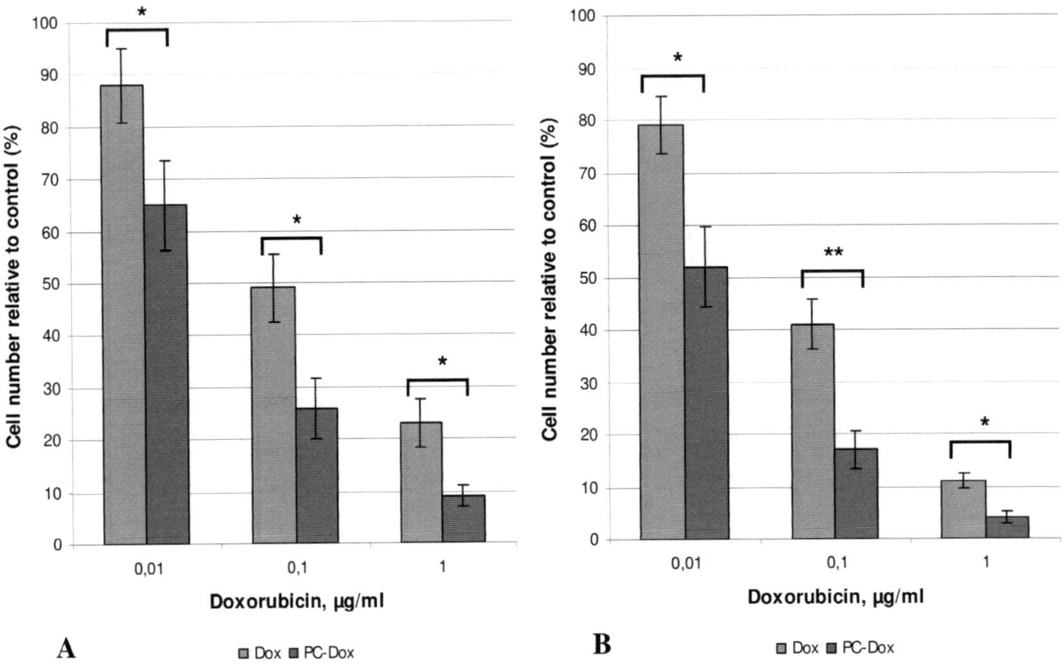

Figure 2. Treatment of murine leukemia cells of L1210 line with free doxorubicin (Dox) and doxorubicin-carrier complexes (PC-Dox) for 24 (A) and 48 (B) hrs (the ordinate notes a decrease in cell number relative to cell number in the control, 100%), n=20. (*P<0.05, **P<0.01).

Higher effectiveness of the antiproliferative and cytotoxic action the PC-Dox comparing with such effects of free Dox was confirmed in other mammalian tumor cell lines.

Fig. 3 presents the results of studying the effect of application of novel polymeric carrier for doxorubicin delivery at treatment of NK/Ly lymphoma in BALB/c mice. Treatment for 8 days with free doxorubicin in daily dose of 0.1 mg per gram of animal weight, led to only a transient stop of tumor growth and prolongation of survival of tumor-bearing mice from medially 21 to 39 days. At the delivery of this drug by the developed carrier, all tumor-bearing mice survived over 60 day observation period, and the treated animals were classified as “cured”. Thus, application of novel polymeric carrier allows increasing the effectiveness of doxorubicin delivery and action towards murine lymphoma, comparing with the action of free form of this drug.

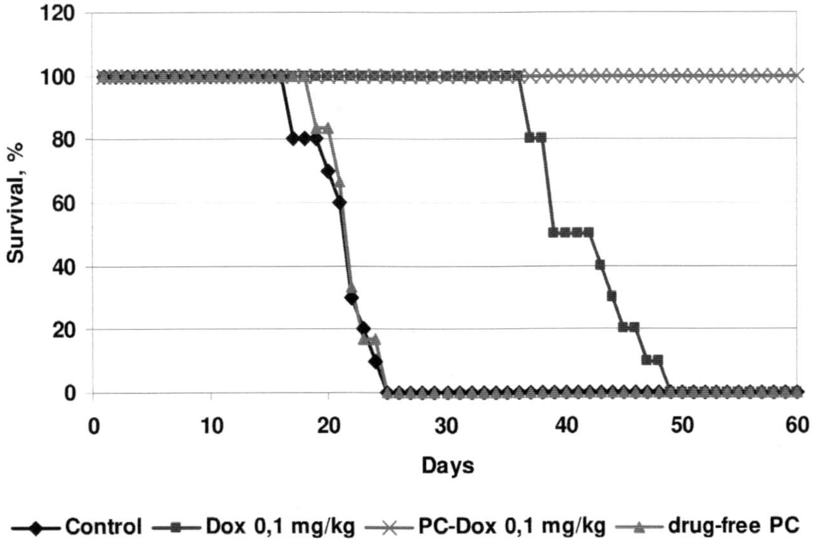

Figure 3. Survival of NK/Ly lymphoma-bearing mice depending on the treatment course. Dox – doxorubicin, PC – polymeric carrier.

Doxorubicin is known to intercalate DNA causing inhibition of its biosynthesis and it also inhibits topoisomerase II activity. That leads to cell stop in G2/M phase of cell cycle [7, 8]. Taking into consideration the above described ability of the nanocarrier to enhance doxorubicin's cytotoxicity, we have compared the impact of free Dox and its complex with nanocarrier on cell cycling in murine L1210 leukemia cells. As one can see (Fig. 4, 5), Dox significantly and dose-dependently affects cell cycling, namely induces L1210 cells' arrest in G2/M phase (from 15.7% in control to 26.8% at 0.05 μg/ml Dox dose). 5 times higher dose (0.25 μg/ml) of this drug caused a complete blocking of cell cycle progression in L1210 cells, resulting in 81.95% cells arrested in G2/M phase (Fig. 4, 5). Dox immobilization on the nanocarrier led to even more significant changes in cell cycling – we observed almost complete inhibition of cell cycle progression (83.56% cells were arrested in G2/M phase) at 0.1 μM of Dox that was immobilized on the nanocarrier. At higher concentrations of Dox (0.25 μg/ml) encapsulated in the nanocarrier, such effect was preserved, although the ratio of G2/M-stopped cells was lower (56%). Such decrease in this ration can be explained by stronger induction of apoptosis by the Dox/PC complex which leads to changes in the pattern of cell cycle phases in treated cells. Thus, the enhancement of cytotoxic action of Dox by its encapsulation in novel polymeric carrier has been confirmed by the results of FACS analysis in treated murine leukemia L1210 cells.

Journal of Physics: Conference Series **429** (2013) 012038 doi:10.1088/1742-6596/429/1/012038

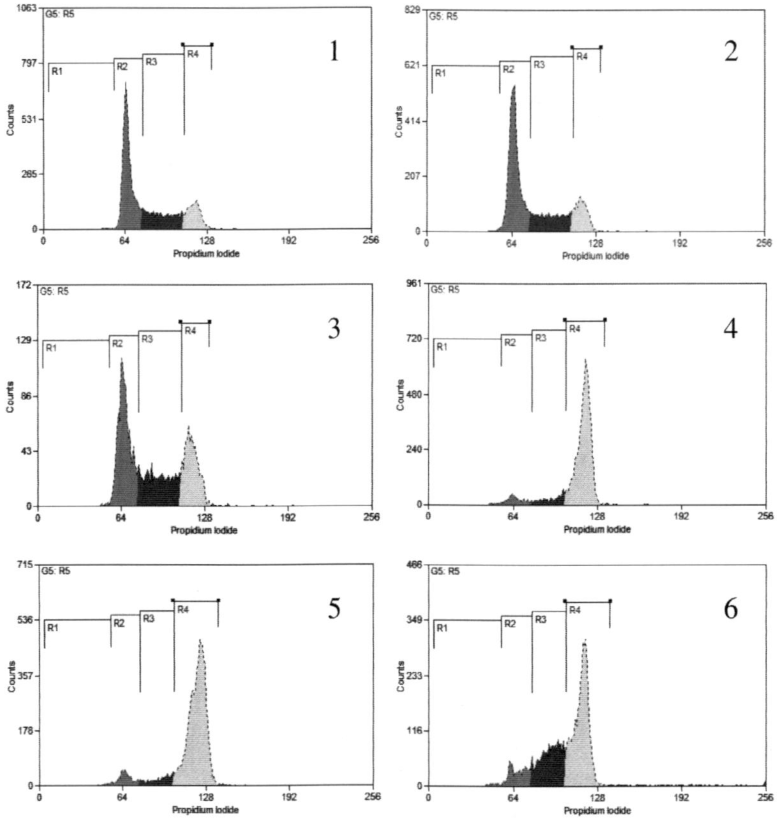

Figure 4. Comparison of impact of free doxorubicin and doxorubicin/carrier complex on cell cycling in murine leukemia L1210 cells (24 h treatment). 1 – control (C); 2 – polymeric carrier (PC); 3 – Dx 0.05 μg/ml; 4 – Dx 0.25 μg/ml; 5 – Dx (0.05 μg/ml)+NC; 6 – Dx (0.25 μg/ml)+NC.

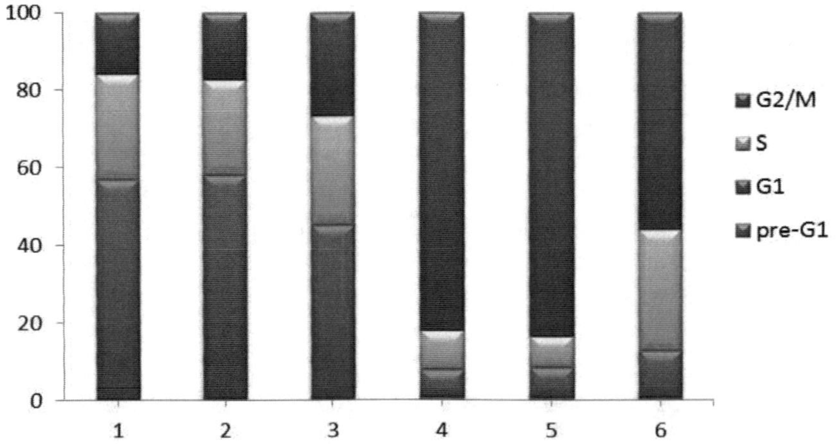

Figure 5. Quantification of the impact of free doxorubicin and doxorubicin/carrier complex on cell cycling in murine leukemia L1210 cells (24 h treatment). 1 – control (C); 2 – polymeric carrier (PC); 3 – Dx 0.05 μg/ml; 4 – Dx 0.25 μg/ml; 5 – Dx (0.05 μg/ml)+NC; 6 – Dx (0.25 μg/ml)+NC.

For studying the mechanisms of enhanced efficiency of the applied drug delivery system at the molecular l evel, Western-blot analysis of the expression of proapoptotic proteins – caspase-3 and PARP-1 (DNA reparation enzyme) was carried out. Murine leukemia L1210 cells were treated for 6 and 12 hrs with the Dox-carrier or free Dox used in 0.1 μg/ml and 1 μg/ml concentrations of Dox. The results of Western-blot analysis of the effector caspase-3 and its protein substrate PARP-1 are shown in the Fig. 6. We have found that the PC-Dox treatment of L1210 cells for 6 h caused activation of the effector caspase-3 and cleavage of PARP-1. Free Dox activated these processes much later – only in 12 hrs of the drug treatment.

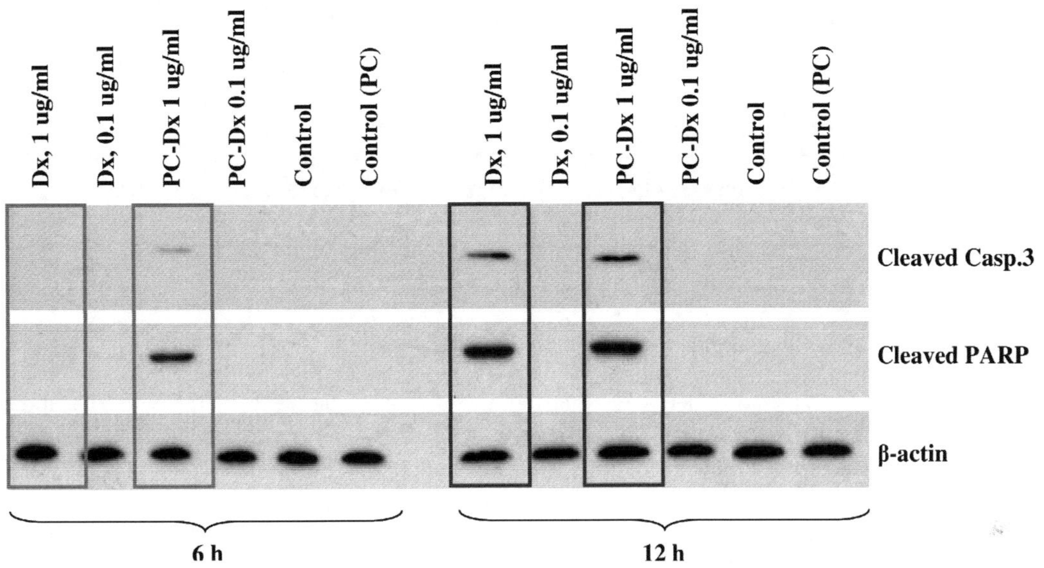

Figure 6. Western-blot analysis of expression of the effector caspase-3 and its protein substrate PARP-1 in murine leukemia L1210 cells treated for 6 h and 12 h with free doxorubicin (Dx) or its complex with the polymeric carrier (PC).

4. Conclusions. Encapsulation of the anticancer drug doxorubicin in novel polymeric nanocomposite enhances delivery of this drug into the targeted mammalian cells, as well as its cytotoxic action towards tumor cells. It also protects of negative side effects of this drug in mice organism. The developed polymeric carriers do not possess their own cytotoxicity *in vitro* and *in vivo.*

References

[1] Ljubimova J Y and Holler E. 2012 *Nanomedicine (Lond)* **7** 467
[2] Plosker G L. 2008 *Drugs* **68** 2535
[3] Riabtseva A *et al* 2012 *Chem. Chem. Technol.* **6** 292
[4] Zaichenko A *et al* 2010 *AIP Conf. Proc.* **1275** 178
[5] Mortelmans K and Zeiger E 2000 *Mutat. Res.* **455** 29
[6] Heffeter P *et al* 2009 *Curr. Cancer Drug Targets* **9** 595
[7] Minotti G *et al* 2004. *Pharmacol Rev.* **56** 185
[8] Park E J *et al* 2012 *PLoS One.* **7** e44990.

Role of water temperature in the fate and transport of zinc oxide nanoparticles in aquatic environment

Seyed Mohammad Majedi[1], Hian Kee Lee[1,2] and Barry C. Kelly[3]

[1]Department of Chemistry, National University of Singapore, 3 Science Drive 3, Singapore 117543, Singapore

[2]National University of Singapore Environmental Research Institute, T-Lab Building #02-01, 5A Engineering Drive 1, Singapore 117411, Singapore

[3]Department of Civil and Environmental Engineering, National University of Singapore, 1 Engineering Drive 2, Singapore 117576, Singapore

Corresponding author's e-mail: chmleehk@nus.edu.sg

Abstract. The influence of water temperature on the aggregation and dissolution kinetics of zinc oxide nanoparticles (ZnO NPs) (mean diameter ~40 nm) was investigated. Samples of 100 mg.L^{-1} ZnO NPs were incubated at 15, 25 and 35 $^{\circ}$C, similar to the surface temperature of cold freshwater, temperate estuarine, and tropical/sub-tropical coastal marine ecosystems, respectively. The natural organic matter (NOM) content, pH, electrolyte type, and ionic strength (IS), were adjusted on the basis of the water chemistries of typical aqueous systems. Specifically, the time-dependent hydrodynamic diameters (HDDs) and sedimentation plots were obtained over the first 3 h and after 24 h using time-resolved dynamic light scattering (TR-DLS) and UV-visible spectroscopy, respectively. The settling distance was further modeled for the aggregates with various HDDs according to the Stokes' sedimentation equation. The dissolution kinetics was studied over the first 12 h and after 48 h in term of percentage of released zinc ions.

The results showed that the HDD increased at elevated temperatures, termed as temperature-induced aggregation, while dissolution was reduced. The aggregation at higher temperatures further hindered the dissolution due to the decrease in the surface area of the NPs. We express this process as "aggregate-suppressed dissolution". The maximum aggregation was reached in the tropical coastal marine environment with the HDD >3 µm, and the released zinc ion of 9.2% was obtained in the cold freshwater. Based on the results, the aggregation rate of 1.57 nm.s^{-1} was estimated for the former, and the dissolution rate of 7.44×10^{-5} mol.L^{-1}.h^{-1} was calculated for the latter, respectively. The predicted values successfully fitted to genuine water samples (<26% deviation). This study provides useful data for environmental risk assessment of exposure of ZnO NPs to water column and benthic organisms.

1. Introduction

Nanosized materials are extensively used in industry and consumer products. They have been reported to undergo transformation in the environment [1]. Several studies have focused on fate and transport of these materials in the environment. Zinc oxide nanoparticles (ZnO NPs) have found several applications among engineered nanoparticles (ENPs). Since environmental occurrence and bioavailability as well as toxicity of ZnO NPs are very high [2,3], their behavior is worthy of attention. Among a number of environmental factors, temperature plays a key role in all physicochemical processes. It may vary significantly with climate geographies from extremely cold poles to very hot deserts.

In aqueous media, microscopic and macroscopic effects of temperature are of utmost importance. Temperature can change physical, chemical, electrical, hydrological and mechanical properties of fluids such as viscosity, density, dielectric constant, refractive index, diffusion coefficient, hydrolysis, etc with change in intermolecular forces such as van der Waals [4]. It also influences colloidal properties and stability of particles. Temperature has been reported to influence aggregation [5], attachment efficiency [6], dissolution [7], and adsorption of ENPs [8].

The current work reports the role of water temperature in the behavior of ZnO NPs under various simulated conditions that are representative of freshwater, estuarine water, and seawater. Temperature levels are set to indicate cold (15 °C), temperate (25 °C), and tropical (35 °C) surface waters. The settling distance is further calculated to predict the extent of exposure of water column and benthic organisms. Finally, the predicted values of synthetic samples are compared with genuine river and coastal marine samples. This investigation can advance our understanding about fate and transport of ENPs in the nature regarding climate conditions.

2. Experimental

2.1. Chemicals

ZnO nanopowder (zincite, particle size: 24 – 71 nm, specific surface area: 15 – 45 m^2/g) was supplied by Alfa Aesar (Ward Hill, MA, U.S.A.). Analytical grade sodium chloride, calcium sulfate, sodium hydroxide, and humic acid (sodium salt), were bought from Sigma-Aldrich (St. Louis, MO, U.S.A.). Nanopure water (18.2 MΩ) was produced with a PURELAB Option-Q water purification system (ELGA LabWater, Marlow, U.K.). All solutions were filtered through 0.2 µm syringe filters (Whatman, Sanford, ME, U.S.A.). The exact humic acid concentration as well as natural organic matter (NOM) concentration were measured with UV–visible spectroscopy at λ = 254 nm using calibration curve of pure humic acid [9]. ZnO nanopowder was dispersed (100 $mg.L^{-1}$) in solutions containing humic acid and salts, followed by pH adjustment with diluted NaOH. Table 1 shows specifications of synthetic freshwater (SFW), estuarine water (SEW), and seawater (SSW), samples. Samples were incubated in the dark, or in amber glass vessels, for 24 and 48 h. Genuine water samples were collected from river and coastal marine in Singapore.

Table 1. Specifications of synthetic water samples

Type	pH	NOM (mg $C.L^{-1}$)	NaCl (mM)	$CaSO_4$ (mM)
SFW	7.5	30	10	-
SEW	8.0	10	50	0.5
SSW	9.0	5	200	1

2.2. Methods

Time resolved dynamic light scattering (TR-DLS) was conducted using Model BIC 90Plus DLS analyzer (Brookhaven Instruments, Holtsville, NY, U.S.A.). Sedimentation plots were obtained by measuring the optical absorbance at λ = 368 nm and recording the change in initial absorbance as a function of time with Model UV-2450 UV–visible spectrophotometer (Shimadzu, Kyoto, Japan). The analyzers were equipped with temperature control accessories for TR analyses. To determine release of zinc ions, samples were first centrifuged at 14000 rpm for 30 min, and filtered through 0.02 μm syringe filters [2,7]. The supernatant was analyzed by inductively coupled plasma optical emission spectroscopy (ICP-OES, Optima 7300 DV, Perkin-Elmer, Waltham, MA, U.S.A.). All experiments were performed in triplicate.

3. Results and discussion

3.1. Aggregation and sedimentation

Synthetic water samples were incubated at different temperatures. Figure 1a shows the aggregation rates of these samples, estimated from linear regression analysis of time-dependent hydrodynamic diameter (HDD) curves within the first 60 min (20 min for the SSW sample). As shown, aggregation rate increased with increasing temperature (temperature-induced aggregation). However, this change was not significant for all samples at the 95% confidence level, based on one-way analysis of variance (ANOVA). Since Brownian motion and collision frequency of the NPs increase at elevated temperatures, the interaction energy barrier between NPs would likely be overcome, and the electric double layer (EDL) compressed, thereby attaching NPs to one another [10]. It is worthy to note that viscosity, density, and dielectric constant of water as well as chemical potential, activity and ion mobility of electrolyte also undergo changes with variation of water temperature [4]. The increased van der Waals attractive forces and reduced EDL repulsive forces at higher temperatures as a result of alteration in physical properties of water, electrolytes, and ZnO NPs, can further be estimated using classical/extended Derjaguin–Landau–Verwey–Overbeek (DLVO/EDLVO) theory. It describes the enhanced propensity of NP attachment.

Figure 1b demonstrates influence of temperature on HDD of synthetic samples after 24 h equilibration. ZnO NPs in SSW sample form aggregates with HDD >3 μm, while other samples were effectively stabilized with NOM. The ANOVA results showed that the ZnO NP aggregation was greatly influenced by temperature in the SFW and SEW samples (significant at $p < 0.025$ and $p < 0.01$, respectively), whereas it did not exhibit significant variation in the SSW sample ($p > 0.05$).

Figure 1c illustrates the aggregation kinetics of these ZnO NP suspensions at 35 °C. ZnO NPs were stabilized in the presence of NOM ≥ 10 mg $C.L^{-1}$, and addition of 50 mM NaCl and 0.5 mM $CaSO_4$ to SEW sample did not markedly enhance the NP aggregation. In contrast, the presence of 200 mM NaCl and 1 mM $CaSO_4$ significantly increased aggregation kinetics in the SSW sample within the first hour of dispersion of ZnO NPs. Critical coagulation concentration (CCC) should be reached at these levels of electrolytes, thereby exerting the highest aggregation rate [10,11].

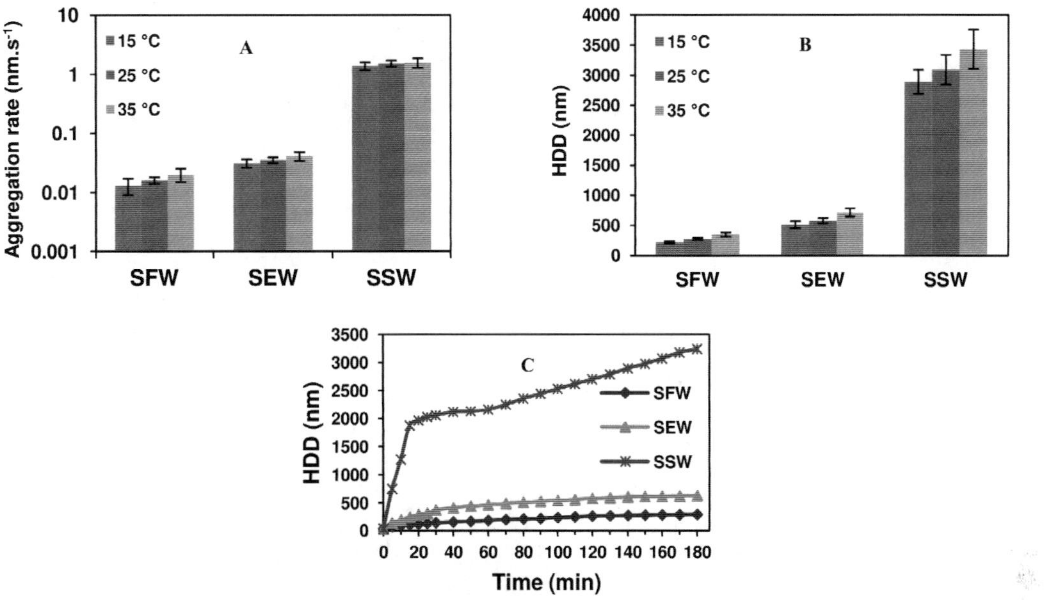

Figure 1. Effects of temperature on aggregation rate (A), and hydrodynamic diameter after 24 h (B), of ZnO NPs in synthetic waters. Aggregation kinetics of ZnO NPs in synthetic waters at 35 $^{\circ}$C (C).

Concomitant with the formation of NP aggregates, they would settle in the water column and increase the risk of exposure to benthic organisms in aquatic systems. They may then undergo several transformations in the sediment [1]. Estimation of settling velocity will enable us to predict NPs exposure time to water column organisms. We further conducted sedimentation experiments and modeled settling distance according to Stokes' sedimentation equation [12]. From aggregation kinetics of synthetic samples (Figure 1c), it was found that the most unstable ZnO NP suspension was in the SSW sample. Hence, the effect of temperature was investigated on this sample.

Figure 2a illustrates sedimentation plots of SSW. While increase in temperature may enhance Brownian motion thereby hindering sedimentation of NPs [6,12], fast formation of large aggregates resulted in efficient removal of the ZnO NPs from the water column, thus decreasing the optical absorbance. Figure 2b shows the calculated Stokes' settling distance of SSW. As expected, longer distance was observed at higher temperature (e.g. 261 mm at 35 $^{\circ}$C). The results were consistent with sharp decrease in absorbance within the initial 20 min due to sedimentation of NPs in the spectrophotometer cuvette (height: 30 mm). The estimated settling distances were 25.6, 31.1, and 33.1 mm, at 15, 25 and 35 $^{\circ}$C, respectively. The settling distance was modeled for different synthetic water samples at 35 $^{\circ}$C and is illustrated in Figure 2c. Stable ZnO NP suspensions (SFW and SEW) were transported 2.8 and 21.4 mm in the water column after 3 h. This further confirmed that NOM >10 mg.L^{-1} effectively stabilized ZnO NPs in water.

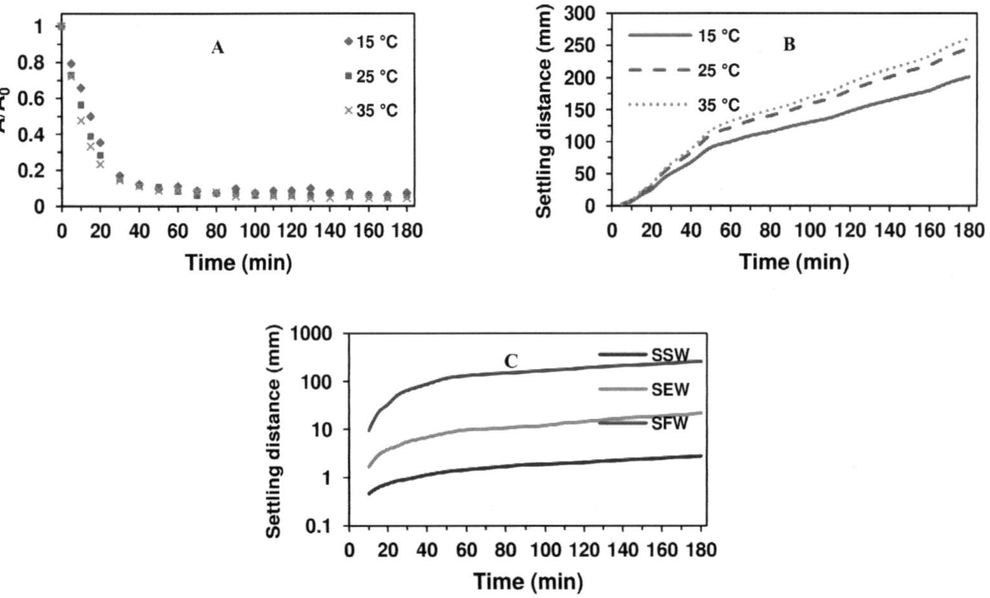

Figure 2. Sedimentation plots (A), and Stokes' settling distance (B), of ZnO NPs in SSW at different temperatures. Stokes' settling distance of ZnO NPs in different synthetic waters at 35 $^{\circ}$C (C).

3.2. Dissolution

The effect of NOM, electrolyte type and concentration as well as incubation temperature were further investigated on dissolution kinetics of ZnO NPs. The high bioavailability and toxicity of ZnO NPs to aquatic organisms have been shown [2,3]. We further simulated real environmental conditions with respect to various temperatures (cold, temperate, and tropical) and water chemistries. Figure 4a depicts the dissolution rate of different synthetic samples within the initial 12 h under various incubation temperature conditions. ZnO has negative enthalpy of dissolution [13]. Hence, increasing temperature would lead to a reduction of its dissolution [7]. The environmental implication is that, if ZnO enters the organism in the form of NPs, body temperature (~37 $^{\circ}$C) will induce aggregation and reduce dissolution, and therefore, decrease bioavailability and toxicity of ZnO NPs. However, since ZnO NP is readily hydrolyzed in water [14,15], it would be less likely to accumulate in form of NPs in the organism at low environmental concentration levels.

As can be seen in Figure 3a, a decrease in dissolution rate was observed for SEW samples. The dissolution rates were 6.36×10^{-5} and 5.15×10^{-5} mol.L^{-1}.h^{-1} at 15 $^{\circ}$C and 35 $^{\circ}$C, respectively. The maximum dissolution rate was given for SFW at 15 $^{\circ}$C (7.44×10^{-5} mol.L^{-1}.h^{-1}). This proved that a high concentration level of NOM (ca. 30 mg C.L^{-1}) and low aggregation rate will enhance the dissolution. The enhanced hydrolysis of the NP surface might be attributed to the surface adsorption of humic acid and weakening the surface oxygen bond as well as providing large surface area due to the formation of small aggregates in the SFW medium [14].

Figure 3b demonstrates total percentage of released zinc ions of the synthetic samples under various incubation temperatures after 48 h equilibration. As expected, the lowest dissolution was observed for SSW while the highest was for SFW at 15 $^{\circ}$C (9.2%). Electrolytes can enhance dissolution of particles according to the Debye–Hückel law [16]. While it was observed for SEW, very high aggregation kinetics of SSW suppressed dissolution of ZnO NPs in SSW. Figure 3c illustrates the dissolution kinetics of synthetic samples. This suppression is observed when dissolution kinetics of SSW is compared with its aggregation kinetics. We term this process as "aggregate-suppressed dissolution".

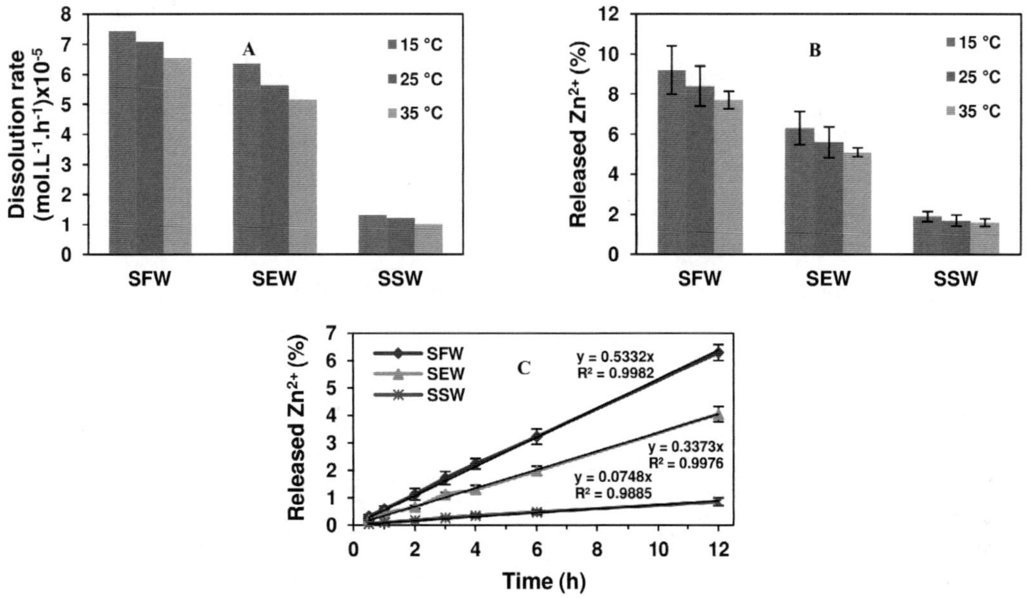

Figure 3. Effects of temperature on dissolution rate (A), and released Zn^{2+} after 48 h (B), of ZnO NPs in synthetic waters. Dissolution kinetics of ZnO NPs in synthetic water samples at 35 °C (C).

3.3. Genuine water samples

To validate anticipation of aggregation and dissolution kinetics of real environmental waters, the results of simulated conditions were compared with those of genuine water samples. ZnO NPs (100 $mg.L^{-1}$) were dispersed in genuine samples and incubated at different temperatures. The specifications of these samples are tabulated in Table 2. The characteristics of genuine river water (RW) fitted well with those of the SEW sample. The zinc contents of river and seawater samples were measured prior to dissolution experiments.

Table 2. Characteristics of genuine water samples

Type	pH	NOM (mg $C.L^{-1}$)	Salinity (ppt)	Zn^{2+} ($\mu g.L^{-1}$)
River	7.6	10.3	6.3	13
Coastal marine	8.1	4.2	25.8	7

Tables 3 and 4 demonstrate the comparison of the predicted aggregation, and dissolution rates under simulated conditions, respectively. Percentages of deviation were in the range 10.8 – 25.5% and 3.8 – 20.5% for aggregation rate and dissolution rate, respectively. The highest deviation of predicted rates for both the aggregation and dissolution of ZnO NPs were observed for genuine RW, possibly due to the effects of various NOMs and electrolytes. Our results showed that the simulated conditions predicted the fate of ZnO NPs very well in the real aquatic environment.

Table 3. Comparison of aggregation rate (nm.s^{-1}) of ZnO NPs in simulated synthetic with genuine samples

Temperature	dr/dt(SEW)	dr/dt (genuine RW)	deviation (%)	dr/dt (SSW)	dr/dt (genuine SW)	deviation (%)
15 °C	0.031	0.038	18.4	1.38	1.62	14.8
25 °C	0.035	0.047	25.5	1.52	1.73	12.1
35 °C	0.041	0.05	18.0	1.57	1.76	10.8

Table 4. Comparison of dissolution rate (× 10^{-5} mol.L^{-1}.h^{-1}) of ZnO NPs in simulated synthetic with genuine samples

Temperature	dC/dt (SEW)	dC/dt (genuine RW)	deviation (%)	dC/dt (SSW)	dC/dt (genuine SW)	deviation (%)
15 °C	6.36	7.11	10.4	1.32	1.19	10.9
25 °C	5.64	6.73	16.2	1.22	1.14	7.0
35 °C	5.15	6.48	20.5	1.01	1.05	3.8

4. Conclusion

ZnO NPs are some of the most widely used ENPs and released, when their fate depends on their physicochemical interactions in the environment. These in turn determine the toxicity of these NPs to a variety of organisms exposed to them. Our study provided some insight into the role of water temperature in the behavior of ZnO NPs and could be extended to other ENPs. The influence of temperature was determined in terms of aggregation, sedimentation, and dissolution of ZnO NPs. While it is costly and labor- and instrument-intensive to investigate the fate of ENPs in various aqueous systems and regions (cold, temperate, and tropical), our results clearly showed that the examination of ENP behavior under simulated conditions allows cost- and time-effective anticipation of their behavior in nature to be made.

Acknowledgement
This work was financially supported by the National Environment Agency, Ministry of the Environment and Water Resources of Singapore, and the National University of Singapore. S M M is grateful to the Agency for Science, Technology and Research, Singapore for the award of a research scholarship.

References

[1] Lowry G V, K B Gregory, S C Apte, and J R Lead 2012 *Environ. Sci. Technol.* **46** 6893
[2] Li M, L Zhu, and D Lin 2011 *Environ. Sci. Technol.* **45** 1977
[3] Blinova I, A Ivask, M Heinlaan, M Mortimer, and A Kahru 2010 *Environ. Pollut.* **158** 41
[4] Goodwin A R, J Sengers, and C J Peters, Eds 2010 *Applied Thermodynamics of Fluids* (Electronic materials) (Cambridge: RSC Publishing)
[5] Misra S K, A Dybowska, D Berhanu, M N Croteau, S N Luoma, A R Boccaccini, and E Valsami-Jones 2012 *Environ. Sci. Technol.* **46** 1216
[6] Zhang W, J Crittenden, K Li, and Y Chen 2012 *Environ. Sci. Technol.* **46** 7054
[7] Reed R B, D A Lander, C P Higgins, P Westerhoff, and J F Ranville 2012 *Environ. Toxicol. Chem.* **31** 93
[8] Dederichs T, M Möller, and O Weichold 2009 *Langmuir* **25** 10501
[9] Saito T, L K Koopal, W H van Riemsdijk, S Nagasaki, and S Tanaka 2004 *Langmuir* **20** 689

[10] Zhou D and A A Keller 2010 *Water Res.* **44** 2948
[11] Li X, J J Lenhart, and H W Walker 2012 *Langmuir* **28** 1095
[12] Liu H H, S Surawanvijit, R Rallo, G Orkoulas, and Y Cohen 2011 *Environ. Sci. Technol.* **45** 9284
[13] Lide D R Ed. 2005 *Handbook of Chemistry and Physics* 86th ed. (Cleveland: Chemical Rubber Company)
[14] Bian S -W, I A Mudunkotuwa, T Rupasinghe, and V H Grassian 2011 *Langmuir* **27** 6059
[15] Mudunkotuwa I A, T Rupasinghe, C -M Wu, and V H Grassian 2012 *Langmuir* **28** 396
[16] Debye P and E Hückel 1923 *Physikalische Zeitschrift* **24** 185

Nanosafe 2012: International Conferences on Safe Production and Use of Nanomaterials IOP Publishing
Journal of Physics: Conference Series **429** (2013) 012040 doi:10.1088/1742-6596/429/1/012040

Characterizing Nanoparticles Reactivity: Structure-Photocatalytic Activity Relationship

J Piella[1,2], N G Bastús[1], E Casals[1], V Puntes[1,2,3]

[1] Institut Català de Nanotecnologia (ICN), ETSE 2nd floor, Campus UAB 08193 Bellaterra / Spain.

[2] Universitat Autònoma de Barcelona (UAB, Campus UAB 08193 Bellaterra / Spain.

[3] Institució Catalana de Recerca i Estudis Avançats (ICREA), 23, Passeig Lluís Companys 08010 Barcelona / Spain.

E-mail: Jordi.piella@icn.cat

Abstract. Nanoparticles are reactive, and their final interactions with the surrounding media are ultimately determined by their reactivity, which in turns depends on the nanoparticles morphology, surface chemistry and environment in which they are embedded. One simple and informative approach for the study of the reactivity of nanoparticles is the determination of their photocatalytic activity. In the present work, we briefly summarize the importance of different parameters such as the size, shape and agglomeration state on the photocatalytic activity of colloidal inorganic nanoparticles. The study of the use of nanoparticles as photocatalyts is relevant not only for its potential applications in environmental remediation issues but also it can provide relevant information about the role of these parameters at the nanoscale.

1. Introduction

The use of engineered nanoparticles (NPs) in photocatalysis has become one of the most investigated research fields during the last decade with potential applications in a variety of environmental remediation processes. The reasons include that currently treatment methods to degrade organic toxins, such as absorption by activated carbon or air stripping, need post-processing steps for the elimination of adsorbed contaminants. Other advanced oxidation technologies, especially those based on the use of strong oxidants, require hydrogen peroxide or ozone, which are expensive chemicals. In this sense, the main advantage of photocatalysis is that there is no further requirement for secondary disposal methods and that the photocatalysts are self-regenerated and can be reused or recycled [1].

Among other inorganic nanoparticles, oxide-based NPs are of specially interest due to their important photocatalytic activity. In some cases their band gap is in the range of UV-Visible light and the electrons required for initiating the photo-oxidation reactions can be easily excited even by sun-light. Moreover, the high reactivity of mater at the nanoscale makes them suitable for photocatalysis. In fact, TiO_2, ZnO and CeO_2 NPs are currently used for photocatalytic remediation technologies [2, 3]. In this context, the arising question is how to extract all this potential in the "real" use of engineered NPs. Recent advances in the synthesis of oxide NPs and the control over their size, shape and agglomeration state set new standards when talking about their uses as photocatalysts and it offers the possibility to even increase the photocatalytic activity of these nanomaterials without the need of dopants or hybrid NPs [4-7].

However, and even taking into account the current effort on the characterization of NPs, there is still a lack of information and of specification regarding the relationship between size, shape and agglomeration state of oxide NPs and their photocatalytic activity.

2. Photocatalytic reaction

The principle of the photocatalytic reaction with oxide NPs is straightforward [2]. Upon adsorption of photons with energy higher than the band gap, electrons are excited from the valence band to the conduction band, creating electron-hole pairs. These charge carriers, in particular the electrons from the conduction band, migrate to the NP surface -where the reactions take place- and facilitate reduction reactions. The capture of holes and electrons by adsorbed species generate a variety of highly reactive radicals and intermediate species such as OH·, O^{2-}, H_2O_2 capable of producing the oxidation of organic matter. In an aqueous media, the kinetic of the photocatalytic reactions mediated by NPs can be followed spectrophotometricaly by studying the degradation of organic dyes such as Rhodamine B (RhB), Methylene Blue or some phenol derivatives after photo-oxidation. These compounds have a characteristic absorption peak in the range of visible light which linearly depends on their concentration. As the reaction evolves the concentration of these compounds decreases and consequently the respective absorption peaks drop (figure 1). Thus, it is possible to follow the kinetics of the photocatalized reaction.

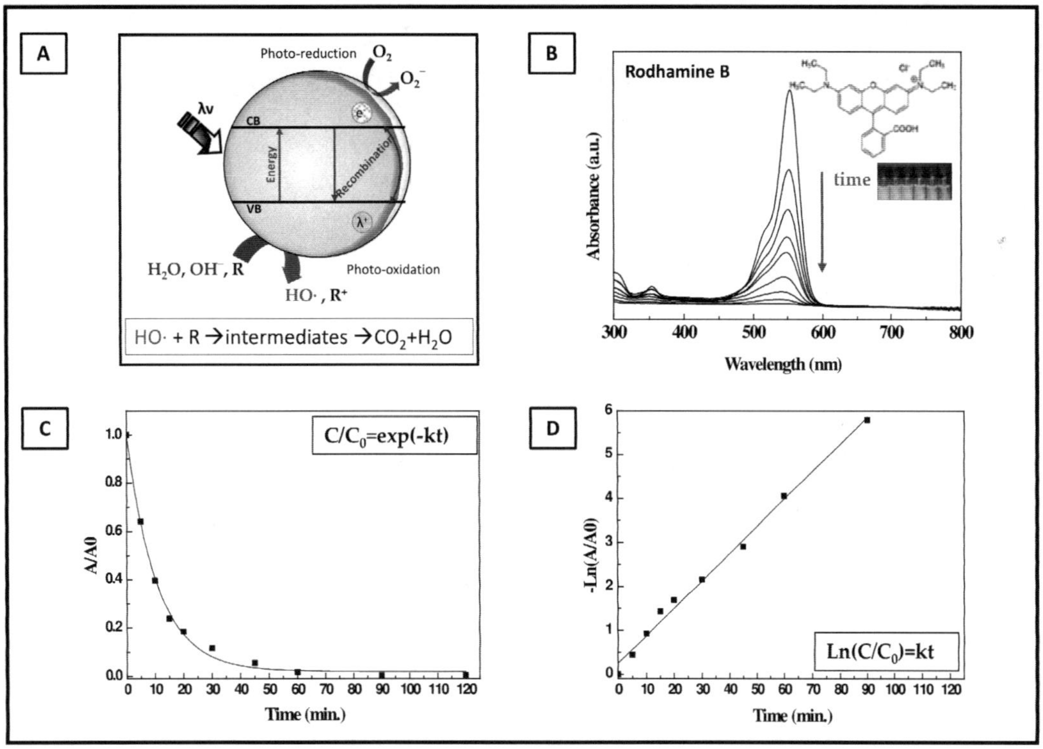

Figure 1. (A) Theoretical electron excitation process of an oxide-based NP. CB: Conduction Band. VB: Valence Band. (B) Temporal evolution of the absorbance of RhB in the presence of Degussa P25 NPs (0.6 mM). The reaction was done under 365 nm light and 8.9 mW/cm^2. (C) Maximum absorbance of the RhB as a function of time. (D) Logarithm of the maximum absorbance of RhB as a function of time.

The photocatalytic degradation using NPs is known to fit in a first ordered kinetic reaction. In the same conditions and normalizing the experiments by the total number of surface atoms, it is possible to compare the photocatalytic activity of two different NPs by comparing the kinetics rate of each reaction.

3. Experimental section

NPs used in this work, CeO_2, TiO_2 anatase, SiO_2 and ZnO, were synthetized in our laboratory following well-known aqueous sol-gel procedures and using tetramethyl ammonium hydroxide (TMAOH) as stabilizer [8-10]. Further purifications were performed by low speed centrifugation to ensure that no residues from these syntheses were in the samples during the experiments. The size and the shape of the NPs were determined by transmission electron microscopy (TEM) and their absorption spectra by UV-Vis spectroscopy.

In the case of CeO_2NPs, different shapes were obtained modifying reaction temperature during the synthesis and reagent concentration. In the case of TiO_2 NPs, the total amount of TMAOH used previous to the purification step leaded to different agglomeration states.

3.1. Preparation of particle suspensions

The concentration of NPs in the solutions were firstly adjusted with miliQ-water and further mixed with an aqueous solution of Rh B until 10 μM. At this concentration, when no degradation occurred, the absorbance was 1 (arbitrary units). Finally, the solutions were kept in dark under stirring for approximately 30 minutes for stabilization.

3.2. Photocatalytic activity

The samples were irradiated with a light source of 365 nm and 8.9 mW/cm^2 at 25°C for a few hours. At different time points, aliquots of 1 mL were extracted and centrifuged at high speed in order to remove the NPs. Following absorbance of Rh B solution was measured. Since the absorbance is directly proportional to non-degradated Rh B this experiment can be considered a direct measurement of the photocatalytic activity of the NPs.

4. Nanoparticles structure related photocatalytic properties

The photocatalytic activity of oxide-based NPs is largely controlled by (i) the light absorption properties, e.g. light absorption coefficient and spectrum, (ii) the energy of the band gap between the valence band and the conduction band, which needs to be below the energy of absorbed light, (iii) the electron-hole recombination rate inside and at the surface of the particle (iv) the absorption of the reactants and finally (v) reduction and oxidation rates on the surface by the electrons and holes [2]. These processes depend on the chemical composition and the nature of the oxide materials rather than just a fact of nanometric size of the material. While photocatalytic TiO_2 anatase, ZnO and CeO_2 NPs show high absorption below 400 nm and a bandgap around 3.2 eV, SiO_2 NPs with less absorbance in this range and a band gap of 1.1 eV do not show any photocatalytic activity (figure 2). Thus, even varying the parameters of SiO_2 particles, the efficiency of these NPs to photodegradate organic compounds is very low.

Crystallinity plays also an important role. The higher the crystallinity the fewer the bulk defects (kinks, steps…) and easier for the electron-hole pairs to reach the surface without recombination. For instance, anatase, rutile, brookite or amorphous TiO_2 NPs display different photocatalytic efficiencies, as well as monocrystalline or polycrystalline structured NPs [11, 12].

However, what ultimately define the effectiveness of a NP as photocatalyst is its intrinsic parameters such as size, shape and agglomeration state. These parameters affect the surface state of a NP and thus the place where the reaction takes place.

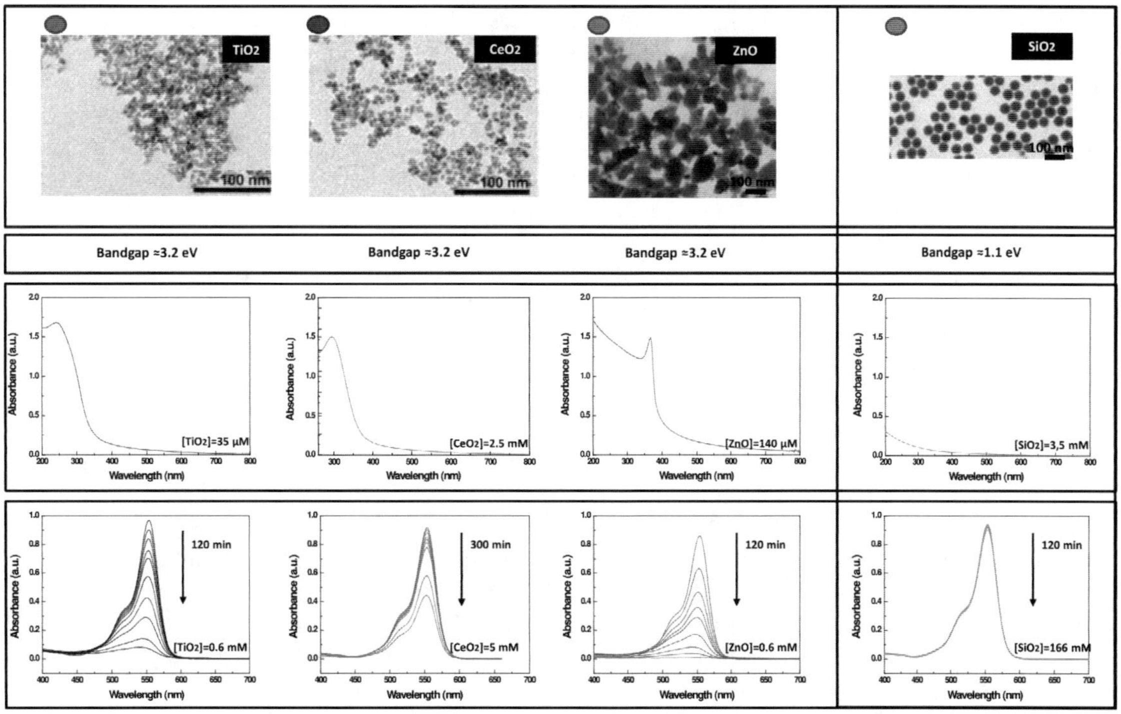

Figure 2. TEM images of different oxide-based NPs (first row), band gaps of NPs (second row), absorbances in the UV-Visible region showing band gaps of NPs (third row) and degradation of Rh B as a function of time under UV-Visible light; 365 nm 8.9 mW/cm (fourth row).

4.1. The role of size in photocatalysis

At the nanoscale, the size of NPs plays an important role not only in determining the efficiency and selectivity of photocatalytic activity, but also in the reactivity of the NP in general. Whereas in bulk materials surface atoms contribute only a relatively small fraction of the total number of atoms, at the nanoscale almost all the atoms of the particle lay close or at the surface. Moreover, when the NP is small enough, the coordination number of the surface atoms is affected by the size. For spherical-shaped NPs, both the curvature radii and the percentage of defects increases as the diameter decreases and surface atoms become less coordinated, which results in a higher reactivity. For larger diameters, the surface of the crystals possesses a larger collection of steps and kinks, independent from the size of the NP, and therefore the reactivity is only determined by the quality of its surface [13]. In this sense, the smaller the size, the higher the photocatalytic activity is. However, other effects can deviate photocatalytic behavior from previous defined predictions, e.g. the surface is a defective site; therefore, the larger the surface area, the faster the recombination.

Quantum effect is also a consequent of the size [2]. In the bulk material, an electron excited by light absorption from the valence band to the conduction band finds a high density of electronic states, where it can exist with different kinetic energies. However, in the case of NPs, as the size decreases below a critical value these electronic states start to discretize and the existence of an electron-hole pair is conditioned by the probability to fit into one of these states. Generally, size quantization is translated into strong changes in the electronic properties of the NP affecting the charge-carrier dynamics and the band structure. In most cases, there is a blue shift in the absorption and an increase on the band gap leading larger redox potentials.

Although, none experiment to correlate all these effects is presented here, the results reported by Wang et al. are indicative of the state of the art of the research, and at the same time, the little progress that has been made so far. They found an optimal size of 10 nm for TiO_2 anatase in the decomposition of chloroform and, interestingly, this activity decreases when the size of the particles used was decreased down do 6 nm [14]. Recently Cerunto et al. investigated methylene blue decomposition in the presence of TiO_2 NPs between 4 nm and 10 nm, founding a maximum efficiency around 6 nm [15]. Unquestionably, size quantization affects the photocatalytic activity of a particle but not in a predictable manner, at least not treating the size as a merely geometric factor. Optimal size conditions for photocatalysis are sought by taking these conclusions into account and may vary from case to case.

4.2. The role of shape in photocatalysis

The photocatalytic activity of NPs itself also depends on the shape [6, 16]. It is known that physicochemical properties of NPs are shape dependent, and for example the absorbance and the band gap vary depending on this parameter. However, if the relation between the physicochemical properties of NPs and size is complex, there is still a greater lack of information related to the shape.

Apart from the quantum effects induced by the shape, there is another important point which has to be taken into account. This is, as in the previous case, the number and the coordination state of the atoms at the surface. Sphere is the form with less surface area per unit of volume and when comparing the same number of total atoms distributed in other shapes such as stamps, belts or others, the number of surface atoms in these forms is always higher than that found in the case of spheres. Besides the number of the surface atoms, there the quality of the surface is also an important parameter. Atoms in flat surfaces such as stamps have both, a higher degree of coordination than in spheres, and less reactivity compared to the atoms on vortex and edges present in the anisotropic shapes [13]. Different shapes also implies different crystal faces exposed to the media and it is known that not all the crystal faces have the same photoreactivity [16].

Without entering into detail, the differences in the physicochemical properties and the variation in the surface state of NPs associated with modifications in their shapes imply necessarily changes in their photocatalytic activity. Figure 3 shows the degradation rate of Rh B in the presence of CeO_2 NPs with different shapes under UV 365 nm light. Even being the specific surface area similar, small 3 nm spherical CeO_2 NPs seemed to be more photoreactive than stamps and belts.

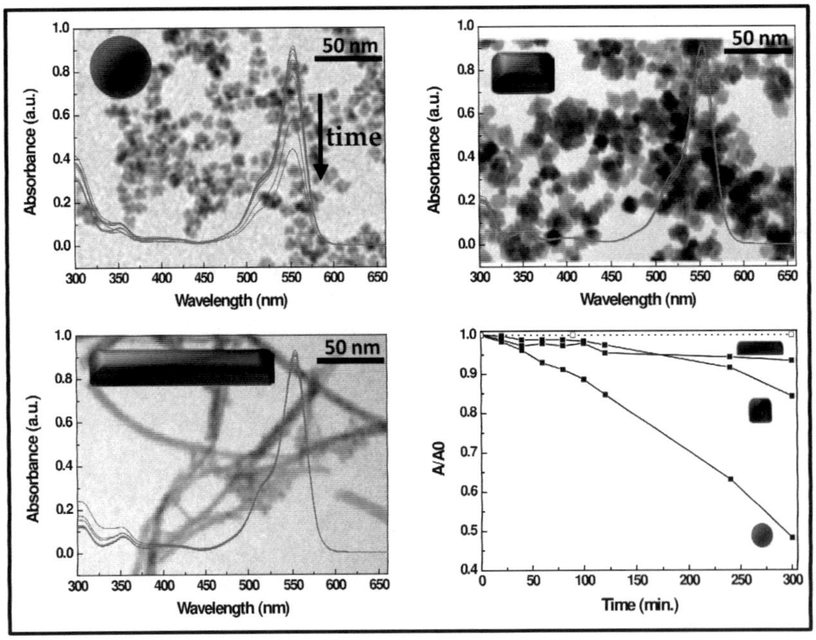

Figure 3. Evaluation of the photocatalytical properties of CeO_2 NPs of different shapes. Temporal evolution of Rh B peak of absorption in the presence of spheres, stamps and belts CeO_2NPs ($C_{0\text{-}CeO2}$=5 mM).

4.3. The role of agglomeration state in photocatalysis

Finally, the agglomeration state of NPs is also an important parameter to be taken into account. The degree of agglomeration has relevant consequences on the final behavior of the NPs which often translates into a loss of their nanometric properties. A direct consequence of agglomeration is the decrease of the free surface, and thus an expected decrease in the photocatalytic activity. Additionally, large sizes difficult the organic molecules to reach the surface. Figure 4 shows two cases of 5-10 nm a TiO_2 anatase NPs dispersed in water; one composed of 100 nm agglomerated and other well dispersed nearly individual particles. The photocatalytic activities of the two samples are different, being the first one less efficient in the photodegradation process than the second one.

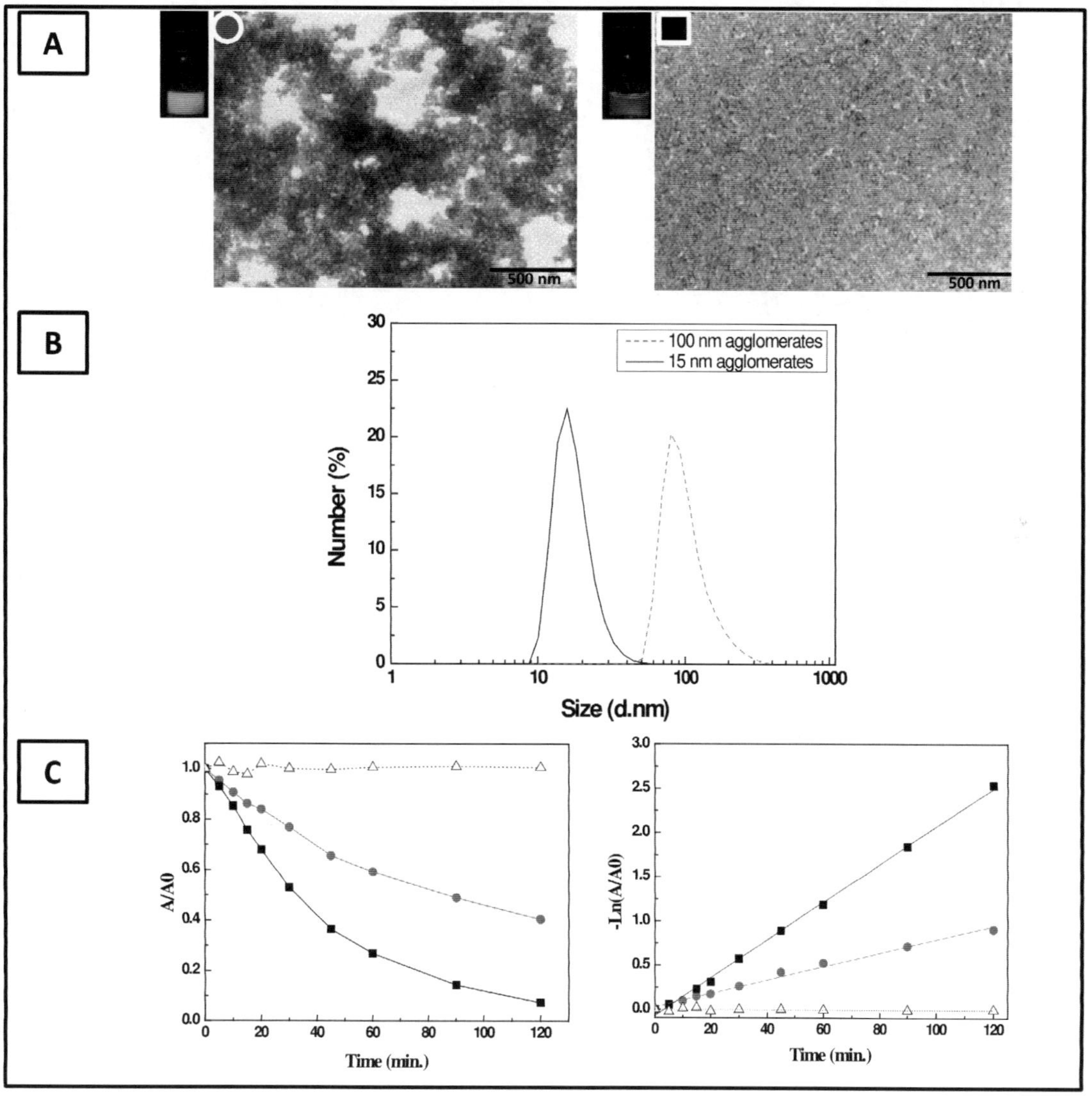

Figure 4. (A) TEM images of: (left) 100 nm and (right) 15 nm agglomerates TiO_2 NPs. (B) Dynamic Light Scattering measurement of both samples. (C) Temporal evolution of the RhB absorbance ($C_{0\ TiO2}$=0.6 mM): △control without NPs, ● 100 nm agglomerates and ■ 15 nm agglomerates.

5. Conclusions

There are many parameters that can affect the photocatalytic activity of NPs. Some of them, size, shape and aggregation have been briefly discussed in the present work.

Taking into account above considerations and the specific cases showed, it is expected for NPs that: (i) small particles are more photocatalytic than larger particles, (ii) atoms on flat surfaces in stamp or belt like shapes have a higher degree of coordination than in spheres, therefore possessing a lower photocatalytic activity compared to those atoms in edges or vortices, (iii) particles lose their nanometric properties and therefore also their efficiency when agglomerates.

However, this is just a general trend and not necessarily has to be always fulfilled. Usually, modifications of physicochemical properties associated to the variation of these parameters are proved for a specific photodegradation process and it is difficult to predict the efficiency of a particular NP as photocatalyst.

Moreover, nowadays there is a huge effort in understanding how NPs interact with the environment, bringing new parameters that are discovered to play an important role in the NPs behavior, e.g. surfactant used, corrosion, dissolution… Hence, the photocatalytic activity of NPs not only depends on the intrinsic properties of the particle but also on the environment in which the NPs are embedded and their evolution over time.

Acknowledgements

The authors acknowledge financial support from the grants Plan Nacional MAT2009-14734-C02-01, Juan de la Cierva Programme, 2009-SGR-776, CNBSS and the EU-FP7 project QNano (GA No. 262163).

References

[1] Hoffmann MR, Martin ST, Choi WY, Bahnemann DW 1995 *Chem Rev.* **95** 69-96.
[2] Beydoun D, Amal R, Low G, McEvoy S 1999 *J Nanopart Res.* **1** 439-58.
[3] Chen X, Mao SS 2007 *Chem Rev.* **107** 2891-959.
[4] Li YF, Liu ZP 2011 *J Am Chem Soc.* **133** 15743-52.
[5] Jassby D, Budarz JF, Wiesner M 2012 *Environ Sci Technol.* **46** 6934-41.
[6] Wu NQ, Wang J, Tafen D, Wang H, Zheng JG, Lewis JP, et al. 2010 *J Am Chem Soc.* **132** 6679-85.
[7] Lakshminarasimhan N, Kim W, Choi W 2008 *J Phys Chem C.* **112** 20451-7.
[8] Zhang F, Jin Q, Chan SW 2004 *Journal of Applied Physics.* **95** 4319-26.
[9] Pottier AS, Cassaignon S, Chaneac C, Villain F, Tronc E, Jolivet JP 2003 *Journal of Materials Chemistry.* **13** 877-82.
[10] Jolivet JP, Cassaignon S, Chaneac C, Chiche D, Tronc E 2008 *J Sol-Gel Sci Technol.* **46** 299-305.
[11] Ohno T, Sarukawa K, Matsumura M 2001 *J Phys Chem B.* **105** 2417-20.
[12] Ohno T, Sarukawa K, Matsumura M 2002 *New J Chem.* **26** 1167-70.
[13] Bastus NG, Casals E, Vazquez-Campos S, Puntes V 2008 *Nanotoxicology.* **2** 99-112.
[14] Wang CC, Zhang ZB, Ying JY 1997 *Nanostruct Mater.* **9** 583-6.
[15] Cernuto G, Masciocchi N, Cervellino A, Colonna GM, Guagliardi A 2011 *J Am Chem Soc.* **133** 3114-9.
[16] D'Arienzo M, Carbajo J, Bahamonde A, Crippa M, Polizzi S, Scotti R, et al. 2011 *J Am Chem Soc.* **133** 17652-61.

Nanosafe 2012: International Conferences on Safe Production and Use of Nanomaterials IOP Publishing
Journal of Physics: Conference Series **429** (2013) 012041 doi:10.1088/1742-6596/429/1/012041

Comparison of effects on crustaceans: carbon nanoparticles and molybdenum compounds nanowires

A Baumerte[1], G Sakale[2], J Zavickis[2], I Putna[1, 3], M Balode[1, 3], A Mrzel[4] and M Knite[2]

[1] University of Latvia, Faculty of Biology- Kronvalda boulevard 4, LV-1010, Riga, Latvia
[2] Riga Technical University, Institute of Technical Physics – Āzenes street 14/24, LV-1048, Riga, Latvia
[3] Latvian Institute of Aquatic Ecology – Daugavgrīvas street 8, LV-1048, Riga, Latvia
[4] Jozef Stefan Institute, Department for Complex Matter – Jamova 39, 1000, Ljubljana, Slovenia

E-mail: anda.baumerte@gmail.com

Abstract. Carbon nanomaterials (CNM) and molybdenum compound nanostructures are materials with various applications yet little is known regarding the toxicity of these nanoparticles in pristine form in aquatic environment. *Daphnia magna* standard acute toxicity test (EN ISO 6341:1996; freshwater) and *Artemia salina* standard acute toxicity test (ArtoxKit standard method; 15 ppt saltwater) were applied to assess the toxicity of non-modified CNM and molybdenum compound nanowires in water. It has been observed that CNM are more toxic in freshwater suspensions and somewhat more toxic than the tested molybdenum compound nanowires.

1. Introduction

Nanomaterial properties may differ from the bulk material as greatly as may their toxic properties and still many new materials provide a challenge for the options of toxicity testing. European Environmental Agency within REACH [1] and the Environmental Protection Agency of the United States of America within the Framework for Metals' Risk Assessment [2] have listed molybdenum compounds in their legislation as a material of (primary-EPA) interest.

The little number of studies addressing molybdenum effects on aquatic crustaceans have demonstrated that the 10% effective concentration (EC_{10}, expressed as measured dissolved molybdenum from added sodium molybdate dihydrate, $Na_2MoO_4 \cdot 2H_2O$) for the most sensitive endpoint is at 62.8–105.6 mg(Mo)L^{-1} for *D. magna* (21 day-reproduction). For *D. magna*, the 48 h LC_{50} value has been determined for sodium molybdate at 2847.5 mg L^{-1} [3]. In addition, by multiple chronic exposure tests, a median hazardous concentration affecting 5% of the species ($HC_{5,\ 50\%}$) of 38.2 mg(Mo)L^{-1} has been derived [1]. Yet the only study addressing molybdenum compounds nanoparticle toxicity describes molybdenum oxide (MoO_3; 150 nm and 30 nm) as moderately toxic to BRL 3A rat liver cells [4].

Carbon nanomaterials (CNM), in their turn, are one of the nanomaterials with the highest number of potential applications, yet still, little of data exist regarding their fate in environment. Carbon black (CB) is produced at rates of several million tons per year, which make it also the most abundant nanoparticle in environment [5]. However, according to the review data provided by the Environment

Journal of Physics: Conference Series **429** (2013) 012041 doi:10.1088/1742-6596/429/1/012041

Canada on CB (CAS RN 1333-86-4) in year 2011, no information existed on CB fate in environment. A few toxicity studies show effects below 0.1–0.5 mg L^{-1}, several indicate moderate toxicity *in vitro* [6]. A number of toxic effects have been observed for the multi-walled carbon nanotubes (MWCNT) *in vitro* and also several tests with aquatic organisms, mainly fish, have been carried out to assess the toxicity of MWCNT in aquatic environment. However, the lowest 48 h EC_{50} for these materials for *D.magna* have been stated at 37.563 mg L^{-1} for 20 000 nm sized CB particles and at 8.723 mg L^{-1} for MWCNT (with LC_{50} being at 61.547 and 22.751 respectively) [7,8].
In the current study, to allow the interpretation of the test results with the respect to the local environment (Baltic Sea region), a local culture of freshwater species *D. magna* was chosen and in addition a saltwater species *Artemia salina.* The salt content in *A. salina medium* was kept maximally similar to that of the Baltic Sea. *D. magna* standard acute toxicity test (modified EN ISO 6341:1996; freshwater) and *A. salina* (ArtoxKit standard method; 15 ppt saltwater) acute toxicity test were applied to assess the toxicity of non-modified CNM and molybdenum nanowires.

2. Materials and methods

As representatives for nanotoxicity tests of CNM were chosen CB nanoparticles Printex XE2 produced by Evonik Degussa GmBH with average diameter of primary CB particle 30 nm, specific surface area 950 m^2/100g, dibutylphtalate absorption 380 ml/100g and MWCNT purchased from Aldrich with outside diameter 40-60 nm, inside diameter 5-10 nm, length 0.5-500 μm. Molybdenum compound nanowires ($Mo_6S_2I_8$) and molybdenum oxide nanowires (MoO_3-x) were used for toxicity testing as well. Scanning electron microscopy images of molybdenum compound nanomaterials are shown in Figures 1 and 2.

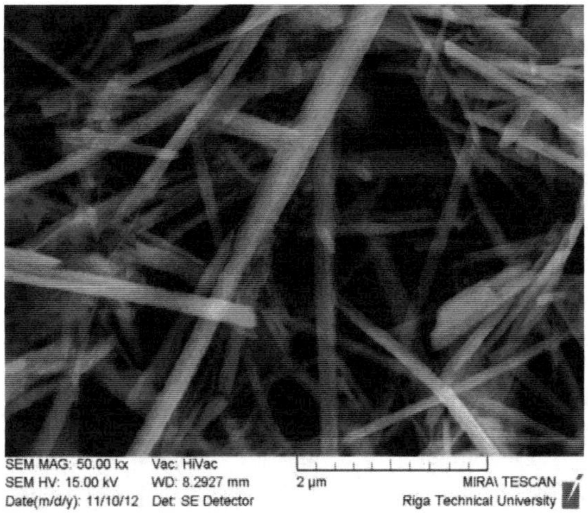

Figure 1. SEM image of $Mo_6S_2I_8$ nanowires in isopropanol (Author: D. Jakovlevs).

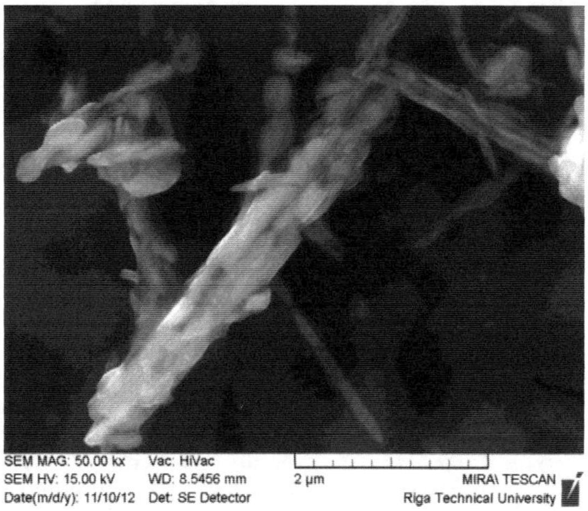

Figure 2. SEM image of MoO_3–x nanowires in isopropanol (Author: D. Jakovlevs).

CNM and molybdenum compound nanowires were sonicated (with power 1 W/ml) in 100 ml *A. salina* and *D. magna* test medium for 5 to 10 minutes to form a stock suspension with total nanoparticle concentration of 10 mg L^{-1}. The stock suspension was diluted then for appropriate test concentrations in the range from 1 mg L^{-1} to 40 mg L^{-1}. The monitoring of air temperature, pH and photo-fixation of test organisms was performed during the test period. The changes of particle solubility in response to varying water salinity were measured, and particle sedimentation was assessed in parallel to the test by photo-fixation of the suspension once per hour for the period of the maximum test duration, i.e. 72 h. Test procedure and conditions are summarised in Table 1. Probit regression analysis was applied for the determination of LC_{50} with regard to the total organisms affected with statistically correct result.

Table 1. *D.magna* and *A.salina* test conditions.

	EN ISO 6341:1996	ArtoxKit
Test species	*Daphnia magna* (younger than 24 h)	*Artemia salina* (A.salina instar II-III)
Source of organisms	culture collection of Latvian Institute of Aquatic Ecology	hatched from dormant eggs; incubation period at 25°C 36-48 h
Test chamber	6 vial plate	24 vial plate
Volume per vial	10 ml	3 ml
Replicates	3	6
Organisms per replicate	5	10
Feeding	-	-
Aeration	-	-
Temperature	20 ± 1 °C	20 ± 1°C
Salinity	0 ppt	15 ppt
Test duration	24 h (48 h)	72 h
Control	test medium	test medium

3. Results and discussion

After 48 h period, **CB** caused the maximum 5.1 % lethality to ***A. salina*** at the concentration of 7.5 mg L^{-1} 10 and 2.4 % lethality at 10 mg L^{-1}. In case of ***D. magna*** the 24 h lethality at 10 mg L^{-1} was 30 % (96.7 % after 48 h).

CNT were susceptible for ***A. salina*** at the concentration of 2.5 mg L^{-1}, causing a maximum 46.7 % lethality, and somewhat more than 20% at 10 mg L^{-1}. For ***D. magna***, 24h lethality at the concentration of 10 mg L^{-1} was 13.3 %.

$\mathbf{Mo_6S_2I_8}$ nanowires caused a maximum of 20 % organism lethality for ***D. magna*** after 48h at 10 mg L^{-1}, yet no lethality was observed after 24 h or by increasing the concentration to 40 mg L^{-1}. ***A. salina*** lethality (48 h) for $\mathbf{Mo_6S_2I_8}$ nanowires was 18.6 % at the concentration of 10 mg L^{-1}.

$\mathbf{MoO_3}$**-x** nanowires turned out to cause 20% lethality after 24h (26.7% after 48h) for ***D.magna***. The percent of affected ***A. salina*** after 48 h were 21.6%, whereas at concentration 30 mg toxicity decreased reaching 13% lethality after 48 h. However, none of the particles caused significant effects to the *A. salina* after 24 h.

The calculated LC_{50} values with regard to the total organisms affected are displayed in Figures 3, 4, 5 and 6. It is believed that this can be considered the most environmentally relevant endpoint, since for some of the materials higher mortality was reached within smaller concentrations. This is believed to be caused largely by the nanoparticle aggregation and suspension properties, such as aggregate size distribution, since we can observe that the smaller particles are ingested by the crustaceans while both smaller and larger particles might adhere to the organism surface.

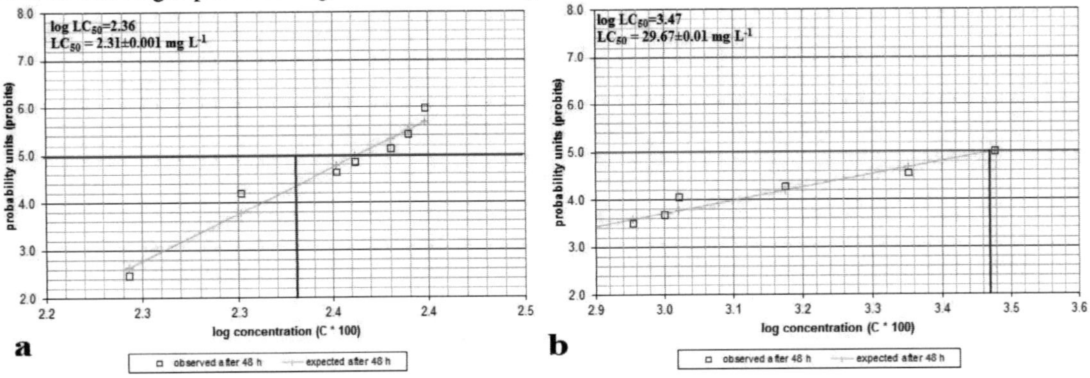

Figure 3. LC_{50} for CB after 48 h incubation period, calculated with regard to the total organisms affected: A – *D .magna*, B – *A.salina*

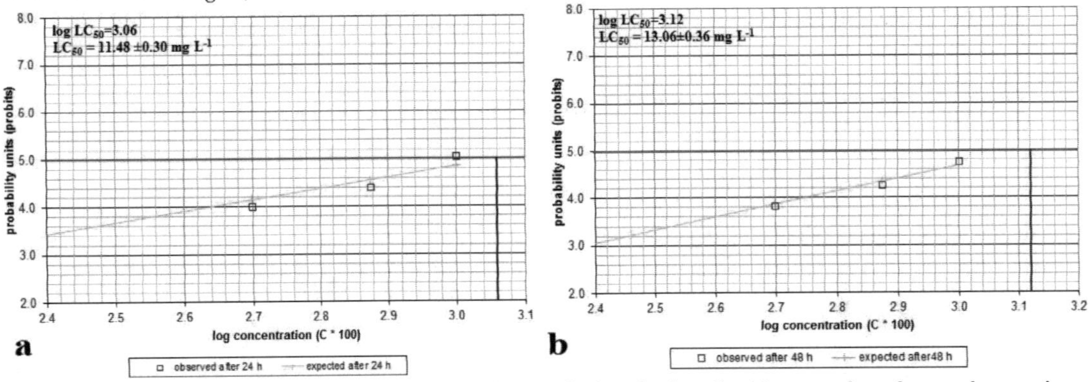

Figure 4. LC_{50} for MWCNT after 48 h incubation period, calculated with regard to the total organisms affected: A – *D .magna*, B – *A.salina*

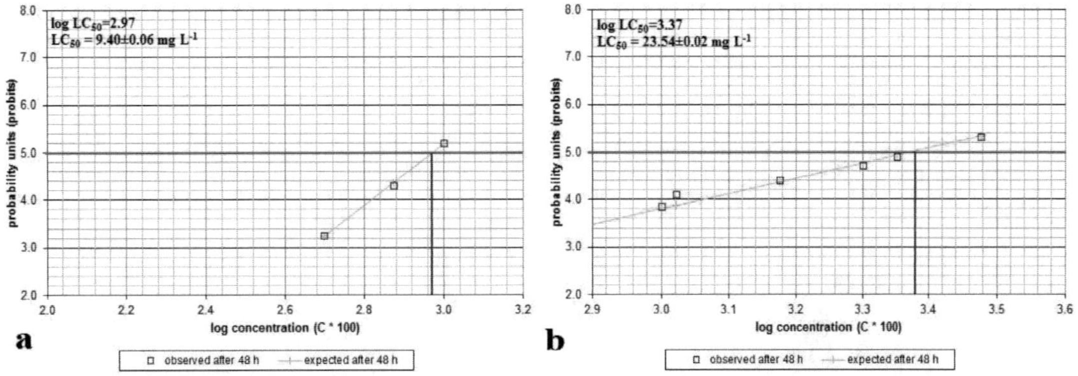

Figure 5. LC_{50} for MoO_3-x nanowires after 48 h incubation period, calculated with regard to the total organisms affected: A – *D .magna,* B – *A.salina*

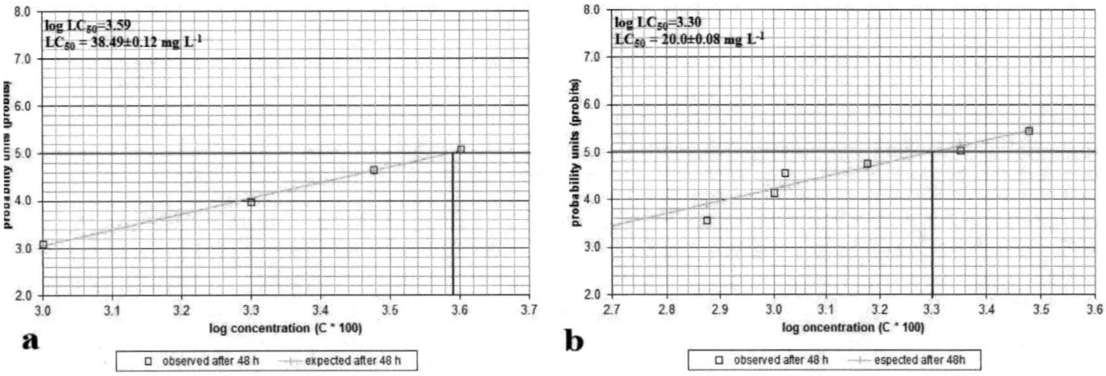

Figure 6. LC_{50} for $Mo_6S_2I_8$ nanowires after 48 h incubation period, calculated with regard to the total organisms affected: A – *D .magna,* B – *A.salina*

Toxic effects were observed for all particles at concentrations 2.5 mg L^{-1}. Such effects ranged from the movement or swimming inhibition to particle aggregate accumulation in the gastro-intestinal tract, particle aggregate adherence to the organism surface as well as, specifically for *D. magna,* – more intensive moulting in particle suspensions (Figures 7, 8 and 9). In 5 mg L^{-1} CNM suspensions, *D. magna* frequently got trapped into large aggregates (Figure 8F) and mortality was caused due to this reason. It is believed that physical interaction of the particle aggregates with organisms may have caused organism lethality as well (Figure 10). Obtained results indicate that for three of the materials 50% lethality was not reached within the water dispersions.

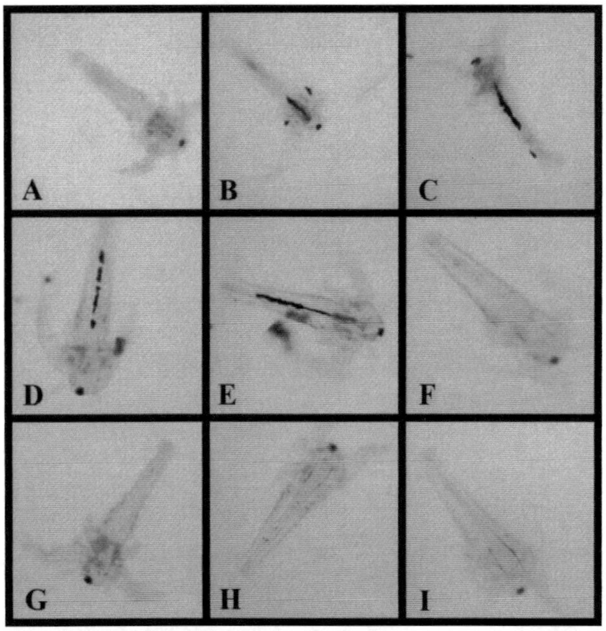

Figure 7. Nanoparticle effects on *Artemia salina*: A – exposed to test medium, B – exposed to 3.5 mg L^{-1} CNT suspension, C - exposed to 5 mg L^{-1} CNT suspension, D - exposed to 7.5 mg L^{-1} CNT suspension, E - exposed to 10 mg L^{-1} CNT suspension; F - exposed to 5 mg L^{-1} MoO_3-x suspension; G - exposed to 7.5 mg L^{-1} MoO_3-x suspension, H - exposed to 10 mg L^{-1} MoO_3-x suspension, I - exposed to 10.5 mg L^{-1} MoO_3-x suspension (Author: A. Baumerte).

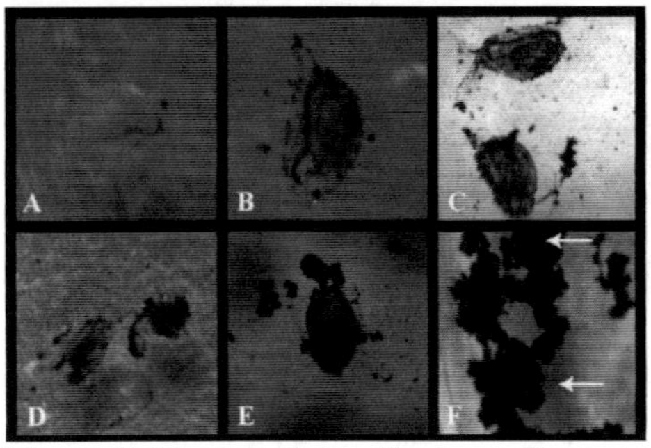

Figure 8. Nanoparticle effects on *Daohnia magna*: A – exposed to test medium, B – exposed to 1 mg L^{-1} CB suspension, C - exposed to 1.75 mg L^{-1} CB suspension, D - exposed to 2 mg L^{-1} CB suspension, E - exposed to 2.5 mg L^{-1} CB suspension; F - exposed to 10 mg L^{-1} CB suspension (Author: A.Baumerte).

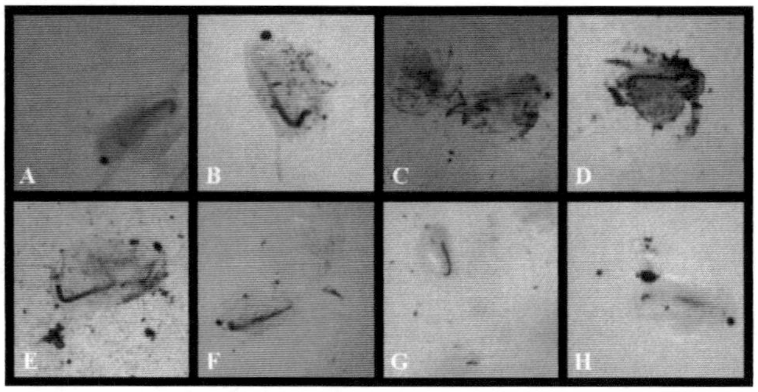

Figure 9. Nanoparticle effects on *Daophnia magna*: A – exposed to test medium, B – exposed to 2.5 mg L^{-1} $Mo_6S_2I_8$ suspension, C - exposed to 5 mg L^{-1} $Mo_6S_2I_8$ suspension, D - exposed to 7.5 mg L^{-1} $Mo_6S_2I_8$ suspension, E - exposed to 10 mg L^{-1} $Mo_6S_2I_8$ suspension; F - exposed to 1.75 mg L^{-1} MoO_3-x suspension; G - exposed to 7.5 mg L^{-1} MoO_3-x suspension, H - exposed to 10 mg L^{-1} MoO_3-x suspension (Author: A. Baumerte).

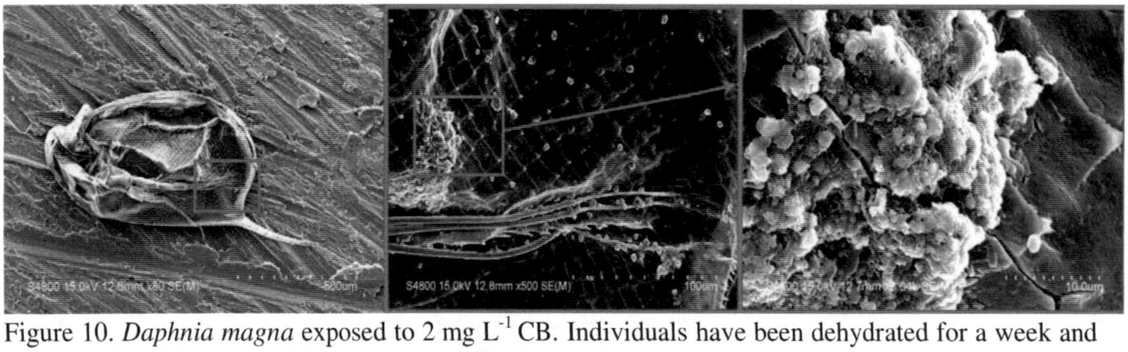

Figure 10. *Daphnia magna* exposed to 2 mg L^{-1} CB. Individuals have been dehydrated for a week and coated with gold prior to taking the pictures (Author: A. Baumerte).

It has been known that CNM are temporally soluble in water. Likewise, in this study we have observed that molybdenum nanowires are partly soluble in water, whereas MoO_3-x dissolve better than $Mo_6S_2I_8$. In addition, the sedimentation tests suggest that the most of the dispersed CNM and molybdenum particles have settled before the test results are read, i.e. before 24h have passed (see Figure 11). Moreover, the particles (particle aggregates) had sedimentated in water suspensions after 6 h for both CNM, yet for the molybdenum nanowires there was an insignificant amount of light-colored aggregates in the vial and no visual time-dependant changes have been observed. According to colloidal theory, it is believed that in the saltwater larger particle aggregates form due to the particle bridging with NaCl salt as a result electrical double layer of the particle is diminished and particles can approach closer to each other (see Figure 12 and 13). Similarly Brant *et al.* has observed for C_{60} fullerene nanoparticles [9, 10]. It has been stated that the stability and therefore the toxicity is affected by the salt content in solution. When the salinity reaches 33% and nanomaterial aggregation occurs [11], toxicity is still observed.

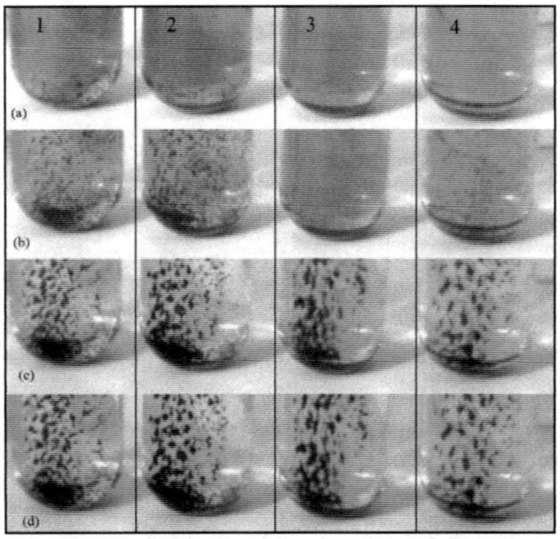

Figure 11. Sedimentation images of CNM: 1 – MWCNT + freshwater, 2 – MWCNT + saltwater, 3 – CB + freshwater, 4 – CB + saltwater; (a) 0.5 h after preparation; (b) 4 h after preparation; (c) 6h after preparation; (d) 12h after preparation (Author: A. Baumert

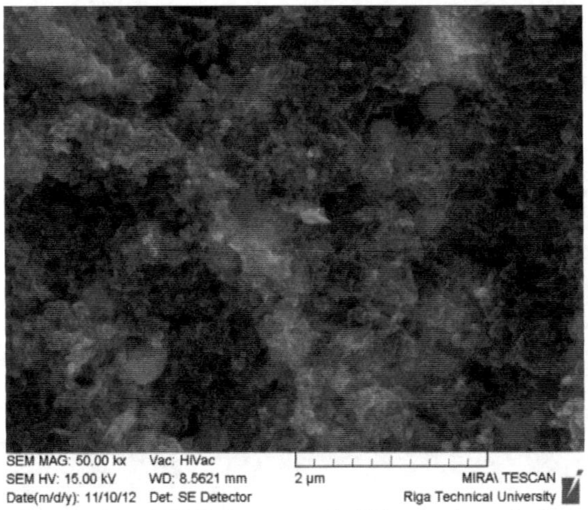

Figure 12. SEM image of CB stock solution, prepared in *A. salina* test medium (15 ppt NaCl) (Author: D. Jakovlevs).

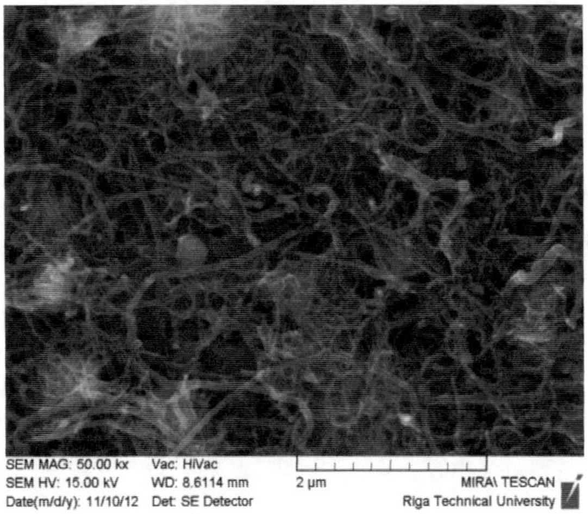

Figure 13. SEM image of CNT stock solution, prepared in *A. salina* test medium (15 ppt NaCl) (Author: D. Jakovlevs).

In conclusion, *D. magna* have showed to be more sensitive to both materials. It is believed that due to the low solubility in water and the particle size, nanoparticle toxicity did not follow dose-dependence. Altogether, MoO_3-x nanowires turned out to be somewhat more toxic than $Mo_6S_2I_8$ nanowires, yet more toxic effects have been observed for CNM. It is suggested that the nanomaterial properties contributing towards the toxicity in environment are complex and also abiotic factors of the environment and nanomaterial physical properties should be accounted for. Further experiments may be needed to explain the toxic effects in interaction with other, including natural, substances; as well as the uptake routes and distribution of nanoparticles (nanoparticle aggregates) within the organisms.

Acknowledgments
The study has been the fruit of collaboration of University of Latvia and the Institute of Technical Physics of Riga Technical University. Study has been financed from ESF grant "Support for Master studies in University of Latvia". The molybdenum nanowires have been provided by the Institute of Technical Physics (in frame of COST action No. MP0902 COINAPO). The test organisms have been provided from the cultures of Latvian Institute of Aquatic Ecology. We are thankful for their technical help support and valuable suggestions to Dr.Dmitrijs Jakovlevs and Dr.biol.Turs Selga.

References

[1] De Schamphelaere K A C, Stubblefield W, Rodriguez P, Vleminckx K and Janssen C R 2010 The chronic toxicity of molybdate to freshwater organisms. I. Generating reliable effects data. *Sci. Total Environ.* **408** (22) pp 5362–5371
[2] Fairbrother A, Wenstel R, Sappington K and Wood W 2007 Framework for Metals Ris k Assessment. *Ecotoxicol. Environ. Saf.* **68**(2) pp 145–227
[3] Diamantino TC, Guilhermino L, Almeida E and Soares M V M A 2000 Toxicity of Sodium Molybdate and Sodium Dichromate to Daphnia magna Straus Evaluated in Acute, Chronic, and Acetylcholinesterase Inhibition Tests. *Ecotoxicology and Environmental Safety* **45**(3) pp 253-259.
[4] Hussain S M, Hess K L, Gearhart J M, Geiss K T and Schlager J J 2005 In vitro toxicity of nanoparticles in BRL 3A rat liver cells. *Toxicology in Vitro* **19** (7) pp 975–983
[5] Lauterwasser C 2005 Small sizes that matter: Opportunities and risks of Nanotechnologies. Allianz Center for Technology and OECD International Futures Programme.
[6] Fent K 2010 *Ecotoxicology of Engineered Nanoparticles* (Nanoparticles in the Water Cycle) eds Frimmel F H, Nießner (Springer-Verlag, Berlin Heidelberg) pp 183-205
[7] Zhu X, Zhu L, Chen Y and Tian S 2009 Acute toxicities of six manufactured nanomaterial suspensions to Daphnia magna. *Nanopart. Res.* **11** pp 67-75
[8] Kahru A and Dubourguier, H C 2010 From Ecotoxicology to Nanoecotoxicology. *Toxicology.* **269**(10) pp 105-119.
[9] Brant J A, Labille J, Bottero J Y and Wiesner M R 2006 Characterizing the impact of preparation method on fullerene cluster structure and chemistry. *Langmuir.* **22** pp 3878–388
[10] Brant J A, Lecoanet H, Hotze M, Wiesner M R 2005 Comparison of electrokinetic properties of colloidal fullerenes (n-C60) formed using two procedures. *Environ. Sci. Technol.* **39** pp 6343–6351
[11] Klaine S J, Alvarez P J J, Batley G E, Fernandes T F, Handy R D, Lyon D Y, Mahendra S, McLaughlin M and Lead J R 2008 Nanomaterials In The Environment: Behaviour, Fate, Bioavailability and Effects. *Environ. Toxicol. Chem.* **27**(9) pp 1825-1851

Metal oxide nanoparticle transport in porous media - an analysis about (un)certainties in environmental research

I Heidmann

Interdisciplinary Research Group for Environmental Studies, University of Koblenz-Landau, Germany

Email: heidmann@uni-landau.de

Abstract. Research about the fate and behavior of engineered nanoparticles in the environment is despite its wide applications still in the early stages. The fast-growing area of nanoparticle research and the high level of uncertainty create a big challenge for describing clearly the recent state of the current scientific knowledge. Therefore, in this study the certain knowledge, the known uncertainties and the identified knowledge gaps concerning mobility of engineered metal oxide nanoparticles in porous media are analyzed. The mobility of nanoparticles is mainly investigated in model laboratory studies under well-defined conditions, which are often not realistic for natural systems. In these model systems, nanoparticles often retain in the pore system due to aggregation and sedimentation. However, under environmental conditions, the presence of natural organic matter may cause stabilization or disaggregation of nanoparticles and favors therefore higher mobility of nanoparticles. Additionally, potential higher mobility of particles using preferential flow paths is not considered. Knowledge of the long-term behavior of nanoparticles concerning disaggregation, dissolution or remobilization in soils under environmental conditions is scarce. Scientific uncertainty itself is rarely mentioned in the research papers. Seldom known methodically uncertainties in nanoparticle characterization are referred to. The uncertainty about the transferability of the results to environmental conditions is discussed more often. Due to the sparse studies concerning natural material or natural pore systems, certain conclusions concerning the mobility of nanoparticles in the soil environment are not possible to drawn.

Keywords: nanoparticles, metal oxide, porous media, soil, environmental research, scientific uncertainty

1. Introduction

"There is a high level of uncertainty in the research concerning the fate and behavior of nanoparticles in the environment" is often stated. Research about the fate and behavior of engineered nanoparticles in the environment is despite its wide applications still in the early stages. As far as is known, no engineered nanoparticle has been directly detected in the environment [1]. Up to now, any risk assessment failure because of the limited knowledge and the huge uncertainties about the expected environmental concentration of engineered nanoparticles in the environment [2]. Scientific uncertainties can be considered from different perspectives. One possibility is to distinguish between (a) the uncertainties related to the basic research like the measuring uncertainty, not yet validated, contradictory, inconsistent, or not reproduce able data and (b) the uncertainties related to applied research questions as e.g. in the risk assessment like transferability of model data or data from model systems to environmental systems, the uncertainties in scaling issues, e.g. spatial heterogeneity or lack of reliable data. Of course, there are several more definitions and perspectives of uncertainties, e.g. in psychology, ethics and social sciences. For this meta-analysis, these two perspectives of scientific uncertainty will be considered.

The fast-growing area of nanoparticle research and the high level of uncertainty create a big challenge for describing clearly the recent state of the current scientific knowledge [3]. One way to analyze the knowledge of a specific topic is to distinguish between certain and uncertain knowledge and knowledge gaps. First descriptions of the uncertainty of the behavior and effects of nanoparticles in the environment includes investigations of the measurement uncertainties in characterizing nanoparticles with different methods [4, 5] and scientific research on environmental, health, and safety risks [6]. Uncertainties have primarily been observed in the areas of test conditions, effect estimations, characterization, and exposure of nanoparticles [6]. However, this list remains descriptive and does not allow for any conclusions about the kind, sources or consequences of these uncertainties.

Metal oxide nanoparticles are because of their expected toxicity and high production volumes one of the nanoparticle classes, from which the main risk for the environment is expected [2]. One way nanoparticles enter the environment is via the industrial and house-hold wastewater to the effluents of sewage plants and through sewage sludge [7-10]. The nanoparticles can then reach aquatic and terrestrial environments [8]. Soils may play a key role for the fate and behavior of nanoparticles, because soils can act as nanoparticle-sink as well as nanoparticle-source, e.g. for groundwater, organisms, and plants [3]. Because one key parameters of the fate of nanoparticles in the environment is their mobility, the meta-analysis was carried out in order to characterize the certain knowledge, the known uncertainties and the knowledge gaps in the field of metal oxide nanoparticle transport in porous media.

2. Material and Methods

To characterize the knowledge about metal oxide nanoparticle transport in porous media, the literature was screened in March 2012 for peer-reviewed research articles dealing with experiments of the topic "transport of metal oxide nanoparticles in porous media". Searches were carried out in Web of Science with search terms like nano*, transport, deposition, porous media, soil. Searching for the terms nano*, (transport or mobility or retention) in the article topic leads together with the term porous media to over 600 results, together with the term soil to around 400 results. Scanning these articles, research papers dealing with metal oxide nanoparticle transport in water saturated porous media like glass beads, sand or soil material and only one paper per author were selected. After this specification, 13 papers were left for the meta analysis, which are shown in Table 1 (next page).

The meta-analysis was carried out concerning the following topics: nanoparticle characteristics (type, coatings, size), porous medium (material, grain size, porosity), experimental conditions (ionic strength, pH, particle concentration, presence of organic substances, Darcy velocity, application time), measurement methods of particle size and effluent concentration, considered models, presented results and drawn conclusions, mentioned or discussed uncertainty related to transport of nanoparticles in the environment.

3. Results and discussion

3.1. Nanoparticle characteristics

The nanoparticle characteristics used in the selected studies are summarized in Table 2. Most studies used titanium dioxide or zinc oxide as studied nanoparticles. In three studies, more than one nanoparticle type was investigated. Titanium dioxide and zinc oxide nanoparticles were chosen because they are the most widely used metal oxide nanoparticles, which are present in a variety of applications and consumer products (e.g. sunscreens and paints) with high production volumes. This is about 3000 respectively 550 t/year TiO_2 and 550 respectively 55 t/year ZnO worldwide and in Europe [11]. It is expected that the ongoing use of both nanoparticles lead to relevant inputs of 1,28 µg TiO_2/kg·year and 0,09 µg ZnO/kg·year (mode values) and a concentration of 0,4 – 4,8 µg TiO_2/kg in soils in Europe [9, 10]. Other metal oxide nanoparticles as aluminum oxide or iron oxide nanoparticles are produced in lower amounts, both about 55 t/year worldwide [11]. According to Som and colleagues, TiO_2 and ZnO are belonging to the most environmental relevant nanoparticles [12].

Table 1: Selected peer-reviewed research articles in alphabetic order dealing with the topic of transport of metal oxide nanoparticles in porous media

first author reference	year	article title	journal
Ben-Moshe [13]	2010	Transport of metal oxide nanoparticles in saturated porous media	Chemosphere
Chen [14]	2011	Transport and Retention of TiO_2 Rutile Nanoparticles in Saturated Porous Media under Low-Ionic-Strength Conditions: Measurements and Mechanisms	Langmuir
Chowdhury [15]	2011	Mechanisms of TiO_2 nanoparticle transport in porous media: Role of solution chemistry, nanoparticle concentration, and flow rate	Journal of Colloid and Interface Science
Fang [16]	2009	Stability of titania nanoparticles in soil suspensions and transport in saturated homogeneous soil columns	Environmental Pollution
Godinez [17]	2011	Aggregation and transport of nano-TiO_2 in saturated porous media: Effects of pH, surfactants and flow velocity	Water Research
Jiang [18]	2012	Transport and deposition of ZnO nanoparticles in saturated porous media	Colloids and Surfaces A
Joo [19]	2009	Influence of Carboxymethyl Cellulose for the Transport of Titanium Dioxide Nanoparticles in Clean Silica and Mineral-Coated Sands	Environmental Science & Technology
Kanel [20]	2011	Influence of pH on the transport of nanoscale zinc oxide in saturated porous media	Journal of Nanoparticle Research
Lecoanet [21]	2004	Laboratory assessment of the mobility of nanomaterials in porous media	Environmental Science & Technology
Li [22]	2011	Transport and deposition of CeO_2 nanoparticles in water-saturated porous media	Water Research
Petosa [23]	2012	Transport of two metal oxide nanoparticles in saturated granular porous media: Role of water chemistry and particle coating	Water Research
Solovitch [24]	2010	Concurrent Aggregation and Deposition of TiO_2 Nanoparticles in a Sandy Porous Media	Environmental Science & Technology
Zhao [25]	2012	Transport and Retention Behavior of ZnO Nanoparticles in two Natural Soils: Effect of Surface Coating and Soil Composition	Journal of Nano Research

About half of the studied nanoparticles were uncoated, bare particles when introduced in the porous system. The other half was coated with various organic polymers, organic acids or surfactants. In technical applications, most nanoparticles are functionalized [26]. However, the characteristics of released nanoparticles are unclear [1]. Released nanoparticles may be embedded into or sorbed to a matrix [26]. They may have their engineered coating and aged coating [27], no coating or will become a substitute coating of e.g. natural substances after released in the environment. Summarized, the surface properties of engineered nanoparticles released and transformed or aged in the environment are still unclear [26, 28, 29]. Coatings may have a tremendous effect via charge or steric stabilization on the stability and therefore on the aggregation behavior and the mobility of nanoparticles [1, 24, 30-33]. Therefore, one important factor of the mobility of nanoparticles, the exact surface properties under environmental conditions, is not known [29].

Table 2: Selected nanoparticle characteristics used in the selected research studies

character	**nanoparticle specification (quantity)**
type	TiO_2 (9), ZnO (5), Fe_2O_3 (1), CuO (1), CeO_2 (1)
coating	without coating (6), CMC (2), SDBS (1), Triton X-100 (1), acetic acid (1), dispersant and polishing additives (1), polymer (1), citrate (1)
primary size	< 10 nm (3); 10-49 nm (11); 50-99 nm (3); > 99 nm (3)
size in suspension	< 100 nm (9); ~100-500 nm (14); ~500-1000 nm(3); >1000 nm (2)

Nanoparticles are commonly defined as particles with at least two or three dimension between 1 and 100 nm (e.g. [1, 34]). The primary sizes of the nanoparticles used in these studies are according to the manufactures or determined by microscopic methods mainly in this range. However, the nanoparticle sizes in the experimental suspensions with ionic strengths in the environmental range were about one order of magnitude bigger. One reason could be the size measurement with DLS, where the hydrodynamic radius is determined. But mainly the bigger size is due to aggregation. These studies are determining therefore transport of aggregated nanoparticles.

3.2. Porous media

Selected characteristics of the porous media are summarized in Table 3. The porous media used are mainly well-defined systems as glass beads and clean industrial sand. Only in two studies, real soil material was used. By using defined mineral surfaces (SiO_2), uniform and defined surfaces are constructed. In environmental systems, a big variety of different mineral surfaces concerning charge and surface functional groups contribute unspecific and specific sorption sites, which affect the retention of nanoparticles in soils. By using clean silica sand or glass beads, these factors are neglected. Additionally, by excluding any natural organic matter from such model systems, an important stabilizer for particles (natural colloids as well as engineered nanoparticles) is not present in these systems, which is always present in natural systems and may increase the mobility of nanoparticles in soils [31, 32, 35-37].

The grain sizes in the studies with glass beads and sand are ranging from fine sand to coarse sand with a respective porosity of the packed material. In such model systems, well-defined pore systems can be established, which allows generally a high reproducibility of the experimental data. By modeling these data, specific transport mechanism could be analyzed. However, in natural system, a great variety of particle sizes and shapes are present. Additionally only sparse data for smaller fractions like silt and clay or mixtures are available; and no data exist with aggregated material, secondary pore systems, or biological aggregates. Cornelis and colleagues assumed a colloid-mediated nanoparticles transport in soils [38], so smaller grain sizes in transport experiments should be taken into account.

Table 3: Selected characteristics of the porous media used in the selected research studies

character	**porous media specification (quantity)**
material	glass beads (2); clean silica sand (9); soil material (2)
average grain size	< 250 μm (2); 250-500 μm (6); 500-750 μm (5); > 750 μm (2)
porosity	< 0,30 (3); 0,30-0,39 (6); 0,40-0,49 (5); > 0,49 (2)

It is uncertain, if results of such model systems can be transferred to natural systems, when such important factors like different mineral compositions, a broad range of grain size and porosity, presence of natural organic matter, colloid-facilitated transport and preferential flow paths are not considered.

3.3. Experimental conditions

A summary of the experimental conditions of the selected studies is shown in Table 4. The ionic strengths used in the experiments are in the board range of the relevant aquatic and terrestrial systems. In some studies also the effects changes in the ionic strength are studied. This is also very probably for natural systems, e.g. in cases of heavy rainfall or flooding events. All studies conducted their experiments in a neutral pH range from 6-8, which is also most common for aquatic and terrestrial systems. Higher and lower pH values are also considered in some studies, which are relevant especially for soils, where pH values can be significant higher or lower. Darcy velocities ranging from below 0,01 up to over 1,00 cm/min, which is in the range which could be expected in natural soil systems.

Table 4: Selected experimental conditions used in the selected research studies

character	**experimental set-up specification (quantity)**
ionic strength	most 0,1 - 100 mM NaCl (some studies used KCl, NaBr, $NaNO_3$), and / or 0,01 - 10 mM $CaCl_2$
pH	< 6 (5); 6-8 (13); > 8 (4)
particle concentration [mg/l]	0-10 (5); 11-50 (8); 51-100 (3); >100 (1)
organic substance	without (10), with soil organic substance (2), Aldrich HA (1)
Darcy velocity [cm/min]	Up to 0,01 (3); 0,11-0,50 (5); 0,51-1,00 (4); > 1,00 (6)
time of NP application [PV]	1-5 (5); 6-10 (5); > 10 (3)

In most studies, the particle concentrations ranges between 5 and 50 mg/l, however, concentrations up to 800 mg/l were also used. The environmentally relevant concentrations for surface water for different nanoparticles are expected in the pg/l to low µg/l range [9, 10, 39]. Considering sewage sludge as a source for nanoparticles in soils, one has to anticipate with a load of mg/kg per year [9]. At high particle concentration, aggregation is favored and a lower mobility would be determined compared with conditions with lower particle concentration. One hint for the aggregation in the used nanoparticle suspension is the considerable higher particle size (in most cases one order of magnitude) in suspension compared to the primary size. Using particle concentrations in the range for the aquatic pathway of nanoparticles to soils is challenging, because the detection methods for most methods are not sufficiently low [9, 10, 28, 39]. However, it is questionable is the results are transferable to natural conditions with several orders of magnitude lower particle concentration.

Organic acid was added to the system only in one study. As discussed in the section 3.2, excluding natural organic matter, an important stabilizer for nanoparticles is omitted. It is assumed that organic molecules (short acids or fulvic and humic acids) will play an important role in stabilizing nanoparticles in environmental systems [31, 32, 35-37].

The period of time of the nanoparticle applications ranged between 1 and mostly 10 pore volumes. This seems an appropriate time for model studies. Only a few studies considered subsequent remobilization steps, which could be relevant for environmental systems, e.g. in case of heavy rainfall or flooding events. No long time behavior is considered.

3.4. Measurement methods

Selected measurement methods used in the studies are presented in Table 5. The effluent concentration was mostly measures by UV-vis or ICP-MS, -AES, or –OES. Measuring with ICP-MS, -AES, or –OES, the total concentration is given as result. No information is available about the distribution between particular and dissolved form. It had been shown that from UV-vis spectra, nanoparticle size and concentration can be determined [40]. However, most studies indicate only the particle concentration in the effluent. No further information of a possible change in particle size or the dissolution of nanoparticles, or a possible aging of nanoparticle coating was given. Additional experiments determining these parameters in a corresponding suspension and in the effluent could strengthen such data and their interpretation, but this is only partly done.

Table 5: Selected measurement methods used in the selected research studies

character	methods (quantity)
effluent NP concentration	UV-vis (7); ICP-AES, ICP-OES and ICP-MS for Zn and Ti (5), Fluorescence Spectroscopy (1)
particle size in suspension	DLS (13), NTA (1), laser diffraction (1)

The particle size in suspension is measured by DLS. With DLS, particle sizes from ~ 3 to 1000 nm in mg/l concentration range can be measured. Though particle size is calculated by calculating their movement from the measurement of the scattered light, many assumptions and simplifications have to be made, coming to a routine measurement result (detailed description see e.g. [39]). One of the assumptions is, that the measured particles are spherical, which should be proofed before the measurement. However, DLS was used for size determination also for non-spherical particles (e.g. [14]). Because DLS provide intensity based results, the bigger particles are overemphasized [34]. This results in difficulties in the interpretation of polydisperse suspension. Therefore giving the polydispersity index together with the particle size would be a help to interpret the results. With NTA, the particles size from 30 up to 1500 nm at lower concentrations (µg/l concentration range and lower) can be determined. However, if the particle concentration in the sample is too high, it have to de diluted, which alters its diffusion behavior and aggregation state [39]. Cross validation and additional specifications like size distribution or polydispersity index could strengthen the data, until better methods are available [34, 41].

3.5. Discussed theories and models

From the selected papers, in 11 studies transport theories and models are discussed and some specifications were calculated, as summarized in Table 6. Even if it is stated that the theories are not able to predict the transport behavior of nanoparticles [18], some specifications mainly from the classical colloid filtration theory were calculated and compared for different experimental conditions or for different nanoparticles. The interpretation of such results is difficult, because the models were not able to completely describe the observed nanoparticle transport in the model systems. The transfer to environmental systems is questionable because of the unconsidered factors as discussed above.

Table 6: Theories and models discussed and calculated specifications in the selected research studies

theories and models	specification (quantity)
classical colloid filtration theory (Yao et al 1971, a.m.)	attachment efficiency (6), deposition rate coefficient (5), L_{max} (4), collector efficiency (3)
DLVO-theory	interaction energy (4)

3.6. Main findings and conclusions

The main findings are summarized in Table 7. If investigated, a dependence of ionic strength, ion valence, pH value, presence of humic acid, surface coatings or surfactants were determined, which could be expected from classical colloid theories like the colloid filtration theory [42, 43] and the DLVO theory. In the conclusions, as transport determining mechanisms are mainly mentioned aggregation followed by DLVO interactions, straining of particles, blocking effects and disaggregation. Further it is concluded that a potential risk for groundwater exist, as well as that bare nanoparticles exhibited a very limited mobility.

Table 7: Main findings and conclusions of selected studies

main results / conclusions	specification (quantity)
mobility/deposition/retention of nanoparticles in porous media depends on	ionic strength (10), ion valence (4), pH (8), presence of humic acid, surface coatings or surfactants (5), nanoparticles characteristics (2)
transport determining mechanisms	aggregation (4), DLVO interactions (3), straining (2), blocking (2), disaggregation (2), deposition (1), ripening (1), pore plugging (1)
potential risk for	groundwater (3)
limited mobility	for bare nanoparticles (3)

3.7. Mentioned uncertainties

The mentioned uncertainties in the selected research papers are summarized in Table 8. Uncertainty itself is rarely mentioned. More often the sources (e.g. lack of analytical methods) or consequences (e.g. questionable transferability) of uncertainties are considered. Only in two studies, uncertainties related to basic research are discussed. These are measuring uncertainties related to the size determination by DLS and NTA and the discussion of reasons of contradicting results of studies with different nanoparticles. The uncertainties related to applied research are mostly discussed in the conclusions or the sections dealing with environmental consequences. Mainly the transferability of the results to environmental conditions is discussed with different foci.

Table 8: Mentioned uncertainties in selected studies

uncertainties	specification (quantity)
measurement uncertainties	size (1)
contradicting results	different characteristics of nanoparticles (1)
transferability to environmental conditions	without specification (4), due to presence of organic molecules (3), preferential pathways (2), high variability (1)
transferability to other nanoparticles	type (2)

3.8. Certain and Uncertain Knowledge and Knowledge Gaps

Certain knowledge about the mobility of metal oxide nanoparticle in porous media is restricted to model systems with a limited number of studies with a limited number of nanoparticles. Generally, nanoparticle concentrations are used, which are much higher than the expected environmental concentrations [9, 10, 33]. The water saturated porous media consist mainly of clean sand without any natural organic matter or other minerals and in in absence of smaller grain sizes or any secondary pores. The identified factors controlling the mobility are ionic strength, ion valence, pH, presence of humic acid, surface coatings or surfactants, which could be expected from classical colloid theories. It is concluded, that various mechanisms as e.g. aggregation and disaggregation, pore plugging and deposition are affecting the dynamic nanoparticle transport, indicating that classical filtration theory often fail to describe completely the transport behavior [14, 15, 17, 18].

The uncertainties in the fields of basic research could be found in the small amount of papers dealing with our topic, in the measurement uncertainty, the sparse characterization of the systems and the fragile validation of the results. Cross validation of particle size determination to validate the data is rarely done. By measuring the effluent concentration only, any possible changes in nanoparticle characteristics as dissolution, aggregation or transformation or aging of coatings are not considered. Because the mobility of different nanoparticles with different sizes and coatings differ significantly [13, 23, 25], comparability from one nanoparticle type to the other type or size or coating is questionable. Therefore, a broad validation of single findings is difficult to date. Defining reference materials and procedures would be helpful in order to compare results.

The uncertainties in the fields of applied research are mostly the transferability of the results to environmental conditions. Because most studies were conducted in the absence of natural organic matter, other minerals than silicium dioxide, smaller grain sizes or secondary pore systems which may affect the retention in soils and sediments, the transferability of these data from model systems to environmental systems is questionable. A good half of the authors mentioned or discussed this uncertainty in their papers.

Major knowledge gaps are the unknown form and the unknown amount of engineered nanoparticles at their release in the environment and their transformation and aging under environmental conditions [1, 26, 33, 44]. The fact, that the surface characteristics, which greatly affect the nanoparticle mobility, are not known for environmental conditions, is discussed very rarely in the selected studies.

3.9. Sources and Consequences of Uncertainties and Knowledge Gaps

The main source of these uncertainties and knowledge gaps is caused by the lack of available methods to separate, characterize and detect engineered nanoparticles in environmental media at environmental concentrations [1, 26, 33, 45]. One of the biggest challenges is to separate and characterize the small amount of engineered nanoparticles in environmental matrices, which contain high amounts of highly heterogenic natural nanoparticles [34].

Because of the lack of analytical methods, the form of nanoparticles at release and the surface properties transformed and aged in the environment are not known [1, 26, 33, 44]. Therefore, one important factor of the mobility of nanoparticles in the environment, the exact surface properties, is still unclear [29]. The characteristics of the chosen bare or coated nanoparticles may be not relevant under environmental conditions.

As consequences of these uncertainties knowledge gaps, any risk assessment, regulation or management concerning engineered nanoparticles based yet on modeling data or studies in model systems, which transferability to environmental conditions is questionable as discussed above. Studies under more natural conditions are challenging to date.

Acknowledgements - The author is grateful for financial support by the Ministry for Education, Sciences, Further Education and Culture of the State of Rhineland-Palatinate (MBWWK), Germany. The author thanks the group Environmental and Soil Chemistry of G.E. Schaumann at the University of Koblenz-Landau for useful discussion, comments and suggestions.

References

1. Klaine, SJ, Alvarez, PJJ, et al. 2008 Nanomaterials in the environment: Behavior, fate, bioavailability, and effects. Environ. Toxicol. Chem., 27 (9) 1825-1851.
2. Aschberger, K, Micheletti, C, et al. 2011 Analysis of currently available data for characterising the risk of engineered nanomaterials to the environment and human health - Lessons learned from four case studies. Environ. Int., 37 (6) 1143-1156.
3. Klaine, SJ, Koelmans, AA, et al. 2012 Paradigms to assess the environmental impact of manufactured nanomaterials. Environ. Toxicol. Chem., 31 (1) 3-14.
4. Kwon, SY, Kim, YG, et al. 2011 Uncertainty analysis of measurements of the size of nanoparticles in aqueous solutions using dynamic light scattering. Metrologia, 48 (5) 417-425.
5. Wagner, R, Moon, R, et al. 2011 Uncertainty quantification in nanomechanical measurements using the atomic force microscope. Nanotechnology, 22 (45).
6. Grieger, KD, Hansen, SF and Baun, A 2009 The known unknowns of nanomaterials: Describing and characterizing uncertainty within environmental, health and safety risks. Nanotoxicology, 3 (3) 1-U17.
7. Benn, TM and Westerhoff, P 2008 Nanoparticle silver released into water from commercially available sock fabrics. Environ. Sci. Technol., 42 (11) 4133-4139.
8. Brar, SK, Verma, M, et al. 2010 Engineered nanoparticles in wastewater and wastewater sludge - Evidence and impacts. Waste Manage., 30 (3) 504-520.
9. Gottschalk, F, Sonderer, T, et al. 2009 Modeled Environmental Concentrations of Engineered Nanomaterials (TiO_2, ZnO, Ag, CNT, Fullerenes) for Different Regions. Environ. Sci. Technol., 43 (24) 9216-9222.
10. Mueller, NC and Nowack, B 2008 Exposure modeling of engineered nanoparticles in the environment. Environ. Sci. Technol., 42 (12) 4447-4453.
11. Piccinno, F, Gottschalk, F, et al. 2012 Industrial production quantities and uses of ten engineered nanomaterials in Europe and the world. J. Nanopart. Res., 14 (9).
12. Som, C, Wick, P, et al. 2011 Environmental and health effects of nanomaterials in nanotextiles and facade coatings. Environ. Int., 37 (6) 1131-1142.
13. Ben-Moshe, T, Dror, I and Berkowitz, B 2010 Transport of metal oxide nanoparticles in saturated porous media. Chemosphere, 81 (3) 387-393.
14. Chen, GX, Liu, XY and Su, CM 2011 Transport and Retention of TiO_2 Rutile Nanoparticles in Saturated Porous Media under Low-Ionic-Strength Conditions: Measurements and Mechanisms. Langmuir, 27 (9) 5393-5402.
15. Chowdhury, I, Hong, Y, et al. 2011 Mechanisms of TiO_2 nanoparticle transport in porous media: Role of solution chemistry, nanoparticle concentration, and flowrate. J. Colloid Interface Sci., 360 (2) 548-555.
16. Fang, J, Shan, XQ, et al. 2009 Stability of titania nanoparticles in soil suspensions and transport in saturated homogeneous soil columns. Environ. Pollut., 157 (4) 1101-1109.
17. Godinez, IG and Darnault, CJG 2011 Aggregation and transport of nano-TiO_2 in saturated porous media: Effects of pH, surfactants and flow velocity. Water Res., 45 (2) 839-851.
18. Jiang, XJ, Tong, MP, et al. 2012 Transport and deposition of ZnO nanoparticles in saturated porous media. Colloids Surf., A, 401 29-37.
19. Joo, SH, Al-Abed, SR and Luxton, T 2009 Influence of Carboxymethyl Cellulose for the Transport of Titanium Dioxide Nanoparticles in Clean Silica and Mineral-Coated Sands. Environ. Sci. Technol., 43 (13) 4954-4959.
20. Kanel, SR and Al-Abed, SR 2011 Influence of pH on the transport of nanoscale zinc oxide in saturated porous media. J. Nanopart. Res., 13 (9) 4035-4047.
21. Lecoanet, HF, Bottero, JY and Wiesner, MR 2004 Laboratory assessment of the mobility of nanomaterials in porous media. Environ. Sci. Technol., 38 (19) 5164-5169.
22. Li, Z, Sahle-Demessie, E, et al. 2011 Transport and deposition of CeO_2 nanoparticles in water-saturated porous media. Water Res., 45 (15) 4409-4418.

23. Petosa, AR, Brennan, SJ, et al. 2012 Transport of two metal oxide nanoparticles in saturated granular porous media: Role of water chemistry and particle coating. Water Res., 46 (4) 1273-1285.
24. Solovitch, N, Labille, J, et al. 2010 Concurrent Aggregation and Deposition of TiO_2 Nanoparticles in a Sandy Porous Media. Environ. Sci. Technol., 44 (13) 4897-4902.
25. Zhao, LJ, Peralta-Videa, JR, et al. 2012 Transport and Retention Behavior of ZnO Nanoparticles in two Natural Soils: Effect of Surface Coating and Soil Composition. J. Nano Res., 17 229-242.
26. Nowack, B and Bucheli, TD 2007 Occurrence, behavior and effects of nanoparticles in the environment. Environ. Pollut., 150 (1) 5-22.
27. Auffan, M, Pedeutour, M, et al. 2010 Structural Degradation at the Surface of a TiO2-Based Nanomaterial Used in Cosmetics. Environ. Sci. Technol., 44 (7) 2689-2694.
28. Farre, M, Sanchis, J and Barcelo, D 2011 Analysis and assessment of the occurrence, the fate and the behavior of nanomaterials in the environment. TrAC, Trends Anal. Chem., 30 (3) 517-527.
29. Mudunkotuwa, IA and Grassian, VH 2011 The devil is in the details (or the surface): impact of surface structure and surface energetics on understanding the behavior of nanomaterials in the environment. J. Environ. Monit., 13 (5) 1135-1144.
30. Domingos, RF, Tufenkji, N and Wilkinson, KJ 2009 Aggregation of Titanium Dioxide Nanoparticles: Role of a Fulvic Acid. Environ. Sci. Technol., 43 (5) 1282-1286.
31. Ghosh, S, Mashayekhi, H, et al. 2010 Colloidal Stability of Al_2O_3 Nanoparticles as Affected by Coating of Structurally Different Humic Acids. Langmuir, 26 (2) 873-879.
32. Jaisi, DP and Elimelech, M 2009 Single-Walled Carbon Nanotubes Exhibit Limited Transport in Soil Columns. Environ. Sci. Technol., 43 (24) 9161-9166.
33. Peralta-Videa, JR, Zhao, LJ, et al. 2011 Nanomaterials and the environment: A review for the biennium 2008-2010. J. Hazard. Mater., 186 (1) 1-15.
34. Tiede, K, Boxall, ABA, et al. 2008 Detection and characterization of engineered nanoparticles in food and the environment. Food Addit. Contam., 25 (7) 795-821.
35. Pelley, AJ and Tufenkji, N 2008 Effect of particle size and natural organic matter on the migration of nano- and microscale latex particles in saturated porous media. J. Colloid Interface Sci., 321 (1) 74-83.
36. Navarro, E, Baun, A, et al. 2008 Environmental behavior and ecotoxicity of engineered nanoparticles to algae, plants, and fungi. Ecotoxicology, 17 (5) 372-386.
37. Keller, AA, Wang, HT, et al. 2010 Stability and Aggregation of Metal Oxide Nanoparticles in Natural Aqueous Matrices. Environ. Sci. Technol., 44 (6) 1962-1967.
38. Cornelis, G, Ryan, B, et al. 2011 Solubility and Batch Retention of CeO_2 Nanoparticles in Soils. Environ. Sci. Technol., 45 (7) 2777-2782.
39. Hassellov, M, Readman, JW, et al. 2008 Nanoparticle analysis and characterization methodologies in environmental risk assessment of engineered nanoparticles. Ecotoxicology, 17 (5) 344-361.
40. Haiss, W, Thanh, NTK, et al. 2007 Determination of size and concentration of gold nanoparticles from UV-Vis spectra. Anal. Chem., 79 (11) 4215-4221.
41. von der Kammer, F, Ferguson, PL, et al. 2012 Analysis of engineered nanomaterials in complex matrices (environment and biota): General considerations and conceptual case studies. Environ. Toxicol. Chem., 31 (1) 32-49.
42. Yao, K-M, Habibian, MT and O'Melia, CR 1971 Water and waste water filtration. Concepts and applications. Environ. Sci. Technol., 5 (11) 1105-1112.
43. Tufenkji, N and Elimelech, M 2004 Correlation equation for predicting single-collector efficiency in physicochemical filtration in saturated porous media. Environ. Sci. Technol., 38 (2) 529-536.
44. Nowack, B, Ranville, JF, et al. 2012 Potential scenarios for nanomaterial release and subsequent alteration in the environment. Environ. Toxicol. Chem., 31 (1) 50-9.
45. Christian, P, Von der Kammer, F, et al. 2008 Nanoparticles: structure, properties, preparation and behaviour in environmental media. Ecotoxicology, 17 (5) 326-343.

Journal of Physics: Conference Series **429** (2013) 012043 doi:10.1088/1742-6596/429/1/012043

Carbon nanotubes enhanced the lead toxicity on the freshwater fish

D S T Martinez[1], O L Alves[1] and E Barbieri[2]

[1]Solid State Chemistry Laboratory (LQES), Institute of Chemistry, University of Campinas – UNICAMP, P. O. Box 6154, 13083-970, Campinas-SP, Brazil.

[2]Instituto de Pesca, APTA – SAA/SP, P.O. Box 61, 11990-000, Cananéia-SP, Brazil.

E-mail: diegostefani.br@gmail.com, oalves@iqm.unicamp.br, edisionbarbieri@yahoo.com.br

Abstract. Carbon nanotubes are promising nanostructures for many applications in materials industry and biotechnology. However, it is mandatory to evaluate their toxicity and environmental implications. We evaluated nitric acid treated multiwalled carbon nanotubes (HNO_3-MWCNT) toxicity in Nile tilapia (*Oreochromis niloticus*) and also the lead (Pb) toxicity modulation after the nanotube interaction. Industrial grade multiwalled carbon nanotubes [Ctube 100, CNT Co. Ltd] were treated with 9M HNO_3 for 12h at 150°C to generate oxygenated groups on the nanotube surface, to improve water dispersion and heavy metal interaction. The HNO_3-treated multiwalled carbon nanotubes were physico-chemically characterized by several techniques [e.g. TEM, FE-SEM, TGA, ζ-potential and Raman spectroscopy]. HNO_3-MWCNT did not show toxicity on Nile tilapia when the concentration ranged from 0.1 to 3.0 mg/L, and the maximum exposure time was 96h. After 24, 48, 72 and 96h the LC50 values of Pb were 1.65, 1.32, 1.10 and 0.99 mg/L, respectively. To evaluate the Pb-nanotube interaction influence on the ecotoxicity, we submitted the Nile tilapia to different concentrations of Pb mixed with a non-toxic concentration of HNO_3-MWCNT (1.0 mg/L). After 24, 48, 72, 96 h the LC50 values of Pb plus nanotubes were: 0.32, 0.25, 0.20, 0.18 mg/L, respectively. These values showed a synergistic effect after Pb-nanotube interaction since Pb toxicity increased over five times. X-ray energy dispersive spectroscopy (EDS) was used to confirm lead adsorption on the carbon nanotube oxidized surface. The exposure of Nile tilapia to Pb plus HNO_3-MWCNT caused both oxygen consumption and ammonium excretion decrease, when compared to the control. Finally, our results show that carbon nanotubes interact with classical pollutants drawing attention to the environmental implications.

1. Introduction

Safe development of nanotechnology depends on the ability to control both health and environmental risks [1].

Carbon nanotubes (CNTs) are among the most promising materials for nanotechnology and their applications demand chemical treatments (e.g. purification and oxidation) of the raw samples. Nitric acid treatment is one of the most used methods for achieving this goal [2]. The surface of CNTs is oxidized during this process, which generates oxygen functional groups such as –OH, –C=O, and –

COOH [3]. Environmental applications have explored these oxygenated surface groups on nanotubes and their large surface area to remove and monitor polar and non polar pollutants in the environment (e.g. PHAs, dioxins, pesticides and heavy metals) [4].

Lead (Pb) is a cumulative and insidious environmental pollutant that induces a broad range of physiological, biochemical and behavioral dysfunctions in living organisms [5]. With the increasing aquatic environment contamination, the potential interaction of Pb with nanomaterials needs to be considered. The CNTs industry is growing very fast and industrial production has already reached several tons a year. Therefore, the investigation of the interaction of classical pollutants with nanotubes is an important issue and requires more attention [6].

Freshwater fish is an important source of protein and lipids to human beings and animals. Nile tilapia (*Oreochromis niloticus*) is not only a very nutritious food, but also an important indicator of environmental contamination and water quality [7,8,9].

We evaluated the nitric acid treated multiwalled carbon nanotubes (HNO_3-MWCNT) ecotoxicity on the Nile tilapia as well as lead (Pb) toxicity modulation after nanotube interaction. This work aims to contribute to the development of a safe nanocarbon-based technology.

2. Experimental

Commercially available chemical vapour deposition (CVD) grown multiwall carbon nanotubes (MWCNT) [*Ctube 100*, CNT Co. Ltd., Incheon – South Korea] were used. $PbCl_2$ (analytical grade) was obtained from Reagen Ltd and used without further purification.

2.1.1. *Nitric acid treatment of multiwalled carbon nanotubes.* The raw MWCNT (1.0 g) was refluxed in HNO_3 9.0 mol/L (200 mL) for 12 h at 150 ºC. After cooling to room temperature, they were filtered through a 0.2 µm PTFE membrane and washed with deionized water until neutral pH of the filtrate was reached. The HNO_3-MWCNT sample was dried in a vaccum system for 24 h.

2.1.2. *Characterization of HNO_3-MWCNT sample.* HNO_3-treated multiwalled carbon nanotubes were physico-chemically characterized by using the following techniques: a) Transmission electronic microscopy [TEM – Zeiss], b) Field emission scanning electronic microscopy [FEG-SEM – FEI], c) thermogravimetric analysis [TGA, SDTQ600 TA Instruments], d) ζ-potential [Malvern Instruments], e) Raman spectroscopy [WiTech].

2.1.3. *HNO_3-MWCNT stock-solution preparation.* 100 mg of HNO_3-MWCNT were dispersed in 200 mL of deionized water and sonicated for 60 min in an ultrasound bath [Cole-parmer 8891].

2.1.4. *Acute ecotoxicity experiments.* The acute toxicity of HNO_3-MWCNT and Pb was evaluated in Nile tilapia (*Oreochromis niloticus*) juveniles cultivated in the laboratory of Fisheries Institute (Cananéia, Sao Paulo state, Brazil). They were exposed to different concentrations of these chemicals for a period of up to 96 h. Groups of 15 individuals were exposed to each of the following concentrations of HNO_3-MWCNT: 0.1, 0.5, 1.0, 2.0, 3.0 and Pb: 0.00, 0.04, 0.4, 0.8, 1.12 and 2.12 mg/L, respectively. After this procedure, we submitted the juveniles to concentrations of 0.04, 0.4, 0.8, 1.12, and 2.12 mg/L Pb mixed with a non-toxic concentration of HNO_3-MWCNT (1.0 mg/L). Prior to this, nanotubes were maintained in contact with Pb for 1 h. Dead fishes were removed from the tanks and counted at 24, 48, 72, and 96 h of exposure. The lethal concentration (LC50 with 95% confidence limits) was calculated by Probit analysis. These experiments were performed in triplicates using glass tanks of 50 L.

2.1.5. *Oxygen consumption and Ammonia excretion measurements.* The respirometer was made in our laboratory, with an acrylic tube and PVC covers. Ten fish were subjected, individually, to oxygen consumption and ammonia excretion measurements in the presence of HNO_3-MWCNT (1.0 mg/L), Pb (1.0 mg/L) and the mixture of nanotubes and Pb (1:1 ratio). Before beginning the experiments, the animals were maintained in the respirometer with continuous water circulation for at least 90 min to attenuate handling stress. The difference between the oxygen and ammonia concentrations determined at the beginning and the end of the confinement was used to calculate consumption and excretion after three hours. Dissolved oxygen was determined according to Winkler Method (Winkler, 1888). Determination of ammonium-nitrogen in the freshwater was based on the phenolhypochlorite method (Solarzano, 1969). Nile tilapia juveniles oxygen-specific consumption and ammonium excretion average were analyzed by using the Tukey's multiple comparisons test ($P<0.05$).

3. Results and Discussion

Due to the economical and ecological importance of freshwater fish and the problems related to water pollution, we evaluated the effects of oxidized multiwalled carbon nanotubes and Pb interaction on Nile tilapia juveniles. Industrial grade multiwalled carbon nanotubes were oxidized with nitric acid, aiming to generate oxygenated groups on the nanotube surface as well as to improve their water dispersion. Fig. 1 shows the nanotubes transmission electronic microscopy (TEM) and scanning electronic microscopy (FEG-SEM) representative images, after the oxidation process with nitric acid (9.0 mol/L) at 150 °C for 12 hours under conventional reflux system. Table 1 shows a summary of the main physico-chemical properties of HNO_3-MWCNT.

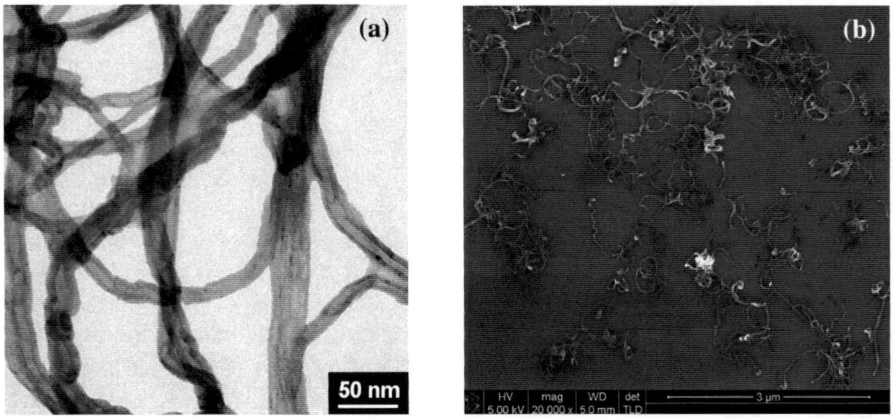

Fig. 1. (a) Transmission electronic microscopy [TEM] and (b) scanning electronic microscopy [FEG-SEM] of acid treated multiwalled carbon nanotubes (HNO_3-MWCNT).

Table 1. Summary of the physico-chemical properties of HNO_3-treated MWCNT used in this work.

Physico-chemical properties	HNO_3-MWCNT	Technique used
Diameter (nm)	10 ~ 40	TEM
Length (μm)	< 5	FEG-SEM
Surface area (m^2/g)	264	BET method
Surface charge (mV)	-27	Zeta-potential
Oxidation temperature (°C)	587	TGA
Iron oxide metallic residue (%)	< 2.0	Analytical microbalance
Defects (I_D/I_G ratio)	1.02	Raman

The effects of HNO_3-MWCNT sample (from 0.1 until 3.0 mg/L) were evaluated according to the acute toxicity assay for Nile tilapia (24, 48, 72 and 96 h). No deaths were observed (data not shown). However, the higher the concentration of metal the fish were exposed to, the higher the observed mortality. After being exposed to Pb, death was first observed at a concentration of 1.12 mg Pb/L in the first 24 h. Mortality rate of 100% was observed after a 24h exposure at concentrations of 2.12 mg Pb/L and after 72 h at a concentration of 1.12 mg Pb/L. The acute toxicity of Pb to Nile tilapia exposed to different concentrations of this metal, expressed as the medium lethal concentration (LC50), were 1.64, 1.32, 1.10 and 0.98 mg/L for 24, 48, 72, and 96h exposure, respectively. Furthermore, we exposed the Nile tilapia to different concentrations of Pb mixed with a non-toxic concentration of HNO_3-MWCNT (1.0 mg/L) to evaluate the influence of Pb-nanotube interaction on the ecotoxicity. After 24, 48, 72, 96 h the LC50 values of Pb plus nanotubes were: 0.32, 0.25, 0.20, 0.18 mg/L, respectively. It was observed an increase over five times of Pb toxicity on the freshwater fish, which demonstrates that there is a synergistic effect after Pb-nanotube interaction (Fig. 2).

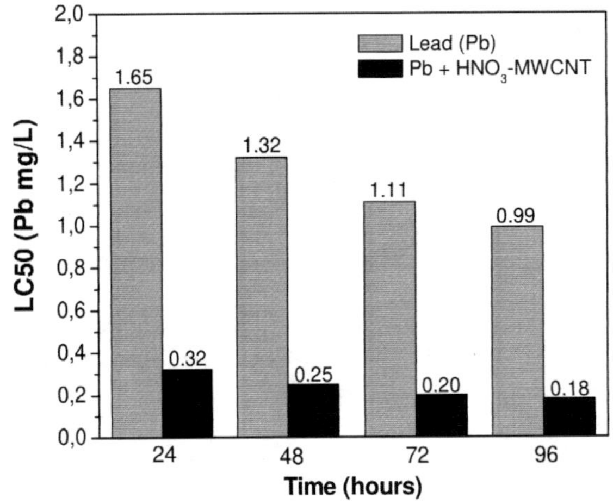

Fig. 2. Modulation of the LC50 toxicity value of lead (Pb) in the freshwater fish, Nile tilapia (*Oreochromis niloticus*) after the interaction with nitric acid treated multiwalled carbon nanotubes (HNO_3-MWCNT). 1.0 mg/L of nanotubes were used during this experiment, a non lethal dose for freshwater fish.

Xu et al. 2011 [10] investigated acute toxicity and synergism of four antifouling biocides (e.g. Irgarol 1051, dichlofluanid, tolylfluanid and Sea-Nine 211) and five heavy metals (e.g. Ni, Pb, Zn, Cd and Cu) using sea urchin embryos (*Glyptocidaris crenularis*). This study reported synergistic effects in almost all 25 mixtures. We also found a synergism effect when we used a mixture of Pb and oxidized multiwalled carbon nanotubes for Nile tilapia.

In order to get further insights into the adsorption of lead on the nanotubes, we performed analysis of HNO_3-MWCNT after incubation in freshwater containing 1.0 mg/L of Pb by using X-ray energy dispersive spectroscopy (EDS) (Fig. 3). We can observe that after the interaction with freshwater containing Pb the nanotubes presented this element on their composition as well as Mg and Ca. Similar results have been reported in the literature [11]. The proposed mechanism to this chemical interaction is the electrostatic force between the lead ions (e.g. Pb^{2+}, positively charged) and the oxygenated groups on the nanotube surface (e.g. –COOH, negatively charged) [12]. However, colloidal chemistry aspects (e.g. adsorption/desorption and aggregation/agglomeration) on the toxicity assessment of this mixture (Pb + nanotubes) should be studied in more detail.

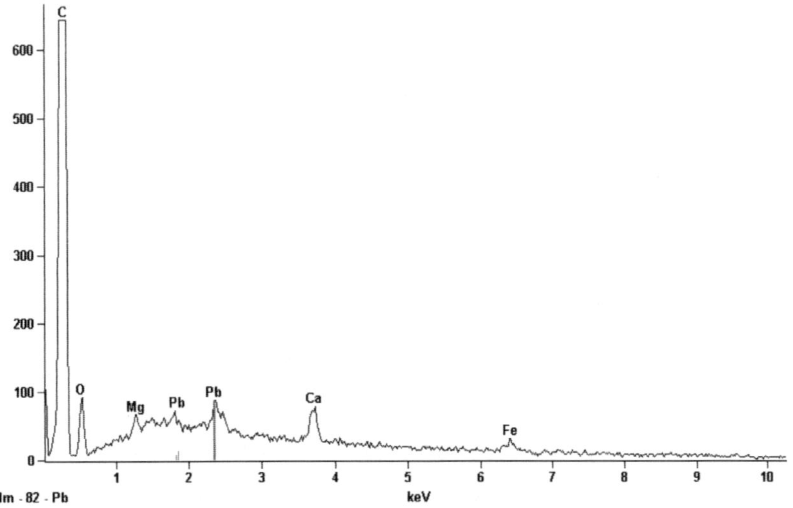

Fig. 3. X-ray energy dispersive spectroscopy (EDS) of HNO_3-MWNCT after incubation in freshwater containing of 1 mg/L of Pb.

Oxygen consumption and ammonia excretion measurements provide a good indicator of animal stress and water quality [13]. We evaluated the effects of Pb, HNO_3-MWCNT and mixture (Pb + HNO_3-MWCNT) at a non lethal concentration (1.0 mg/L) on these physiological parameters after three hours of exposure.

The specific oxygen consumption decreased in the presence of Pb, carbon nanotubes and the mixture of Pb-nanotubes treatments. The average of specific oxygen consumption of fish exposed to Pb, HNO_3-MWCNT and mixture (Pb + HNO_3-MWCNT) at a fixed concentration (1.0 mg/L) were 0.32, 0.52, 0.05 mL oxygen/g/min, respectively. These values represent a metabolic level decrease of 74%, 58% and 95% when compared to the control (Fig. 4).

Using the Tukey ($p<0.05$) statistical test, it was verified that the averages of the specific ammonia excretion in the groups treated with nanotubes and Pb were not different when compared to the control (Fig. 5). However, the ammonia excretion decreased considering the mixture (Pb + HNO_3-MWCNT). Thus, we found a clear synergistic effect when the mixture was evaluated. Our future work will focus on both the acute effects of these nanotubes on Nile tilapia at other biological levels such as histological and biochemical, and chronic effects on metabolism, molting and growth rates, which are also very important for the toxicological effects.

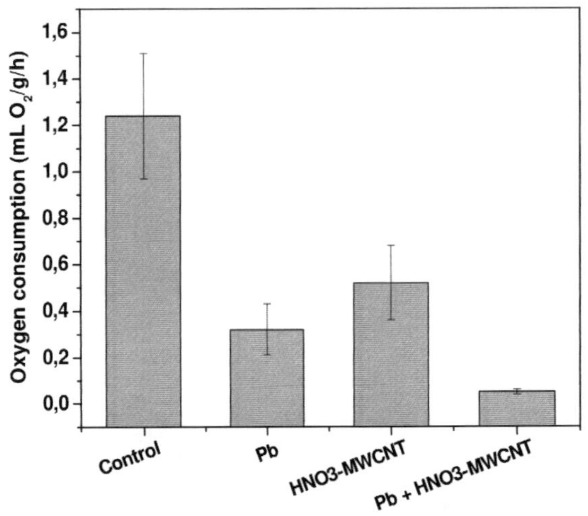

Fig. 4. Average of oxygen consumption in the freshwater fish, Nile tilapia (*Oreochromis niloticus*) after exposure to nitric acid treated multiwalled carbon nanotubes (HNO_3-MWCNT), Pb and mixture (Pb + HNO_3-MWCNT).

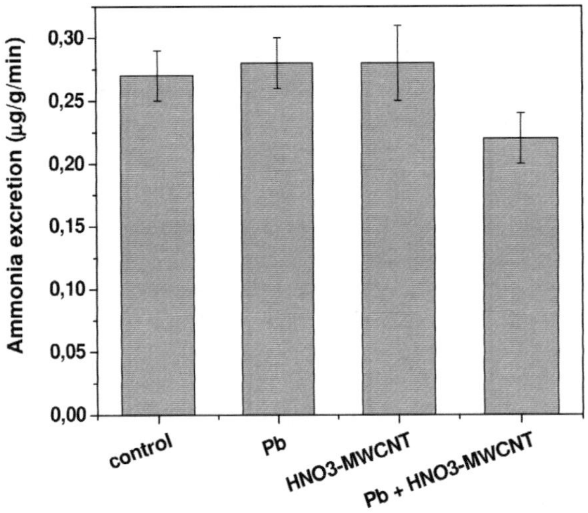

Fig. 5. Average of ammonia excretion in the freshwater fish, Nile tilapia (*Oreochromis niloticus*) after the exposure with nitric acid treated multiwalled carbon nanotubes (HNO_3-MWCNT), Pb and mixture (Pb + HNO_3-MWCNT).

4. Conclusion

The oxidized multiwall carbon nanotubes (HNO_3-MWCNT) enhanced the lead acute toxicity on the freshwater fish Nile tilapia (*Oreochromis niloticus*). An increase over five times on the LC50 value of lead after the nanotube interaction was observed. The carbon nanotubes studied did not show acute toxicity on the freshwater fish up to 3.0 mg/L. Furthermore, the nanotubes decreased the oxygen consumption and ammonia excretion after lead interaction. Since the industrial production of carbon nanotubes has already reached tons per year, this work draw attention for the implications of carbon nanomaterials released in the aquatic environment and their interaction with classical pollutants.

Acknowledgments

The authors are grateful to CNPq and FAPESP for financial support. This is a contribution of the National Institute of Science, Technology and Innovation on Complex Functional Materials (Inomat) and the Brazillian Nanotoxicology Network (CiGeNanotox).

5. References

[1] Stéfani D, Paula A J, Vaz, B G, Silva, R A, Andrade, N F, Justo, G Z; Ferreira, C V, Filho, A. G., Eberlin, M. N., Alves, O. L 2011. Structural and proactive safety aspects of oxidation debris from multiwalled carbon nanotubes *Journal of Hazardous Materials* **189** 391-396

[2] Datsyuk V, Kalyva M, Papagelis K, Parthenios J, Tasis D, Siokou A, Kallitsis I, Galiotis C **2008**. Chemical oxidation of multiwalled carbon nanotubes *Carbon* **46 (6)** 833-840

[3] Marques, R. R. N.; Machado, B. F.; Faria, J. L.; Silva, A. M. T., Controlled generation of oxygen functionalities on the surface of Single-Walled Carbon Nanotubes by HNO_3 hydrothermal oxidation *Carbon* **48 (5)**, 1515-1523

[4] Wu W H, Chen W, Lin D H, Yang K 2012. Influence of Surface Oxidation of Multiwalled Carbon Nanotubes on the Adsorption Affinity and Capacity of Polar and Nonpolar Organic Compounds in Aqueous Phase *Environmental Science & Technology* **46 (10)** 5446-5454

[5] Dai W, Liu S X, Fu L L, Du H H, Xu Z R 2012. Lead (Pb) accumulation, oxidative stress and DNA damage induced by dietary Pb in tilapia (Oreochromis niloticus) *Aquaculture Research* **43 (2)** 208-214

[6] Umbuzeiro G A, Coluci V R, Honorio J G, Giro R, Moraes D A, Lage A S G, Mazzei J L, Felzenszwalb I, Souza Filho A G, Stéfani D, Alves O L 2011. Understanding the interaction of multi-walled carbon nanotubes with mutagenic organic pollutants using computational modeling and biological experiments *TRAC Trends in Analytical Chemistry* **30 (3)** 437-446

[7] Navarro R D, Navarro F K S P, Ribeiro O P, Ferreira W M, Pereira M M, Seixas J T 2012. Quality of polyunsaturated fatty acids in Nile tilapias (Oreochromis niloticus) fed with vitamin E supplementation *Food Chemistry* **134 (1)** 215-218

[8] Kaya H, Akbulut M, Celik E S, Yilmaz S, Aydin F, Duysak M 2011. Effects of lead nitrate on haematological and immunological parameters in the tilapia (Oreochromis mossambicus, L. 1758) *Toxicology Letters* **205** S128-S132

[9] Rashed M N 2011. Cadmium and lead levels in fish (Tilapia nilotica) tissues as biological indicator for lake water pollution. *Environmental Monitoring Assessment* **68 (1)** 75-89

[10] Wang X 2011. Acute toxicity and synergism of binary mixtures of antifouling biocides with heavy metals to embryos of sea urchin Glyptocidaris crenularis. *Human & Experimental Toxicology* **30 (8)** 1009-1021

[11] Li Y H, Wang S G, Wei J Q, Zhang X F, Xu C L, Luan Z K, Wu D H, Wei B Q 2002 Lead adsorption on carbon nanotubes *Chemical Physics Letters* **357 (4)** 263-266

[12] Yu X Y, Luo T, Zhang Y X, Jia Y, Zhu B J, Fu X C, Liu J H, Huang X J 2011. Adsorption of Lead(II) on O-2-Plasma-Oxidized Multiwalled Carbon Nanotubes: Thermodynamics, Kinetics, and Desorption *ACS Applied Materials and Interfaces* **3 (7)** 2585-2593

[13] Barbieri E, Paes E T 2011. *The* use of oxygen consumption and ammonium excretion to evaluate the toxicity of cadmium on Farfantepenaeus paulensis with respect to salinity *Chemosphere* **84(1)** 9-16

Nanosafe 2012: International Conferences on Safe Production and Use of Nanomaterials IOP Publishing
Journal of Physics: Conference Series **429** (2013) 012044 doi:10.1088/1742-6596/429/1/012044

Monitoring migration and transformation of nanomaterials in polymeric composites during accelerated aging

G Vilar[1,2] *, E Fernández-Rosas[1,2] *, V Puntes[2,3], V Jamier[2], L Aubouy[1,2], S Vázquez-Campos[1,2]

[1] *LEITAT Technological Center, C/ Innovació 2, 08225 Terrassa, Barcelona, Spain*
[2] *CNBSS, Centre for NanoBioSafety and Sustainability, Campus de la UAB, Edifici CM7 08193 Bellaterra (Barcelona), Spain*
[3] *Catalan Institute of Nanotechnology (ICN), Campus UAB, 08193 Bellaterra, Spain*
* *These authors have contributed equally to this work.*
E-mail: svazquez@leitat.org

Abstract. The incorporation of small amounts of nanoadditives in polymeric compounds can introduce new mechanical, physical, electrical, magnetic, thermal and/or optical properties. The properties of these advanced materials have enabled new applications in several industrial sectors (electronics, automotive, textile...). In particular, for the nanomaterials (NM) described in this work, multi-walled carbon nanotubes (MWCNT) and silicon dioxide nanoparticles (SiO_2 NP), the following properties have been described: MWCNT act as nucleating agents in thermoplastics, and change viscosity, affecting dispersion, orientation, and therefore mechanical, thermal, and electrical properties; and SiO_2 NP act as flame retardant and display improved electrical and mechanical properties.

The work described here is focused on the evaluation of the migration and transformation of NM included in polymer nanocomposites (NC) during accelerated climatic ageing. To this aim, we generated polyamide 6 (PA6) NC with different degree of compatibility between the NM and the polymeric matrix. These NC were submitted to accelerated aging conditions to simulate outdoor conditions (simulation of the use phase of the polymeric NC). The NC contain as nanofillers MWCNT and SiO_2 NP with different surface properties to influence the compatibility with the polymeric matrix. The generated NC were evaluated by scanning electron microscopy (SEM), transmission electron microscopy (TEM) with Energy-dispersive X-ray spectroscopy (EDX), thermogravimetry (TGA) and differential scanning calorimetry (DSC) before and after the aging process, to monitor the compatibility of the NM with the matrix: dispersion within the matrix, migration during aging, and modification of the polymer properties.

The dispersion of SiO_2 NP in the NC depended on their compatibility with the matrix. However, independently of their compatibility with the matrix, SiO_2 NP were aggregated at the end of the accelerated aging process. In addition, degradation of the matrix and migration of nanoparticles to the surface was observed as well in the different types of SiO_2 NP aged NC. Oppositely, compatibilized MWCNT ($MWCNT_{MB}$) decreased the degradation of the polymer. Nevertheless, the nanomaterial migrated likewise to the surface during the ageing process.

In order to evaluate the possible changes in the structure of nanomaterials due to the aging process, NM were extracted from the polymer by calcination. The nanomaterials extracted were analyzed by TGA, Fourier transform infrared spectroscopy (FT-IR), BET and TEM and its properties compared with calcinated raw NM. SiO_2 hydrophilic nanoparticles were not affected by the aging process. However, both types of MWCNT were affected by the aging of the NC.

1. Introduction

Polymer materials are a growing area in all the industrial sectors because of their properties and ease processing. The addition of small amounts of nanoparticles to polymers leads to new and/or improved properties for these materials such as mechanical, physical, electrical, magnetic, thermal or optical properties. These NC present interesting applications in most of the industrial sectors, in particular in the electronic, automotive, textile and mining sectors.

To assess the potential value of NC, it is important to determine which nanomaterials can be effectively integrated into NC and what new or improved properties this enables. Then, it is important to determine the effectiveness of dispersion of the nanomaterial in the matrix and how this affected the structure of the polymer to enable the desired property. Polyamides are highly used engineering thermoplastic materials due to their good performance /cost balance. Moreover, PA offers interesting chemical and mechanical properties such as rigidity, chemical resistance, abrasion resistance and barrier properties.

The incorporation of the NM inside of polymeric matrix contributes to an improvement of the polymer properties or adding new properties. Nanofillers are only a few nanometers in dimension offering large surface area per volume and novel properties at this scale. The addition of small amounts of these nanofillers leads to NC materials with enhanced or novel properties, which in some cases allows the use of these materials in new sectors. In this work, SiO_2 NP and MWCNT have been selected to generate PA6 polymer NC. SiO_2 NP can offer interesting mechanical and thermal properties such as mechanical resistance, anti scratching and anti abrasion and anti corrosion.[1-2] On the other hand, MWCNT act as nucleating agents [3] leading to new crystallographic morphologies of the polymers which offer enhanced mechanical, thermal and electrical properties compare to the starting material.[4-7]

The improvement of the material properties by the addition of nanofiller is achieved if the interaction between nanomaterials and the matrix is optimal (homogenous dispersion of the nanofiller inside the matrix). The big differences in the properties of polymer and nanomaterials can often cause phase separation, which it could affect the properties of the resulting NC. [8-9]

Several methods have been studied to enhance the compatibility between the polymer and nanomaterial. The most frequently method used to enhance the compatibility between NM and polymer and therefore to improve the dispersion of them in the polymeric matrix is the nanoparticle surface chemical modification.

Regarding SiO_2 NP, chemical surface modification is the most commonly used method. In this case, silane coupling agents were used as surface modifier agents. The general structure of silane coupling can be represented as $RSiX_3$, where X represents the hydrolyzable groups and R corresponds to organo groups with different functional groups which will improve the interaction between SiO_2 NP and polymer. Different polymeric silane coupling agents has been developed such as aminopropyl methydiethoxysilane (APMDES), vinyltriethoxysilane (VTES), dimethyldichlorosilane (DDS), among others. [10-16]

Commonly, carbon nanotubes have been used as reinforcing materials by their incorporation into a polymeric matrix. However, optimal properties are difficult to achieve due to the deficiencies in nanotubes dispersion and alignment. The state of CNT dispersion within the diluted composites is influenced by processing conditions and compatibility between CNT and polymer matrix. It was demonstrated by recent scientific papers that satisfactory carbon nanotubes dispersion can be obtained using different strategies including physical and chemical methods. Physical methods include carbon nanotubes direct mixing using a mechanical force while chemical methods are based on carbon nanotube surface modifications such as surfactant or polymer wrapping technology. [17-18] Master batches, thermoplastic polymers containing high loading of CNTs (15-20%), have recently became widely used in the melt preparation of CNT based polymer NC. For industrial application of melt mixing extrusion technique, the use of master batch materials is preferred to improve CNT dispersibility in the polymer matrix.[19-24] Additionally, master batches offer a dust-free environment and consequently no safety-risk concerns are considered.

Nowadays, different strategies to recycle the NM or NC are being studied because of the added value of the nanofillers. These strategies can be distributed in two big sectors: mechanical and chemical recycling. Mechanical recycling is less destructive method for the recovery and reuse of NM-polymer mixture. However the chemical recycling methods can be very interesting because it is possible to separate the NM from polymeric matrix. The last can be calcinated or dissolved with the corresponding solvent depending on the polymer used.

This work describes the evaluation of the potential release of nanomaterials from polymeric NC during their use in outdoor applications. The importance of NP dispersion and compatibility with the matrix in NC materials are key issues to avoid release of NM to the environment. Size, shape and composition of the nanofillers are found to influence the potential release of SiO_2 NP and MWCNT from PA6 NC.

2. Materials & methods

MWCNT (NC7000, 90% pure) and MWCNT master batches with a 15% of nanofiller content (PLASTICYL™ PA1503) were obtained from Nanocyl™. Two different types of nanosized SiO_2 were acquired from PlasmaChem to get different degrees of compatibility between NM and polymer: non-modified SiO_2 (hydrophilic; PL-SiOF) and SiO_2-octyl (hydrophobic; PL-SiOF-OS, 99.8% pure).

2.1. Extrusion and Injection of the NC

The polymer used, PA6, is an aliphatic hygroscopic homopolymer. Thus, it was dried with a dehumidifier Wittmann PDC (Portable Drying Conveyor) before extrusion/injection treatment to prevent the hydrolytic degradation and alteration of its properties. To ensure maximum dispersion and homogenization of the NP within the polymeric matrix, first NM were mixed with PA6 and extruded in a co-rotating twin screw extruder (TSE20, Brabender) to obtain a pellet with a 3% of nanofiller content. NC pellets (3% nanofiller content) were injected to obtain standard test specimens in a 100 tons clamping force injection machine (TM110, Battenfeld).

2.2. Accelerated weathering process

In order to study the climate effect on the different NC, accelerated aging was performed in a climatic chamber (Suntest XXL+, Atlas). Conditions chosen were based on ISO 4892/06 (table 1):

Ageing Conditions	
Irradiation (at 300-400 nm):	0.50 W/m^2
Internal filter:	borosilicate
External filter:	borosilicate
BST temperature:	65ºC (±3ºC)
Relative humidity:	50% (±5%)
Wetting/cycle:	1 ± 0.5 min
Drying:	29 min
Irradiation:	continuous
Total exposure:	1000 h

Table 1. Ageing conditions based on ISO 4892/06

After aging, the characterization of all the samples was performed using TGA, FT-IR, DSC, TEM, SEM and EDX techniques.

The weight loss of the samples was analyzed, both qualitatively and quantitatively, by Hi-Res TGA (Q500, TA instruments), applying a temperature increase of 10°C/min from 30-1000°C and a flow rate of air at 50 mL/min. Weight loss 160°C was considered due to hydration of the sample and discarded for analysis. DSC was used to determine melting and crystallization temperatures (T^a_m & T^a_c respectively), glass transitions and the estimation of the crystallization kinetics in each NC injected. mDSC (Q20, TA instruments) with a RCS module was used for these analyses. The conditions selected were temperature programming of 10ºC/min from 30-250ºC, and a flow rate of air at 60mL/min. Functional groups present in the samples were identified using a FT-IR spectrophotometer (IR Affinity-1 8400, Shimadzu). Finally, compatibility between NM and the polymeric host and migration of the nanofiller were determined using SEM (MERLIN FE-SEM, Zeiss) and TEM (Jeol 2010, JEOL Ltd). For TEM microscopy: 2 mm^2 cubes of non-aged and aged specimen tests were cut and included in Epon resin. After resin polymerization (60 °C, 48 h), sections with a thickness of 50 nm were cut with an ultramicrotome and placed on TEM grids (Formvar carbon-coated Cu grids, EMS).EDX in TEM cuts was used to determine the atomic composition of the sample.

2.3. Calcination process
The NM were extracted from non aged and aged NC by calcination process. Calcination of the nanomaterial took place at 410 °C for 3h 30 min. The material recovered was characterized by TGA, FT-IR, TEM and EDX.

3. Results & discussion
3.1. Alteration of the NC properties by the nanofiller nature and compatibility
It has been widely described that the quality, useful life and properties of the polymers can be significantly modified adding only a small amount of nanofiller. However, the same NM can modify differently the final NC properties depending on its dispersibility and interaction within the matrix. The objective of the present study is to analyze the different behaviour of PA6 NC filled with a 3% of compatibilized and non-compatibilized MWCNT or SiO_2 NP, to identify the modifications caused on the mechanical and structural properties, as well as to evaluate the possible aggregation, migration or release of the NM during the fabrication and use of the NC.

It is accepted that an homogeneous dispersion of the nanofillers within the matrix contributes in the improvement of the physical, chemical and mechanical properties of the NC. PA6 is essentially a hydrophilic polymer. Thus, beforehand, unmodified SiO_2 NP (with naturally formed –OH groups on its surface) and $MWCNT_{MB}$ (additived to obtain a high-loaded and homogeneous mixture of MWCNT) were expected to be more compatible with the matrix. Contrarily, hydrophobic SiO_2 NP, with octyl groups on NP surface, and pristine MWCNT, would be more likely to show worse compatibility with the hydrophilic polymeric matrix (PA6), and consequently resulting in NC with worse properties.

In order to study the different dispersion and structural properties provided by the NM and compare its modification between compatibilized and non-compatibilized samples, TEM and DSC analysis were performed. Ultrafine TEM sections of the inner part of each PA6 NC after injection showed a uniform distribution of the nanofiller in all the cases excepting for NC including hydrophobic SiO_2 NP, which formed aggregates due to their incompatibility with the matrix (figure 1). Contrarily to the hydrophilic SiO_2 NC, this aggregation resulted in an increase of the melting and crystallization enthalpies (table 2). Both MWCNT were homogeneously distributed within the matrix, though compatibilized $MWCNT_{MB}$ favours flame retardancy and PA6 crystallinity. Thus, a good compatibility NM-matrix guarantees the maintenance or improvement of the NC properties respecting the original polymer, which cannot be said for the non-compatibilized NM.

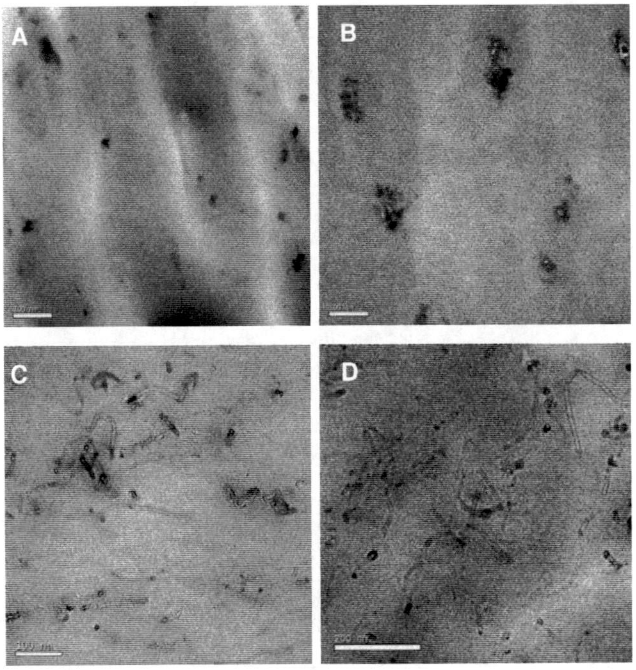

Figure 1. **Section of specimen tests with a 3% of a) hydrophilic SiO_2, b) hydrophobic SiO_2, c) MWCNT and d) $MWCNT_{MB}$.**

Table 2. Values of DSC and percentage of crystallinity

NC	midpoint T^a_c (°C)	ΔH_{cc} (J/g)	midpoint T^a_m (°C)	ΔH_{cm} (J/g)	crystallinity (%)
PA6	191,52	69,02	218,82	75,43	3,36 ± 0,02%
PA6 aged	191,02	79,14	214,51	87,95	2,86 ± 0,00%
PA6/SiO_2 hydrophobic	191,51	75,11	217,71	80,71	1,47 ± 0,57%
PA6/SiO_2 hydrophobic aged	191,565	74,86	217,04	78,83	1,04 ± 0,17%
PA6/SiO_2 hydrophilic	189,61	60,40	216,91	66,68	1,65 ± 0,01%
PA6/SiO_2 hydrophilic aged	190,94	69,47	216,06	73,64	1,09 ± 0,17%
PA6/MWCNT	196,09/209,62	57,95	220,04	60,47	0,66 ± 0,06%
PA6/MWCNT aged	244,765	71,38	261,08	84,99	3,57 ± 0,01%
PA6/$MWCNT_{MB}$	245,14	72,67	260,52	76,29	1,50 ± 0,13%
PA6/$MWCNT_{MB}$ aged	196,95/207,21	73,95	220,27	82,08	2,13 ± 0,08%

3.2. Influence of NM nature and compatibilization on NC aging

Apart from the production stage, the present work simulated usage of the NC by accelerated aging in controlled conditions to determine possible changes in this phase of their life cycle. The use of NC can change its physicochemical properties, cause its degradation, and affect the distribution or release of the nanofillers.

In order to determine the alteration of the NC due to aging, TGA, TEM and DSC analysis were performed. The TGA showed that the weight loss of organic mater (o.m.) is higher in all the aged NC excepting PA6/$MWCNT_{MB}$. It indicates the appearance of degradation products from the PA6 that are consumed at higher temperatures; the NC with compatibilized MWCNT is protected by the uniformly dispersed nanofiller, avoiding the mentioned degradation (table 3). It is also supported by TEM imaging, where PA6/$MWCNT_{MB}$ samples do not show detectable differences in NM dispersion and PA6 structure after aging, but it causes the fragmentation of all the other polymeric matrices (figures 2 & 3). In addition, regarding SiO_2 NC, TEM imaging shows that aging causes the aggregation of compatibilized and well-dispersed SiO_2 NP; this changes the properties of the NC, making them more similar to those observed in the non-compatibilized SiO_2 NC.

DSC characterization (table 2) shows that the different nature of SiO_2 NP and MWCNT affected differently the crystallinization of PA6. Compatibilized SiO_2 NP reduced the crystallinization temperature, though this effect was lost during the NC aging. Crystallinity, however, was reduced heavily in both compatibilized and non-compatibilized SiO_2 NC. Regarding MWCNT NC, the nanofiller had the opposite effect on PA6. Crystallinization started at higher temperatures, and this effect was much more evident when the nanotubes had been compatibilized. This is due to the nucleating effect that this nanomaterial causes in the process. When PA6-MWCNT had not been compatibilized, interactions between both components of the NC is worst, the nucleating effect is less notorious and occurs in two steps, which indicates two reordinations of the atoms. After aging, both MWCNT samples show an increase in the crystallinity values, specially the compatibilized sample. Even both nanofillers have different effect on the polymer, it can be said that an increase of the compatibility between NM and matrix causes a decrease of the PA6 crystallinity. However, the nucleating effect of MWCNT leads to an earlier crystallinization.

Table 3. Thermal degradation (TGA) of polymer and NC

	non-aged		aged	
NC	**Weight loss 700°C (o.m.)**	**i.m.%**	**Weight loss 700°C (o.m.)**	**i.m.%**
PA6	99.14%	---	97.78%	---
PA6/SiO_2 hydrophobic	97.87%	1.21%	97.00%	0.81%
PA6/SiO_2 hydrophilic	97.49%	1.81%	96.75%	1.14%
NC	**Weight loss 600°C (o.m.)**	**i.m.%**	**Weight loss 600°C (o.m.)**	**i.m.%**
PA6	98.98%	---	97.02%	---
PA6/MWCNT	95.35%	3.70%	94.09%	1.97%
PA6/$MWCNT_{MB}$	94.16%	4.92%	95.70%	2.21%

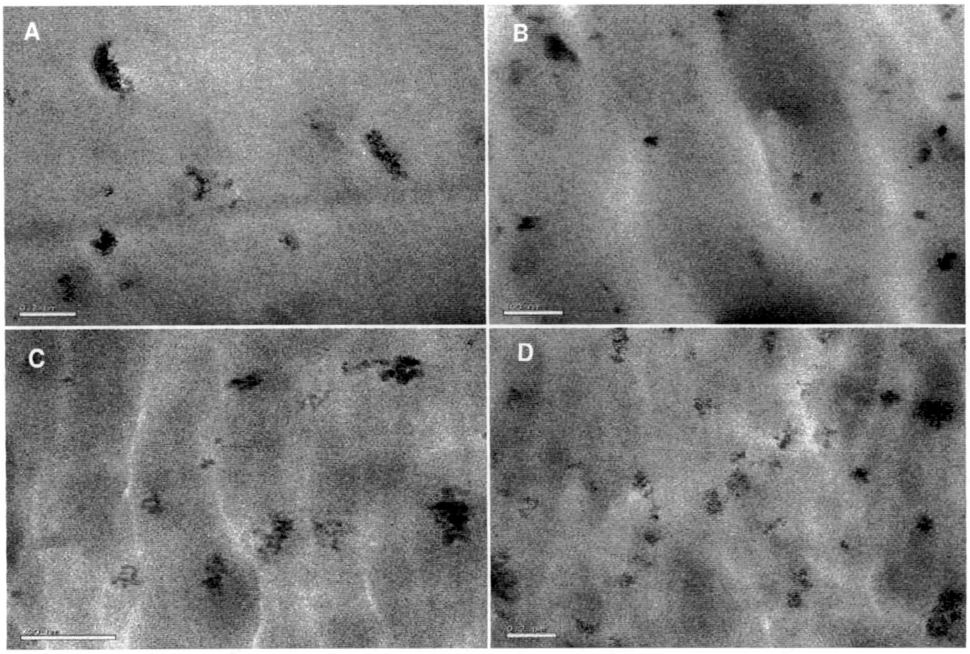

Figure 2. Image TEM of non compatibilized (A-C) compatibilized (B-D) non aged (A-B) and aged (C-D) SiO_2 NC.

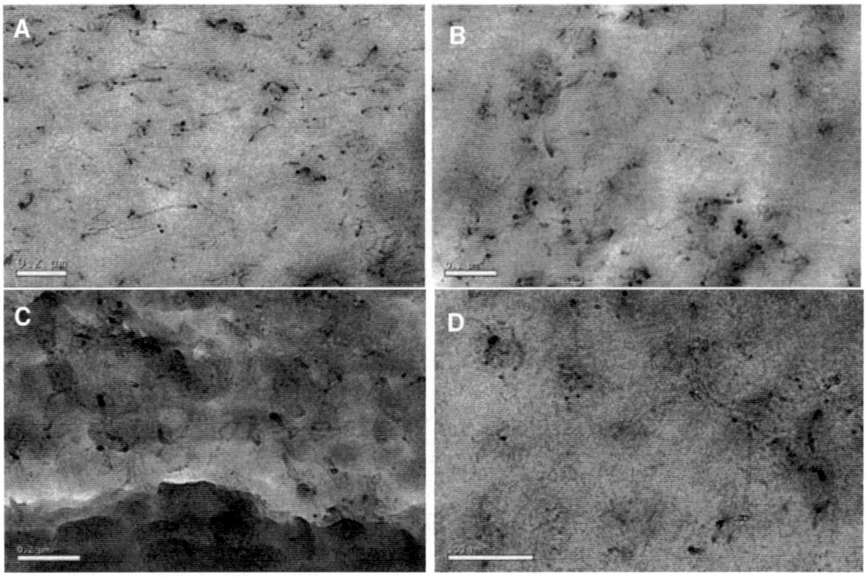

Figure 3. Image TEM of non compatibilized (A-C) compatibilized (B-D) non aged (A-B) and aged (C-D) MWCNT NC.

The degradation occurred in the NC during the simulation of its use (accelerated aging) is also observable in its surface by SEM-Merlin imaging. This technique allowed us to observe the external part of the NC without any extra treatment. After injection, SiO_2 NP and MWCNT were observable embedded in the matrix. However, after aging, both types of nanofillers were clearly on the surface, exposed to the environment (figure 4). Altogether with TGA data, which indicate a loss of i.m. during aging, results indicate that (being NM compatibilized or not) the degradation of PA6 causes the release of SiO_2 NP and MWCNT to the environment.

SEM images show same degradation degree of the matrix in its surface for both SiO_2 NC, as well as for non-compatibilized MWCNT NC. In these cases, it leads to the aggregation of the NM. It is important to mention that the aggregation of the non-compatibilized MWCNT was not observable in TEM, most probably due to a more accused degradation of PA6 in the surface. $MWCNT_{MB}$ samples seem to be more protected by the nanofiller, and there is no evidence of degradation or migration.

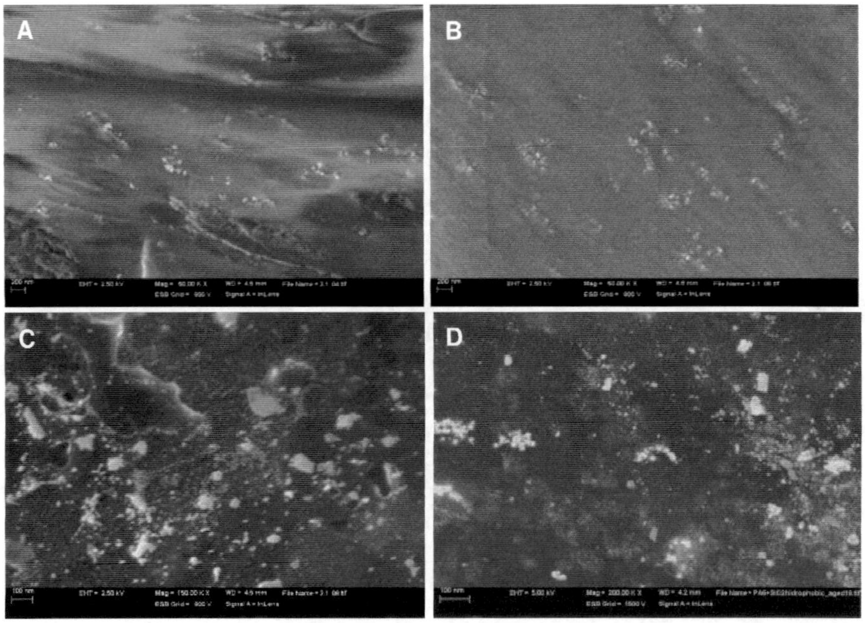

Figure 4. Image SEM of non compatibilized (A-C) compatibilized (B-D) non aged (A-B) and aged (C-D) SiO_2 NC.

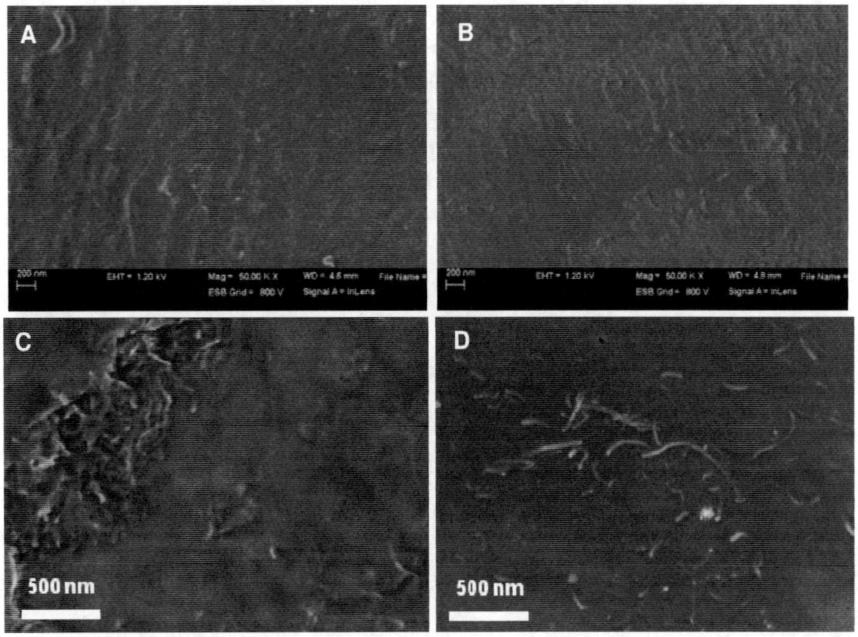

Figure 5. Image TEM of non compatibilized (A-C) compatibilized (B-D) non aged (A-B) and aged (C-D) MWCNT NC.

3.3 Extraction of the NM for recycling by calcination

In recent years, there has been an increasing interest in the use of NM to improve interesting properties of different polymers. Therefore, it is important to know what to do with these NM at the end of their life. There are different options; NC can be recycled (mechanical recycling) or it could be better recycling the NM or depending on their physicochemical properties and their toxicity the NM can be disposed. It is important to consider that the physicochemical properties could change during their useful life and therefore, their toxicity can also be affected. When this happen and in those cases that these new properties are not interesting, the NM have to be disposed.

In this project, calcination was chosen as a method to recover the NM from NC. A previous study about calcinations of raw NM was performed in order to determine the changes which could be due to the extraction process. Calcination conditions were 410°C during 3h 30 min. The temperature was determined according to the data obtained by TGA therefore, it was considered the temperature in which polymer is burned and NP do not suffer any change. Calcinated NM were characterized by FT-IR and TGA and neither of the two analyses showed any differences between NM before and after calcination except for the case of hydrophobic SiO_2 whose octyl groups are burned by calcination.

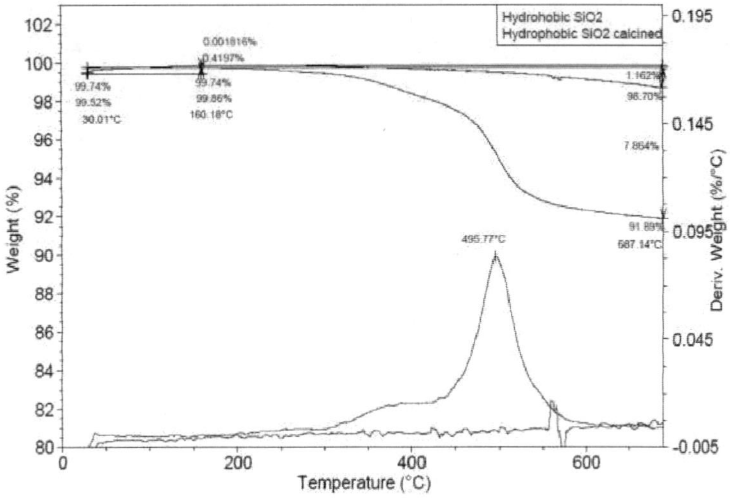

Figure 6. TGAs comparison between calcined and non calcined hydrophobic SiO_2.

Figure 6 shows a comparison between NM before and after calcination, and it can be observed that the loss of weight corresponding to the functional group of NM cannot be observed after calcination. Nevertheless, it was decided to calcinate all NC to recover NM because non functionalized SiO_2 NP can also be used in the synthesis of NC.

3.4. SiO_2 NP recovered

According to FT-IR analyses it cannot be observed any difference between NM recovered from compatibilized and non compatibilized NC. Recovered material exhibits bands which are associated to the vibrations corresponding to SiO_2 NP (1055 cm^{-1} and 806 cm^{-1} assigned at Si-O st and Si-O bd, respectively) and remaining polymer (1599 cm^{-1} and 3360 cm^{-1}). Moreover, TGA results are consistent with those obtained by FT-IR. The first analyses showed a weight loss at ~600 °C which corresponds to remaining polymer. However, a curious behavior was detected in the case of hydrophobic SiO_2 NC. Figure 7 shows a comparison between hydrophobic SiO_2 NP non-calcinated and extracted material from aged and non aged SiO_2 NC. It can be observed that the weight loss at 496 °C, which is attributed at the octyl groups, cannot be observed in the case of extracted material from non aged SiO_2 NC indicating that these groups have been calcinated during processing but it is not so in the case of the material recovered from aged SiO_2 NC. This fact indicates that the degraded matrix is interacting with octyl groups in a way that protects these groups from degradation.

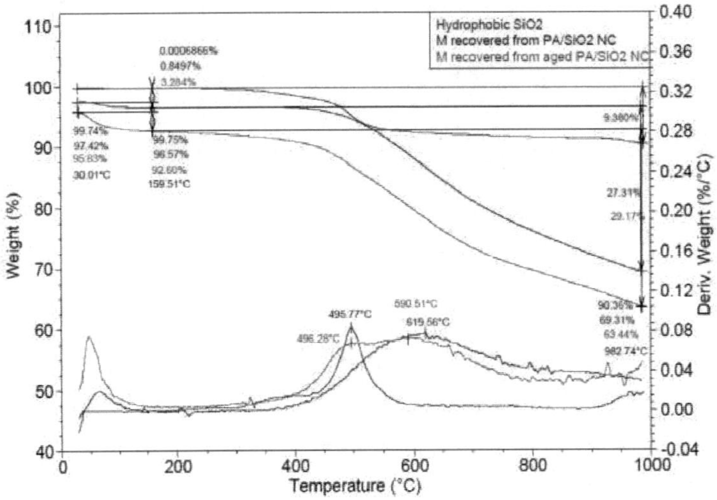

Figure 7. TGA comparison between hydrophobic SiO_2 and NM recovered from aged and non aged PA/SiO_2 NC.

The TEM images obtained from recovered material support these last results. NM recovered and some amount of polymer can be observed before and after aging processes, but in the last case, the amount of polymer remaining is higher. Their atomic composition was determined by EDX (figure 8).

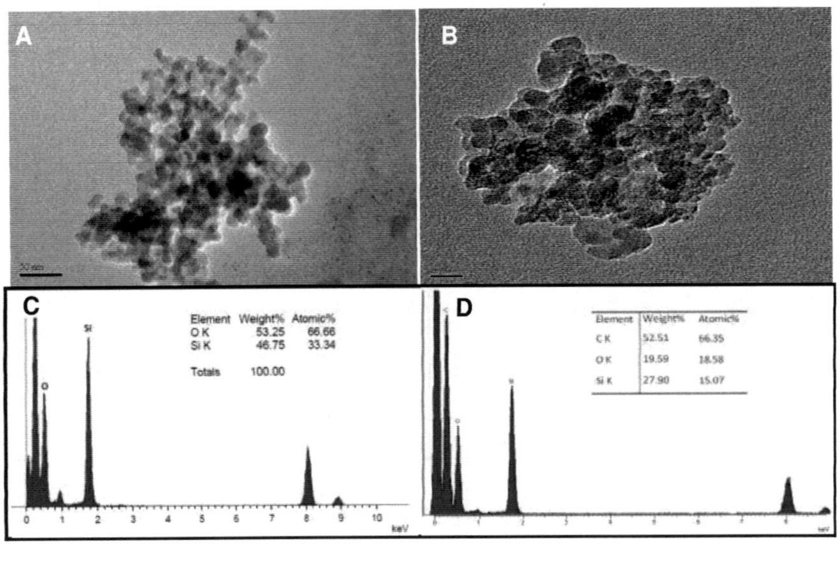

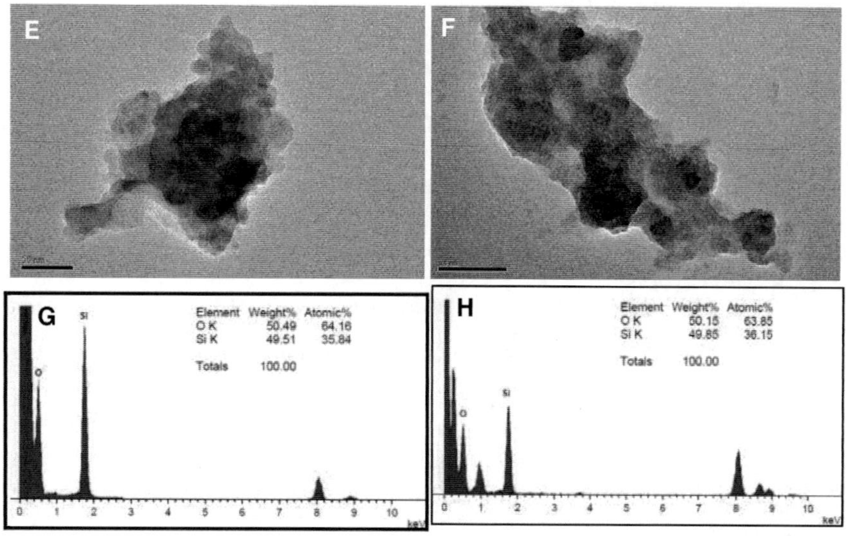

Figure 8. TEM images of material recovered from non-compatibilized (A-E) compatibilized (B-F) non-aged (A-B) and aged (E-F) NC. EDX spectrum of material recovered from non-compatibilized (C-G) compatibilized (D-H) non-aged (C-D) and aged (G-H) NC

3.5. MWCNT NM recovered

According with FT-IR and TGA, it is concluded that the material recovered from both NC contains MWCNT and polymer remaining. On the other hand, TEM analysis showed that the recovered material before the aging process consisted of MWCNT which are inserted inside the polymeric matrix and there are a little amount of them that are free. However, this tendency is different when the NC were aged. In these cases, it can be observed a big amount of MWCNT free and a small amount inserted in the polymeric matrix. Therefore, it can be concluded that the release of MWCNT from polymeric matrix is easier when the last has been affected by the aging process (figure 9).

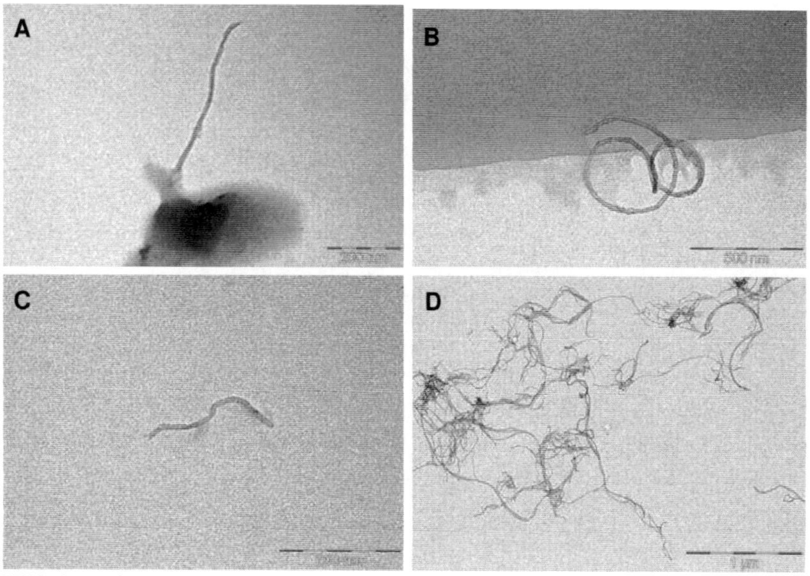

Figure 9. Image TEM of material recovered from non-compatibilized (A-C) compatibilized (B-D) non-aged (A, B) and aged (C, D) NC

4. Conclusions

Different PA6 NC were prepared using compatibilized and non compatibilized NM. The NM chosen were SiO_2 NP and MWCNT whose surface was modified in order to obtain different compatibilities between NM and polymer. Hence, hydrophobic SiO_2 NP (with octyl groups in their surface), hydrophilic SiO_2 NP, MWCNT in powder form and MWCN master batch were used.

Specimen NC were obtained by extrusion and injection processes and were aged by accelerated aging processes (outdoors simulation processes) in climatic chambers. Analysis of NC properties before and after ageing demonstrated that release of NM from polymer NC depends mainly on polymer degradation. The process of polymer degradation in NC materials under climatic conditions can be reduced by the inclusion of well-dispersed MWCNT. However in the case of SiO_2 NP degradation of the polymer is not dependent on the dispersibility of the nanofiller in the matrix. Therefore, the size, shape and chemical composition of the nanofiller was shown to influence the release of NM to the environment.

Calcination is the method proposed in this work to recover the nanofillers from NC for recycling. A preliminary study was performed to identify possible modifications in NM structure due to the method of the extraction; any modifications could be observed, to except of hydrophobic SiO_2 which lost their octyl groups. Therefore, NM can be recovered from NC with a small amount of polymer, and reused.

5. Acknowledgements

This work has been supported by the Generalitat de Catalunya, Departament d'Innovació, Universitat i Empresa, and by the Spanish Government, Ministerio de Ciencia e Innovación.

6. References

[1] Ouabbas Y, Chamayou A, Galet L, Baron M, Thomas G, Grosseau G, Guilhot B 2009 *Powder Technol.* **190** 200
[2] Xue L, Li J, Fu J, Han Y 2009 *Colloids Surf. A Physicochem. Eng. Asp.* **3308** 15
[3] Assouline E, Lustiger A, Barber A H, Cooper C A, Klein E, Wachtel E, Wagner H D 2003 *J. Polym. Sci. Part B: Polym. Phys.* **41** 520
[4] Wong K K, Shi S Q, Lau K T 2007 *Key Eng. Mater.* **334-335** 705
[5] Zhang W D, Shen L, Phang I Y, Liu T 2004 *Macromolecules* **37** 256
[6] Park O K, Jeevananda T, Kim N H, Kim S, Lee J H 2009 *Scr. Mater.* **60** 551
[7] Sulong A B, Muhamad N, Sahari J, Ramli R, Md Deros B, Park J 2009 *Eur. J. of Sci. Res.* **29** 13
[8] Solarski S, Ferreira M, Devaux E 2008 *Polym. Degrad. Stabil.* **93** 707
[9] Su S, Wilkie C A 2004 *Polym. Degrad. Stabil.* **83** 347
[10] Yoshinaga K, Shimada J, Nishida H, Komatsu M 1999 J. *Colloid Interface Sci.* **214** 180
[11] Yoshinaga K, Tani Y, Tanaka Y 2002 *Colloid Polym. Sci.* **280** 85
[12] Kang S, Hong S I, Choe C R, Park M, Rim S, Kim J 2001 *Polymer* **42** 879
[13] Sugimoto H, Daimatsu K, Nakanishi E, Ogasawara Y, Yasumura T, Inomata K 2006 *Polymer* **47** 3754
[14] Su Y H, Wei T Y, Hsu C H, Liu Y L, Sun Y M, Lai J Y 2006 *Desalination* **200** 656
[15] Su Y H, Liu Y L, Sun Y M, Lai J Y, Wang D M, Gao Y, Liu B J, Guiver M D 2007 *Membr. Sci.* **296** 21
[16] Moncada E, Quijada R, Retuert J 2007 *Nanotechnology* **18** 335606
[17] Mackay M E, Tuteja A, Duxbury P M, Hawker C J, Van Horn B, Guan Z, Chen G, Krishnan R S 2006 *Science* **311** 1740
[18] Abdalla M, Dean D, Robinson P, Nyairo E 2008 *Polymer* **49** 3310
[19] Willmow T, Pötschke P, Pegel S, Haussler L, Kretzschmar B 2008 *Polymer* **49** 3500
[20] Pegel S, Pöstchke P, Petzold G, Alig I, Dudkin S M, Lellinger D 2008 *Polymer* **49** 974
[21] Hong J, Lee J, Hong C K, Shim S E 2010 *Curr. Appl. Phys.* **10** 359
[22] Pöstchke P, Bhattacharyya A R, Janke A 2003 *Polymer* **44** 8061
[23] Pöstchke P, Bhattacharyya A R, Janke A 2004 *Carbon* **42** 965
[24] Meincke O, Kaempfer D, Weickmann H, Frierich C, Vathauer M, Warth H 2004 *Polymer* **45** 739

Journal of Physics: Conference Series **429** (2013) 012045 doi:10.1088/1742-6596/429/1/012045

Nanoparticle release from nanocomposites due to mechanical treatment at two stages of the life-cycle

Daniel Göhler, André Nogowski, Petra Fiala and Michael Stintz[1]

Research Group Mechanical Process Engineering, Institute of Process Engineering and Environmental Technology, Technische Universität Dresden, Münchner Platz 3, D-01062 Dresden, Germany.

E-mail: Michael.Stintz@tu-dresden.de

Abstract. Currently, an increase of published studies regarding nano-object release into the air can be observed. In the absence of a harmonized methodology only a few studies provide quantitative data for further use. In this study, an appropriate methodology was adopted and partly enhanced for the characterization of the nanoparticle release from coatings and composites containing nanoscaled pigments. Systematic analyses at two stages of the life-cycle (non-weathered/arterially weathered) were performed for three simulated treatment processes – sanding, dynamic friction and wind erosion. Results show that the sample material, the sample composition, the sample condition (e.g. weathering state) and the type of mechanical treatment have a strong effect on the particle release. SEM, TEM and EDX analyses revealed that none of the released particles were free nanoscaled pigments.

1. Introduction

The ability and the quantity of nanoparticle release into air from engineered nano-composites due to mechanical treatment can be determined systematically by exposure studies in laboratory for estimating potential risk to health, safety and environment [1]. Currently, no harmonized methodology for nanoparticle release investigations in laboratory exists, whereby the comparability and transferability of most of such studies is very limited. This is firstly due to the huge variety of possible release scenarios and the multitude of materials and secondly results from limitations of the measurement technology and from physical constrains of aerosols. Figure 1 illustrates the complexity of the experimental design for analyses of particle release into the air, which is necessary for the determination of qualitative and quantitative data regarding an exposure assessment.

Save for a few exceptions [e.g. 2, 3], most laboratory studies [e.g. 4-8] regarding the release of engineered nano-objects into the air from nano-composites due to mechanical treatment processes have shown, that engineered nano-objects are firmly embedded in released wear/swarf particles [1]. The exceptions can maybe explained by artifacts from sample preparation and/or conditioning (e.g. incomplete wetting, incomplete or no surface modification, state of dispersion). In general, we believe that a release of isolated nano-objects from well-prepared nanocomposites requires a prior chemical or thermal degradation of the matrix material.

External impacts like sunlight, moisture or temperature fluctuations may promote the degradation of the matrix material and eventually expose embedded nanoparticle additives or nanoscaled pigments.

[1] To whom any correspondence should be addressed.

In turn, this is probably associated with a higher release risk as shown in [9] for the transition of TiO_2 from facades into the aquatic environment.

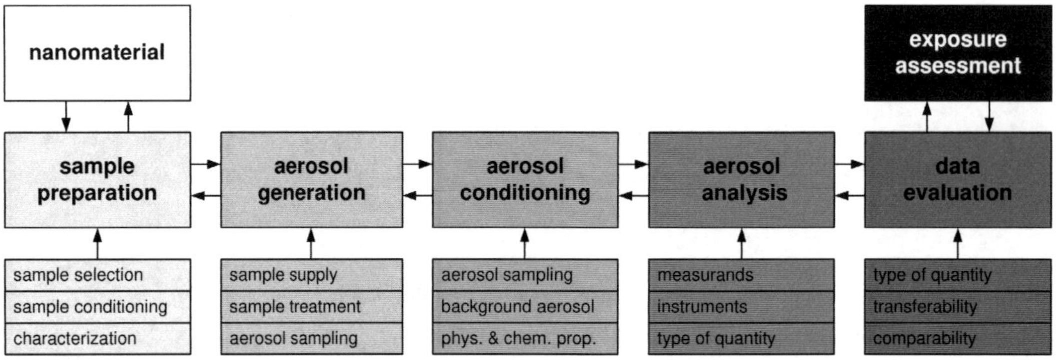

Figure 1. Selected main and secondary stages of the experimental design for the characterization of particle release into the air.

This study addresses the effect of weathering on the (nano)-particle release from two different nano-composites (coatings and composites) that are exposed to sanding, dynamic friction and wind erosion. A recently published method [4] was adapted and partly enhanced to quantify the particle release.

2. Materials and Methods

2.1. Sample selection, conditioning and characterization

Nanoparticle release was analyzed for a total of 22 different sample systems based on two different matrix-materials (11 acrylate coatings, 11 polypropylene composites). The matrices were filled with 10 different pigments, which varied in particle size, morphology and material. Apart from two (TD, CT), all pigments have dimensions in the nanoscale. In addition, a pigment-free coating (AL) and a pigment free composite (PP) were examined. The coatings were applied on aluminum-substrates by a defined squeegee procedure, while the PP-composites fabrication was accomplished by injection molding.

Table 1. Sample description: Identification key of the sample systems, employed pigments, their intensity weighted harmonic mean size (x_{PCS}), polydispersity index (PI) and BET surface area.

sample ID [-]		CIGPN[a] [-]	pigment material [-]	pigment content [wt.-%]		x_{PCS}[b] [nm]	PI[b] [-]	BET surface area [$m^2 \cdot g^{-1}$]	
coating	composite			coating	composite			coating	composite
AL-CB	PP-CB	black 7	carbon black	2.0	1.0	110	0.13	320	120
AL-PC	PP-PC	green 7	phthalocyanine green	1.0	0.2[d]	212	0.28	47	37
AL-TD	PP-TD	white 6	titania	10.0	1.0	311	0.13	n.a.	n.a.
AL-TT	PP-TT	white 6	titania, tansparent	0.7	1.0	153	0.26	110	160
AL-BS	PP-BS	white 21	barium sulfate	5.0	1.0	303	0.13	30	30
AL-TE	PP-TE	red 101	iron oxide, transparent	1.9	0.2[d]	85	0.25	85	n.a.
AL-RP	PP-RP	red 122	quinacridon	2.8	0.2[d]	227	0.27	78	78
AL-AO	PP-AO	white 24	aluminum oxide	7.2	1.0	202	0.16	100	100
AL[c]	PP[c]	-	-	-	-	98	0.06	-	-
AL-CT	PP-CT	brown 24	chrom.-antimony-titan.	15.0	1.0	829	0.25	2	2
AL-KB	PP-KB	blue 28	cobalt blue	10.0	1.0	597	0.29	9	n.a.

[a] CIGPN ... color index generic pigment name
[b] data valid for the coating-pigments, obtained by DLS (HPPS5002, Malvern Inst. GmbH, Herrenberg, Germany)
[c] reference systems: polyacrylate dispersion (AL) for coatings and polypropylene (PP) for composites
[d] white balance with 0.8 wt.-% TiO_2

For each sample system, ten individual subsamples were produced. Four of them were conditioned by artificial weathering in a weathering device (Model Xenotest Beta LM, Atlas Material Testing Technology GmbH, Linsengericht, Germany) by filtered xenon-arc radiation according to ISO 11341:2004 to simulate a natural weather impact for a time span between 2 and 5 years. The AL-coatings were weathered in two phases. In the first phase the samples have experienced an alternating illumination (bright and dark cycle's á 120 min) for 1000 h, while the second phase lasted for 1500 h and consisted of repeated dry (102 min illumination) and wet phases (18 min irrigation). The PP-composites were weathered for 1000 h similar to the second phase of the coatings. Surface conditions of the samples were analyzed before, during and after the artificial weathering procedure by specular gloss measurements in accordance with DIN EN ISO 2813:1999 (see figure 2), color difference measurements in accordance with DIN 6174:2007 and scanning electron microscopy (SEM) analyses.

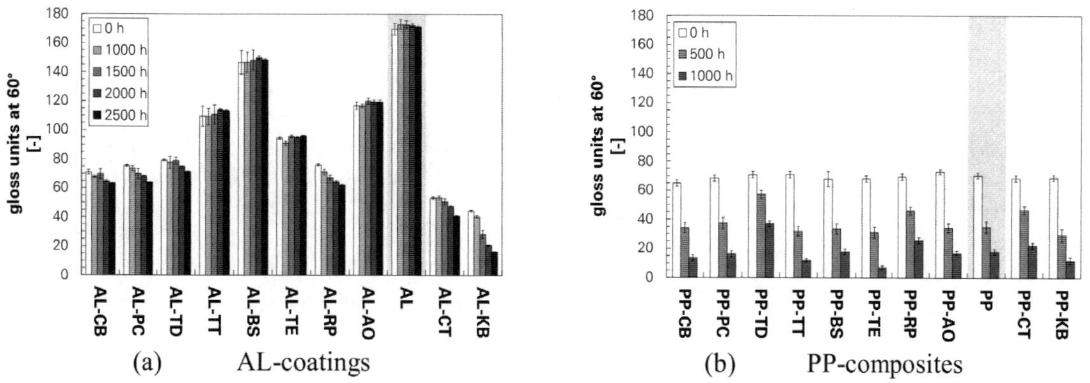

Figure 2. Gloss data of the AL-coatings and the PP-composites before, during and after the artificial weathering process; reference samples are highlighted.

2.2. Experimental details

2.2.1. Sample treatment. The prepared samples were exposed to sanding, dynamic friction and wind erosion, which were simulated by an extended version of the experimental apparatus described in [4]. Within the apparatus the samples were supplied by a translation sample carrier beneath the acting tools, which were encapsulated for sampling of all released particles as close as possible to the particle source. The used treatment process parameters are given in table 2.

The parameters of the sanding procedure for the AL-coatings were adopted from [4] except for the abrasive paper graining. Finer abrasive paper was used to achieve a cutting depth < 5 μm, thus to

Table 2. Process parameters of the simulated treatment processes.

process parameter		sanding process		dynamic friction		wind erosion	
		coatings	composites	coatings	composites	coatings	composites
stressed area	[cm^2]	10.4	10.4	12.7	12.7	1.6	1.6
sample speed	[$mm \cdot min^{-1}$]	5.0	5.0	5.0	5.0	5.0	5.0
maximum face velocity[a]	[$m \cdot s^{-1}$]	4.9	4.9	1.5	1.5	30	30
normal force	[N]	0.5	0.5	13	13	-	-
contact pressure	[kPa]			84	84	-	-
abrasive paper graining	[-]	P1200	P240	-	-	-	-
rotational velocity	[$m \cdot s^{-1}$]	1.83	0.73	-	-	-	-
cutting velocity ratio	[-]	366	146	-	-	-	-
power[b]	[W]	1.3	0.5	< 0.078	< 0.103	-	-

[a] in the case of sanding and dynamic friction parallel velocity between the lower part of the sampling hood and the sample surface, in the case of wind erosion parallel velocity between nozzle inlet and sample surface.
[b] cutting power for sanding, friction power for dynamic friction

ensure that sanding occurs within the weathered region of the weathered samples. Preliminary investigations have shown that the sanding parameters used for the AL-coatings led to a thermal particle generation in the case of the PP-composites, which was indicated by high number concentrations and average particle sizes below 30 nm. Accordingly, the sanding parameters for the PP-composites were adjusted as shown in table 2. Since the PP-composites showed larger weathering depths as the AL-coatings, the cutting depths remained within the weathered region.

Wind erosion was simulated by a stainless steel suction nozzle with an inner diameter of 2 mm, while surface stressing by dynamic friction was realized by a flat-lying and circular elastic polyurethane body (diameter of 14 mm) with a depth of roughness of Rz = 7.4 µm (mean roughness index Ra = 1.3 µm, total height of roughness Rt = 13.1 µm).

2.2.2. Measuring, sampling and conditioning instruments. The instruments employed for sampling, conditioning and characterization are listed in table 3. Detailed information on the operation principles can be gathered from the cited references. Target measurands for the (nano)-particle release characterization are the particle number concentration (PNC), the number weighted particle size distribution (PSD) and the material composition of the released particles. Both, PNC and PSD, can be detected by means of conventional aerosol measurement techniques, while the determination of the material composition requires additional analyses. Thus, electrostatic precipitated aerosol particles were subsequently analysed by scanning electron microscopy (SEM, Model Gemini 982, Karl Zeiss AG, Jena, Germany), transmission electron microscopy (TEM, Model Tecnai 20, FEI Company, Hillsboro, USA) and energy dispersive X-ray spectroscopy (EDX). The used dilution units DDS 560 and DIL 556 realised well-defined (non-classifying) dilution of aerosols.

Table 3. Operated measuring, sampling and conditioning instruments for the aerosol characterization.

Instrument	Model; manufacturer	metric; equivalent diameter	size range; time resolution[a]; aerosol flow rate[b]	reference
Condensation Particle Counter (CPC)	Model 3022A; TSI Inc., Shoreview, USA	PNC	6 nm - < 10 µm; 1 s; 0.3 L·min^{-1}	[10]
Diffusion Filter Battery (DB) + CPC	Model 3022A; TSI Inc., Shoreview, USA	PNC	> 100 nm; 1 s; 0.3 L·min^{-1}	[11]
Engine Exhaust Particle Sizer (EEPS)	Model 3090; TSI Inc., Shoreview, USA	PNC, PSD; electrical mobility diameter	5.6 nm – 560 nm 0.1 s 10.0 L·min^{-1}	[12]
Aerodynamic Particle Sizer (APS)	Model 3321; TSI Inc., Shoreview, USA	PNC, PSD; aerodynamic particle diameter	0.5 µm – 20 µm; 80 s; 5.0 L·min^{-1}	[13]
Electrical Aerosol Detector (EAD)	Model 3070A; TSI Inc., Shoreview, USA	aerosol length	-; 1.0 s; 2.6 L·min^{-1}	[14]
Electrostatic Precipitator (ESP)	Prototype of Model 3089; TSI Inc., Shoreview, USA	sampling	> 5 nm; > 100 s; 0.3 L·min^{-1}	[15]
Dynamic Dilution System (DDS)	DDS 560; Topas GmbH, Dresden, Germany	dilution 1:10	-; -; 0.6 L·min^{-1}	
Dilution Unit (DIL)	DIL 556; Topas GmbH, Dresden, Germany	dilution 1:10	-; -; 5.0 L·min^{-1}	
Mass Flowmeter (MFM)	Model 41221; TSI Inc., Shoreview, USA	flow rate		

[a] adjusted time resolution
[b] adjusted aerosol flow rate

2.2.3. Experimental setups. Depending on the applied stress and the associated PNC two different experimental setups (see figures 3 and 4) were installed for the (nano)-particle release characterization. Both experimental setups were operated in a laminar flow bench (Model LF-VM-K0615; Steag Laminarflow Prozesstechnik GmbH, Pliezhausen, Germany) to achieve a nearly particle free environment with PNC < 0.01 cm^{-3}. Each experimental setup can be divided in three main sections, i.e. aerosol generation, aerosol conditioning and aerosol characterization.

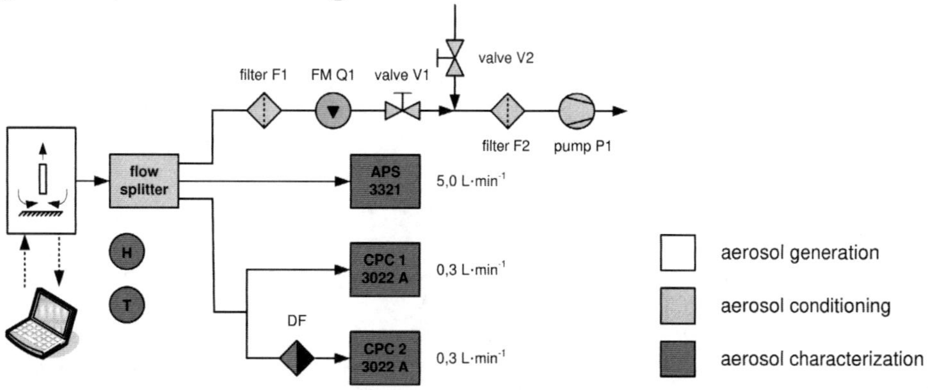

Figure 3. Schematic diagram of the experimental setup used for the (nano)-particle release characterization from coatings and composites due to wind erosion and dynamic friction.

Previous investigations [e.g. 4, 5] have shown that sanding lead to a considerable release of particles. Thus the experimental setup shown in figure 4 consisted of two dilution stages (adjustable predilution for the complete aerosol flow and device specific dilution) with known dilution factors to avoid changes in the PSD (e.g. by coagulation) and to enable best measurement conditions (e.g. single particle count mode of the CPC).

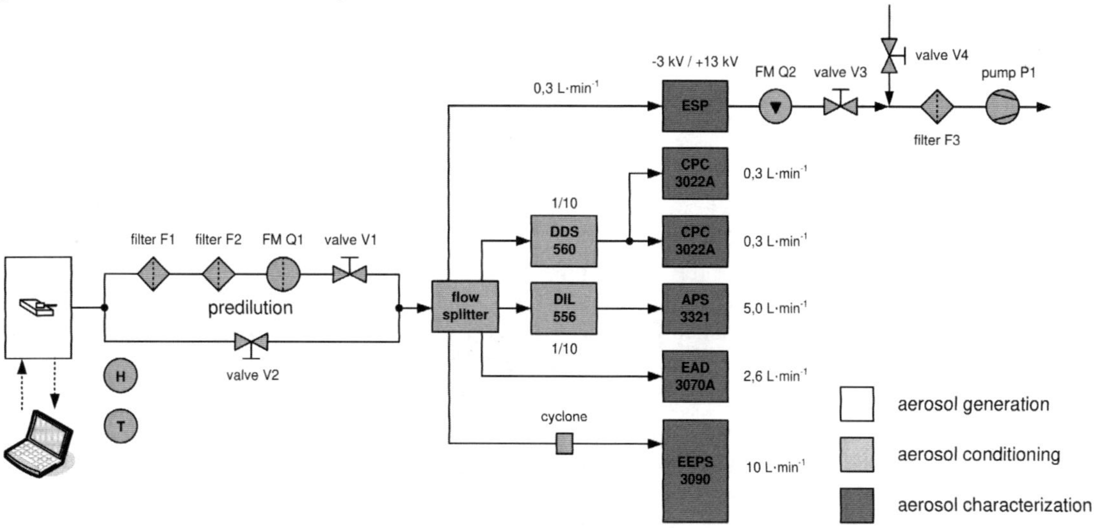

Figure 4. Schematic diagram of the experimental setup used for the (nano)-particle release characterization from coatings and composites due to sanding.

2.2.4. Experimental procedure. Each sample was treated 5 times (2 × sanding, 2 × dynamic friction, 1 × wind erosion) at different local positions. Hence, a total of 8-12 analyses for sanding and dynamic friction and 4-6 analyses in the case of wind erosion were performed for each condition of a sample system.

To monitor the whole release process (including startup and shutdown), the total measurement time of one sanding test was 100 s and in the cases of dynamic friction and wind erosion 90 s. Within this time, measurement instruments were successively switched on for data recording. The actual treatment process was started after all instruments were running for ≥ 10 s (lead time for zero background characterization) and lasted for 16 s. Subsequent the aerosol analyses were continued for ≥ 30 s (lag time for collecting all released particles).

2.3. Data evaluation

In order to draw meaningful conclusions about the (nano)-particle release, it is necessary to distinguish between the process concentration (PNC based on the simulated treatment and the suction flow rate at the particle source), the measuring concentration (PNC after aerosol conditioning or far away from the source) and the true concentration. In the literature, it is common to show the measured process concentration. But without additional sampling parameters the data can be severely misinterpreted by underestimation.

A reasonable approach is the determination of absolute release numbers from the measurement data and their relation to suitable process parameters (e.g. surface area or sample mass) to enable a consistent emission characterization, which is the basis for the estimation of particle-immission into the working space or atmosphere. This approach was already used for the data evaluation in [4, 6, 16] and is now an inherent part of ISO/TS 12025:2012.

The released number of particles $n(x,\Delta t)$ in a time interval $t_1 - t_2$ of a size fraction $x_i - x_j$ can be determined for a closed experimental system by equation (1), where Q_{tot} is the total volumetric air flow rate and f the used total dilution ratio.

$$n(x,\Delta t) = Q_{tot} \cdot f \cdot \int_{t_1}^{t_2} \int_{x_i}^{x_j} q_0(\xi,\tau) c_{n,tot}(\tau) d\xi d\tau \qquad (1)$$

The measured PNC in each size class $c_n(\xi,\tau)$ are available as discrete data. Numerical integration without additional steps gives equation (2), which was used in this study and in previous works [4, 16] for the data evaluation.

$$n_k(\Delta t) \approx Q_{tot} \cdot f \cdot \Delta t \cdot \sum_{i=a}^{b} \left(\sum_{k=c}^{d} c_{n,k,i} \right) \qquad (2)$$

In equation (2) Δt is the time between to measurement points. The running index k refers to the size class, while i is the running index of the time course.

Table 4 shows the fractional release numbers that are used in this study. Note that fractional release numbers can be in principle established for each size channel (high statistical uncertainty) or over the whole PSD (low information content) for each operated instrument.

The fractional release numbers can be used for comparison within one treatment procedure and are independent from the adjusted aerosol conditioning. In order to compare different treatment processes, the process characteristic fractional numbers of released particles have to be related to a suitable parameter of the process. This study dealt with surface treatment of coatings and composites, thus the corresponding stressed areas (see table 2) of the simulated treatment processes were used.

Assuming ideal mixing without particle coagulation or particle losses a simple way to convert the

Table 4. Evaluated fractions of released particles with annotations.

release fraction	device	annotation
$x \leq 100$ nm	EEPS or CPC-CPC+DB[a]	number of released nanoparticles (5.6 nm ≤ x ≤ 100 nm)
$x < 10$ µm	CPC	total particle release (6 nm ≤ x < 10 µm), most robust results
$x \geq 1$ µm	APS	number of released particles in the micrometre size (1 µm ≤ x ≤ 20 µm)

[a] based on non-ideal separation and particle deposition, the DB-principle can lead to an overestimation of n(x ≤ 100 nm)

resulting fractional release numbers in a model concentration is the relation to an appropriate air column volume [4], as performed for this study. The usage of dispersion models is also conceivable.

3. Results

Figure 5 shows the determined numbers of released nanoparticles for the AL-coatings (left) and the PP-composites (right).

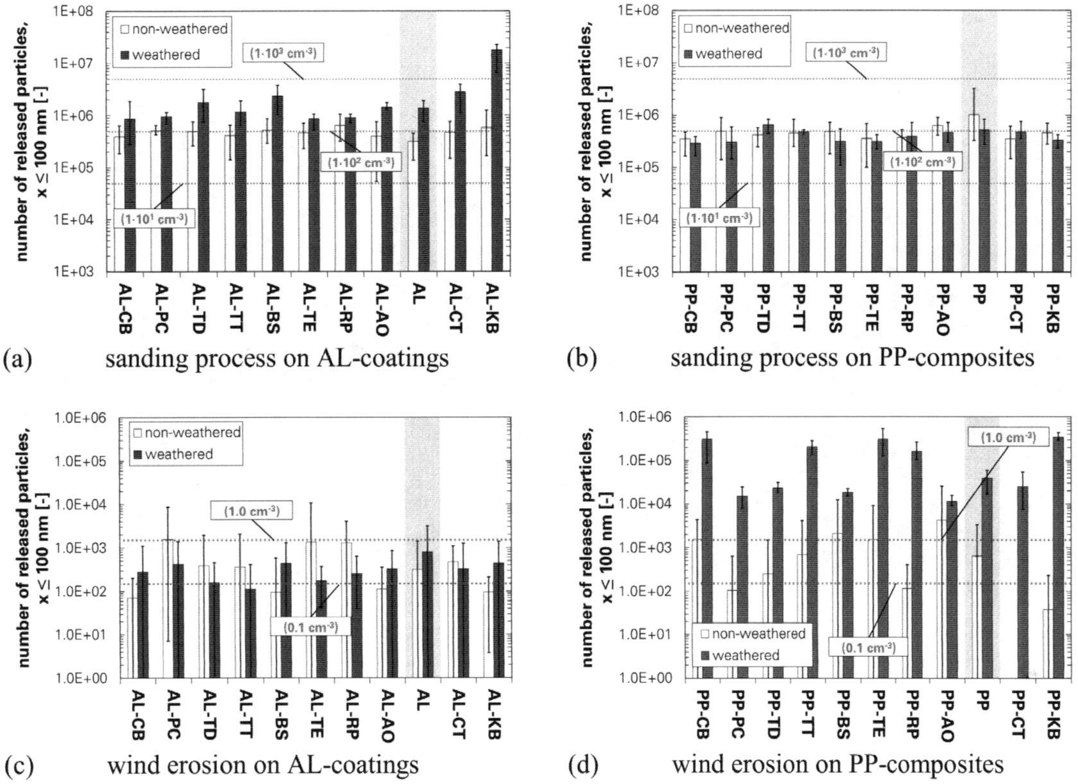

(a) sanding process on AL-coatings (b) sanding process on PP-composites

(c) wind erosion on AL-coatings (d) wind erosion on PP-composites

Figure 5. Experimental determined nanoparticle release numbers of the simulated treatment processes (sanding and wind erosion); error bars = extrema; dotted lines and values in brackets = process concentration at particle source.

3.1. Impact of pigmentation

The comparison between the non-pigmented samples with the pigmented ones for both the non-weathered and the artificial weathering state (figures 5 and 6) lead to the conclusion that the pigmentation has a non-negligible impact on the particle release. For example, the total particle release ($x < 10$ μm) due to sanding of the non-weathered coatings tends to increase with growing pigment size x_{PCS} (table 1), but it should be noted that for the examined samples the pigment size positively correlates with solid content (see table 1), too.

Extensive SEM-analyses of the contact materials and electrostatically precipitated aerosol particles showed no indication for free nanoscaled pigments. Independently performed TEM- and EDX-Analyses have shown firmly embedded pigments. Pigments with dimension significantly larger than the nanoscale (TD, CT, KB) were recovered during SEM-analyses on the contact materials (abradants for dynamic friction and sanding) that can maybe explained by shearing off due to their large contact surface.

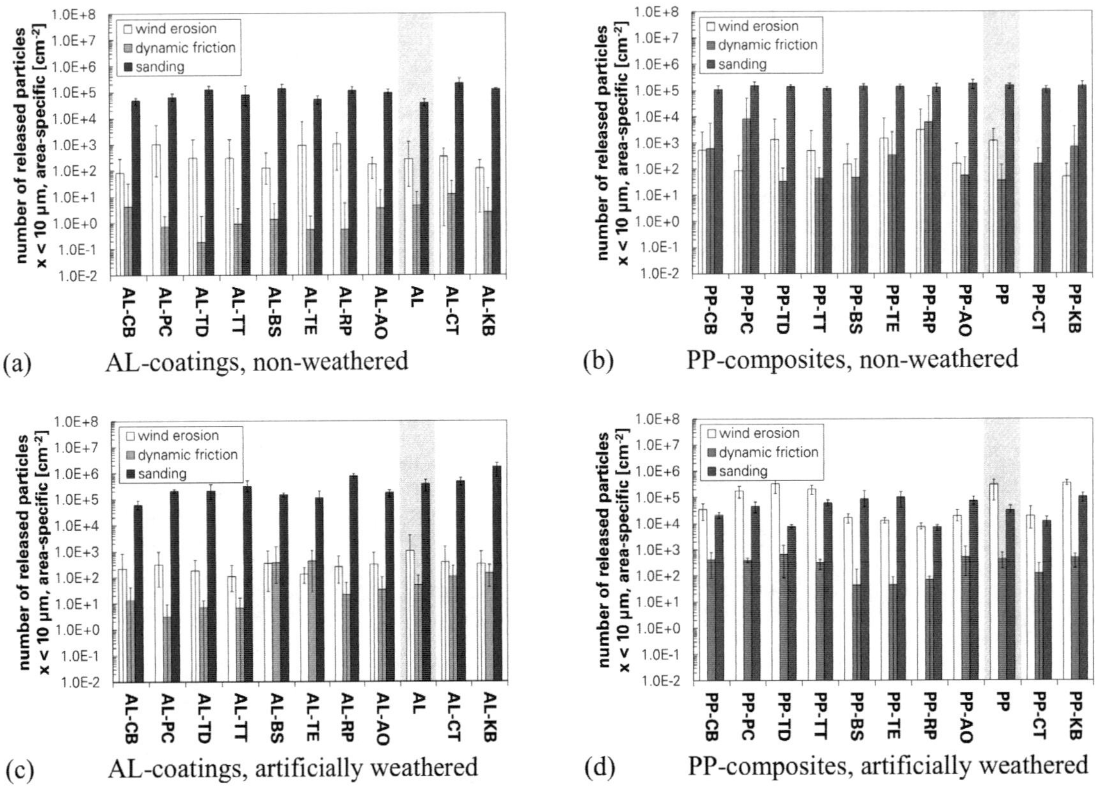

Figure 6. Cross-process comparison of the AL-coatings and PP-composites by the area specific number of released particles < 10 µm; error bars = extrema.

3.2. Impact of artificial weathering

3.2.1. Sanding. Results in the case of sanding have shown more often than not a significant increase in the total particle release ($x < 10$ µm) for coatings after they had been artificially weathered, whereas the reverse effect was observed for PP-composites. The artificial weathering of the coatings led also to a significant increase in the nanoparticle release ($x \leq 100$ nm). No significant differences in the nanoparticle release ($x \leq 100$ nm) were detected for the PP-composites. Both, coatings and composites, have shown a significant weathering-induced increase in the release of micron particles ($x \geq 1$ µm).

3.2.2. Dynamic friction. The artificial weathering led in the case of dynamic friction to a significant increase in the release $x \geq 1$ µm for the AL-coatings and the PP-composites. The release fractions $x < 10$ µm and $x \leq 100$ nm showed only for the coatings the tendency to increase, whereas no significant differences were determined for the PP-composites.

3.2.3. Wind erosion. Wind erosion led in the majority of cases to a significant increase in the particle release $x < 10$ µm, $x \leq 100$ nm and $x \geq 1$µm for the artificially weathered PP-composites in comparison to the non-weathered counterparts. No significant differences in each release fraction resulted from wind erosion of the coatings.

3.3. Comparison of the treatment processes and data transferability

Figure 6 shows for example the resulting data for x < 10 µm for the non-weathered (a, b) and weathered (c, d) samples.

In the case of the analysed coatings (non-weathered and artificially weathered), the most particles were released by sanding followed by wind erosion. The lowest numbers of released particles were observed for dynamic friction. The non-weathered PP-composites showed a similar behaviour, although no significant difference between wind erosion and dynamic friction could be observed. However, wind erosion of the weathered PP-composites led in most cases to higher particle release numbers than the other simulated processes.

Model concentrations of x < 10 µm for selected air columns are given in table 5. For the purpose of comparison, measured urban PNC (x ≤ 0.8 µm) in a street canyon ranged from $2.5 \cdot 10^4$ cm^{-3} - $5.5 \cdot 10^4$ cm^{-3} in the night and on weekend and up to $1.5 \cdot 10^5$ cm^{-3} during traffic rush hour on weekdays [17]. Comparable concentrations yielded only the simulated sanding processes. Natural flow velocities in the magnitude as simulated in this study accompany with significantly higher air volumes and thus would result in negligible low concentrations.

Table 5. Model concentrations for x < 10 µm based on the simulated processes; estimation based on the maxima values of the measured data.

process	example	air column height	model PNC for x < 10 µm			
			coatings		composites	
			non-weathered	weathered	non-weathered	weathered
[-]	[-]	[cm]	[cm^{-3}]	[cm^{-3}]	[cm^{-3}]	[cm^{-3}]
sanding	processing of semi-finished parts; re-preparation of surfaces	10	$\leq 3.3 \cdot 10^4$	$\leq 2.3 \cdot 10^5$	$\leq 2.5 \cdot 10^4$	$\leq 1.8 \cdot 10^4$
		50	$\leq 6.7 \cdot 10^3$	$\leq 4.7 \cdot 10^4$	$\leq 4.9 \cdot 10^3$	$\leq 3.6 \cdot 10^3$
		200	$\leq 1.7 \cdot 10^3$	$\leq 1.2 \cdot 10^4$	$\leq 1.2 \cdot 10^3$	$\leq 8.9 \cdot 10^2$
dynamic friction	handling of objects (e.g. toys, pencils, …)	10	$\leq 3.9 \cdot 10^0$	$\leq 1.5 \cdot 10^2$	$\leq 6.0 \cdot 10^3$	$\leq 1.5 \cdot 10^2$
		50	$\leq 7.7 \cdot 10^{-1}$	$\leq 3.0 \cdot 10^1$	$\leq 1.2 \cdot 10^3$	$\leq 2.9 \cdot 10^1$
		200	$\leq 1.9 \cdot 10^{-1}$	$\leq 7.6 \cdot 10^0$	$\leq 3.0 \cdot 10^2$	$\leq 7.3 \cdot 10^0$
wind erosion	facades erosion; moved vehicles	10	$\leq 7.4 \cdot 10^2$	$\leq 4.0 \cdot 10^2$	$\leq 1.9 \cdot 10^3$	$\leq 5.7 \cdot 10^4$
		50	$\leq 1.5 \cdot 10^2$	$\leq 8.1 \cdot 10^1$	$\leq 3.7 \cdot 10^2$	$\leq 1.1 \cdot 10^4$
		200	$\leq 3.7 \cdot 10^1$	$\leq 2.0 \cdot 10^1$	$\leq 9.3 \cdot 10^1$	$\leq 2.9 \cdot 10^3$

4. Summary and conclusions

A recently published analysis method [4] was used and partly enhanced for the qualitative and quantitative characterization of the (nano)-particle release from AL-coatings and PP-composites. Systematic analyses were performed for artificially weathered samples based on three simulated treatment processes – sanding, dynamic friction and wind erosion.

Immediate results of the methodology are total and size-fractionated number concentrations and number weighted particle size distributions. With attention to the process conditions, these data were converted in process- and fractional numbers of released particles for the size ranges x ≤ 100 nm, x < 10 µm and x ≥ 1 µm. Combined with sufficient reproducibility, the evaluation allows comparison of non-pigmented reference samples with pigmented ones and also of non-weathered samples with weathered ones.

4.1. Nanoparticle release results

Investigations have shown that the sample material, the sample composition, the sample condition (e.g. weathering state) and the type of mechanical treatment have a strong effect on the particle release. For example artificial weathered coatings showed in the case of sanding a higher (nano)-particle release as the non-weathered samples. In contrast, the artificially weathering of PP-composites led to a reduced release.

Despite occasional high numbers of released (nano)-particles, no free nano-pigments were observed during independently performed SEM- and TEM-analyses of electrostatic precipitated aerosol particles. SEM-analyses of used abradants for sanding and dynamic friction showed the presence of pigment particles with dimensions significantly larger than the nanoscale for both the non-weathered and weathered samples.

4.2. Methodology

Particle concentration, particle size and particle material are the three main measurands which are necessary for exposure assessment. Despite the increasing amount of published studies, there is still a need for a harmonized methodology of investigations on the (nano)-particle release into the air, especially in the field of experimental implementation and data evaluation. Most published studies have only a qualitative character, while exposure assessment requires additional quantitative data to enable cross-process comparison and data transferability.

As demonstrated in this study by analysis of millions of particles, particle size and number concentration can be determined statistically reliable with commercial aerosol measurement instruments. In contrast, the representativeness of the particle nature identification by SEM and TEM image analysis is limited by the small number of analyzed particles.

Acknowledgement

This work was supported by the German Federal Ministry of Economics and Technology (BMWi) under grant 01FS10004. The pigments and matrix materials were provided by the companies of the Verband der Mineralfarbenindustrie e.V. (VdMi), Frankfurt am Main, Germany and PlasticsEurope Deutschland e.V. Frankfurt am Main, Germany. Thank goes to the Bruchsaler Farbenfabrik GmbH & Co. KG, Bruchsal, Germany and the Orion Engineered Carbons GmbH, Hanau-Wolfgang, Germany for the professional fabrication of the AL-coatings and the PP-composites, the Institute of Coating Systems and Painting Technology of the Fraunhofer Institute for Manufacturing Engineering and Automation (IPA), Stuttgart, Germany for the accomplishment of the artificial sample weathering and surface characterization and Bayer Technology Services GmbH (BTS), Leverkusen, Germany for the comprehensive TEM- and EDX-analyses.

References

[1] Kuhlbusch T A J, Asbach C, Fissan H, Göhler D and Stintz M 2011 *Part. Fibre Toxicol.* **8** 22
[2] Bello D, Wardle B L, Zhang J, Yamamoto N, Santeufemio C, Hallock M and Virji M A 2010 *Int. J. Occup. Environ. Health* **16** 434-50
[3] Schlagenhauf L, Chu B, Buha J, Nüesch F A and Wang J 2012 *Environ. Sci. Technol.* **46** 7366-72
[4] Göhler D, Stintz M, Hillemann L and Vorbau M 2010 *Ann. Occup. Hyg.* **54** 615-24
[5] Koponen I K, Jensen K A, Schneider T 2011 *J. Expo. Sci. Env. Epid.* **21** 408-18
[6] Vorbau M, Hillemann L and Stintz M 2009 *J. Aerosol Sci.* **40** 209-17
[7] Bello D, Wardle B L, Yamamoto N, deVilloria R G, Garcia E J, Hart A J, Ahn K, Ellenbecker M J and Hallock M 2009 *J. Nanopart. Res.* **11** 231-49
[8] Wohlleben *et al.* 2011 *Small* **7** 2384-95
[9] Kaegi R *et al.* 2008 *Environ. Pollut.* **156** 233-39
[10] McMurry P H 2000 *Aerosol Sci. Technol.* **33** 297-322
[11] Cheng Y and Yeh H 1980 *J. Aerosol Sci.* **11** 313-20
[12] Biskos G, Reavell K and Collings N 2005 *Aerosol Sci. Technol.* **39** 527-41
[13] Wilson J C and Liu B Y 1980 *J. Aerosol Sci.* **11** 139-50
[14] Li L, Chen D R and Tsai P J 2009 *J. Electrostat.* **67** 765-73
[15] Dixkens J and Fissan H 1999 *Aerosol Sci. Technol.* **30** 438-53
[16] Stintz M and Göhler D 2012 *Technical Proc. of the 2012 NSTI Nanotechnology Conf. and Expo* (NSTI-Nanotech 2012, vol. 1) pp 525-526
[17] Wehner B, Birmili W, Gnauk T and Wiedensohler A 2002 *Atmos. Environ.* **36** 2215-23

On Nanoparticles Release from Polymer Nanocomposites for Applications in Lightweight Automotive Components

S Sachse[1], L Gendre[1], F Silva[1], H Zhu[2], A Leszczyńska[3], K Pielichowski[3], V Ermini[4], and J Njuguna[1]

[1] Department of Environmental Science and Technology, Cranfield University, Bedfordshire MK43 0AL, UK
[2] Cranfield Health, Cranfield University, Bedfordshire MK43 0AL, UK
[3] Department of Chemistry and Technology of Polymers, Cracow University of Technology, 31-155 Krakow, Poland

E-mail: j.njuguna@cranfield.ac.uk

Abstract. Nano and micro reinforced glass fibre-polymer composites have been manufactured to investigate different effect such as filler type, filler size and matrix materials on the particle emission during low velocity impact test. Nano and micro- silica, as well as nanoclay reinforced crash cones were prepared with a two step extrusion process and final injection moulding of the structures. The addition of secondary filler into the glass-fibre reinforced polymer composites had significant influence on the mechanical behaviour of the material as well as on the particle emission. In general, nano and ultrafine airborne particles were emitted from all investigated materials. However, composite filled with nanoclay emitted higher amounts of particles than those filled with nano and microsilica. One reason for the increase of particle emission of the nanoclay filled composites was the change of the failure behaviour of the matrix. However, similar results of particle emission were obtained for both nano and microsilica fillers, which in general did not vary significantly from the results obtained from traditionally reinforced glass fibres polymer composites.

1. Introduction

Material used for crashworthy structural application, traditionally have been metals, due to their plastic deformation characteristics, which enable them to absorb impact energy in a controlled manner. Polymer composite materials, contrasting metals, do not typically exhibit plastic deformation. However their stress- strain relationships may show signs of other types of nonlinearities, but they are superior to metals for specific energy absorption. Nanoreinforced polymers have focused the attention, because of their potential to exhibit impressive enhancements of material properties compared to the pure polymers [8]. Therefore, polymer nanocomposites are manufactured commercially for diverse engineering applications and are used in many economic sectors.

However, the increasing research, production and utilisation of nanomaterials nurture questions on their fate and behaviour. It is likely that, once nanocomposites undergo mechanical and thermal stress situations nanoparticles will be detached as free nanofillers or hybrids particles into the air. Therefore, the understanding of the exposure mechanism is crucial for the assessment of the potential environmental and health risks.

A present various research groups have investigated the release of nanoparticles during different mechanical stress situations such as shredding [9], drilling [10, 11, 2], sanding [7, 4], impact [12] and abrasion [1,18, 16, 13, 5, 6] of nanocoating and nanocomposites. In general, only very low amounts of nanoparticle were measured and no free pristine nanoparticles were detected. These findings raise the question on the release-ability of nanomaterials and the compatibility of the applied methods.

In order to release pristine nanofillers from a matrix, huge accelerations are necessary to generate forces able to compete with van der Waals forces, for example through instantaneous shocks [15]. However, larger nano-objects including carbon nanotubes (CNT) or nanofibres are expected to be removed under lower acceleration from the matrix surfaces. Hence, the aim of this study is to simulate mechanical shocks, through low velocity impact on nanocomposite cones and to evaluate the particle released from the structures.

2. Experiments

2.1. Material selection and preparation of test samples
Polyamide 6 (Tarnamid T30, Azoty Tarnow, Poland) and polypropylene (Moplen HP500J, Basell Polyolefins) were used as matrix material for all composites. As fillers the following materials were utilised (5mass%); surface modified montmorillonite (DELLITE® 43B and 72T, Laviosa) and fumed nanosilica particles (AEROSIL® R 974 and 200, Degussa, Evonik Industries). Nanofillers and polymer materials were compounded via extrusion to a master batch (Granulate 1 in Figure 1). Additionally, polyamide with 30% glass fibres and 30% glass spheres (MM-PA I 1F30 & MM-PA I 1K30, MacroMass Verkaufs AG, Germany) and polypropylene 30% glass fibres and 25% glass spheres (MM-PP BI 24 and MM-PP HE25, MacroMass Verkaufs AG, Germany) were used as secondary filler. Further, glass fibres for PP and PA6 (Thermoflow 636/672, Johns Manville Fibres) compounding were integrated. This approach of melt blending was chosen to maximise nano and micro particle distribution throughout the polymeric matrix, as well as the ease of granulates handling. The composition of the manufactured crash cone was as follows: glass fibre content was kept at 30 mass% for all composites, while nanofiller loading accounted 1.6 mass% and microfiller loading varied between 9 mass% for the PP matrix and 10.7 mass% for the PA6 matrix.

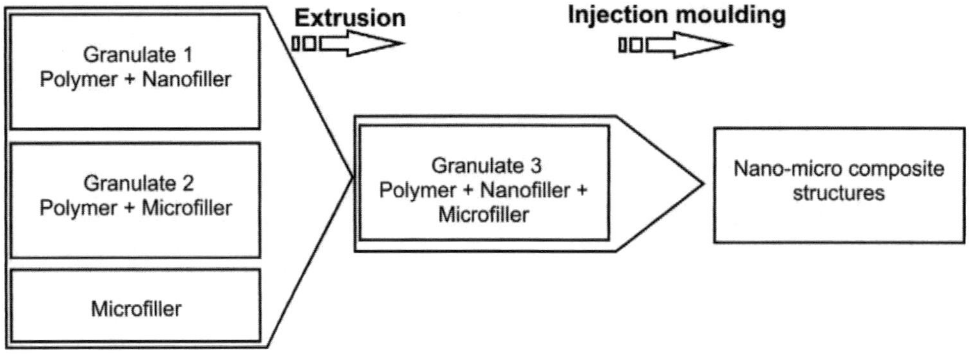

Figure 1: Manufacturing of test specimen

Exploiting this recipe, fibre reinforced nanocomposites granulate were obtained and then utilised for injection moulding of crash cones and tensile bars for mechanical testing. **Figure 2** shows the dimensions of the manufactured crash cones.

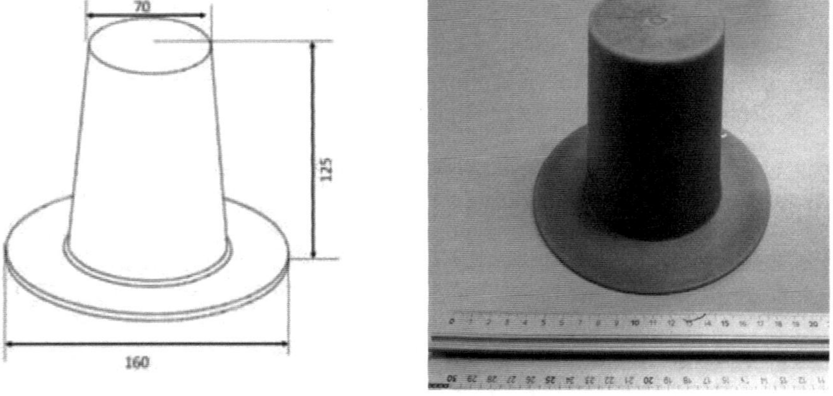

Figure 2: Crash cone dimentions (left) and injected moulded crash cones (right)

2.2. Measurement of airborne particle number, concentration and size distribution

Crash cones were impacted utilising a high energy capacity drop tower rig. This machine permits impact testing at up to 8 m/s velocity and maximum falling weight of 300 kg. External devices for measuring of load and displacement were used for external acquisition system. The low velocity impact tests were conducted at three different velocities 4.4, 6.2 and 7.7 m/s corresponding to 1, 2 and 3 m drop heights. The tests were performed by direct impact of the falling beam. The impactor mass of 54 kg was constant in all tests, giving the overall impact energies of 520 J, 1050 J and 1580 J. The load was measured using a 200 kN load cell which was located underneath the cones.

To characterize the physical properties of particles generated during these impact test, crash cones were placed into a crash chamber, as illustrated in Figure 3. The particle emissions were measured using a Condensation Particle Counter "CPC" 5.403 with Classifier "Vienna"-DMA 5.5-U (SMPS+C, Grimm Aerosol, Germany). SMPS+C measured sub-micrometer particles generated during impact process over a particle size range of 5.6-1083 nm and a particle size resolution of 32 channels in total. Prior to the measurements, the chamber was purged with laboratory air for about 20 min. Each sampling cycle comprised a 60 min background air monitoring in the chamber, and a 60 min post-impact period. The experiment was repeated 2 times for each material composition.

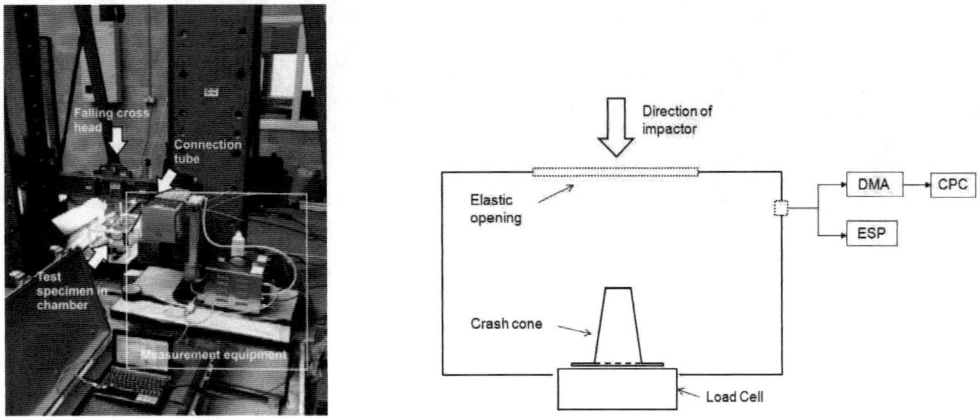

Figure 3: Experimental set up (left) and chamber set up (right)

3. Results and Discussion

3.1. Effect of matrix type on particle emission

For low velocity impact test, the matrix material was found as an important parameter on the particle emission. In general PA6 generated 4-8 times more airborne particles than PP based composites. A pronounced difference in energy absorption between the composites made of PP and PA matrices was observed as shown in Figure 4. All PP composites absorbed more impact energy compared to the PA6 composites. It was discovered that all PP composites fail in a progressive and stable manner, whereas the PA6 composite in a brittle and unstable condition. This behaviour can be directly associated with the mechanical properties of the matrix/fibre interaction. In general, PP is a ductile material with relatively low strength, whereas PA6 is a brittle material which possessed a high strength.

The brittleness of the material causes larger cracks and fragmentation, reducing the energy absorption capabilities of the material. In case of PP nanocomposites the matrix cracks and failure were localised in a close proximity of the impact point and they were no crack propagation along the structure. Delamintation and debonding of the fibres increase the effectiveness of the energy absorption, and at the same time avoiding the weakening of the non-crashed section of the structure. As impact on PP based composite only caused localised failure and not complete failure of the composite, it can be assumed that the emission rate of particles during impacting can be influenced, through the matrix materials.

Figure 4: Fracture behaviour of manufactured nanocomposites (a) PP+GF/nanosilica, (b) PP+GF/nanoclay, (c) PP+GF/microsilica, (d) PP+GF, (e) PA6+GF/nanosilica, (f) PA6+GF/nanoclay, (g) PA6+GF/microsilica, (h) PA6+GF

3.2. Effect of nanofiller type on particle emission

A significant influence of the nanofiller type used on the particle emission was noted for the low velocity impact test. Nanoclay filled PP and PA6 composite generated the most particles throughout all impact tests. The cause of this behaviour can be explained by the fracture behaviour of the composite cones during impact test. The incorporation of nanosilica particles did not increase the impact strength of the material, however changed the fracture mechanism. The brittleness of the material was significantly reduced, which was observed as an increase in elongation to brake, determined in the tensile tests. However, the opposite behaviour was observed for the nanoclay nanocomposites. In this case the impact strength of the material was increased but at the cost of reduced ductility. Hence, the strain reached the maximum allowable limit and the crack propagated along the structure, leading to a complete failure of the structure. As a result the energy absorption capability of the material remained on the same level, in spite of the increase in quasi-static strength and stiffness.

Damages due to low-velocity impact events weaken the structure of composite materials, due to a continuous service load. In every material, energy is absorbed in two ways (i) deformation of material and (ii) fracture of material. As material deformation is limited to the material properties, the fracture of the material in smaller pieces is the direction towards higher energy absorption capacities. In recent studies, it has been reported that the impact behaviour and related properties (energy absorption capacity) of polymer matrices can be engineered by adding nano-scale fillers. For example, rigid nano-sized particles, carbon nanotubes and clay nanoplatelets have been used [14]. Figure **5** shows the basic principle behind the increase in fracture pieces due to nanofiller and hence a higher degree of energy absorption capacity. In a micro and macro filled composite cracks will propagate from one filler particle to the other. The same principle is valid for nanocomposites, however the distance between particles is reduced and therefore the material breaks in smaller pieces.

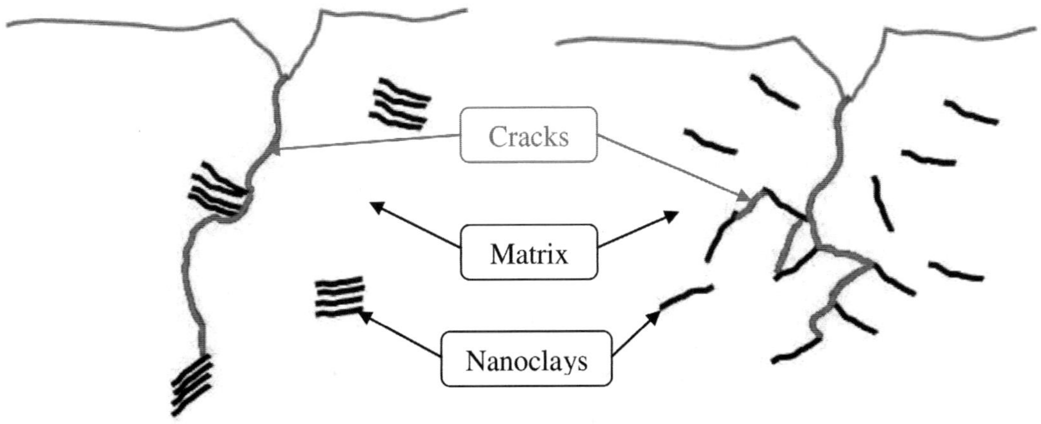

Figure 5: Difference in crack propagation in nanoclay composites; left - Intercalated nanoclay in matrix and right - Exfoliated nanoclay in matrix

3.3. Effect of filler size on particle emission

As shown in the literature [17, 3] and by the results obtained by this study, the utilisation of nanofiller over microcfiller in an identical matrix material will lead higher degree of property enhancement. The tensile modulus differed by 0.31 GPa by using nanosilica instead of microsilica particles in PP matrix. Additionally, the specific energy absorbed during impact test increased from 20.7 kJ to 22.6 kJ by using nanosilica alternative to microsilica particles. Hence, it can be assumed that these changes in properties will directly affect the quantity and properties of the release particles.

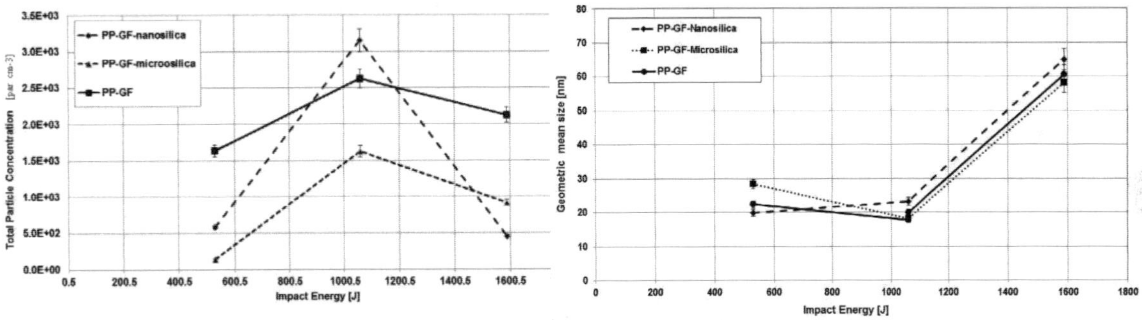

Figure 6: Comparison plot of nanosilica and microsilica fillers on particle emission, left-total particle concentration vs impact energy, right- geometric mean particle size vs impact energy

However, comparing the quantity and geometric mean size of particles release during different impact test (Figure 6) it was observed that there was a clear pattern. While there was a maximum peak of particles concentration for the impact at 1050J energy level, the particles geometric mean size was at a minimum peak. It was further observed that the quantity of particles release was significantly lower for the 530J and 1560J impact, however, particle size increased significantly depending on the impact energy. The geometric particle size increased from approx. 25 nm for the 530 J to approx. 60 nm for the 1560 J impact. Nevertheless, this behaviour was observed for all PP and PA6 reinforced with conventional glass fibre, and the composites having nanosilica or microsilica as a secondary reinforcement Therefore, it can be concluded that the particle size of the secondary reinforcement is not necessarily affecting the particle emission during impact testing, even though it influences the mechanical properties of the composite material, notably.

4. Conclusions

Nano and micro reinforced crash cones have been manufactured through a two step extrusion process and final injection moulding of the nanoreinforced granulates. Quasi-static mechanical properties and crushing behaviour of the various polymer composites were studied. It was shown that addition of secondary filler into the glass-fibre reinforced polymer composites had significant influence on the mechanical behaviour of the material. It could be shown that by varying of the secondary fillers it is possible to change the micro-mechanism of a crash and therefore control the energy absorption characteristics of the composite.

Particulate emissions were evaluated based on size-resolved from various silica based composites during impacting process. Physical characterization of the number concentration and size distribution of sub-micron particles from 5.6 to 512 nm was carried out, for the different composites. In general, nano and ultrafine airborne particles were emitted from all investigated materials. However, composite filled with nanoclay emitted higher amounts of particles than those filled with nano and microsilica. One reason for the increase of particle emission of the nanoclay filled composites was the change of the failure behaviour of the matrix. Nanoclay induced a transition from ductile to brittle fracture. Brittle material behaviour results in fracture of material in many pieces and a low deformation, and hence more particles were generated.

However similar results, of particle emission where obtained for both nano and microsilica fillers, which in general did not vary significantly from the results obtained from traditionally reinforced glass fibres polymer composites.

Acknowledgment

This work was funded by the European Commission (FP7 NEPHH Project- CP-FP7; Project Reference No. 228536 - 2. Additionally, the authors would like to thank the EPSRC Equipment Pool for the utilisation of the high speed cameras.

References

[1] Bello D, Wardle B L, Yamamoto N, Guzman de Villoria R, Garcia E J, Hart A J, Ahn K, Ellenbecker M J and Hallock M D 2009 Exposure to nanoscale particles and fibers during machining of hybrid advanced composites containing carbon nanotubes *Journal of Nanoparticle Research* **11** 231

[2] Bello D, Wardle B L, Zhang J, Yamamoto N, Santeufemio C, Hallock M and Virji M A 2010 Characterization of exposures to nanoscale particles and fibers during solid core drilling of hybrid CNT advanced composites Int*ernational Journal of Occupational and Environmental Health* **16** 434

[3] Fu S Y, Feng X Q, Lauke B and Mai Y W 2008 Effects of particlesize, particle/matrixinterfaceadhesion and particleloading on mechanicalproperties of particulate-polymer composites *Composites Part B: Engineering* **39** 933

[4] Gohler D, Stintz M, Hillemann L and Vorbau M 2010 Characterization of Nanoparticle Release from Surface Coatings by the Simulation of a Sanding Process *The Annals of Occupational Hygiene* **54** 615

[5] Golanski L, Gaborieau A, Guiot A, Uzu G, Chatenet J and Tardif F 2011 Characterization of abrasion-induced nanoparticle release from paints into liquids and air *Journal of Physics: Conference Series* **304** 012062

[6] Guiot A, Golanski L and Tardif F 2009 Measurement of nanoparticle removal by abrasion *Journal of Physics: Conference Series* **170** 012014

[7] Koponen I K, Jensen K A and Schneider T 2009 Sanding dust from nanoparticle-containing paints: Physical characterisation *Journal of Physics: Conference Series* **151** 012048

[8] Njuguna J, Peña I, Zhu H, Rocks S A, Blázquez M and Desai S A 2009 Opportunities and environmental health challenges facing integration of polymer NanoComposites technologies for automotive applications, *International Journal of Applied Polymers and Technologies* **1(2-3)** 113.

[9] Raynor P C, Cebula J I, Spangenberger J S, Olson B A, Dasch J M and D'Arcy J B 2012 Assessing Potential Nanoparticle Release During Nanocomposite Shredding Using Direct-Reading Instruments *Journal of Occupational and Environmental Hygiene* **9** 1

[10] Sachse S, Silva F, Irfan A, Zhu H, Pielichowski K, Leszczynska A, Blazquez M, Kazmina O, Kuzmenko O and Njuguna J 2012 Physical characteristics of nanoparticles emitted during drilling of silica based polyamide 6 nanocomposites *IOP Conference Series: Materials Science and Engineering* **40** 012012

[11] Sachse S, Silva F, Irfan A, Zhu H, Pielichowski K, Leszczynska A, Blazquez M, Kazmina O, Kuzmenko O, and Njuguna J 2012 The Effect of Nanoclay on Dust generation during Drilling of PA6 Nanocomposites *Journal of Nanomaterials* Article ID 189386

[12] Sachse S, Silva F, Irfan A and Zhu H 2010 Morphology studies of nanodust generated from polyurethane/nanoclay nanofoams following mechanical fracture *Journal of Nanostructured Polymers and Nanocomposites* **7** 5

[13] Schlagenhauf L, Chu B T T, Buha J, Nueesch F and Wang J 2012 Release of Carbon Nanotubes from an Epoxy-Based Nanocomposite during an Abrasion Process *Environmental Science & Technology* **46 (13)** 7366

[14] Sun L, Gibson R F, Gordaninejad F and Suhr J 2009 Energy absorption capability of nanocomposites: A review *Composites Science and Technology* **69 (14)** 2392

[15] Tardif F, Malarde M, Golanski L, Guiot A and Pras M 2011 Nanotechnology 2011: Advanced Materials, CNTs, Particles, Films and Composites *Proceedings of the nanotech 2011 conference, Boston* 495

[16] Vorbau M, Hillemann L and Stintz M 2009 Method for the characterization of the abrasion induced nanoparticle release into air from surface coatings *Journal of Aerosol Science* **40** 209

[17] Wacharawichanant S, Thongyai S, Phutthaphan A and Eiamsam-ang A 2008 Effect of particle sizes of zinc oxide on mechanical, thermal and morphological properties of polyoxymethylene/zinc oxide nanocomposites *Polymer Testing* **27** (8) 971

[18] Wohlleben W, Brill S, Meier M W, Mertler M, Cox G, Hirth S, von Vacano B, Strauss V, Treumann S, Wiench K, Ma-Hock L and Landsiedel R 2011 On the Lifecycle of Nanocomposites: Comparing Released Fragments and their In-Vivo Hazards from Three Release Mechanisms and Four Nanocomposites *Small* 2384

Behaviour of nanoparticles during high temperature treatment (Incineration type)

S Derrough [1], G Raffin [2], D Locatelli [1], P Nobile [1], C Durand [1]
[1]CEA, DRT, NanoSafety Platform (PNS), F-38054 Grenoble
[2] Institut des Sciences Analytiques (ISA), CNRS-SCA, F-69100 Villeurbanne

E-mail: samir.derrough@cea.fr

Abstract. The treatment of waste containing nanoparticles (NP) will become a matter of first importance being given the increasing production and use of engineered NP. At present no specific end of life treatment is planned for such waste and most of the time it follows the path of conventional waste in incineration plants. The study of the behavior of NP at high temperature may help to define dedicated procedures and eventually lead to new regulations. This work deals with the set up of an incineration mounting at a laboratory scale. This assembly tested on NP samples shows significant results and interesting trends.

1. Introduction

In a global context of reduction and follow-up of the atmospheric particle emissions, limiting the release of NP when submitted to waste incineration is a non negligible element. This subject is growing while the implementation of end of life solutions under control are put in place, as a necessary continuation of the growth in the manufacturing and use of engineered NP.

Nowadays waste containing nanomaterials are most of the time treated by elevated temperature (from 850°C to 1100°C) like classical waste in incineration plants. There is not much information about the behaviour of NP during incineration to establish which fraction stays in the slag and which percentage becomes airborne and is released outside the plant. However, some studies have reported the problem [1] and on-line analysis were performed [2-3] to characterize concentration and size distribution of NP released from municipal waste incinerators.

Measuring and especially understanding NP release when submitted to end of life high temperature processes is one of the topics of the NanoSafety Platform at CEA Grenoble. The first step of this fundamental study has been to develop a specific bench test including a laboratory furnace connected at its exhaust to NP analysis tools. Different NP samples were subjected in this way to incineration tests to validate our assembly.

2. Experimental setup and methods

2.1. Samples

The choice was made to focus the study on starting samples entirely at a nanometric scale. Another option would have been to conduct the incineration tests on waste containing NP. However, the NP concentration in waste being usually very low (around one part per thousand), we were not sure to be sensitive enough to measure any release of NP (the ones included from the start) when heating such samples. Indeed, our goal was to avoid measuring NP potentially released from the macroscopic part of the waste. Three samples were selected: silver, tin and nickel. For each incineration trial, around 600 mg of NP are used.

2.2. Furnace

The incineration tests are performed using a benchtop tubular furnace. As seen on Fig. 1, the furnace is made of different parts with the most important element of the assembly being the quartz tube. On the right side of this tube, the NP sample is being held in a crucible waiting to be introduced in the central heating part once the chosen temperature is reached. The incineration trials are conducted at 850°C and 1100°C which are the usual temperatures used in waste incineration. On the opposite end of the quartz tube, all the emitted particles are collected and measured using various on-line particle sizing techniques described in next section. It is important to point out that after the sample being positioned in its ceramic crucible and waiting to be carried in the central heating portion, the furnace is isolated from the outside environment. Clean air filtered with an HEPA filter is then sent all along the tube at an adequate flow rate during the rest of the experiment. By this manner, no NP are added from the outside while the flow rates needed by the analytical equipments are matched. The experiment is then started when the background noise count is showing no NP along the furnace.

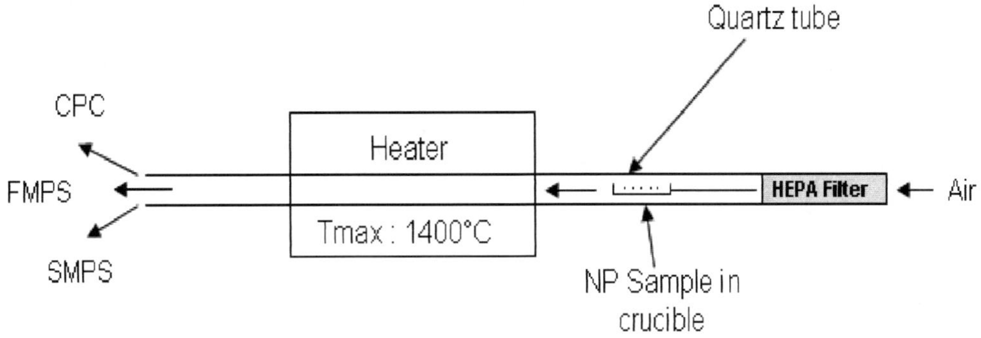

Figure 1. Schematic of the furnace

2.3. Analytical equipments

To measure NP liable to be released during incineration, we used several techniques to characterize particles aerosol in terms of volume concentration and size distribution:

Particles counting is performed using a water-based Condensation Particle Sizer (CPC 3785) manufactured by TSI Inc [4] operating on the principle of enlarging small particles using a condensation technique to a size allowing optical detection. The CPC detects particles in the size range from 5 nm to 3 µm and provides measurements over a particle concentration range from 0 to 10^7 p.cm^{-3} with time resolutions in the order of one second.

We also used a GRIMM 5.416 Scanning Mobility Particle Sizer (SMPS) [5] consisting of a particle neutralizer (Am241), a short-column Differential Mobility Analyzer (DMA) and a CPC. In such a configuration, the SMPS provides the size distribution of NP in the range 5-350 nm up to a concentration of 10^7 p.cm^{-3} with time resolution of around three minutes.

Finally, we performed real-time measurements by using a Fast Mobility Particle Sizer (FMPS) [6]. The FMPS 3091 from TSI measures the particle mobility diameter in the range 5.6-560 nm with a time resolution of 1 s this time.

Journal of Physics: Conference Series **429** (2013) 012047 doi:10.1088/1742-6596/429/1/012047

For solid samples characterization, before and after heating, we used a Scanning Electronic Microscopy (SEM) 5500 from HITACHI. The high resolution of this tool allows the visualization of the nanoparticles for size measurement and shape determination.

3. Results

An example of CPC profile recorded when a NP sample is introduced and kept in the hot furnace is given on Fig. 2. We can first notice that before the test (when the furnace is empty), no particle emission is measured. This point proves that our setting equipped with adequate filters is in conformity with our need of a background noise count close to zero before any heating of the sample. The second point is that a substantial emission of NP is measured when the sample is heated. As seen on Fig. 2, at the maximum of emission, a concentration around 1.5 10^5 p.cm^{-3} is reached. It is important to notice that the amount of sample being heated and the sizing of our furnace lead to timescales of few minutes. For this reason, the SMPS having a too long sampling time was replaced by the FMPS very early in our study. Finally, we see that the particle emission is not an endless process. The concentration value goes back to the same level as before the start of the experiment while the sample is not yet removed from the heating section. Our setup being ready to use and giving consistent results, we extended the tests to NP comparisons.

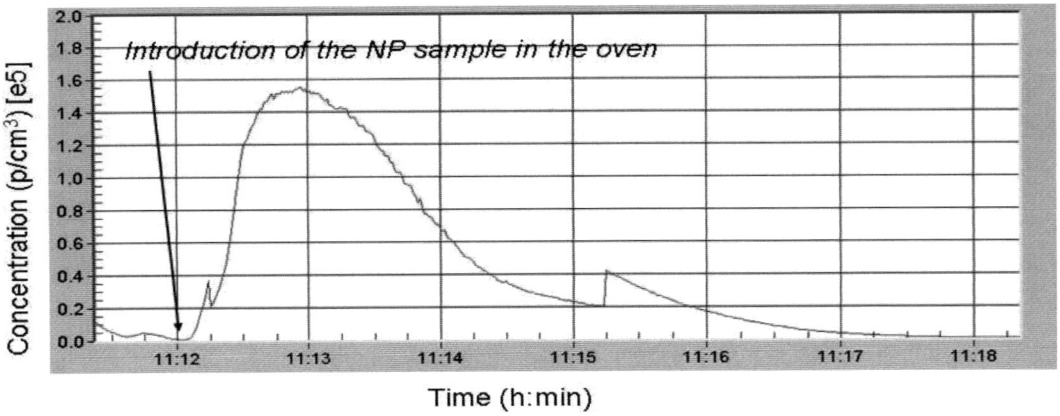

Figure 2. CPC data – NP sample heated at 850°C

For nickel and tin NP, the number of particles emitted per cm^3, at 850°C and 1100°C, is reported on Fig 3. The first information coming out from those data is that the number of particles released from the sample is significantly higher when the temperature is set at 1100°C. The maximum of concentration peaks are around 7 times higher at 1100°C compared to 850°C. One can also see that the process of emission is much shorter when the temperature is low. For both two samples, the duration of emission at 850°C is around one third of the 1100°C one.

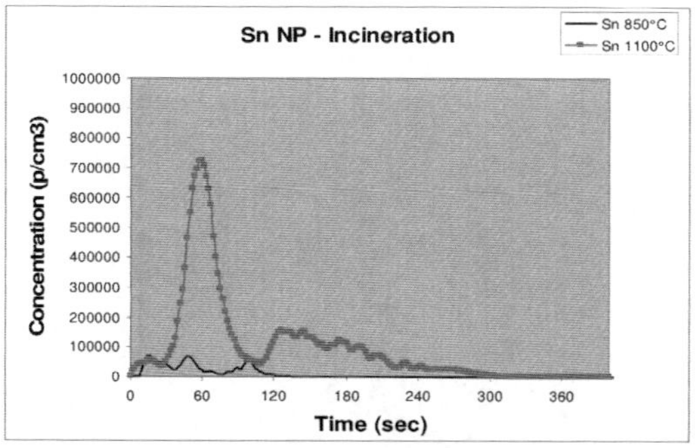

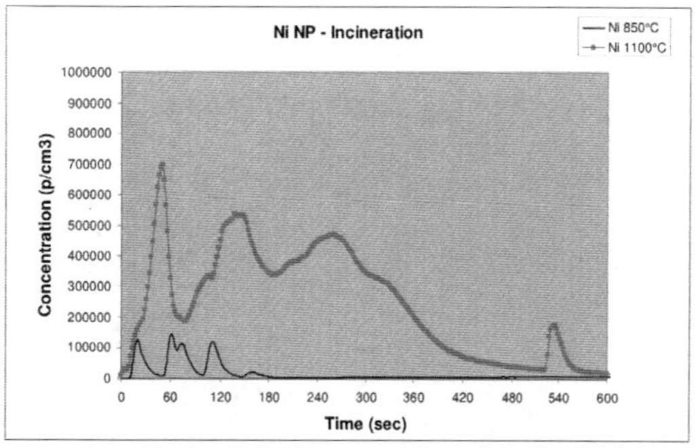

Figure 3. FMPS data – Tin and Nickel NP samples heated at 850°C and 1100°C

For silver NP, the number of particles emitted per cm^3 at 850°C and 1100°C is reported on Fig 4. For this sample, the temperature level seems to have a very low impact on the emission process as the results are very close. Although, the number of particles emitted at 1100°C is slightly higher than at 850°C, the length of the emission processes is very similar.

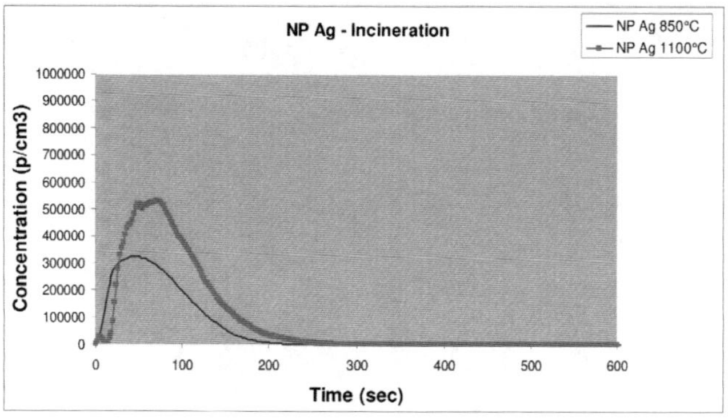

Figure 4. FMPS data – Silver NP samples heated at 850°C and 1100°C

Concerning the size of the released particles, as seen on Fig. 5 for tin NP, we can associate small size particles to low incineration temperature and higher sizes to high temperature. Indeed, very small NP of diameter lower than 50 nm are released at 850°C while at 1100°C, NP of diameter 100 nm and higher are measured. Although the results are not presented here, the same behaviour is found for nickel. One explanation is that at 850°C, the smallest "elementary" particles from the starting NP sample are released whereas at 1100°C, high size NP and aggregates are likely to be emitted. This assumption is illustrated on figure 6, on a Tin NP starting sample.

Journal of Physics: Conference Series **429** (2013) 012047 doi:10.1088/1742-6596/429/1/012047

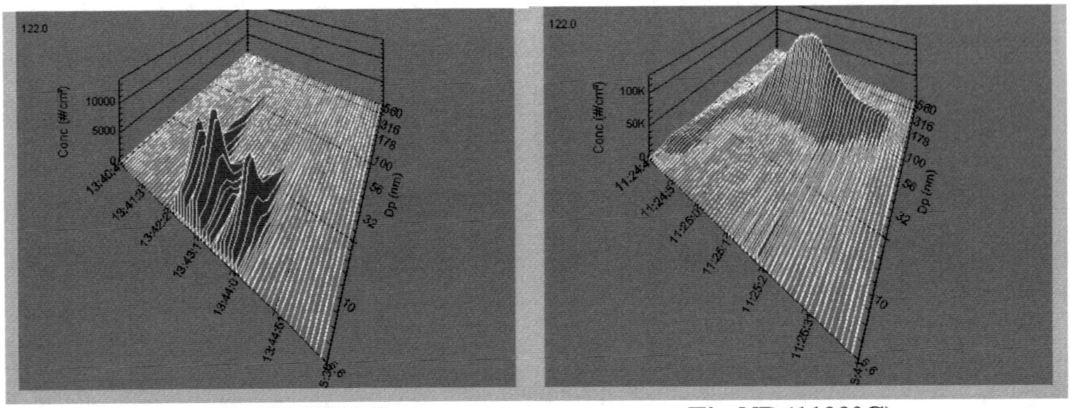

Tin NP (850°C) **Tin NP (1100°C)**

Figure 5. FMPS data (Sizes vs. time) – Tin NP heated at 850°C and 1100°C

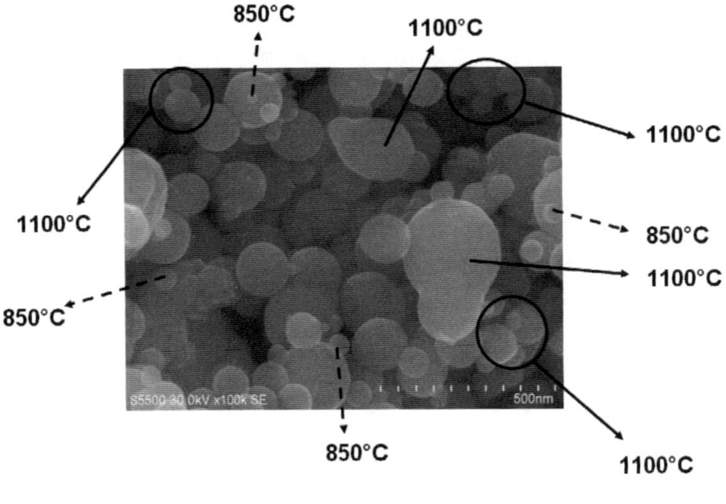

Figure 6. SEM Picture- Initial Sn NP- Potential NP release at 850°C and 1100°C

For silver NP (see Fig. 7), the behaviour is similar with tin and nickel at 850°C. At 1100°C, the behaviour is somewhat different as NP close to 100nm are first released before a size decrease.

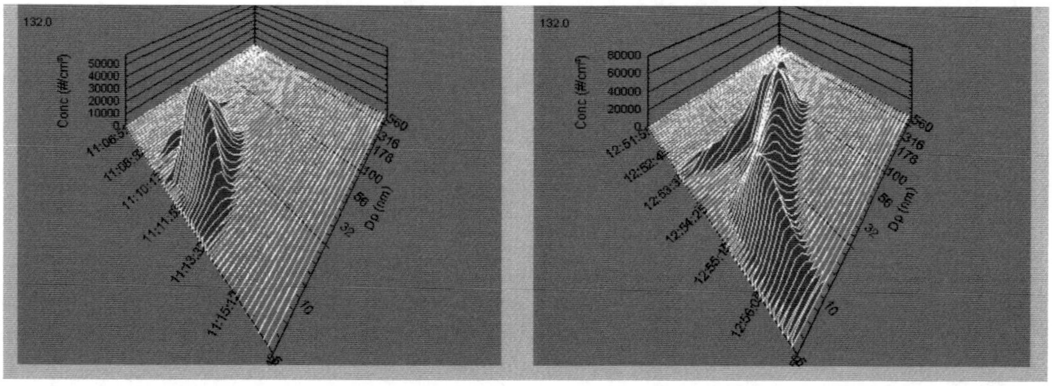

Silver NP (850°C) **Silver NP (1100°C)**

Figure 7. FMPS data (Sizes vs. time) – Silver NP heated at 850°C and 1100°C

This specific behavior of the silver sample has to be linked to the fact that melting starts even at 850°C, which is not the case for other samples. Indeed, a SEM picture of the silver residue, collected from the crucible after a 850°C heating (see Fig.8), shows that melting already occurred. This point highlights the fact that physical properties of the samples at the incineration temperature are parameters of first importance for NP release.

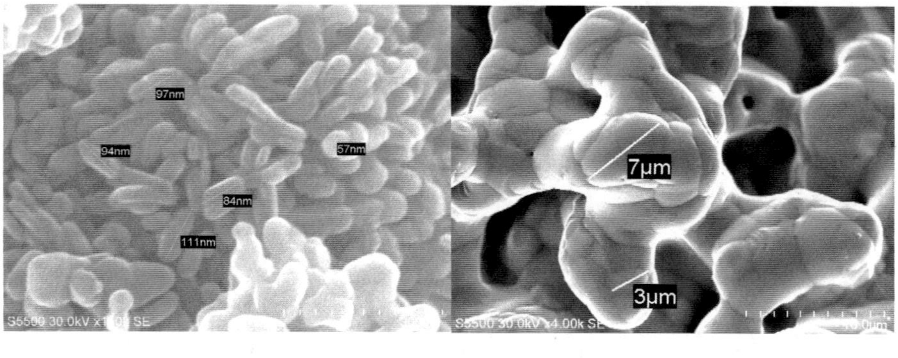

Figure 8. SEM Pictures – Silver NP before heating (left) and residue collected from the crucible after heating at 850°C (right)

4. Conclusion

The study of the incineration of NP at a laboratory scale is possible. We were able to develop an installation and a protocol that gives interesting results. Preliminary data show that release of NP from an initial sample at a "nano" scale is significant and measurable with our mounting. Differences between samples and global trends were identified. For instance, correlations were established between incineration temperatures, concentrations and sizes of NP released. Moreover, the position of the incineration temperature with respect to the melting temperature of the initial NP sample seems to be one of the critical points that control the properties of NP release. Future studies will deal with adjustments that can be undertaken with our mounting and our incineration methodology. In addition this study will be extended to other type of NP samples.

References

[1] N. C. Mueller, B. Nowack, (2008), Exposure Modeling of Engineered Nanoparticles in the Environment, Environ. Sci. Technol., 42 (12), pp 4447–4453

[2] J. Maguhn, E. Karg, A. Kettrup, R. Zimmermann, (2003), On-line Analysis of the Size Distribution of Fine and Ultrafine Aerosol Particles in Flue and Stack Gas of a Municipal Waste IncinerationPlant: Effects of Dynamic Process Control Measures and Emission Reduction Devices, Environ. Sci. Technol., 37 (20), pp 4761–4770

[3] G. Buonanno, G. Ficco, L. Stabile, (2009), Size distribution and number concentration of particles at the stack of a municipal waste incinerator, Waste Management, Vol. 29, Issue 2, Pages 749–755

[4] M. Hermann, B. Wehner, O. Bischof, H.-S. Han, T. Krinke, W. Liu, A. Zerrath, A. Wiedensohler, (2007), Particle counting efficiencies of new TSI condensation particle counters, Journal of Aerosol Science, vol. 38, Issue 6, pp 674-682

[5] S. C. Wang, R. C. Flagan, (1990), Scanning Electrical Mobility Spectrometer, Aerosol Science and Technology, vol. 13, pp 230-240

[6] T. Johanson, R. Caldow, A. Pöcher, A. Mirme, D. Kittelson, (2004), A New Electrical Mobility Particle Sizer Spectrometer for Engine Exhaust Particle Measurement, Society of Automotive Engineers, Technical Paper Series 2004-01-1341

Exploring release and recovery of nanomaterials from commercial polymeric nanocomposites

Martí Busquets-Fité[1,2,6], Elisabet Fernandez[3], Gemma Janer[3], Gemma Vilar[3], Socorro Vázquez-Campos[3], R Zanasca[4], C Citterio[4], L Mercante[4], Víctor Puntes[1,2,5,6,7]

[1]Catalan Institute of Nanotechnology (ICN), Campus UAB, 08193 Bellaterra, Spain
[2]Universitat Autònoma de Barcelona (UAB), Campus UAB, 08193 Bellaterra, Spain
[3]LEITAT Technological Center. C/ Innovació 2, 08225 Terrassa, Barcelona, Spain
[4]LATI Industria Termoplastici S.p.A. , Vedano Olona, Italy
[5]Institució Catalana de Recerca i Estudis Avançats (ICREA), Passeig Lluís Companys, 23, 08010 Barcelona, Spain
[6]Centre d'Investigacions en Nanociència i Nanotecnologia CIN2 (ICN-CSIC), Campus UAB, 08193 Bellaterra, Spain.
[7]Author to whom any correspondance should be addressed

E-mail: victor.puntes@icn.cat

Abstract. Much concern has been raised about the risks associated with the broad use of polymers containing nanomaterials. Much is known about degradation and aging of polymers and nanomaterials independently, but very few studies have been done in order to understand degradation of polymeric nanocomposites containing nanomaterials and the fate of these nanomaterials, which may occur in suffering many processes such as migration, release and physicochemical modifications. Throughout the UE funded FP7 project NANOPOLYTOX, studies on the migration, release and alteration of mechanical properties of commercial nanocomposites due to ageing and weathering have been performed along with studies on the feasibility of recovery and recycling of the nanomaterials. The project includes the use as model nanocomposites of Polyamide-6 (PA), Polypropylene (PP) and Ethyl Vinyl Acetate (EVA) as polymeric matrix filled with a 3% in mass of a set of selected broadly used nanomaterials; from inorganic metal oxides nanoparticles (SiO2, TiO2 and ZnO) to multi-walled carbon nanotubes (MWCNT) and Nanoclays. These model nanocomposites were then treated under accelerated ageing conditions in climatic chamber. To determine the degree of degradation of the whole nanocomposite and possible processes of migration, release and modification of the nanofillers, nanocomposites were characterized by different techniques. Additionally, recovery of the nanomaterials from the polymeric matrix was addressed, being successfully achieved for PA and PP based nanocomposites. In the case of PA, dissolution of the polymeric matrix using formic acid and further centrifugation steps was the chosen approach, while for PP based nanocomposites calcination was performed.

1. Introduction The first section in your paper

Continuous increase on production of nanoparticles and other nanomaterials during last decades [1] has yield to extensive studies and characterization of their properties at the many stages of their life cycle [2-6]. The same happened for the most broadly produced commercial polymers. Nevertheless, when both are combined forming nanocomposites most of the studies and characterization effort has been focused on the emerging and enhanced mechanical properties of these new nanocomposites [7-

13], and little attention has been drawn to the evolution and processes suffered by the nanomaterials used as fillers [14-15]. This is because the nanofillers have been often considered to be trapped and stabilized into the polymeric matrix, and so the consequences of its interaction with the outside environment almost neglected. This inactivity of the nanomaterials used as nanofillers could not be so categorical and there could be release of material containing nanofillers (microsized or nanosized pieces of composites or even isolated or aggregated nanomaterial). Consequently, if there is release of material, potential exposure of nanomaterial has to be addressed and (combined with toxicological and (eco)toxicological data) risk for human health and environment must be determined.

In addition, if nanocomposites are going to be (and already are in some specific fields such as in automotive industry) produced at large scales, it is also crucial to study how to deal with them when they get to the end of their life in terms of both recycling and disposal.

All these aspects concerning nanocomposites, amongst other aspects of their life cycle, such as studies of potential migration and agglomeration/aggregation processes of the nanofillers due to aging, recycling of the nanocomposites by mechanical processes, extraction and reuse of the nanofillers, toxicological and ecotoxicological studies, life cycle assessment (LCA) studies and others are addressed by the project NANOPOLYTOX (funded by the European Community's Seventh Framework Program FP7/2007-2013).

NANOPOLYTOX is a preeminently industrial orientated project where 18 nanocomposites were generated by the combination of three of the more commercially produced polymers: Polyamide-6 (PA), Polypropylene (PP) and Ethyl Vinyl Acetate (EVA) with a 3% in mass charge of 6 largely produced nanomaterials: 3 metal oxide nanoparticles (SiO2, TiO2 and ZnO); multi-walled carbon nanotubes (MWCNT) and and two variations of organomodified montmorillonite (MMT) nanoclays (the quaternary ammonium modifiers dimethyl dihydrogenated tallow ammonium (CECA, France) and dimethyl benzylhydrogenated tallow ammonium (Stepan, USA)). It is important to note that in order to achieved good compatibility between the nanomaterials and the polymeric host, a prior tailored functionalization of the nanomaterials surface was performed. As an example, to make metal oxide nanoparticles (MOx) surface compatible with the highly hydrophobic polarity of polypropylene (PP), they were conjugated with either octyl or propyl capped groups.

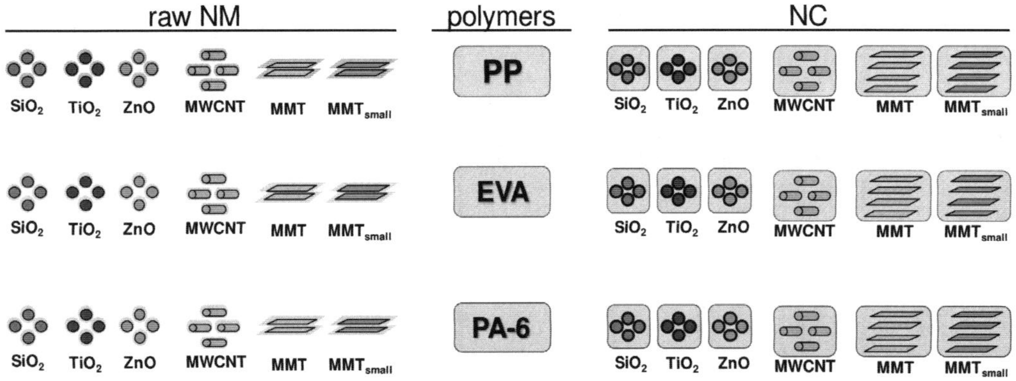

Figure1. General Scheme of the composites produced and studied within NANOPOLYTOX project (left column, under Nanocompoistes (NC) label) prepared combining three broadly produced polymers (central column; PP,EVA and PA-6) and broadly used used nanomaterials (labeled as raw NM)

An accelerated aging and weathering process was then applied to each nanocomposite inside aging chambers to study the evolution of both the whole composite and the nanofillers throughout the aging process and possible release of material. As mentioned, a substantial part of the project deals also with the development of recycling, recovery and disposal processes.

In this article we present our results on the release due accelerated aging processes and on the extaction and recovery of nanomaterials from both PA and PP based nanocomposites are presented. Analysis on the data on the composites containing nanoclays is still not finalized so it will be presented in future work.

2. Experimental procedures

2.1 Conditions of the accelerated aging process
Each nanocomposite underwent an accelerated aging and weathering process equivalent to 10 years of normal aging following the normative *UNE-EN ISO 4892-2/06, and* consisting of 1000 hours of weathering cycles under UV light and simulated heavy rain episodes. The thecnical details are as follows:

Continuos UV irradiation (340nm): 0,50 W/m^2

Black Strandard Temperature (BST) conditions: 65 ± 3°C

Relative Humidity: 50 ± 5%

Wetting cycle: 1± 0,5 min

Drying: 29 mins

Filters: Borosilicate

2.2 Procedures of extraction and recovery of nanofillers

2.2.1 Extraction and recovery of nanomaterials from Polyamide-6 based composites In order to recover the nanomaterials incorporated to the PA polymeric matrix, the chosen approximation was the dissolution of the mentioned polymer using concentrated formic acid (HCOOH). The method was optimized to maximize the amount of polymer that can be dissolved at once. It consisted on the dissolution of PA-based composite pieces in formic acid 95%, and the recovery of the nanomaterials after several series of washing and centrifuging processes with fresh formic acid 95% and ultrapure water. A final drying step under vacuum was needed to obtain the recovered nanomaterial as dry powder. The optimized lab-scale extraction and recovery method is as follows:

- **Dissolution Step:** Cut the composites into small pieces (0.5g-2g) and dissolve between 7g and 9g of the composite into 40mL of formic acid, using 50mL centrifuge tubes (made of PP). Complete dissolution takes between 2 and 4 days depending on the quantity of composite and the regularity of the shaking of the tubes. Three shaking operations, consisting of few seconds shaking and an upside down turn of the tubes, per day are enough for a complete dissolutions in 48h.

- **Centrifugation and washing steps:** Once the composite is fully dissolved, centrifuge the sample (the longest and under more times gravity force, xg or g's). Decant the dissolved polymer. Washed and centrifuged the obtained solid twice with 10mL of fresh formic acid and then again twice with 10mL of ultrapure water. The reason of using fresh formic acid for the first two washings is to get rid of polyamide residues, because if the pH of the media increases while there is still a relevant concentration of polyamide it automatically precipitates again into solid polyamide, compromising the purity of our recovered material. As was said before, each kind of nanocomposite requires different centrifugation times and speeds. The optimal numbers are shown in the table below in **Table1.**

- **Drying Step:** Dry out carefully the recovered and washed material under vacuum conditions.

Table 1. Optimal centrifugation conditions for each type of nanomaterial. G's stands for times gravity force

	1st Centrifugation (dissolved polymer)		Formic acid washings+ centrifugation 2 x 10mls		mQ washings + centrifugation 2 x 10mls	
Type of NM	**mins**	**g's**	**mins**	**g's**	**mins**	**g's**
Metal oxides	15	7000	5	6000	5	6000
Clays	20	10000	5	9000	5	9000
MWCNTs	40	13000	10	11000	10	11000

2.2.2 Extraction and recovery of nanomaterials from Polypropylene based composites The mildest methods attempted to dissolve PP were too aggressive for nanoparticles and implied an important affectation of its functionalization and inorganic structure. For that reason, dissolution approach was dismissed and replaced for calcination approach. Conditions used were 430°C for 3h and 30min, selected to both achieve a virtually complete degradation of PP into CO_2 and allow further characterization of the recovered material using thermogravimetry (TGA).

3. Results and discussion Another section of your paper

3.1 Studies on the release during accelerated aging process

To study the release of material due to the accelerated aging and weathering process, the aging waters were collected and lyophilized. The values of released mass, without determining yet its nanometric or micrometric nature or composition, can be seen in Figure2 in % of mass released found on the aging waters respect the initial mass of composite. The values themselves give no information as the percentage of mass released may vary depending on the duration of the aging process and the shape (or exposed surface) of the sample, but they are valuable to compare between composites and extract some information. EVA based nanocomposites show much less loss due to aging than PA or PP, One may think that is because of a higher resistance to degradation, but a closer evaluation of the TEM and SEM imaging data of the microtomo cuts of the nanocomposites before and after undergoing the aging process reveals that the ones with higher release values were the ones with more clearly observable aggregates and segregates of nanomaterial, of even few microns in the worst cases (PA6-ZnO-OH one). This is most probably due to a not totally efficient conjugation of the surface process, and thus, a decrease of compatibility between nanofillers and polymeric matrix. In the case of EVA composites we can attribute the lower release values to the fact that the metal oxide nanoparticles no dot require an extra conjugation step to make their surface compatible with the polarity of the polymeric matrix host, because they already are compatible.

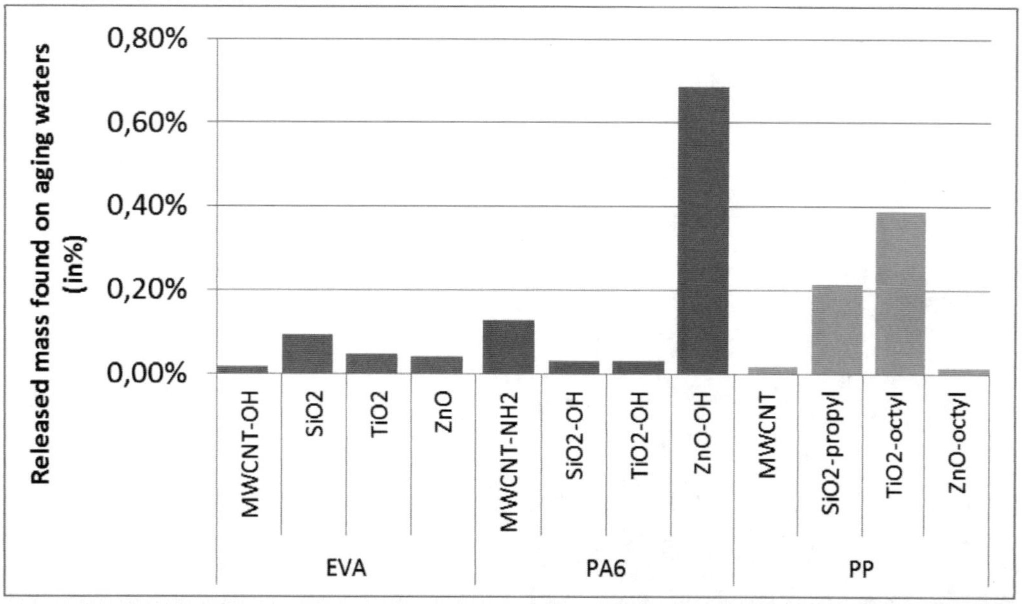

Figure2. Graphic in bars showing the percentage of released material collected in the aging waters during the accelerated aging and weathering process. In % respect the initial total mass of composite.

Characterization of the material found in the aging water was limited by the amount of material recovered, as some techniques require higher amounts than the actually found material. Nevertheless, TEM imaging using EDX to identify the composition (even quantitatively) gave 2 different kinds of behaviors in terms of nanomaterial content on the material found on the lyophilized aging waters. The extreme cases were the PP-SiO2-Propyl and PP-MWCNT composites were in the first case many SiO2 nanoparticles were found by TEM and confirmed by EDX to be the 19,9% in mass of the total material (7 times higher content than the original 3% charge of all composites) while no hints of MWCNT were found and only few pieces of polymer were observed (see Fig.Y). This observations match with both release due to degradation and goodness of the mixing observed by TEM and SEM of the composites microTomo cuts, in the sense that when the compatibility is not well achieved, there was obseeved more release of material during aging and there is more presence of nanomaterial in to the released material.

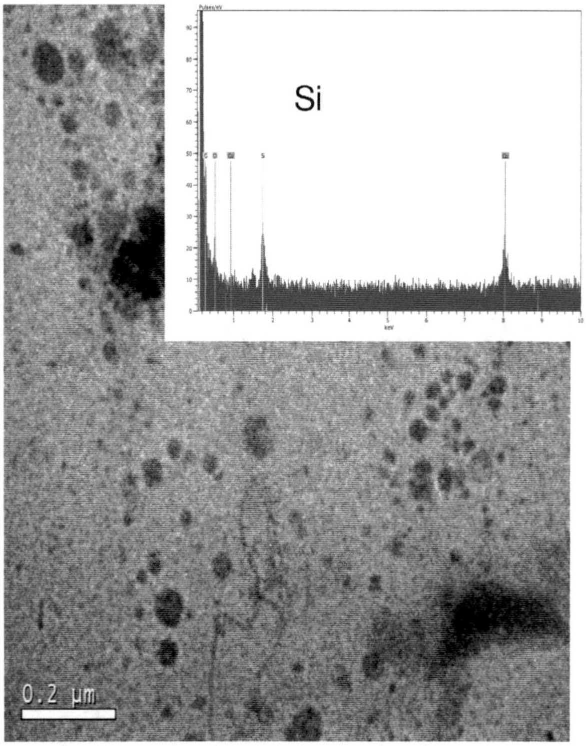

Figure3. Background: TEM image of material recovered from PP-SiO2-Propyl nanocomposite.**Inset:** EDX spectra of the recovered material from PP-SiO2-Propyl nanocomposites, where quantification showed a 19.9% in mass presence of SiO2 on the released material

Further analysis and collection of bigger amounts of aging waters will be done to corroborate the results presented here. If confirmed, it would be possible to categorize the samples into composites releasing nanomaterials and composites not releasing them. Then exposure matters, and further (eco)toxicological assays would only have to be addressed for the ones where release and exposures is actually taking place.

3.2 Discussion on the extraction and recovery of nanomaterials

3.2.1 Discussion on the extraction and recovery of nanomaterialsfrom Polyamide-6 based composites

After the whole dissolution plus centrifugation and washing process, powder-like material was recovered in yields between 43% and 60% respecting to the original 3% charge of the nanomaterial (see **Fig4.**). Characterization of the recovered materials showed a huge change in some surface properties such as BET surface area, which combined with TGA spectra and TEM imaging indicates an incomplete removal of PA residues. Nevertheless, ICP-MS values of the metalic elements (Ti in case of TiO_2 and so on) are not in any case statistically significantly lower.

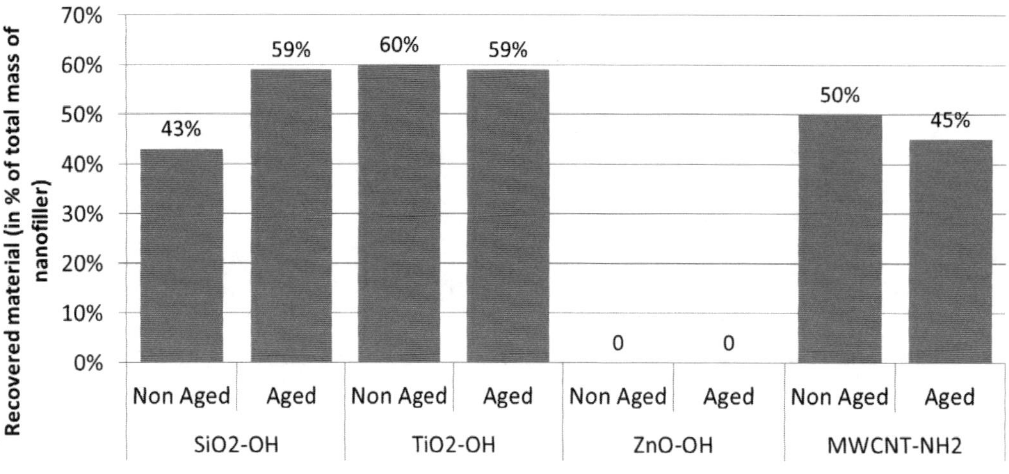

Figure4. Yields of recovery of extracted and recovered materials from PA based nanocomposites

In the case of the composites containing ZnO, the developed dissolution with formic acid method is not valid due to the high acidity of the media and the well-known high reactivity towards ionization (Zn^{2+}) of ZnO nanoparticles. Then is not possible to recover them in nanoparticulate solid form.
Another remarkable consequence of this dissolution approach to extract and recover nanoparticles is that, even with the aggressiveness of the procedure, the surface conjugation is kept almost unaltered. This is clearly seen by FTIR spectroscopy, where all the peaks corresponding to the functional groups (–OH and Si-O-C vibration bands) are still present (**see Fig.5-B**)

A

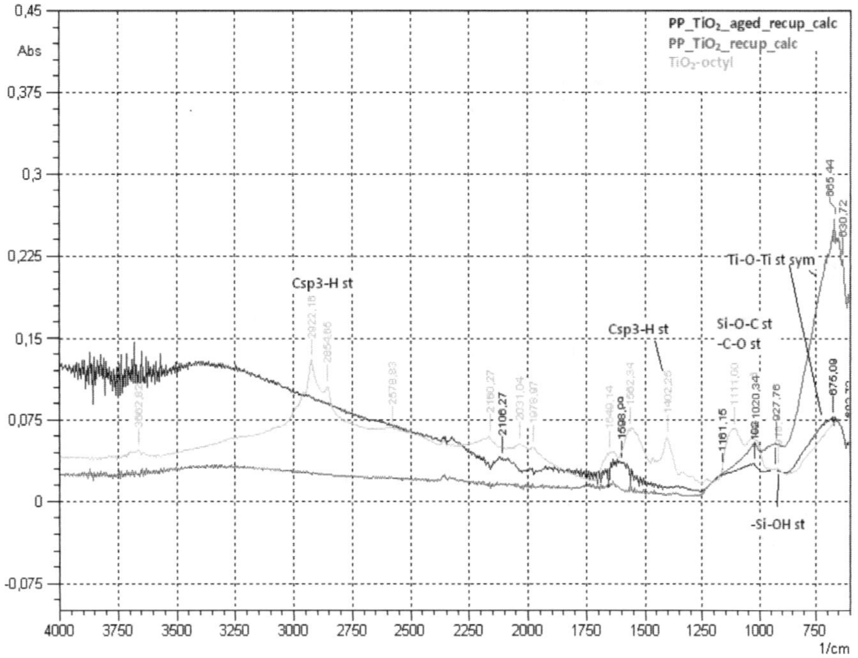

B

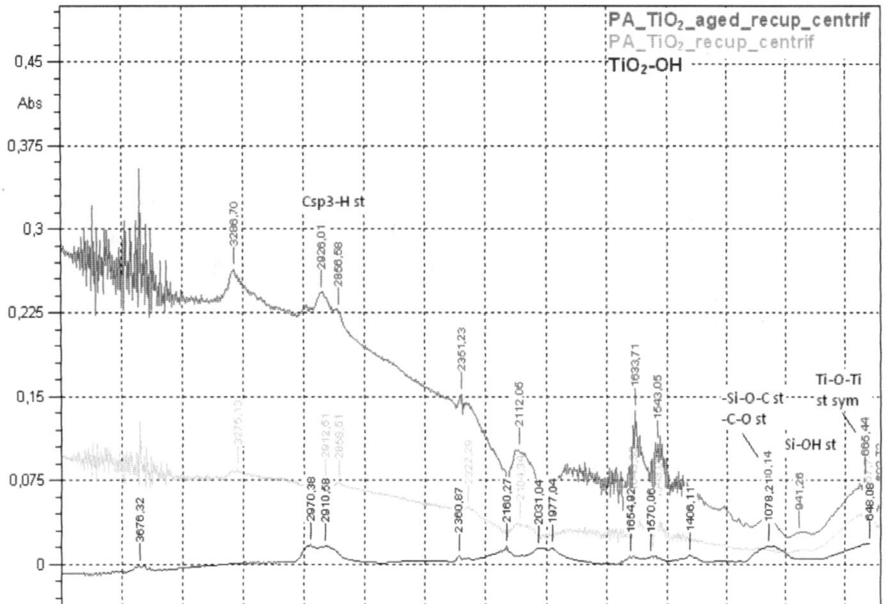

Figure.5 A) The figure shows a comparison among FTIR spectrum of extracted material from PP+TiO_2 aged and non aged nanocomposites and TiO_2-octyl. The FTIR analysis indicates the presence of TiO_2 NP characteristic bands (Ti-O-Ti symmetric stretching at 680 cm^{-1}), whilst the bands corresponding to the functional group (C_{sp}^3-H stretching at 2922 and 2855 cm^{-1}, C_{sp}^3-H bending at 1402 $cm^{-1)}$ cannot be observed, indicating its loss. Even so, the band corresponding to silica (-Si-OH and Si-O-C stretching mode at 928 cm^{-1} and 1020 cm^{-1} respectively) used to attach the octyl group to TiO_2 nanoparticles can be observed indicating that it is not eliminated during the calcination process. **B)** Comparison between the FTIR spectra of starting nanomaterial TiO_2-OH and the extracted materials from PA (non-aged/aged nanocomposites). The FTIR analysis indicates the presence of the TiO_2 NP characteristic band (Ti-O-Ti symmetric stretching at 665 cm^{-1}) in all the samples. Additionally, the bands corresponding to the silica (-Si-OH and Si-O-C stretching modes at 928 cm^{-1} and 1020 cm^{-1} respectively) used to attach the –OH group to the TiO_2 nanoparticles surface, and the bands corresponding to the functional group (C_{sp}^3-H stretching at 2922 and 2855 cm^{-1}, C_{sp}^3-H bending at 1402 cm^{-1}) can be seen as well. This fact implies that the functional groups were not eliminated by the extraction process.

Maintaining the surface conjugation and at the same time a small residual polyamide could turn out to be a huge advantage by the time to recycle the recovered material to produce new PA based nanocomposites as it could be avoided an extra conjugation step to add functional groups into the surface and the material could be treated as a master batch for the production of new PA composites. During the dissolution process significant amounts of dissolved PA in highly acidic formic acid was produced. However, far from being treated as waste, it can be used to recover and reuse again the PA. Actually, it is a very commonly used industrial way to recycle PA from old polymers.

3.2.2 Discussion on the extraction and recovery of nanomaterials from Polypropylene based composites Very high yields of recovered material from calcinations of nanocomposites were obtained (see Fig.6) of above 95% for all of them but for the PP-MWCNT aged one. Yields were corrected taking into account that calcinations of PP alone gave a total residue of 0.41% of the initial composite mass.

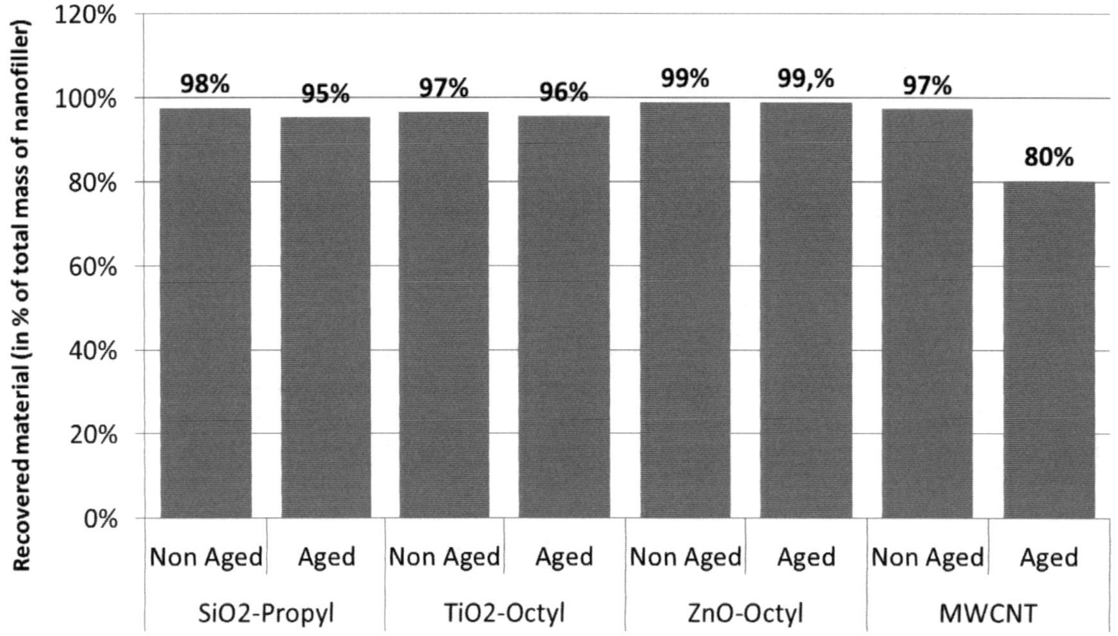

Figure 6. Yields of recovery of extracted and recovered materials from PP based nanocomposites

It is remarkable that even under such thermal conditions (430ºC) for 3h 30min, no sintering phenomena was observed by TEM imaging. This may be due to the low charge of nanofillers (3%) and the presence of the PP polymeric matrix, that keeps the particles mostly isolated and separated one from each other while degrading, preventing them from being in contact and, thus, from suffer sintering.

Regarding the conjugation of the surface, in contrast with the material extracted and recovered from PA composites following a dissolution approach, there is an almost complete lost of the surface modifiers during the calcination process, observable in Fig5-B, where the peaks corresponding to the -octyl modifier are almost not present in both spectra of aged and non-aged calcinated PP+TiO_2-octyl nanocomposites. This may then be a drawback if we intend to reuse these recovered nanomaterials to be extruded again into a PP based nanocomposites, as for that purpose an extra conjugation step would be mandatory.

4. Conclusions

Regarding release of material due to accelerated aging: Release of material has been found for all the tested composites, but exposure, and so potential risk, only would have to be addressed for the ones were nanomaterial rather than only polymer has been found. Then it turns out that a good functionalization of nanomaterials surface to make them compatible with the polarity of the polymeric host is a key factor to determine the degree of degration upon aging of the resulting composite Further (eco)toxicity results will determine their hazard and risk, but released material is low, under 0.7% of mass in the worst case and less than 0.05% of the orginal mass in the majority of the testes nanocomposites

Regarding extraction and recovery methods: Calcination of PP nanocomposites gave higher yields of recovery (≈100%) than PA dissolution (40-60%), though the morphology and distribution of the organic core remains almost the same in both cases. PA preserves the organic surface modifier, which confers compatibility between the nanofillers and the polymeric host, and keeps a certain amount of PA residue, which combined make the recovered material a perfectly suitable master batch to be recycled again into a new nanocomposite. On the other side, calcination of PP yields to nanomaterials with an almost complete loss of the functional groups, and would need to undergo an extra surface conjugation step in order to be used again to produce new PP nanocomposites.

Acknowledgements

The authors acknowledge financial support from the grant 2009-SGR-776 from the Catalan Government, and from the FP7 project NanoPolyTox funded by the European Commission under Grant Agreement No.247899. We thank Dr. Cecilia Lopez for assistance preparing this contribution.

References

[1] Hendren C O, Mesnard X, Dröge J and Wiesner M R 2011 *Environmental Science & Technology* 2011 **7** 2562-2569
[2] Adams L K, Lyon D Y; Alvarez P J 2006 **40** 3527– 3532
[3] Klaine S *J Environ. Toxicol. Chem.* 2008 **27** 1825– 51
[4] Nel A 2006 *Science* **311** 622– 627
[5] Davis J M 2007 *J. Nanosci. Nanotechnol.* **7** 402– 409
[6] Wiesner M R 2006 *Environ. Sci. Technol.* **40** 4336– 4345
[7] Usuki A, Kawasumi M, Kojima Y, Okada A, Kurauchi T and Kamigaito Osami 1993 *Journal of Materials Research* **8** 1174.
[8] Usuki A, Kojima Y, Kawasumi M, Okada A, Fukushima Y, Kurauchi T and Kamigaito Osami 1993 *Journal of Materials Research* **8** 1179.
[9] Bhattacharyya, A R, Sreekumar T V, Liu T, Kumar S, Ericson L M, Hauge R H and Smalley R E 2003 *Polymer* **8** 2373-2377
[10] Assouline E., Lustiger A., Barber A H., Cooper C A., Klein E, Wachtel E. and Wagner H D 2003 *J. Polym. Sci., Part B: Polym. Phys.* **5** 520–527.
[11] Kearns J C and Shambaugh R L 2002 Journal *of Applied Polymer Science* **8** 2079–2084.
[12] Moore E M, Ortiz D L, Marla V T, Shambaugh R L and Grady B P 2004 *Journal of Applied Polymer Science* **6** 2926–2933.
[13] Masenelli-Varlot K, Reynaud E et al. 2002 Journal *of Polymer Science Part B: Polymer Physics* **3** 272-283.
[14] Wohlleben W, Brill S et al. 2011 Small **7** 2384-2395.
[15] Reijnders L 2009 **94** 873-876.

Potential release of carbon nanotubes from their composites during grinding

I Ogura[1,2], M Kotake[2], M Shigeta[2], M Uejima[2], K Saito[2], N Hashimoto[2] and A Kishimoto[1,2]

[1] National Institute of Advanced Industrial Science and Technology (AIST), 16-1 Onogawa, Tsukuba, Ibaraki 305-8569, Japan
[2] Technology Research Association for Single Wall Carbon Nanotubes (TASC), 1-1-1 Higashi, Tsukuba, Ibaraki, 305-8565, Japan

E-mail: i-ogura@aist.go.jp

Abstract. We investigated the particle release caused by the grinding of polystyrene-based composites with and without single-wall carbon nanotubes (CNTs). In the results of real-time aerosol monitoring, considerable increases in the number concentration of nano-sized aerosol particles were observed during the grinding of both CNT-containing and CNT-free polystyrene. When a thermodenuder was used, the number of released nanoparticles was reduced by over 99.9%, indicating that the nanoparticles were presumably volatile particles released by the friction heat produced by grinding the composite. In an electron microscopic analysis of the aerosol particles, micron-sized particles with protruding fibers (probably CNTs) were observed, whereas free-standing CNTs were not observed.

1. Introduction

Carbon nanotubes (CNTs) have unique properties that make them an interesting prospect as a filler material in composites. CNT composites are expected to be used in a wide range of industrial applications and consumer products. However, there is still a concern about the potential impact of CNTs on health. To perform a risk assessment of CNT composites, it is important to know the potential release of CNTs throughout the life cycle.

The potential release of multi-wall CNTs and carbon nanofibers (CNFs) from composite materials as a result of cutting, drilling, sanding, grinding, and UV-light weathering has been investigated [1-9]. However, few studies have investigated composites with single-wall CNTs.

In this study, we investigated the particle release caused by the machining of polystyrene-based composites with and without single-wall CNTs (i.e., CNT-containing and CNT-free).

2. Methods

We utilized polystyrene-based composites containing single-wall CNTs at 0 and 5wt%. These CNTs had a tube diameter of approximately 3 nm and were synthesized using a water-assisted chemical vapor deposition (CVD) method [10]. The composites were ground using a microgrinder in a conductive/antistatic plastic box that had an air supply opening with a high efficiency particulate air (HEPA) filter (see figure 1). Before the test, purified air was passed through the box to eliminate background particles. The aerosol particles released into the box by the grinding process were measured using real-time aerosol monitoring instruments such as a scanning mobility particle sizer

(SMPS, model 3936L72, TSI Inc., USA) and a condensation particle counter (CPC, model 3007, TSI Inc., USA). Moreover, a thermodenuder [11], which was designed to remove volatile and semivolatile compounds from sample aerosol particles by heating the sample aerosol and adsorbing the volatile compounds in active charcoal, was used to distinguish the CNT-containing particles released by the grinding process from volatile particles potentially released by the friction heat produced by grinding the composite.

In addition, aerosol particles were collected on Nuclepore membrane filters (25-mm diameter, 0.080-μm pore size, Nomura Micro Science Co., Ltd., Japan). The particles collected on the Nuclepore membrane filters and those deposited on the inner surface of the box were observed using a field-emission scanning electron microscope (SEM, S-4300, Hitachi High-Technologies Corporation, Japan) under an accelerating voltage of 3 kV. Prior to the SEM observations, the samples mounted on SEM grids were coated with platinum-palladium (approximately 2 nm) to avoid image charging.

Conductive silicone tubing was used to transmit the particles to the measuring instruments.

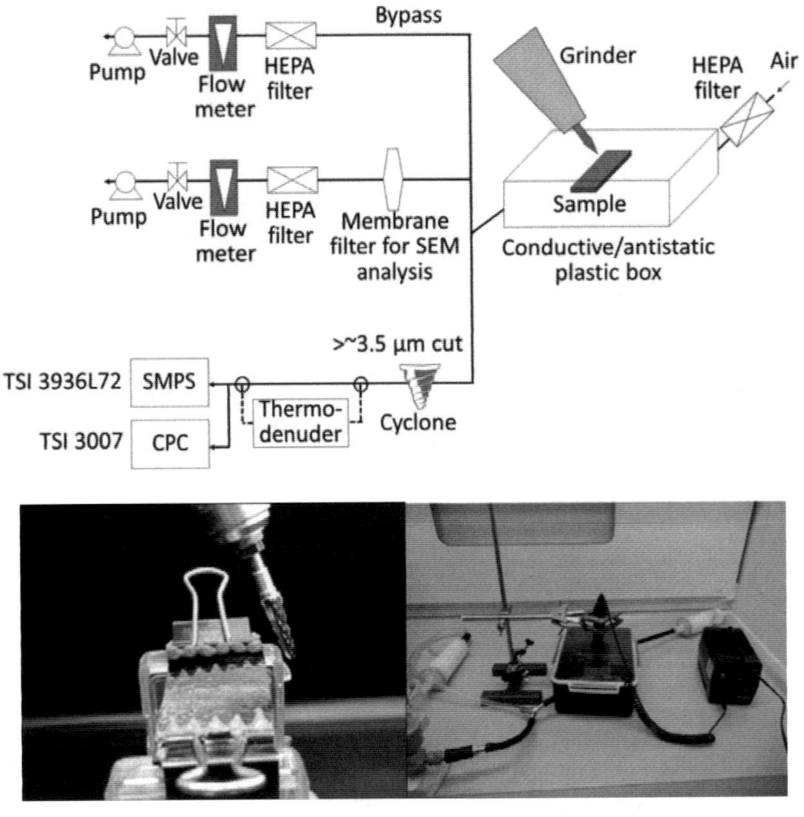

Figure 1. Schematic of experimental setup.

3. Results and discussion

In the results for the real-time aerosol monitoring, considerable increases in the number concentration of nano-sized aerosol particles were observed during the grinding of the CNT-containing polystyrene (figure 2). However, similar increases were also observed when CNT-free polystyrene was ground (figure 2). When the thermodenuder was used, the number of released nanoparticles was reduced by over 99.9% (figure 3), indicating that the nanoparticles were presumably volatile particles released by the friction heat produced by grinding the composite.

SEM images of the particles released into the air and those deposited on the inner surface of the box when the CNT-containing and CNT-free polystyrene were ground are shown in figures 4–7,

Journal of Physics: Conference Series **429** (2013) 012049 doi:10.1088/1742-6596/429/1/012049

respectively. For the aerosol particles collected during the grinding of the CNT-containing polystyrene, micron-sized aerosol particles with protruding fibers (probably CNTs) were observed (see figure 4), whereas free-standing CNTs were not observed. The particles deposited during the grinding of the CNT-containing polystyrene had a fluffy, fibrous surface (see figure 6).

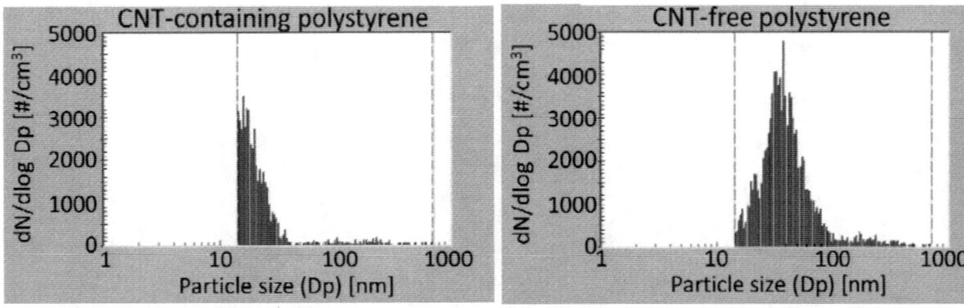

Figure 2. Number-based particle size distributions of released particles when CNT-containing and CNT-free polystyrene were ground. The particle sizes are expressed in terms of the electrical mobility diameter.

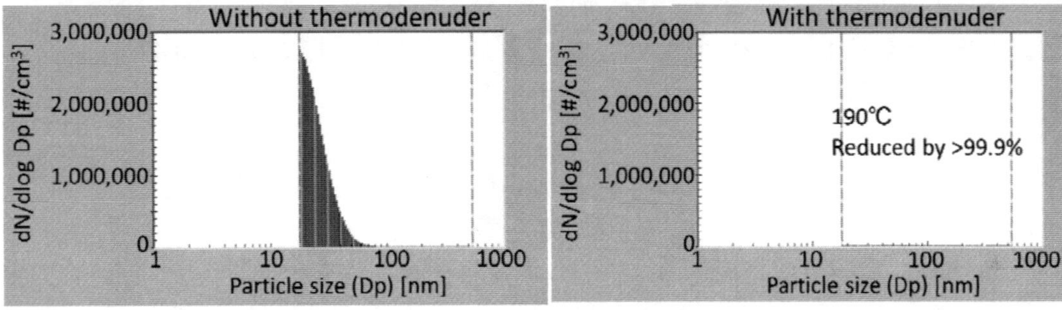

Figure 3. Number-based particle size distributions of released particles measured by SMPS with/without thermodenuder when CNT-containing polystyrene was ground. The particle sizes are expressed in terms of the electrical mobility diameter.

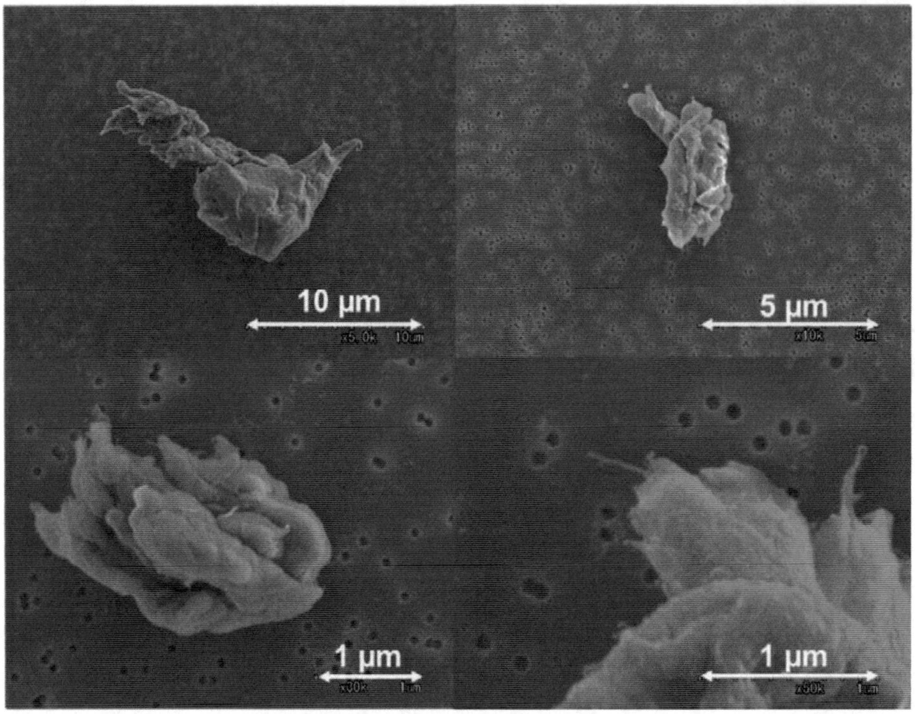

Figure 4. SEM images of particles released into air when CNT-containing polystyrene was ground.

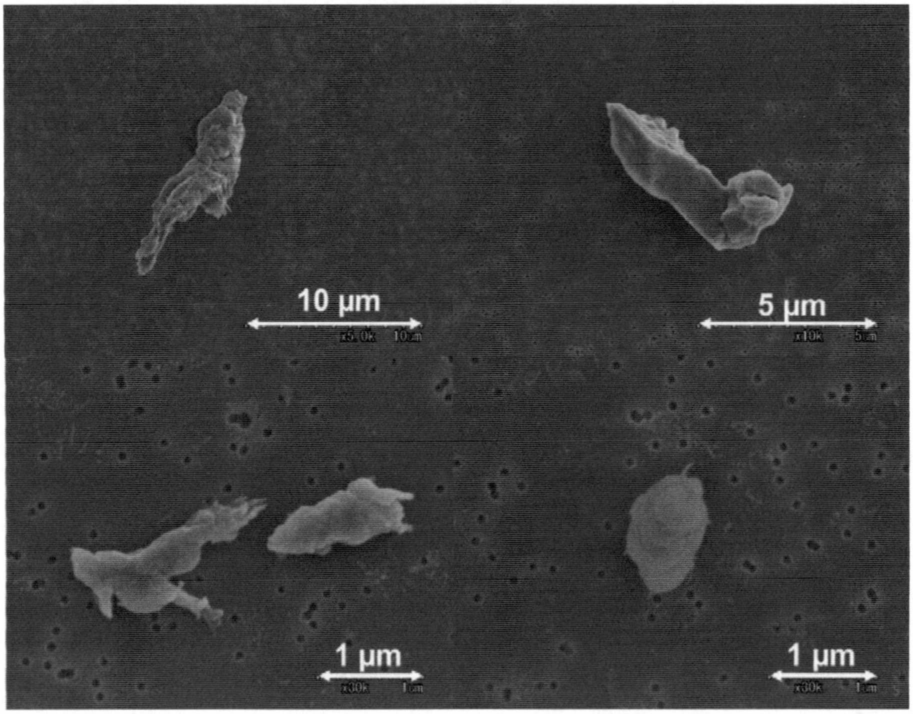

Figure 5. SEM images of particles released into air when CNT-free polystyrene was ground.

Figure 6. SEM images of particles deposited on inner surface of box when CNT-containing polystyrene was ground.

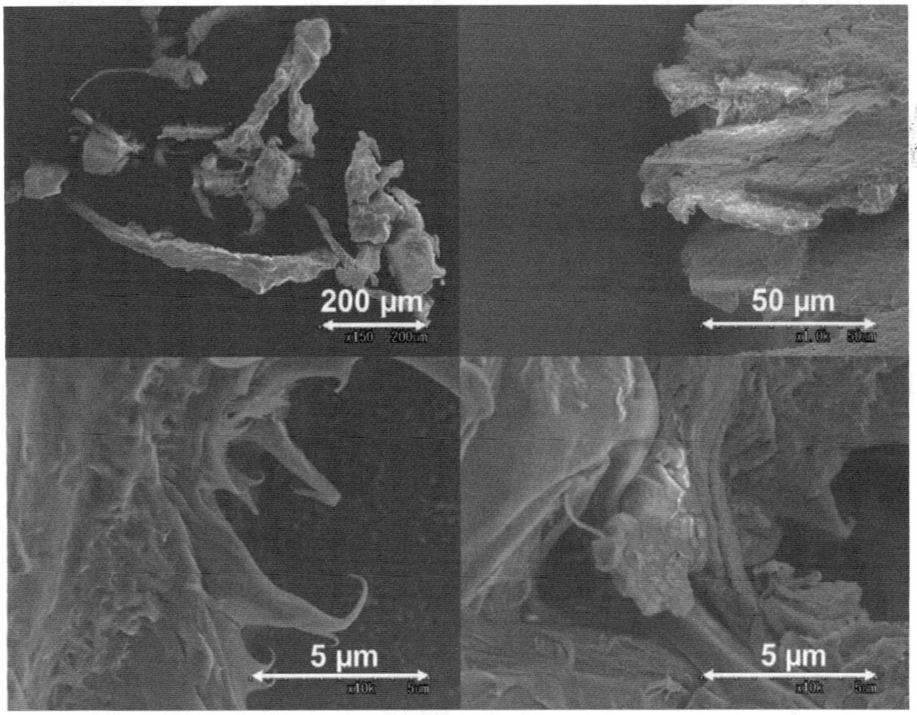

Figure 7. SEM images of particles deposited on inner surface of box when CNT-free polystyrene was ground.

Acknowledgment
This work was funded by the New Energy and Industrial Technology Development Organization of Japan (NEDO) under a grant for "Innovative carbon nanotubes composite materials project toward achieving a low-carbon society" (no. P10024).

References
[1] Bello D, Wardie B L, Yamamoto N, deVilloria R G, Garcia E J, Hart A J, Ahn K, Ellenbecker M J and Hallock M 2009 Exposure to nanoscale particles and fibres during machining of hybrid advanced composites containing carbon nanotubes *J. Nanopart. Res.* **11** 231–49
[2] Bello D, Wardle B L, Zhang J, Yamamoto N, Santeufemio C, Hallock M and Virji M A 2010 Characterization of exposures to nanoscale particles and fibers during solid core drilling of hybrid carbon nanotube advanced composites *Int. J. Occup. Environ. Health* **16** 434–50
[3] Cena L G and Peters T M 2011 Characterization and control of airborne particles emitted during production of epoxy/carbon nanotube nanocomposites *J. Occup. Environ. Hyg.* **8** 86–92
[4] Gupta A, Gaspar D J, Yost M G, Gross G M, Rempes P E, Clark M L and Martin J C 2006 Evaluating the potential for release of carbon nanotubes and subsequent occupational exposure during processing of a nanocomposite *Nanotechnology occupational and environmental health and safety (Cincinnati, OH, USA, 4–7 December 2006)* pp 110-111
[5] Methner M M, Birch M E, Evans D E; Ku B K, Crouch K and Hoover M D 2007 Identification and characterization of potential sources of worker exposure to carbon nanofibers during polymer composite laboratory operations *J. Occup. Environ. Hyg.* **4** D125–30
[6] Methner M, Crawford C and Geraci C 2012 Evaluation of the potential airborne release of carbon nanofibers during the preparation, grinding, and cutting of epoxy-based nanocomposite material *J. Occup. Environ. Hyg.* **9** 308–18
[7] Nguyen T, Pellegrin B, Bernard C, Gu X, Gorham J M, Stutzman P, Stanley D, Shapiro A, Byrd E, Hettenhouser R and Chin J 2011 Fate of nanoparticles during life cycle of polymer nanocomposites *Journal of Physics: Conference Series* **304** 012060
[8] Schlagenhauf L, Chu B T T, Buha J, Nüesch F, Wang J 2012 Release of carbon nanotubes from an epoxy-based nanocomposite during an abrasion process *Environ. Sci. Technol.* **46** 7366–72
[9] Wohlleben W, Brill S, Meier M W, Mertler M, Cox G, Hirth S, von Vacano B, Strauss V, Treumann S, Wiench K, Ma-Hock L and Landsiedel R 2011 On the lifecycle of nanocomposites: comparing released fragments and their in-vivo hazards from three release mechanisms and four nanocomposites *Small* **7** 2384–95
[10] Hata K, Futaba D N, Mizuno K, Namai T, Yumura M and Iijima S 2004 Water-assisted highly efficient synthesis of impurity-free single-walled carbon nanotubes *Science* **306** 1362–64
[11] Ogura I, Saito K and Kishimoto A 2012 Development and evaluation of thermodenuder for separation of background volatile aerosol particles from airborne nanomaterials *SENN2012 - Int. Congress on Safety of Engineered Nanoparticles and Nanotechnologies (Helsinki, Finland, 28–31 October 2012)* p 65

Towards large scale aligned carbon nanotube composites: an industrial safe-by-design and sustainable approach

P Boulanger[1], L Belkadi[1], J Descarpentries[1], D Porterat[1], E Hibert[1], A Brouzes[1], M Mille[1], S Patel[1], M Pinault[1], C Reynaud1, M Mayne-L'Hermite[1], and JM Decamps[2]

[1]: CEA, IRAMIS, SPAM, Laboratoire Francis Perrin (CNRS URA 2453), 91191 Gif sur Yvette, France

[2]: Annealsys : Bâtiment T2, PIT de la Pompignane, Rue de la Vieille Poste, 34055 Montpellier

E-mail: pascal.boulanger@cea.fr

Abstract. We present the main results demonstrating the feasibility of high surface (> A4 format size) semi-industrial fabrication of composites embedding VACNT in organic matrices. The process of growing VACNT exhibits several advantages regarding safety issues: integrating *de facto* a safe collecting procedure on the substrate, avoiding additional preparation steps and simplifying handling and protection by impregnation into a matrix. The following steps of the overall process: VACNT carpet functionalization, alignment control and impregnation, can be processed on-line in a closed and safe continuous process and lead to dramatically reduced direct nanotube exposure for workers and users. This project opens the route to a continuous, roll-to-roll, safer, cost-effective and green industrial process to manufacture composites with controlled and aligned greener « black » carbon nanotubes.

1. Introduction

Since their structural identification by Ijima in 1991, Carbon NanoTubes (CNTs) have remained an exciting material over many years. Their properties are exceptional: stronger than steel, better thermally conductive than diamond, better electrical conductive than copper, blackest material on earth, easy chemical functionalized, faster and frictionless molecule transportation [1]. Industrial production capacity of

dispersed carbon nanotube powder has boomed since 2005, mainly conditioned in masterbatches. The coming of industrial production of Vertically Aligned Carbon Nanotubes carpets/arrays/forests (VACNT) is still at its first stages and is generally considered as the second generation of safer carbon nanotube production [2][3]. Among the different growth processes, CVD methods are the most promising to produce with a high growth rate array/forest or carpet of well-aligned CNTs. The Vertically Aligned CNTs (VACNTs) topology has demonstrated promising properties by combining CNT individual properties and specific arrangement [4].
Recent papers are investigating the embedding of polymers or other matrix materials such as ceramics in the free space between the CNTs in order to fabricate nanostructured multifunctional composites and materials [5][6].

2. Front-end process: VACNT Synthesis by Aerosol Assisted CVD

From an industrial point of view, the main challenges are the control of carbon nanotube characteristics and structure through a safe and low cost process applicable to large surfaces. In this context, Catalytic Chemical Vapour Deposition (CCVD) appears to be a versatile process with great potential. In order to grown VACNT carpet, a carbon source and a catalyst is needed (often Iron, Cobalt or Nickel although other catalysts have been promoted). Two families of CCVD process can be distinguished according to the way the catalyst is engineered:

- The pre-deposition method, which consists of preparing a substrate out of the furnace where nanotubes grow and then synthetize VACNT. Once the catalyst is pre-deposited, the substrate is then submitted to a gas flow of carbon sources. This method has been for years developed for electronic application as it allows a lower synthesis temperature and the patterning of the growth through the patterning of the catalyst layer [7][8]. This method has the disadvantage of a quite slow growth rate and a limitation in the height of the resulting carpets. The addition of water molecules, in a very low quantity, has revolutionized the method leading to a so-called "super-growth" method [9] that can be up-scaled to large surface substrates [10].
- The « co-injection » method which consists of feeding simultaneously and continuously both catalyst and carbon sources in a single step simpler process. Especially, the Aerosol-Assisted CCVD process developed at CEA [11] offers a fast [12] and continuous growth of VACNT directly on various substrates [5][14] with control of nanotube characteristics as density, diameter, length and doping [13].

Although some aspects of the growth process is still not understood, the industrial stakes are now to propose schemes that could be extended for larger surfaces at mastered costs in safer conditions. In this context, aerosol assisted growth, as illustrated in fig. 1.a, reveals to be the more economical and the safer one.

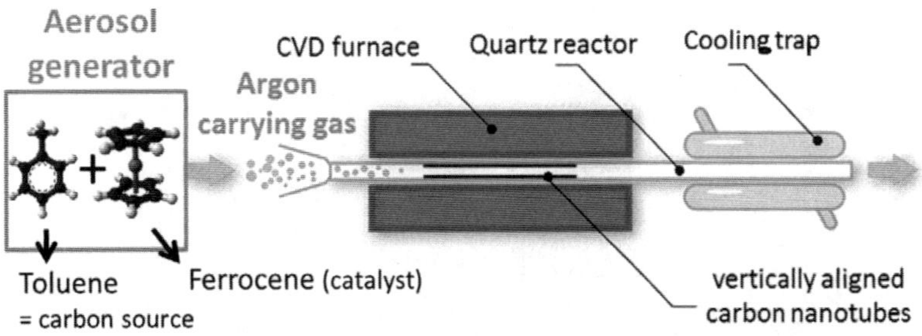

Figure 1: Aerosol Assisted CCVD process basics and typical horizontal furnaces configurations

The synthesis standard procedure has been described previously [11][12]. Briefly, the precursor mixture is composed of 2.5 wt% of ferrocene dissolved in toluene, a composition that was found to be an optimum in terms of growth rate and catalyst efficiency [13]. The resulting solution is nebulized through an injection generator (Kemstream Company, France) associated to an evaporator maintained at 200°C. The temperature is typically fixed at 850°C. The aerosol is carried out by a gas flow of 1 $L.min^{-1}$ of argon. Moreover we have demonstrated, in laboratory furnace configurations (small and short reactors (external diameter of 1,9 cm and length of 45 cm)), that the process is robust and allows a large control of VACNT carpet characteristics such as:

- Nanotube length ranges from few microns to several millimeters and depends directly on the synthesis duration [5][14] ;
- Nanotube density [3] ranges from 10^9 to more than 10^{10} CNTs per cm^2, according to the iron concentration;
- Nanotube diameter can be reduced by the use of additional gas (H_2) [16];
- At last, nanotube structure can be changed by the use of different additional gas (NH_3 etc ..) [17].

Growth is also possible on many other substrates (Quartz, Si, Stainless steel, …) [12] or even flexible fiber substrates [5][15] by adding in the same process a pre-deposition step of catalyst barrier layer.

3. Scale-up of VACNT growth through Aerosol Assisted-CCVD process

3.1. First step: growth of VACNT in laboratory equipment in horizontal configuration

Syntheses were performed in a laboratory set-up equipped with a quite large reactor (55 mm internal diameter), which is introduced in a 1 m long tubular and horizontal furnace.

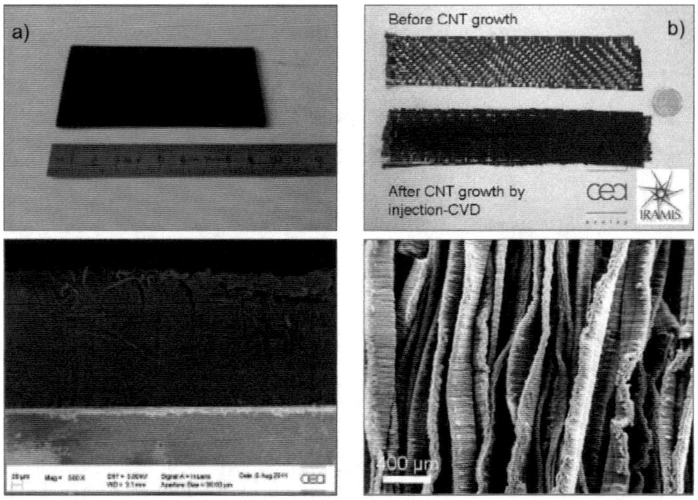

Figure 2 : VACNT growth on intermediate substrate in an horizontal furnace. (a) VACNT on quartz substrat with STM view of the carpet size, (b) on carbon fiber cloths with STM typical arrangement in aligned ribbons

Squared or rectangular flat quartz or silicon or metal substrates of 5x10 cm2 were placed in the reactor and synthesis is operated at 850°C from precursor solutions of ferrocene (2,5 wt%) in toluene. In addition, growth was also performed on woven carbon fibers exhibiting sizes of 5 x 30 cm. In the case of metal substrates and carbon fiber cloth, a previous treatment is performed consisting in depositing a ceramic oxide barrier layer in order to prevent catalyst diffusion in the substrate [15].
Figure 2 shows the growth on both quartz substrates (a) and carbon fiber cloths (b), with respectively SEM observations. Both substrates are homogeneously covered by a dark layer, which corresponds of carpet of aligned CNT perpendicularly aligned to the quartz substrate surface or to aligned CNT arranged as ribbons along the carbon fiber axis. Therefore, with this quite large experimental set-up, it is possible to homogeneously cover quite large substrates, which are interesting for R&D studies.

3.2. Second step: growth of VACNT in commercial equipment, transposition to vertical configuration

Synthesis on circular silicon wafers of 100 mm diameter (78.5 cm^2 surface) was performed in a vertical CVD set-up (MC-100 commercial equipment, ANNEALSYS company, Montpellier, France), keeping the substrate in a rotating motion. This set-up is usually used for metals, alloys, oxide and nitride compound layer deposition.
We obtain a homogeneous covering of the whole Si substrate by a black layer (figure 3 a)). SEM and TEM observations indicate that this layer is composed of well aligned

and clean carbon nanotube carpets (figure 3 b)) made of multi-walled carbon nanotubes (Figure 3 c)) as in the horizontal configuration. Alignment has been checked by SEM (Figure 3 b) and CNT length is homogeneous all along the substrate surface (figure 4).

This successful synthesis demonstrates that we obtain VACNT of the same characteristics that in horizontal lab furnace, while both temperature and gaz flow is different. Moreover, VACNT growth on moving substrates (rotating), opens the road to roll-to-roll VACNT industrial process.

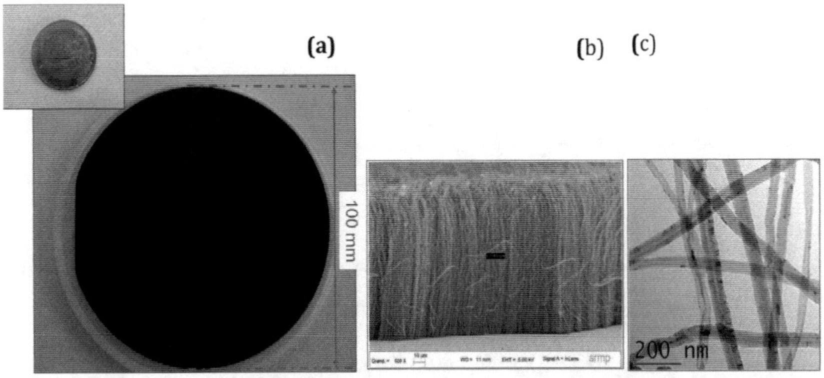

Figure 3: a) 100 mm Si substrate homogeneously covered by VACNT, b) SEM characterisation of the VACNT carpet c) TEM images of multiwall nanotubes

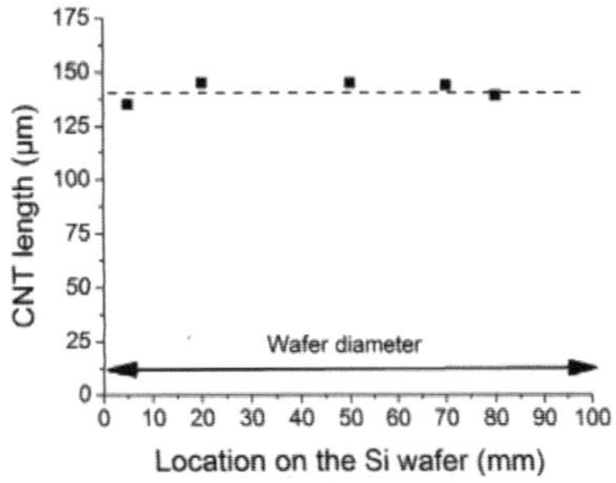

Figure 4 : Homogeneousness of the VACNT carpet's height (nanotubes length) all along the substrate)

3.3 Design of a new semi-industrial batch-to-batch pilot-plant

We designed a new pilot furnace to allow the semi-continuous production of 12 inches wafer scale or A4 format sheets, to up-scale the growth process for larger surfaces and industrial scales. A prototype equipment was therefore built in a vertical configuration for production of VACNT on large substrates such as silicon wafers exhibiting a diameter of 300 mm (706.8 cm2 surface).

Synthesis are performed from toluene/ferrocene precursor solution, in Ar atmosphere at pressures varying between up to 1000 mbar and at temperatures between 800 and 850 °C on immobilized silicon substrates. The prototype equipment based on vertical configuration and on immobilized substrate enables the homogeneous formation of VA-MWCNT carpets covering homogeneously the whole surface of the Si substrate (Figure 5 (a)) similar to the ones obtained both in horizontal configurations on laboratory equipment and at intermediate scale.

SEM observations (using a SEM-FEG Zeiss Ultra 55 electron microscope) performed on the cross section of the Si substrates at 11 cm from the edge of the Si wafer (Figure 5 (b-c)) reveal the formation of a carpet of well aligned CNT exhibiting a thickness of 200µm for a synthesis duration of x min using a 2,5 wt.% solution of ferrocene dissolved in toluene. After dispersion in alcohol of the CNT carpet collected on the Si substrate at the same position (11cm from the edge of the substrate), CNT morphology and size are obtained from transmission electron microscopy (TEM Philips CM 12) and confirm that carpet is composed of multi-walled CNT with a mean external diameter of x nm and exhibit the cleanliness of the samples with very few by products (Figure 5 (d)).

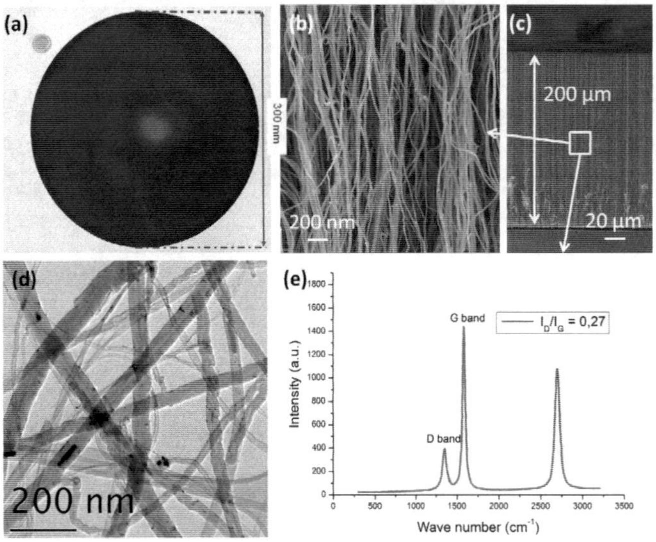

Figure 5: Large surface VACNT carpets, (a) 30 mm in diameter Si substrate homogeneously covered by VACNT, (b) SEM picture of alignment (c) SEM picture of the carpet side (d) TEM characterization showing clean Multi-wall CNT, (e) Raman Spectroscopy of CNTs

Moreover, carbon structure has been studied by Raman spectroscopy (Renishaw Invia Reflex, l=532 nm) at the same position (11cm from the edge of the substrate). Recorded spectrum (Figure 5 (e)) exhibit the 2 strong D and G bands (at ~ 1354 cm-1 and ~ 1582 cm-1 respectively) commonly observed for CNT structures. The resulting ID/IG intensity ratio is about 0.27 which is lower than the one reported for aligned MWCNT raw samples synthesized by CVD [1, 2] which is about 0.4 and is significantly lower than the ones generally reported for MWCNTs produced by CVD

[18]. Thus these analyses confirm the good structural quality of the well aligned multi-walled CNT carpet obtained at the surface of the large Si substrate in our prototype equipment.

4. Back-end process: impregnation, and membrane thinning

Our approach is to use and keep vertically aligned carbon nanotubes carpets in order to evaluate their properties in applications such as:

- Energy system electrodes: the use of VACNTs is electricity storage devices such as ultra-caps, batteries or solar systems is of great interest [19];
- The fact that VACNT can be grown on flexible substrates such as carbon fibers [15], is of great interest for lighten composite material and make them electrically and thermally conductive [16]
- Other exceptional examples are given in the field of environment where one will benefit from both chemical and nanofluidic properties of nanotubes in applications ranging from CO2 sequestration and water desalination [22].

We developed several processes to engineer raw VACNT and fabricate an intermediate product such as an electrode, a composite material or a membrane. For each process, we adopted classical industrial processes, coming from either mechanical or microelectronics industries that could lead to a continuous implementation.

Functionalization, coating or impregnation of VACNT carpet are processes to treat individual nanotubes external surfaces and to fill the space in between the nanotubes to fabricate a dense composite material made of aligned nanotubes in a polymeric matrix.

- In particular, diazonium chemistry in aqueous media was shown to be a powerful tool to graft nanotubes [23]. The grafting phase is obtained by pure chemical protonation of diazonium that serves as a seed for polymer growth. Thickness is easily controlled by the treatment duration and the process is easy to handle and eco-friendly.

The objectives of the coating can be manifold:

- coating can protect from individual nanotubes release during the process [20];
- Functionalization to add a functionality to the composite;
- Allow the removal of substrate by maintaining nanotubes;
- The coating can also be useful as a sub layer in order to improve the link between the impregnation matrix and carbon nanotubes.

Impregnation step consists of filling, without any porosity, the space in between the nanotubes by different polymerization methods. Several polymers have been tested, from PS (polystyrene), PMMA (Poly-methyl methacrylate), epoxydes, PET and other specialties polymers. It has been proven that we can impregnate thick carpets of more than 1 mm in thickness without any detectable porosity or crack (figure 6.a).

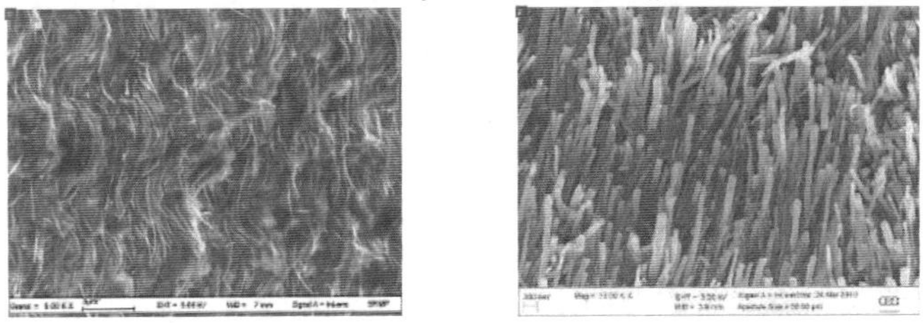

Figure 6: Complete impregnation of a thick VACNT carpet
(a) side impregnation 1 mm nanotubes with of epoxy (b) top of the carpet

At the end of the process, Chemical Mechanical Polishing (CMP) was used to thin the composite, laser ablation to engineer the carpet [24] and plasma treatment was used to open nanotubes caps (figure 6.b) in a membrane fabrication context. At the end we were able to extend and chain all the steps of the process continuously as illustrated in figure 7, leading to the fabrication of the current biggest VACNT membrane.

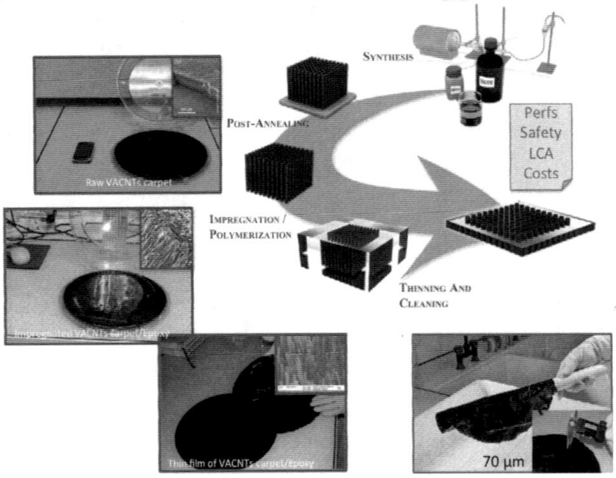

Figure 7: global process illustration allowing us to realize
the current biggest VACNT membrane (70 microns of thickness)

5. Sustainability and Safe-by design approach

We adapted the process to replace chemical species such as toluene, which is recognize as CMR, by alternative solvents coming from green sources. We tested three different green oils:

Source	Chemical Formula	Density (gml^{-1})	Molecular weigth ($g .mol^{-1}$)	Vaporisation temperature
Toluene	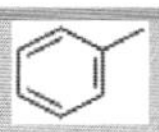 C_7H_8	0.867	92.14	111 °C
Camphor oil	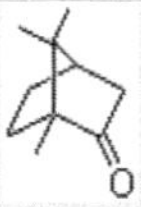 $C_{10}H_{16}O$	0.992	152.23	204 ºC
Palm oil	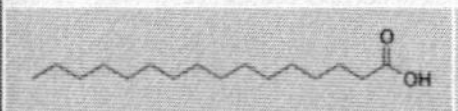 $C_{16}H_{32}O_2$	0,8527	256	350 °C
Eucalyptus Oil	 $C_{10}H_{18}O$	0.913	154.25	175 ºC

Table 1: Alternative green Carbon sources for VACNTs

Figure 8 illustrates the surprisingly good quality of nanotubes obtained by these sources, in particular Camphor Oils, which is already observed in case of dispersed nanotubes [25]. We tested the same mix of ferrocene in these oils (2,5 %w) and proceeded as in a classical synthesis. In this case VANCT growth rate has been observed as higher than with Toluene/ferrocene reaching up to 200 microns/min.

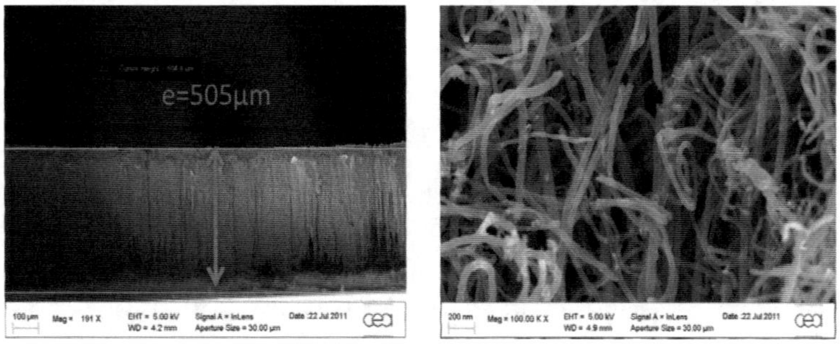

Figure 8: VANCT made from Camphor oil, Alignment and cleanliness observed by SEM

As already mentioned, one of our motivation is to be able to design a complete fabrication process starting from chemical products to final product without any manipulation, transportation or manual treatment, thus avoiding any exposure (expect of course in the case of maintenance).

In this objective, it appears that AACVD is certainly the safest process as it exhibits many advantages in regards to more classical fluidized-bed-type dispersed nanotubes processes that multiply steps to produce final clean nanotubes.

- Nanotubes in carpets are Non Volatile, Homogeneous in length, clean (less than 4 % catalyst content)
- The process is operated in classical room condition (no clean room needed), at ambient pressure, with the use of liquid precursor (easy to handle).
- The fact that the growth occurs on a substrate, allows a direct collection, no sorting neither cleaning steps are needed
- Nanotubes in carpets are easier to store, to manipulate and finally to disperse in case we want to get dispersed nanotubes with this process
- We keep the alignment all along the process
- VACNT carpets can then be directly coated or impregnated to encapsulate CNTs, reducing any further direct exposure.

In order to evaluate and improve safety aspects, we collected residues during the CMP thinning phase. Indeed, by principle, all other phases of the global process, growth, functionalization, impregnation or laser steps do not release any residues. For the thinning step, we never observed by SEM any individual NTCs alone, they are always encapsulated in big polymers particles and often in bundles (« big » particles > 500 microns in Figure 10).

5. Perspectives and Conclusions

We present the main results obtained within the framework of the French program NaWaA4, which aimed at demonstrating the feasibility of high surface (> A4 format size) industrial fabrication of composites embedding VACNT in organic matrices. The process of growing VACNT exhibits several advantages regarding safety issues: integrating de facto a safe collecting procedure on the substrate, avoiding additional preparation steps and simplifying handling and protection by impregnation into a matrix. The following steps of the overall process: VACNT carpet functionalization, alignment control and impregnation, can be processed on-line in a closed and safe continuous process and lead to dramatically reduce direct nanotube exposure for employees and users. We collected sample and residues all along the process and found only large composite particles (of size > 100 nm) made of nanotubes into polymer, which are less toxic than volatile CNT powders. This project opens the route to a continuous, roll-to-roll, safer, cost-effective and green industrial process to manufacture composites with controlled and aligned greener « black » carbon nanotubes.

Acknowledgments.

This work has been supported by the French National Agency (ANR) in the frame of its technological Research NANO-INNOV/RT program (NaWaA4, project N°ANR-09-NIRT-05). Within the framework of the project, the author wants to thanks :

- Julien Cambedouzou, Denis Petermann, Pascale Launois from Université Paris Sud/Laboratoire de Physique des Solides
- Sebastien Pacchini and Christina Villeneuve from CNRS/LAAS (toulouse)
- Sebastien Roussel from Pegas-tech
- Jacques Cinquin, Stéphane Bechtel and Amélie Darfeuille from EADS-IW
- Nicolas Debski for Raman Analysis
- Thomas Gabard for synthesis on pilot

This work has also been supported by the SAPHIR European project (SAfe and Integrated Nanomanufacturing). Authors want to thank Frédéric Schuster for fruitful discussions and CEA's Materials transverse program financial support.

References

[1] M. Endo, M. Strano, P. Ajayan, « Potential applications of carbon nanotubes », in Carbon nanotubes, Topics Applied Physics 111, Springer Verlag, 2008
[2] BCC market research Nanotubes 2010
[3] Q. Zhang et al. « Advances in Carbon nanotubes mass production » ChemSusChem, 4, 864-889, 2011
[4] Y. Lan, Y. Wang, Z.F. ren, « Physics and applications of aligned carbon nanotubes », in Advances in Physics, 60, 4, 553-670, 2011
[5] M. Delmas, et al. Proceeding euro CVD (2009)
[6] F. Harris, "Carbon nanotubes composites", Int. Mat. Review, 49, 1, 2004
[7] Singh C, Shaffer MSP, Koziol KKK, Kinloch IA, Windle AH. Towards the production of large-scale aligned carbon nanotubes. Chemical Physics Letters 2003;372(5-6):860-5.
[8] Singh C, Shaffer MSP, Windle AH. Production of controlled architectures of aligned carbon nanotubes by an injection chemical vapour deposition method. Carbon 2003;41(2):359-68.
[9] K. Hata et al, Science. 306 (2004)
[10] Yasuda et al, ACS nano (2009)
[11] M. Mayne et al. Chemical Physics Letters 338 (2001).
[12] M. Pinault et al, Nano Lett., 2005, 5 (12)
[13] Mayne-L'Hermite M, Armand X, Porterat D, Reynaud C. Pyrolysis of mixed aerosols: a versatile CVD-based process to produce high yields of clean and long aligned carbon nanotubes. In Electrochemical Society Proceedings: Chemical Vapor Deposition-XVI and EUROCVD-14, ed M Allendorf, et al. 2003; pp 549–556.
[14] C. Castro et al, Carbon, 48 (2010)
[15] M. Delmas et al Nanotechnology 2012

[16] C. Castro Thesis, Univ Paris XI, 2010

[17] M.H. Rümmeli et al. Nanoscale Research Letters, 6 (2011)

[18] Tan P, An L, Liu L, Guo Z, Czerw R, Carroll DL, et al. Probing the phonon dispersion relations of graphite from the double-resonance process of Stokes and anti-Stokes Raman scatterings in multiwalled carbon nanotubes. Physical Review B 2002; 66(24): 245-410

[19] C. Candelaria et al, Nano Energy, 1, 195-200, 2012

[20] S. Patel Thesis, univ Paris XI, 2012

[21] S. Patel et al. « Secure Process for elaboration of organic composites reinforced with aligned carbon nanotubes grown on carbon fibers », Nanosafe 2012 (submitted)

[22] Hinds, B. J.; Chopra, N.; Rantell, T.; Andrews, R.; Gavalas, V.; Bachas, L. G, "Aligned Multiwalled Carbon Nanotube Membranes", Science 303 (2004) 62.

[23] L. Tessier, J. Chancolon, P.-J. Alet, A. Trenggono, M. Mayne-L'Hermite, G. Deniau, P. Jégou, S. Palacin, Grafting organic polymer films on surfaces of carbon nanotubes by surface electroinitiated emulsion polymerization, Phys. Stat. Sol. (a) 205, 1412–1418 (2008)

[24] P. Boulanger, O. Sublemontier, O. Gobert, C. Cornaggia, R. Ladret, M. Pinault, M. Mayne-l'Hermite, « Patterning Vertically Aligned Carbon Nanotubes Carpets by Lateral Femtoseconde Laser Ablation », LMP Proceeding, Takamatsu, Japan, 2011

[25] M. Kumar, Y Ando, JNN, 10, 3739-3758, 2010

Plasma polymerized thin coating as a protective layer of carbon nanotubes grafted on carbon fibers

A Einig[1,2], P Rumeau[2], S Desrousseaux[3], Y Magga,[1] J B Bai[1]

[1]Laboratoire MSSMat, CNRS UMR 8579, Ecole Centrale Paris, 92290 Châtenay-Malabry, France
[2]Institut Français du Textile et de l'Habillement, Avenue Guy de Collongue, 69134 Ecully, France
[3]Laboratoire Chimie et Sécurité des Nanomatériaux, LITEN/DTNM/LCSN, Commissariat à l'Energie Atomique, 17 rue des Martyrs, 38054 Grenoble Cedex 9, France

Corresponding author: antinea.einig@ecp.fr

Abstract. Nanoparticles addition is widely studied to improve properties of carbon fiber reinforced composites. Here, hybrid carbon fiber results from grafting of carbon nanotubes (CNT) by Chemical Vapor Deposition (CVD) on the carbon fiber for mechanical reinforcement and conductive properties. Both tows and woven fabrics made of the hybrid fibers are added to the matrix for composite processing. However handling hybrid fibers may induce unwilling health risk due to eventual CNT release and a protective layer is required. A thin coating layer is deposited homogeneously by low pressure plasma polymerization of an organic monomer without modifying the morphology and the organization of grafted CNTs. The polymeric layer effect on the electrical behavior of hybrid fiber is assessed by conductivity measurements. Its influence on the mechanical properties is also studied regarding the interface adhesion between fiber and matrix. The protective role of layer is demonstrated by means of friction constraints applied to the hybrid fiber.

1. Introduction

In extreme conditions, such as for aeronautic applications, composites need to get higher properties than conventional materials as improved tensile strength and conductivity while being high performance lightweight structure. One of the key points of their properties is the interface between the fibers and matrix, which can be modified by nanoparticles addition. A possible solution is to graft carbon nanotubes (CNT) on the surface of carbon fibers (CF) to induce better interfacial properties in the composite [1]. Furthermore CNT have both high electric and thermal conductivities [2] and associated to carbon fibers they can bring more charges dissipation efficiency in the final composite [3]. During textile processing, fibers undergo different types of mechanical stresses before being incorporated into the matrix. In this step, nanoparticles can be released and breathed by operators [4] with unknown consequences on health. Within our study, the approach is to protect CNT with a thin polymeric layer on hybrid fibers. Deposit is made by plasma polymerization [5] under vacuum of an acrylic acid monomer leading to a plasma polymerized acrylic acid (ppAA). This process keeps confined grafted CNT in the vessel under vacuum instead of atmosphere and avoids direct exposure.

Layer of ppAA is also used to coat different substrates like PS spheres [6] and CNT [7]. It improves interface behavior of carbon fiber in composites [8].

2. Experimental

2.1. Plasma reactor and experimental conditions

Deposition process is carried out in a stainless steel reactor of 86 dm^3 as shown in Fig. 1. The microwave plasma source of 2.45 GHz is coupled to a surfaguide crossed by a quartz tube. The working pressure in the vessel is maintained under primary vacuum $5\text{-}20.10^{-2}$ mbar by Roots pump and can be isolated by a valve. The monomer, acrylic acid (Acros Organic 99.5%, extra pure, stabilized) is injected by a mass flow controller for liquid (Bronkhorst Liqui-flow®) at 100% full scale. Simultaneously co-gas are introduced at 47 sccm for Argon (99.999%, Air Products) and 73 sccm for Oxygen (99.999%, Air Products). Pressure in the vessel is maintained at 10.10^{-2} mbar and deposition time is 1 minute.

Carbon nanotubes are grafted to carbon fibers (Toray T700 GC 12k) by CVD process in a quartz reaction tube [9,10]. Unsizing of fibers is carried out by removing sizing at the same temperature used for CVD process.

CNT grafted carbon fibers and unsized carbon fibers are both positioned onto a home-made rotating drum (radius 10cm) inside the vessel with its axis at a distance of 37cm from the plasma source and with a rotating speed of 3 rpm.

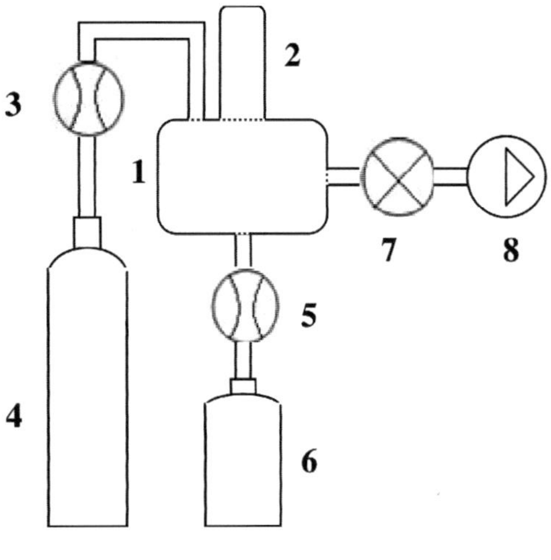

Figure 1. Schematic drawing of plasma reactor (1. vessel 2. microwaves source 3. co-gas flow meter 4. co-gas 5. flow meter monomer 6. monomer 7. valve 8.pump).

2.2. Materials

Fiber morphology has been studied by Scanning Electron Microscopy Leo Gemini 1530 with Field Emission Gun. CNT morphology was observed by Transmission Electron Microscopy Philips CM20 Ultra-Twin at 200 kV. After ultrasonic treatment of ppAA coated hybrid fibers, released CNTs were collected and placed onto copper grids with holy-carbon film for TEM analysis.

Surface composition of ppAA treated unsized fibers were analysed by X-Ray Photoelectron Spectroscopy PHI Quantera SXM using monochromatic Al Ka X-rays (1486.6 eV) with spot diameter of 200µm. Standard take-off angle was 45° giving analysis depth in the range of 3-10nm. Survey spectra were recorded with a pass energy of 280 eV and carbon 1s high resolution spectra were recorded with a pass energy of 55 eV.

Carbon fiber tows conductivity is determined by standard NF EN 1149-2. A circular electrode is put on fibers with a controlled load and an auxiliary one under the same fibers. Both are connected to an electrometer (6517 Keithley). Conductivity perpendicular to the fibers direction was measured in this case.

Friction test is carried out on coated and uncoated hybrid carbon fibers tows. Fibers under fixed load are placed around mobile Teflon cylinder (turning on its axis in 480 rpm) and move on 60cm for 2 min. The setup is inside a glove box with a hole connected to an Electrical Low Pressure Impactor (ELPI Dekati).

3. Results and discussions

3.1. Characterization of ppAA coating on hybrid fiber

Morphology of grafted CNTs on carbon fibers before and after ppAA plasma treatment are shown in Fig. 2 and 3. Carbon nanotubes cover entirely carbon fiber as shown in Fig. 2 and hide its surface. CNT diameters are in the range of 8-12 nm and their length between 0,5-1μm. After plasma treatment as shown in Fig. 3, CNT's density and their morphology are unchanged. Layer of ppAA overlays uniformly the hybrid fiber in combination with a possible slight agglomeration of CNT tips. The ppAA tends to wrap CNTs as shown in Fig. 4 with a thickness in this case of approximately 9nm. XPS C1s spectra of ppAA is decomposed into two peaks corresponding to bonding energy of C-O (286.7eV) and O-C=O (288.8eV) (Fig. 5). Percentage of ester groups or carboxylic acid groups in Table 1 suggests a high retention of carboxylic acid groups from the monomer of acrylic acid [11,12]. The presence of C-O in layer results from fragmented forms of monomer occurring during plasma polymerization [13].

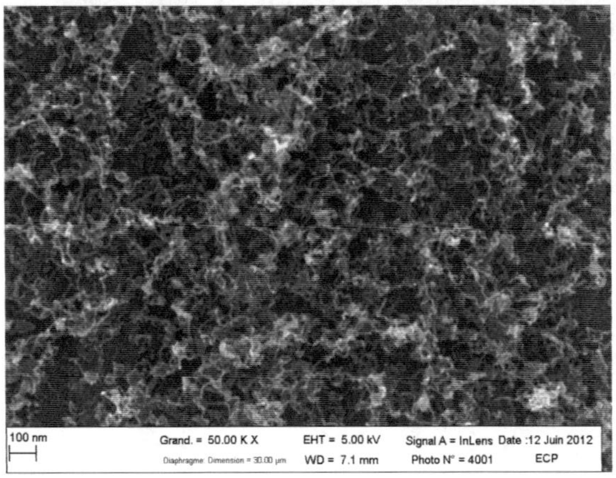

Figure 2. SEM micrograph of CNT grafted carbon fibers .

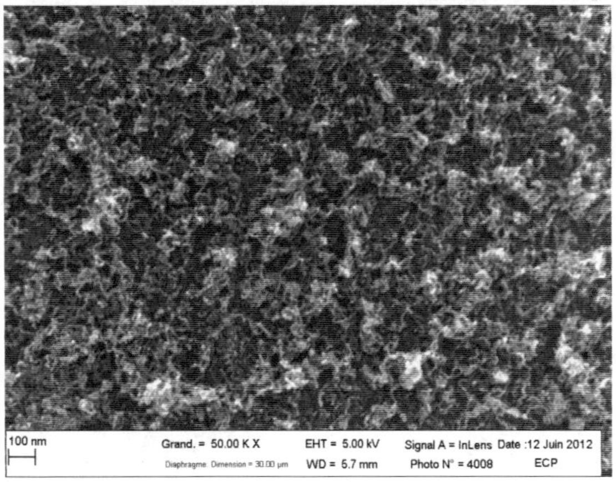

Figure 3. SEM micrograph of CNT grafted carbon fibers after ppAA plasma treatment.

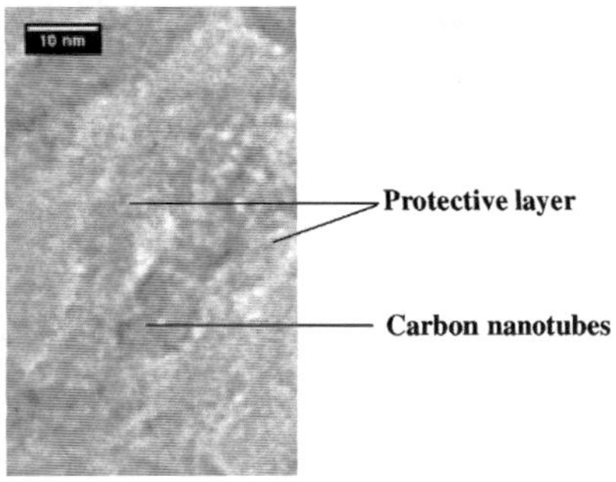

Figure 4. TEM micrograph of CNT coated by ppAA layer.

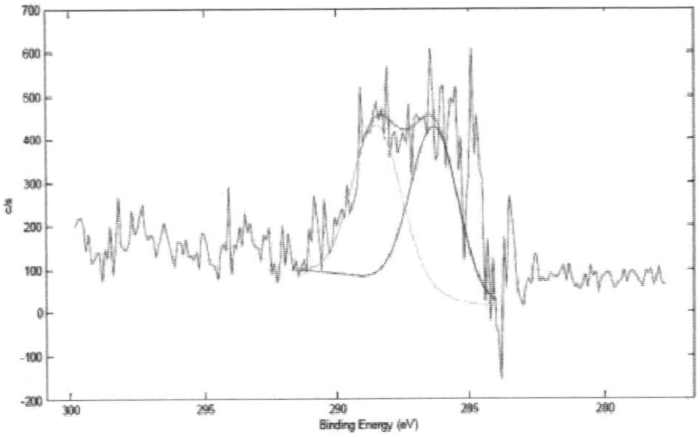

Figure 5. XPS C1s spectra from ppAA layer.

Table 1. Chemical bond concentration of ppAA layer by XPS.

% C-O *(286.7eV)*	% O-C=O *(288.8eV)*
56	44

3.2. Layer effect on the hybrid fiber properties

The layer is supposed to modify the fiber surface as well as the fiber behavior at macroscopic scale. In order to quantify its effect, influence of ppAA coating on conductivity and fiber/matrix interface is studied on hybrid fibers.

The conductivity is one evidence of coating effect in electrical performance of fibers among others[14,15]. Grafting CNTs on carbon fibers highly increase the conductivity (X 10 000), as shown in Fig. 6, compared to that of the raw fiber. It results from the combination of unsizing and CNTs grafting of the carbon fiber and occurring in CVD process. After plasma coating of ppAA, hybrid fiber conductivity slightly decreases in comparison with CNT grafted carbon fiber but still has a much higher conductivity than the pristine one.

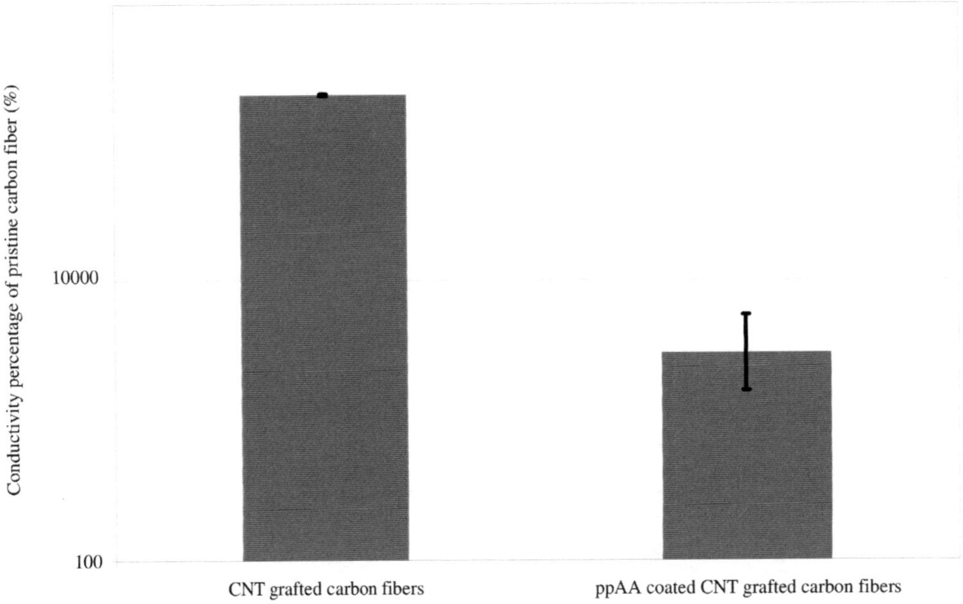

Figure 6. Conductivity evolution of CNT grafted CF and plasma treated CNT grafted CF compared to pristine carbon fiber.

Fragmentation samples were prepared with a single fiber extracted from the tows and embedded longitudinally in the epoxy matrix (resin WWA and hardening agent WWB4 from RESOLTECH) cured while 15 hours at 60°C. Dogbone shape specimen is loaded at 0.1 $mm.min^{-1}$ up to 1.8 mm displacement. The number of fragments is then counted by exposing specimen to transmission light (Fig. 6). Failure number in fragmentation test is an indication of the cohesive interface between matrix and fiber. Grafting CNT on the fiber seems to damage this interface compared to the pristine fiber, with a lower failure number. Concerning plasma coated hybrid fiber failure number increases and overtakes pristine fiber value. ppAA layer balances lost of adhesion at fiber/matrix interface observed after CNT grafting, and improves adhesion properties compared to the as received fiber. ppAA layer promotes bonds with epoxy matrix resin which improve level of adhesion of hybrid fiber [8,16].

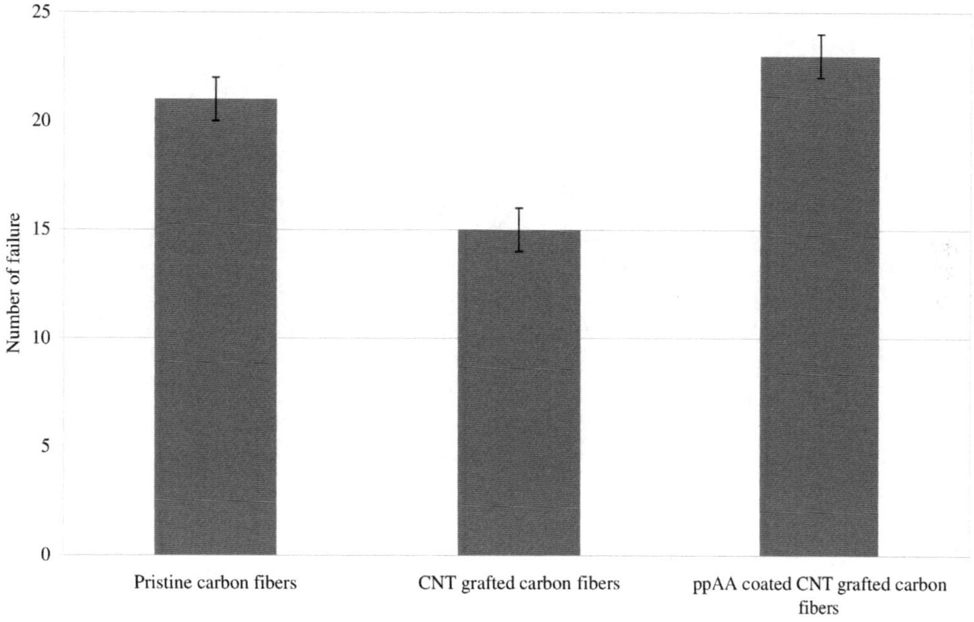

Figure 7. Results of fragmentation test for CNT grafted CF and plasma treated CNT grafted CF compared to pristine carbon fiber.

3.3. Protective layer effect measurement

Electrical Low Pressure Impactor (ELPI Dekati) gives a real-time measurement of aerodynamic diameter particles distribution and concentration [17] emitted during friction test as shown in Fig. 7 and 8. For hybrid fibers, most of the particles sizes are below 28 nm with 3.10^3 $particles.cm^{-3}$. After ppAA plasma treatment on hybrid fibers, same amount of particles was detected but with a significant change of the particle size distribution showing a majority of particles in the ranges of [94-155nm] and of [155-260nm] referring to histograms of in Fig. 8. ppAA layer increases released particles size above nanoscale (i.e > 100 nm) certainly because of embedding CNTs blocks and avoiding releasing of single grafted CNTs. It reduces drastically the concentrations of particles size below 100nm when frictions are applied to hybrid fiber.

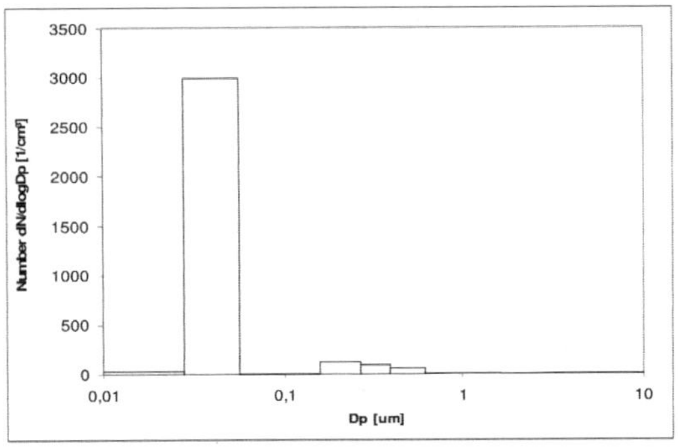

Figure 7. Released particles size distribution under friction test measured by ELPI for CNT grafted carbon fiber.

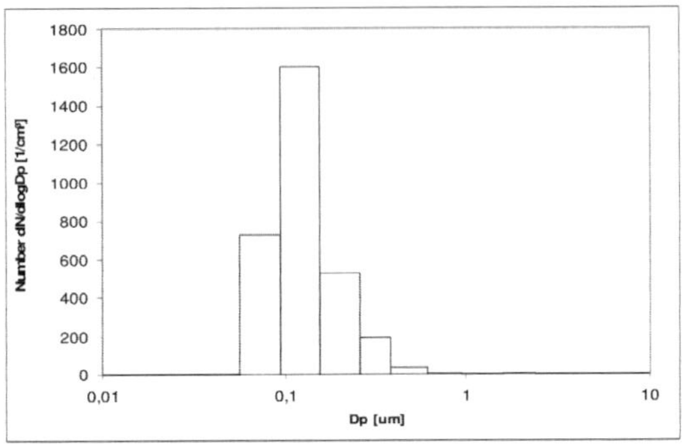

Figure 8. Released particles size distribution under friction test measured by ELPI for ppAA coated CNT grafted carbon fiber.

4. Conclusions

The plasma polymerized acrylic acid (ppAA) coating, with a thickness of about 10 nm, wraps uniformly carbon nanotubes grafted on carbon fibers with no modification of hybrid fiber aspect and more specifically, no morphology modification of carbon nanotubes. ppAA polymeric layer slightly reduces hybrid fibers conductivity which still remains clearly higher than in the case of the pristine fibers. Considering interfacial properties, ppAA contributes to the compatibility with epoxy and induce improvement of adhesive properties between hybrid fibers and epoxy. The protective role of the ppAA layer is demonstrated under friction test. ppAA coating prevents single CNT releasing probably by embedding CNTs blocks. It induces a decreasing of the particles concentrations and a shift of the particles diameter distribution beyond nanoscale (i.e. >100 nm).

Acknowledgments

The authors thank Mr. Jacques Maguin for technical support at IFTH. The authors are grateful to Mrs. F. Garnier for performing SEM and Dr. P. Haghi-Ashtiani for performing TEM observation at Ecole Centrale Paris. The authors want also thank Mr. A. Guiot for ELPI measurements for friction test at the Laboratoire Chimie et Sécurité des Nanomatériaux, CEA Grenoble.
This work was financially supported by national ANR-PROCOM project.

References

[1] Qian H, Bismarck A, Greenhalgh ES, Shaffer MSP. Carbon nanotube grafted carbon fibres: A study of wetting and fibre fragmentation. *Composites Part A: Applied Science and Manufacturing*. 2010 sept;**41**(9):1107-14.

[2] Dominiczak M, Otubo L, Alamarguy D, Houzé F, Volz S, Noël S, et al. Evaluation of the nanotube intrinsic resistance across the tip-carbon nanotube-metal substrate junction by Atomic Force Microscopy. *Nanoscale Research Letters*. 2011;**6**(1):1-10.

[3] Allaoui A, Hoa SV, Evesque P, Bai J. Electronic transport in carbon nanotube tangles under compression: The role of contact resistance. *Scripta Materialia*. 2009 sept;**61**(6):628-31.

[4] Cena LG, Peters TM. Characterization and Control of Airborne Particles Emitted During Production of Epoxy/Carbon Nanotube Nanocomposites. *Journal of Occupational and Environmental Hygiene*. 2011;**8**(2):86-92.

[5] Yasuda H. *Plasma polymerization*. Academic Press; 1985.

[6] Jafari R, Tatoulian M, Arefi-Khonsari F. Improvement of the stability of plasma polymerized acrylic acid coating deposited on PS beads in a fluidized bed reactor. *Reactive and Functional Polymers*. 2011 avr;**71**(4):520-4.

[7] Zhao J, Shi D, Lian J. Small angle light scattering study of improved dispersion of carbon nanofibers in water by plasma treatment. *Carbon*. 2009 août;**47**(10):2329-36.

[8] Kettle AP, Beck AJ, O'Toole L, Jones FR, Short RD. Plasma polymerisation for molecular engineering of carbon-fibre surfaces for optimised composites. *Composites Science and Technology*. 1997;**57**(8):1023-32.

[9] Zhao Z-G, Ci L-J, Cheng H-M, Bai J-B. The growth of multi-walled carbon nanotubes with different morphologies on carbon fibers. *Carbon*. 2005 janv;**43**(3):663-5.

[10] Li H, He D, Li T, Genestoux M, Bai J. Chemical kinetics of catalytic chemical vapor deposition of an acetylene/xylene mixture for improved carbon nanotube production. *Carbon* 2010 déc;**48**(15):4330-42.

[11] Morent R, De Geyter N, Van Vlierberghe S, Vanderleyden E, Dubruel P, Leys C, et al. Deposition of Polyacrylic Acid Films by Means of an Atmospheric Pressure Dielectric Barrier Discharge. *Plasma Chemistry and Plasma Processing*. 2009;**29**(2):103-17.

[12] Zanini S, Ziano R, Riccardi C. Stable Poly(Acrylic Acid) Films from Acrylic Acid/Argon Plasmas: Influence of the Mixture Composition and the Reactor Geometry on the Thin Films Chemical Structures. *Plasma Chemistry and Plasma Processing*. 2009;**29**(6):535-47.

[13] D'Agostino R. Advanced Plasma Technology. John Wiley & Sons; 2008.

[14] Tzeng S-S, Chang F-Y. Electrical resistivity of electroless nickel coated carbon fibers. *Thin Solid Films*. 2001 juin 1;**388**(1–2):143-9.

[15] Athanasopoulos N, Kostopoulos V. Prediction and experimental validation of the electrical conductivity of dry carbon fiber unidirectional layers. *Composites Part B: Engineering*. 2011 sept;**42**(6):1578-87.

[16] Lopattananon N, Kettle AP, Tripathi D, Beck AJ, Duval E, France RM, et al. Interface molecular engineering of carbon-fiber composites. *Composites Part A: Applied Science and Manufacturing*. 1999 janv;**30**(1):49-57.

[17] Marjamäki M, Keskinen J, Chen D-R, Pui DYH. Performance evaluation of the electrical low-pressure impactor (ELPI). *Journal of Aerosol Science*. 2000 févr;**31**(2):249-61.

Silica-coating as protective shell for the risk management of nanoparticles

D Gardini[1], M Blosi[1], C Delpivo[1], S Ortelli[2] and A L Costa[1]

[1] CNR-ISTEC - National Research Council of Italy, Institute of Science and Technology for Ceramics, Via Granarolo, 64, I-48018, Faenza, Italy
[2] Centro Interdipartimentale di Ricerca Industriale Meccanica Avanzata e Materiali, Viale Risorgimento 2, I-40136, Bologna, Italy

E-mail: anna.costa@istec.cnr.it

Abstract. Nanoparticles (NPs) surface functionalization with silica (SiO_2) has attracted high attention due to hydrophilicity, biocompatibility, chemical and thermal stability of silica. The present work is addressed to the production and characterization of SiO_2-coatings on titanium dioxide (TiO_2) and silver (Ag) NPs dispersed in aqueous solutions (commercial nanosols) with the aim to manage the potential risk that such NPs could generate in occupational exposure scenarios. A colloidal approach, based on principles of heterocoagulation in which opposite charged NPs are forced to coagulate in hierarchical structures, imposed by their relative size and weight ratio, was followed. The results were compared with a chemical approach, based on nucleation of silica phase from silica precursor solutions on the surfaces of TiO_2 or Ag NPs seeds. In order to increase the adhesion of silica on TiO_2 and Ag surfaces, heterocoagulated sols were spray-dried and subsequently redispersed in water to check the feasibility of such approach at industrial level. Physicochemical properties such as zeta potential, electrical conductivity, particle-size distribution, specific surface area and morphology of the samples produced with different SiO_2:TiO_2 and SiO_2:Ag weight ratios were collected and compared. Indirect evidences of silica coating were obtained.

1. Introduction

The present work is performed on behalf of SANOWORK (FP7-NMP4-SL-2011-280716) project that addresses the following objectives.

- Promoting, developing and implementing "DESIGN OPTION BASED" risk remediation strategies.
- Integrating such strategies within MANUFACTURING PROCESSING LINES.
- Implementing EXPOSURE assessment methodologies in the WORKPLACES.
- Performing a RISK analysis BEFORE and AFTER the application of the proposed strategies.
- Analyzing COST / BENEFIT on the basis of the risk analysis results and nanomaterials (NMs) performance.

The development of "DESIGN OPTION BASED" risk remediation strategies is a challenging and beyond the state of the art objective aimed to design out risks rather than address them when they occur.

The main strategy developed by SANOWORK project is based on surface NMs engineering through inorganic/organic surface coating by playing with mechanism and kinetics of self assembled monolayer (SAM) formation [1]. SAMs on NPs, including colloids and nanocrystals, stabilize the reactive surface of the particles and represent a simple way to control the surface reactivity at the particle-solvent interface. Adsorbate molecules adsorb readily because they lower the surface free-energy of the substrate and are stable due to the strong specific adsorption of the "head groups." These bonds create monolayers that are more stable than the physisorbed bonds of Langmuir–Blodgett films [2] and still remain bonded to the surface even when the solid substrate is removed from the solution. The colloidal functionalization of nanoparticles by electrosteric self-assembling is particularly attractive [3] due to the ease of preparation, the tunability of solid/liquid dispersion state and of surface properties via modification of adsorbed layer molecular structure and functions, and the use of adsorbed layers as building blocks in more complex structures, e.g. for 'grafting' additional layers to a surface. Within possible coatings, those based on silica (SiO_2) have attracted high attention due to their hydrophilicity, biocompatibility, chemical and thermal stability, even in aqueous media.

The present work is addressed to the production and characterization of SiO_2-coatings on titanium dioxide (TiO_2) and silver (Ag) NPs dispersed in aqueous solutions (commercial nanosols). A colloidal approach, based on principles of heterocoagulation in which opposite charged NPs are forced to coagulate in hierarchical structures, imposed by their relative size and weight ratio, was followed. The results were compared with a chemical approach, based on nucleation of silica phase from silica precursor solutions on the surfaces of TiO_2 or Ag NPs seeds.

Physicochemical properties such as zeta potential, electrical conductivity, particle-size distribution, specific surface area and morphology of the samples produced with different SiO_2:TiO_2 and SiO_2:Ag weight ratios were collected and compared.

2. Experimental

2.1. Materials

Titanium dioxide (TiO_2) sol (NAMA41, 6 wt%, 40 nm), silver (Ag) sol (NAMA39, 4 wt%, 125 nm), both from Colorobbia Italia, and silica (SiO_2) sol (LUDOX HS-40, 40 wt%, 20 nm) from Grace Davison were used for heterocoagulation process. For the silica coating of titanium dioxide and silver NPs by chemical synthesis (according to Stöber method), tetraethyl orthosilicate (TEOS) from Sigma-Aldrich as precursor of silicon and ammonium hydroxide (NH4OH) solution at 30 wt% from Carlo Erba as basifying agent were used.

2.2. Methods of production of heterocoagulated and synthesized sols

The commercial sols were diluted with water (TiO_2 and SiO_2 sols to 3 wt% and Ag sol to 2 wt%) and, in the case of SiO_2 sol, acidified by cationic exchange on resin till pH 4. After such treatments, the titania and silica sols, as well as the silver and silica sol, were mixed in well-defined ratios (indicated in Figure 1 as SiO_2:TiO_2 and SiO_2:Ag weight ratios) and ball milled (BM) for 24 hours with alumina spheres as milling media.

In addition to the heterocoagulated (HC) sols some synthesized sols were also prepared by diluting the TiO_2 and Ag commercial sols, adding TEOS in well-defined amount in basic environment and leaving the reaction to occur for 24 hours. After that time the reaction systems were washed three times with ethanol. In Figure 2 are shown the sols produced in this way.

2.3. Heterocoagulated sols treatments

In order to increase the adhesion of silica NPs on the surfaces of TiO_2 and Ag NPs, the heterocoagulated sols were spray-dried in counterflow with a stream of hot air (at 220 °C) through a nozzle of 500 μm of diameter (Figure 3).

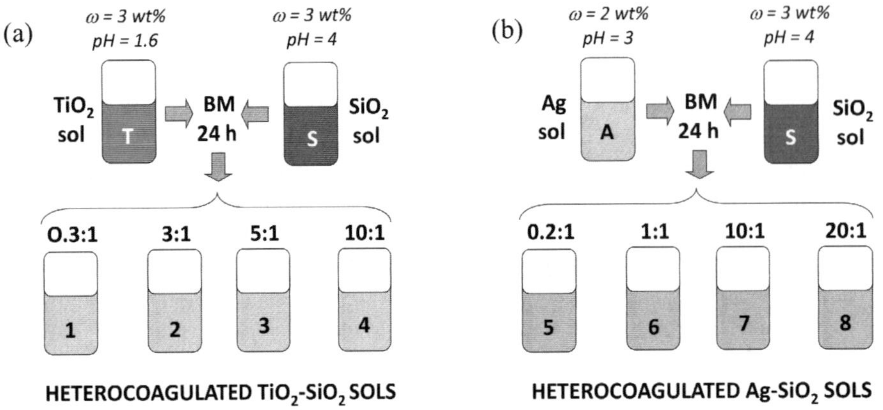

Figure 1. Diagrams showing the method of production of heterocoagulated (a) TiO_2-SiO_2 and (b) Ag-SiO_2 sols. SiO_2:TiO_2 and SiO_2:Ag weight ratios and nomenclature used are indicated.

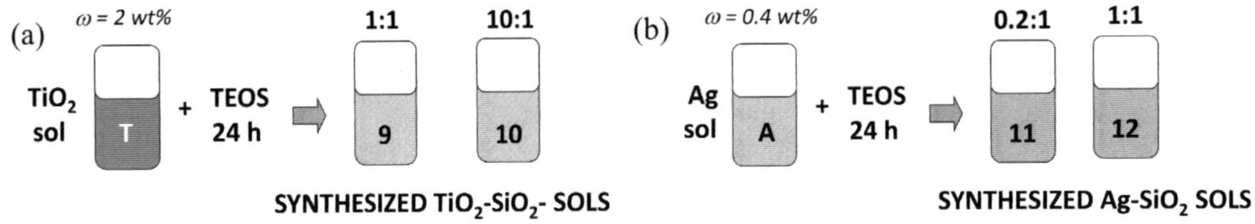

Figure 2. Diagrams showing the method of production of synthesized (a) TiO_2-SiO_2 and (b) Ag-SiO_2 sols. SiO_2:TiO_2 and SiO_2:Ag weight ratios and nomenclature used are indicated.

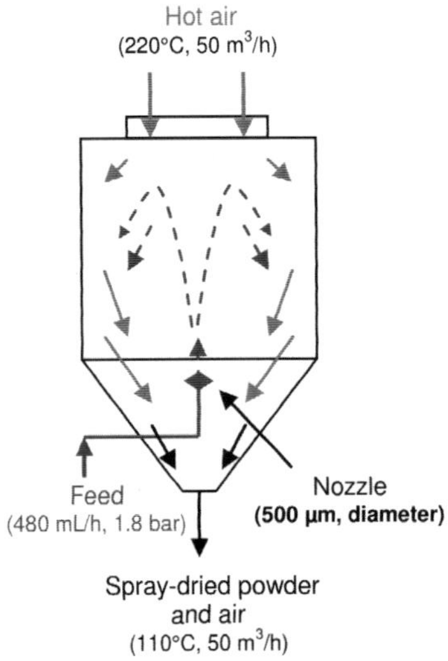

Figure 3. Spray-drying technique applied to heterocoagulated sols.

The spray-dried powders were subsequently redispersed in water by ball milling or ultrasonication with a solid content of 0.5 wt%. The spray-dried powders and the corresponding redispersed sols were indicated with the name of the starting sols followed by SD or RED, respectively. Redispersed sols were compared with the starting ones to study the effect of spray-drying on the silica coating.

2.4. Measurements
The particle-size distribution of sols was measured by dynamic light scattering (DLS) (ZetaSizer, Malvern). The zeta potential of the particle in suspensions was determined by electroacoustic method (AcoustoSizer II, Colloidal Dynamics). Surface area of spray-dried powders was measured by nitrogen adsorption (BET) (Sorpty 1750, Carlo Erba). Powder size and shape were measured by scanning electron microscopy (SEM) (Stereoscan 360, Cambridge Instruments). The electrical conductivity of sols was measured with a conductimeter (AMEL 134, AMEL).

3. Results and discussion
In order to get information about the occurrence of silica coating on titania or silver NPs surfaces, particle-size distribution by dynamic light scattering (DLS) and zeta potential by electroacoustic measurements on the heterocoagulated sols were carried out. Moreover, the spray-dried powders obtained from such sols were observed by SEM before and after redispersion in water. Zeta potential and electric conductivity of the redispersed sols were also measured.

3.1. Particle-size distribution
The dynamic light scattering (DLS) spectra of commercial and heterocoagulated sols are reported in Figure 4. Titania sol (T) has an average hydrodynamic diameter of 32 nm, whilst silica sol spectra (S) is characterized by two peaks at 20 and 330 nm. Silver sol spectrum (A) shows a peak at 125 nm. As comparison, STEM analysis performed on Ag sols shows a diameter size distribution between 20-60 nm, confirming an overestimate of real Ag diameter due to the presence of organic capping agent.

The DLS pattern of TiO_2-SiO_2 sol 2 (SiO_2:TiO_2 weight ratio 3:1) shows essentially a peak at 297 nm. Ag:SiO_2 sols 5 (SiO_2:Ag weight ratios 0.2:1) and 6 (SiO_2:Ag weight ratios 1:1) show peaks at 220 and 190 nm, respectively. Basing on these results, the presence of silica seems to strongly affect only the size of TiO_2 (Figure 4).

The different behavior of the two systems could be due to different affinity between oxide (TiO_2) and metal (Ag) surface towards silica, but also to the presence of a capping agent (polyvinyl pyrrolidone, PVP) on Ag surface that could hide the increase of Ag particle diameter expected by the formation of silica coating, as briefly schematized in Figure 5.

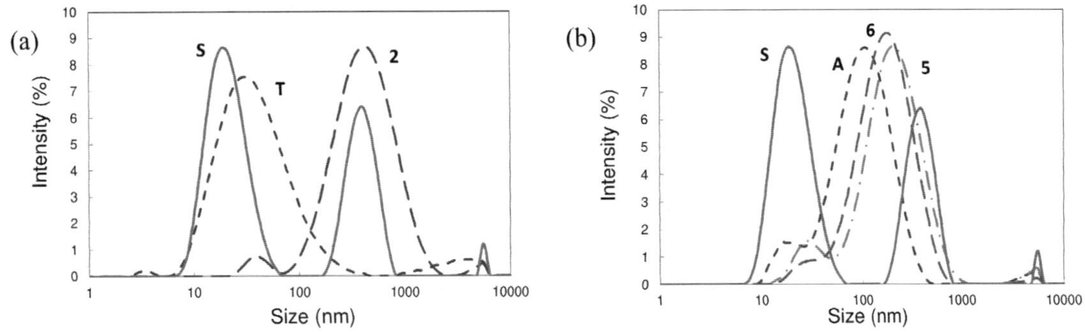

Figure 4. DLS pattern for (a) TiO_2-SiO_2 sol 2 and (b) Ag-SiO_2 sols 5 and 6 compared with the DLS pattern for the commercial titania, silica and silver sols.

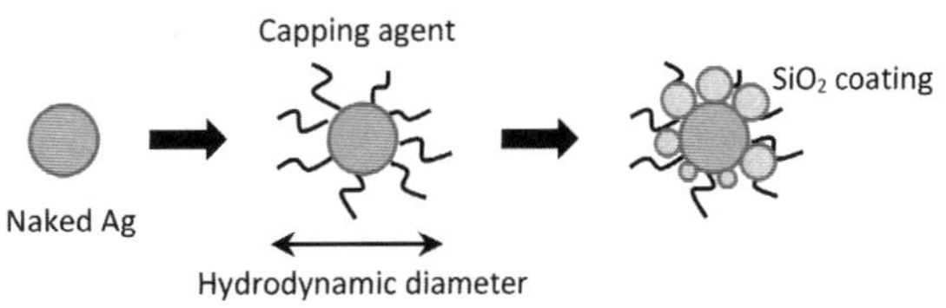

Figure 5. Possible explanation of missed increasing of hydrodynamic size for SiO_2-coated Ag NPs.

3.2. Zeta potential measurements

Zeta potential of heterocoagulated and redispersed sols are plotted in Figure 6. By increasing the weight ratio of silica with respect to titania, the zeta potential value decreases from the value of titania sol towards that one of silica sol (-23 mV); the same occurs for the silver sols (Figure 6a). However, considering that the resulting zeta potential is a combination of the signals arising from the two NPs dispersed, it is difficult to ascribe the decrease of zeta potential only to the formation of silica coating.

As far as the redispersed sols are concerned (Figure 6b), a first evidence is that zeta potential values are systematically lower than the ones of the heterocoagulated starting sols. Such a result could be explained by agglomeration occurring during spray-drying that decreases the number of acid and basic surface sites available for protons exchange with water. Furthermore, it can be seen that the addition of small amount of silica abruptly decreases the zeta potential towards the value of silica at the same pH condition (~ 2). This allows to identify the minimum amount of silica that should completely coat the surfaces ($SiO_2:TiO_2 \sim 3$, $SiO_2:Ag \sim 1$), corresponding to the onset of plateau characterized by small variations of zeta potential.

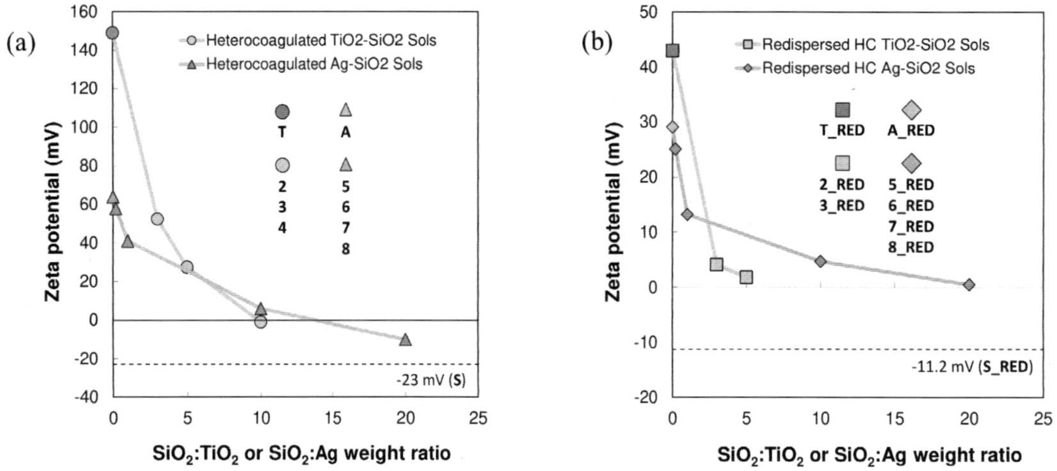

Figure 6. Zeta potential of (a) heterocoagulated and (b) redispersed sols.

3.3. Morphologic characteristics of spray-dried powders

The size and morphology of the powders obtained from the commercial and heterocoagulated sols after spray-drying were observed by scanning electron microscopy. The spray-dried powder obtained from the commercial titania sol is micrometric in size and characterized by irregular shapes and hollows. However, the specific surface area (from BET measurements) is 237 m^2/g proving that the micrometric particles are actually nanostructured. On the other hand, the spray-dried powders obtained from the commercial silica are micrometric in size, as well, with high specific surface areas (201 m^2/g), but their shape is much more regular being circular with a hole in the middle. The spray-dried powders obtained from the heterocoagulated SiO_2-TiO_2 sols show regular morphologies, like spheres or hollowed spheres, high BET specific surface area (255 and 177 m^2/g), with smooth surfaces, similar to the spray-dried silica powders, suggesting the probable formation of silica coating (Figure 7).

The spray-dried powders from the commercial silver sol are spherical with holes due to the evaporation of organics. The spray-dried powders from the heterocoagulated SiO_2-Ag sols have irregular shapes with average diameter that increases with the SiO_2:Ag weight ratio.

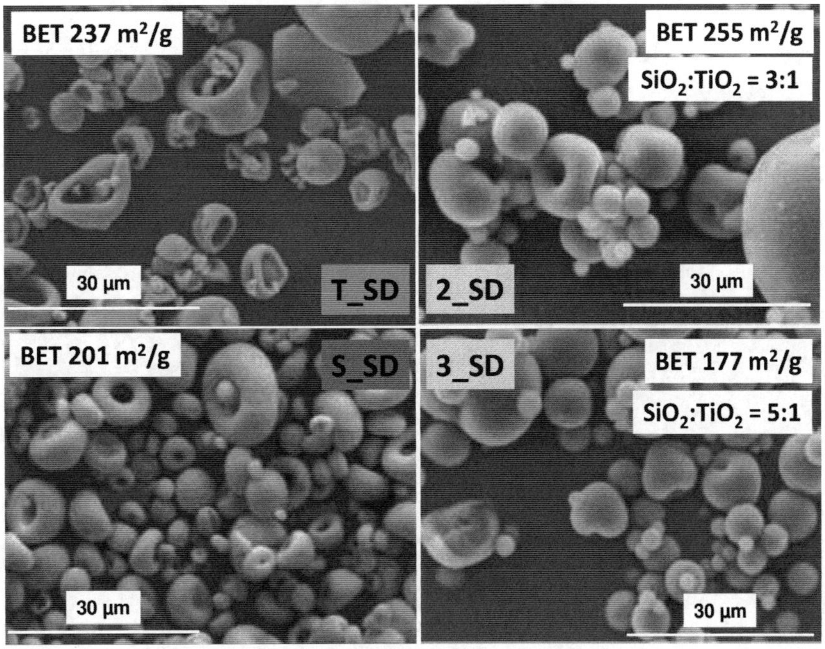

Figure 7. SEM images of spray-dried powders from (a) titania sol (T_SD), (b) TiO_2-SiO_2 sol with a SiO_2:TiO_2 weight ratio 3:1 (2_SD), (c) TiO_2-SiO_2 sol with a SiO_2:TiO_2 weight ratio 5:1 (3_SD), and (d) silica sol (S_SD). The BET specific surface areas are also indicated.

3.4. Electrical conductivity of heterocoagulated and redispersed sols

The electrical conductivities of heterocoagulated sols result higher than those of redispersed sols, due to the spray-drying treatment, which probably causes the immobilization of the salts present in the sols on the particle surfaces (Figure 8). The good fitting between experimental and weighted mean conductivity data and the very low value of conductivity measured for the spray-dried silica sol (0.04 mS/cm) make difficult to establish if the observed decrease as a function of silica content is due to the formation of a silica coating or to the presence of free silica nanoparticles.

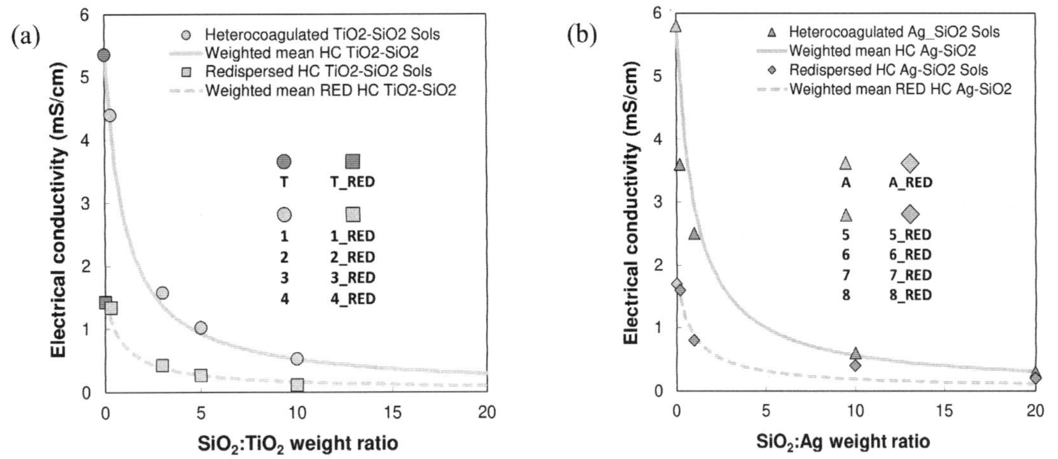

Figure 8. Electrical conductivity of HC and redispersed (a) TiO_2-SiO_2 and (b) Ag-SiO_2 sols.

3.5. Size and shape of redispersed powders
SEM images show that in order to get well-dispersed sols after redispersion in water of the spray-dried powders a long ball milling treatment (100 h) is required for TiO_2-SiO_2 sols, while a simple ultrasonication treatment (10 min) is enough for Ag-SiO_2 sols (Figure 9).

Figure 9. SEM images of redispersed TiO_2-SiO_2 3_RED sol after (a) 10 min of ultrasonication and (b) 100 hours of ball milling and (c) redispersed Ag-SiO_2 5_RED sol after 10 min of ultrasonication.

The redispersion of the spray-dried powders represents a fundamental goal. In fact, spray-drying provides highly reactive and nanostructured micrometric powders, which are more stable over time with respect to nanosols, more easily stored and transported. For these reasons to be successful in maintaining the same characteristics of the starting sol with the redispersed powder is a great achievement.

3.6. Synthesized sols
The TiO_2-SiO_2 sols obtained by nucleating silica from precursors on titania NPs surfaces are characterized by spherical aggregates with submicrometric primary particles, as observed by SEM on dried samples from sol 9. By increasing the silica content (sol 10) irregular aggregates with primary particles lower than 100 nm are observed (Figure 10). For the SiO_2-Ag sols spherical and regular shapes with nanometric structure were observed.

With respect to the corresponding heterocoagulated and redispersed sols, the electrical conductivity of chemically synthesized samples show an abrupt decrease. Particularly the measured conductivities are 0.07 and 0.09 mS/cm for the TiO_2 samples, and 0.04 mS/cm for both the Ag samples. These values, very close to the conductivity of the spray-dried and redispersed silica (0.04 mS/cm), may represent an indirect evidence of the presence of more compact silica layer, which limits the ion release by the particle surfaces.

Conclusions
The main strategy developed by SANOWORK project is based on surface NMs engineering through inorganic/organic surface coating by playing with mechanism and kinetics of self assembled monolayer (SAM) formation. The inorganic coating of titania and silver aqueous commercial nanosols with silica has been performed by using a colloidal approach, based on principles of heterocoagulation, in which opposite charged NPs are forced to coagulate in hierarchical structures, imposed by their relative size and weight ratio. In order to optimize the silica amount and evaluating the silica effect, several SiO_2:TiO_2 and SiO_2:Ag weight ratios have been considered. Moreover, the results were compared with those arising from a chemical approach, based on nucleation of silica phase from silica precursor solutions on the surfaces of TiO_2 or Ag NPs seeds.

Figure 10. SEM images of redispersed (a) TiO_2-SiO_2 9 sol, (b) TiO_2-SiO_2 10 sol, (c) Ag-SiO_2 11 sol and (d) 12 Ag-SiO_2 sol.

The experimental evidences collected till now on the samples produced with the colloidal and chemical approaches do not allow to certainty state that a SiO_2-coating on TiO_2 and Ag occurred, but some clues are in that direction. Further investigation are in progress. However, some preliminary cytotoxicity tests proved that the spray-drying process, particularly when coupled with silica, provides an abrupt decrease of nanosilver hazard potential. Considering that the successfully achieved redispersion of nanosilver spray-dried powders represents a process improvement for this kind of nanomaterials, further investigation are in progress in order to identify key factors responsible of biological activities.

Acknowledgments
The research leading to these results has received funding from the European Community's Seventh Framework Programme (FP7/2007-2013) under Grant Agreement n. 280716.

References
[1] Schwartz D K 2001 Mechanisms and kinetics of self-assembled monolayer formation, *Annu. Rev. Phys. Chem.* **52** 107
[2] Kaifer A 2001 *Supramolecular Electrochemistry* (Coral Gables: Wiley VCH) pp 191–3
[3] Schreiber F 2000 Structure and growth of self-assembling monolayers, *Prog. Surf. Sci. 65* 151–257

Development and optimization of a secure injection CVD process to grow aligned carbon nanotubes on large substrates

S Patel[1], Y Magga[1], L Belkady[1], E Hibert[1], D Porterat[1], P Boulanger[1], M Pinault[1], M Mayne-L'Hermite[1]

[1]CEA, IRAMIS, SPAM, Laboratoire Francis Perrin (CEA-URA 2453), 91191 Gif-Sur-Yvette, France

E-mail: martine.mayne@cea.fr

Abstract. Growth of aligned carbon nanotubes (CNT) by a secure injection CVD process on large quartz substrates and carbon fiber (CF) cloths has been reported in this study. Adjustments of CVD setup and synthesis parameters have been achieved to control the CNT growth in terms of homogeneous covering of the substrates and to tailor the length of CNT. Two parameters have been optimized such as the precursor feeding rate per reactor surface unit and the carrier gas flow rate.

1. Introduction

Because of their properties, carbon nanotubes (CNT) have opened window for research on materials improvement. They are potential materials for sensors, electrodes, membranes or composites [1–4]. Recently, lots of studies deal with the introduction of CNT in polymer matrix composites reinforced with carbon fiber (CF) in order to enhance mechanical and electrical properties [5–24]. In general, CNT are dispersed in matrix, then the fibers are impregnated with this modified matrix leading to forming composite materials which are compatible for a further industrial processes [7–9], [11], [13], [15], [16], [20], [22]. However, it is often reported that the presence of CNT increases the viscosity of the polymer inducing a change in the composite process elaboration and a non homogeneous impregnation [20]. To avoid this problem, another route to elaborate composite is to integrate CNT directly on the reinforcement materials. It can be either a deposition of CNT or a direct growth of CNT on the carbon fiber cloth. In this latter case, injection CVD (Chemical Vapour Deposition) process is the process mostly used for a direct growth of CNT. It has been demonstrated that a growth of aligned dense and long CNT is possible in the case of a treatment on carbon fiber performed previously to the growth of CNT [10], [25], [26] . In general, this previous treatment is achieved by a process which is different from CVD process such as PVD (Physical Vapour Deposition) processing two different reactors, thus involving intermediate handling phases which are not compatible in terms of secure process development [10], [26]. Recently, a process consisting in performing the pretreatment in the same CVD reactor through two successive steps has been developed for the growth of aligned CNT [25]. Nevertheless, a current problem concerns the achievement of a homogeneous growth of CNT on large substrates such as flexible carbon fiber clothes. Few papers deal with the growth of CNT on substrates exhibiting sizes over 10 cm. We consider that this size constraint is one of the reasons why there are only few studies on mechanical properties of composites (interlaminar shear strength) reinforced with carbon fiber clothes and more papers exhibiting mechanical properties results obtained on individual carbon fibers (interfacial laminar shear strength) [5], [6], [18], [21], [24]. The CVD process used for the growth of aligned CNT is well-known for flat quartz substrates with sizes above 10 mm* 10 mm. Therefore, it is of particular importance to develop and adjust such processes in order to grow aligned CNT on large fiber cloths and to further study the effect of CNT on the properties of composite materials resulting from aligned CNT covered fiber cloth impregnated with organic matrixes.

In this context, the present paper is focused on the adjustment of the DLI-CVD process to grow aligned CNT on large quartz substrates and CF cloths allowing standard mechanical tests on the final composites. Thus, we have investigated the characteristics of aligned CNT grown in quite big reactor and furnace exhibiting sizes over 1 m in length and 80 mm in diameter. From this study, we have pointed out that two synthesis parameters need to be optimized such as the precursor feeding rate per reactor surface unit and the carrier gas flow rate.

2. Experimental

The aligned CNT synthesis is based on a catalytic chemical vapor deposition process which consists in a catalytic decomposition of a liquid hydrocarbon-based aerosol [27–30]. The aerosol is generated by an injection system (Qualiflow-Jipelec) coupled with an evaporator maintained at 200 °C. The precursor mixture, made of ferrocene (Acros, 99,9%) dissolved in toluene (Merck, 99.9%) is filled in a tank and the resulting solution is nebulized through the injection system which feeds a cylindrical quartz reactor (55 mm internal diameter) introduced in a 1 m long tubular and horizontal furnace. The aerosol is carried by argon flow in the reactor and, the argon carrier gas feeding is separated from the precursor solution tank resulting in a precursor aerosol feeding rate of the reactor independent of the argon flow rate. The growth mechanisms have been detailed previously [27–30] and the set-up has been adjusted for an efficient growth of aligned CNT on carbon fiber [25]. It consists in depositing a silica based sub-layer previously to the CNT growth. This is performed with the same CVD process and in the same CVD reactor, which reduces considerably all intermediate handling steps of materials which is a real challenge for the development of a secure CNT growth process. In order to perform such synthesis, the evaporator is fitted with another injection system similar to the one for the growth of CNT. This injection system is connected with another tank containing a solution of tetraethylorthosilicate namely TEOS, (Aldrich, 99%) (figure 1).

The temperature furnace profile is illustrated in the figure 2 which indicates that the isothermal area (+/- 20°C) starts at 20 cm and finishes at 80 cm from the entrance of the furnace. Our previous results show that the best growth rate of CNT and catalyst efficiency is obtained for a solution of ferrocene diluted in toluene (2.5 wt. %), and pyrolysed at 850°C.

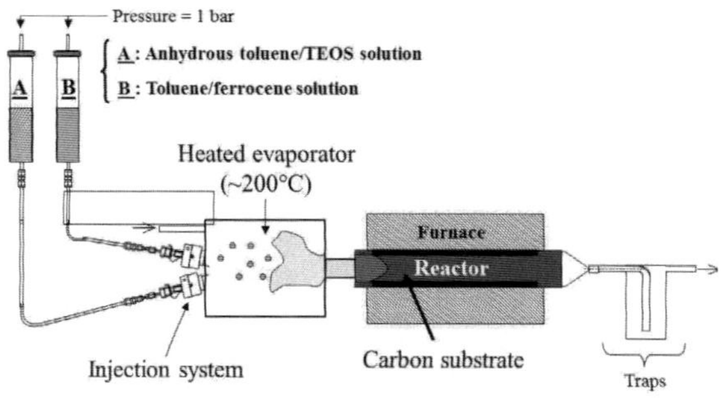

Figure 1. Experimental setup for both SiO2-based layer deposition and CNT growth [25].

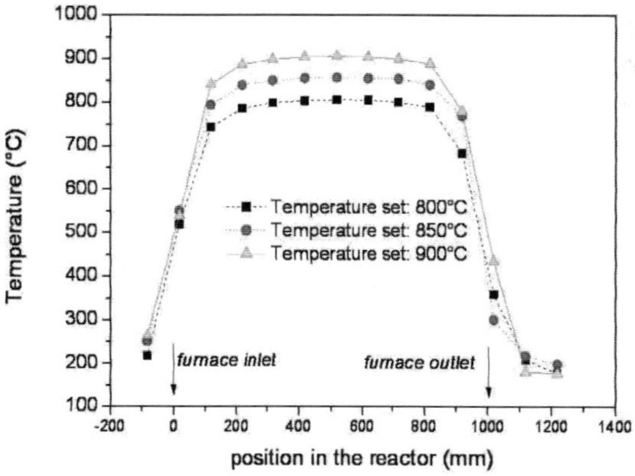

Figure 2. Temperature furnace profile.

Characterizations of synthesized materials are performed with Scanning Electron Microscopy SEM (Carl Zeiss Ultra 55) to observe the morphology and to measure the length of aligned CNT carpets on quartz substrates as well as on carbon fiber cloths. TGA analyses are carried out under air up to 1000°C with a heating rate of 10°C/min in order to assess the iron content in CNT which, in particular, corresponds to iron contained in catalytic particles generating CNT.

3. Results and discussion

2.1 Growth of aligned carbon nanotubes in a long reactor

Castro *et al.* [30] reported studies on synthesis of aligned CNT growth in a short quartz reactor and studied catalyst particle location as well as CNT length along this reactor after CNT growth. Iron-based catalyst nanoparticles are formed by homogeneous nucleation in the gas phase at a location in the reactor where the temperature is optimal for the ferrocene decomposition (826°C), which happens to be before the isothermal area. From this point, nanoparticles carried by the gas flow are deposited gradually on the reactor walls and their surface density decreases. They give rise to aligned CNT growth with efficiency strongly dependent on the temperature profile. CNT growth is optimal in the isothermal area at the temperature chosen to optimize the carbon precursor catalytic decomposition. This dependence explains the inhomogeneity of CNT products along the reactor especially in terms of thickness [30].

Taking into account the aim of the present study which is to grow homogeneously aligned CNT carpet on large fibre cloths, it is important to check, first, that the same phenomena are observed on a longer reactor device same as the one reported in [25]. Therefore, the same procedure as the one reported in [30] has been performed involving the collection of powdered samples on the internal walls of the reactor by decomposing the reactor in 8 distinct areas of 50 mm or 10 mm long each (no substrates were placed in the reactor). The injection of the precursors (solution of ferrocene (2.5 wt.%) diluted in toluene) is performed at 800°C with a feeding rate of 1,5 g/min and with a argon flow rate of 2,5 l/min during 75 min. The aligned CNT carpet thickness (e.g. CNT mean length) have been measured by SEM and the iron content has been measured by TGA for all samples resulting from the collection in each reactor area.

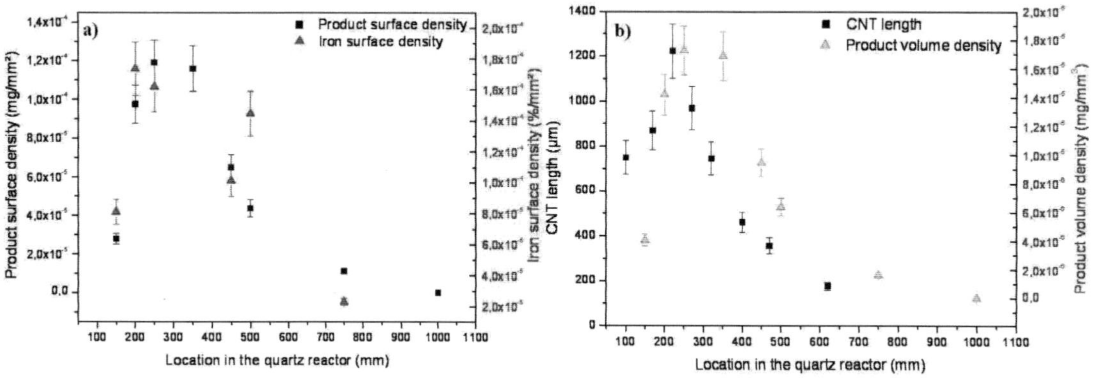

Figure 3. Characteristics of the synthesis products as a function of the location in the quartz reactor (a) surface density of total products and iron, (b) CNT carpet thickness and product volume density (error bar corresponds to the measurement error).

Figure 3 reports the changes of product surface density together with iron surface density (figure 3(a)) as well as the CNT length together with the product volume density (figure 3(b)) both as a function of the location in the quartz reactor. A correlated profile between all these data is shown and the trend is very similar to the one reported by Castro *et al.* [30]. Indeed, the CNT mean length and the iron surface density, as well as both product volume and surface density, increase to reach their maximum at a location of 250 mm corresponding to a 800 +/- 20 °C temperature (beginning of the isothermal area). This temperature is only slightly lower than the decomposition temperature of ferrocene (826 °C) [30], and we can assume, similarly to our previous study [30], that in this area ferrocene is decomposed and numerous iron-based catalyst nanoparticles are suddenly nucleating by homogeneous nucleation. After the 250 mm location, a decrease of CNT mean length and iron surface density, as well as both product volume and surface density is observed. As a consequence, these results indicate that the phenomena reported by Castro *et al.* [30] in a short reactor are confirmed in a long reactor and that the germination of iron-based catalyst particles occurs always at the same temperature whatever the CVD device and the programmed temperature. Therefore, the growth of aligned CNT can be extended on a much longer area of the internal walls when using a long reactor as compared to a short one.

Besides, it was important to check if these phenomena are similarly occurring on substrates placed in this long reactor since the aim is to synthesize aligned CNT on large and flexible carbon fiber cloths. Therefore, we introduced either several short quartz substrates (20) with sizes of 10 mm*10 mm or 3 large quartz substrates with sizes of 100 mm*50 mm, which width corresponds to the quartz reactor diameter. We observe CNT carpet present on the different substrates put in the reactor and those collected from inner reactor walls for the 8 different areas defined previously along the reactor. First, on substrates we observe the same results mentioned previously namely the decrease of the CNT length along the reactor because of the impoverishment of catalytic particles along the reactor. Moreover, in presence of substrates, CNT length measured from CNT carpet collected from reactor wall in the same area varies from 140 µm to 1.37 mm, depending on the area collected in the circumference of the reactor (figure 4). We suppose that it is the consequence of a disruption of the flow in the reactor due to the presence of the substrates.

In other words, two main phenomena have to be considered for a homogeneous growth of CNT on large substrates which are the impoverishment of catalytic particles along the reactor and the disruption of the gas flow in the reactor. Thus, it is necessary to change the reactor entrance configuration to allow a large homogeneous growth on substrates and powder of CNT growing on the quartz reactor wall to limit the disruption and to limit the rapid reduction of catalytic particles along the reactor.

Figure 4. SEM observations of CNT carpet collected from the inner quartz reactor wall.

2.2 Growth of aligned carbon nanotubes on large substrates in a long reactor

2.2.1 Set-up synthesis modifications. To avoid the flow disruption generated by the presence of substrates, the reactor entrance was modified in order to localize the precursor flow only above the substrates. This leads to a decrease of reaction volume by a factor of 2 as compared to the initial configuration of the reactor. This modification plays a role on two parameters: the precursor content per surface unit of the reaction chamber increases as compared to the initial configuration of the reactor which should limit the decrease of iron surface density along the reactor and the precursor flow along the reactor should be less turbulent with low disruption of the flow.

2.2.2 Growth of CNT on large quartz substrates. Before testing this new configuration on carbon fiber cloths, we performed preliminary tests on large quartz substrates. The growth of CNT has been performed on three large quartz substrates (100 mm * 50 mm) placed along the reactor (figure 5) at a temperature of 850 °C during 5 min and with an argon carrier flow rate of 3 l/min and precursor feeding rate of 1 g/min. The temperature was increased as compared to the previous part of the study in order to improve the catalytic yield as reported in the past [31]. The first substrate, located between 40 to 140 mm from the furnace inlet, contains only a thin layer of nanotubes deposited at around 120 mm from the entrance of the furnace which corresponds to a 800°C temperature and is related to the decomposition of ferrocene. This is consistent with the results reported previously, even if the configuration of the reactor has been changed. The part of the quartz substrate located near the reactor entrance is heated at temperatures which are not enough high to achieve the ferrocene decomposition. The second and the third substrates are located in the isothermal area of the furnace (850°C) allowing at the growth of aligned CNT. SEM observations show that on these two substrates the length of CNT is similar, namely around 370 µm +/- 10 µm. As a consequence, this set-up modification enables a homogeneous covering of large flat quartz substrates on an area of 200 mm x 50 mm. This is most probably due to both the precursor content per surface unit of the reaction chamber which is increased as compared to the initial configuration of the reactor and to the precursor flow which is less disrupted, validating thus the set-up modifications adopted.

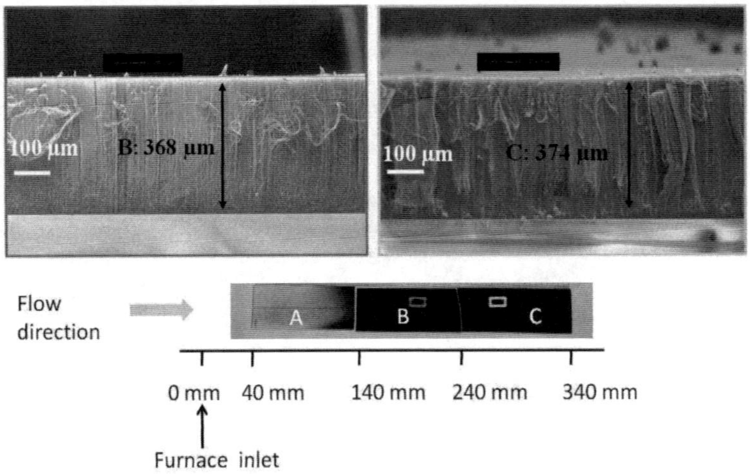

Figure 5. SEM observations of CNT grown on substrates after CVD process adjustment.

2.2.3 Growth of CNT on large carbon fiber cloths. The previous results indicate that the set-up modification enables to get homogeneous CNT growth on flat quartz substrates along a 200 mm long reaction area of the reactor, which means that the CNT synthesis parameters and configuration are optimized. Regarding porous and flexible CF cloth substrates, they exhibit a higher specific surface area as compared to flat substrate, therefore CNT growth conditions need to be adjusted as compared to flat quartz substrates. Moreover, an efficient growth of CNT on CF cloth in terms of alignment and density needs a pre-deposition of ceramic sub-layer on CF as presented in the experimental section. In fact, the feeding rate of ceramic sub-layer precursor (TEOS in this study) and the flow rate of carrier gas have to be adjusted to achieve a homogeneous covering of aligned CNT along the porous CNT cloth (figure 6). Figure 6 shows the influence of carrier gas flow rate which varies from 1 l/min to 10 l/min. For 3 l/min, we have slightly improved the surface covering of CNT on CF cloth as compared to the other flow rates but this is not enough to cover the whole cloth surface. The TEOS feeding rate is then adjusted to 0.1 g/min which involves a homogeneous covering of CF cloth with aligned CNT, as shown in figure 6(c).

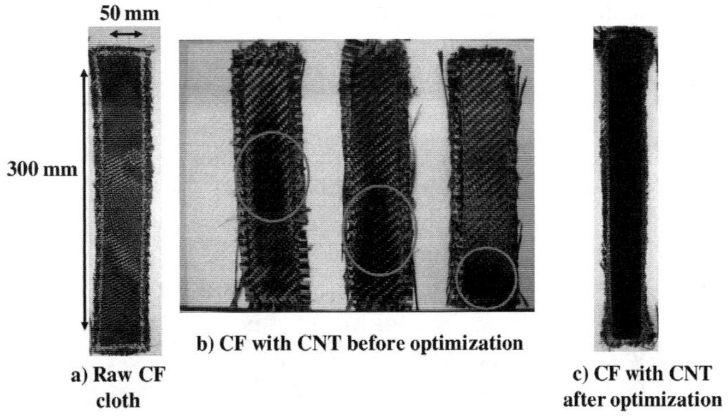

Figure 6. Pictures of carbon cloth (a) raw cloth, (b) cloth without optimization showing CNT on small surface of the carbon fiber cloth, (c) cloth with homogenous growth of CNT on the whole carbon fiber cloth.

The length and the morphology of CNT can be tailored by injection duration of toluene/ferrocene mixture. It is possible to grow either entangled CNT with high density covering the whole CF of the cloth for 3 min precursor injection or long CNT (50 µm mean length), dense and aligned perpendicularly to the CF axis along the CF cloth (200 mm * 50 mm) for 5 min precursor injection as illustrated on figure 7.

Figure 7. (a) CF cloth with high density of entangled and short CNT covering CF (3 min precursor injection duration) and (b) CF cloth with long dense and aligned CNT covering CF (5 min precursor injection duration).

4. Conclusion

In conclusion, we have demonstrated that the trends regarding vertically aligned CNT growth in small sized CVD reactor are confirmed on larger CVD reactor. Moreover, we have adjusted and optimized the CVD equipment and synthesis parameters in order to achieve a homogeneous covering of large substrates (quartz substrate or carbon fibre cloth with size of 200 mm * 50 mm) by aligned CNT. The length of CNT can be tailored in order to answer to different applications such as membranes or composites. Moreover, growth of CNT on large substrate opens the route to the elaboration of composite which sizes are compatible with the measurement of mechanical properties such as the interlaminar shear strength.

Acknowledgements
This work has been mainly performed in the framework of the ANR PROCOM project coordinated by EADS IW. We, therefore, gratefully acknowledge the main industrial partners: S. Bechtel (EADS IW) and S. Lambour (Porcher Industry) for providing and preparing the primary materials (carbon fibre cloths mainly) and for technical information regarding requirements for the elemental materials necessary for the elaboration. We are very thankful to F. Schuster (CEA-Saclay) for fruitful discussion in this research area. We also thank the ANR agency (Matériaux et Procédés) for financial support in the PROCOM project, and the Région Ile de France for financial support for the SEM equipment.

5. References

[1] Ren X, Chen C, Nagatsu M, Wang X 2011 *Chemical Engineering Journal*, **170** 395-410
[2] Herrera-Herrera A V, González-Curbelo M A, Hernández-Borges J, Rodríguez-Delgado M A 2012 *Analytica Chimica Acta* **734** 1-30
[3] Gohier A, Chancolon J, Chenevier P, Porterat D, Mayne-L'Hermite M, Reynaud C 2011 *Nanotechnology* **22** 105501
[4] de las Casas C and Li W 2012 *Journal of Power Sources* **208** 74-85
[5] Agnihotri P, Basu S, Kar K K 2011 *Carbon* **49** 3098-3106.
[6] An F, Lu C, Li Y, Guo J, Lu X, Lu H, He S, Yang Y 2012 *Materials & Design* **33** 197-202
[7] Aldajah S and Haik Y 2012 *Materials & Design* **34** 379-383
[8] Ashrafi B, Guan J, Mirjalili V, Zhang Y, Chun L, Hubert P, Simard B, Kingston C T, Bourne O, Johnston A 2011 *Composites Science and Technology* **71** 1569-1578
[9] De Greef N, Gorbatikh L, Lomov S V, Verpoest I 2011 *Composites Part A: Applied Science and Manufacturing* **42** 1635-1644
[10] de Resende V G, Antunes E F, de Oliveira Lobo A, Oliveira D A L, Trava-Airoldi V J, Corat E J 2012 *Carbon* **48** 3655-3658
[11] Godara A, Mezzo L, Luizi F, Warrier A, Lomov S Van Vuure A W, Gorbatikh L, Moldenars P, Verpoest I 2009 *Carbon* **47** 2914-2923
[12] Hung K H, Kuo W S, Ko T H, Tzeng S S, Yan C F 2009 *Composites Part A: Applied Science and Manufacturing* **40** 1299-1304
[13] Kim M T, Rhee K Y, Lee J H, Hui D, Lau A K T 2011 *Composites Part B: Engineering* **42** 1257-1261
[14] Kepple K L, Sanborn G P, Lacasse P A, Gruenberg K M, Ready W J 2008 *Carbon* **46** 2026-2033
[15] Lee S B, Choi O, Lee W, Yi J W, Kim B S, Byun J H, Yoon M K, Fong H, Thostenson E T, Chou T W 2011 *Composites Part A: Applied Science and Manufacturing* **42**, 337-344
[16] Li J and Zhang L Q 2011 *Materials Science and Technology* **27** 252-256
[17] Lomov S V, Gorbatikh L, Kotanjac Z, Koissin V, Houlle M, Rochez O, Karahan M, Mezzo L, Verpoest I 2011 *Composites Science and Technology* **71** 315-325
[18] Lv P, Feng Y Y, Zhang P, Chen H M, Zhao N, Feng W 2011 *Carbon* **49** 4665-4673
[19] Mathur R B, Chatterjee S, Singh B P 2008 *Composites Science and Technology* **68** 1608-1615
[20] Reia da Costa E F, Skordos A A, Partridge I K, Rezai A 2012 *Composites Part A: Applied Science and Manufacturing* **43** 593-602
[21] Sager R J, Klein P J, Lagoudas D C, Zhang Q, Liu J, Dai L, Baur J W 2009 *Composites Science and Technology* **69** 898-904
[22] Siddiqui N A, Khan S U, Ma P C, Li C Y, Kim J K 2011 *Composites Part A: Applied Science and Manufacturing* **42** 1412-1420

[23] Sharma S P and Lakkad S C 2011 *Composites Part A: Applied Science and Manufacturing* **42** 8-15
[24] Thostenson E T, Li W Z, Wang D Z, Ren Z F, Chou T W 2002 *Journal of Applied Physics* **91** 6034
[25] Delmas M, Pinault M, Patel S, Porterat D, Reynaud C, Mayne-L'Hermite M 2012 *Nanotechnology* **23** 105604
[26] Patton S T, Zhang Q, Qu L, Dai L, Voevodin A A, Baur J 2009 *Journal of Applied Physics* **106** 104313
[27] Pinault M, Mayne-L'Hermite M, Reynaud C, Beyssac O, Rouzaud J N, Clinard C 2004 *Diamond and Related Materials* **13** 1266-1269
[28] Pinault M, Mayne-L'Hermite M, Reynaud C, Pichot V, Launois P, Ballutaud D 2005 *Carbon* **43** 2968-2976
[29] Pinault M, Pichot V, Khodja H, Launois P, Reynaud C, Mayne-L'Hermite M 2005 *Nano letters* **5** 2394-8
[30] Castro C, Pinault M, Coste-Leconte S, Porterat D, Bendiad N, Reynaud C, Mayne-L'Hermite M 2010 *Carbon* **48** 3807-3816
[31] Mayne-L'Hermite M, Armand X, Porterat D, Reynaud C 2003 *Proceeding of chemical vapour deposition-XVI and EuroCVD-14*, edited by Allendorf M, Maury F, Teyssandier F, 2003–2008, p. 549

Journal of Physics: Conference Series **429** (2013) 012054 doi:10.1088/1742-6596/429/1/012054

From Safe Nanomanufacturing to Nanosafe-by-Design processes

F Schuster[1] and F Lomello[1]

[1] CEA Cross-Cutting program on Advanced Materials Saclay, 91191 Gif-sur-Yvette, France

E-mail: frederic.schuster@cea.fr

Abstract. Industrial needs in terms of multifunctional components are increasing. Many sectors are concerned, from the integrated direct nanoparticles production to the emerging combinations which include the metal matrix composites (MMC), ductile ceramics and ceramic matrix composites, polymer matrix composites (PMC) for bulk application and advanced surface coatings in the fields of automotive, aerospace, energy production and building applications. Moreover, domains with a planetary impact such as environmental issues, as well as aspects for instance health (toxicity) and hazard assessment (ignition and explosion severity) were also taken into account. Nanotechnologies play an important role in promoting innovation in design and realization of multifunctional products for the future, either by improving usual products or creating new functions and/or new products. Nevertheless, this huge evolution in terms of materials could only be promoted by increasing the *social acceptance* and by acting on the different main technological and economic challenges and developing *safe oriented processes.* Nowadays, a huge number of developments of nanoparticles are potentially industrial up-scalable. However, some doubts exist about the handling's safety of the current technologies. For these reasons, the main purpose was to develop a *self-monitored* automation in the production line coupling different techniques in order to simplify processes such as *in-situ* growth nanoparticles into a nanostructured matrix, over different substrates and/or the nanopowders synthesis, functionalization, dry or wet *safe* recovery system, granulation, consolidation in *single-step*, by monitoring at real time the processing parameters such as powder stoichiometry. With the aim of assuring the traceability of the product during the whole life, starting from the conception and including the R&D, the distribution and the use were also considered. The optimization in terms of processing, recovery and conditioning, permits to increase its versatility, thus leading to the increase of the added value to the production. This review put in evidence -by different examples- the progress achieved since the beginning of *Nanosafe integrated initiative* developed by CEA and its partners as well as promoting, firstly the safety at work place, and then extending to an integrated and controlled *in-situ* production by of *Nanosafe-by-Design controlled processes.*

1. *Objectives of the R&D*

Nowadays, the nanotechnologies applied to the materials science are a strategic issue for almost all the industrial sectors.

CEA took profit on its safety culture knowledge inherited from the nuclear field, in order to carry out multidisciplinary scientific studies regarding the risk assessment related to human health, the environment, final workers and costumers at all stages of the life cycle of nanotechnology. This life cycle involves the starting at the point of conception and including R&D, manufacturing and distribution.

The first step was to verify and quantify the current exposition risks (*Nanosafe*), followed by reassuring the *safe nanomanufacturing (SAPHIR- Safe, Integrated & controlled production of high-tech multifunctional materials and their recycling)* and then evolving up to a *single step* production and *in-situ* integration of nano-objects without any handling - *Nanosafe-by-design* processes.

This allowed the development of *self-monitorated*, *integrated* and *controlled* production of high-tech multifunctional nanostructured products.
Furthermore, the principal aim is to produce objects at a unit cost equal or slightly lower than those of existing objects especially for mass production. The industrial demonstration platforms enabled to assess both the technical and economic effectiveness of the concept of the *factory for nano's*. The combinatorial approach will also be a powerful tool to select the best candidates reducing time-spent on nanostructured products preparation and thus time-to-market.
The different developed tools, techniques and technological demonstrators are resumed in the figure below.
In order to achieve the main aim, *Safe Nanomanufacturing* tools and *Nanosafe-by-Design* processes were developed. In the following lines, the examples regarding the safe processes are given.

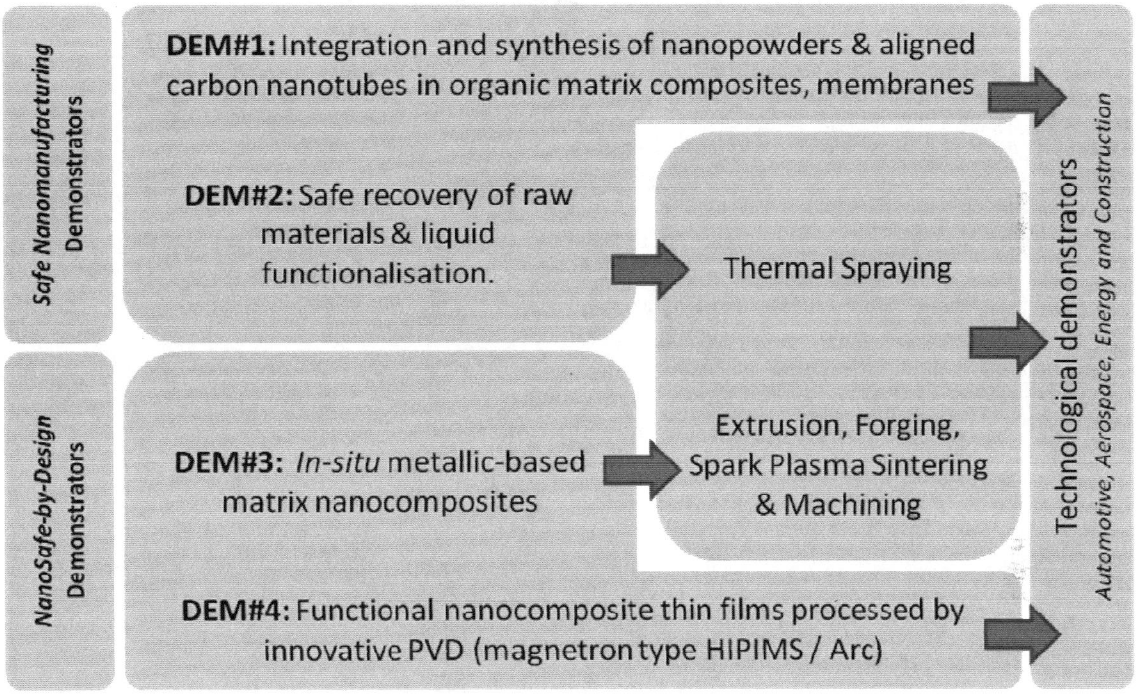

Figure 1. The *Safe Nanomanufacturing* and *Nanosafe-by-Design* processes.

2. *Introduction: some definitions regarding nanopowders*

During the 80s, the nanopowders production was initiated on large scale and the attention was focused on nanopowders processing.
In the following decade, the effort was emphasized to develop reproducible processing methods for manufacturing nanopowders into useful components taking profit on its *new* properties.
During the second part of the 90s many significant advances in the production of nanometric objects were achieved as they present new interesting properties. However, the handling of nanopowders and the possible industrial production created new challenges.
Generally, nanomaterials could be classified into different groups as shown in the figure below.

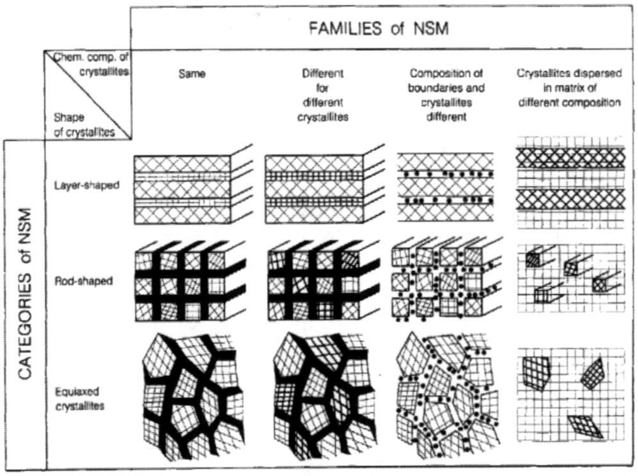

Figure 2. Schema identifying all the nanomaterials as a function of the chemical composition and shape and crystallites [1].

According to *Gleiter* [1]], nanomaterials are classified into twelve groups depending on their shape (dimensionality) and chemical composition of the structural elements. On one hand, as a function of crystallites shape, three categories of nanocrystalline materials could be distinguished: layer-shaped crystallites, rod-shaped crystallites with layer thickness or rod diameters in the order of a few nanometers and nanocrystalline materials composed of the equiaxed nanometer-sized crystallites.
On the other hand, depending on the chemical composition of the crystallites, the three categories of nanocrystalline materials could be divided into four families.

2.1. Brief description and properties regarding the nanomaterials

Nanomaterials and in particular nanopowders are composed of individual particles having sizes between 1 to 100 nm according to cf. standard ISO/TS 27687. The particles in nanopowders are smaller than the wavelength of the visible light. Nanopowders are characterized of having an extremely high surface area to volume ratio [2].
Specific surface area of the particle is related to the average particle size, based on a geometrical consideration. In case of spherical particles, the formula [3] is given below.

$$SSA_{sphere} = \frac{SA_{sphere}}{m_{sphere}} = \frac{4\pi r^2}{\rho.V_{sphere}} = \frac{3}{\rho.r} \quad (1)$$

where SSA is the specific surface area (generally measured in m^2/g), SSA is the surface area, m is the mass and V is the volume. Generally, the rougher the particle surface, the greater surface area regardless of the particle size. The equation underestimates the surface associated with the surface texture.
Regarding the new properties, size effects on nanomaterials, for example in case of Ni_3Al crystallites dispersed in a NiAl matrix lead having a different of flow stress as a function of the precipitates size (Fig. 3 a). Similarly, quantum size effects are observed allowing a blue shift in the absorption and luminescence spectra as it is corroborated for instance when CdS nanoparticles when dispersed in an aqueous solution.

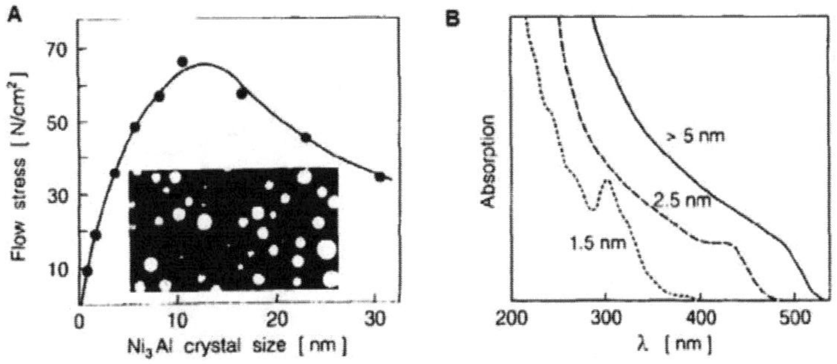

Figure 3. Flow stress of Ni-13 at% Ni alloys with a dispersion of Ni_3Al (a) & Absorption Spectrum of CdS particles (b) as a function of the particle size [1].

When light quantum is absorbed, an electron is transferred from the valence band into the conduction band, leaving behind a positive hole. In small particles, the wave functions of the electron and the hole are confined to the particle volume.

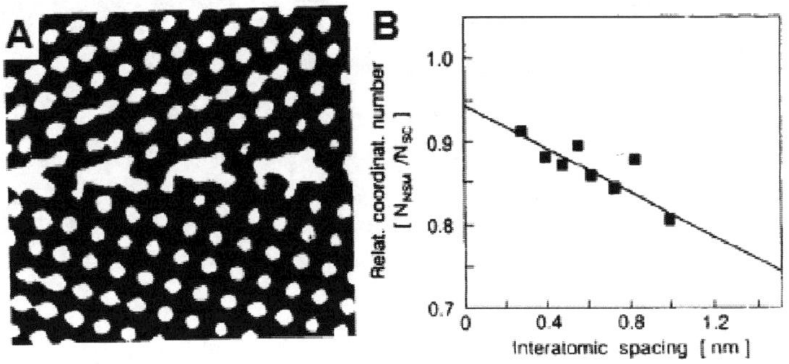

Figure 4. (a) Atomic structure in the core of grain boundary between two NiO tilted relative to one another by 36.9° about a common [1 0 0] and (b) Coordination number for nanocrystalline Pd (12 nm crystal size) relative to the single crystalline as a function of the interatomic spacing [1].

Another characteristic of nanomaterials is their incoherent interfaces which introduce misfit between the crystallites. This is the responsible for modifying locally the atomic microstructure by reducing the atomic density and the coordination [1]. The relative coordination number for interatomic spacing is less than one. This fact is thought to be the responsible of changing the properties in nanostructured materials (Fig. 4).

The third aspect, which differs the nanosized powders from the classical, micronic ones is the morphological metastability related to fine grain size. This metastability in nanocrystalline materials could be divided into three categories: compositional (extended solution ranges), structural or topological (alternate crystal structure or amorphous phases) and morphological (finely divided structures) [4].

However, the influence in some fields are not clear, for example in the mechanical field it has not been proved yet the benefits in keeping the nanoregime [5].

2.2. Aspects related to nanopowders

The potentialities of nanostructured powders can be affected by some drawbacks, first of all by their inherent agglomeration, due to *Van der Waals* forces [6]. Therefore, in order to produce dense

materials, with tailored microstructure, it is important to control all the elaboration steps, beginning with the dispersion.
Most nanocrystalline powders are not composed of single nanometer sized particles. Such particles known as crystallites are bonded together to form agglomerates or aggregates as it is illustrated in Fig. 5. The distinction between the agglomerate and the aggregate is usually based on the degree of bonding.
Soft agglomerates in which the particles are bonded by weak Van der Waals and hard agglomerates or aggregates in which particles show necks between adjacent particles [8].

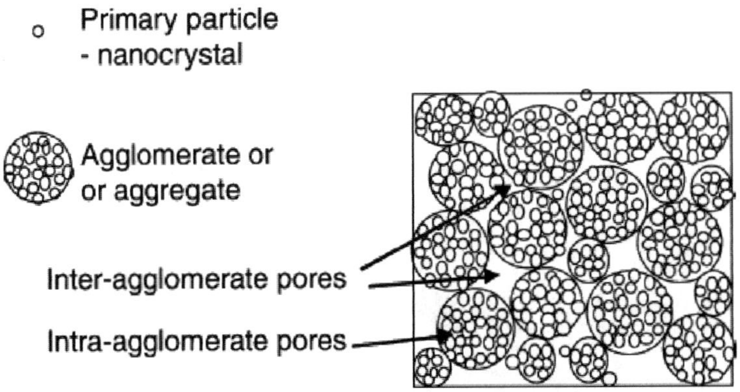

Figure 5. Schematic illustration of nanocrystallites and agglomerates/aggregates [7].

Those agglomerates and/or aggregates, could present two different types of pores; inter-agglomerate pores (micronic) which coexist with smaller inter-crystallite pores (nanometric).

3. *Risk evaluation tools during the entire products' life-cycle: development of secure industrial systems and workplaces*

3.1. *Quantification of individual protections: efficiency of personal protective equipment*

It is thought that the exposition of workers to nanoparticles will be increased in a near future. Indeed, the National Science Foundation in the USA has estimated an overall market for nanoproducts exceeding 1 trillion of dollars in 2015.
As a consequence, the potential human exposure to nanoparticles can occur by inhalation, dermal contact and ingestion during processing should be considered. Actually, for non developed fibrous media, the maximum penetrating particle size (MPPS) is about 200-300 nm [8].
In many cases, nanoparticles under 200 nm tends to re-agglomerate by the *Van der Waals* forces. Although, the electrostatic filters allow reducing the MPPS down to 50-60 nm. These filters are fabricated by fibers electrically charged by corona or triboelectric effect.
In addition, *Balazy et al.* reported a MPPS near 40 nm for N95 electrostatic filters exposed to a NaCl monodispersed aerosol [9].
Golanski et al. tested FPP3 electrostatic filters (according the EN 149) with graphite and TiO_2 aerosols [10,11]. The researchers employed a nanoparticle penetration device, as shown in the figure below.
The *through diffusion bench* was designed so as to reproduce the conditions of workers' clothing. In the upper cell, the concentration of nanoparticles is maintained constant. A differential pressure of few megabars is kept between the two parts. The particle flow was accurately measured in order to estimate the efficiency of the protective clothing. Finally, the ^{85}Kr was employed with the aim of neutralizing the aerosol into the test filter or the diffusion cell

Journal of Physics: Conference Series **429** (2013) 012054 doi:10.1088/1742-6596/429/1/012054

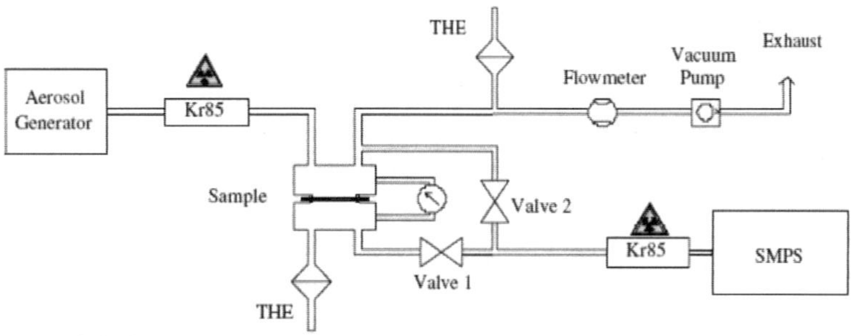

Figure 6. Schematic diagram of the nanoparticle penetration device [11].

On one hand, the graphite particles were generated by spark plasma discharge between two graphite electrodes with an average distribution of around 35 nm. On the other hand, TiO_2 and Pt particles are produced by condensation from supersaturated vapour obtained with an average distribution of 40 and 10 nm, respectively.

Firstly, the authors tested the efficiency of the HEPA filters. The results regarding the penetration are presented in Fig. 7. The HEPA filter shows higher efficiency for graphite nanoaerosols, if compared with TiO_2 and Pt, for particles ranging from 10 to 20 nm. A difference of one order of magnitude is observed for a given particle diameter. TiO_2 nanoaerosols show the highest penetration through the filter: the maximum penetration for 19 nm particles is around 0.2%.

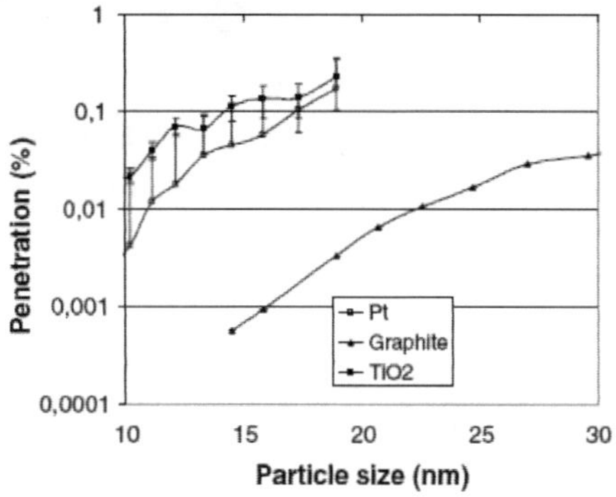

Figure 7. Evolution of graphite, TiO_2 and Pt nanoparticle penetration for the same HEPA fibrous filter as a function of particle size (nm) for a face velocity of 5.3 cm/s [11].

An increase of fibrous filter efficiencies inversely proportional particle size is predicted by the classical filtration model, considering the Brownian diffusion and the *Van der Waals* forces [12].

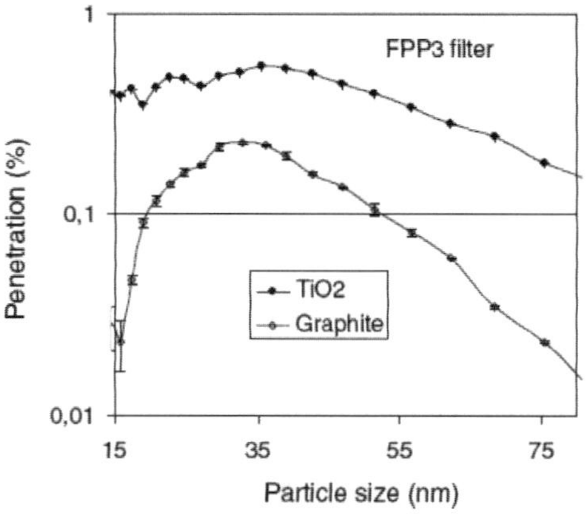

Figure 8. Penetration as a function of the particle size of FPP3 electrostatic filters [11].

The efficiency of an FPP3 electrostatic filter was investigated for TiO_2 and graphite particle diameters ranging between 15 and 75 nm at the flow speed of v = 5.3 cm/s. The authors reported MPPS values around 35 nm in case of TiO_2, five times higher if compared with graphite as shown in Fig. 8.
A shift of the MPPS toward smaller particles is consistent with the theory: the fiber charge has an effect on the particle deposition [8]. A difference in terms of efficiency between different materials was noted. The electrostatic forces in the graphite could enhance the nanoparticles' capture in comparison with TiO_2 nanoparticles.
Subsequently, the diffusion through the protective clothing was also measured. The results concerning the efficiencies of cotton, polyester and polyethylene are shown in Fig. 9.

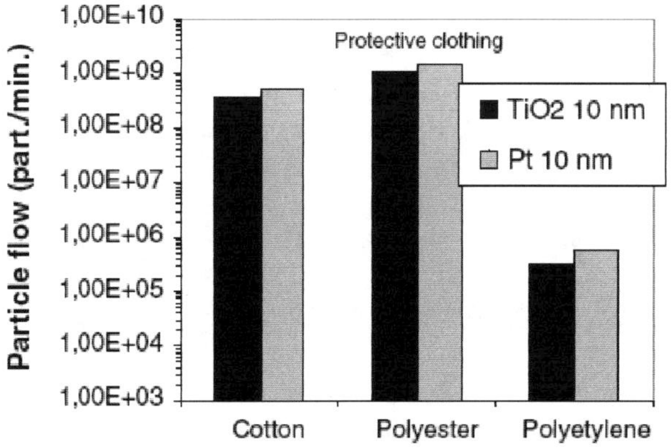

Figure 9. Evolution of particle flow for TiO_2 and Pt nanoparticles as measured by through diffusion technique [11].

The results indicated that cotton and polyester present almost the same efficiency when tested with TiO_2 and Pt particles with mean geometric diameter of around 10 nm. Non-woven fabrics (air-tight materials) present the best efficiency against Pt and TiO_2 nanoparticles centered around 10 nm.
These results are in good agreement with those published by *Golanski et al.*, who reported similar results for graphite nanoparticles of 40-80 nm [10].
Finally, the authors analyzed three types of gloves i.e. nitrile, latex and neoprene with thicknesses 100, 150 and 700 μm.

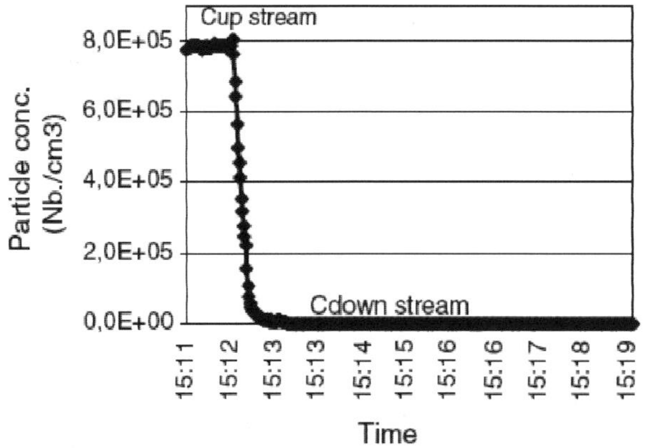

Figure 10. The number of particles as a function of time for a nitrile exposed to TiO_2 [11].

No TiO_2 penetration through nitrile glove after exposing for 5 minutes is observed, as shown in Fig. 10. Furthermore, no TiO_2 and Pt penetration were found when the gloves are exposed at

10^5-10^6 particles/cm^3. As a conclusion, nitrile, latex and neoprene gloves are efficient for TiO_2 and Pt around 10 nm.

3.2. Monitoring by means of the Laser-Induced Breakdown Spectroscopy (LIBS)

Amodeo et al. studied the particle physicochemical characteristics by means of a spectral imaging-aided *Laser-Induced Breakdown Spectroscopy* (LIBS) coupled with a Scanning Mobility Particle Sizer (SMPS) [13]. They employed the technique for detecting sodium chloride and metallic particles ranging from 40 nm to 1 μm.

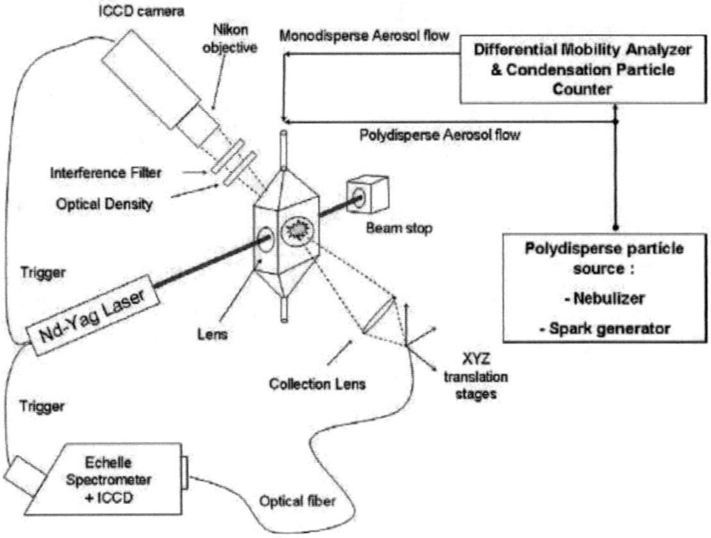

Figure 11. Experimental set-up of the second device [13].

As shown in Fig. 11, the general experimental set-up combines the LIBS technique and the SMPS. Two nanoparticle generators were used during our experiments. An atomizer and a spark generator were utilized for particle generation. With both generators, a flow of clean air was obtained connecting the laboratory compressed air line to a filtered air supply unit.

Particle sizes, number and concentration were measured using a Differential Mobility Analyzer (DMA) combined with a particle counter (CPC). On one hand, the DMA connected with a particle counter allowed to measure the granulometry and total mass concentration. On the other hand, the particle number involving polydispersed flows was detected by another particle counter - designed for counting the particles whose sizes were inferior to 10 nm.

Time-resolved plasma imagery on laser-induced sparks, generated in air subsequent to single shots was recorded, in order to determine temporal evolution of plasma characteristics such as size, shape and lifetime. An example is given in the figure below.

Shortly after the end of the laser pulse, the cone-shaped plasma expands rapidly longitudinally in the radial direction between 0 and 20 ns. From 1500 ns to 4500 ns time delay, the plasma expansion slows down, stabilizes in shape and plasma dimensions are approximately 4.5 mm long and 2.5 mm wide. The plasma lifetime value was estimated to be more than 60 μs.

In all cases three regions within the laser induced plasma are distinguished.

The size effects do not occur during LIBS measurements on particles with sizes ranging from 60 nm to 500 nm, due to a span incomplete vaporization.

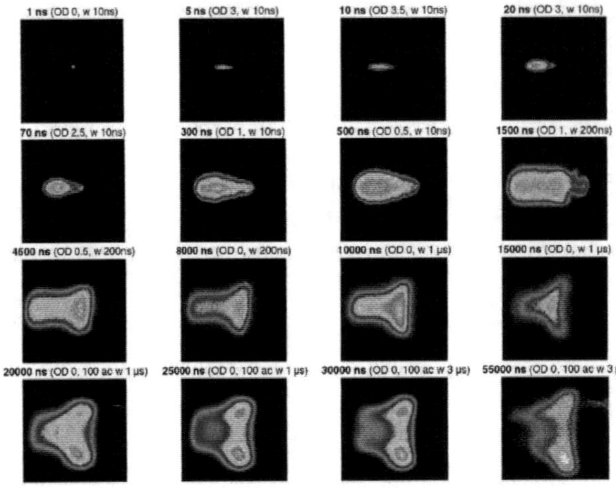

Figure 12. Laser-induced plasma spatio-temporal evolution as a function of time delay using laser shots of 120 mJ – Above the images are indicated the delays, optical densities (OD) and Gate widths. [13].

Carranza and Hahn reported that an incomplete vaporization of micrometric-sized particles even for high laser irradiance. The authors confirmed the existence of an upper particle size of 2.1 µm beyond which, the vaporization was not completed thus leading a LIBS signal saturation for short time delays closer to vaporization times [14,15].

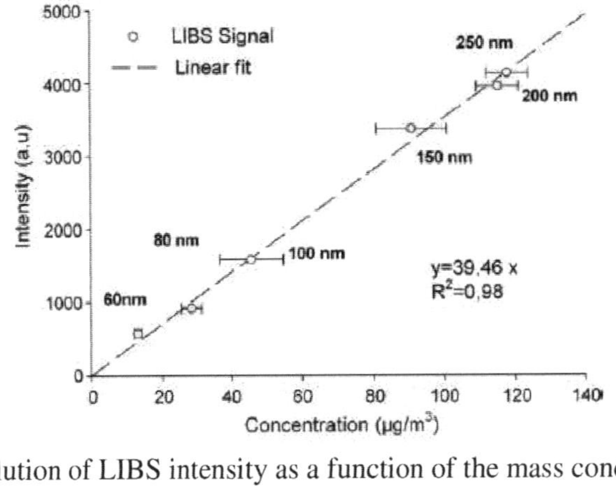

Figure 13. Evolution of LIBS intensity as a function of the mass concentration [13].

The LIBS intensity is linear as a function of the mass concentration as it is reported in Fig. 13. The authors concluded that no particle size effect is observed. As reported in literature, the dependency on LIBS signal and the size effects were detected by employing carbonaceous nanoparticles in the 30-100 nm at early time delays (5-10 µs) closer to vaporization times [15].

With the aim of characterizing the particle size *in-situ*, the LIBS technique was successfully coupled with the transmission electron microscopy (TEM) in a pilot production of carbon nanotubes (CNT) - by the catalytic chemical vapor deposition technique (CCVD)-, as reported by *R'mili et al.* – Fig. 14 [16].

Two TEM samplers were connected in order to measure the particle size and the morphology. The two TEM samplers were designed by Technical Research of Finland [17].

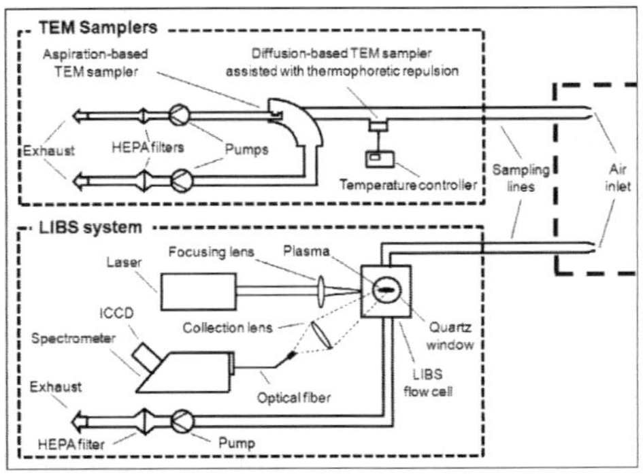

Figure 14. Experimental set up of TEM and LIBS systems [16].

The first sampler is based on aspiration through holey TEM grids and collects particles without particle size selection. The second sampler works on diffusion assisted with *thermophoretic* repulsion as shown in the figure above. In the latter the temperature gradient of few degrees, between the TEM grid and the sampling pipe, helps to repel the micrometric particles.
The authors were successfully able to separate by the thermophoretic –by repulsion- the carbon nanotubes as it is compared in Fig. 15. Moreover, LIBS measurements presented detection inconvenience, as in the previous case due to lower-upper particle size limits. In their study, the authors claimed that no efficient results were obtained over the 1-80 μm range, due to an incomplete vaporization.

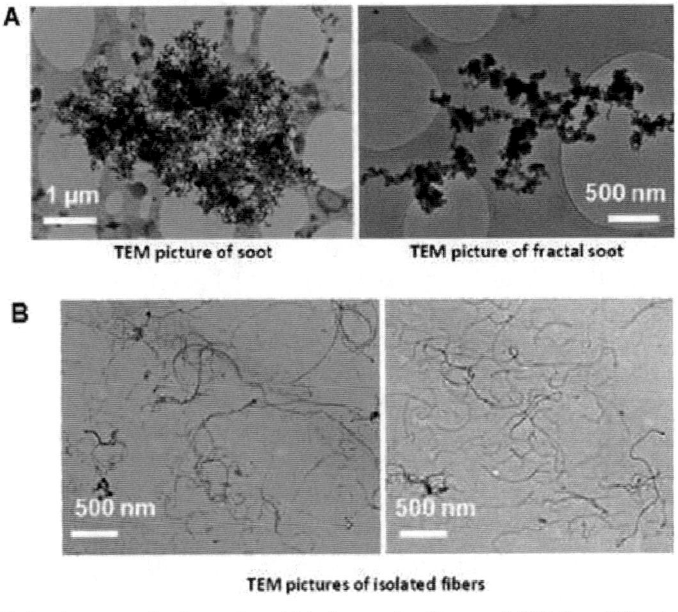

Figure 15. TEM micrographs by using (A) the aspiration and (B) the diffusion assisted with thermophoretic repulsion sampler [16].

3.3. NanoBadge

The NanoBadge was developed during *Nanosafe project* in order to measure the personal exposure as well as the indoor and outdoor air quality assessment. The two generations are presented in the Fig. 16.

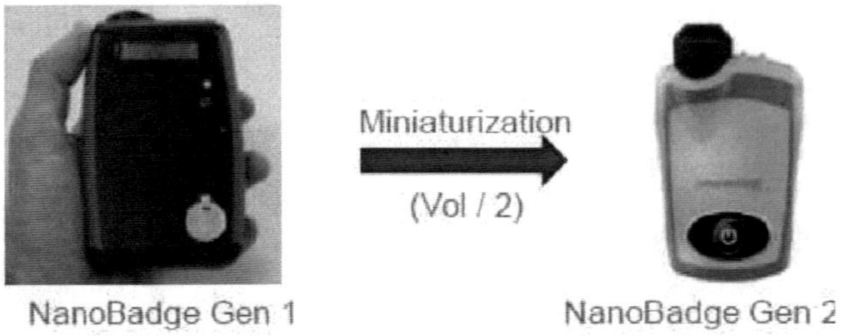

Figure 16. Two Nanobadge generations.

It is a compact solution for measuring, the inhalation exposure to engineered nanomaterials in the worker zone, in order to determine the particle mass concentration. The workplace exposure to airborne nanoparticles and their aggregates are quantified by means of elemental quantification and mass concentration in the respiratory fraction (alveolar).

The NanoBadge sampler collects particles with cut-off aerodynamic diameter < 4 μm with 50 % efficiency in order to fit the alveolar fraction (Fig. 17). The inlet size selection stage is a home-made impactor designed specifically for NanoBadge. The nanoparticles are then collected on the filter downstream.

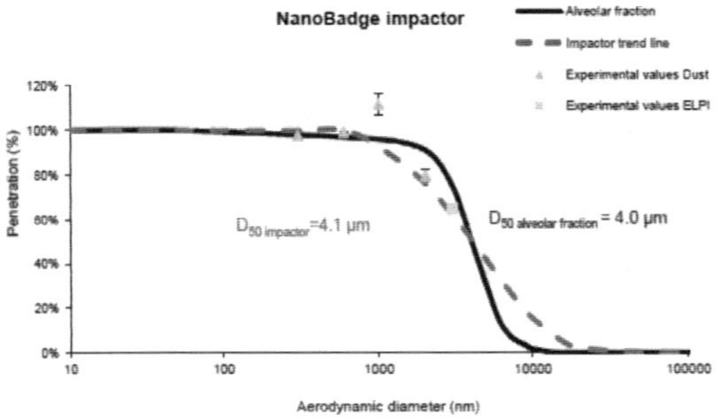

Figure 17. Particle size selection prior to collection.

In parallel, additional data such as user identity, timestamp, flow rate, events and errors are also recorded. Subsequently, an *ex-situ* analysis is made extracting the filter from the cassette. All the possible characterization techniques are contained in the figure below.

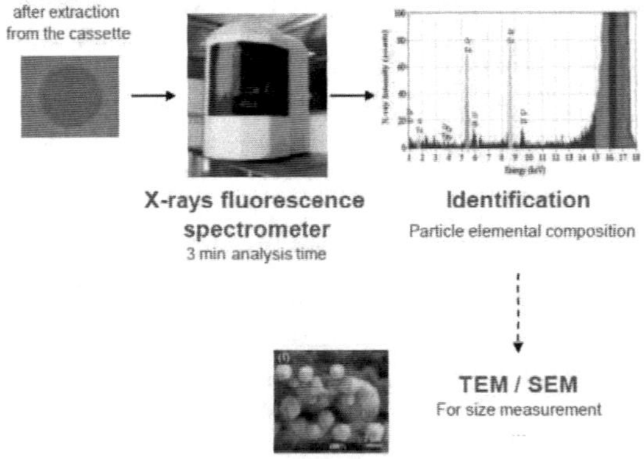

Figure 18. *Ex -situ* elemental analysis of particle collected by the NanoBadge.

3.4. Health and hazard assessment: in vitro toxicity tests

During the *Nanosafe* project, the toxic effect of 24 different nanoparticles with similar equivalent spherical diameter and various elemental compositions were evaluated on 2 human pulmonary cell lines: A549 and THP-1 [18]. A secondary aim was to elaborate a generic experimental set-up that would allow the rapid screening of cytotoxic effect of nanoparticles. By comparing 2 cytotoxicity assays (MTT and Neutral Red) and by analyzing 2 time points (3 and 24 hours) for each cell type and nanoparticle. The data treatment allowed the TC50 (Toxic Concentration 50 i.e. nanoparticle concentration inducing 50% cell mortality) calculation.
The MTT assay on THP-1 cells exposed for 24 hours allowed assessing the cytotoxic effect of one type of nanoparticles. Results revealed that Copper- and Zinc-based nanoparticles were the most toxic. Titania, Alumina, Ceria and Zirconia-based nanoparticles show moderate toxicity, and no toxicity was observed for Tungsten Carbide. No correlation between cytotoxicity and the physico-chemical characteristics such as the equivalent spherical diameter and/or specific surface area was found.

3.4.1. Development of a physiologically based kinetic model for ^{99m}Tc-labelled carbon nanoparticles – In Silico computer model

Particulate air pollution is associated with respiratory and cardiovascular morbidity and mortality as it was reported by *Péry et al.* [19]. The researchers proposed a model which simulates the translocation of nanopowders into the bloodstream in humans.
The authors based the model Technegas® distribution data on 5 healthy male volunteers from 24 to 47 years old reported by *Nemmar et al.* [20]. The male volunteers were exposed to 100 MBq in 3 to 5 acts of breathing - intake of nanoparticles occurs initially through breathing. Technegas® is an aerosol suspension of ^{99m}Tc-labelled carbon nanoparticles (5 to 10 nm) produced in high purity argon.
The blood radioactivity data was measured at 1, 5, 10, 20, 30, 45 and 60 min in order to observe the concentration evolution. A gamma-camera enables to identify the radioactivity distribution at the liver, stomach, urine and blood.
Based on these results, *Péry et al.* [19] developed a parameter estimation in the Bayesian context with the Markov chain Monte Carlo (MCMC) techniques. The authors developed a physiologically based pharmacokinetic (PBPK) model, adapted to imaging data analysis of *Nemmar et al.* which simplifies the mechanisms of absorption, distribution and elimination of technetium-labelled nanoparticles.
The authors distinguished and modeled three different fractions of nanoparticles according to the particle sizes: small ^{99m}Tc bonded particles (*i.e.* able to translocate), large ^{99m}Tc bonded particles (*i.e.* unable to translocate) and free ^{99m}Tc. On one hand, large nanoparticles are supposed to remain at constant concentration in the respiratory tract, lungs and the stomach. On the contrary, nanoparticles in the lung are transferred to blood by diffusion. Furthermore, the free ^{99m}Tc is able to translocate from the lung into the gastrointestinal tract. These phenomena were taken into account in the model.
Generally, nanoparticles diffuse into the various organs excepting the brain, since experimental measurements do not confirm the of ^{99m}Tc presence. The elimination is supposed to occur by means of the renal elimination (urine). Nonetheless, during the time of exposure, the renal elimination of particle bound-^{99m}Tc was not observed, since only free ^{99m}Tc was found in urine.

3.4.2. Rapid screening *in vitro* tests

The screening of nanoparticles toxic effects in humans was studied and a device was developed that it is able to detect them. These nanoparticles are thought to be able to enter into the human body by the lung-blood barrier. The translocation of nanoparticles by the in vitro model implies increased risks of toxicity, because the internal organs are also exposed to nanomaterials due to the blood stream.

The developed system consists in a microchip containing an internal biological system and a data detector. The nanoparticles screen is based on an in vitro translocation test using a model of pulmonary epithelium, i.e. a watertight layer of epithelial cells.

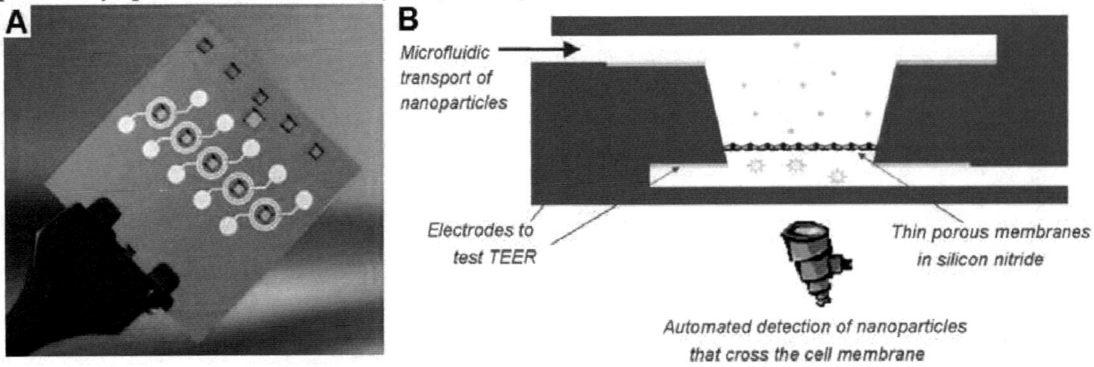

Figure 19. (a) Photo & (b) schematic view of the chip.

A chip contains an internal gap in which the epithelial cells (CaLu-3 cells) are cultured forming a tight layer that simulates the human lung epithelium. The microfluidic system feeds the cell culture media with the nanoparticles representing the interior of the lung.
Nanoparticles cross into the lower compartment allowing the detection by measuring the electrical resistance on the on-chip electrodes ('TransEpithelial Electrical Resistance' or 'TEER'). The silicon chip is fabricated using standard photolithography and silicon microfabrication techniques. The fluidic layers were casted in PDMS using a microfabricated master in silicon and SU8 photoresist.
With the aim of calibrating the device, the researchers employed four elemental standards: Cd, Te, Au, and Ti were used. These elements were chosen because of the possibility of analyzing nanoparticles such as Cd/Te, Au and TiO_2. For instance in Fig. 20, a test done with a mixture of TiO_2, Au (with sizes of 20nm) and CdTe nanoparticles of 590 nm is presented.

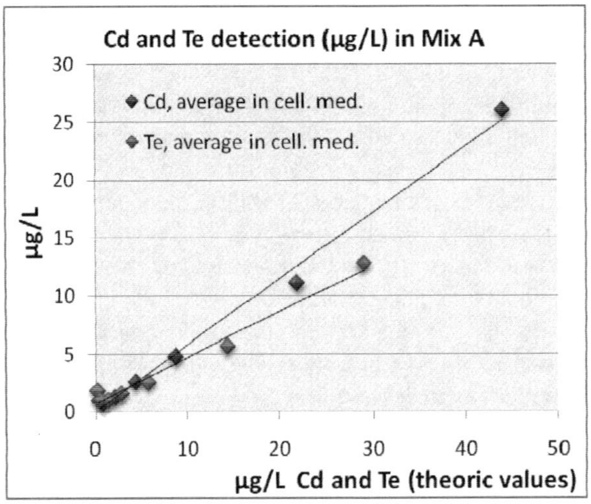

Figure 20. Calibration curves for Cd and Te detection.

As illustrated in Fig. 20, the system allows to quantify the nanoparticles detection in a culture medium. Similar results were obtained employing different powder mixtures.

3.5. Flammability and explosivity risks: nanopowders & multiwall carbon nanotubes

Many powders in suspension in air tend to ignite, burn and propagate as the powder nature influences the explosion propensity [21]. This phenomenon depends on the concentration, the degree of agglomeration and turbulence.
Several studies were devoted to the micro-sized powders [21]. However, the study of flammability hazard was extended to nano-sized powders in the *EU Nanosafe 2* project.
Bouillard et al. compared and tried to establish some correlation by using carbonaceous nanoparticles, aluminium particles as well as multiwall carbon nanotubes (MWNTs) [22,23].
The authors centred the attention on the dependencies between the particle size, the ignition temperature and minimal explosive concentration (MEC), since nanopowders are thought to present higher potential inflammation and explosion risks.
Generally, nanopowders are heavily agglomerated due to the *Van der Waals* forces as mentioned in the previous section. The fact of presenting voids in the bulk state increases the required storage volumes. According to the *Geldart's* classification, nanopowders belong to the cohesive type since their difference in terms of the particle density ρ_s and the fluid density ρ_g [24]. For the sake of clarity in Fig. 21 the different groups are presented.

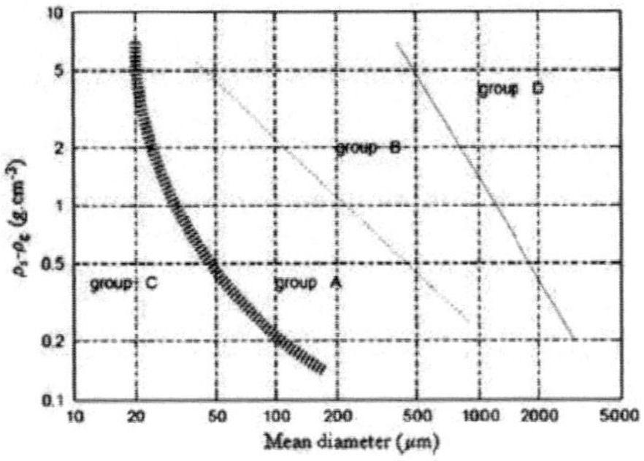

Figure 21. *Geldart's* classification [24].

As shown in Fig. 21, they present low bulk density that involves a low thermal conductivity which enables a higher self-heating and ignition risks.
Bouillard et al. claimed that nanopowders present two combustion regimes kinetically controlled (chemical) for small sized ones i.e. nanopowders and diffusion controlled for large sized one (micro). Both regimes for carbon (a) and aluminium (b) are presented in Fig. 22.

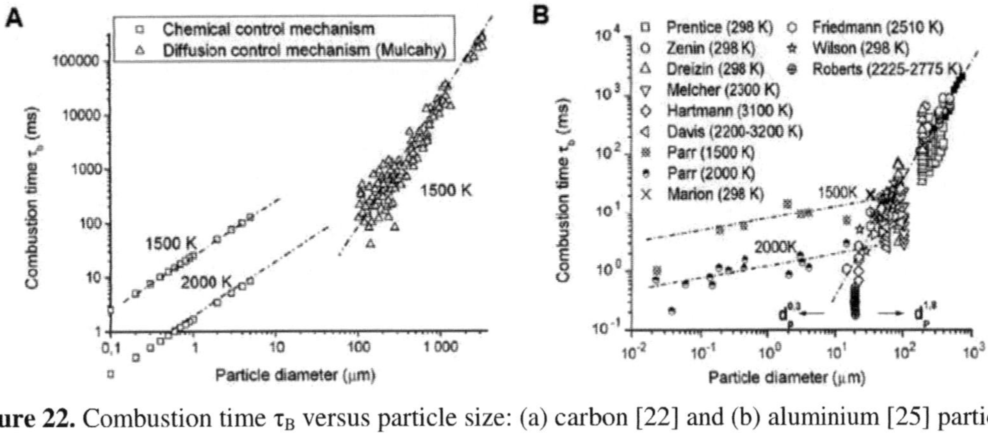

Figure 22. Combustion time τ_B versus particle size: (a) carbon [22] and (b) aluminium [25] particles.

Carbonaceous nanoparticles present well separated the two combustion mechanisms as it is shown in Fig. 22 (a). Indeed, the phenomenon was corroborated in case of the aluminium in Fig. 22 (b), in good agreement with the research of *Huang et al.*[25].
In addition, *Bouillard et al.* claimed that aluminium oxidation at low temperatures (1000°C<) is essentially ruled by diffusion mechanisms.
Furthermore, their work and the observation of *Huang et al.* [25] put in evidence how the particle size affects the ignition temperature. As a matter of fact, the decreasing of the particle size is accompanied by a reduction of the ignition temperature as it is shown in Fig. 23.

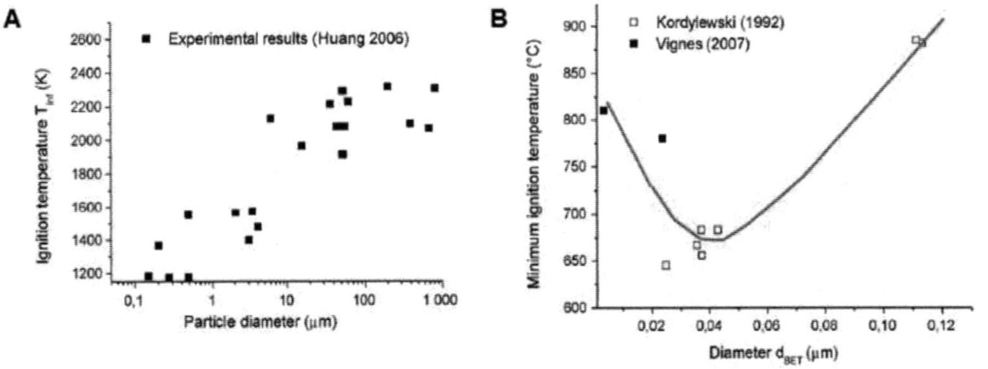

Figure 23. Ignition temperature versus particle diameter for (a) aluminium [25] and (b) carbonaceous particles [26,27].

Nonetheless, by comparing figures 23 (a & b) a minimum ignition temperature is observed in nanosized agglomerated carbonaceous such as black carbon and MWNTs Fig. 23 (b). In fact, the agglomeration and/or aggregation in nanoparticles oblige counting additional energy required to oxidize the powder. This fact explains the higher MIT observed in the nanosized range Fig. 23 (b).
The minimum explosible concentration (MEC) and the minimum ignition energy (MIE) as a function of the particle size are also important factors to be taken into account- Fig. 24. On one hand, many studies were devoted to the minimal explosive concentrations. *Bouillard et al.* found that no significant differences regarding the minimum explosible concentration exists between the nanopowders' types.

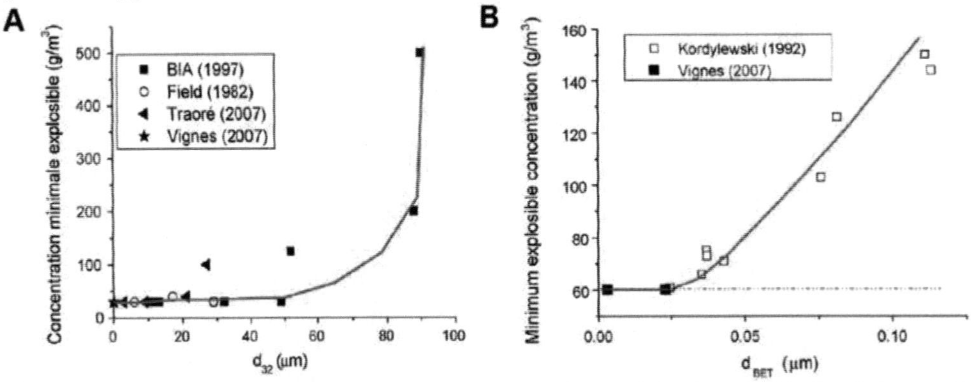

Figure 24. Minimal explosible concentration versus the particle size of (A) aluminium and (B) carbonaceous nanoparticles [22].

An almost constant MEC is noted at low particle size range being 30 gm^{-3} in the case of aluminium.

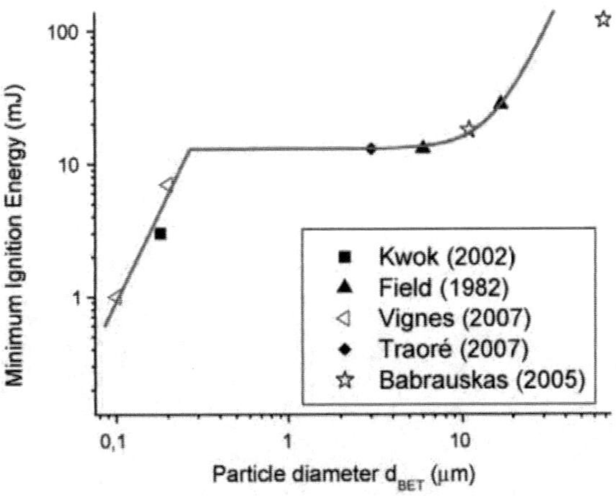

Figure 25. MIE as a function of the particle size [27,28,29,30,31].

On the other hand, a second factor is the minimum ignition energy (MIE), which is determined by the energy required for a given cloud of particles to sustain ignition and in a medium at an initial temperature T_u, while the burning temperature is at the final temperature T_{ad} (the flame adiabatic temperature).
As reviewed by the authors from the information found in literature, the two combustion regimes the kinetically and the diffusion controlled regimes are distinguished in Fig. 25. The former depends on the d_p^3 and the latter $d_p^{3/2}$, respectively. In aluminium particles, a transition in a range of 400-2000 nm is noted. No experimental data was found by the authors for the carbonaceous nanoparticles.
Bouillard et al. have shown that the particle diameter influences the explosion severity in terms of maximum pressure. Indeed, the maximum pressure rise is determined by the particles concentration and the specific surface area (SSA). In Fig. X the explosion pressure is presented as a function of both factors corresponding to different aluminium nanopowders.

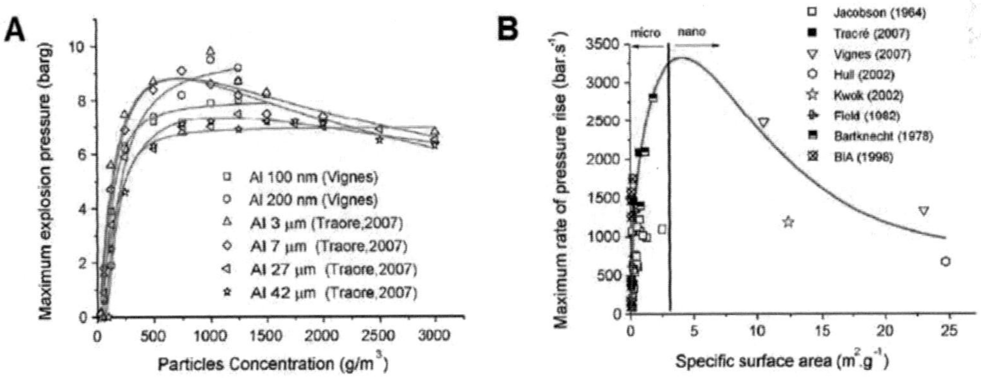

Figure 26. The explosion Pressure as a function of the aluminium nanoparticles concentration (a) [27,30,32] and the SSA (b) [27,28,27,30,33,34,35,36].

As a conclusion, the combination of both parameters defines the maximum explosion pressure. However, the SSA is not the only factor to be considered. The chemical elements of the compound should be pondered. From their study, *Bouillard et al.* claimed that MWNTs and black carbon have higher SSA. MWNTs are heavily agglomerated if compared with black carbon as it is shown in Fig. 27, thus hindering the risk explosivity. Nevertheless, the contribution of particle size is lower if compared with the parameters as shown before.

The degree of agglomeration is the principal parameter to be controlled in order to increase the safety regarding nanopowders' handling. As a conclusion, the carbon nanopowders tend to explode while metallic powders have the propensity to ignite in air (self-heating).

Vignes et al. focused the attention on the kinetics of combustion of MWCNTs with the purpose of clarifying the global risk assessment regarding the production and storage inside *the EU Nanosafe 2* project in reply to the needs of the European directive 1999/92/EC [37,38]. The different parameters were measured by means of thermal gravimetric analysis (TGA) and differential scanning calorimetry (DSC). The results indicated that the single step reaction is accompanied by a cylindrical shrinking.

Figure 27. SEM and TEM micrograph of MWCNTs at different magnifications [38].

Furthermore, the authors determined the activation energies by the Kissinger method being 150 kJ mol^{-1} [39]. The calculated value was higher if compared with SWCNTs [40]. The authors confirmed that the explosion severity of these materials is limited. In fact, the authors performed the tests as well as adiabatic calculation, as presented in Fig. 28. By comparing the black carbon and the MWCNT behaviour, it is possible to observe a lower maximum explosion pressure in the latter one.

Although, good storage facilities should be required in order to avoid self-heating and dust explosions as well as ignition sources such as hot surfaces, electric sparks and smouldering combustion

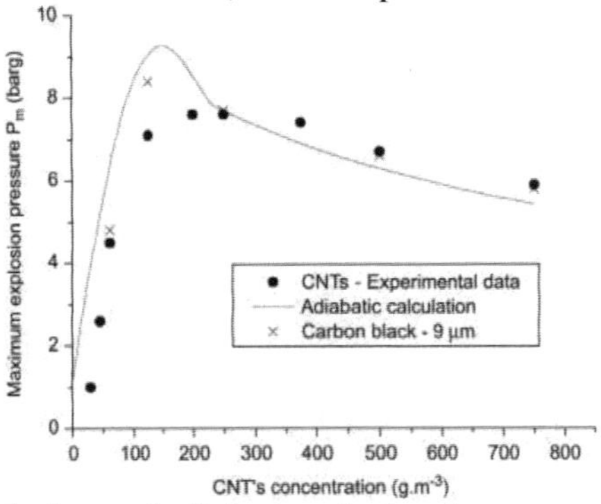

Figure 28. Maximum explosion and adiabatic calculated pressure as a function of the CNTs concentration [38].

3.6. Releaseability of nano fillers from different nanomaterials during the entire products' life-cycle

In the *Nanosafe* project, CEA developed a specific bench abrasion test using normalized Taber equipment – according the ASTM C1353-07 method- in order to quantify the nanomaterials before their market introduction in terms of their constitutive nanoparticle release-ability.
Firstly, *Guiot et al.* focused the attention on the calibration and the determination of the nanoparticles release of PET layer coated with a PVC layer and nanoclays as fillers [41]. As shown in the figure below, the scanning mobility particle size (SMPS) is coupled with a condensation particle counter (CPC) so as to permit the particle measurement in a range of 5-1000 nm. These tests revealed a release of 50 nm nanoparticles corresponding to the nanofillers size.
In a second publication, *Golanski et al.* reported the abrasion results regarding four types of paints containing nano-TiO_2, calcium carbonate and a TiO_2 pigment [42]. The paints were deposited on PVD substrates.
In this particular research, the abrasion tests were done adding a liquid solution on the coating in order to generate a dynamic friction. The *Taber 5131 standard tool* was employed. During the tests, the resulting nanoparticles were collected on filters for SEM analysis. At the same time, an Electrical Low Pressure Impactor (ELPI) was coupled with the aim of determining the number size distribution in the range of 7 nm-10 μm.
No nanoparticle release was measured during wet abrasion tests by the ELPI granulometer, only agglomerates in their matrix of around 100-300 nm were detected. This fact was confirmed by SEM observations.
In a third study, the nanoparticles release was measured in polycarbonate (PC), epoxy, and PA11 polymers containing carbon nanotubes (CNT) up to 4 % wt as fillers in polymers, paints, and fabrics are mechanically solicited [43].
Especially in the case of paints, hard abrasion tests were performed in non-optimized paint containing SiO_2 (12 nm) nanoparticles up to 35 %wt.
On one hand, an abrasive paper was rubbed on the sample to generate a dynamic friction on the surface of the sample. On the other hand, the Taber Abrader was employed for the abrasion tests in air. A bench test with several techniques disposed in a clean glove box was designed for these specific experiments – Fig. 29.

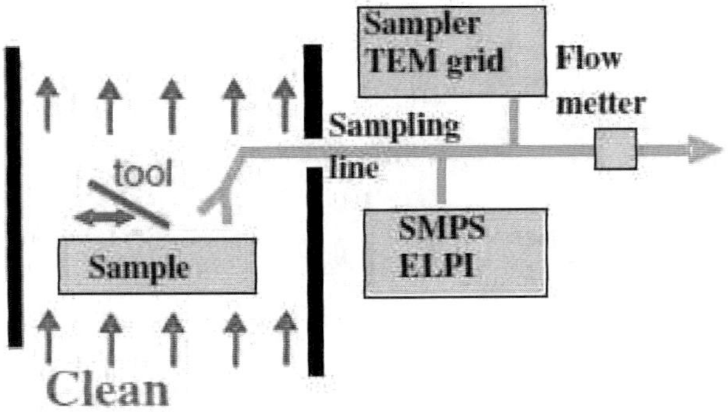

Figure 29. Schematic view of the bench test [43].

As in the previous examples, the debris finished in a sampler- a transmission electron microscopy (TEM) grid.
Firstly, an important release of nanopowders was detected in the nanometric range for polymers PA6, PC containing CNT. The release of CNT from epoxy was only observed when instantaneous shocks induce the CNT separation from the matrix thus overcoming the *Van der Waals* forces. A very low release of CNTs was only detected in samples showing the better dispersion of CNT. The lower CNT agglomerates population in the matrix reduced probability of aerosolizing and releasing.

Secondly, nanoparticles release when induced by scratching of two commercial fabrics made of PVC containing nanoparticles. An important release of nanoparticles was detected in the nanometric range indicating the removal efficiency of the metallic rake.
The release of nanoparticles by abrasion in different materials was tested. A very low release of micro- and sub-micronic agglomerates of nano-SiO_2 was measured in some products.
As a conclusion, the authors agreed that it is still necessary to pursue the studies regarding the nanoparticle release in order to assure *nanosafety*.

4. *Safe Nanomanufacturing 2-D & 3-D processes: Development of Safe Nanomanufacturing generic tools*

As it was illustrated in Fig. 1, the different *Safe Nanomanufacturing generic tools* were integrated into the existing technologies. The second aim was to keep the nanoparticles properties during the new different processes.
Finally, after the optimization of the actual processes it is was possible to obtain new advanced multifunctional properties.

4.1. Direct liquid recovery and granulation by CEA

In order to avoid the dissemination of nanoparticles after synthesis, a liquid recovery system was developed. The liquid recovery allows a significant reduction of nanoparticle emissions during powder recovery and bagging operations, thus increasing the safety for the workers. CEA was the responsible for developing the suitable equipment with the aim of collecting the TiO_2 nanoparticles in a liquid medium after a laser pyrolysis synthesis [44]. These nanoparticles were produced by decomposing a mixture of titanium isopropoxide (TIP) and Argon.
Indeed, it was demonstrated that is possible to recover and collect the nanoparticles in water medium- i.e. *one-step* process. The Liquid Recovery system is presented in the figure below.

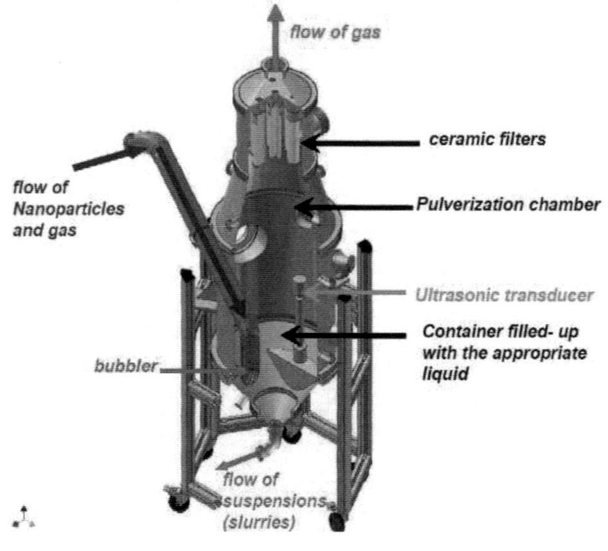

Figure 30. Liquid Recovery System.

4.2. Granulation, liquid phase functionalization & dispersion by GEA Niro/SU

The *SAPHIR* consortium centered the efforts on the development of versatile nanoparticles production routes allowing the integration into safe modules. The principal aim was to package the nanoparticles in its dry form or r*edispersible nanoparticles retaining their properties.* Dispersing the particles as a concentrated suspension reduces the health hazards associated with the respiratory intake [45].

Different approaches were developed in order partially-agglomerate the nanoparticles. New technologies, such as spray drying and freeze granulation, allow to achieve the safe production of free flowing (low dusty granules) from suspensions of nanoparticles.

GEA Niro focused the attention on developing safe processing routes for the production of non-dusty micron-sized granules so as to obtain a safe handling of nanopowders. The GEA Niro Test Centre in Copenhagen has an important experience in producing nanoparticles' granules employing different materials, i.e. SiC, SiO_2, TiO_2, Hydroxyapatite and carbon nanotubes.

With the aim of characterizing the performance, the Drying Kinetic Analyzer™ was employed, since this apparatus enables to determine drying behaviour. The GEA Niro conducted the investigation using a commercial Degussa Aeroxide TiO_2 P25 which possess a specific surface area of 50 m^2/g dispersed in an aqueous solution of Dispex N40 and 2-Octananole as foam reducer (37 wt. % of solid content).

The atomization of the slurry was conducted by means of a GEA Niro two fluid nozzle coupled with a Malvern Spraytec (Malvern Instruments) in order to characterize the kinetic measurements and morphology. Both devices are shown in Fig. 31.

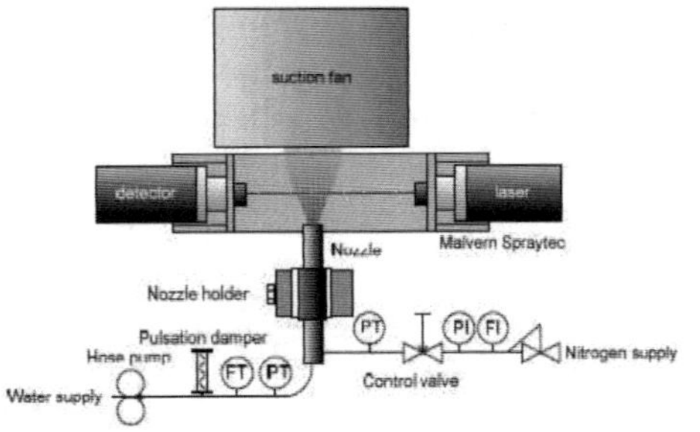

Figure 31. – Schematic illustration of the GEA Niro set-up. [46]

A Mobile Minor™ developed by the GEA Niro was used in four different configurations, as it is reported in the figure below.

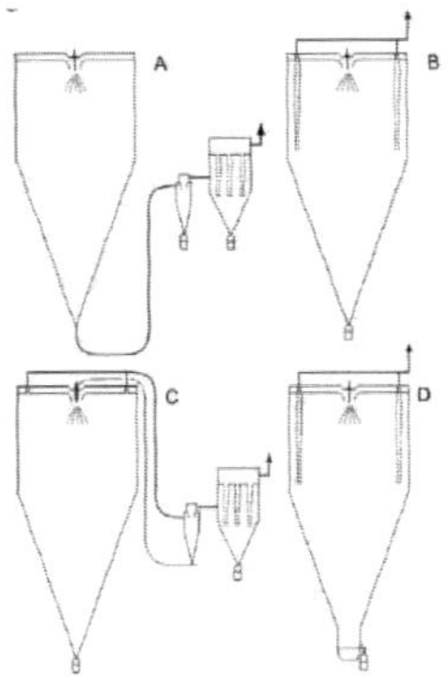

Figure 32. - Different configurations of the Mobile Minor: A: SD-MM configuration with product outlets from cyclone and a bag filter; B: SD-IF-MM configuration with bag filters placed inside the drying chamber and with product outlet under the chamber; C: SD configuration with fine return from the cyclone to the atomization zone and main powder outlet under the chamber & D: IFD-MM

configuration with bag filters placed inside the drying chamber and with powder outlet from the integrated fluid bed [46].

After spray drying, an important volume reduction of ≈ 45% was observed. By analyzing the spraying performance, it was observed that the droplet size is dependent on the liquid/atomization gas flow rate as it is shown in Fig. 33. The ratio between the atomization mass flow rate and the liquid mass flow rate (ALR) is frequently used to quantify the performance of two fluid nozzles.

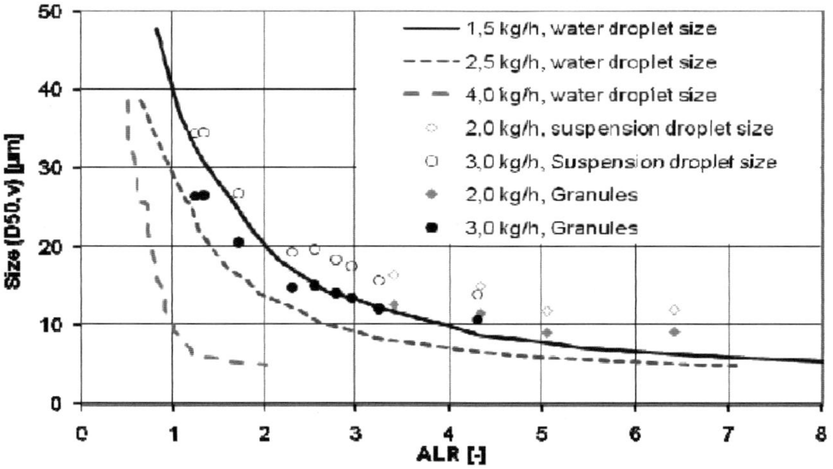

Figure 33. Suspension droplet sizes calculated from the measured granule sizes using the volumetric reduction at the A configuration determined by means of Drying Kinetic Analyzer™ [46].

In fact, different configurations allow to vary the granules' morphology from D_{50} equal 26-43 µm. From all the tested configurations, the B solution (SD-IF-MM) was the most suitable as it presents the lower shear and less system complexity.

In a second work, the same authors evaluated the effect of the different dispersants employing several techniques such as rheology, particle size and electrokinetic measurements [47]. Different commercial polymeric dispersing agents were tested and one commercial titania the P-25 is produced by hydrolysis of titanium tetrachloride in a H_2/O_2 flame at 985 °C [48,49].

From these dispersing agents, only the Dispex N40 (polycarboxylate) was able to stabilize the primary aggregates of the titania nanoparticles at 2.4 wt. % lower content and over a relatively large pH-range. The carboxylate has a great affinity with the titania surface [50]. These aggregates exhibited lower sizes if compared with the initial powder agglomeration.

It is well known that higher surface charge may present a stronger and longer range electrostatic repulsion which is related to the low viscosity and long-term stability of the TiO_2 suspensions at low ionic strength. The Dispex N40 is believed to create steric interactions due to the observed slurry behaviour.

The nanotitania suspension was dispersed by Dispex N40 in an aqueous medium and then spray dried. In order to quantify the influence on granulometry distribution after redispersion, the spray dried was submitted to ultrasonication. Indeed, a similar granulometry (in volume) was measured in the spray dried nanopowder. After redispersion the spray dried powder exhibited an average particle size of 160 nm if compared with the raw powder 180 nm – Fig. 34.

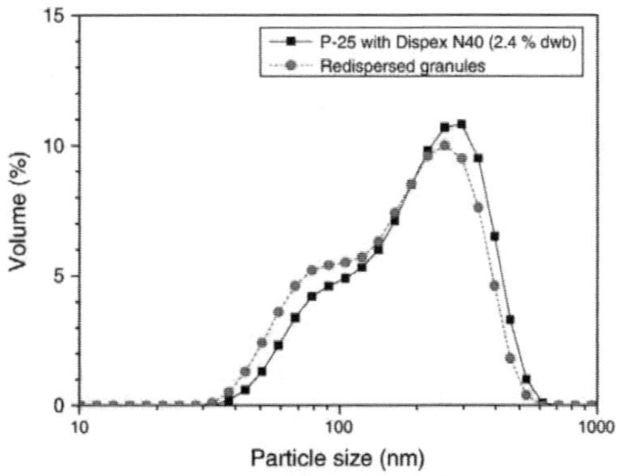

Figure 34. Particle size distribution (in volume) of the P-25 with Dispex N40 and the redispersed granules [47].

After spray drying, the powder was submitted to FTIR measurements. The FTIR spectra revealed that the powder contained polymeric chains imputed to the dispersing agent. This fact is favorable, since the dispersing agent could act as a binder even at dry state [51,52].

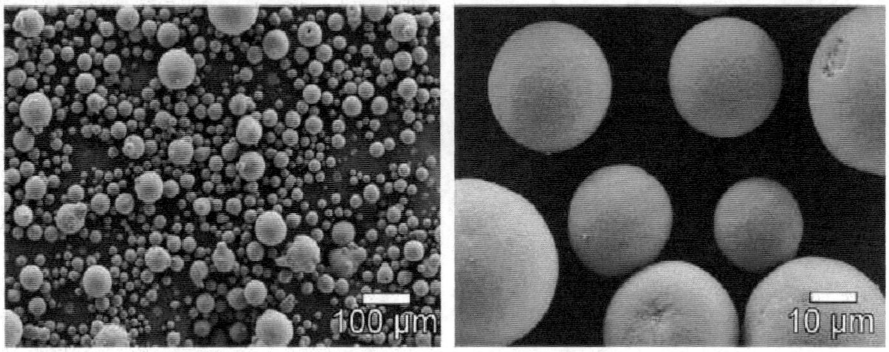

Figure 35. Spherical titania granulated powder after spray drying process [47].

The final granulated powder analyzed by SEM is shown in the figure above (Fig. 35).

4.3. Development of a complete monitoring strategy at workplace

The Safe monitored system was one of the most important objectives of the *SAPHIR* project. The main target was to control the size and the chemical composition during synthesis.

Two different on-line characterization techniques, i.e. scanning mobility particle sizer (SMPS 3936, TSI) and the long path multiwavelength extinction sensor (LPME, Wizard Zahoranski), were proposed as promising *Nanosafe safety tools* for the existing laser pyrolysis production lines at CEA Saclay [53]. The first solution consisted in an APTL coupled with an aerosol dilution system with the aim of allowing the SMPS measurements.

A fixed tunnel Venturi vacuum generator and a customized ejector type diluter were tested. A data acquisition system was especially developed for the LPME technique. The preliminary tests have been performed during the carbon nanoparticles production, in which the particle size distribution was analyzed by means of the LPME and the SMPS.

However, these *on-line measuring techniques* were not tested in the laser pyrolysis production line. Finally, this system was replaced by a nanoparticle impactor, provided by NanoMoudi, MSP. A thermophoretic precipitator was installed in order to collect the SiC aggregates/agglomerated particles on TEM grids with the aim of performing the *ex-situ* primary particle size distribution analysis. The analysis of the raw data of SiC nanoparticles produced in CEA laser pyrolysis line has been performed by employing fractal development theory.

In order to achieve this objective, a complete monitoring strategy was built up and tested by the partners on the CEA Pyrolysis platform, as shown in the following Fig. 36.

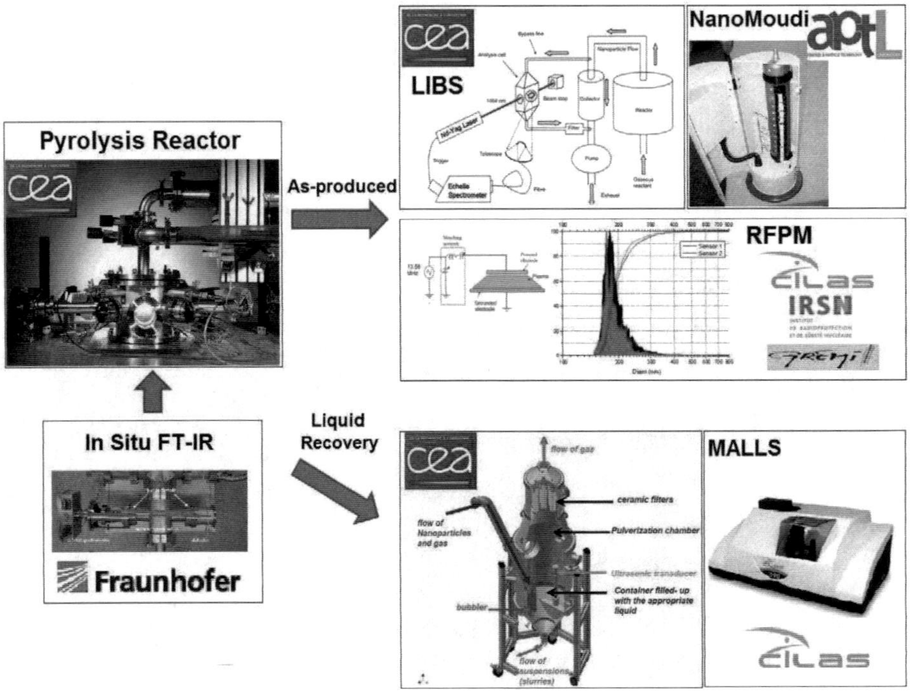

Figure 36. Integration of the different *Nanosafe safety tools* on the CEA Pyrolysis Platform.

As shown in Fig. 36, many techniques such as LIBS, RFPM, Nanomoudi and MALLS were employed.

The LIBS permits to monitor different vapour nanopowders synthesis (combustion, laser pyrolysis and plasma) and estimate the chemical composition at the *workplace*. For this purpose, temporarily-resolved emission spectroscopy has been successfully employed for controlling the Si/C_x stoichiometry as reviewed by *Amodeo et al.* [54].

LIBS method assures the real time stoichiometry control during the production of different materials such as solid, liquid, gas and aerosols [54,55,56,57,58,59,60]. In fact, the method has been already tested in the steel production [61,62]. LIBS has been integrated into the laser pyrolysis existing equipment at the French Atomic Energy Commission. The technique is thought to be a promising candidate for workplace surveillance in industries producing nanosized-based materials.

Laser pyrolysis is known to be an interesting technique for producing various carbide nanopowders [63]. This latter already developed for large production rates (more than 2 kg/hour for SiC) has been extensively studied since its development for the synthesis of SiC [64]. The gaseous process, using the decomposition of silane mixed with a gaseous carbon precursor as acetylene or ethylene, allows to control the final grain size and degree of crystallization [65].

The synthesis was analyzed by the temporarily-resolved emission spectroscopy. In Fig. 37, the experimental set-up is shown.

Journal of Physics: Conference Series **429** (2013) 012054 doi:10.1088/1742-6596/429/1/012054

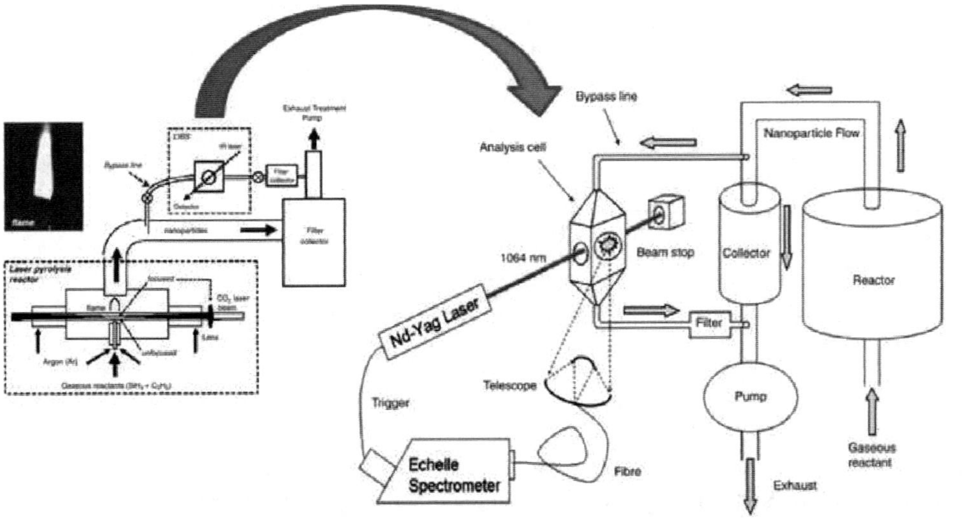

Figure 37. Schematic illustration of the LIBS unit integrated into the laser pyrolysis reactor [54].

As it is shown, an analysis cell mounted at a by-pass line of the up-stream permits to record the data. A Q-switched Nd-YAG laser at 20 Hz is focused on that cell in which a LIBS signal (originating from the plasma created in the argon-nanoparticles mixture) is collected by an *Echelle* spectrometer equipped with a fast intensified CCD camera. The elemental composition is then obtained through optical emission spectroscopy analysis.
The plasma at a local thermal equilibrium state (LTE) allows to calculate the stoichiometry S from the intensity and the density ratios recorded during the LIBS spectroscopy measurement. Essentially, as claimed by the authors, stoichiometry determination consists of measuring the ratio of the total number of atoms.
In Fig. 38, a typical recorded spectrum during the vaporization of SiC nanoparticles is presented (time delay 4.2 μs, ICCD gate width 2 μs). It is composed of a carbon line at 247.85 nm and six Si lines.

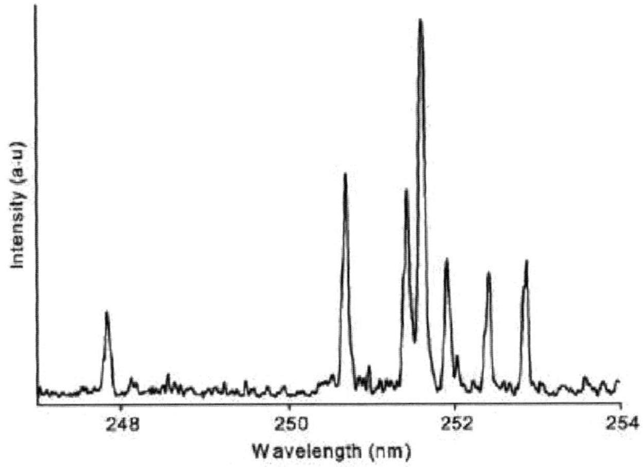

Figure 38. A typical recorded spectrum during SiC synthesis [54].

In their study, the authors focused the attention on the 247 nm neutral carbon line and the Si 288.157 nm, as they have a better signal to the background ratio if compared with the other detected transitions. The typical spectroscopic characteristics are presented in the following Table I.

Table I – From reference [54].

Wavelengths λ, transition probabilities $A_{ij} \times 10^7$, upper level energies E_i, lower level energies E_j, statistical weights of the upper states g_i, of the Si and C selected lines for stoichiometry calculation

Element	λ (nm)	$A_{ij} \cdot 10^7$ (s^{-1})	E_i (eV)	E_j (eV)	g_j	Z	T (K)
Si	288.16	18.9	5.082	0.780	3	16.87	14,800
C	247.85	3.4	7.684	2.684	3	11.51	14,800

Z are the partition functions of Si and C in the neutral state calculated for temperature value T of 14,800 K.

The electronic temperature was calculated by using Boltzmann plot method. It is built by using the following formula.

$$ln\left[\frac{I_{nm}\lambda_{nm}}{g_n A_{nm}}\right] = -\frac{E_n}{kT_e} + C \qquad (2)$$

n and m being the upper and lower levels of the selected transition respectively, I_{nm} the integrated intensity of the transition, λ_{nm} its wavelength, g_n statistical weight of the upper state level, A_{nm} transition probability, En energy of the upper state level, Te the electronic temperature, k Boltzmann constant, and C a constant. For this experience, it is assumed that the laser-induced plasmas are closer to the LTE condition [66]. Despite the good results, the technique presents some inconveniences related to the limitation for determining the electronic temperature.

Secondly, RFPM employs the low pressure and power plasma RF levitating the particles in plasma by the electrostatic force which counteracts the gravitational one, thus enabling the trapping of particles. The presence of nanoparticles within the plasma modifies the plasma properties allowing the size determination. The nanoparticles arranged in layers of around 10 μm depending on their density.

Thirdly, such as multi-angle laser light scattering (MALLS) was coupled to the laser pyrolysis. It allows to determine the size distribution by means of optical sensors that record the scattered light (in the range of 20-405 nm). A mathematical model determines the final particle size distribution.

Furthermore, a differential mobility spectrometer (DMS) and a mass aerodynamic diameter with a rotating micro-orifice impactor NanoMoudi (MSP) were also tested. On one hand, the DMS measures the particle size by quantifying the electrical mobility of the charge on the particle and its aerodynamic drag. On the other hand, the nano-range cascade impactor NanoMoudi (MSP) collects mass-based size fractioned particles due to their aerodynamic diameter-in a range equal to 10 μm down to 10 nm-.

Finally, the FT-IR in transmission and emission allowed the *in-situ* monitoring. A laser diode spectroscopy (LDS) was mounted in order to detect the critical impurities such as oxygen.

The most promising results are summarized in the following lines:

- Chemical composition & stoichiometric *in-situ* monitoring by Laser Induced Breakdown Spectroscopy (LIBS) technology.
- Particle size on-line real time monitoring by means of various techniques for gas and liquid. The Radio Frequency Plasma Monitoring (RFPM) technology exhibited the best results if compared with the multi-angle sampling of the scattered intensity (MALLS).
- Real-time Workplace monitoring was also investigated by using the LIBS technology. This technique allows to analyze the nanoparticles - in an air medium - in order to quantify the *risk* assessment of *particles inhalation-* .

These *Nanosafe-by-Design in-situ* safety tools allow to evaluate the inherent risks of nanopowders production.

4.4. Safe Nanomanufacturing processes

Safe processing techniques were integrated into the existing processes. As a result, the *safe processed nanopowders* – presented in the section before- were employed with the aim of producing new nanocomposites. The principal goal was to demonstrate the production feasibility and the improved properties of the nanostructured materials if compared with conventional materials.
These *Safe Nanomanufacturing processes* were divided into the 3-D (bulk) high temperature and 2-D (surface) processes according their architectures.

4.4.1. 3-D (Bulk) Processes

4.4.1.1. Monolithic ceramic-based matrix nanocomposites

SiC nanopowders synthesized by laser pyrolysis were consolidated by *Spark Plasma Sintering*. Grain sliding is proposed to be responsible for the fast densification due to the strain rate of around 10^{-3} s^{-1}. Rapid grain growth is observed in a narrow temperature range, which can occur even in a porous body with a relative density below 80%. The early grain growth inhibits grain sliding which allows the creation of large intergranular pores, thus impeding further densification.

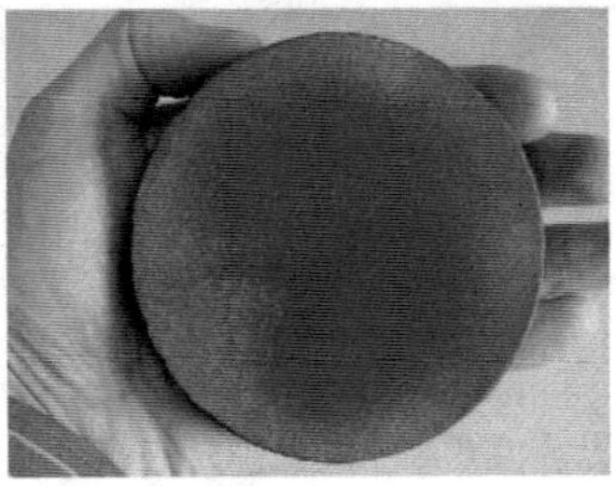

Figure 39. SiC pellet produced by *Spark Plasma Sintering.*

A large nano-SiC pellet of 100 mm. was produced keeping a nanometric and homogeneous microstructure, as shown in the SEM micrographs (Fig. 40). Scanning Electron Microscopy (SEM) revealed that the microstructure is mainly composed of grains of around 200-400 nm whereas a second fraction of grains lower than 50 nm is present. The porosity seems to be limited and their size is < 100 nm (closed porosity). However, X-Ray Diffraction confirmed the presence only of the β-SiC phase in good agreement with the absence of lengthened grains.

Figure 40. SEM micrographs of the final pellet.

Pure SiC nanoparticles produced by laser pyrolysis at pilot scale were employed. Prior sintering, the pellets were compacted by Cold Isostatic Pressing (CIP) in order to increase the green density up to 45% of theoretical density (TD).
The optimized conditions for the sintering were at temperature of 1890°C, an applied force of 100 MPa using a dwell time of 5 minutes. The final density reached was 98% of T.D. [67].
The same SiC nanopowders were employed in order to produce SiC nanocomposites reinforced with *Cfs (carbon fibers). Electrophoretic infiltration* (EPI) coupled with *Spark Plasma Sintering* were employed in order to evaluate an alternative processing route. This technology allows to produce *Cf*/SiC nanocomposites with lower associated costs if compared with other techniques such as the *Chemical Vapour Infiltration* (CVI).
The EPI parameters were evaluated with nanopowders in the system SiCB. The excellent SiCB nanopowders' suspensions let an acceptable infiltration in the green state. Finally, SPS parameters were determined for each nature of the tested fibres. An example of the final microstructure is presented in Fig. 41.

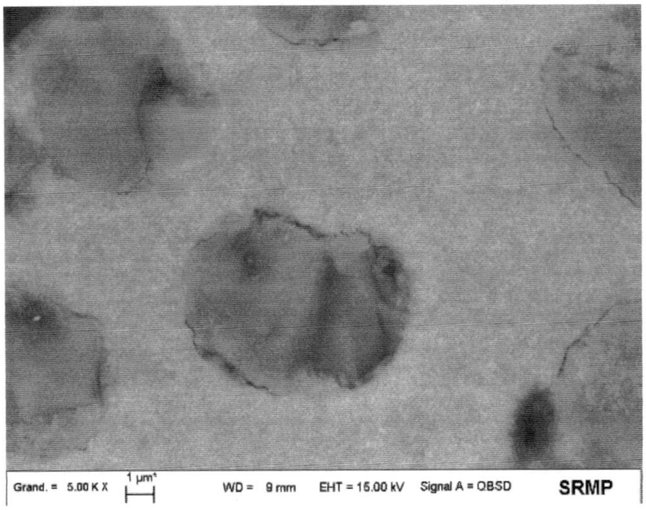

Figure 41. SEM micrograph of the Cf embedded in a nano-SiC matrix.

The dynamic grain sliding was studied in this project by *Herrmann et al.* [68]. Two nitride nanopowders α- and β-Si_3N_4 nanopowders' mixtures were prepared by mechanical alloying and by radio frequency inductively coupled nitrogen plasma. The nanopowders were subsequently sintered by *Spark Plasma Sintering*. The consolidation parameters were established in order to avoid the reprecipitation which induces the grain coarsening. The aim of the study was to achieve the dynamic grain sliding by a liquid forming additive (the β-Si_3N_4 nanoparticles are coated by a glassy phase) under mechanical pressure. The authors claimed that is possible to sinter at relatively low temperature of 1400°C for 5 min with an applied force of 75 MPa.
The final average grain size was around 52 ± 10 nm which lead to obtain interesting hardness Hv_{10} values of around 18.7 GPa.
A third study developed during the project, concerned the processing of Ca-α-SiAlON ceramics prepared by mechanical alloying from Si_3N_4, AlN and CaH_2 precursors [69]. Samples were consolidated by hot pressing in a N_2 atmosphere. In their work, the researchers varied the CaH_2 content in order to obtain different microstructures and Ca-α- and Ca-β-SiAlON contents. By increasing the CaH_2 it is possible to obtain more quantity of elongated Ca-α-SiAlON grains.

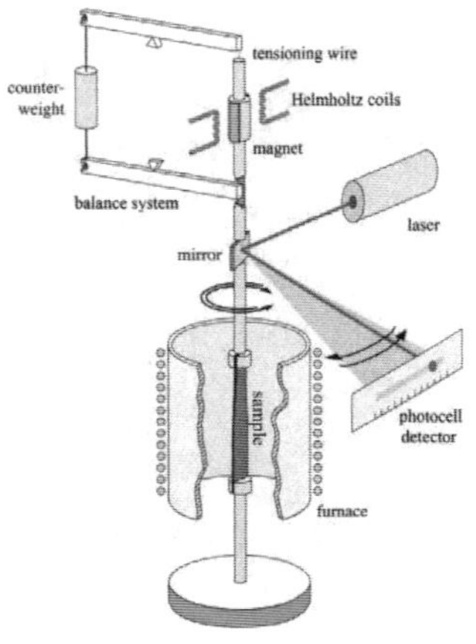

Figure 42. A schematic illustration of the forced torsion pendulum [69].

For the mechanical characterization, mechanical spectroscopy measurements were carried out in an inverted forced torsion pendulum working in subresonant mode. In Fig. 42, the schematic illustration of the device is shown.
In this study, by applying an oscillation of 8 MPa at 1 Hz, the sample deformation (torsion angle) by a displacement of a laser beam spot on an optical photocell was measured. The excitation signal is measured by means of a frequency response analyzer. The response parameters phase lag (tan (φ) or mechanical loss) and the ratio between the excitation signal and the strain signal (proportional to the shear modulus G) is recorded.
As a result, the authors demonstrated the importance of having an equiaxed microstructure since it increases the internal friction thus impeding the displacement.
In contrast, at high temperatures (>1400 K) an α-relaxation due to the formation of a glassy phase, usually found at grain boundaries in SiAlON. The maximum restoring force intensity (that counteracts the applied load) depends on the glassy phase as well as on the grain elasticity.
In this context, the specimen consisting of equiaxed elongated grains exhibited the lower peak height in comparison with a sample of lower glassy phase content.

4.4.1.2. MWCNTs reinforced polymeric, metal and ceramic nanocomposites

CNTs mechanical and electrical properties are remarkable in comparison with other known materials. These materials are envisaged to reinforce polymers with the aim of increasing their strength as reported by several authors [70]. Composite samples containing acrylonitrile-butadiene-styrene ABS matrix reinforced by Arkema MWCNTs were produced by blending and injection molding at the *Centro Ricerche Plast-Optica*. Three carbon contents i.e. 3, 7 and 15 wt. % were employed [71,72]. A micrograph corresponding to the ABS 15 wt. % MWCNTs is presented in Fig. 43.
In their study, mechanical spectroscopy was performed by means of the device presented in section *monolithic ceramic-based matrix nanocomposites* at temperatures varying from 100 to 430 K.
Mari et al. identified two relaxations at 189 and 381 K (1 Hz) ascribed to the glass transition of the polybutadiene and the styrene–acrylonitrile (SAN) phases, which are believed to provoke a first β-transition involving the movement of side groups and a second α-transition related to the chain sliding. The interaction of the MWCNTs with the SAN phase enables the softening of the polymer at higher temperatures. Furthermore, no elastic modulus reinforcing effect was measured - probably due to the method. Indeed, during the tests only shear modulus is measured. A processing inconvenience regarding the reduction of the adhesion between the two constituents could explain this phenomenon.

Figure 43. SEM micrograph of the ABS 15 wt. % MWCNTs microstructure [71].

MWCNTs reinforced magnesium and magnesium- 23.5 wt. % nickel nanocomposites were processed by mechanical alloying [73]. The consolidation was achieved by means of hot pressing at 600°C for the magnesium and 430°C under a pressure 50 MPa for 30 min. Prior to mechanical spectroscopy, samples were annealed in a 0.4 MPa hydrogen atmosphere at 620 K thinking about the final application of the hydrogen storage capability. These samples were compared with the pure magnesium.

As in the previous case, *Schaller et al.* indexed two relaxation peaks at ≈190 and ≈330 K (≈6 kHz) in the magnesium- 23.5 wt. % nickel. The authors reported that H atoms destroy the crystalline order in the Mg_2Ni phase of the hexagonal structures, precipitating as Mg_2NiH_4 hydrides. Hydrogen occupies the tetrahedral interstitial sites within the hexagonal structure. This fact is increased in 25 orders in magnesium- 23.5 wt. % sample containing magnesium 23.5 wt. %. Nevertheless, no hydrogenation was detected in a sample which contained pure magnesium.

Finally, the study was extended to production of MWCNTs reinforced nanoceramics. For this purpose, nanosized ZrO_2 (Y-stabilized) with 0, 0.5, 1.5, 3 and 5 wt% CNT contents were consolidated by SPS. The characterization was conducted by means of the mechanical spectroscopy.

The measurements were performed at 10^{-3} Pa either as a function of temperature up to 1600 K at a fixed frequency or as a function of frequency (10^{-4} to 10 Hz) at a fixed temperature.

Schaller et al. processed polycrystalline 3Y-TZP mixed either with CNTs or SiC_w via attrition milling with the aim of producing ceramic nanocomposites [74]. Samples compacted by cold uniaxial pressing at 100 MPa were naturally sintered in argon atmosphere.

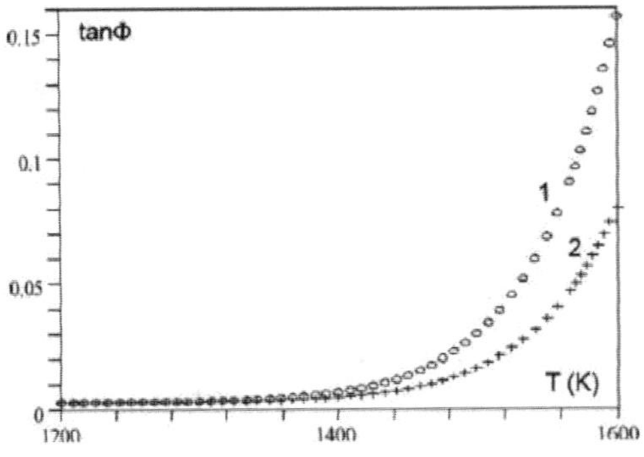

Figure 44. Mechanical-loss (tan φ) spectra for two zirconia samples: pure (curve 1) and reinforced with carbon nanotubes (curve 2) [74].

In Fig. 44, shows the theoretical mechanical-loss spectra of the pure zirconia and the composite. Indeed, an exponential increase of the mechanical loss (tan φ) is observed when the temperature increases in both samples. It was corroborated that the MWCNTs improve the restoring forces, since the pinning effect on the GB sliding was measured. This allows to obtain a better final creep resistance (in factor 2) in the sample containing 1.5 wt. % MWCNTs, as less angular displacements were measured.

The higher creep resistance was attributed to the presence of MWCNTs pinning grain boundaries at high temperatures.

Moreover, *Schaller et al.* produced 3Y-TZP/MWCNTs composites varying as in the previous case the MWCNTs content [75]. In contrast, in the second study, samples were consolidated by *Spark Plasma Sintering*. The different microstructures of the 3Y-TZP/MWCNTs composites are shown in Fig. 45.

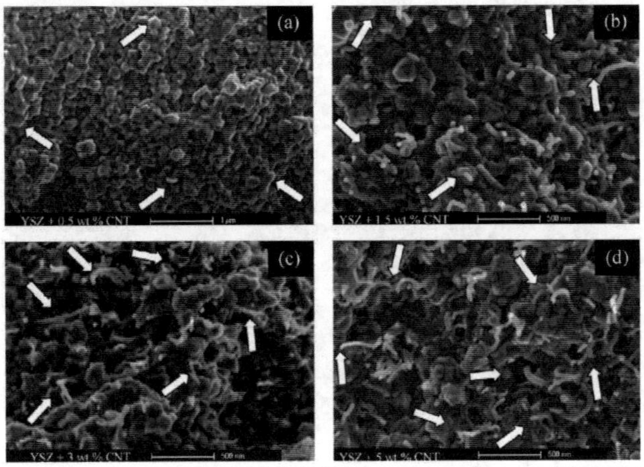

Figure 45. SEM micrographs of the 3Y-TZP/MWCNTs composites: (a) 0.5, (b) 1.5, (c) 3 and 5 wt. % (the arrows indicated the MWCNTs position) [74].

The authors measured an important increase of toughness K_{Ic} by means of single edge V-notch bean (SEVNB), being 6.1 ± 0.15 for monolithic zirconia and 11 MPa.m$^{1/2}$ for 5 wt. % CNT reinforced zirconia composite. Furthermore, a significant enhancement of electrical conductivity (14 order of magnitude) is observed with 5 wt.% MWCNT addition due to the percolation network through the insulating zirconia matrix [76].

As reported in the previous study, the increasing of MWCNTs content allows to enhance the mechanical resistance at high temperatures as presented in Fig. 46.

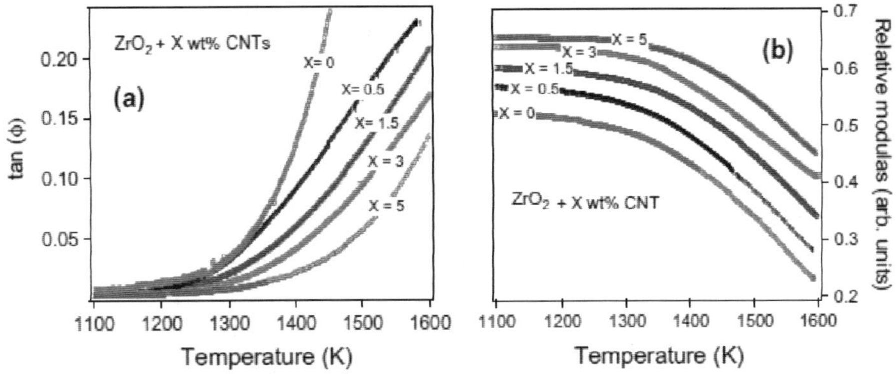

Figure 46. Mechanical-loss (tan φ) spectra and the dynamic shear modulus G/G_{ref} as a function of the MWCNTs content [76].

4.4.1.3. Aluminium-based matrix composites

The principal aim was to develop nanoparticle reinforced Al-based Metal Matrix Composite. In this context, Al2009-15%SiC (nano) powders have been milled and consolidated by SPS, HIP and extrusion.
The latter was selected as the forming preindustrial demonstrators. The final bars exhibited a dimension of about 13 mm (diameter) and 1500 mm (length) – Fig. 47. The bars were consolidated at CEA and Mecachrome.

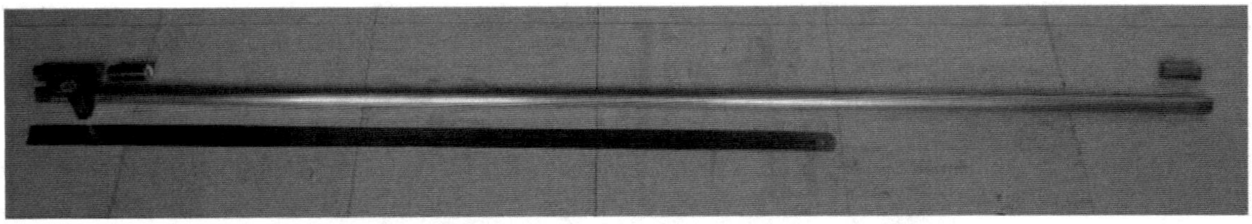

Figure 47. Al2009-15%SiC preindustrial extruded bar.

The mechanical properties of the bars were measured by the final user (EADS). The results were compared with the commercially available Al + 15% micro-SiC (SiC mean size 6 μm). A strong improvement of the yield and tensile strength were measured. Indeed, the yield strength (R 0.2) was increased in >2 order of magnitude from 350 to 747 MPa. Nevertheless, the strain to failure was reduced taking into account the final applications' tolerances.
On the contrary, the fatigue behavior was improved by employing the nano-SiC dispersion in the Al matrix. As shown in Fig. 48, the Al nano-SiC sample exhibited higher fatigue strength if compared to a commercial micro-MMC (in a factor 1.6).

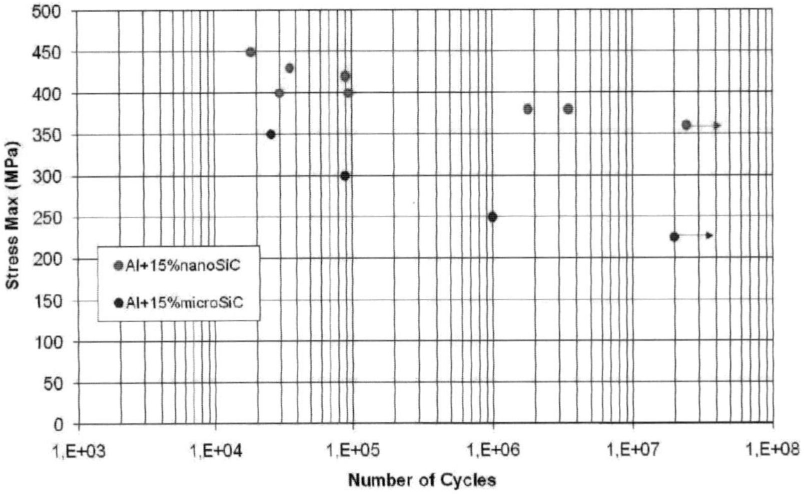

Figure 48. Comparison of the S-N curves between the Al micro/nano SiC samples.

4.4.1.4. Direct growth of aligned MWCNTs on substrates without manipulation

Aligned multi-walled carbon nanotubes (MWCNT) carpets can be successfully synthesized on both carbon fibres and any metal substrates compatible with the injection-CVD process. This injection-CVD process involves two stages, including fibre pretreatment by depositing a SiO_2-based sub-layer from an organometallic precursor followed by CNT growth from toluene/ferrocene precursor mixture as illustrated in Fig. 49 (b).

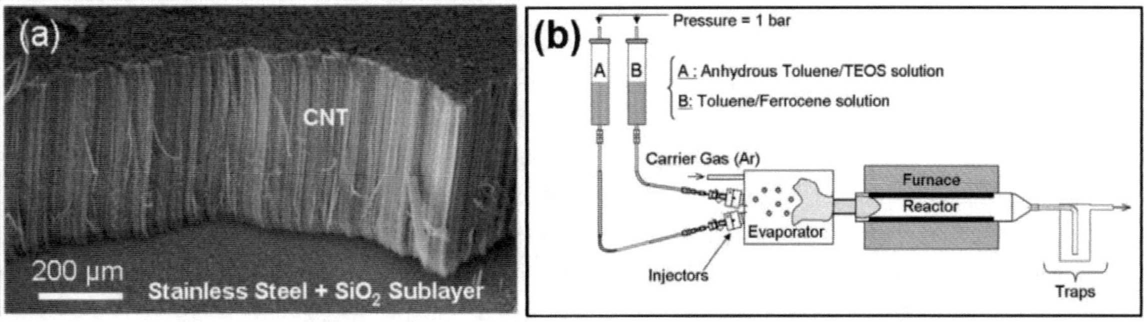

Figure 49. (a) FEG-SEM micrograph of CNTs grown on SiO_2-based coated stainless steel substrate & (b) injection-CVD experimental set-up [77].

Nowadays, this technique was *up-scaled* into an injection-CVD demonstration for the safe production of aligned carbon nanotubes in an A4 format Carbon Fiber Fabric.

4.4.1.5. Complete integrated PEM fuel cell bipolar plate

ZBT developed an industrial scale-up production of highly filled carbon compounds for bipolar plate materials at high temperatures (up to 200°C) PEM Fuel Cells – Fig. 50 (a) [78]. The bipolar plate production is achieved by compression and injection moulding lead to achieve a substantially increase of the performances in terms of electrical conductivity. The process of compression moulding enables a higher filling load due to the perfect alignment of graphite and spherical flakes or the carbon nanotubes within an injection moulded plate - Kraus Maffei KM500- .
A handling of the carbon based compounds was assured by feeding the fillers in an enclosed environment via bag-dumpers. The same quality of compound material could be achieved on a twin screw extruder. A fully robotized plate removal from injection moulding was developed in order to prevent the operator exposition to CNTs.

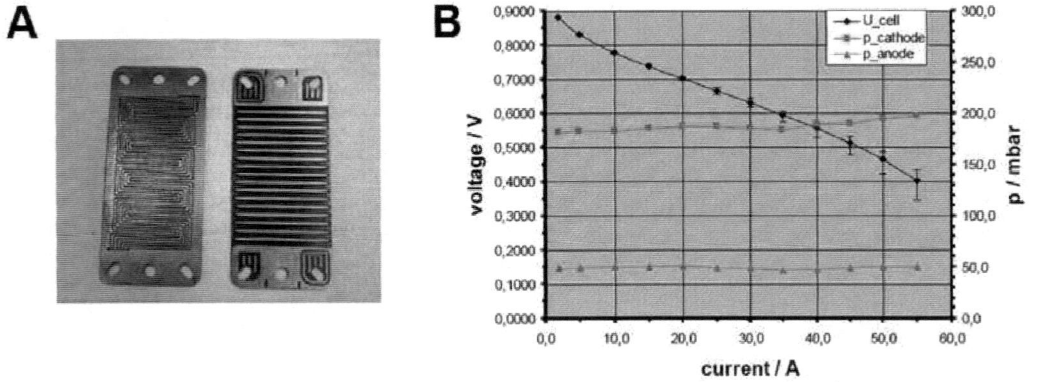

Figure 50. (a) Injection molded CNT-based bipolar plates & (b) Polarization curve of the 5-cell PEM stack at 70°C under a cathodic/anodic pressure of 1.2/1.04 bars.

A 5-cells stack PEM cell containing CNTs was assembled. After testing in humidified hydrogen and air, an interesting behaviour was observed. The polarization curves at 70 °C are presented in Fig. 50 (b). The power output was about 125 W at 0.5 V single cell voltage and a current of 50 A.

Journal of Physics: Conference Series **429** (2013) 012054 doi:10.1088/1742-6596/429/1/012054

As a conclusion, among all the carbon based nanoparticles, the CNTs exhibited the best operational performances and a better rheological behaviour. The latter leads to facilitate the fabrication.

4.4.2. 2-D (Surface) Processes

4.4.2.1. Photocatalytic activity of the spray dried nanopowders

Initially, the main purpose was to develop technical solutions for the industrial demonstration of functionalized flat products, like antibacterial or self cleaning panels and/or anti-corrosion panels. The different coatings were evaluated by the industrial partner Arcelor.
On one hand, the first aim was to integrate the spray dried nanoparticles into the existing solution as they allow the fabrication of transparent coatings. On the other hand, they also permit to employ crystalline phases such as TiO_2 anatase which possesses interesting photocatalytic activity. The latter is difficult to obtain at low temperatures.
Moreover, the nanoparticles should have a direct contact with the air molecules as well as to be cohesive, adherent, flexible and reliable. In addition, the coatings require a second phase in order to serve as a matrix, without avoiding the contact between TiO_2 and atmosphere.
The application of the photocatalytic coatings was for indoor air purification devices as illustrated in Fig. 51.

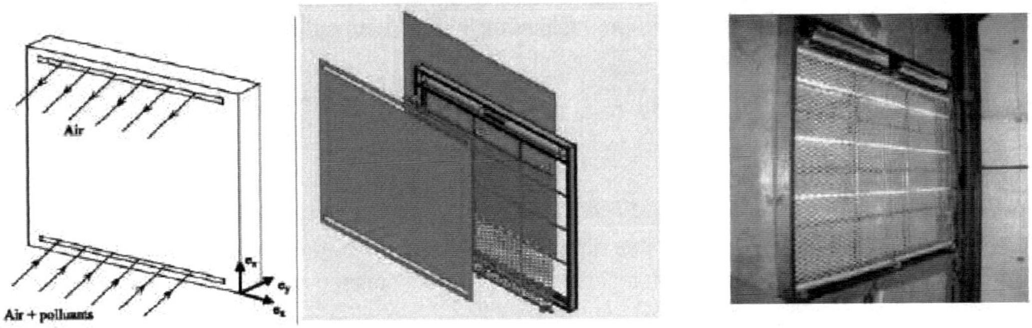

Figure 51. Air purification wall.

The Fraunhofer IWS tested different thermal spray technologies which showed interesting photocatalytic characteristics. In fact, the suspension deposited by HVOF kept the active TiO_2 anatase phase despite the high temperatures involved [79,80]. The researchers from Fraunhofer IWS employed a TiO_2 Degussa P25 composed of mainly anatase (80 vol. %, 25 nm) and rutile. The powder was spray dried by using the IFD Minor Plant (Niro A/S, Denmark) in an $Ar–H_2$ and $Ar–H_2–He$ plasma.
Suspensions were injected into the Atmospheric Plasma-Sprayed (APS) and High Velocity Oxy-fuel (HVOF) flame. In fact, it was demonstrated that it is possible to limit the surface overheating at 150°C on substrates such as hot dip galvanized steel.
Moreover, the HVOF technology was adapted for the desired applications with a robotized arm. It is well known, that this technique exhibits better results in *3-D* pieces.
IWS employed different substrates such as stainless steel and hot dip galvanized steel in which it was subsequently measured a high/constant photo catalytic activity. In their study, the photocatalytic efficiency was evaluated by means of decolouration of the pink dye Rhodamine B (performed at LABEIN Tecnalia) and degradation of gaseous acetaldehyde (Arcelor Mittal Liège R&D).

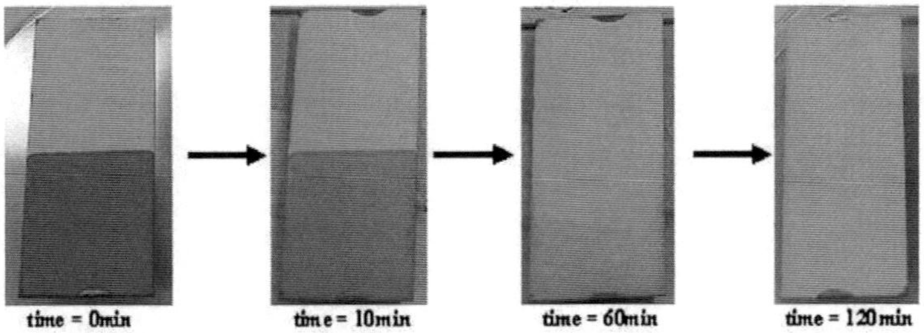

Figure 52. Photocatalytic decolouration of pink dye Rhodamine B on TiO_2 coating deposited by APS as a function of the irradiation time [79].

As shown in Fig. 52, during UV irradiation the pink colour continuously decreases in intensity. After 120 min, the UV light leads to fully deteriorate the Rhodamine B in coating deposited by APS. In contrast, uncoated steel completed the degradation without UV irradiation after 96 hours. For the sake of clarity, the comparison between the different titanium oxide sprayed coatings regarding the photocatalytic decolouration of the pink dye Rhodamine B is presented in Fig. 54.

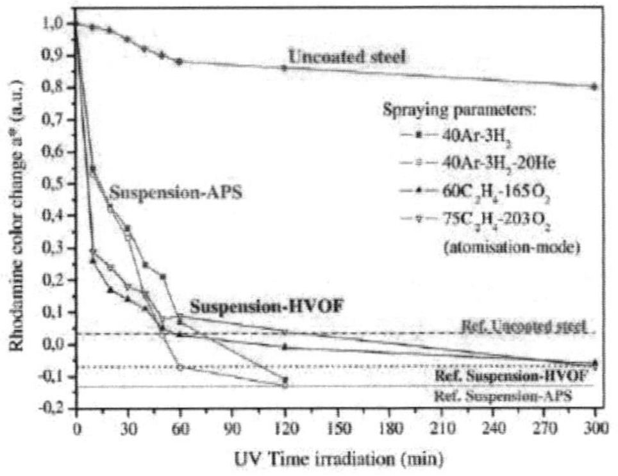

Figure 53. Normalised data for photocatalytic degradation of Rhodamine B as a function of UV irradiation time [79].

A second photocatalytic test revealed new encouraging results – Fig. 54. Indeed, the samples processed by HVOF using non spray dried powder showed a very low degradation (10%) of the pollutant. In contrast, spray dried samples exhibited a final degradation of 140 min and 80 min in the case of HVOF and APS samples, respectively.

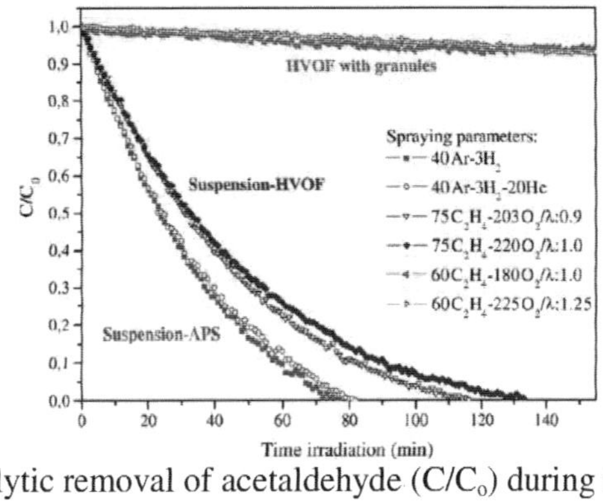

Figure 54. Photocatalytic removal of acetaldehyde (C/C_0) during the UV irradiation [79].

This phenomenon is attributed to the higher anatase content in the spray dried samples (>65 vol.%) if compared with the non-spray dried HVOF samples (15.9 vol.%).
But if the deformability is promising, the adhesion of the coatings is still low, even if it can be improved by an optimization of the surface temperature or by pretreatment of the substrate.

5. Nanosafe-by-design processes

The *Nanosafe-by-Design processes* concern the development of new *2-D* and *3-D* nanostructured materials. The main purpose is to generate and integrate *in-situ one-step* nano-functional objects, avoiding any risk of dispersal into the environment. In addition, it leads to *redesign* the current industrial value chains since some processing steps could be removed and/or skipped.
In the following lines, the *Nanosafe-by design processes* developed by CEA and other examples found in literature are presented.

5.1. 2-D Surface Processes

5.1.1. Nanostructured thin films processed by physical vapour deposition

The production of nanostructured superlattice hard coatings with different periods could be easily up-scaled into industrial cathodic arc reactors only by varying the substrate rotation as shown by *Ducros et al.* [81,82]. In the following Fig. 55, the TEM cross section of these coatings is shown.

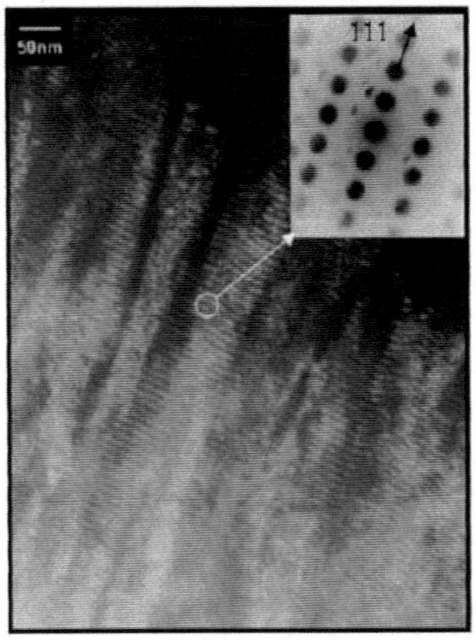

Figure 55. TEM images of the cross-section of TiN/AlTiN [81].

The authors measured important hardness values ≈ 55 GPa and good elastic properties. Furthermore, authors demonstrated that the nanocoatings on cutting tools were the optimal solution for reducing the forces during machining.
Nanolayered superlattice coatings are well known due to their enhanced mechanical properties if compared with single-phase coatings. These are often called the fourth generation of hard coatings when hardness is >40 GPa [83,84].
Many authors concentrated the studies around the superlattice coatings after the publication of *Koehler* [85]. The author suggested that by creating an alternated architecture, it would be possible to create an efficient dislocation barrier. The second condition is to decrease layer thicknesses with the aim of avoiding the dislocation generation (≈ 100 atomic layers or less) and the dislocation pile-up.

The strength improvement is explained by applying the formula $Q=(G_A-G_B)/(G_A+G_B)$ where G_A and G_B are the shear moduli of the superlattice single constituents [86]. The second explanation of the inhibition of dislocation is attributed to the layer coherency strains due to the increase of interface area which allows to glide throughout the layers [87].
Another interesting example of multilayered nanocomposites was given by *Kamath et al.* [88]. TiAlCN/VCN multilayered coatings were prepared combining the one operating in high-power impulse magnetron sputtering (HIPIMS) with the other in unbalance magnetron sputtering (UBMS) mode. These coatings possess low friction (μ=0.42) and high hardness ($N_{K\ 0.25}$=2800), which enables an increase of the tool life 4 times more, when aerospace grade Al 7010–T7651 alloy is dry machined. The XTEM microstructural observation (Fig. 56) revealed that the period was 2.2 nm.

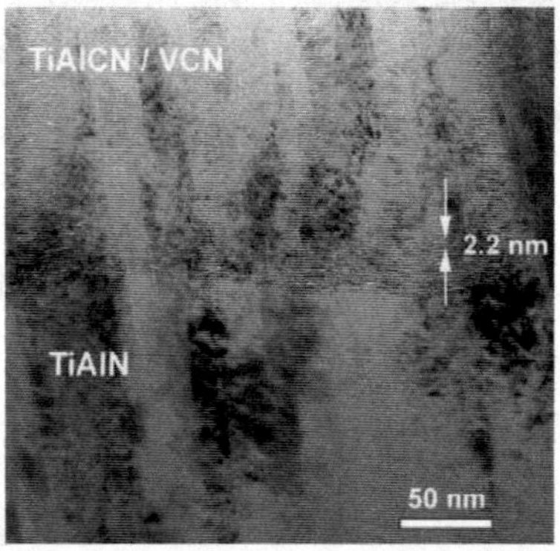

Figure 56. XTEM cross section of TiAlCN/VCN [88].

5.1.2. Antibacterial titania-based nanostructured thin films processed by DLI-MOCVD

TiO_2–Cu and TiO_2-Ag nanocomposite films were grown by pulsed direct liquid injection-metalorganic chemical vapor deposition (DLI-MOCVD) on different substrates at 683 K [89,90]. The aim of the study was to develop new antibacterial surfaces. As a matter of fact, it is well known that the addition of metals (Cu, Ag, Au, Co, Fe, etc.) to anatase TiO_2 improve the photocatalytic activity [91,92,93].
The two metalorganic precursors as Cu/Ag and Ti sources were employed in a conventional MOCVD reactor: copper bis (2,2,6,6-tetramethyl-3,5-heptadionate) ($Cu(TMHD)_2$) or (silver pivalate $(CH_3)_3CC(O)OAg$ (namely AgPiv) and titanium tetra-iso-propoxide (TTIP). The precursors were introduced through injectors to a flash vaporization chamber and then the reactive vapors followed the deposition zone under low pressure using N_2. A schematic view of the reactor is shown in Fig. 57.

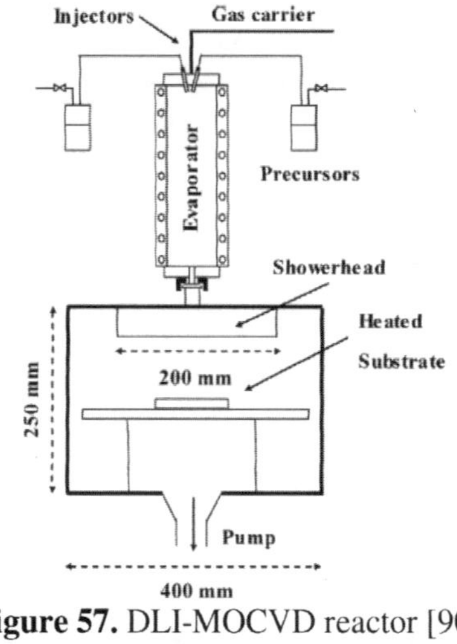

Figure 57. DLI-MOCVD reactor [90].

Journal of Physics: Conference Series **429** (2013) 012054 doi:10.1088/1742-6596/429/1/012054

After deposition, the TEM microstructural observation revealed that the microstructure is composed of a TiO_2 matrix with crystallites of about 50-150 nm embedding Ag or Cu dispersoids (indicated with an arrow in the image), as is illustrated in Fig. 58.

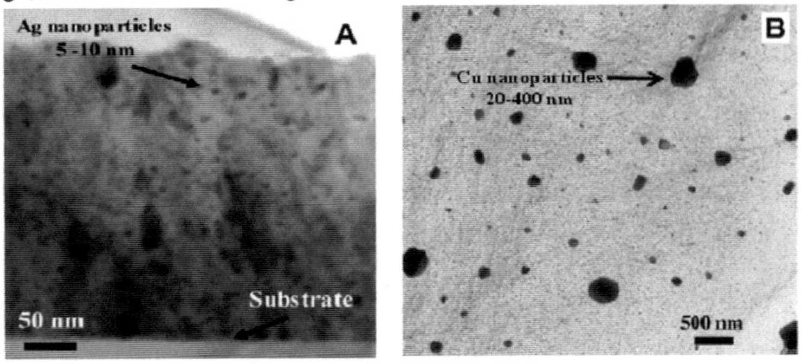

Figure 58. Bright field TEM cross-section image of (a) TiO_2–Ag & (b) TiO_2-Cu [89,90].

In their work, the authors were capable to alter the metal content inside the TiO_2 matrix as a function of the precursor content. This fact indicated that is possible to control the anti-bacterial activity without UV irradiation only varying the metal content determined by secondary ion mass spectroscopy (SIMS) as shown in the following Fig. 59.

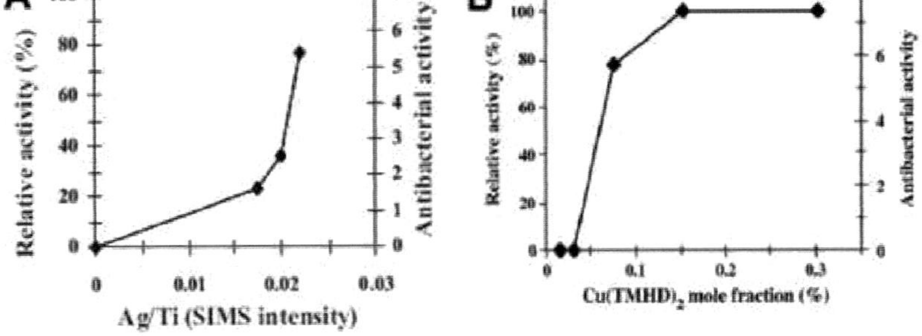

Figure 59. Anti-bacterial activities against Gram-positive *Staphylococcus aureus* as a function of the metal content: (a) TiO_2–Cu & (b) TiO_2-Ag [89,90].

In the case of TiO_2-Cu, the authors reported a slight decrease from 100% to 70 % of the bactericidal property which only changes after 5 months of storage.
A second advantage of TiO_2-Ag is the photo-induced self-cleaning behaviour. With the aim of evaluating the photocatalytic behaviour, the degradation of Orange G in aqueous solution (10^{-5} mol L^{-1}) under UV irradiation (365 nm) was studied. In Fig. 60, the photocatalytic behaviour is reported. As a conclusion, in this particular case the silver content above >0.1 substantially affected the photocatalytic activation of the films.

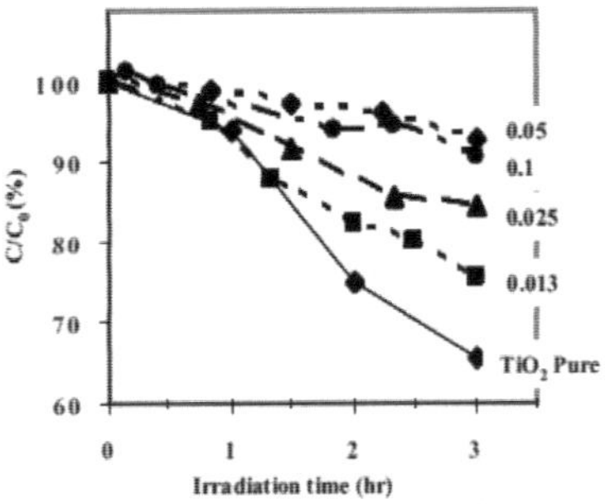

Figure 60. Photocatalytic decomposition kinetics of Orange G under UV light [90].

5.2. Hybridization of different the physical - and chemical- deposition methods

5.2.1. Cathodic arc coupled with plasma enhanced chemical vapour deposition

The hybridization of the physical (PVD) and a plasma assisted chemical vapour processes (PECVD) lead to produce in *one step* TiN-α-Si_3N_4 nanocomposites, as reported by *Bendavid et al.* [94]. The experimental set up is shown in the following Fig. 61.

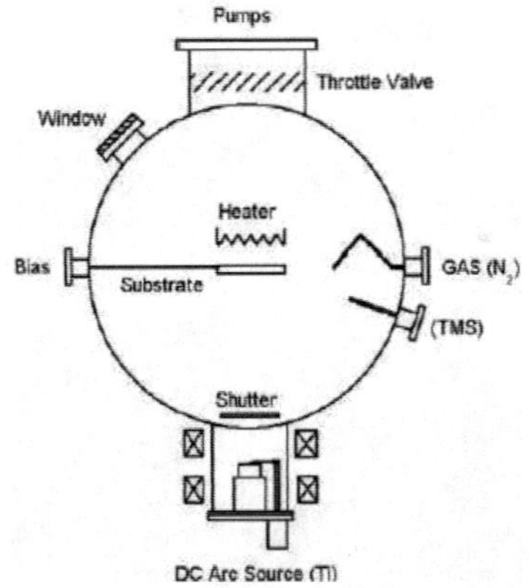

Figure 61. Experimental set-up of the PVD and CVD hybridation technologies [94].

The nanocomposite TiN-Si_3N_4 thin films are formed by the reactive deposition of Ti produced from a classical cathodic arc and the Si from a liquid precursor, the tetramethylsilane (TMS). This technique consists of combining both the cathodic arc and chemical vapour deposition technologies employing a nitrogen atmosphere.

In their study, the researchers varied the TMS flow rate on the structure which lead to obtain different mechanical properties. The film was composed of TiN crystallites surrounded by amorphous α-Si_3N_4. By increasing TMS flow, a decrease in the TiN crystalline size was measured passing from 33 nm to 4 nm - Fig. 62 (a). Indeed, the Si content enables to obtain different hardness values, being 41 GPa when the Si content was 5%, corresponding to a pressure of TMS of 0.8 Pa.

Hardness enhancement was ascribed to the crystallite size evolution when the presence of the amorphous α-Si_3N_4 phase is increased Fig. 62 (b). This increase is the result of the creation of additional interfaces between both phases which hinders dislocation movement allowing the energy dissipation and crack deflection.

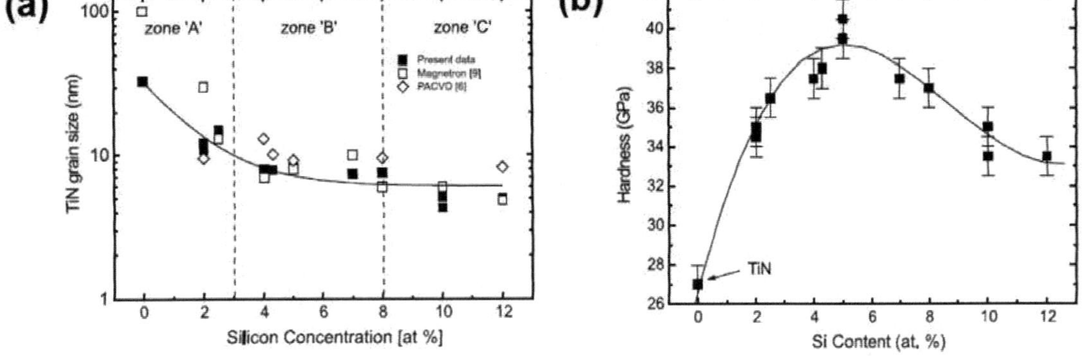

Figure 62. (a) Hardness and (b) Grain size evolution as a function of the Si content in TiN-Si_3N_4 films [94].

The decreasing from the maximum value - when the Si content evolves - is ascribed by the authors to some stress relaxation.
The microstructural morphology was analyzed by high resolution transmission electron microscopy (HRTEM) analysis – Fig. 63, which confirmed the existence of well orientated nc-TiN/α-Si_3N_4 (≈10 nm grains) microstructure in the case of the sample containing 5 at. % Si.

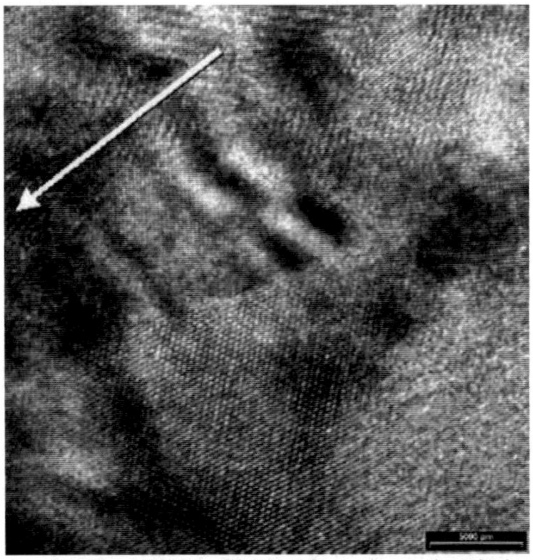

Figure 63. HRTEM micrograph of nc-TiN/α-Si_3N_4 (the orientation is indicated with an arrow [94].

An interesting phenomenon reported by the authors, is that in samples containing more than 4 at. % of silicon, their hardness is retained after annealing at 500°C in vacuum. This fact is ascribed to the percolative nature of the Si_3N_4 phase segregation, which creates one monolayer protection around TiN crystal allowing no changes in terms of grain sizes [95].

5.2.2. Laser Pyrolysis coupled with magnetron sputtering

Leconte at al. at CEA developed a system that combines two nanoparticles processes for producing Si Q-Dots embedded in a SiO_2 matrix [96]. The purpose was to create a one *single-step process* for applications related to photovoltaic solar cells. The technique has an interesting versatility regarding the large size range (2-100 nm) and different possible chemical compositions. It is a safe process, since it presents no worker exposure to nanoparticles.
As a matter of fact, similar films were processed by sol-gel and spin coating. However, they present limitations in terms of chemical composition, film thickness and a process step complexity.
The single step process consists in forming Si quantum dots by laser pyrolysis in the expansion chamber. Subsequently, a supersonic expansion in the intermediate chamber enables to increase the particles' energy.
Finally, these particles arrive to the deposition chamber in which the supersonic jet is destroyed but keeping their energy (2 keV). The quantum dots are deposited on a substrate. Simultaneously, the magnetron sputtering leads to create the SiO_2 matrix. The following schematic illustration resumes the process - Fig. 64.

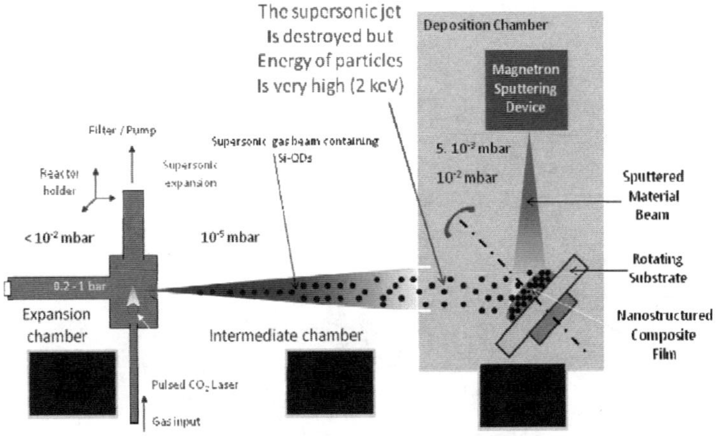

Figure 64. Schematic illustration of the system.

5.2.3. Hybridization of supersonic cluster beam combined with cathodic arc deposition

Piazzoni et al. produced nanocomposite thin films consisting of a titanium nitride matrix with embedded molybdenum disulphide fullerene-like (IFLN) nanoparticles [97].
The nanocomposite layer was deposited by combining an industrial cathodic arc evaporator (for TiN) with a pulsed microplasma cluster source (PMCS) for the production of supersonic MoS_2 cluster beams.
In Fig. 65, TEM images of multilayer MoS_2-TiN sample deposited on Si substrates are presented. Two phases of IFLN-MoS_2 nanoparticles of 10–50 nm and TiN matrix were observed.

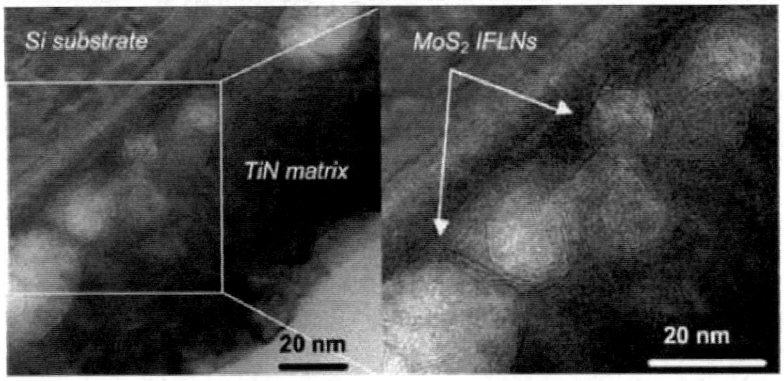

Figure 65. TEM micrograph of a MoS_2-TiN nanocomposite film [97].

5.3. 3-D Processes

5.3.1. Oxide dispersed strengthened steels four fuel cladding pins (ODS)

The objective was to produce a demonstrator of a fuel cladding pin made of ODS steel by mechanical alloying for the RNR nuclear reactor (Gen IV) [98]. Two different consolidation routes were proposed with the aim of producing steel-based matrix composites with dispersed nanometric Y_2O_3 reinforcements: i.e. hot extrusion and *Spark Plasma Sintering* [99].
The optimization of consolidation route was successfully achieved by means of *Spark Plasma Sintering*. Different samples diameters 60 & 100 mm. were sintered. An optimal reproducibility in terms of final density was obtained.

Interesting results regarding the microstructure were obtained. Indeed, in sample processed by extrusion, the microstructure was mainly composed of fine grain elongated along the extrusion direction.
After extrusion, the tube was machined up to the right dimensions. Subsequently, an annealing step (1050°C for 1 hour) was employed in order to reduce the strain to the desired hardness (380 HV). A further cold rolling step allows to increase the hardness up to 400HV.
However, intermediate heat treatments on the tube at 1200°C during 1 hour are required to reach the recrystallization state. The dimensions of the finished tube are 1500mm in length, 10.7mm in diameter and 0.5mm of internal thickness as shown in the figure below.

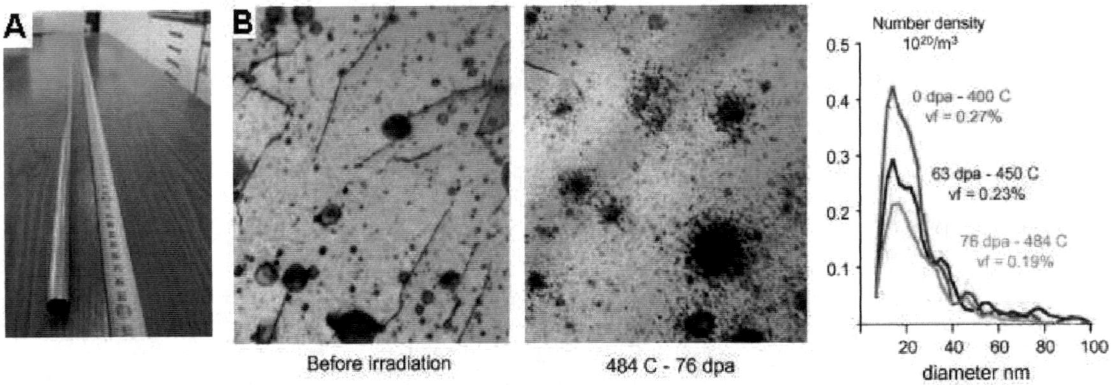

Figure 66. (A) First prototype of ODS steel and (B) the evolution of the oxides' distribution after irradiation tests [99].

From the impact tests, an important decrease in terms of the ductile-brittle transition temperature (-150°C) was measured – if compared with a PM 2000 industrial ODS steel (+100°C). The encouraging tensile tests revealed that the 18Cr ODS steel behaviour is quite similar between in the 20-750°C range if compared with the industrial ODS steels.

5.3.2. Oxide dispersed strengthened alloys by EBPVD

Srinivasan et al. processed NiCr alloys reinforced with nanoscale yttria and alumina dispersoids by electron beam physical vapor deposition (EBPVD) [100].
For this purpose, targets containing master alloys of Ni-20 wt% Cr (NiCr) and Y_2O_3/Al_2O_3 were evaporated in order to produce the nanocomposite-ODS sheets. These sheets were processed at the International Center for Electron Beam Technology (ICEBT), Ukraine.
With this technique it is possible to fabricate ODS alloys in a *one-step process*. In their work, the produced samples NiCr-4% Y_2O_3 and NiCr-4% Al_2O_3 alloys were submitted to a TEM microstructural observation, as presented in Fig. 67.
As shown in the figure above, it is possible to process ODS alloys containing 50–200 nm grains of the NiCr matrix with well dispersed of Y_2O_3 and Al_2O_3 nanoparticles of 5–20 nm.
A second study employing the NiCr-Y_2O_3 nanocomposites revealed that it is possible to tailor the mechanical properties [101]. The ultimate tensile strength and elongation percentage measured in the as-processed ODS nanocomposite are 1060 MPa and 0.07.
By aging the sample at 800°C for 3 h, it is possible to increase the ultimate tensile strength up to 1220 MPa due to the formation of $Cr_{23}C_6$. Whereas samples heat treated at 1100°C for 0.5 h or at 800°C for 16 h exhibited an ultimate tensile strength reduction down to 960 MPa. The same phenomenon was corroborated in the sample treated at 1100°C for 3 h in which a diminution down to 930 MPa was measured.

On the contrary, the latter treatment allows the formation $Y_3Al_5O_{12}$ (YAG) precipitates which are believed to enhance the final ductility.

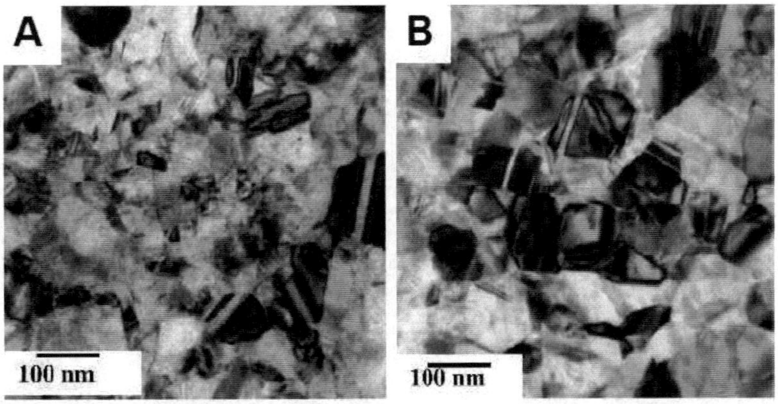

Figure 67. TEM micrographs of (A) NiCr-4% Y_2O_3 and (B) NiCr-4% Al_2O_3 [100].

Another interesting study regarding the mechanical properties was reported by *He et al.* [102]. Indeed, the researchers submitted a similar ODS thin film of 0.1 mm by hot isostatic pressing (HIP). The treatment was performed at 1100°C, with an applied pressure of 1100°C for 1 h.
After the treatment, no significant differences regarding the microstructure were found. Only dislocations located at the grains and along the grain boundaries were observed. As a result, the treatment provokes an increasing of tensile strength and elongation evolving from 725 MPa and 0.49 to 1230 MPa and 0.92, respectively.

5.3.3. Thermal plasma in-situ synthesis of TiC-Al(Ti) nanocomposite

Another example of *single-step* process is the TiC-Al (Ti) nanopowders' production by thermal plasma technology. As reported by *Tong et al.*, the process consists in vaporizing Al (sizes ranging 45 to 375 µm), Ti (45 µm) and CH_4 in a thermal plasma plume followed by a supersaturated product gases quenching [103].
The thermal plasma unit is composed of a power source and plasma torch, powder processing system (including reactor, condenser and filter), starting material feeder and data acquisition system – Fig. 68.

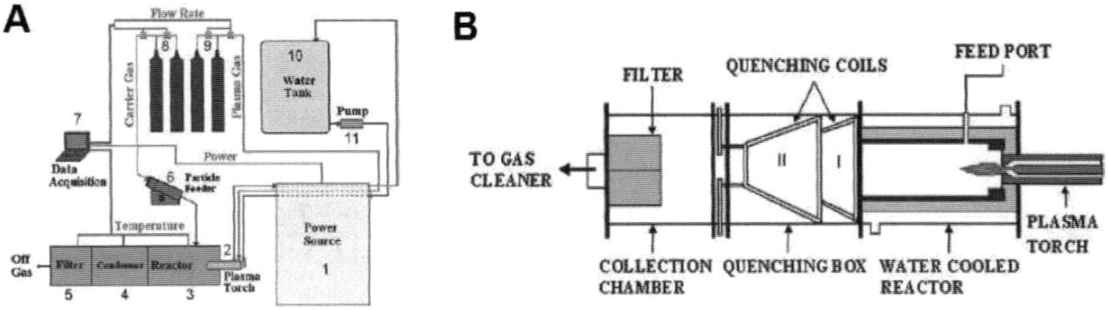

Figure 68. (a) Schematic illustration of the thermal plasma and (b) powder processing systems [103].

First, the reinforcement particles with a high melting point are nucleated from the supersaturated gas phase during shock quenching (Fig. 69). Along the nanopowders production, a particle size was

measured. In fact, the nanoparticles collected in the coil I/II and the filter zone confirmed this phenomenon.
For the sake of clarity the SEM observations performed with powders collected in the filter zone are presented in Fig. 69.

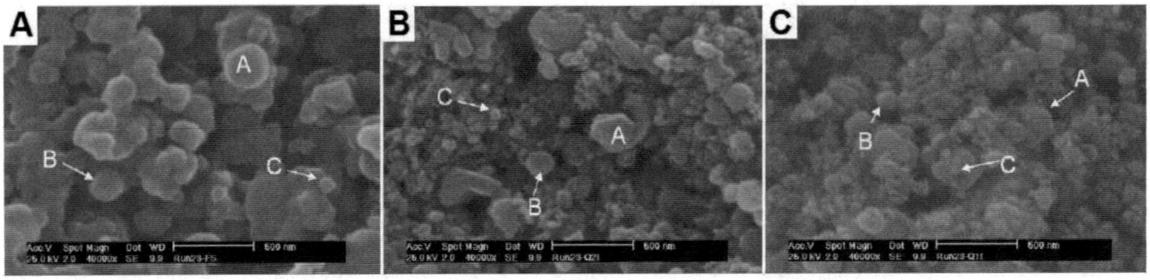

Figure 69. SEM micrographs of particles collected at the: (a) filter zone, (b) quenching zone 2 and (c) quenching zone 1 (A-B-C: indicates typical particles) [103].

The condensed ceramic particles act as heterogeneous nucleation seeds for the condensation of the metallic phase. The condensed metallic phase grows and encapsulates ceramic reinforcement particles in order to form composite powders with grain sizes >100 nm. Subsequently, nanopowders are collected, pressed and sintered with the aim of producing the Al (Ti) alloy matrix composites with a uniform distribution of TiC reinforcements.

5.3.4. Reactive mechanical milling of TiC/Ti-Al nanocomposites

TiC/Ti-Al nanocomposite powders were synthesized in *one step* by mechanical alloying from a mixture consisting of elemental Ti (30 μm), Al (16 μm), and graphite powder (30 μm) - $Ti_{50}Al_{25}C_{25}$ - as reported by *Gu et al.* [104]. During the mechanical alloying in an argon atmosphere, Ti and Al form a primary hcp solid-solution, whereas the TiC is formed by a rapid interdiffusion and an abrupt reaction between the Ti and C elements ($Ti + C = TiC$).

The authors found that by varying the milling time, it is possible to observe a refinement of the crystallite size accompanied with an increase of lattice strain as shown in Fig. 70 (a). The lattice strain directly influences the hardness of powder particles – Fig. 70 (b).

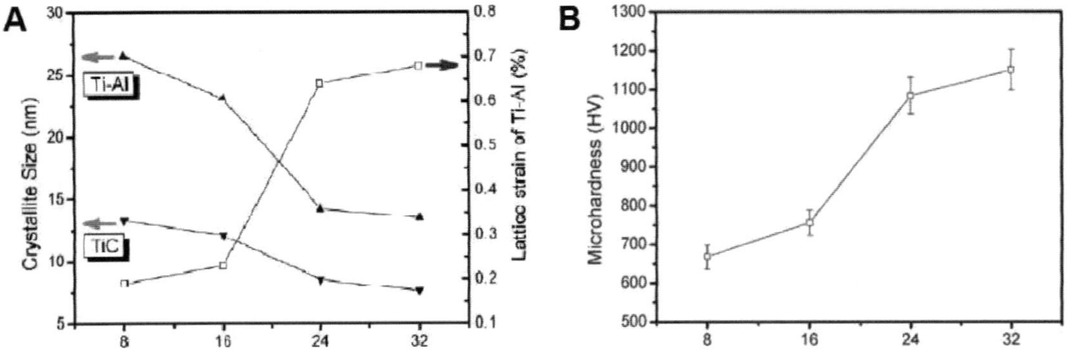

Figure 70. Average crystallite size and lattice strain and microhardness as a function of milling time [104].

In order to explain the mechanism for the formation of *in-situ* TiC/Ti-Al nanocomposites, the authors proposed the following a→d steps as presented in Fig. 71.

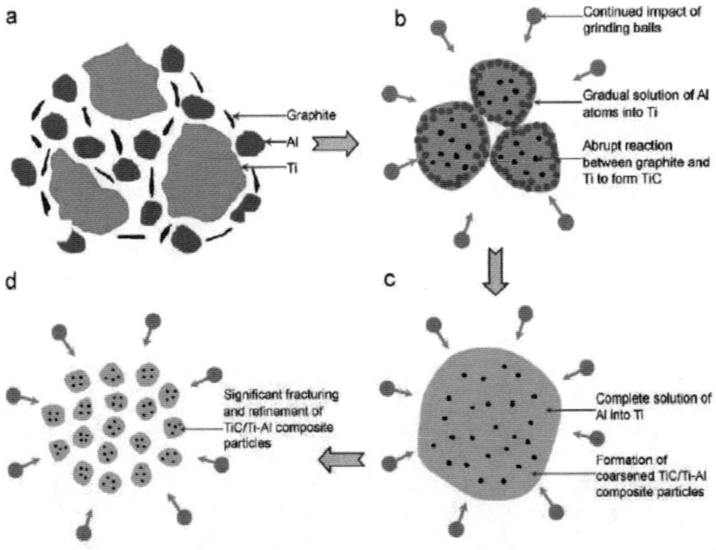

Figure 71. Mechanism for the formation of *in-situ* TiC/Ti-Al nanocomposites [104].

TEM and EDX-spectroscopy allowed identifying the TiC nanoparticles within the Ti-Al matrix. Indeed, in the MET micrograph - Fig. 72 (a)- two zones A and B were analyzed.
By observing the matrix (zone A), on Al-Ti with an equal atomic proportion is measured by EDX - Fig. 72. On the contrary, the zone B with ≈8 nm revealed the formation of the stoichiometric TiC reinforcements, since the atomic ratio of the C element and the residual Ti element is near 1:1.

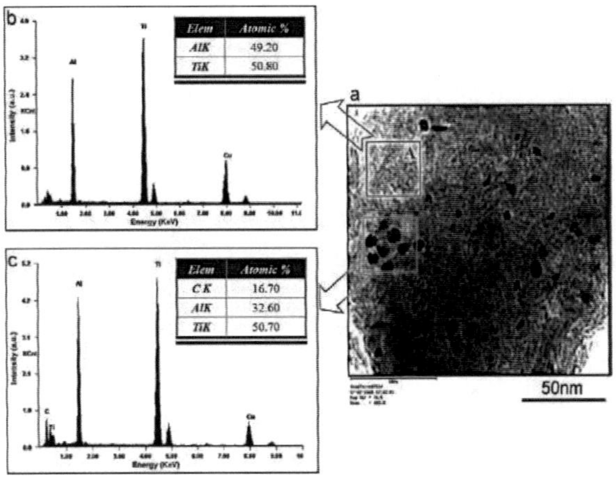

Figure 72. TEM micrograph on the 32 hours milled nanopowders (Cu is due to the copper-mesh grid) [104].

5.3.5. Nanocomposites processed in-situ by solid state reaction

Surface doping of two commercial alpha alumina powders was performed in order to prepare an Al_2O_3-5 vol.% YAG ($Y_3Al_5O_{12}$) composites. These materials were composed of micronic or highly sub-micronic alumina particles and very fine YAG grains, yielded by *in-situ* reaction as reported by several researchers [105,106,107].

Two different α-alumina powders were dispersed in aqueous suspensions by ball-milling and then doped with an yttrium chloride solution. A spray drying step was carried out by atomization in order to avoid the dopant segregation. Fast heat treatments were carried out in order to partially induce the crystallization of yttrium-aluminates on the alumina particles surface, minimizing the crystallites growth and the aggregates formation during heating. Subsequently, the powders were re-dispersed by ball milling and slip casted in porous moulds.

Samples were uniaxially pressed with the aim of increasing the green density and then consolidated by three sintering routes: Pressure-less Sintering, Hot-Pressing (HP) and *Spark Plasma Sintering* (SPS). The complete formation of the YAG phase was only yield after sintering.

The final microstructures were analyzed by field emission scanning electron microscopy (FESEM) are shown below– Fig. 73. They correspond to the commercial alumina powder characterized by the finer crystallite size (150 nm).

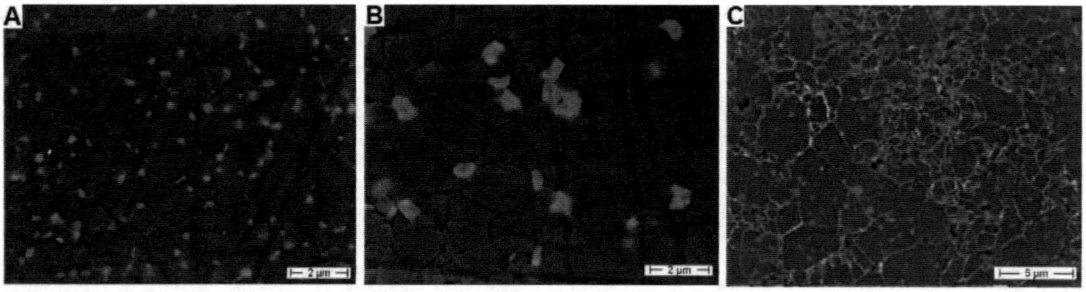

Figure 73. ESEM micrographs of polished surfaces: (a) Pressure-less Sintered, (b) Hot Pressed and (c) *Spark Plasma Sintered* [107].

As shown in Fig. 74, the microstructures are composed of equiaxed alumina grains with submicronic YAG grains (brighter in the micrographs) located at inter- and intra-granular positions (along/inside grain boundaries).

As reported by the authors, the YAG formation allowed the enhancing of the creep behaviour in these composites, since it is believed that Y^{3+} reduces the grain boundary diffusivity [108].

6. *Conclusion*

In this review, many examples regarding the *Safe-Nanomanufacturing* and *Nanosafe-by-Design* are given. As a matter of fact, ten years ago the CEA and its partners began the development of an integrated multidisciplinary *nanosafety* approach.

As a consequence, different generic tools were set-up in the frame of three European projects: *Nanosafe, Nanosafe 2* and *SAPHIR*. In these projects, the risks associated with the *nanomanufacturing* were not considered as drawbacks.

On the contrary, they were the *driving force* for developing new advanced measurement, monitoring and *Safe-Nanomanufacturing* tools. These tools were designed taking into consideration the entire products' life cycle: nanostructured materials or nano-objects.

The latter are believed to be the reaction intermediaries, which allow the microstructure and final properties tailoring. Finally, as reviewed in the last section, the *Nanosafe-by-Design* processes are thought to overcome the associated risks of dispersal into the environment. Furthermore, by implementing these processes, the *nanoproducts' manufacturers* will be able to skip some processes in their value chains.

References

[1] Gleiter H 1995 *Nanostruc. Mater.* **6** 3

[2] Mayo M 1996 Int. Mater. Rev. **41** 85

[3] Choe G M and Kurihara L K 2002 *Chemical Synthesis of Processing of Nanostructured Powders and Films. In:* Nanostructured Materials: processing, Properties and Applications ed KOCH C C (New York: William Andrew Publishing)

[4] Groza J R 1999 *Nanostruc. Mater.* **12** 987

[5] Zhan G D, Kuntz J, Wan J, Garay J and Mukherjee A K 2002 *J. Am. Ceram. Soc.* **86** 200

[6] Niihara K 1991 *The Centennial Memorial Issue of the Ceramic Society of Japan* **99** 974

[7] Bowen P, Carry C, Luxembourg D and Hofmann H 2005 *Powder Technol.* **157** 100

[8] Bazin B H, Les nanoparticules: un enjeu majeur pour la santé au travail 2007 (Les Ulis: EDP Sciences)

[9] Balazy A, Podgorski A and Gradon L 2004 *J. Aerosol Sci.* S967–S968

[10] Golanski L, Guiot A, Rouillon F, Pocachard J and Tardif F 2009 *Hum. Exp. Toxicol.* **28** 353

[11] Golanski L, Guiot A and Tardif F 2010 *J. Nanopart. Res.* **12** 83

[12] Hinds W C 1999 *Aerosol technology: properties, behaviour and measurement of airborne particles* (New York: Wiley-Interscience)

[13] Amodeo T, Dutouquet C, Le Bihan O, Attoui M and Frejafon E 2009 *Spectrochim. Acta Part B* **64** 1141

[14] Carranza J E and Hahn D W 2002 *Anal. Chem.* **74** 5450

[15] Hohreiter V and Hahn D W 2005 *Anal. Chem.* **77** 1118

[16] R'mili B, Duotuquet C, Sirven J B, Aguerre-Chariol O and Frejafon E 2011 *J. Nanopart. Res.* **13** 563

[17] Lyyränen J, Backman U, Tapper U, Auvinen A and Jokiniemi J 2009 *J. Phys.: Conf. Ser.* **170** 012011.

[18] Lanone S, Rogerieux F. Geys J, Dupont A, Maillot-Marechal E, Boczkowski J, Lacroix G and Hoet P 2010 *Part. Fibre Toxicol.* **6** 14

[19] Péry A R R, Brochot C, Hoet P. H M, Nemmar A and Bois F Y 2009 *Inhal. Toxicol.* **13** 1009

[20] Nemmar A, Hoet P H M, Vanquickenborne,B, Dinsdale D, Thomeer M, Hoylaerts M F, Vanbilloen H, Mortelmans L and Nemery B 2002 *Circulation* **105** 411

[21] Pritchard D 2004 Literature review – Explosion Hazards Associated with Nanopowders (London: Health and Safety Laboratory)

[22] Bouillard J, Vignes A, Dufaud O, Perrin L and Thomas D 2010 *J. Hazard. Mater.***181** 873

[23] Dufaud O, Vignes A, Henry A, Perrin L and Bouillard J 2011 *J. Phys.: Conf. Ser.* **304** 012076

[24] D. Geldart, Powder Technology 7 (1973) 285-292.

[25] Huang Y, Risha G, Yang V and Yetter R 2007 *Proceedings of the Combustion Institute 31* 2001

[26] Korylewski W and Seweryniak M 1992 *Journal of the Thermodynamics and Combustion Committee of the Polish Academy of Science* **12** 153

[27] Vignes A 2009 *Evaluation of nanopowders flammability and explosivity: an essential step towards risk assessment* PhD Thesis Institute Polytechnique de Lorraine

[28] Kwok Q, Fouchar R, Turcotte A, Light P, Bowes R and Jones D 2002 *Propell. Expos. Pyrot.* **27** 229

[29] Field P 1982 Dust Explosions (Amsterdam: Elsevier)

[30] Traoré M 2007 *Dust and hybrid mixtures.* PhD. Thesis, Institute Polytechnique de Lorraine.

[31] Babrauskas V 2003 *Ignition Handbook: Principles and Applications to fire safety engineering, fire investigation, risk management and forensic science* (Issaquah WA: Fire Science Publishers)

[32] Dufaud O, Traoré M, Perrin L, Chazelet S and Thomas D 2010 *J. Loss. Prevent Proc.* **23** 226

[33] Jacobson M, Cooper A R, Nagy J 1964 *Explosibility of Metal Powders* (Pittburg: U.S. Bureau of Mines)

[34] *Combustion and explosion characteristics of dusts* 1997 (Sankt Augustin: HVBG)

[35] Bartknecht W 1990 *Dust Explosion: Course, Prevention, Protection* (Berlin: Springer-Verlag)

[36] Hull M 2002 *Powder Metall.* **45** 8

[37] European directive 1999/92/EC. Directive on minimum requirement for improving the safety and health protection of works potentially at risks from explosive atmospheres. 15th Individual Directive within the meaning of Article 16 of Directive 89/391/EEC.

[38] Vignes A, Dufaud O, Perrin L, Thomas D, Bouillard J, Janès A and Vallières C 2009 *Chem. Eng. Sci.* **64** 4210

[39] Kissinger H E 1957 *Anal. Chem.* **29** 1702

[40] Illekova E and Csomorova K 2005 *J. Therm. Anal. Calorim.* **80** 103

[41] Guiot A, Golanski L and Tardif F 2009 *J. Phys.: Conf. Ser.* **170** 012014

[42] Golanski L, Gaborieau A, Guiot A, Uzu G, Chatenet J and Tardif F 2011 *J. Phys.: Conf. Ser.* **304** 012062

[43] Golanski L, Guiot A, Pras M, Malarde M, Tardif F 2012 *J. Nanopart. Res.* **14** 962

[44] Sentein C, Guizard B, Giraud S, Yé C and Ténégal F 2009 *J. Phys.: Conf. Ser.* **170** 012013

[45] Lockwood D J and Webster T J 2009 *Safety of Nanoparticles: From Manufacturing to Medical Applications* (New York: Springer New York)

[46] Lindeløv J S and Wahlberg M 2009 *J. Phys.: Conf. Ser.* **170** 012027

[47] Faure B, Lindeløv J S, Wahlberg M, Adkins N, Jackson P and Bergströn L 2010 *Powder Technol.* **203** 384

[48] Schubert U, Hüsing N 2005 *Formation of Solids from the Gas Phase, Synthesis of Inorganic Materials*, 2nd ed. (Berlin: Wiley-VCH)

[49] Roth P 2007 *Proc. Combust. Inst.* **31** 1773

[50] Foissy A, El Attar A and Lamarche J M 1983 *J. Colloid Interface Sci.* **96** 275

[51] Tanaka S, Chia-Pin C, Kato Z and Uematsu K 2007 *J. Eur. Ceram. Soc.* **27** 873

[52] Chaubal M V and Popescu C 2008 *Pharm. Res.* **25** 2302

[53] Wattieaux G, Lecerf P, Meyer L, Boufendi L, Leconte Y, Sublemontier O, Herlin N, Asimakopoulou A, Tsakis A, Daskalos M, Konstandopoulos A G and Ouf F X 2011 *J. Phys.: Conf. Ser.* **304** 012021

[54] Amodeo T, Dutouquet C, Tenegal F, Guizard B, Maskot H, Le Bihan O and Fréjafron E 2008 *Spectrochim. Acta Part B* **63** 1183

[55] Fichet P, Mauchien P and Moulin C 1999 *Appl. Spectroscopy* **53** 1111

[56] Fichet P, Mauchien P, Wagner J F and Moulin C 2001 *Anal. Chem. Acta* **429** 269

[57] Dudragne L, Adam P and Amoureaux J 1998 *Appl. Spectroscopy* **52** 1321

[58] Cremers D A and Radziemski L J 1983 *Anal. Chem.* **55** 1252

[59] Hahn D W and Lunden M M 2000 *Aerosol Sci. Tech.* **33** 30

[60] Radziemski L J, Loree T R, Cremers D A and Hoffman N M 1983 *Anal. Chem.* **55** 1246

[61] Noll R, Bette H, Brysch A, Kraushaar M, Mönch I, Peter L and Sturm V 2001 *Spectrochim. Acta Part B* **56** 637

[62] Gruber J, Heitz J, Strasser H, Bäuerle D and Ramaseder N 2001 Spectrochim. Acta Part B **56** 685

[63] Lomello F, Bonnefont G, Leconte Y, Herlin-Boime N and Fantozzi G 2012 J. Eur. Ceram. Soc. **32** 633

[64] Cannon W R, Danforth S C, Flint J H, Hagerty J S and Marra R A 1982 J. Am. Ceram. Soc. **65** 324

[65] Cauchetier M, Croix O and Luce M 1988 Adv. Ceram. Mater. **3** 548

[66] Cremers D A and Radziemski L J 2006 *Handbook of Laser-induced Breakdown Spectroscopy* (Chischester: John Wiley & Sons)

[67] Mengeot C, Guizard B, Poissonnet S, Boulanger L, Le Flem M, Guillard F and Ténégal F 2007 *Mater. Tech.* 95 289

[68] Herrmann M, Shen Z, Schulz I and Jancar B 2010 *J. Mater. Res.* **25** 2354

[69] Mazaheri M, Mari D, Schaller R, Cai Y, Esmaeilzadeh S and Shen Z 2011 *J. Am. Ceram. Soc.* **94** 1536

[70] Coleman J N, Khan U and Gun'ko Y 2006 *Adv. Mater.* **18** 689

[71] Mari D, Schaller R 2009 *Mater. Sci. Eng. A* **521-522** (2009) 255

[72] Mari D, Schaller R and Mazaheri M 2011 *J. Phys.: Conf. Ser.* **304** 012085

[73] Schaller R, Mari D, Marques dos Santos S, Tkalcec I and Carreno-Morelli E, 2009 *Mater. Sci. Eng. A* **521-522** 147

[74] Schaller R and Ionascu C 2009 *Mater. Sci. Eng. A* **521-522** 217

[75] Mazaheri M., Mari D, Schaller R, Bonnefont G and Fantozzi G 2011 *J. Eur. Ceram. Soc.* **31** 2691

[76] Mazaheri M, Mari D, Hesavi Z R, Schaller R and Fantozzi G 2011 *Compos. Sci. Technol.* **71** 939

[77] Delmas M, Pinault M, Patel S, Porterat D, Reynaud C and Mayne-L'Hermite M 2012 *Nanotechnology* **23** 105604

[78] Derieth T Bandlamudi G, Beckhaus P, Kreuz C, Mahlendorf F and Heinzel A 2008 *J. New Mat. Elect. Syst.* **11** 21

[79] Toma F L, Berger L M, Jacquet D, Wicky D, Villaluenga I, de Miguel Y R and Lindeløv J S 2009 *Surf. Coat. Technol.* **203** 2150

[80] Toma F L, Berger LM, Stahr C C, Naumann T and Langner S 2010 J. *Therm. Spray Techn.* **19** 262

[81] Ducros C, Cayron C and Sanchette F 2006 *Surf. Coat. Technol.* **201** 136

[82] Ducros C and Sanchette F 2006 *Surf. Coat. Technol.* **201** 1045

[83] Zhou Y, Asaki R, Soe W H, Yamamoto R, Chen R and Iwabuchi A 1999 *Wear* **236** 150

[84] Hovsepian P E, Lewis D B and Münz W D 2000 *Surf. Coat. Technol.* **133-134** 166

[85] Koehler J S 1970 *Phys. Rev.* **B2** 547

[86] Long Y, Giuliani F, Lloyd S J, Molina-Aldareguia J, Barber Z H and Clegg W J 2006 *Compos. Part. B – Eng* **37** 542

[87] Cahn J W 1963 *Acta Mat.* **11** 1274

[88] Kamath G, Ehiasarian A P, Purandare Y and Hovsepian P E 2011 *Surf. Coat. Technol.* **205** 2823

[89] Mungkalasiri J, Bedel L, Emieux F, Doré J, Renaud F N R , Maury F 2009 *Surf. Coat. Technol.* **204** 887

[90] Mungkalasiri J, Bedel L, Emieux F, Doré J, Renaud F N R , Sarantopoulos C, Maury F 2010 *Chem. Vap. Deposition* **16** 35

[91] Zhang W, Chen Y, Yu S, Chen S, Yin Y 2008 *Thin Solid Films* **516** 4690

[92] Arabatzis I M , Stergiopoulos T, Bernard M C, Labou D, Neophytides S G, Falaras P 2003 *Appl. Catal. B Environ.* **42** 187

[93] Bugyi L, Berko A, Ovari L, Kiss A M, Kiss J 2008 *Surf. Sci.* **602** 1650

[94] Bendavid A, Martin P J, Cairney J, Hoffman M and Fischer-Cripps A C 2005 *Appl. Phys.* A **81** 151

[95] Veprek S, Nesladek P, Niederhofer A and Glatz F 2003 *Nanostruc. Mater.* **10** 679

[96] Leconte Y, Paquez X and Sublemontier O 2012 Patent: Device for synthesising a nanostructured composite material and associated method. *Commissariat à l'énergie atomique et aux énergies alternatives* WO 2012/110914.

[97] Piazonni C, Blomqvist M, Podesta' A, Bardizza G, Bonati M, Piseri P, Milani P, Davies C, Hatto P, Ducati C, Sedláčková K and Radnóczi G 2008 *Appl. Phys.* A **90** 101

[98] A Technology Roadmap for Generation IV Nuclear Energy Systems 2002 Issued by the U.S. DOE Nuclear Energy Research Advisory Committee and the Generation IV International Forum

[99] Dubuisson P, de Carlan Y, Garat V and Blat M 2012 *J. Nucl. Mater.* **428** 6

[100] Srinivasan D and Subramanian P R 2007 *J. Mater. Res.* **22** 68

[101] Chen S, Qu S, Liang J and Han J 2011 *Rare Metals* **30** 76

[102] He X D, Xin Y, Li M W, Sun Y 2009 *J. Alloy. Comp.* **467** 347

[103] Tong L and Reedy R G 2006 *Metall. Mater. Trans. B* **37B** 531

[104] Gu D and Shen Y 2009 *Adv. Eng. Mater.* **11** 573

[105] Palmero P, Naglieri V, Chevalier J, Fantozzi G and Montanaro L 2009 *J. Eur. Ceram. Soc.* **29** 59

[106] Lomello F 2010 *Optimization of nanostructured oxide-based powders by surface modification* PhD Thesis Politecnico di Torino/Institut National de Sciences Appliquées de Lyon

[107] Palmero P, Fantozzi G, Lomello F, Bonnefont G and Montanaro L 2012 *Ceram. Int.* **38** 433

[108] Yoshida H, Ikuhara Y and Sakuma T 1998 *J. Mater. Res.* **13** 2597

Evaluating performance of containment equipment designed for handling manufactured nanomaterials by use of nanoparticle tracer

Sébastien Artous[1], Pierre Bombardier[2], Samir Derrough[1], Dominique Locatelli[1], Pierre Nobile[1], Catherine Durand[1]

[1] CEA-Grenoble, NanoSafety Platform, 17 avenue des Martyrs, F-38054, GRENOBLE Cedex 9

[2] FAURE QEI, 27 bd des Alpes, F-38240 MEYLAN

E-mail: sebastien.artous@cea.fr

Abstract. The implementation in many products of manufactured nanoparticles is in strong growth and raises new questions. For this purpose, the CEA - NanoSafety Platform is developing various research topics for health and safety, environment and nanoparticles exposure in professional activities. The working group Nano-CERT/MTD, driven by INERIS, federates actors of the sector: experts, research organizations, industrial users and manufacturers of collective protection. The main activity of this group is to establish specific guidelines or a voluntary certification of collective protection, at a national scale, but with the possibility of a further extension at an European level. The group aims to establish an experimental protocol of certification to characterize collective protections for workers faced with nanomaterials potential exposure. The NanoSafety Platform provides in this presentation a method of collective protection characterization based on the developments in nanoparticles metrology and on the study of existing standards and practices in related areas (chemicals, dust, microbiological and nuclear). This study presents the results obtained during the experimental characterization of a potential nanoparticles transfer in a prototype laboratory fume hood by the use of a particulate tracer of sodium-fluorescein. The efficiency of the equipment and more specifically the efficiency of dynamical air barrier is evaluated, with the experimental results, by calculating the backward diffusion coefficient.

1. Introduction

Nanoparticles are present in various and numerous fields and activities (medicine, cosmetology, building materials,...). As a result, the number of workers in contact with nanoparticles is increasing. As collective protections are the first one to be implemented, before individual protections, it becomes necessary to be able to test their efficiency for nanoparticles.

The CEA - NanoSafety Platform (NSP) located in Grenoble works on all matters of safety and security related to the handling of nanomaterials. It conducts in parallel actions of R & D and services missions of measurement and characterisation, expertise, training and intervention in accidental situations. It employs nearly 150 professionals, researchers and experts, whose jobs are to provide a strong support to researchers and partners using nanotechnology, while maintaining the highest level knowledge and skills in the field of nanosafety.

In this objective the NanoSafety Platform is developing various research topics for health and safety, environment and nanoparticles exposure in professional activities. One of this topic is to determine the best available method to characterize protection techniques for workers faced with nanomaterials exposure. For this purpose the NSP joined the working group Nano-CERT/MTD, driven by INERIS. The working group brings together various stakeholders, experts, research organizations, industrial users and manufacturers of collective protection.

The main activity of this group is to establish specific guidelines for a voluntary certification of collective protections, at a national scale, but with the possibility of a further extension at a European level. The first certifications with an experimental evaluating of performance of containment equipment by a particulate tracer, are expected for 2013.

The NSP provides a method of collective protections characterization based on the use of a particulate tracer of sodium-fluorescein. This study presents the results obtained during the experimental characterization of a potential nanoparticles transfer in a prototype laboratory fume hood. The efficiency of the equipment and more specifically the efficiency of dynamical air barrier are evaluated, with the experimental results, by calculating the backward diffusion coefficient.

2. Principle of determination of Aeraulic parameters with a tracer

2.1. Extraction airflow

The principle of the method is based on the dilution of a tracer injected at a constant flow rate in the pipe. The determination of mass balance (equation 1) allows to calculate the extraction airflow of the equipment.

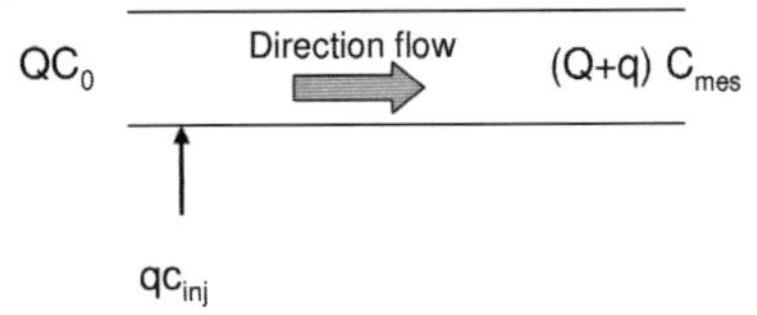

$$Q = \frac{qc_{inj}}{C_{mes} - C_0} \quad \text{with } q<<Q \qquad (1)$$

With :
Q the flow in the pipe in m^3/h
q the flow of tracer in m^3/h
C_{INJ} injected volume concentration of tracer
C_0 volume concentration of tracer in ambient air
C_{mes} the volume concentration of the tracer at the measurement point

2.2. Air renewal

The residence time distribution (RTD) of a chemical reactor is a probability distribution function that describes the amount of time a fluid element could spend inside a reactor or, for this study, in the containment equipment. The RTD characterizes the mixing and flow in the equipment and allows comparing the behavior of real equipment to his ideal models.

The theory of residence time distributions generally begins with three assumptions:

- the reactor or equipment is at steady-state,
- flow where no random event happens at the macroscopic scale,
- the fluid is incompressible.

If ventilation is assumed to be homogeneous, the equipment is considered equivalent to a perfectly stirred reactor. The theoretical system response (here the equipment) to an inverse step experiment is resulting in equation 2.

$$\ln\left(\frac{C(t)}{C_0}\right) = -\frac{1}{\tau}.t = -R.t \qquad (2)$$

The plot of ln((C-C0) / (C *-C0)) = f (t) gives a straight line whose slope represents, except for the sign, the renewal rate R. The renewal rate R (equation 3) is equal to the inverse of renewal time τ (equation 4).

$$R\left(h^{-1}\right)=\frac{Q\left(m^3h^{-1}\right)}{V\left(m^3\right)} \qquad (3) \qquad\qquad \tau(h)=\frac{1}{R} \qquad (4)$$

With:
Q the flow in the pipe m^3/h
C the concentration at the instant t
C* the equilibrium concentration
C_0 the concentration of tracer in ambient air
V the volume of the equipment

2.3. Backward diffusion coefficient
The backward diffusion coefficient K is defined as the ratio of the concentration at the point of measurement "m" (plane of the fume hood, respiratory tract or at the level of a nanoparticulate detection device...) on the flow of pollutant. This coefficient is used to characterize the potential transfer of pollutant used in the equipment.

$$K=\frac{C_m}{q_s} \qquad (5)$$

With:
C_m the concentration at the point of measurement m g/m^3
q_s the flow of pollutant or tracer at its source S in g/h

3. Materials and method

3.1. Containment equipment
The prototype of containment equipment of FAURE QEI company has two configurations. The first without the front panel is called "fume hood mode". The second with the front panel in place, is called "glove box mode". A removable filtration / ventilation block allows to filtrate air with a HEPA filter and set the extraction airflow (Figure 1-a). The exhaust of equipment is driven in a laboratory fume hood. The *Figure 1*-b shows the localization of the measurement points for the determination of backward diffusion coefficients. The sampler used for gas and particulate tracer is defined by the French standard NF EN 14175 [1].

For determining the air renewal, the injection takes place at the entrance to the equipment using a cable entry at the side and by using various injection sites in order to homogenize the concentration inside the equipment. Measuring the gas concentration is carried out downstream of the filtration unit. Measuring by counting or by sampling on a filter for fluorescein is performed immediately upstream of the filtration unit.

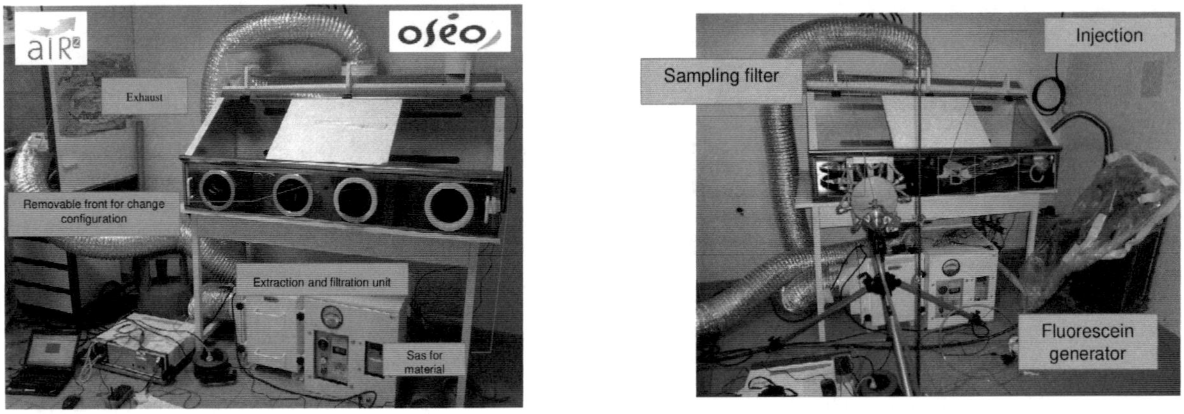

a b

Figure 1. FAURE QEI containment equipment and localization of measurement point for backward diffusion coefficients

1.1. Method

3.1.1. The helium tracer

The injection of helium is controlled by a mass flow meter ANALYT MTC-358 (Figure 2). The localization of injection is adjusted according to the desired information (extraction flowrate, air renewal, backward diffusion coefficient). The concentration of tracer gas is measured by a mass spectrometer adixen - type ASM 102S. The sampling point is adjusted according to the desired information (extraction flowrate, air renewal, backward diffusion coefficient).

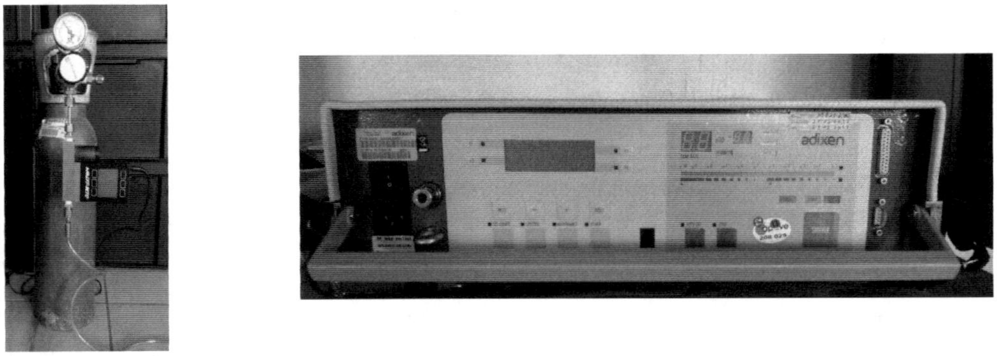

Figure 2. mass flow meter and mass spectrometer

3.1.2. The sodium fluorescein tracer

The principle of measurement is to inject a solid aerosol of sodium fluorescein. The number median diameter of aerosol is 60 nm (Figure 3). The generator is described in the French standard NFX44-011 [2]. The injection point is adjusted according to the desired information (extraction flowrate, air renewal, backward diffusion coefficient). The mass flow rate of generator is of the order of 30 mg/h.

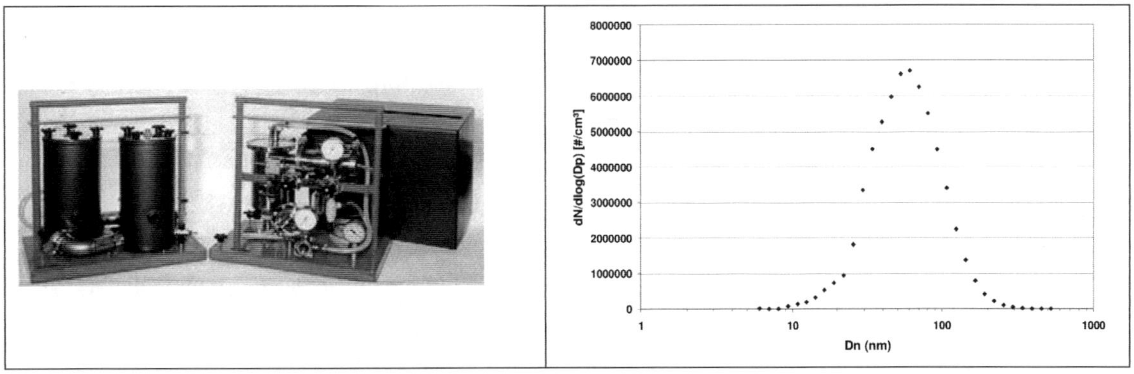

Figure 3. Fluorescein generator and size distribution obtained by FMPS of TSI

4. Results

4.1. Aeraulic characterization

The figure 4-a shows the curves of RTD obtained for two extraction airflows in fume hood mode and in glove box mode. The set of curves shows that the renewal is homogeneous in fume hood mode and in glove box mode. The exploitation of results gives renewal time around 20 s in glove box mode. For the fume hood mode the renewal time is around 4 s with an extraction airflow at 450 m^3/h and around 7 s with a extraction airflow at 280 m^3/h. The values of renewal and extraction airflow allow to calculate the volume of equipment. The result is of the order of 0.5 m^3.

The figure 4-b shows the curves of RTD obtained in fume hood mode and in glove box mode with helium and Fluorescein tracer. The set of curves shows that the renewal is similar with a nanoparticulate tracer and a gaseous tracer. This result is interesting for the characterization of equipment. In fact, several equipments have an exhaust in the room. Gas is not retained by the filters and is sent back in the room. Thus it's not possible to characterize the aeraulic with a gaseous tracer.

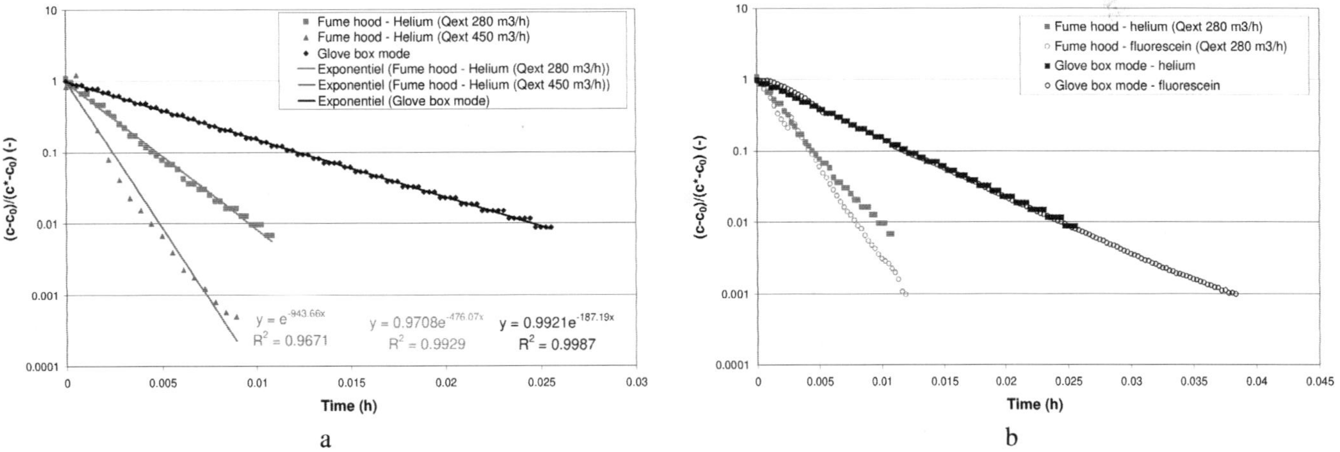

Figure 4. RTD of containment equipment

The table 1 summarizes the results obtained when measuring RTD for different modes of equipment and for the two types of tracer. The values obtained with a gaseous or a particulate tracer, for a same configuration of the equipment (fume hood mode or glove box mode) are similar.

Table 1. Synthesis of determination of aeraulic parameters

Mode	Tracer	Extraction airflow (m^3/h)	Hourly renewal (R/h)	Renewal time (s)	Volume of equipment (m^3)
Fume hood (280 m^3/h)	helium	280	539 et 476	6.6 et 7.7	0.52 et 0.6
	fluorescein	250	537 et 581	6.7 et 6.2	0.43 et 0.47
Fume hood (450 m^3/h)	Helium	450	944	4.6 et 3.8	0.57 et 0.48
Glove box	helium	102	187 et 203	17 et 19	0.49
	fluorescein	-	188 et 195	19 et 20	-

4.2. Backward diffusion coefficient

The Figure 5-a shows the backward diffusion coefficients obtained in fume hood mode for two extraction airflows (280 m^3/h and 450 m^3/h) for localizations A and B (Figure 1-b). The lower the value of backward diffusion coefficient at the measurement point is, the more efficient is the equipment.

The results show that helium tracer method is near from the detection limit. The consequence is that there's no significant difference between values of backward diffusion coefficients between extraction airflow at 280 m^3/h and extraction airflow at 450 m3/h. For the extraction airflow at 280 m^3/h, the values of backward diffusion coefficients obtain with the gas or particulate tracer are similar. The results show that, for this operating point, it's possible to use a tracer gas to account of the behavior of a nanoparticle pollutant.

The Figure 5-b compares the backward diffusion coefficients obtained in fume hood mode for two extraction airflow and in glove box mode with the fluorescein tracer. The results show that the glove box mode is 100 times more efficient than the fume hood mode. Finally the low limit detection of fluorescein allows determining the decrease of the backward diffusion coefficients with the increase of extraction airflow in fume hood mode.

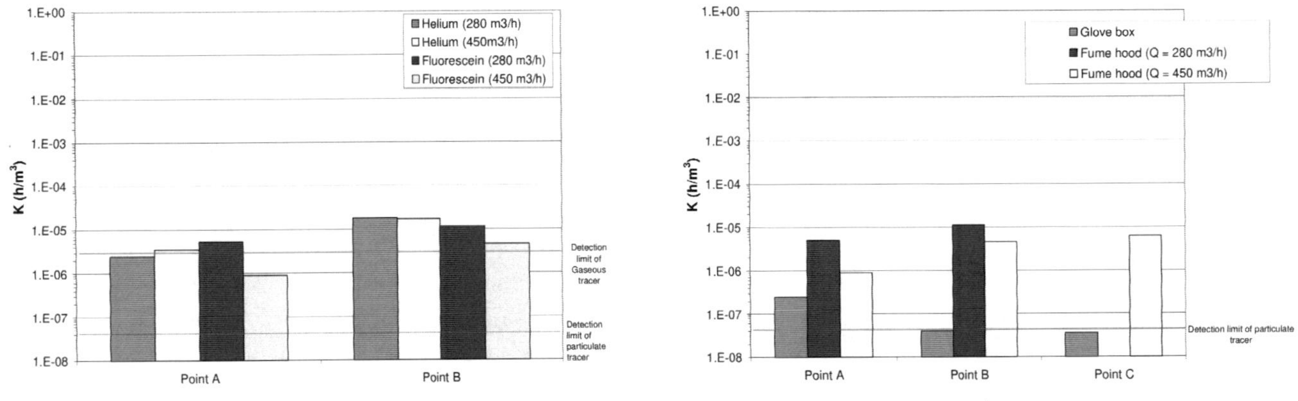

a b

Figure 5. Backward diffusion coefficients

5. Conclusion

This study shows the importance of carrying out tests for characterizing ventilation (extraction airflow, RTD, renewal) because these parameters have a significant influence on the performance of containment equipment. In addition, the study shows that this characterization can be performed both with a gas or nanoparticulate tracer. The feasibility of study the renewal of equipement using a nanoparticle tracer, is particularly advantageous when the extraction airflow is in the room where the equipment is located.

These measures have to be completed by measurements of coefficients either gaseous or particulate tracers in the plane of equipment. The values of backward diffusion show that a gas tracer can account of the behavior of a nanoparticulate pollutant if the backward diffusion is high enough to be above the detection limit of the method of gas analysis. However, the analytical method with a nanoparticulate tracer of fluorescein is more sensitive and allows to differentiate changes in the operating point of the equipment

In conclusion the method using a solid aerosol fluorescein is robust, reliable and sensitive. Moreover, sodium fluorescein does not present any danger to humans until today.

Acknowledgments
We want to thank all the persons that have helped us in this work, and particularly those from CEA / LITEN / LCSN lab.

References

[1] Août 2004, French standard, Sorbonnes - Partie 3 : méthodes d'essai de type, NF EN 14175-3, AFNOR
[2] Mai 1972, French standard, Méthode de mesure de l'efficacité des filtres au moyen d'un aérosol d'uranine (Fluorescéine), NF X44-011, AFNOR

Experimental evaluation of the resistance of nitrile rubber protective gloves against TiO_2 nanoparticles in water under conditions simulating occupational use

L Vinches[1], P Dolez[3,1], K J Wilkinson[2], S Hallé[1]

[1] École de technologie supérieure - 1100 Notre-Dame Ouest, Montréal, QC H3C 1K3 Canada
[2] Dept. de chimie, Université de Montréal - C.P. 6128, succ. Centre-ville, Montréal, QC H3C 3J7 Canada
[3] CTT Group - 3000 avenue Boullé, Saint-Hyacinthe, QC J2S 1H9 Canada

E-mail: ludwig.vinches.1@ens.etsmtl.ca

Abstract – Manufactured nanoparticles (NPs), including titanium dioxide nanoparticles ($nTiO_2$), now enter in the formulation of several commercial products. Some studies have been carried out to assess the resistance of protective gloves against NPs, they have generally not considered the conditions prevailing in occupational settings.
This study was designed to evaluate the behavior of protective gloves materials against NPs in solution under conditions simulating occupational use. Mechanical deformations, simulating those produced by flexing the hand, were applied to nitrile rubber glove samples in contact with $nTiO_2$ in water. The first analysis showed that $nTiO_2$ solution penetrates some of the materials after prolonged dynamic deformations. These results were partly attributed to modifications in the physical and mechanical properties of protective gloves materials that were induced by repetitive mechanical deformations and/or the presence of the colloidal solution.

1. Introduction

The global market for nanotechnologies should represent more 2.5 trillion dollars in 2015 [1] and the National Science Foundation has estimated that the number of nano-products manufactured around the world will double every three years until 2020. At the same time, the number of workers and researchers in the industry will reach 6 million [2]. Titanium dioxide nanoparticles ($nTiO_2$) are present in commercial products such as paints, varnishes, sunscreens, etc. [3, 4] and represent a new potential source of hazard. Indeed, some concerns exist about their potential noxious effects on health. For example, studies conducted on rats and mice, exposed to 250-nm TiO_2 pigment particles and 20-nm $nTiO_2$, observed a greater occurrence of pulmonary effects [5]. In addition, the observed inflammatory response was higher with $nTiO_2$ [6, 7]. In 2007, a small increase in the number of cancers among workers in contact with $nTiO_2$ has been reported [8]. Based on these results, the International Agency for Research on Cancer (IARC) has classified nanosized titanium dioxide in the 2B-group, as being possibly carcinogenic to humans [9]. Furthermore, some studies have shown that the skin is not an impervious membrane against NPs. The penetration, through epidermal layers, is possible when the skin presents small lesions [10], after repeated flexion [11] or even intact [12]. In response to these studies and the IARC classification, several agencies have recommended application of the precautionary principle (e.g.[13]). While this implies the use of protective gloves, the number of studies on dermal protection against NPs, and especially on gloves, is scarce.

Whereas the penetration of 30 and 80 nm graphite NPs has been reported for nitrile, vinyl, latex and neoprene commercial glove samples [14], no penetration was observed for the same glove models for 40 nm graphite, 10 nm TiO_2 or Pt nanoparticles [15, 16]. For his part, Ahn *et al.* exposed nitrile and latex glove samples to nanoclays and alumina nanopowder [17]. Scanning electron microscope (SEM) observations revealed the accumulation of the NPs in micrometer-size pores at the surface of the gloves material.

In occupational settings, protective gloves are subjected to mechanical deformations, for example biaxial strains at interphalangeal joints which may also affect the penetration of NPs [18]. Moreover, the presence of a liquid carrier facilitates the penetration of NPs because elastomers are sensitive to solvents and induce a significant swelling [18, 19].

This paper reports some results on the penetration of $nTiO_2$ in water through protective gloves. The tested protective glove materials consisted of nitrile rubber. It also analyses the effect of contact of the glove materials with the colloidal solution liquid carrier as well as of dynamic biaxial deformations.

2. Materials

2. 1. Protective gloves

Two models of protective gloves were selected for this study: disposable nitrile rubber gloves with 100 and 200 μm thickness. In this paper, the corresponding results are identified, respectively, as NBR-100, NBR-200. All the samples were taken from the palm section of the gloves.

2. 2. Nanoparticles

Nanoparticle suspension in water used for this study contained 15 nm, anatase titanium dioxide nanoparticles (15% w/w). It was obtained from Nanostructured & Amorphous Materials, Inc. (Houston, TX). Particle sizes of 21 ± 2 nm were measured by fluorescence correlation spectrometry (FCS) [20].

3. Methods

3. 1. NP penetration experimental setup

An experimental test setup has been developed for the purpose of this study (Figure 1). It has been designed to expose glove samples to colloidal solution while simultaneously subjecting them to biaxial constraints simulating deformations produced by flexing the hand. The test system includes two chambers, an exposure chamber and a sampling chamber, as well as all components are made of ultrahigh molecular weight polyethylene loaded with black carbon in order to limit the adsorption of $nTiO_2$. Both chambers are separated by the sample. Colloidal solution is introduced in the exposure chamber and placed into contact with the external surface of the glove sample. As shown in Figure 1, the test setup is also equipped with a probe mounted on an electrical system for deforming the sample. The system includes a 200 N load cell and is computer controlled. The whole setup is enclosed in a glove box in order to ensure operator safety during assembly, dismounting and clean-up operations, as well as during tests.

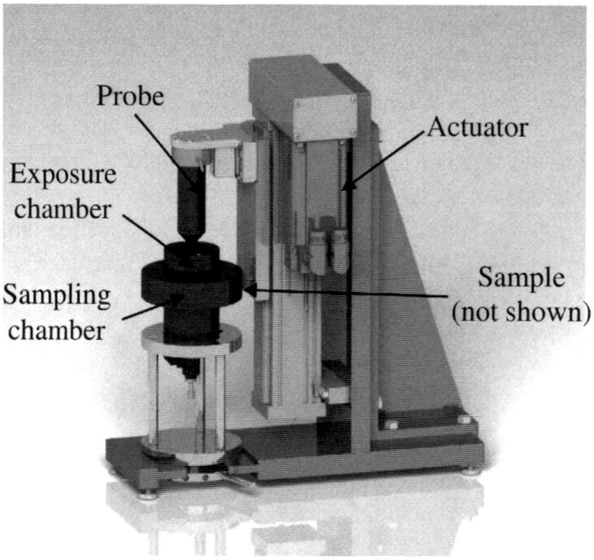

Figure 1. Isometric view of the test setup.

As illustrated in Figure 2(a), the time profile of the sample deformations consists in 50% out-of-plane deformations applied every 1 or 5 minutes. The probe head used in this study is shown in Figure 2(b) and corresponds to a conical-spherical geometry which simulate biaxial deformations (BD) produced by flexing hand [21].

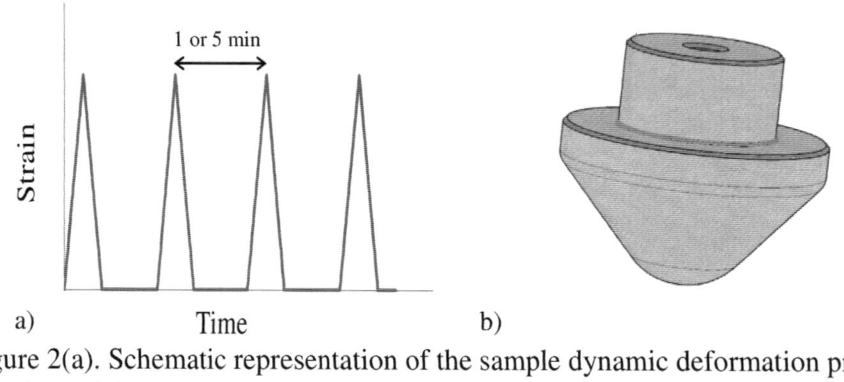

Figure 2(a). Schematic representation of the sample dynamic deformation program and Figure 2(b) Schematic representation of the conical-spherical probe head.

3. 2. NP sampling protocol and detection methods

$nTiO_2$ sample collection was facilitated using a sampling solution (ultra-high purity water with 1% nitric acid) which was placed in the sampling chamber during setup. Before the beginning of each test, the absence of contamination of the sampling chamber by $nTiO_2$ was verified by analysing a control sample that was produced by rinsing the sampling chamber with the sampling solution. If a trace of contamination in the test control was detected, the result of the following test was discarded. Titanium concentrations were analysed by ICP-MS (Inductively coupled plasma mass spectrometry, Varian 820).

3. 3. Chemical composition of the glove surface and quantification of the characteristic features.

Energy dispersive spectroscopy (EDS) analysis was conducted with Oxford INCA X-sight 200 to perform the quantitative analysis of the chemical elements present in both glove materials. Moreover, some characteristic features (micrometer-size pores) can be observed on the glove surface as already reported in the literature [17]. The surface morphology of five samples for each elastomer, was analysed by scanning electron microscopy (SEM, Hitachi S3600N – V_{acc} = 15 kV – magnification × 1000). The quantification of the surface area of these features was performed using image processing software.

3. 4. Strain energy
In order to follow the variation of the strain energy corresponding to the applied 50% dynamic deformation during tests, stress-strain curves were plotted. An estimation of the relative strain energy was carried out by computing the area under the curves using the trapezoidal method.

3. 5. Swelling measurements
Length change measurements were performed on glove samples in order to evaluate the effect of the $nTiO_2$ solution. Indeed, elastomers may be sensitive to swelling in some solvents. Rectangular samples (5 × 60 mm) were taken from the palm section of the gloves. To obtain statistically significant data, triplicate measurements were carried out for each solution. The samples were immersed in the $nTiO_2$ in water. At regular intervals, samples were removed from the liquid and their length measured using a calliper (±0.01 mm). The length change data were computed using Eq. (1):

$$\Delta L(t) = \frac{L_t - L_0}{L_0} \qquad (1)$$

with L_t being the length at t time and L_0 the corresponding value before immersion.

4. Results and discussion

4. 1. Characterization of the $nTiO_2$ in water and the protective gloves surface.
A series of experiments were performed to characterize the nanoparticle solution. Analysis of the aqueous $nTiO_2$ stock solution (following dilution to 10 mg L^{-1}) by fluorescence correlation spectroscopy (FCS) gave a hydrodynamic diameter for the particles of 21 ± 2 nm. Thermogravimetric analyses were carried out to estimate the $nTiO_2$ mass fractions. The mass fraction obtained for $nTiO_2$ in water is 14.3 ± 0.8 %. These results confirmed the manufacturer data. In addition, FT-IR analyses were performed to bring out the presence of additives such as stabilizing agents. Figure 3 compares FT-IR spectra of $nTiO_2$ in water and ultra-high purity water. An additional peak can be observed for $nTiO_2$ in water. It may be due to the presence of stabilizing agents.

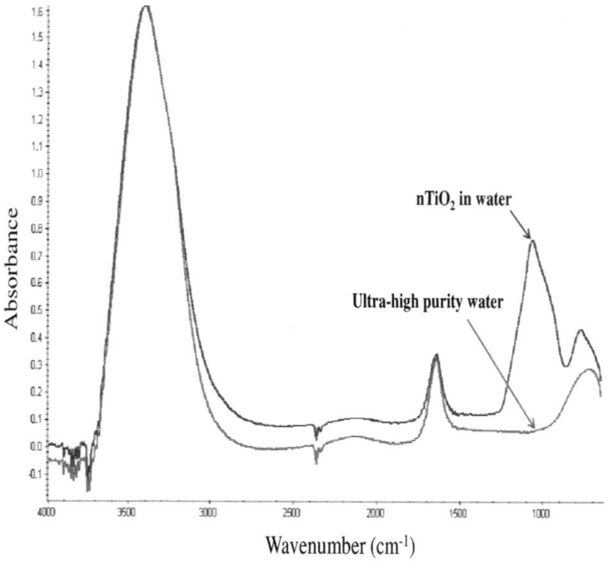

Figure 3. FT-IR spectra of nTiO2 in water (purple) and in ultra-high purity water (red).

Figure 4 shows the micrometer-size pores which can be observed on the outer surface of both nitrile rubber gloves. The same features were observed on the inner surface. Table 1 compares the mass fraction of the chemical elements for both nitrile rubber gloves. Significant differences exist between the two models of protective gloves but both contain titanium.

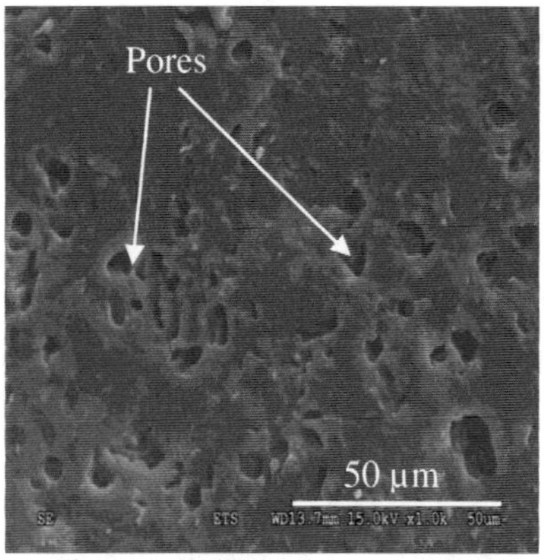

Chemical elements	NBR-100 Mass fraction (%)	NBR-200 Mass fraction (%)
C	85.64	90.27
O	10.41	5.44
S	1.24	1.20
Ca	1.85	0.71
Si	0.20	0.38
Al	-	0.35
Ti	0.44	0.73
Zn	-	1.13

Figure 4. SEM images of the native outer surfaces both nitrile rubber gloves and Table 1 Chemical composition of glove materials obtained by EDS analysis.

4. 2. NP penetration: ICP-MS analyses.

Glove samples (NBR-100) were exposed to the $nTiO_2$ in water while being simultaneously subjected to a 50% biaxial deformation every 5 minutes for up to 7 h. In order to determine if the titanium concentration is solely due to the penetration of the $nTiO_2$ solution, the same tests were repeated without NPs. The results are shown in Figure 5. For nitrile gloves, a clear increase in titanium concentration in the sampling solutions can be observed after 5 and 7h of deformations as a result of the presence of the $nTiO_2$ in the sampling chamber. These results indicate the possible penetration of $nTiO_2$ in water through nitrile gloves after deformations.

These first results have been obtained with conditions simulating soft occupational use (a deformation every 5 minutes up to 7h, for a total of 85 deformations). In typical occupation use, the number of flexions of the hand per minute is probably more important and the wearing time of protective glove does not exceed 3 hours.

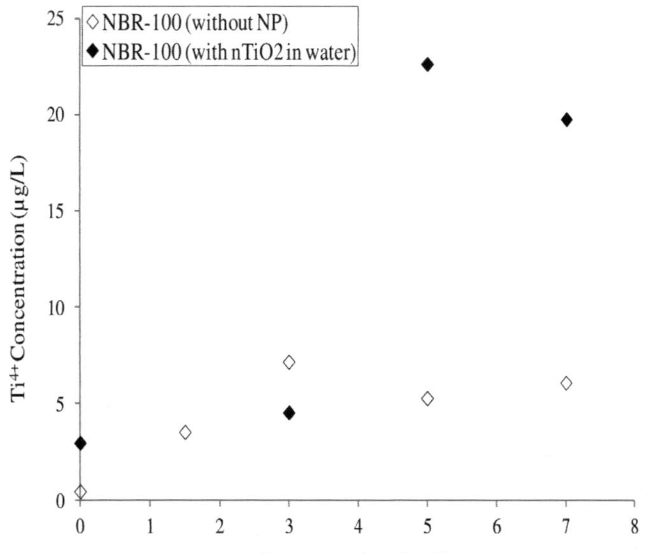

Figure 4. Variation in titanium concentration in the sampling solutions as a function of test duration for NBR-100 samples exposed simultaneously to 50% biaxial deformations and $nTiO_2$ in water.

4. 3. Mechanical and physical phenomena facilitating the penetration of $nTiO_2$ solution.

In order to investigate the cause of the NP penetration through the nitrile gloves, the swelling of nitrile rubber immersed in $nTiO_2$ solution has been studied. In addition, some mechanical tests were performed to study the behaviour of the gloves material.

Swelling measurements were performed by recording the length change of both nitrile rubber samples after immersion in aqueous $nTiO_2$. Figure 6 displays a gradual length change increase for the two glove materials, indicating significant penetration of $nTiO_2$ solution. Nonetheless, a larger swelling ratio was recorded for NBR-100. For example, for an immersion time of 3 hours, the length change ratio for the NBR-100 was 5% and 2% for NBR-200. It should be noted that maximum was not attained for either of the two nitrile rubber gloves after 3 hours of immersion. A comparison of the length change behaviour was therefore performed over a longer period. A 9% gain in length was observed for the NBR-100 after 20 hours of immersion and a 12% after 8 days of immersion for NBR-200. This difference in behaviour between the two nitrile rubber gloves can be attributed to their difference in chemical composition which induces a different affinity to the colloidal solution. The difference in swelling behaviour between both kinds of protective gloves may well be one of the primary reasons that the aqueous $nTiO_2$ was able to more easily penetrate the NBR.

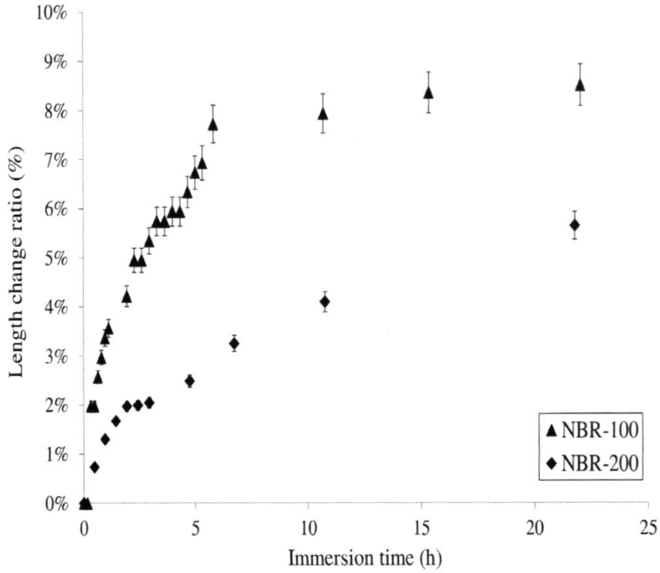

Figure 6. Variation in length change ratio as a function of immersion time in aqueous $nTiO_2$ for both nitrile rubber samples.

On the other hand, the sample glove surface morphology was characterized following exposure to the dynamic biaxial deformations (BD) without $nTiO_2$ in water. Figure 7(a) displays the variation of the surface features as a function of the number of deformations for NBR-100. In the case of the inner surface, no significant effect is recorded. By cons, the outer surface seems to be strongly affected by mechanical deformations. Indeed, an increase in pore surface area is observed after only 30 BD and more than three-fold after 180 BD. The deterioration of the materials could be attributed to abrasion of the sample surface by the probe [22]. This hypothesis is supported by the reduction in surface features observed when the mechanical deformations are combined with exposure to $nTiO_2$ in water which plays the role of a lubricating agent as shown in Figure 7(b). Such an observation could explain the jump in the Ti concentration in the sampling solution that was measured for NBR-100 after this number of deformations.

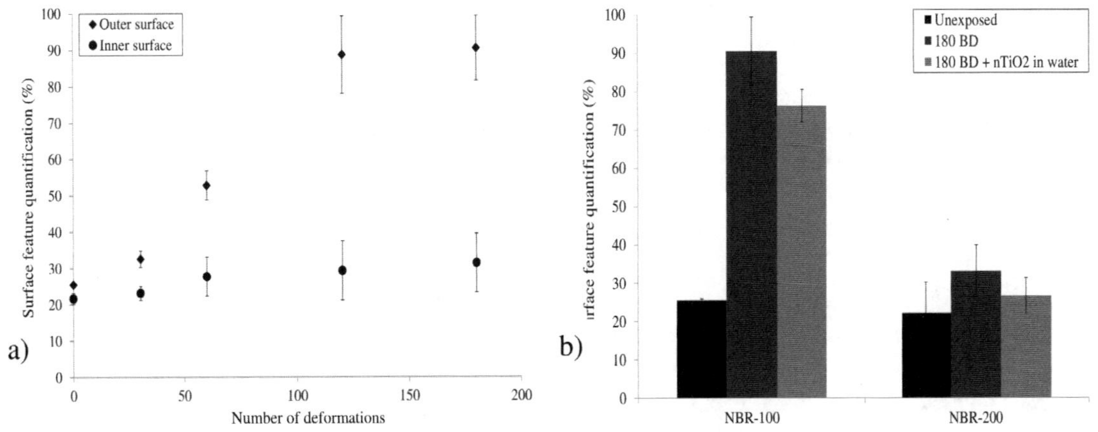

Figure 7(a). Variation of the surface features as a function of the number of BD for NBR-100 and Figure 7(b) Effect 180 BD combined or not with aqueous $nTiO_2$ on the outer surface features for gloves material samples.

A second mechanical test was to measure the relative strain energy during each deformation. As shown in Figure 8, an important decrease of the relative strain energy was observed especially during the first deformations and reached a tray. This phenomenon must be due to the Mullins effect. Indeed, some polymers chains have reached their limit of extensibility and break during the first deformations, reducing the energy required to deform the sample for the next ones [23]. The variation of the relative strain energy for NBR-100 and NBR-200 without NP was comparable. With aqueous $nTiO_2$, the energy required to deform the sample is less important because the elastomer membrane is weakened by its swelling [24]. As shown before, NBR-100 swells more than NBR-200 for a same immersion time. This explains the difference between both plateaus obtained with $nTiO_2$ in water. It is observed that the strain energy is low when the swelling of the sample is important.

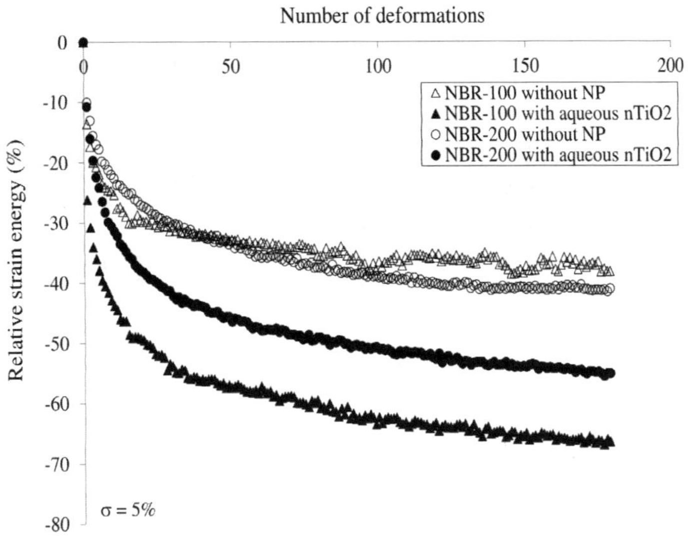

Figure 8. Variation of the relative strain energy as a function of the number of deformations for gloves material samples in contact or not with $nTiO_2$ in water.

5. Conclusion

This paper has investigated the penetration of $nTiO_2$ in water through nitrile protective gloves that were subjected to BD, corresponding to soft conditions of occupational use. The penetration of $nTiO_2$ was recorded for disposable nitrile gloves exposed to aqueous $nTiO_2$ after 5 and 7 hours. This penetration of the NPs through the protective material was attributed in part to a degradation of its physical and mechanical properties following repetitive mechanical deformations and contact with aqueous $nTiO_2$. Moreover, to simulate realistic work conditions encountered by protective gloves, new tests were performed. They consist in one 50%-BD applied per minute during the experiment duration (3 hours corresponding to 180 BD). These conditions increase the physical degradation of the glove samples which should further facilitate the penetration of NPs through protective gloves. In conclusion, the results presented show that great care must be exercised when selecting protective gloves for the handling of NPs. Even if more work is needed to confirm these worrying results, it is however already possible to recommend a frequent replacement of disposable gloves in case of exposure to NPs, especially when exposure of the NPs occurs in the liquid phase.

Acknowledgements

The authors would like to acknowledge the contribution of G. Perron, M. Ben Salah, J. De Santa Barbara (École de technologie supérieure) to the project as well as the contribution of Madjid Hadioui (Université de Montréal).

References

[1] Maynard, A., *Nanotechnology and occupational health*, 2005, EPA, June 13th 2005.
[2] Roco, M.C., C.A. Mirkin, and M.C. Hersam, *Nanotechnology Research Directions for Societal Needs in 2020 - Retrospective and Outlook*, S.P. Reports, Editor 2010, WTEC Study on Nanotechnology Research Directions: Berlin and Boston. p. 610.
[3] Hervé-Bazin, B., *Les nanoparticules: Un enjeu majeur pour la santé au travail ?* 2007: EDP Sciences. 704.
[4] Robichaud, C.O., et al., *Estimates of upper bounds and trends in nano-TiO_2 production as a basis for exposure assessment.* Environmental Science and Technology, 2009. **43**: p. 4227-4233.
[5] Höhr, D., et al., *The surface area rather than the surface coating determines the acute inflammatory response after instillation of fine and ultrafine TiO_2 in the rat.* International Journal of Hygiene and Environmental Health, 2002. **205**(3): p. 239-244.
[6] Warheit, D.B., et al., *Pulmonary toxicity study in rats with three forms of ultrafine-TiO_2 particles: Differential responses related to surface properties.* Toxicology, 2007. **230**(1): p. 90-104.
[7] Warheit, D.B., et al., *Pulmonary Instillation Studies with Nanoscale TiO_2 Rods and Dots in Rats: Toxicity Is not Dependent upon Particle Size and Surface Area.* Toxicological Sciences, 2006. **91**(1): p. 227-236.

[8] CCHST. *Basic Information on Titanium Dioxide*. 2007; Available from: http://www.cchst.ca/oshanswers/chemicals/chem_profiles/titanium_dioxide/basic_td.html.

[9] IARC, *Monographs on the evaluation of carcinogenic risks to humans - carbon black, titanium dioxide and talc*, 2010, World health organization: Lyon.

[10] Lahmani, M., F. Marano, and P. Houdy, *Les nanosciences - Nanotoxicologie et nanoéthique*. Vol. 4. 2010. 608.

[11] Rouse, J.G., et al., *Effects of mechanical flexion on the penetration of fullerene amino acid-derivatized peptide nanoparticles through skin.* Nano Letters, 2007. **7**(1): p. 155-160.

[12] Ryman-Rasmussen, J.P., J.E. Riviere, and N.A. Monteiro-Riviere, *Penetration of intact skin by quantum dots with diverse physicochemical properties.* Toxicol. Sci., 2006. **91**(1): p. 159-165.

[13] OECD, *Comparison of guidance on selection of skin protective equipment and respirators for use in the workplace: Manufactured nanomaterials*, in *Series on the safety of manufactured nanomaterials*, Editor 2009. p. 25.

[14] Golanski, L., A. Guiot, and F. Tardif. *Are conventional protective devices such as fibrous filter media, respirator cartridges, protective clothing and gloves also efficient for nanoaerosols ?* 2008; Available from: http://www.nanosafe.org/home/liblocal/docs/Dissemination%20report/DR1_s.pdf.

[15] Golanski, L., et al., *Experimental evaluation of personal protection devices against graphite nanoaerosols: fibrous filter media, masks, protective clothing, and gloves.* Human and Experimental Toxicology, 2009. **28**(6-7): p. 353-359.

[16] Golanski, L., A. Guoit, and F. Tardif, *Experimental evaluation of individual protection devices against different types of nanoaerosols: graphite, TiO_2 and Pt.* Journal of Physics: Conference Series, 2009. **170**.

[17] Ahn, K. and M.J. Ellenbecker, *Dermal and respiratory protection in handling nanomaterials at the center for high-rate nanomanufacturing (CHN)*, in *AIHce Conference*2006: Chicago, IL.

[18] Vinches, L., P. Dolez, and T. Vu-Khanh, *Study on the penetration of TiO_2 nanoparticles through nitrile and butyl protective gloves*, in *International Conference on Nanotechnology : Fundamentals and Applications*2011: Ottawa.

[19] Nohile, C., P.I. Dolez, and T. Vu-Khanh, *Parameters controlling the swelling of butyl rubber by solvents.* Journal of Applied Polymer Science, 2008. **110**(6): p. 3926-33.

[20] Domingos, R., N. Tufenkji, and K.J. Wilkinson, *Aggregation of titanium dioxide nanoparticles: role of natural organic matter.* Environ. Sci. Technol., 2009. **43**: p. 1282-1286.

[21] Harrabi, L., et al., *Evaluation of the flexibility of protective gloves.* International Journal of Occupational Safety and Ergonomics, 2008. **14**(1): p. 61-68.

[22] Dolez, P., et al., *Development of a test method for protective gloves against nanoparticles in conditions simulating occupational use.* Journal of Physics: Conference Series, 2011. **304**(1): p. 012066.

[23] Bueche, F., *Molecular basis for the mullins effect.* Journal of Applied Polymer Science, 1960. **4**(10): p. 107-114.

[24] Vinches, L., et al., *Swelling of elastomers in solutions of TiO_2 nanoparticles.* International Scholarly Research Network – Polymer Science, 2012.

Release characteristics of single-wall carbon nanotubes during manufacturing and handling

I Ogura[1,2], M Kotake[2], N Hashimoto[2], K Gotoh[3] and A Kishimoto[1,2]

[1] National Institute of Advanced Industrial Science and Technology (AIST), 16-1 Onogawa, Tsukuba, Ibaraki 305-8569, Japan
[2] Technology Research Association for Single Wall Carbon Nanotubes (TASC), 16-1 Onogawa, Tsukuba, Ibaraki, 305-8569, Japan
[3] Okayama University, 3-1-1 Tsushima-naka, Okayama, 700-8530, Japan

E-mail: i-ogura@aist.go.jp

Abstract. We investigated the release characteristics of single-wall carbon nanotubes (CNTs) synthesized by a pilot-scale plant. In addition to on-site aerosol measurements at the pilot-scale plant where the CNTs were synthesized, harvested, and packed, we conducted dustiness tests by vortex shaking and by transferring CNTs from one bowl to another. In the results of the on-site aerosol measurements, slight increases in the concentration were observed by aerosol monitoring instruments in the enclosure where CNTs were harvested and packed. In filter samples collected in this enclosure, micron-sized CNT clusters were observed by electron microscopy analysis. For samples collected outside the enclosure or during other processes, no CNTs were observed. The concentrations of elemental carbon at all locations were lower than the proposed occupational exposure limits of CNTs. The results of the dustiness tests revealed that submicron-sized particles were dominant in the number concentration measured by aerosol monitoring instruments, whereas micron-sized CNT clusters were mainly observed by electron microscopy analysis. The results of dustiness tests indicate that these CNTs have a low release characteristic. The lower drop impact of CNT clusters due to their lower bulk density resulted in lower CNT release from falling CNTs.

1. Introduction

Although carbon nanotubes (CNTs) have unique properties that make them potentially useful in many applications, there is growing concern regarding their potential impact on health. At present, workers (and researchers) involved in CNT manufacture and handling are potentially exposed to elevated levels of CNTs, mainly through inhalation. Recently, occupational exposure limits of airborne CNTs have been proposed [1-5]. Information on airborne CNT release, such as the magnitude, size distribution, and morphology of airborne CNTs, is required in order to evaluate and control CNT exposure.

In a previous study [6], we conducted on-site aerosol measurements at a research laboratory where a relatively small amount of single-wall CNTs were synthesized and handled. Recently, a pilot-scale plant has started the production of CNTs at a rate of 600 g/day, and the supply of CNT samples has accelerated the development of new applications. In this study, we investigated the release characteristics of single-wall CNTs synthesized by the pilot-scale plant. In addition to on-site aerosol measurements at the pilot-scale plant, we conducted dustiness (emission) tests by vortex shaking [7-8] and by transferring CNTs from one bowl to another.

2. Methods

2.1. Materials

CNTs were synthesized using the water-assisted chemical vapor deposition (CVD) method, which resulted in the massive growth of superdense and vertically aligned nanotube forests with heights of up to a few millimeters [9]. These can be easily separated from the catalysts on the substrate, providing nanotube material with a carbon purity of 99.9%. The outer diameter of the CNTs is approximately 3 nm. The specific surface area (using the Brunauer, Emmett, and Teller (BET) N_2 adsorption method) is around 1000 m^2/g.

2.2. On-site aerosol measurements at the pilot-scale plant

We performed on-site aerosol measurements at the pilot-scale plant where CNTs were synthesized, harvested, and packed.

CNTs were synthesized using a continuous reactor in which a series of synthesis processes were carried out along a conveyer belt. After the synthesis of CNTs, substrates (50 × 50 cm in size) with a few grams of CNTs emerged from the reactor along the conveyer belt. Subsequently, these were placed into a case by an automatic operation. The outlet of the reactor and its surroundings (including the conveyer belt and case) were enclosed, and there were local ventilation exhaust ports to purge air from the enclosure. During the on-site aerosol measurements, the case containing the substrates with synthesized CNTs was taken out from the enclosure twice, after 27 and 7 substrates of CNTs were synthesized.

In another enclosure, the synthesized CNTs were harvested from the substrates and were placed in a plastic bag by an automatic operation. During the on-site aerosol measurements, after 27 substrates of CNTs were harvested, the enclosure was opened once and the bag was sealed and taken out from the enclosure. Subsequently, 17 substrates of CNTs were harvested.

During the synthesizing, harvesting, and packing processes, the aerosol particles inside the enclosures, outside the enclosures (at a distance of approximately 50 cm from the opening of the enclosure), and at the center of the room (at a distance of several meters from the opening of the enclosures) were measured using three sets of portable real-time aerosol monitoring instruments. These included a condensation particle counter (CPC, model 3007, TSI Inc., USA, number concentration of aerosol particles with the size range between ~10 to >1000 nm), an optical particle sizer (OPS, model 3330, TSI Inc., USA, number concentration of aerosol particles with the size range between 300 to 10000 nm in optical diameter, 16 channels), a light-scattering laser photometer (LP, DUSTTRAK II, model 8530, TSI Inc., USA, mass concentration of aerosol particles) with a cyclone for respirable dust (GS-3 Multiple-inlet Cyclone; 4 μm particles are cut by 50% at 2.75 L/min; SKC Inc., USA), and a black carbon monitor (BC, microAeth, model AE51, AethLabs, USA, mass concentration of aerosol black carbon). The sampling flow rates of the CPC, OPS, LP, and BC were 0.7, 1.0, 2.75, and 0.1 L/min, respectively. The relative responses of the LP and BC to the mass concentration of the CNTs synthesized in this plant were 1.2 and 0.23, respectively [10].

In addition to real-time aerosol monitoring, the aerosol particles were collected on quartz fiber filters (37-mm diameter, 2500 QAT-UP, PALL Corporation, USA) with/without a cyclone for respirable dust (GS-3 Multiple-inlet Cyclone). The quantities of elemental carbon in respirable and total dust were measured by a carbon analysis (see section 2.6). The sampling flow rates of the respirable and total dust sampling were set at 2.75 and 3.0 L/min, respectively.

Aerosol particles were also collected on Nuclepore membrane filters (25-mm diameter, 0.080-μm pore size, Nomura Micro Science Co., Ltd., Japan) with a stainless steel filter holder at a sampling flow rate of 0.5 L/min for scanning electron microscopy (SEM) analysis.

2.3. Dustiness tests by vortex shaking

Dustiness tests were conducted by vortex shaking [7-8]. Approximately 1 cm^3 of nanomaterial was placed in a test tube having an outer diameter of 25 mm and a length of 20 cm and agitated using a laboratory vortex shaker (Cole-Parmer Instrument Co., Vortex Genie 2 Shaker), as shown in figure 1. Aerosol particles released during agitation with clean air passing through the test tube were continuously measured. Conductive silicone tubing was used to transmit the particles to the measuring instruments.

The aerosol particles were measured using the CPC, OPS, LP, a scanning mobility particle sizer (SMPS, model 3936L72, TSI Inc., USA, number concentration of aerosol particles with the size range between 14 to 400 nm in electrical mobility diameter, 64 channels/decade), and an aerodynamic particle sizer (APS, model 3321, TSI Inc., USA, number concentration of aerosol particles with the size range between 500 to 20000 nm in aerodynamic diameter, 32 channels/decade). The sampling flow rates of the CPC, OPS, LP, SMPS, and APS were 0.7, 1.0, 2.5, 1.0, and 5.0 L/min, respectively. The sheath flow rate of the SMPS was 3 L/min. The SMPS up-scan time and retrace time were 160 s and 15 s, respectively, and the SMPS data were saved at 3-min intervals by one scan. A cyclone (SKC aluminum cyclone; 4 μm particles are cut by 50% at 2.5 L/min; SKC Inc., USA) was connected to the inlet of the LP to obtain the concentration of respirable dust.

The aerosol particles were also collected on Nuclepore membrane filters at a sampling flow rate of 0.3 L/min for SEM analysis.

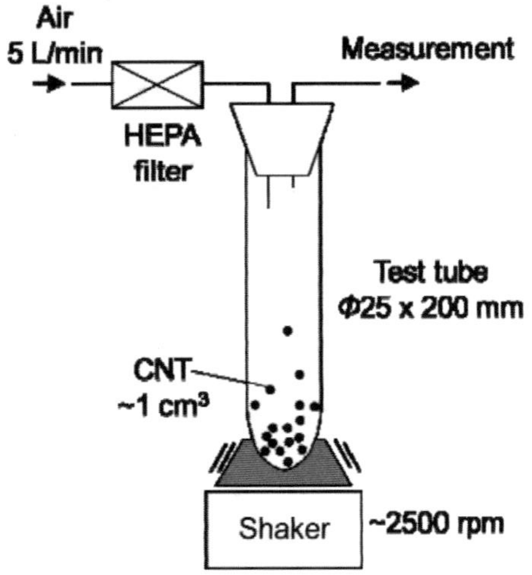

Figure 1. Dustiness tests by vortex shaking.

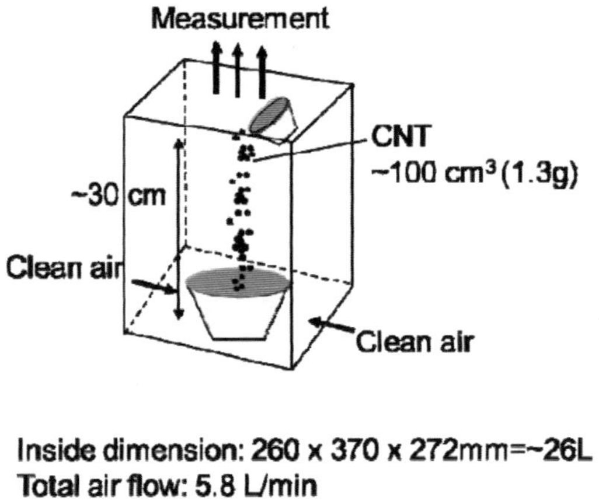

Figure 2. Dustiness tests by transferring CNTs.

2.4. Dustiness tests by transferring CNTs

In an acrylic glove box with an inflow of air through a high efficiency particulate air (HEPA) filter, we repeatedly transferred approximately 100 cm^3 of CNTs from one bowl to another, as shown in figure 2. The majority of the inner surface of the glove box was lined with aluminum foil for static protection.

The aerosol particles were measured using the CPC, OPS, LP, and SMPS. The sampling flow rates of the CPC, OPS, LP, and SMPS were 0.7, 1.0, 2.75, and 1.0 L/min, respectively. A cyclone (GS-3 Multiple-inlet Cyclone) was connected to the inlet of the LP to obtain the concentration of respirable dust. Conductive silicone tubing was used to transmit the particles to the measuring instruments.

Aerosol particles were also collected on Nuclepore membrane filters at a sampling flow rate of 0.3L/min for SEM analysis.

2.5. SEM analysis

The aerosol particles collected on the Nuclepore filters were observed using a field-emission scanning electron microscope (SEM, S-4300, Hitachi High-Technologies Corporation, Japan) under an accelerating voltage of 3 kV. Prior to sample collection, the filters were coated with platinum-palladium (approximately 2 nm) to avoid image charging.

2.6. *Carbon analysis*

The CNT mass on the quartz fiber filter was quantified as a quantity of elemental carbon using an organic and elemental carbon (OC/EC) analysis instrument (CAA-202M-D, Sunset Laboratory Inc., USA) based on the NIOSH 5040 method (however, optical pyrolysis correction was not used).

3. Results and discussion

3.1. *On-site aerosol measurements at the pilot-scale plant*

While ethylene gas was being used for the synthesis of CNTs, unreacted ethylene gas in the exhaust was burned in the upper portion of the reactor. During that time, increases in the number concentrations were observed by the CPCs at all locations. These increases could have been a result of the particles emitted by a combustion process, such as soot and condensation particles, as combustion processes are commonly accompanied by the emission of nanoparticles.

Slight increases in the concentration of aerosol particles in the enclosures were observed by the aerosol monitoring instruments when the first substrate with synthesized CNTs emerged from the reactor (<2 particles/cm^3 with the OPS and <2 μg/m^3 with the LP) and when CNTs were being harvested and packed (<30 particles/cm^3 with the CPC, <2 particles/cm^3 with the OPS, <10 μg/m^3 with the LP, and <1 μg/m^3 with the BC). No distinct increase associated with the work was observed in the concentration of aerosol particles outside the enclosures.

In filter samples collected in the enclosure during the harvesting process, elemental carbon was significantly detected (but at levels less than the quantitation limit) through carbon analysis (table 1), and micron-sized CNT clusters were observed by SEM analysis (figure 3). For samples collected outside the enclosures or during other processes, elemental carbon was not significantly detected and no CNTs were observed. The concentrations of elemental carbon at all locations were lower than the occupational exposure limits of airborne CNTs proposed by the National Institute for Occupational Safety and Health (NIOSH), US (7 μg/m^3) [2] and the National Institute of Advanced Industrial Science and Technology (AIST), Japan (30 μg/m^3) [3].

Table 1. Mass concentration of elemental carbon at the pilot-scale plant (μg/m^3).

	Respirable dust [a]	Total dust [a]
Synthesizing CNTs (inside enclosure)	<2.3	<2.1
Synthesizing CNTs (outside enclosure)	<2.3	<2.1
Harvesting and packing CNTs (inside enclosure)	<1.2	(2.1)
Harvesting and packing CNTs (outside enclosure)	<1.2	<1.1
The center of the room	<0.56	<0.55

[a] The values preceded by '<' are the concentrations that are below the detection limit. The values in parentheses are the concentrations that are below the quantitation limit.

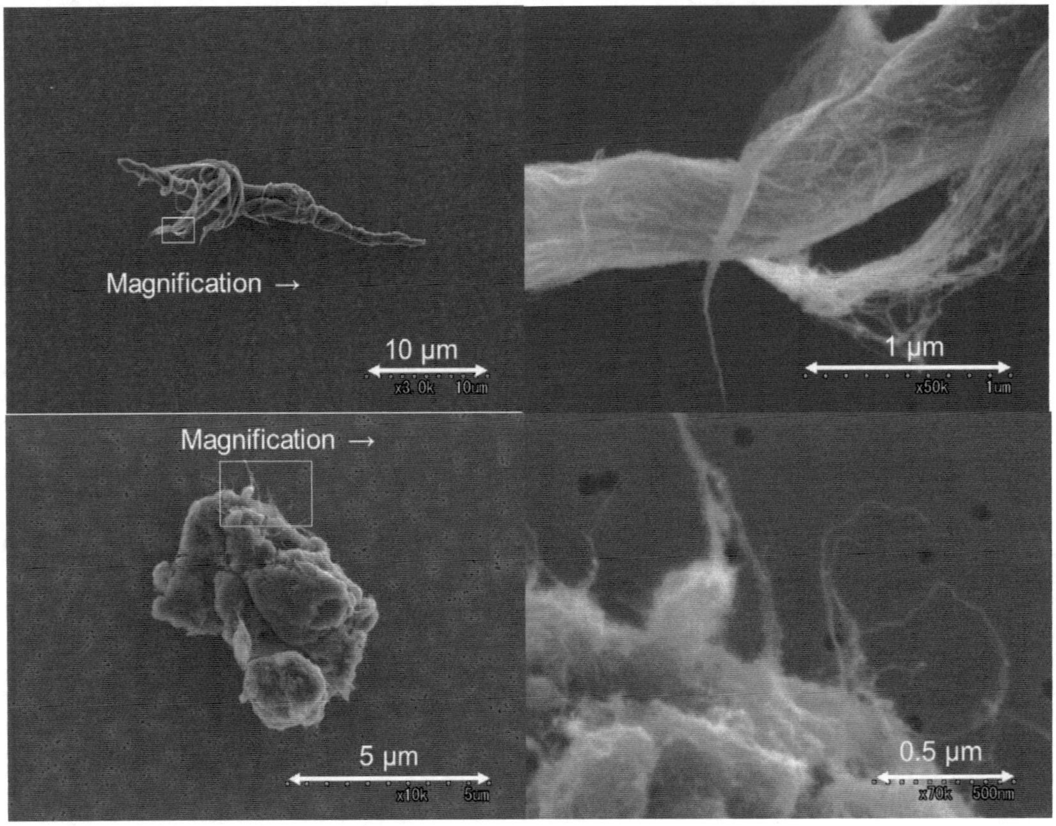

Figure 3. SEM images of aerosol particles collected in the enclosure where CNTs were harvested and packed.

3.2. Dustiness tests by vortex shaking

The number concentration of particles, with a diameter of approximately 10–>1,000 nm, as measured by the CPC, was 2.1 particles/cm^3, whereas the mass concentration of respirable dust measured by the LP was 1.6 μg/m^3 (mean values from 1 min to 31 min after commencement of agitation, $n = 3$). These concentrations were considerably lower than those of most other nanomaterials [8], indicating that these CNTs have a lower release characteristic.

The number-based particle size distribution for emitted particles is shown in figure 4. Submicron-sized particles were dominant in the number concentration measured by aerosol monitoring instruments.

SEM images of emitted particles in dustiness tests are shown in figure 5. Micron-sized CNT clusters were mainly observed.

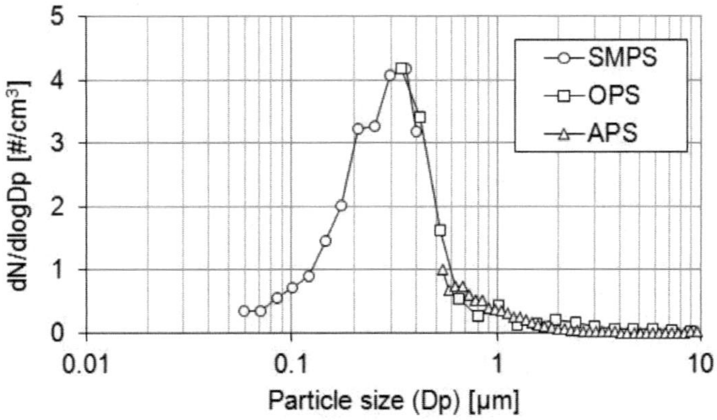

Figure 4. Number-based particle size distribution for emitted particles in dustiness tests by vortex shaking. Mean values from 1 min to 31 min after commencement of agitation (n = 3). The particle sizes are expressed in terms of electrical mobility diameter for the SMPS result, optical diameter for the OPS result, and aerodynamic diameter for the APS result.

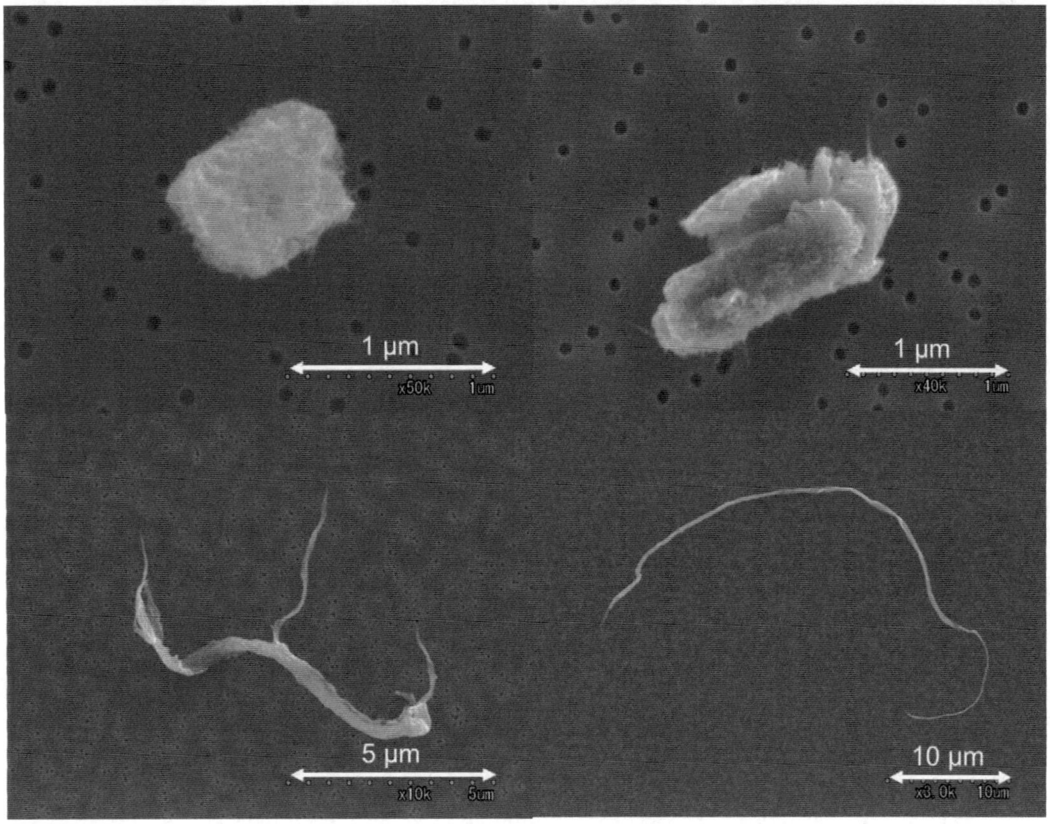

Figure 5. SEM images of emitted particles in dustiness tests by vortex shaking.

3.3. Dustiness tests by transferring CNTs

Figure 6 shows the changes in the concentrations during the transfer of CNTs, as measured by the CPC, OPS, and LP. Only slight increases in the concentration of aerosol particles were observed (<10 particles/cm^3 with the CPC, <1 particles/cm^3 with the OPS, and <10 μg/m^3 with the LP). The lower drop impact of CNT clusters due to their lower bulk density resulted in lower CNT release from falling CNTs. For the LP result, a gradual elevation of the concentration was seen. The fact that the elevated concentration was not recovered even when a HEPA filter was connected to the inlet of the LP explained that this elevation was caused by the zero-point shifting of the LD response.

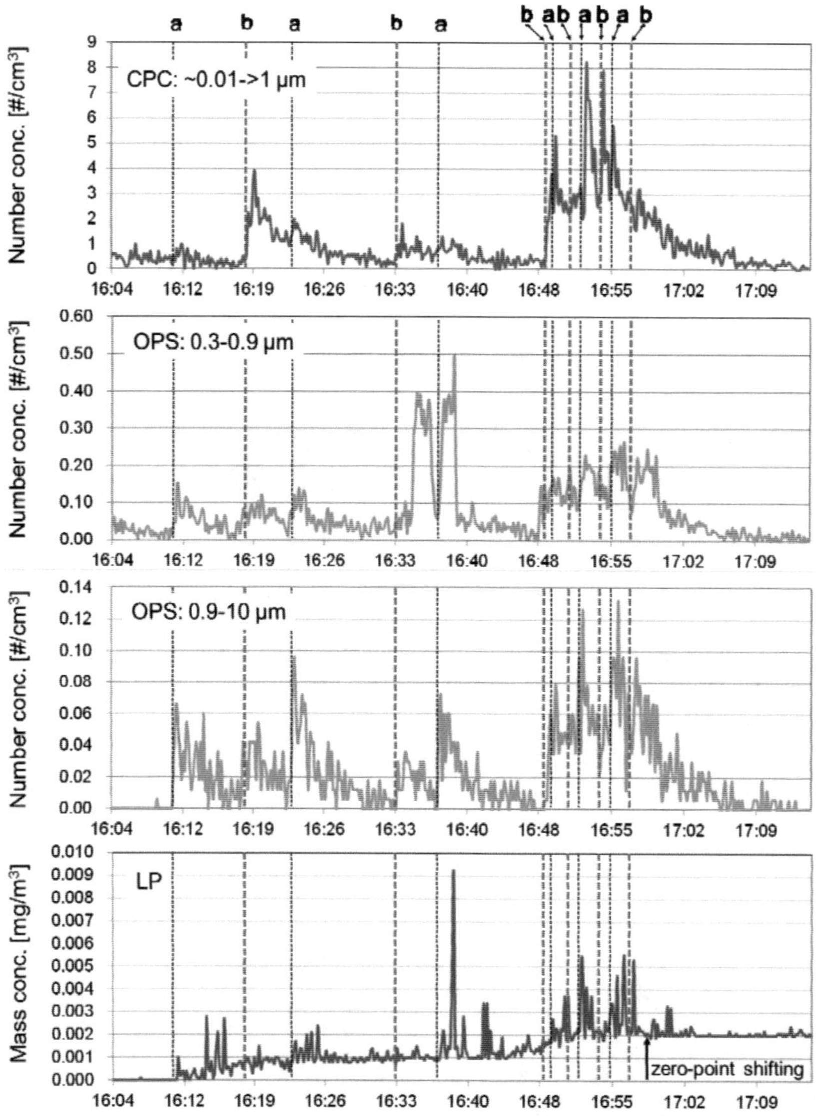

Figure 6. Changes in the concentrations of emitted particles during the transfer of CNTs.

[a] CNTs were dropped from one bowl at a height of approximately 30 cm to another
[b] CNTs were poured directly from one bowl to another

The number-based particle size distribution for emitted particles during the transfer of CNTs, as measured by the OPS, together with those from harvesting and packing and vortex shaking, are shown in figure 7. The patterns of particle size distributions measured by the OPS were similar to each other. Meanwhile, the reliable particle size distribution could not be obtained from the SMPS data during the transfer of CNTs due to the low concentration.

For the filter sample collected during the transfer of CNTs, as with those collected during harvesting and packing and vortex shaking, micron-sized CNT clusters were observed by SEM analysis.

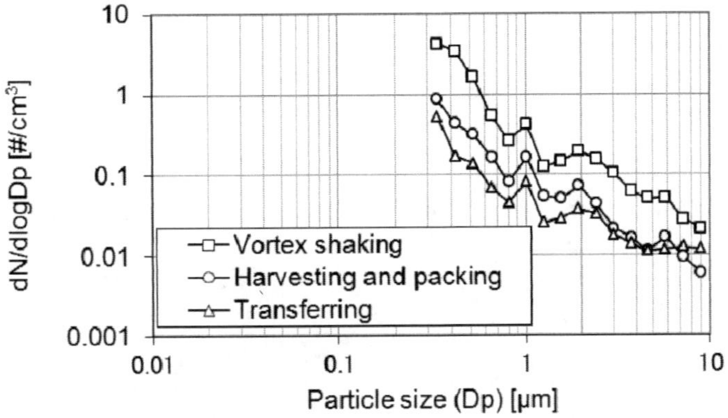

Figure 7. Number-based particle size distributions for emitted particles in harvesting and packing, vortex shaking, and transferring of CNTs measured by OPS. The particle sizes are expressed in terms of optical diameter. For the harvesting and packing results, background concentration has been subtracted.

4. Conclusions

We investigated the release characteristics of single-wall CNTs synthesized by the pilot-scale plant. In the results of on-site aerosol measurements, an obvious release of CNTs was only seen in the enclosure where CNTs were harvested and packed. For the filter sample collected in this enclosure, micron-sized CNT clusters were observed by SEM analysis. The concentrations of elemental carbon at all locations including the inside and outside of the enclosures were lower than the occupational exposure limits of airborne CNTs proposed by US NIOSH (7 $\mu g/m^3$) and AIST, Japan (30 $\mu g/m^3$).

The results of dustiness tests revealed that submicron-sized particles were dominant in the number concentration measured by aerosol monitoring instruments, whereas micron-sized CNT clusters were mainly observed by SEM analysis. The results of dustiness tests indicate that the CNTs have a lower release characteristic. The lower drop impact of CNT clusters due to their lower bulk density resulted in lower CNT release from falling CNTs.

Acknowledgment

This work was funded by the New Energy and Industrial Technology Development Organization of Japan (NEDO) under a grant for "Innovative carbon nanotubes composite materials project toward achieving a low-carbon society" (no. P10024).

References

[1] EC 2010 *Engineered Nanoparticles: Review of Health and Environmental Safety (ENRHES). Project final report* http://ihcp.jrc.ec.europa.eu/whats-new/enhres-final-report

[2] NIOSH 2010 NIOSH *Current Intelligence Bulletin, Occupational Exposure to Carbon Nanotubes and Nanofibers (November 2010 Draft)*

http://www.cdc.gov/niosh/docket/review/docket161A/default.html

[3] Nakanishi J (ed) 2011 *Risk Assessment of Manufactured Nanomaterials: Carbon Nanotubes. Final report issued on 17 August 2011.* NEDO project (P06041) "Research and Development of Nanoparticle Characterization Methods" http://www.aist-riss.jp/main/modules/product/nano_rad.html?ml_lang=en

[4] Pauluhn J 2010 Multi-walled carbon nanotubes (Baytubes®): approach for derivation of occupational exposure limit *Regu.l Toxico.l Pharmacol.* **57** 78–89

[5] Luizi F 2009 Responsible care and nanomaterials - case study Nanocyl *European Responsible Care Conference (Prague, Czech Republic, 21–23 October 2009)* http://www.cefic.org/Documents/ResponsibleCare/04_Nanocyl.pdf

[6] Ogura I, Sakurai H, Mizuno K and Gamo M 2011 Release potential of single-wall carbon nanotubes produced by super-growth method during manufacturing and handling *J. Nanopart. Res.* **13** 1265–80

[7] Maynard A D, Baron P A, Foley M, Shvedova A A, Kisin E R and Castranova V 2004 Exposure to carbon nanotube material: aerosol release during the handling of unrefined single-walled carbon nanotube material *J. Toxicol. Environ. Health* A **67** 87–107

[8] Ogura I, Kotake M, Sakurai H and Gamo M 2012 *Emission and Exposure Assessment of Manufactured Nanomaterials. English Version. (26 October 2012).* NEDO project (P06041) "Research and Development of Nanoparticle Characterization Methods" http://www.aist-riss.jp/main/modules/product/nano_rad.html?ml_lang=en

[9] Hata K, Futaba D N, Mizuno K, Namai T, Yumura M and Iijima S 2004 Water-assisted highly efficient synthesis of impurity-free single-walled carbon nanotubes *Science* **306** 1362–64

[10] Hashimoto N, Ogura I and Kishimoto A 2012 Evaluating usefulness and responses of portable aerosol measuring instruments when measuring carbon nanotubes *SENN2012 - Int. Congress on Safety of Engineered Nanoparticles and Nanotechnologies (Helsinki, Finland, 28–31 October 2012)* p 53

Mapping the Use of Engineered NM in Quebec's Industries and Research Laboratories

Claude Ostiguy[1,2,] Claude Emond[2], Inès Dossa[2,3], Yasmina Malki[3], Chantale Boily[3], David Roughley[3], Anton Plavski[2] and Charles-Anica Endo[3]

(1) Institut de recherche Robert-Sauvé en santé et en sécurité du travail – 505 West Blvd. De Maisonneuve, Montréal, Qc, Canada H3A 3C2,
(2) University of Montreal, Department of Environmental and Occupational Health, PO Box 6128 Main Station, Montréal, Qc, Canada,
(3) AGY Consulting – 3 Place du Commerce, suite 500, Verdun, Qc, Canada, H3E1H7

Corresponding author's: e-mail: Charles-a.endo@agyconsulting.com,

Abstract. Engineered NanoMaterials (NM) offer an opportunity to develop a wide variety of new products with unique properties but many studies have shown potential OHS risks specific to NM. Addressing these risks requires knowledge about release of NM into the workplaces. This research aimed to map the state of nanotechnology OHS practices in Quebec through a questionnaire following a first contact by telephone when possible and by compiling the type and volumes of NM used as well as gathering information related to the working conditions and OHS aspects. This survey was conducted among 1310 Quebec industries and 653 researchers working in different specialties potentially involved in the development/production/distribution/integration of NM and use of NM containing products. Overall, 90 questionnaires, including 51 from the industries, were completed. These showed that NM are mainly used into the powder form, in many different sectors and deserve a wide range of markets. The prevention measures implemented vary widely from a workplace to another but about one third of the participants report that they have implemented NP adapted prevention measures but they remain worried on some specific operations. More than 50% of the participants request more information about the safe laboratory/plant design, toxicity, regulation, good work practices and prevention measures, efficiency of personal protective equipment and environmental impacts.

1. Introduction

Nanotechnology (NT) is developing rapidly and new products containing nanomaterials (NM) are being marketed every week. Extrapolating the existing data, it can be estimated that around 1800 products containing NM are currently commercially available [1]. In 2009, these products were worth 224 billion US dollars and contained for one billion dollars of NM [2]. The development of these NM, their industrial large-scale production, distribution, and their integration into different products followed by machining of some NM containing products already results in a potential occupational exposure to NM of hundreds of thousands of workers worldwide. It is estimated that by 2015, up to 10% of workers into the manufacturing sector could be involved in nanotechnologies (NT) [3]. Research in NT is very active in Québec and NanoQuébec [4] promotes the industrial implementation of new applications. Consequently, for the last 5-10 years, many new companies are entering this field. This trend should continue for several years. One direct impact from this displacement from research laboratories to industrial production is that the number of potentially exposed workers as well as the volume of NM handled substantially increases. Another impact is that the level of knowledge about the specific potential risks and the prevention measures to implement with NM can be lower in the industry, especially in small and medium size companies, compared to research laboratories or large companies.

The toxicity of NM has been widely studied and several literature reviews are now available [5-9]. Many toxicological studies on different substances have demonstrated toxic pulmonary effects including inflammation, fibrosis and granuloma. Health effects have also been reported on various organs and biological systems including the circulatory, cellular, cutaneous, renal and reproductive systems. It is found that in general, a NM will usually be more toxic than the same chemical substance with a larger dimension. The literature gives us very little information specific to NM relating to their physical hazards such as fires or explosions but NM can be much more reactive than larger particles [10, 11]. Overall, it is recognized that NM remains an important source of concern in OHS. In fact, not only does the diversity of commercially available chemical products of nanometric dimensions continues to increase, but also, the information available about the hazards specific to each of these substances is still very fragmentary.

Addressing the risks of NM requires knowledge about the occupational exposure. However, in Québec, this information is not currently available and risk assessment lacks of quantitative data. Consequently, the essential first step is to identify the different workplaces where workers could potentially be exposed to NM. These include the development (research laboratories), the industrial production, distribution, and integration of NM into final products and the use of these products into production or assembly lines (Ex. CNT containing composites used in the production of transportation vehicles). *The project aims to 1)-_ identify and document different workplaces using NM and, 2)-map the state of nanotechnology OHS practices in Quebec as well as gather specific information related to the working conditions and needs for additional information.*

2. Methods

2.1. Nanoparticle Definition

Engineered NM are designed with specific properties. Engineered NM encompasses nano-objects and nanostructured materials. The former are defined as materials with one (nanoplate), two (nanofibre), or three external dimensions (nanoparticle) at nanoscale level (i.e., between approximately 1 and 100 nm). Examples of nanostructured materials are nanocomposites composed of nano-objects embedded in a solid matrix or nano-objects bonded together in simple random assemblies as in aggregates and agglomerates or ordered as in crystals of fullerenes or carbon nanotubes (CNT) based on ISO/TS 80004-137 [12, 13]. In this paper, NM definition excludes undesired pollutants of nanometric dimensions (ultrafine dusts) which originate from natural (fire fumes, pollens…) or anthropomorphic (diesel or welding fumes…) sources of Nanoparticles.

2.2. *Questionnaire.*

A questionnaire adapted to the specific objectives of the study has been elaborated from a literature review covering previous international or local surveys [14-24]. The questionnaire was validated by a small group of 3 researchers and 3 representatives from the industries directly involved in NT. The questionnaire was available in French and in English.

The 17 questions aimed to document potential exposures to NM in different workplaces. It targeted the size of the total workforce and the number of people directly involved in NT, their education level, the NM involved, and their potential release into the work environment, their quantities, origins and physical states, the already developed or targeted markets, the OHS procedures and prevention measures already implemented and the additional information/knowledge and support deemed useful to improve OHS aspects regarding NM.

2.3. *Study population.*

Previous international surveys have been used as a starting point to identify the economic sectors where NM has been documented [14-24]. A total number of 1310 companies have been identified and cover the activity sectors previously identified among which chemistry, aeronautics, rubber, plastic manufacturing, pharmaceutics and neutriceutics, photonics, foundries and metal transformation, composites, textiles, inks and pigments, paints, cosmetics, and electronics. The proportion of companies in each sector is not representative of the Quebec industry.

Researchers in universities (n=653) and research centers were identified through publicly available information (NanoQuébec web site, lists of research granting agencies, etc.). Research is performed on a very wide diversity of topics including physics, chemistry, mechanics, industrial engineering, agro-foods, mines, metallurgy, materials, software and information technologies, biology, biochemistry, biotechnology, bio-products, wood and forest sciences, construction, environment, pharmaceutics and pharmacology, telecommunications, etc…

2.4. Survey design and sampling

Two different approaches were used. The companies were contacted (once or twice if no answer) by phone to explain the project, document if they work with NM and, if so, invite them to fill up the secured questionnaire available on the web (Survey Monkey™). The academic researchers were not contacted by phone. The companies which confirmed their use of NM and agreed to fill-up the questionnaire, the companies which did not return our phone calls and the academic researchers received, by email, an invitation to fill-up the questionnaire. The email contained all the required information about the objectives of the project, the research team, the ethical aspects and a personal confidential identification code. Two recall emails were sent to the participants who did not complete the survey.

2.5. Data Analysis

The data collected from the English and the French versions of the questionnaire filled on Survey Monkey™ have been integrated in the same file. The data have been analysed using the SPSS® version 17, and Microsoft Excel®, version 2010. Using the SPSS® software, tables were produced to group participants by sectors for specific questions. Microsoft Excel® was used to produce Graphs.

3. Results

3.1. Description of the participants

A total of 2001 participants were selected including 1310 from the industries, 38 from CEGEPs (technical schools programs) and 653 from universities and research centers. Globally, 90 participants completed the questionnaire (Table 1). In the industrial sector, 64% employ less than 50 workers and 6% more than 500. Nanotechnology represents less than 10% of the activities in 75% of the industrial plants, suggesting that the NM are used mainly in already established industries. The industrial sector covers the production of NM and the required equipment, importation, exportation, distribution, and use/integration of NM as well as suppliers of raw materials already integrating NM and consultants. Among the participants involved in NT, the level of education is higher than in the other industries as 30% of the industries have at least one employee with a master degree or higher. Only 8% of the industries (mainly producers of NM) report that NM involves more than 60% of their activities. Very few researchers are involved only in nanotechnology. In fact, NT accounts for less than 10% of the activities for 53% of the researchers and more than 60% for only 13% of them.

The industrial sector has developed its market in several sectors including health, biotechnology, pharmaceutics, chemicals and chemical processes, equipment, composite materials, sport goods, transportation, aeronautics, electronics, construction, consumer products, military, energy, safety, mines and the environment, agriculture and food, and information and communication technologies. Research is very active in the fields of nanochemistry, bionanotechnology, medical diagnostic, nanophotonics, nanoelectronics and nanocellulose. Some academic researchers are also involved in OHS, and environment.

In the province of Québec, the results clearly suggest that NM are currently used mainly into powder form in the industries as well as in the academic research sector. Carbon black, nanoclays, nanocellulose, and carbon nanotubes are the most frequently used NM. Most laboratories use less than one kilogram per year and only seven industries reported that they use more than 100 kgs per year.

3.2. Occupational health and safety practices

Many operations and work practices can release some NM in the work environment. Among these, the employers are mainly concerned by weighing, dry transfer, and waste management. Processing, dry packaging, reactors and mixers operations, and dismantling and cleaning of equipment are also of concern to many industrials. Many employers report they have implemented satisfactory control measures but some are still looking for improved control measures, especially for dismantling and cleaning operations, and waste management.

Surprisingly, more academic researchers seem concerned by the potential release of NM in their labs environment despite the fact that more industrials completed the survey. Many academic researchers have implemented satisfactory control measures but some are still concerned and are seeking for new or improved methods for all the operations proposed in the survey.

About 50% of the industries and 60 % of the academic researchers have implemented some specific OHS practices to handle NM in the workplace. The most frequently reported actions by the industry are OHS directives given to their employees, information on risks, training of workers, prevention/information on safe practices, and availability of a health professional. The academic researchers report, in decreasing frequency information on risks, training of workers, OHS directives, information/prevention programs on safe practices, and availability of a health professional. Figure 1 illustrates the prevention measures implemented in the industry and in the academic research laboratories. Ventilation, personal protective equipment and detailed work procedures are the most frequently reported specific prevention measures reported. About one third of the participants reported that prevention measures are not applicable.

The last question of the survey covered the additional information and knowledge deemed useful to improve OHS regarding NM. Figure 2 clearly shows that more information/knowledge is considered valuable on all aspects of OHS. In general, a larger proportion of researchers would like to get more information than the industry.

4. Discussion

The aim of this survey was 1) to develop a questionnaire specific to NM and to establish the mapping of NM in Québec based on public and private industry lists available.as well as 2) to gather information about workplaces ,work practices and prevention measures related to NP. Based on an initial literature review, the survey questionnaire has been developed and adapted to reflect Québec's reality. Two experimental approaches have been used in the survey. Some companies (n=650) have been directly contacted by phone and 29% of those who confirmed their use of NM have answered, on a voluntary basis, to the questionnaire sent by email. This proportion is higher than the 19% of the international ICON study published in 2006 [18, 19] using a similar approach but lower than the 58% of the Swiss study [23] where the participation to such study is mandatory. The other companies and the researchers have received an invitation by email only. Respectively 6.40% and 6.33% of the industries and the researchers completed the survey without being previously contacted. This result suggests that a preliminary contact promotes the level of response but some people not involved in NT possibly concluded that it was not useful to complete the survey. Considering that 51 industries involved in NT completed the survey and assuming a response ratio of 29%, this preliminary survey would suggest that, at least, 175 industries are involved in NT in Québec. Far more than 200 research teams are currently running research projects on NT according to NanoQuébec. The minimal current workforce potentially exposed to NM in Québec is estimated to be more than 2 000 people.

The methodology of this study could be improved in some ways. First, the industries could be selected on a totally random basis, covering all the activity sectors and be proportional to the importance of each economic sector. Every participant should first be contacted by phone and it could be offered to complete the survey by phone or in their workplace. The participant should be contacted again for quality-control when some answers are incomplete or seem inadequate. For example, a few industries mentioned that they already have products on the markets, gave complete indications on the NM they use but report that no employee are involved in NT or handling NM. Finally, working in collaboration with associations (Ex. industry associations) could increase the participation to this voluntary survey. It is not surprising to see that most of the industries involved in NT are small or medium size industries. This reflects the Québec industrial profile [25]. NT represents more than 10% of the activities for 53% of the companies involved in this survey. Many companies developed their own applications, have a research laboratory and have already marketed some products. Also, some factories have the capacity of large size production of carbon nanotubes, nanometric size metals, metal oxides, ceramics, and nanocellulose. Considering the industry and research in universities and research centers, the total manpower involved in NT is estimated to be larger than 2000 spread in 175 industries and more than 200 research teams.

The survey suggests that NM are used mostly into the powder form, not only in the industry but also in most research laboratories. Considering OHS, the industries use much larger quantities of NM but seem generally less concerned about the potential release of NM in their workplace or have implemented adequate prevention measures. A detailed study of the results suggests that the most concerned industries about the safety of their employees are the industries which have their own research laboratories. In academic research laboratories as well as in the industries which produce NM, some participants are concerned and are looking for efficient prevention measures especially for dismantling and cleaning of equipment and NM waste management. OHS practices in industry and research are quite similar but follow-ups seem more regular for the industries. Some prevention measures are already implemented in most workplaces but surprisingly many people report that this is not applicable in their workplace. Globally, most participants would appreciate to have additional information on the OHS and environmental aspects.

5. Conclusion

Despite the limitations of such a targeted survey, it appears clearly that Québec is deeply involved in research in NT, has an important NM production capacity (nanocellulose, carbon nanotubes, nanometric metals, metal oxide and ceramics) and many workplaces have already implemented good work practices. New products are being continuously developed, shifting from the laboratory to the industry. Consequently, an increasing number of workers are potentially exposed to NM and handle much larger quantities. Some industries and laboratories are still concerned about the safety of some specific operations (dismantling and cleaning of reactors, waste management) but most participants request more information about all OHS aspects (design, engineering, administrative measures and personal protective equipment).

6. References

[1] Woodrow Wilson Center for Scholars, 2013. *Consulted January 15, 2013.* http://www.nanotechproject.org/inventories/consumer/
[2] Bradley L, 2010. Nanotech's evolving environmental, health and safety landscape: the regulations. NanoSafe Conference, Grenoble, France, November 16, 2010.
[3] INRS, 2012. Nanomatériaux – Prévention des risques dans les laboratoires. ED6115, Institut national de recherche et de sécurité, France, 56 p.
[4] NanoQuébec web site, 2013. http://www.nanoquebec.ca/fr/index.php
[5] Stone, V., Hankin, S., Aitken, R., Aschberger, K., Baun, A., Christensen, F., Fernandes, T., Hansen, S.F., Hartmann, N.B., Hutchison, G., Johnston, H., Micheletti, C., Peters, S., Ross, B., Sokull-Kluettgen, B., Stark, D., Tran, L. 2009, ENRHES - Engineered nanomaterials: review of health and environmental safety, EC contract number 218433 448 p.
[6] Ostiguy C, B, Roberge C Woods, B Soucy, G Lapointe, L Ménard, 2010. Les nanoparticules de synthèse: connaissances actuelles sur les risques et les mesures de prévention en SST, Seconde édition, IRSST, Études et recherches / Rapport R-646, 159 p.
[7] S Sharifi, S Behzadi, S Laurent, L Forrest, P Stroeve, M Mahmoudi, 2012. Toxicity of NM. Critical Review. Chem Soc Rev 41:2323-2343.
[8] AA Shvedova, An Pietroiusti, B Fadeel, VE Kagan, 2012. Mechanisms of carbon nanotubes-induced toxicity : focus on oxidative stress. Review article. Tox Appl Pharm 261:121-133.
[9] Reijnders L, 2012. Human health hazards of persistent inorganic and carbon nanomaterials. Review. J Mater Sci 47:5061-5073.
[10] Eckhoff R.K., 2011. Are enhanced dust explosion hazards to be foreseen in production, processing and handling of powders consisting of nano-particles? Journal of Physics: Conference Series 304 (2011) 012075, 20 p.
[11] Li Q., Lin B., Li W., Zhai C., Zhu C., 2011. Explosion characteristics of nano-aluminum powder-air mistures in 20L spherical vessels. Powder Technology 212:303-309.
[12] International Standard Organisation (ISO): ISO/TC 229 Nanotechnologies. Nanotechnologies – Terminology and Definitions for Nanomaterials. ISO/TS 27687:2008. Geneva: ISO Central Secretariat, 2008.
[13] Canadian Standards Association, 2012. Nanotechnologies — Exposure control program for engineered NM in occupational settings. CSA Z12885-12. Mississauga CSA, 2012.
[14] NanoQuébec (2007). Profil de la nanotechnologie au Québec. Raport interne # 28; Les petits executants R-D au Québec : rapport d'enquête, Institut de la Statistique du Québec.
[15] StatCan (2007). Aperçu et examen des résultats de l'enquête pilote sur la nanotechnologie au Canada. Ottawa, Section des enquêtes, de l'innovation et de l'information electronique (DSIIE); Ministère de l'industrie.
[16] ISQ (2010). Les petits exécutants R-D au Québec : rapport d'enquête, Institut de la statistique du

Québec, Gouvernement du Québec.
[17] Hock, B., D. Kassel, E. March and D. Netherton (2006). Survey of U.S. nanotechnology executives, Small Times Magazine and The University of Massachusetts Lowell.
[18] ICON, G. Gerritzen, L.-C. Huang, K. Killpack, M. Mircheva and J. Conti (2006a). A Review of Current Practices in the Nanotechnology Industry. Phase Two Report: Survey of current practices in the nanotechnology workplace. Santa Barbara, University of California.
[19] ICON, G. Gerritzen, L.-C. Huang, K. Killpack, M. Mircheva and J. Conti (2006b). Review of Safety Practices in the Nanotechnology Industry.
[20] Linberg, J. (2007). Project on Emerging Technologies A Survey on Environmental Health and Safety Risk Management Information Needs And Practices Among Nanotechnology Firms in the Massachusetts Region. Massachussetts
[21] Mehta, M. (2009). 2009 NCMS Study of Nanotechnology in the U.S. Manufacturing Industry. Ann Arbor, Michigan, National Center for Manufacturing Sciences (NCMS).
[22] CORDIS-FP7 (2010). NanoFATE Nanoparticle Fate Assessment and Toxicity in the Environment. Bruxelles: Union Européenne, European Commission.
[23] Schmid, K., B. Danuser and M. Riediker (2010). "Nanoparticle usage and protection measures in the manufacturing industry-a representative survey." J Occup Environ Hyg 7(4): 224-232.
[24] Song, T.-T. (2010). Taiwan Position on Nano EHS. NanoTechnology Research Center, Industrial Technology Research Institute. Singapore, Lee KuanYew School of Public Policy, NUS.

1) Table 1: Survey population and level of participation.

Category of participants	Features	Number of participants
Industry	**Participants targeted**	**1310**
	Closed or non functional phone number	**129**
	Reached by phone	**650/1181 (55.0 %)**
	Confirmed the use of NM and agreed to answer the survey	**48/650 (7,5 %)**
	Number of invitations sent by email	**607/1181 (51,8 %)**
	Participants who filled the questionnaire	**51 (8,4 % of 607). Only 29,16% (14/48) of the participants who confirmed the use of NM and agreed to fill the questionnaire**
Universities and research centers	**Participants targeted**	**653**
	Emails who did not reach their recipient	**37**
	Participants who filled the questionnaire	**39/616 (6,3 %)**
Researchers in CEGEPs	**Participants who filled the questionnaire**	**0/38**

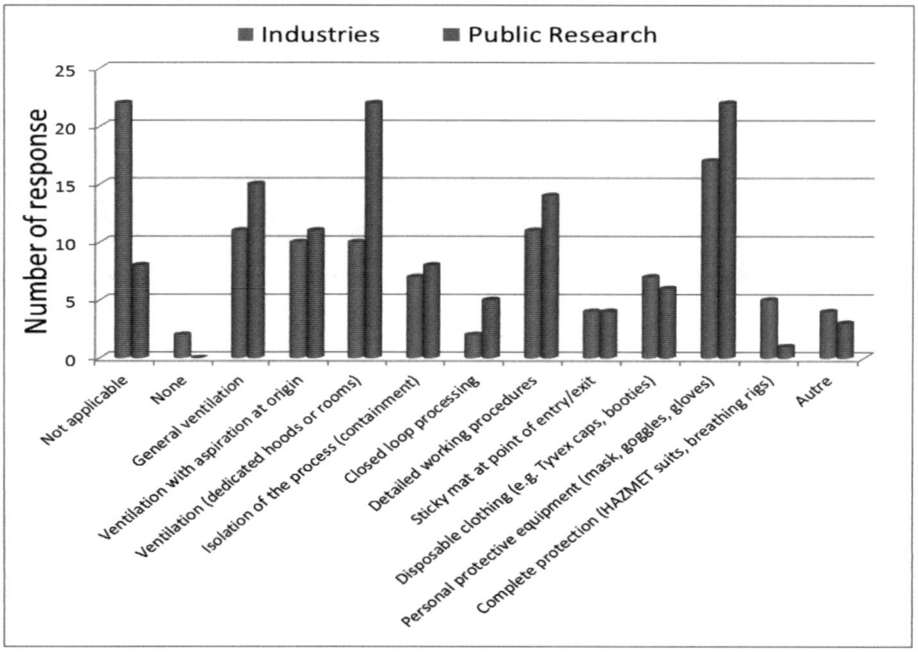

Figure 1: Protection methods used by the participants during the manipulation of nanoparticles or their aggregates. A comparison between Industry and Public research organizations.

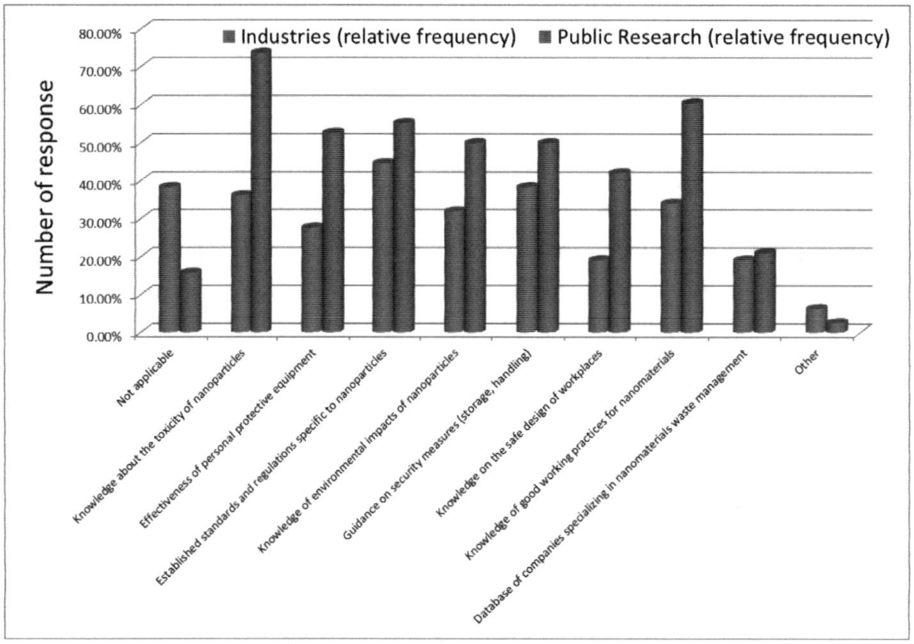

Figure 2. Information/knowledge deemed useful to improve OHS regarding NM. A comparison between Industry and Public research organizations.

The social context of nanotechnology and regulating its uncertainty: A nanotechnologist approach

Vincent Jamier[1], Ignasi Gispert[2], Victor Puntes[3]

[1] Centre for NanoBioSafety and Sustainability (CNBSS), Campus UAB, 08193 Bellaterra (Barcelona), Spain
[2] Institute of Law and Technology-UAB, Campus UAB, 08193 Bellaterra (Barcelona), Spain
[3] Catalan Institute of Nanotechnology (ICN), Campus UAB, 08193 Bellaterra (Barcelona), Spain. Institut Català de Recerca i Estudis Avançats (ICREA), Barcelona, Spain.

Vincent.jamier@cnbss.eu

Abstract. With an increasing number of consumer's products containing nanomaterials (NMs), Nanotechnology is already a reality on the market despite it may only be the emerging part of an iceberg. The potential exponential increase of nanotechnology uses is starting to alert major public bodies due to the uncertainty inherent of every emerging technology. According to the number of publications on safety/toxicity of NMs, major concern seems to be focused on NMs that may be released as free nano-objects. From a nanotechnologist point of view, major identified aspects on inorganic nano-objects are: i) Corrosion of metallic nano-objects, ii) Aggregation to sizes that trigger immune exacerbation, iii) Presence of toxic bystanders, iv) Biological membrane/barrier perturbation, v) Protein association & protein alteration. While the description of phenomena occurring at the nano-scale has not been classified yet, added to the contradictory studies on "unknown" samples, regulatory agencies have no clear idea on what should be considered "nano" or which NMs might pose a threat for health, safety or the environment. In addition to provide full and accessible information on new nano-product, it would seem advisable to plan research and innovation in a responsible and precautionary manner in order to be pro-active towards (responsive) NMs regulation.

1. Introduction

With 1317 consumers' products already registered on the database of the Project on Emerging Nanotechnology (**Figure 1**), and little less on European Consumers' Organisation (**Figure 2**), the use of Nanotechnology is a reality on the market despite that number may represent only the emerging part of an iceberg. As every emerging technology where the balance between risks and benefits is not clearly identified, the exponential increase of nanotechnology entering the consumer market is starting to alert major Non-Governmental Organisation (NGO's), other organization and regulatory agencies due to the uncertainty inherent to novelty.

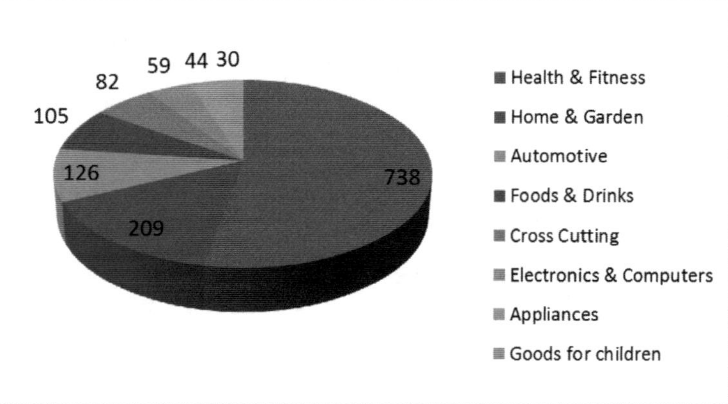

Figure 1. Worldwide inventory of NMs in consumers' products and their repartition from www.nanotechproject.org

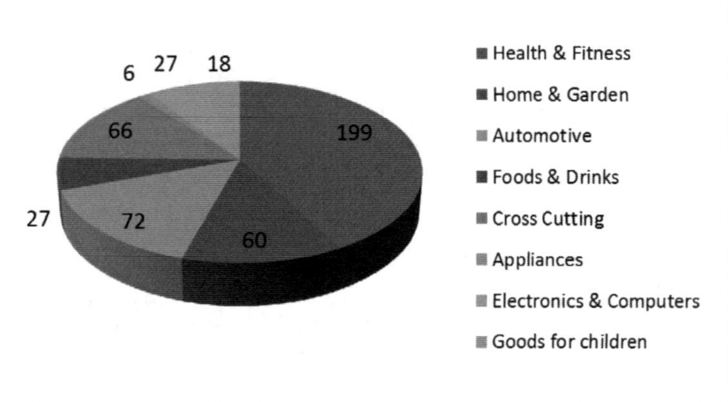

Figure 2. European inventory of NMs in consumers' products and their repartition from the European Consumers' Organisation (www.beuc.org) in 2010.

By definition, Nanotechnology is a technology that permits the manipulation of matter on an atomic and molecular scale, generally with materials with at least one dimension sized from 1 to 100 nanometres. At contrary of other emerging technologies, the nanotechnology cannot be easily defined as an independent technology, it is a field that cover most of the technology already existing (*i.e.* Physics, Chemistry, Biology, Electronic, etc.). With a significant number of publication covering the safety/toxicity of NMs, it seems that a majority of the health concerns related to nanotechnology are focused on nano-objects, which may enter the environment *via* the air or water and be inhaled or ingested. Probably for this reason, there is much less concerns in the development of nano-electronics, probably because these electronic components are less likely to be in direct contact with the consumer. Similarly as many toxic components of electronic apparatus, the isolation of the toxic product from consumer's exposure allow its safe use (or at least, risk is limited to the production -worker's H&S and final disposal).

2. Identification of major concerns

Exposition to such nano-objects has made of importance the nanotoxicology field. It is known that, at contrary of their analogue "bulk" or "molecular" material, the toxicology of nano-objects could not only be established through the evaluation of mass composition doses (*i.e.* the same mass of nano-object will not have the same toxicity depending on particle size, shape,[1] surface state, stabilization process, etc.). Here, the major question is why a nanoparticle (NP) of gold may be different from another NP of gold. Now, it is becoming more and more evident that the precise determination of physico-chemical properties of the investigated nano-object is primordial to be further correlated with others[2] and to understand what make a nano-object less toxic than another one. Here, we can cite some parameters of importance: mean size, shape (if controlled), size range, surface state, surrounding coating (naked, electrostatic or steric), concentration (in solution), cristallinity. Moreover, the instability and short-lived evolution of NPs is also another aspect to be considered as it is common to any of the inorganic NPs, which may often be considered as evolving objects. This makes their toxicity evaluation more complex to understand. This behaviour can be principally attributed to their reduced size, where the surface energy is of great importance. Thus, in their search to minimize this energy, they have the tendency to interact with their environment and/or themselves by aggregation or by dissolution toward more stable phases, and engender the consequent associated concerns.

2.1. Corrosion of metallic nano-objects

Following a simple physical process describes by the Gibbs-Thompson effect and the Noyes-Whitney equation, NPs will dissolve as their surface energy increases (and therefore their size decreases and/or shape evolve toward spherical forms). Due to their reduced size, high surface-to-mass ratio, high radii of curvature, and the corresponding low coordinated atoms at the surface, dissolution is enhanced by chemical reactions with its surrounding or simply themself if it is thermodynamically or kinetically favourable. The last is often the case happening while stocking colloidal solutions for a certain amount of time, morphological modifications are occurring through the physical Ostwald ripening phenomenon. As the concentration of ions increases moderately after the purification and, unless free atoms are removed from the equilibrium, the following consequences should be devised: i) ion concentration remains constant while the size distribution of NPs broadens, ii) NPs shape is modified most probably via selective etching, iii) the number of NPs decreases as the smaller ones are completely dissolved.

2.2. Aggregation to sizes that trigger immune exacerbation

Competitively, NPs have also the tendency to aggregate following the Derjaguin and Lanadau, Verwey and Overbeek (DLVO) rules. It is known that the aggregabillity of NP suspensions depends on parameters such as the surface charge or coating and on the medium in which they are dispersed (*i.e.* highly conductive cell culture medium, physiological buffers, serum and plasma). For example, it has been reported that dramatic change in the state of aggregation, dispersability and charge (through zeta-potential measures) of NPs is occurring while transferring from a buffered aqueous solution to commonly used cell culture medium.[3] This is even more evident while transferring nanopowder (with "naked" surface coating) to aqueous solutions.[4] Those cases enhance the production of agglomerates that precipitate/deposit by weight accordingly to the Stokes Law. As NPs are destabilized in biological media, their special physicochemical properties are progressively/partially lost and may not be more consider as "nano" (neither properties nor the dynamics are any longer the same). Of course, NPs aggregation is a time-dependent process which may takes from a couple of second to months, and this has to be taken in consideration when performing biological tests, in particular in toxicological study attempting to investigate a particular type of NPs, ending up evaluating its micrometric aggregates (some time a source of confusion).

2.3. Presence of toxic bystanders

Under realistic conditions, NPs will frequently come in contact with cells as a component of a complex mixture. While most bystander substances will not affect the interaction between NPs and cells, negative synergistic mechanisms may also occur, including weakening of barrier cell layers, changes in membrane permeability, and changes of the cell activation status by bacterial compounds. Those bystander substances often originate from NP formulations or from ubiquitous bacterial fragments that will directly stimulate immune receptors. It is therefore of interest to understand to what extent the presence of contaminants affects the formation of a bio-corona, and how transcription of key genes is affected by bystander substances. In another hand, sterilisation of NPs may not sufficient to eliminate the bioactive contamination where bacterial fragments by triggering the defence reaction by binding to pattern recognition receptors (*i.e.* Toll-like receptors, Nucleotide Oligomerization Domain receptors, etc.) and enhancing immune reactions. Controlling and understanding such contamination is of key importance, as it has been shown that contamination with lipopolysaccharide can be fully responsible for the "inflammatory" effect of some NPs, leading to false positive results.[5] In some cases, the association of bacterial fragments bound to NPs may also change the effects of each ones in unexpected synergetic ways. Therefore, two key questions should be addressed: i) which assays are suitable to rule out contamination with bacterial substances? ii) to what extent can NPs exacerbate the level of "danger" of bacterial compounds? While bacteria are recognized by the human innate immune system through the repetitive patterns of their moieties (*i.e.* flagellin, CpG DNA), such patterns may be disrupted by NPs (altering the capacity of bacteria to induce usual inflammatory reaction). Emerging from such patterns, one may ask how NPs can modify the current immune-stimulating effects of bacterial fragments (*e.g.* antigen organization influences B cell tolerance [6]).

2.4. Biological membrane/barrier perturbation

NPs may cross barriers like the cell membrane, skin and mucosal barriers in order to fully make their biological impact. Biological membranes have a major significance in mechanisms of toxicity both because of the function as a cell barrier (plasma membrane) and because they are involved in many important physiological mechanisms. Any disruption of the membrane's structure by NP binding will lead to a modification of its barrier properties and physiological processes associated with them. It is important to tackle the present lack of understanding on the interaction of NP with bio-membranes and its mechanistic significance in NPs' bioactivity and toxicity. This interaction is fundamental to assess cell toxicity and translocation (transcytosis) potential.

2.5. Protein association & protein alteration

For many of the previous physicochemical reactivity reasons, the surface of NPs readily interacts with other particles, and in particular with biological molecules, such as proteins. Proteins are the predominant way for communicating to the exterior (*i.e.* living entities or environment) of the living cell. While hydrophobicity and surface charge have been the factors taken in account to explain protein adsorption to functionalized inorganic surfaces,[7] the interaction with NPs surface is greater as protein and NPs are in the same size range and therefore similarly distributed in solution. Those old concepts have been more recently grouped under the well-known name of Protein Corona formation (that may evolve also with time[8]). Major interests have been focused on this formation because the proteins forming the corona remain associated with nanoparticles in *in vitro* and *in vivo* exposure, thereby conferring their biological identity to the NP-Protein Corona complex and determining the interactions between NPs and the host. Despite this association with proteins may indeed biocompatibilize NPs (*i.e.* albuminization of potentially toxic drugs, such as Abraxan©), there is the potential of creating a "Trojan horse" effect, enhancing the accumulation or vectorization of the toxic components of NPs that will slowly release their toxic cargo. Additionally, it has been postulated that the creation of this protein layer may trigger the depletion of specific protein; however, the amounts of NPs or the NP selectivity for a determined property should be much larger than the normal used NP concentrations and observed NP specificity.

Concluding, we do know that the NPs interact with biological systems and molecules in a non-very specific manner and that the existing protective barriers in the body designed to avoid invasion are normally effective preventing NP uptake and distribution. However, if these barrier are crossed (as by intravenous injection), then the NPs have the ability of both, activate the immune system, leading to NP phagocytosis and/or inflammation, or to interfere in the biomolecular machinery what can have long term undesired effects or negative effects in the case of disease. The subtle but prolonged effects that NPs may produce after repeated and chronic exposure are still a subject of research. This knowledge already allow us to initially classify NPs and NP features regarding they hazard, as, for instance, the use of NPs that may aggregate into macroparticles, specially anisotropic, that results in chronic inflammation are subject of concern, as much as metallic semiconductor NPs that leaches toxic ions as CdSe NPs, or NP degradation (corrosion) effects that results in the induction of generation of reactive oxygen species (free radicals) that have been correlated with the REDOX potential of the material in the NP. This knowledge, combined with strategies for reducing exposure, by embedding and employing the NPs embedded in solid or liquid matrices, allows for the first time the safety by design approach, and the most likely, materials will be classified in families depending of their size and morphological ranges and other properties as valence state, electronegativity and surface state. It is therefore expected that in the future the toxicity of a NP will be easily predicted by the knowledge of which NP features result in an increased toxicity at the nanoscale. For advanced and more complex materials, protocols to assess their potential toxicity and strategies to decrease the exposure and dispersabilty of those NPs will be determined.

3. Regulating uncertainty: a nanotechnologist approach

Taking in consideration those scientific concerns, it is understandable that uncertainty dominates the field of nanotoxicology (aren't we aiming to investigate the unknown?). By waiting a scientific clear answer to generate a general regulatory approach, it is evident that we may now be facing a "paralysis by analysis" situation.[9] The description of phenomena occurring at the nanoscale has not been classified yet, thus while there is contradictory studies on "unknown" samples regulators agencies have no idea where "nano" starts or even if it exists, and have no precise strategy to address these issues yet (except a case by case approach). To answer social concerns the development of Nanotechnology need to provide full and accessible information on new product, establish the safety at the first stage of nanomaterial design, attempt to restrict the risk at the production site (where safety monitoring is easily controllable) and control their dispersions in the environment.

3.1. General regulation framework

There are no specific regulations for nanotechnologies or NMs at EU level. Instead, the manufacture, use and disposal of NMs are covered, at least in principle, by a complex set of existing regulatory regimes: i) at a "vertical" level with regulation on Chemical Products (under REACH), pharmaceutical products, pesticide and biocide; Medicines and medical devices, cosmetic, novel foods, food additive, food contact materials and etc. ii) at a "horizontal" level with regulation related to Health and safety of workers and also environmental legislation (like air quality, water, waste, soil protection, IPPC, Seveso II) as well as with environmental liability, product liability or general product safety.

3.1.1. Effectiveness and Regulatory gaps. Regulatory regimes are designed (almost by definition) to handle regulatory concerns existing at the time of promulgation and, as a consequence, emerging technologies often exacerbate or create regulatory gaps.[10] In other words, regulatory systems are badly equipped to respond to NMs. In fact, it could be seen as logical to expect that the emergence of nanotechnology creates regulatory gaps. For instance, the Monash Centre for Regulatory Studies[11] has reported six main types of regulatory gaps for the families of NMs and applications available or likely to become available in Australia before 2017:

- the issue of whether changing the particle size of a substance creates a "new" substance;
- regulatory triggers based on mass or volume;
- regulatory triggers that assume knowledge of risk;
- regulatory triggers reliant on risk assessment protocols;
- regulations that contains exceptions for R&D;
- regulatory triggers reliant on international or external documents that do not themselves address the risk of NMs.

The appraisal as to how important those gaps are is contested: while the Commission considers that "current legislation covers in principle the potential health, safety and environmental risks in relation to NMs",[12] others, like Carl Schlyter (the responsible for the European Parliament resolution on regulatory aspects of NMs) clearly highlighted the limitation of such system when stating that

"(Current) rules are about as effective in addressing nanotechnology as trying to catch plankton with cod fishing net".[13]

3.1.2. Regulation explicitly "Nano". Regardless of the assessment that the Commission or other institutions, organizations and scientific literature have done on the degree of adequacy of the legal framework, the fact is that since 2008 we have seen an increasing regulatory activity related with NMs: basically, nano specific Recommendations (horizontal measures) and Sectorial Regulations (within the above explained framework of non-specific nano regulation). The nano specific Recommendations are the Code of Conduct for responsible research in Nanotechnology and Nanoscience,[14] and more recently, the Recommendation on the definition of nanomaterial.[15] On the Sectorial Regulation regulatory processes has been developed in different areas such as REACH (excluding carbon and graphite from the Annex IV),[16] food, food packaging and food contact materials,[17-19] cosmetic,[20] hazardous substances in electrical and electronic equipment (postponing possible nano regulation at Recital 16),[21] and biocides.[22] Generally speaking, sectorial regulation identifies the need to differentiate nanomaterial from "bulk" material, the need to identify them (labelling) and the need for a case by case approach. It is therefore not surprising that the recent second appraisal on NMs regulation from the Commission endorses the case by case approach.[23, 24] What remains an open question is whether the case-by-case approach is efficient and can be handled with the exponential increase of NMs (or discovery of already existing NMs) in the marketplace.

3.2. Beyond actual framework

From a legal standpoint, the lack of scientific certitude on the possible effects of NM over health and safety must be seen as a trigger for the implementation of preventive measures, economically reasonable, to avoid, or as a default mitigate, the potential risks associated with its use. Moreover, the existence of regulatory gaps and the existence of open legal concepts that imposes general legal obligations imply that companies no longer satisfy their obligations by complying with (non-existent nano-) standards, but by outlining a proactive behaviour in the context of responsive regulation.[25] The regulators' interest in opening up the regulatory process to all stakeholders and the need for a regulatory framework that is flexible (adaptation to uncertainty and new scientific data) is shaping a new regulatory landscape for NMs where the hard law mechanisms[1] (traditional command-and-control measures) will be increasingly complemented with the so called soft law mechanisms (those non legally binding but that might have certain indirect legal effect).[26] Among those soft law mechanisms for NMs we can mention: mandatory/voluntary reporting schemes, Codes of Conduct, Risk Management Systems or Schemes and Guidelines and Standards (such as ISO, etc.). Overall, it can be said that voluntary measures share similar principles and actions (due respect for precaution; priority on safety; raise/consider stakeholder awareness; inclusive approach) and goals (build trust and confidence in the technology; promoting health and environmental safety; gathering information).[27]

3.3. Principles of precaution

A precautionary approach is not only the cornerstone of responsive regulation, but also the basic strategy toward the use of NMs in industry, as shown in **Figure 3**.

---

[1] Following Senden, soft law mechanisms can be defined as rules of conduct that are laid down in instruments which have not been attributed legally binding force as such, but nevertheless may have certain (indirect) legal effects, and that are aimed at and may produce practical effects.

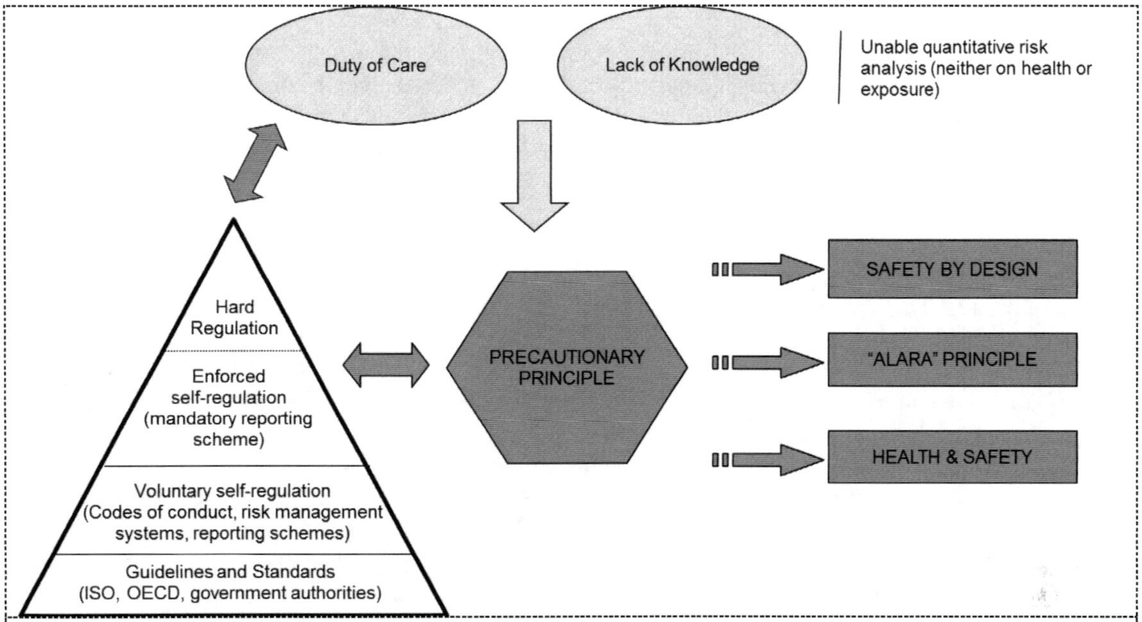

Figure 3. Precautionary approach in Industry[2].

3.3.1. Safety-by-design. The integration of Safety within the design of NMs should not be limited to the investigation of the possible advert effect of used NMs but should be considered in the first step of the material design. It means to evaluate from the actual data available which nanomaterial is less likely to possess toxic properties with the desired final properties. In this search, it is necessary to cover the wide spectrum of available scientific investigation (however the polemic subject and judgement of the publication quality) and reduce by all mean the exposure to consumers as well as establishing a road guideline of the full lifecycle of the NM from cradle to grave assessment. To permit a such safety-by-design there is a need to stopping the "seek and try" evaluation techniques and start the effective and potential "bottom up design" where the properties (and principally safety features) could be controlled from the design of the material directly from the nano-scale as well as help to calibrate/understand the risk and benefit of advanced NMs. A strong inventory database of properties/parameters may therefore grow supported by high throughput toxicity screening of engineered NMs.[28]

3.3.2. "ALARA (As Low As Reasonably Achievable)" Principle and H&S. Adoption of preventive and protective measures similar to those used to control exposure to hazardous (principally aerosols one) is necessary to protect workers manipulating NMs. The main measures can be regrouped in three area: i) the technical measures such as substitution of substances, processes, equipment, isolation or enclosure of the process, local exhaust; ii) the organisational measures such as cleanness of the work area by using wet sweeping systems (and vacuum systems equipped with high efficiency filters), reduction of exposure time and limit the number of workers exposed; iii) the personal protective equipment such as the use of clothing chemical protective, gloves, glasses with side protection.

[2] With integration of the regulatory pyramid from Mantovanni, E., Porcari, A. "Hard e Soft Regulation per le Nanotecnolgie" 4a Conferenza Nazionale del Programma NIC. Milano 2 Decembre 2012.

4. Conclusion

Similarly as the two approaches on nano-object production, the bottom-up and top-down approach, regulation of nanotechnology needs similar processes, like we have seen in the "regulatory pyramid" of **Figure 3**. The presence and use of naturally transformed, accidental and engineered nano-objects (such as processed proteins, ultrafine particles and powders grinded to the nano-scale) is not new. Such "low tech" nanotechnology has been surrounding us or used since centuries for example pigmentation in cosmetics (since the pharaons time) and in ceramics processes as well as their possible adverse effects is well estimated allowing us to use a body of knowledge that while it has not been collected with a nanosafety perspective, it can be extremely helpful to frame the risks and limits of exposure to NPs. This is linking knowledge from microparticles toxicity (sarcoidosis, asbestosis, etc.) and metal compounds toxicity.

It is important that regulatory bodies are able to cover the actual use of NPs that was unnoticed, as in food additives, pigments, catalysis and consumer products as talc or polymeric materials and also the use of next and advanced generation of new NPs which will be each time more artificial than already existing as natural, unintended and unnoticed ones. For that it may be needed to establish complex matrices of calibrated properties (mean size with low size range, shape, surface state, stabilization process, interaction behaviour) covering the toxicology including majors concerns (corrosion of metallic nano-objects, aggregation to sizes that trigger immune exacerbation, presence of toxic bystanders, biological membrane/barrier perturbation, protein association & protein alteration) of existing and foreseen upcoming NPs. Nanoinformatics, to deal with the existing and proposed plethora of useful NPs and their evolution in different matrices of different complexity will surely be a determining tool for NP risk assessment.

For being able to do so the regulatory system must be focused on risk (avoiding regulation based on hazard) and the burden of proof must be shifted from public administration to NMs producers (in line with a responsive regulation and following the REACH model). From the research and innovation stage to marketplace new strategies needs to be implemented, based on selecting those low hazard NMs and designing NP and NPs that minimize exposure during the whole life cycle (so responsible research and innovation strategies). The regulatory system, accordingly, should promote that strategy, but, for doing so, regulation cannot rely solely on command-and-control measures. To complement them, soft law mechanisms are playing and increasing important role in nanotechnology regulation. All stakeholders might need to analyse how they are integrating soft law mechanisms into their organizations and how they are shifting towards pro-active behaviour for legal compliance.

References

[1] L.C. Stoehr, E. Gonzalez, A. Stampfl, E. Casals, A. Duschl, V. Puntes, G.J. Oostingh, Shape matters: effects of silver nanospheres and wires on human alveolar epithelial cells, Part Fibre Toxicol, 8 (2011).

[2] P. Rivera Gil, G. Oberdorster, A. Elder, V. Puntes, W.J. Parak, Correlating Physico-Chemical with Toxicological Properties of Nanoparticles: The Present and the Future, Acs Nano, 4 (2010) 5527-5531.

[3] E. Casals, T. Pfaller, A. Duschl, G.J. Oostingh, V.F. Puntes, Hardening of the nanoparticle-protein corona in metal (Au, Ag) and oxide (Fe3O4, CoO, and CeO2) nanoparticles, Small, 7 (2011) 3479-3486.

[4] E. Casals, E. Gonzalez, V.F. Puntes, Reactivity of inorganic nanoparticles in biological environments: insights into nanotoxicology mechanisms, J. Phys. D: Appl. Phys., 45 (2012) 443001.

[5] G.J. Oostingh, E. Casals, P. Italiani, R. Colognato, R. Stritzinger, J. Ponti, T. Pfaller, Y. Kohl, D. Ooms, F. Favilli, H. Leppens, D. Lucchesi, F. Rossi, I. Nelissen, H. Thielecke, V.F. Puntes, A. Duschl, D. Boraschi, Problems and challenges in the development and validation of human cell-based assays to determine nanoparticle-induced immunomodulatory effects, Part Fibre Toxicol, 8 (2011) 8.

[6] M.F. Bachmann, U.H. Rohrer, T.M. Kundig, K. Burki, H. Hengartner, R.M. Zinkernagel, The influence of antigen organization on B cell responsiveness, Science, 262 (1993) 1448-1451.
[7] S. Goy-Lopez, J. Juarez, M. Alatorre-Meda, E. Casals, V.F. Puntes, P. Taboada, V. Mosquera, Physicochemical characteristics of protein-NP bioconjugates: the role of particle curvature and solution conditions on human serum albumin conformation and fibrillogenesis inhibition, Langmuir : the ACS journal of surfaces and colloids, 28 (2012) 9113-9126.
[8] E. Casals, T. Pfaller, A. Duschl, G.J. Oostingh, V. Puntes, Time Evolution of the Nanoparticle Protein Corona, Acs Nano, 4 (2010) 3623-3632.
[9] S.F. Hansen, A. Baun, When enough is enough, Nat Nanotechnol, 7 (2012) 409-411.
[10] G.N. Mandel, Regulating Emerging Technologies, Law, Innovation & Technology, 1 (2009) 75-87.
[11] D. Ludlow, D. Bowman, G. Hodge, A Review of Possible Impacts of Nanotechnology on Australia's Regulatory Framework, in, 2007.
[12] Communication of the EC of 17/06/08 on Regulatory Aspects of Nanomaterials (COM(2008)366).
[13] On the explanatory statement from the Resolution. Available at http://www.europarl.europa.eu/oeil/FindByProcnum.do?lang=en&procnum=INI/2008/2208. (Accessed February 2011). in.
[14] COMMISSION RECOMMENDATION of 07/02/2008 on a code of conduct for responsible nanosciences and nanotechnologies research, in.
[15] COMMISSION RECOMMENDATION of 18.10.2011 on the definition of nanomaterial (2011/696/EU) OJ L275/38 of 20.11.2011, in.
[16] COMMISSION REGULATION 987/2008 of 08.10.2008. Carbon and graphite are excluded from REACH Annex IV. OJ L268/14, in.
[17] REGULATION (EC) 1333/2008 of 16.12.2008 on food additives. Article 12 OJ L345/16, in.
[18] Commission Regulation (EU) No 10/2011 of 14 January 2011 on plastic materials and articles intended to come into contact with food (OJ L12, 15.1.2011, page 1), in.
[19] Regulation (EU) No 1169/2011 of the European Parliament and of the Council of 25 October 2011 on the provision of food information to consumers, amending Regulations (EC) No 1924/2006 and (EC) No 1925/2006 of the European Parliament and of the Council, and repealing Commission Directive 87/250/EEC, Council Directive 90/496/EEC, Commission Directive 1999/10/EC, Directive 2000/13/EC of the European Parliament and of the Council, Commission Directives 2002/67/EC and 2008/5/EC and Commission Regulation (EC) No 608/2004 (OJ L304, 22.11.2011, page 18), in.
[20] Regulation (EC) No 1223/2009 of the European Parliament and of the Council of 30 November 2009 on cosmetic products (Text with EEA relevance). in.
[21] DIRECTIVE 2011/65/EU OF THE EUROPEAN PARLIAMENT AND OF THE COUNCIL of 8 June 2011 on the restriction of the use of certain hazardous substances in electrical and electronic equipment, in.
[22] Regulation (EU) No 528/2012 of the European Parliament and of the Council of 22 May 2012 concerning the making available on the market and use of biocidal products, in.
[23] EUROPEAN COMMISSION - PRESS RELEASE of the 03/10/12 - Nanomaterials: Case by case safety approach for breakthrough technology, in.
[24] Communication of the EC of 03/10/12 on the Second Regulatory Review on Nanomaterials (COM(2012) 572 final), in.
[25] M. Fuhr, K. Bizer, REACH as a paradigm shift in chemical policy - responsive regulation and behavioural models, Journal of Cleaner Production, 15 (2007) 327-334.
[26] L. Senden, Soft Law in European Community Law, Hart Publishing, 2004.
[27] E. Mantovani, The role of voluntary measures in the governance of nanotechnologies: the case of the European Code of Conduct, in: AIRI/Nanotec IT. Euronanoforum 2011, Budapest, 2011.

[28] S. George, T.A. Xia, R. Rallo, Y. Zhao, Z.X. Ji, S.J. Lin, X. Wang, H.Y. Zhang, B. France, D. Schoenfeld, R. Damoiseaux, R. Liu, S. Lin, K.A. Bradley, Y. Cohen, A.E. Nal, Use of a High-Throughput Screening Approach Coupled with In Vivo Zebrafish Embryo Screening To Develop Hazard Ranking for Engineered Nanomaterials, Acs Nano, 5 (2011) 1805-1817.

Journal of Physics: Conference Series **429** (2013) 012060 doi:10.1088/1742-6596/429/1/012060

Latest research results on the effects of nanomaterials on humans and the environment: DaNa - Knowledge Base Nanomaterials

C Marquardt[1], D Kühnel[2], V Richter[3], H F Krug[4], B. Mathes[5], C. Steinbach[5] and K Nau[1]

[1]Karlsruhe Institute of Technology (KIT), Eggenstein-Leopoldshafen, Germany
[2]UFZ-Helmholtz Centre for Environmental Research, Leipzig, Germany
[3]Fraunhofer-Institute for Ceramic Technologies and Systems (IKTS), Dresden, Germany
[4]Empa - Swiss Federal Laboratories for Materials Science and Technology, St. Gallen, Switzerland
[5]DECHEMA Society for Chemical Engineering and Biotechnology e.V., Frankfurt/Main, Germany

E-mail : clarissa.marquardt@kit.edu

Abstract. Nanotechnology is considered one of the key technologies of the 21st century. The success of this fascinating technology is based on its versatility. It will bring about fundamental changes of basic research as well as of many sectors of industry and also of daily life from electronics to the health care system. However, consumers often miss reliable and understandable information on nanomaterials and all aspects of this versatile technology. A huge body of data on the potential hazards of nanoobjects towards human and environmental health already exists, but is either not easily accessible for a broad audience or presented unprocessable for non-experts. But risk communication is an essential and thus integral component of risk management. For that purpose, the DaNa-Project aims at filling this gap by collecting and evaluating scientific results in an interdisciplinary approach with scientists from different research areas, such as human and environmental toxicology, biology, physics, chemistry, and sociology. Research findings from the field of human and environmental nanotoxicology are being prepared and presented together with material properties and possible applications for interested laymen and stakeholders. For the evaluation of literature a "Literature Criteria Checklist" has been developed as well as a Standard Operation Procedure template (SOP) based on careful scientific practice.

1 Introduction

Nanomaterials (in this article meaning "any material containing nanoobjects") comprise a large variety of substances which come in many modifications. A huge amount of data has been published in recent years on the toxicology of nanomaterials, yet it is difficult to get a general overview on the effects of a specific material. Furthermore, the information contained in the scientific literature is hardly assessable for consumers interested in health aspects of nanomaterials. Hence, a knowledge base was created providing a unique combination of material science-based information on nanomaterials and nano-applications combined with toxicological data. This was achieved by collecting scientific results from publications, research projects and latest news on human and environmental toxicity in the field of nanotoxicology. This state of knowledge has been wrapped up in a knowledge base and presented in a transparent understandable format.

In an interdisciplinary approach, the DaNa-project team assembles scientists from different research areas, such as human and environmental toxicology, biology, physics chemistry and sociology, who create this transparency and present the research on nanomaterials and their effects on humans and the environment in a public orientated way.

Therefore the *www.nanoobjects.info* knowledge base provides easily comprehensible information regarding production, application and toxicology of various nanomaterials. The introduction of all current nanotoxicology-related projects funded by the German Federal Ministry of Education and Research (BMBF)[1] is the second goal of DaNa. Likewise, these project results are being collected and presented recipient-friendly.

2 DaNa – general information on nanoobjects

Our aim is to provide a tool for the websites' visitors enabling them to make responsible decisions on the opportunities and risks of nanotechnology. The presented information can be accessed via three different routes:

Knowledge base - providing information about specific nanomaterials and applications
Glossary - providing definitions of all relevant terms
FAQ Section - collection of questions and answers, providing the opportunity to ask questions to the DaNa experts

The content of all published articles is presented in three levels of complexity according to the needs and interests of different recipient groups:

Simple:	interested citizens/consumers (public oriented)
Detailed:	journalists, stakeholders, scientists from other fields
Complex:	scientists from related fields, regulators

In order to complete the picture the website also provides additional information about current developments in research, news, press releases, events and conferences related to the field of nanotechnology.

2.1 DaNa knowledge base

The core and thus most important part of the DaNa web platform is its knowledge base. It contains information on the actual state of knowledge being assembled from project results (funded by the German Federal Ministry of Education and Research, BMBF), analysis of scientific publications, reports, and latest news on human and environmental toxicology. Literature providing the basis of the nanomaterials knowledge base is selected in accordance with quality criteria that have been acknowledged worldwide within the scientific community. A "Literature Criteria Checklist"[2] to evaluate publications has been developed by the DaNa expert team and made publicly available (see also section 2.2).

The provided detailed information regarding specific nanomaterials can be accessed either directly via the menu toolbar or by using the integrated application-based database, the so-called "slot-machine". This database contains only information about actual applications with nanomaterials. You choose a certain application, for example lacquer and plastics additive, and you get the corresponding material(s), here cerium dioxide and many others, with brief information on exposure, uptake and behaviour (indicated by red highlighting). Starting with a specific material, you get the corresponding applications as well as facts about e.g. its behaviour in the body and in the environment. Using this playful approach a wider range of appeal for the public is generated and interested visitors are then navigated to the detailed information provided for each nanomaterial.

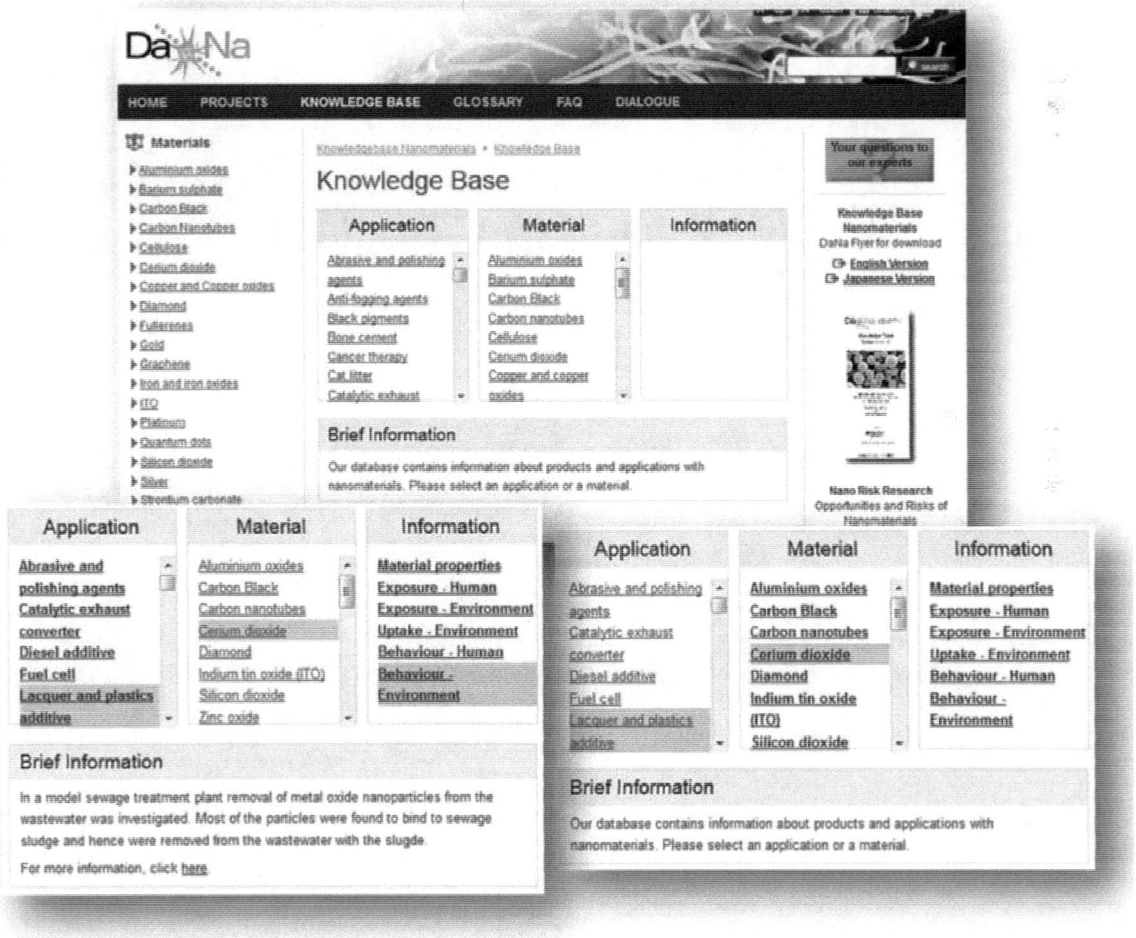

Figure 1 : Slot-machine: an application-based approach to map connections between application of nanoproducts, material and outcome/effects [3].

The core part of the material knowledge base is structured according to the specific nanomaterials, e.g. nano-scaled cerium dioxide, titanium dioxide or quantum dots. Each section provides general information on the nanomaterials production, usage and material properties. Furthermore, paragraphs

on human toxicology summarize the available data regarding exposure, uptake and behaviour in the body and epidemiology. Regarding exposure to nanomaterials the section takes into account the *in vivo* and *in vitro* situation whereas the uptake section focuses on the different uptake routes for nanoobjects (skin, lung, gastro-intestinal tract). Behavioural aspects are addressed relating to cells or the blood-brain barrier.
Regarding nanomaterials in the environment, the knowledge base provides information on environmental exposure and behaviour of particles in aquatic and terrestrial surroundings as well as uptake and potential risks for environmental organisms (terrestrial and aquatic, e.g. fish, daphnia, worms, plants, bacteria).

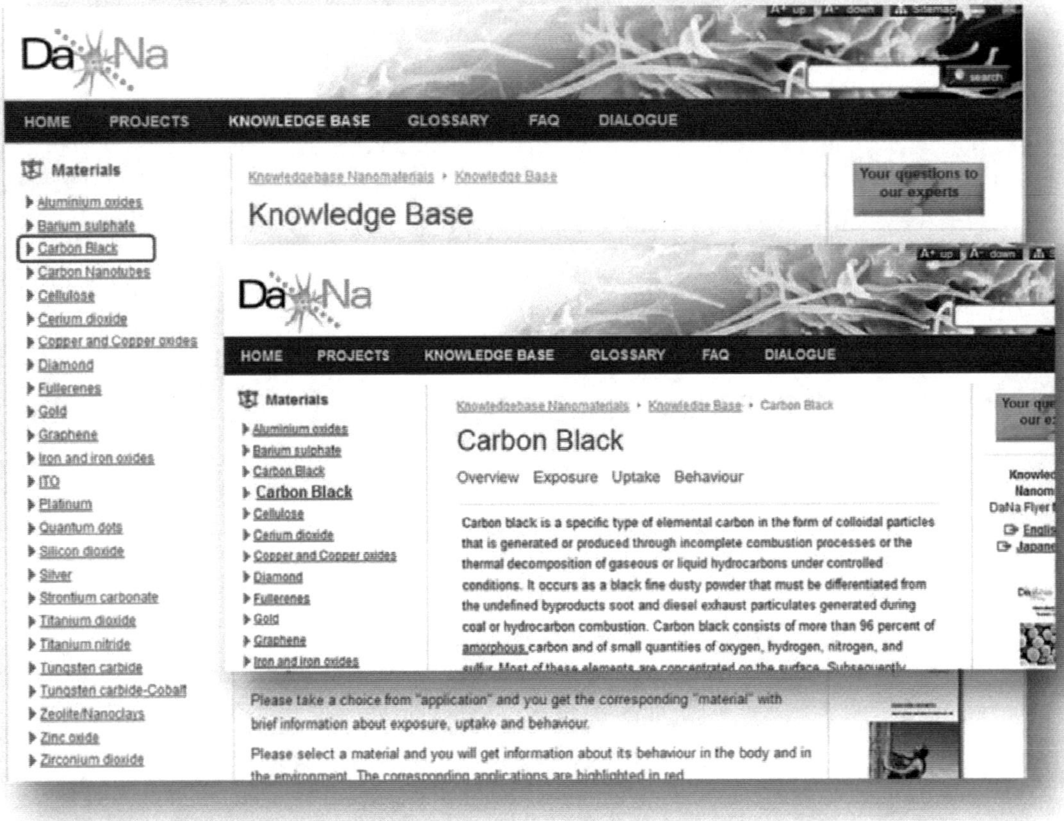

Figure 2 :The Dana knowledge base presents a constantly growing number of nano-scaled materials, e.g. Carbon Black, with information on their application and toxicity aspects [3, 4].

2.2 *Preparation of published data – DaNa Criteria Checklist*

Our claim is to evaluate all data regarding their scientific value before publishing them on the website. But toxicological assessment of nanoobjects is quite a new scientific discipline and poses many pitfalls. For example, the testing methodology has basically been adopted from the testing of chemicals which is not always appropriate to analyse nanoobjects. Furthermore methods and procedures are often not yet standardized or normed creating difficulties for interpreting and comparing of data. In this context a study published by Hristozov *et al* 2012 [5] stated that in case of nano-TiO_2 only one third of all published data from various chemical databases is suitable for a thorough risk assessment. Taking into account the large amount of publications in scientific literature

their assumption was that currently the majority of published data/information deals with (eco)toxicity and less with exposure and risk estimation. Thus there is a need for a tool to facilitate this decision process.
Therefore a methodology of selection, acquisition and evaluation of scientific publications was developed, including the definition of mandatory and desirable assessment criteria. A set of general (e.g. statistics, standardisation), biological (e.g. appropriate controls, dosing procedure) and physico-chemical (e.g. particle size and surface area, impurities) parameters was defined [2]. The list of criteria is accessible to all interested scientists via the DaNa website [2].

2.3 DaNa Standard Operating procedures (SOP)
In accordance with the DaNa Criteria Catalogue the need for a common format to share, compare and describe methodologies specific for nanomaterials has clearly been shown. Therefore Standard Operation Procedures (SOPs) should be a basis for future projects and publications. On the basis of careful scientific practice, the DaNa project team has compiled a SOP template to fill in which is also available via the website [6].

2.4 DaNa Frequently Asked Questions (FAQ) & Glossary
Most important questions on nanomaterials and nanotechnology are answered in this FAQ section. It also provides the opportunity to interact with the DaNa expert team directly by submitting questions via email. Answers will be given by an expert in the addressed scientific field and published on the website if considered to be of public relevance.

Due to the novelty of many aspects in nanotechnology, topic related specific terms need further explanation to improve the comprehensiveness of the articles. Therefore a glossary facilitates quick and easy location of relevant information in the articles. In alphabetic order, short descriptions of terms often used on the DaNa website like `nanoobject`, `zeta potential` or `agglomerate` are given and linked correspondingly.

Figure 3. General information on nanoobjects presented on the DaNa website. Both the FAQ section and the Glossary provide easy access to the topic of nanotechnology and its related research [7, 8].

2.5 DaNa project presentations

Another important goal of the DaNa project is to give an overview of all nanotoxicology-related projects currently funded by the BMBF (German Federal Ministry for Education and Research) them being part of the funding actions NanoCare and NanoNature. It provides a presentation platform for the projects to illustrate with a short description the main scope and aims of each project. Furthermore project findings such as reports, publications etc. can be made available for the public also promoting exchange between the different project groups. Dissemination of project findings via the website are further supported by annual cluster meetings of all researchers involved in these projects

Analogous to the literature used in the knowledge base all project-generated data will be evaluated with regard to their scientific value prior to publishing them in the DaNa-knowledge base and will be presented in a comprehensible way.

Figure 4. The "Projects" section lists and introduces shortly all relevant projects of the NanoCare and NanoNature cluster. The main scope of each project is presented as well as all corresponding project partners, exemplarily shown here for the project "NanoGEM" [9, 10].

3 Summary and outlook

The information provided by the DaNa website offers an easy access to the public on the topic of nanomaterials and toxicological questions for humans and the environment. Constant actualisation of the DaNa website has been achieved by continuous extensions of the knowledge base with additional materials and glossary with further relevant terms, respectively. So far the knowledge base evaluated information on 25 different nanomaterials. By 2011, over 60.000 visitors from more than 100 different countries retrieved information from the DaNa webpage and for 2012 visitor numbers are expected to be even higher. Worldwide appreciation of the website is also reflected by the international background of the sites' visitors, the top five countries of origin being Germany, USA, China, India and the United Kingdom. The acceptance of downloadable material (Criteria Checklist and SOP) is also validated by increasing download numbers. A first SOP derived from one of the BMBF-funded projects about "Detection and semi-Quantification of Endotoxin Contaminations in Nanoparticle Suspensions" [11] is already online available.

The project section is likewise extended according to the progress of each project. To improve the synergism between related projects, an annual cluster meeting of all project participants has been established. Sociological aspects regarding nanotechnology have also been addressed by the DaNa Project team in terms of a series of focus group discussions with advanced level pupils about nanotechnology and environmental protection – chances and risks. They will enable the pupils to form their own knowledge-based opinion on chances and risks associated with nanotechnological processes in the environmental sector. Cooperation and exchange with other dissemination organizations (databases, projects, websites) in the field of nanotechnology on national and international level are further future DaNa tasks. In turn, we will also contribute to relevant international activities (e.g. OECD programme). Citation of the DaNa website as single national database in the recent publication of the European Commissions' "Commission Staff Working Paper"[12] and in the commentary "Focusing the research efforts"[13] also clearly demonstrates its importance and the quality of the provided data respectively. Finally, our comprehensive data collection will assist by defining knowledge gaps and hence further research needs.

Acknowledgement

We thank the German Federal Ministry for Education and Research (BMBF) for funding the DaNa project (Acquisition, evaluation and public-oriented presentation of society-relevant data and findings relating to nanomaterials; grant # 03X0075).

References

[1] BMBF website : http://www.bmbf.de/en/index.php
[2] http://www.nanopartikel.info/files/content/dana/Dokumente/Form_Selected%20Literature.pdf
[3] http://www.nanopartikel.info/cms/Wissensbasis
[4] http://www.nanopartikel.info/cms/lang/en/Wissensbasis/CarbonBlack
[5] Hristozov D R, Gottardo S, Critto A and Marcomini A 2012 *Nanotoxicology* 1-19.
[6] http://www.nanopartikel.info/files/content/dana/Dokumente/DaNa_SOP%20template2012_online.pdf
[7] http://www.nanopartikel.info/cms/lang/en/FAQ
[8] http://www.nanopartikel.info/cms/lang/en/glossar#letter_408
[9] http://www.nanopartikel.info/cms/lang/en/Projekte
[10] http://www.nanopartikel.info/cms/lang/en/Projekte/nanogem
[11] http://www.nanopartikel.info/files/content/dana/Dokumente/Projekte/SOPs/Nanokon%20SOP%202%202%202_gesch%C3%BCtzt.pdf
[12] European Commission. COMMISSION STAFF WORKING PAPER - Types and uses of nanomaterials, including safety aspects SWD(2012) 288 final. 2012.
[13] Schrurs F and Lison D 2012 *Nat Nanotechnol* **7** 546-8.

NanoScan SMPS - A Novel, Portable Nanoparticle Sizing and Counting Instrument

Torsten Tritscher[1], Michael Beeston[1], Axel F. Zerrath[2], Sherrie Elzey[2], Thomas J. Krinke[1], Eric Filimundi[1], Oliver F. Bischof[1]

[1] TSI GmbH, Particle Instruments, Neuköllner Str. 4, 52068 Aachen, Germany

[2] TSI Incorporated, 500 Cardigan Road, Shoreview, MN 55126, USA

E-mail: torsten.tritscher@tsi.com

Abstract. A novel instrument, the NanoScan SMPS (TSI 3910), is introduced here that provides the capability of employing well-established Scanning Mobility Particle Sizing (SMPS) technology in a format that can be easily utilized in the field. The use of a non-radioactive unipolar charger, radial Differential Mobility Analyzer (rDMA) and isopropanol-based Condensation Particle Counter (CPC) removes transport and application restrictions on occasion encountered with current SMPS technology. The NanoScan SMPS was compared initially to a TSI component SMPS reference system using several sizes of monodisperse silver particles. A further comparison was completed with polydisperse NaCl and Emery Oil particles. A linearity test for the instrument was performed by comparison to the ultra-fine butanol CPC (TSI 3776) as concentration reference. Subsequently, in-vehicle measurements during a freeway journey between two German cities were made to provide an example of everyday exposure to nanoparticles while driving. Coupling continuous in-vehicle measurements with GPS technology enabled an easy on-route characterization of particle size distributions, allowing source appointment of measurements, identification of evident hotspots as well as a variety of other sources of interest. Finally, the NanoScan SMPS was applied to different processes within a production test facility. Here direct comparisons between ventilation (background) and in the breathing zone were made. These measurements demonstrated the variation between the nature of the nanoparticles in the ventilation system from those within the breathing zone. During all tests the NanoScan SMPS was found to provide data that compares very well with that of the reference SMPS. It also showed a clear linearity with the CPC measurement up to concentrations as high as 10^6 particles/cm^3. In conclusion, the NanoScan SMPS was found to be a reliable tool for nanoparticle sizing and counting in applications where current SMPS instruments would prove to be challenging.

Keywords: SMPS, CPC, particle measurement, in-vehicle, exposure monitoring, ventilation, breathing Zone

1. Introduction

There are numerous sources of nanoparticles, including a large variety of anthropogenic origin such as engineered nanoparticles and particles from combustion and industrial processes. However, nanoparticles are also ubiquitous in ambient air, both indoors and outdoors. The size distribution, number concentration, and surface area of such nanoparticles represent key parameters in the determination of their risk, as has been identified by many recent health studies. Oberdörster et al., 2005 reported a high rate of pulmonary deposition of nanoparticles, and their ability to travel from lung to systemic sites as well as their high inflammation potential. Particles in the nanometer range are more biologically active due to their greater surface area per mass (Gurr et al., 2005). However, only few specific types of particles are studied in depth (e.g. asbestos) that allow an estimation of their potential health impacts (Sanchez et al., 2009). Measuring exposure to nanoparticles has typically been conducted as academic research using high-end instrumentation and sophisticated software. For many applications, e.g. in the field of indoor air quality and occupational health and safety, this may not be necessary. Hence, the ability to characterize the properties of airborne nanoparticles to a similar level of accuracy and resolution similar to that of the current state-of-the-art laboratory-based research instruments is at the fore-front of the current air quality measurement needs.

This paper introduces a novel instrument and provides the validation of it through direct comparison to the standard SMPS technique. It will also offer insight into the application of the NanoScan SMPS to the measurement of aerosols within environments that would be challenging to apply the component SMPS that is frequently used in the laboratory environment.

The application of the Scanning Mobility Particle Sizing (SMPS) technology to the air quality monitoring field by researchers has provided a level of detail that has become standard within the community in terms of sizing accuracy and detail of the information delivered. Hence, taking this technology from the laboratory into the field is a desirable next step. With regards to portable sizing, the current state-of-the-art is the application of advanced optical measurement technology like the one of the Optical Particle Sizer (OPS, TSI 3330). Despite providing detailed size distribution information, the absence of particle growth technology within the instruments limits it sizing capability to particles larger than 300 nm. Hence, it does not provide the resolution and small sizing capability that has become customary with laboratory-based research equipment such as the component SMPS. The NanoScan SMPS (TSI 3910) was designed to provide size resolution and accuracy in the field close to the one researchers have grown accustomed to within the laboratory. It utilizes the proven technology of electrical mobility sizing and particle detection. In the instrument nanoparticles are charged, classified by their size and subsequently counted in a particle counter. The latter magnifies these small particles by condensing liquid onto their surface to a size that is detectable in a laser optic assembly. The number distribution is based on a mobility equivalent diameter.

The NanoScan SMPS has been designed for multiple applications such as monitoring of industrial process emissions and workplaces, mobile studies, in-vehicle and outdoor air quality measurements. Number size distributions of nano-size airborne particles in the range from 10 to 420 nm can be obtained. The size range can be extended to coarse particles with an additional compact, portable instrument, the Optical Particle Sizer (OPS). Together these instruments can cover the full range from 10 nm to 10 µm. The instrument includes a built-in computer, is battery-powered and contains no hazardous substances (neither butanol, nor radioactivity) that could restrict its use to designated areas.

There are two key features of the instrument that enable its portable use: The first is a patented opposed flow unipolar diffusion charger (Medved et al., 2000), which works without any radioactivity and thus avoids related transport issues. The second is the incorporation of a radial Differential Mobility Analyzer (rDMA) in the instrument. The radial DMA technology was introduced by Pourpix and Daval (1990), followed by work done by Zhang et al., 1995 and Fissan et al., 1998. As opposed to the cylindrical DMA developed by Knutson and Whitby (1975), the sheath and sample air flow radially inwards between two flat electrodes in this DMA type.

2. Instrumentation: Working Principle and Features

Prior to discussing the internal workings of the NanoScan SMPS, the exterior of the instrument is introduced here in Figure 1A, along with a visualisation of the data output given on the on-board computer's color-touch screen (Figure 1B).

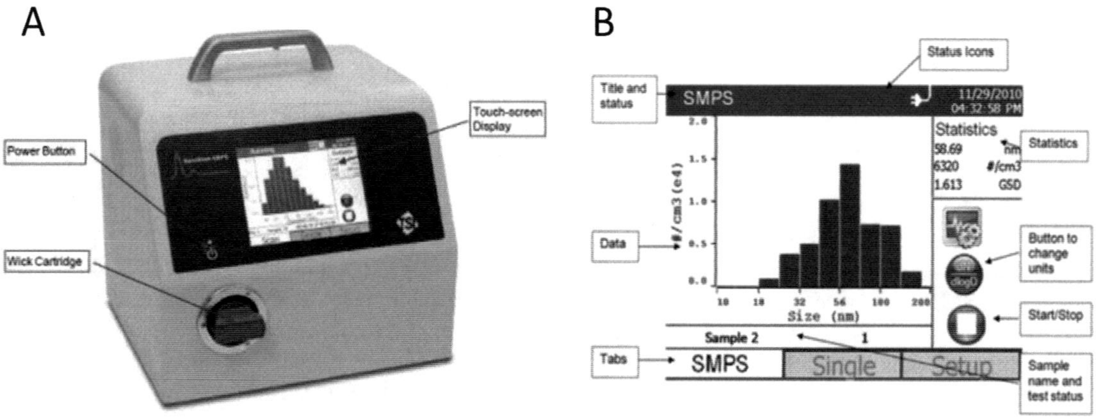

Figure 1: A) The NanoScan SMPS (left) and B) the instrument's color touch screen, showing the output of a measurement in SMPS mode (right).

The internal operation of the NanoScan SMPS is outlined in Figure 2. The four main components of the instrument are its inlet cyclone, the opposed flow unipolar diffusion charger, the radial DMA and the isopropanol CPC.

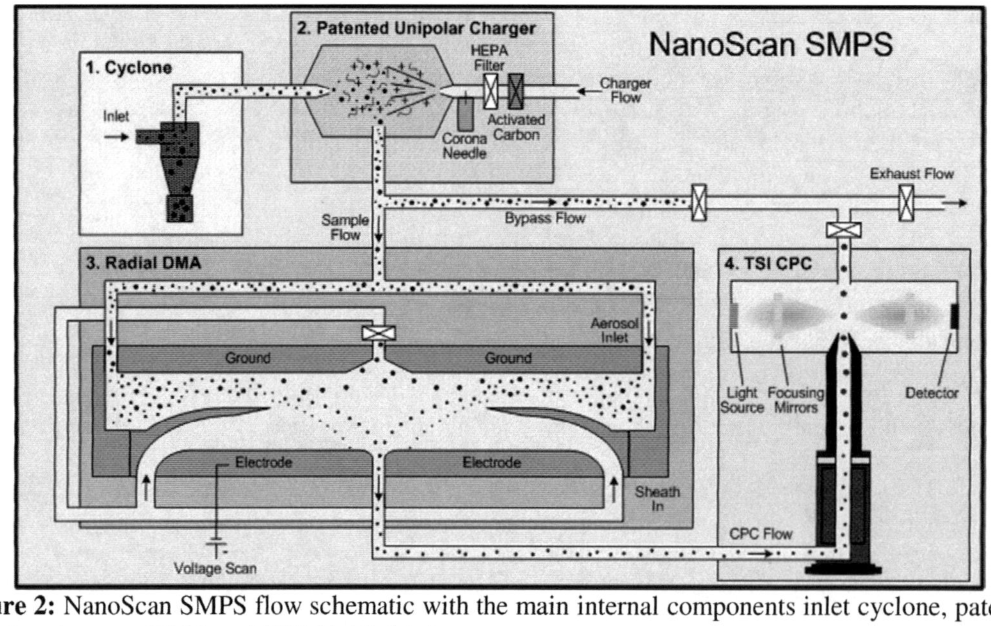

Figure 2: NanoScan SMPS flow schematic with the main internal components inlet cyclone, patented unipolar charger, rDMA and CPC highlighted.

On entering the instrument, the aerosol flow (0.75 L/min) is pre-conditioned to remove large particles by the cyclone which has a cut point (D_{50}) of 550 nm. The cyclone is crucial for long-term use of the instrument as it prevents the issue of plugging that can occur within an impactor. After that the polydisperse aerosol enters the patented unipolar charger, which is a corona jet type charger (Medved et al., 2000). This unipolar charger works through the introduction of an opposed charger flow that is filtered with both an active carbon filter and a HEPA filter. After passing over the charger needle, a jet of positive ions flows into the field-free mixing chamber, allowing the interaction of ions with the polydisperse sample flow using the opposed flow technique, which improves mixing and charge repeatability. After flow splitting, 0.25 L/min of the sample flow enters the radial DMA. The remaining flow enters as particle-free sheath air from the bottom outside edge of the rDMA into a circular channel. The polydisperse sample flow is introduced tangentially through an inlet channel at the top. The top plate of the rDMA is at ground, and the bottom plate is at a high negative voltage, thus creating an electric field. More details on the radial DMA technique can be found in Zhang et al., 1995 and Fissan et al., 1998. The size classified aerosol exits out of the bottom center port of the rDMA through the 'monodisperse outlet' and continues to the isopropanol-based Condensation Particle Counter (CPC). This CPC is based on the handheld CPC design (TSI 3007) that was introduced several years ago and is well-described in Hämeri et al. (2002). The excess flow exits through the top center port of the rDMA.

During the course of a measurement the DMA's voltage is ramped up to scan the entire size range. During every 60 second sampling time there is a 45 second up-scan in which the measurement occurs, and a 15 second down-scan (retrace). This describes the size distribution measurement mode of the NanoScan SMPS (SMPS, see Figure 1B). In the second mode, the so-called Single mode, the instrument monitors the aerosol concentration at a single electrical mobility diameter with a resolution of 1 second. The mobility diameter can be set by the operator to particle sizes between 10 and 420 nm. This single mode enables for instance detection and tracing of a specific engineered nanomaterial of a known size.

The NanoScan SMPS can run on battery power for several hours and its on-board data storage is sufficient for 8 to 10 days of continuous operation. The USB storage drive option can be used to export the data. Data acquisition can be done conveniently via a computer with dedicated measurement software. This is especially useful when the instruments runs in its extended mode on A/C power and with an external working fluid reservoir connected. Further details can be found on the TSI webpage/specification sheet.

3 Results and Discussion

3.1. NanoScan SMPS Verification and Validation Tests

A direct comparison was made between the NanoScan SMPS and two research-grade SMPS systems commonly used in the laboratory (TSI 3034 and TSI 3936L76). The proven technology of the SMPS was used as the reference in several applications. The tests involved challenging the NanoScan SMPS with both monodisperse and polydisperse aerosols. The monodisperse aerosol consisted of silver (Ag) particles generated by a furnace according to Scheibel and Porstendörfer (1983). The corresponding data are shown as upper graphs in Figure 3. The polydisperse aerosols were made from NaCl and Emery Oil (EO, poly-alpha-olefin, PAO 4 cSt). EO generated by an Electrospray Aerosol Generator (TSI 3480) provides spherical particles of synthetic lube oil. Buffer and EO solutions were prepared according to TSI Application note EM-004, 2006, TSI Inc., Shoreview, MN, USA.

Results of the comparison of the NanoScan SMPS with the reference SMPS involving both monodisperse (Ag) and polydisperse (NaCl & EO) aerosols are shown in Figure 3. The data agree very well for the smaller sizes (13.3 and 23.7 nm), although the coarser resolution of the NanoScan SMPS is noticeable. For the monodisperse 133 nm silver particles, the mode agrees as well but a broadening is visible in the results of the NanoScan SMPS. For the polydisperse comparison and validation, NaCl and EO were chosen. Despite the lesser 13 channel resolution of the NanoScan SMPS, the measurements reveal an excellent correlation to the 64 channel/decade resolution of the component SMPS utilized for the two data sets (see lower panels in Figure 3).

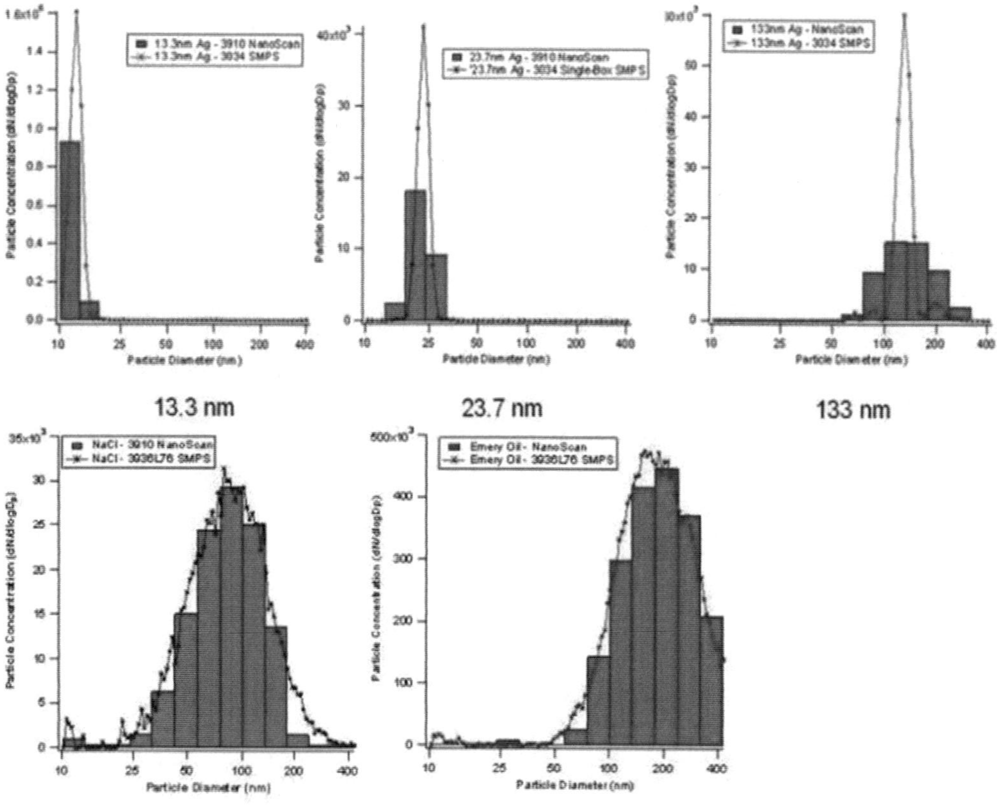

Figure 3: Upper graphs: Parallel NanoScan SMPS and SMPS (TSI 3034) scans of classified Ag aerosol at 13.3, 23.7 and 133 nm. Lower graphs: Scans of polydisperse NaCl and Emery Oil.

The concentration linearity of the NanoScan SMPS was then verified by comparison to an ultrafine butanol CPC (TSI 3776). The results highlighted during the presentation (not shown here) demonstrated a clear concentration linearity of the NanoScan SMPS up to 10^6 particles/cm³. In addition the NanoScan SMPS was also successfully compared to a Fast Mobility Particle Sizer (FMPS, TSI 3091) during measurements of a workplace aerosol originating from soldering.

3.2. In-Vehicle Measurements –Mobile Study

Due to the ever increasing amount of time spent within vehicles, they represent an important location with regards to workplace exposure. Hence, a mobile study of in-vehicle measurements was conducted through the continuous measurement of in-vehicle air during a journey from Cologne to Bremen in Germany (380 km distance). The NanoScan SMPS was placed on the front seat of the car with its sampling port open to sample the cabin air of the vehicle. The measurements were completed by a GPS tracker to enable the accurate correlation of data reported by the instrument with the sampling environment. Figure 4 summarizes the data set, showing the total number concentration plotted on a Google Earth map.

Figure 4: The total particle number concentration during the mobile study shown in Google maps. Warmer colors indicate higher particle concentrations.

Results for the total number concentration and the geometric mean diameter of the measurements calculated from individual recorded number size distributions are displayed in Figure 5A and 5B. Note that figure 5A shows the same total concentration data as Figure 4.

As can be seen from Figure 5A, the variation in the total number concentration measured along the route highlights a number of distinct high particle number hot spots such as in Munster (see Figure 5, iii) where a concentration of 227,444 particle/cm^3 was recorded. It is clear from the size distribution measured that the majority of the particles within this situation are very small, with diameters smaller than 20 nm. The largest particles were found at a construction site at the Lohne-Dinklage location. Here the mode of the distribution is at 86.6 nm (see Figure 5B). The results taken at Bremervörde are the final example of the different nature of particle size distributions along the route. Here the mode of the distribution is at 36.5 nm and particle concentration is low.

With regards to the geometric mean diameter variation over the course of the drive it is clear that there appears to be some degree of correlation between the total number concentration and particle size along the route. This is indicated by the areas shown as hotspots in terms of total number concentration, which display particle size distributions with a geometric mode that is less than 20 nm. In contrast the sites with lower total number concentration exhibit a geometric mean that is approximately 40 to 50 nm in diameter. This implies that sites with higher total number concentration are influenced to a greater degree by aerosols generated by combustion processes as indicated by Figure 5i, whereas those areas demonstrating a lower total number concentration are influenced to a greater degree by aerosols generated by mechanical processes such as construction, as indicated by Figure 5iii. This study demonstrates the value of such data but also shows that nanoparticle concentrations at common "work" places such as inside a vehicle can be very high.

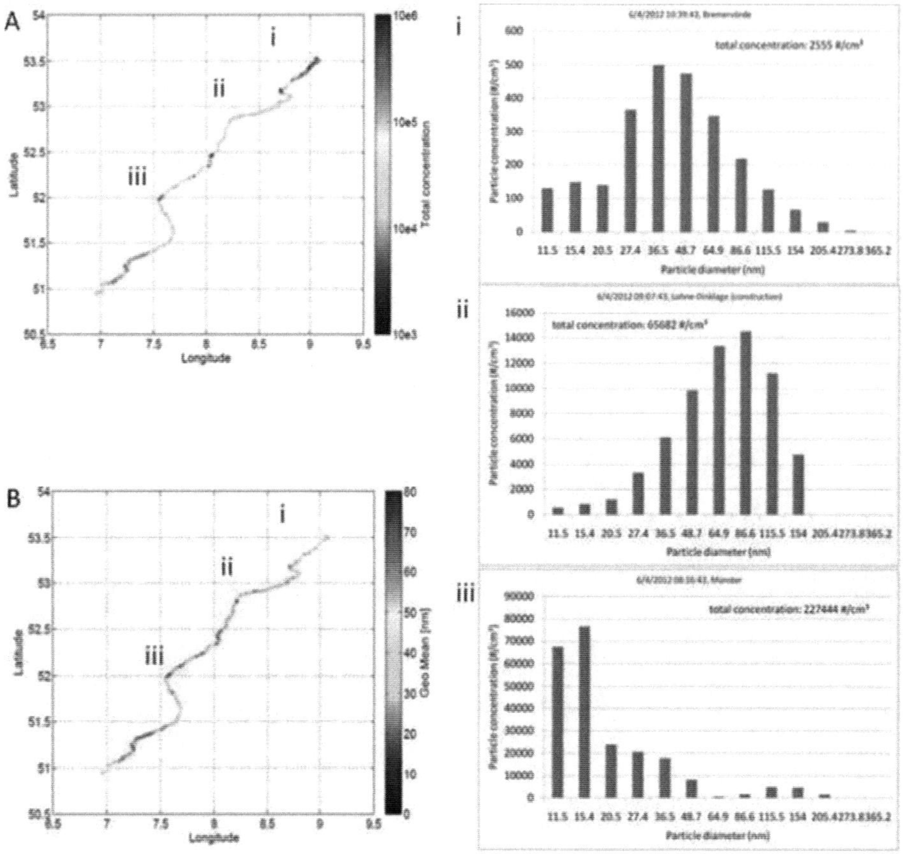

Figure 5: The total particle concentration (A) and geometric mean diameter in nm (B) of the aerosol in the car driving from Cologne to Bremen in Germany. The graphs to the right show the size distributions with 13 channels of resolution at: Bremervörde (i); a construction site in Lohne-Dinklage (ii); and Munster (iii). Their approximate location is indicated on the maps of panels A and B.

3.3. Indoor Air Monitoring

Nanoparticles at indoor workplaces represent an important area of environmental monitoring. Hence, the NanoScan SMPS was also tested in a production facility. Utilizing the NanoScan SMPS, an assessment of worker exposure was made during several work scenarios. These ranged from material tests like abrasion and extrudability to measurements done during milling of the same material. The background particle concentrations in indoor air and exposure levels during the material processing were determined. Background particles were measured at each site and at several sampling points within the rooms. Exposure to nanoparticles was measured in the breathing zone. The background data recorded from the ventilation diffusers supplying fresh air to the rooms were measured, along with the breathing zone during all processes.

The particle size distributions measured at a number of different risk areas inside an indoor production facility are summarized in Figure 6. Different processes result in the generation of nanoparticles which pose a known risk within a work place. Hence, concrete measures could be taken within the facility to limit their impact on the employees. One such measure is the use of enclosures and glove cabinets. Figure 6B shows data from the NanoScan SMPS while the efficiency of the enclosure is challenged.

With regards to the data shown in Figure 6A, background particles were measured from the ventilation diffusers that supply fresh air to the laboratories and at several sampling points within these rooms. By averaging the data from all locations studied, it was found that the total number concentration in the fresh air supplied through diffusers from the ventilation system was around 6,100 particles/cm^3 in the nanoparticle size range (< 0.5 μm). In contrast the total number concentration of particles in the breathing zone during the various processes was found to be twice that, with concentrations on the order of 12,000 particles/cm^3. The dominant peaks in the size distribution within the breathing zone were determined to be approximately 15 and 37 nm. In the background signal a third peak was evident at about 116 nm, which is a particle size not uncommon for indoor aerosols.

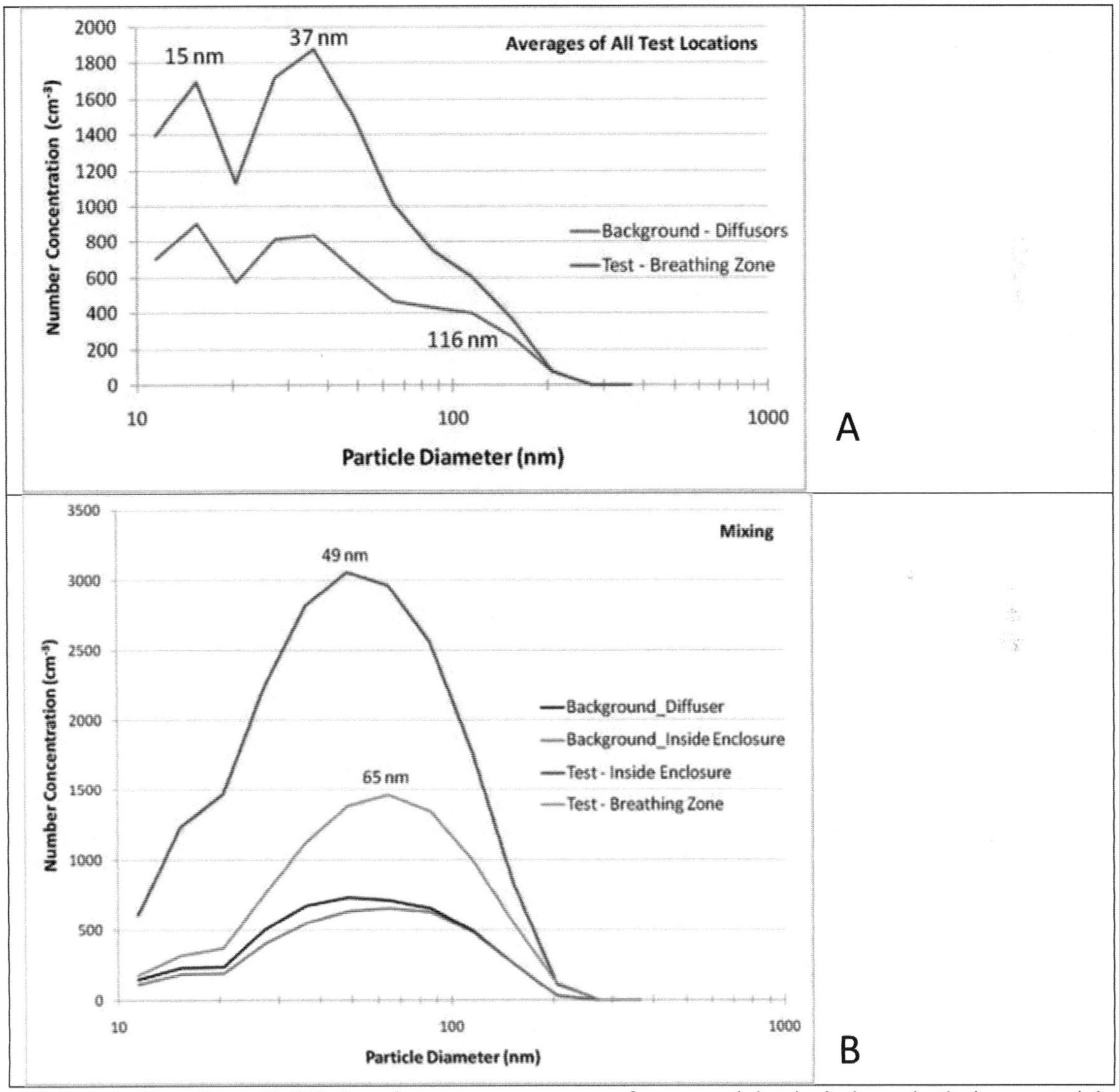

Figure 6: NanoScan SMPS data from measurements of nanoparticles in indoor air during material production at a test facility. (A) The blue line represents data from the ventilation diffuser system and corresponds to background. The red line represents data taken in the breathing zone during the material processing. (B) NanoScan SMPS scans from the emissions generated during a composite mixing process, aimed at testing the efficiency of the enclosure.

The efficiency of enclosures employed as a safety measure is highlighted from the analysis of the measurements during a composite mixing process, which took place in a ventilated enclosure (Figure 6B). It is clear that the particle counts in the room air measured at the ventilation diffuser were equal to the background measurements in the enclosure. The total number concentration was determined to be on the order of 4,700 particles/cm^3. As the material is processed within the enclosure, the number concentration within the enclosure increased by a factor of four to about 20,000 particles/cm^3. Therefore it is evident that the enclosure was able to capture the majority of the particles emitted. Despite this fact about 8,600 particles/cm^3 were still measured in the breathing zone, which is approximately twice that of the background.

4. Conclusions

This paper described the new NanoScan SMPS (TSI 3910) and highlighted its capabilities as a reliable instrument for real-time nanoparticle measurements within the range from 10 to 420 nm. The NanoScan SMPS provides the means to employ SMPS technology in a format that can be easily utilized in the field. The incorporation of a unipolar charger, rDMA and isopropanol-based CPC removes transport and application restrictions currently in place for SMPS technology. The NanoScan SMPS demonstrated concentration linearity up to 10^6 particles /cm^3. It provided good comparability to the laboratory-based reference SMPS systems when challenged with monodisperse silver and polydisperse NaCl and EO particles. Two case studies with the NanoScan SMPS demonstrated the wide range of possible applications and ease of use of this portable instrument. Coupling continuous in-vehicle measurements with GPS technology enabled an easy characterization of the particle size distributions while driving from Cologne to Bremen in Germany. This allowed source apportionment based on measurements, highlighting distinct hotspots and sources of interest. It also allowed an assessment of the exposure to nanoparticles in the vehicle as a workplace. Finally measurements within a production test facility showed the variation between the nature of the nanoparticles in the ventilation diffusers compared to those detected in the breathing zone.

5. References

Fissan, H.; Pöcher, A.; Neumann, S.; Boulaud, D. and M. Pourprix, Analytical and empirical transfer functions of a simplified spectromètre de mobilité electrique circulaire (SMEC) for nano particles. Journal of Aerosol Science 29, 3: 289-293, 1998.

Gurr, J.R., Wang, A.S.S., Chenb, C-H. and K.Y. Jan, Ultrafine titanium dioxide particles in the absence of photoactivation can induce oxidative damage to human bronchial epithelial cells. Toxicology, 213: 66-73, 2005.

Hämeria, K., Koponen, I.K., Aalto, P.P., and M. Kulmala, The particle detection efficiency of the TSI-3007 condensation particle counter. Journal of Aerosol Science 33, 1463–1469, 2002.

Kaufman, S.L. and F.D. Dorman, Aerosol Charge Adjusting Apparatus Employing a Corona Discharge. US Patent 6,544,484 April 8, 2003.

Kinney, P., Pui, D., Mullholland, G. and N. Bryer, Use of the Electrostatic Classification Method to Size 0.1 μm SRM Particles – A Feasibility Study. Journal of NIST 96: 147, 1991.

Knutson, E. and K. Whitby, Aerosol Classification by Electric Mobility: Apparatus, Theory, and Applications. Journal of Aerosol Science, 6: 443-451, 1975.

Medved, A., Dorman, F., Kaufman, S.L. and A. Pöcher, A New Corona-based Charger for Aerosol Particles. Journal of Aerosol Science, 31:S616-S61, 2000.

Mullholand, G., Bryner, N. and C. Croarkin, Measurement of the 100 nm NIST SRM 1963 by Differential Mobility Analysis. Aerosol Science & Technology, 31: 39-55, 1999.

Oberdörster G, Oberdörster E and J. Oberdörster, Nanotoxicology: an emerging discipline evolving from studies of ultrafine particles. Environ Health Perspect, 11: 823-839, 2005.

Pourpix, M. and J. Daval, Electrostatic precipitation of aerosol on wafers, a new mobility spectrometer. In 'Aerosols: Science, Industry, Health and Environment', Pro. of the Third Int. Aerosol Conference, 24–27 Sept., 1990, Kyoto, Japan, Vol. II, edited by S. Masuda and K. Takahashi, Pergamon Press, 1990.

Sanchez V.C., Pietruska J.R., Miselis N.R., Hurt R.H., and A.B. Kane, Biopersistence and potential adverse health impacts of fibrous nanomaterials: What have we learned from asbestos? Wiley Interdiscip Rev Nanomed Nanobiotechnol., 1: 511-529, 2009.

Scheibel, H. G., and J. Porstendörfer, Generation of monodisperse Ag- and NaCl aerosols with particle diameters between 2 and 300 nm. Journal of Aerosol Science 14, 113-126, 1983.

TSI Inc., Generating Emery Oil Aerosol for Calibration. TSI Application note EM-004. TSI Inc., Shoreview, MN, USA. Available at
http://www.tsi.com/uploadedFiles/ Site Root/Products/Literature/Application Notes/EM-004-Generating Emery Oil Aerosol-5001326A.pdf, 2006.

Zhang S.H., Akutsu Y., Russell L.M., Flagan R.C. and J.H. Seinfeld, Radial Differential Mobility Analyser. Aerosol Science & Technology, 23: 357-375, 1995.

Acknowledgments
The authors like to acknowledge the input and support by further TSI colleagues. We thank also Ines Engel for her help with creating the map-based plots for the in-vehicle study.

Journal of Physics: Conference Series **429** (2013) 012062 doi:10.1088/1742-6596/429/1/012062

Overview of Risk Management for Engineered Nanomaterials

PA Schulte[1], CL Geraci[1], LL Hodson[1], RD Zumwalde[1], ED Kuempel[1], V Murashov[1], KF Martinez[1], DS Heidel[1]

[1]National Institute for Occupational Safety and Heath
4676 Columbia Parkway, MS-C14
Cincinnati, OH 45226

PSchulte@cdc.gov

Abstract. Occupational exposure to engineered nanomaterials (ENMs) is considered a new and challenging occurrence. Preliminary information from laboratory studies indicates that workers exposed to some kinds of ENMs could be at risk of adverse health effects. To protect the nanomaterial workforce, a precautionary risk management approach is warranted and given the newness of ENMs and emergence of nanotechnology, a naturalistic view of risk management is useful. Employers have the primary responsibility for providing a safe and healthy workplace. This is achieved by identifying and managing risks which include recognition of hazards, assessing exposures, characterizing actual risk, and implementing measures to control those risks. Following traditional risk management models for nanomaterials is challenging because of uncertainties about the nature of hazards, issues in exposure assessment, questions about appropriate control methods, and lack of occupational exposure limits (OELs) or nano-specific regulations. In the absence of OELs specific for nanomaterials, a precautionary approach has been recommended in many countries. The precautionary approach entails minimizing exposures by using engineering controls and personal protective equipment (PPE). Generally, risk management utilizes the hierarchy of controls. Ideally, risk management for nanomaterials should be part of an enterprise-wide risk management program or system and this should include both risk control and a medical surveillance program that assesses the frequency of adverse effects among groups of workers exposed to nanomaterials. In some cases, the medical surveillance could include medical screening of individual workers to detect early signs of work-related illnesses. All medical surveillance should be used to assess the effectiveness of risk management; however, medical surveillance should be considered as a second line of defense to ensure that implemented risk management practices are effective.

1. Introduction

The products of nanotechnology are relatively new, generally coming into commerce the last 10-15 years [1]. To date, there are more than 1,600 nanotechnology-enabled products in commerce [1-3]. Each nanomaterial and the products that contain them are developed and produced by workers. Workers are the first people in society to be exposed to new technologies and materials such as those arising from nanotechnology. Moreover, if there is to be vast societal benefit from nanotechnology, society needs to be assured that it is being developed responsibly. Attention to the safety and health of workers is the foundation of responsible development that ultimately delivers benefit to society while protecting human health and engendering public trust [4,5].

Risk management (RM) is arguably the most critical step in the protection of workers. Driven by hazard, exposure, and risk information, risk management involves evaluating the extent of risks, and deciding on the most appropriate exposure control measures. The role of RM in nanotechnology may be best considered by taking what can be termed a "naturalistic" view where RM is treated as an evolving set of guidance and control concepts and part of a larger system because it is early in the development of the technology. Employers and workers want RM guidance because there are still vast uncertainties about hazards, exposures, and risks. Because of these uncertainties, hazards, exposures, and risk management approaches should be seen as parts of a dynamic system—one that will be changing, and one where RM approaches and guidance will need to be continuously evaluated, improved, and verified as risk information becomes more substantial.

A naturalistic view of RM for ENMs is illustrated in Figure 1. RM is part of a dynamic iterative system that involves societal and workplace level efforts. At the societal level, two overarching principles prevail. First, is that workers have a right to a safe and healthy workplace and a right to know about potential hazards. Second, employers have the responsibility to provide a safe and healthy workplace and keep the workers safe (workers have the responsibility to cooperate with employers in this regard).

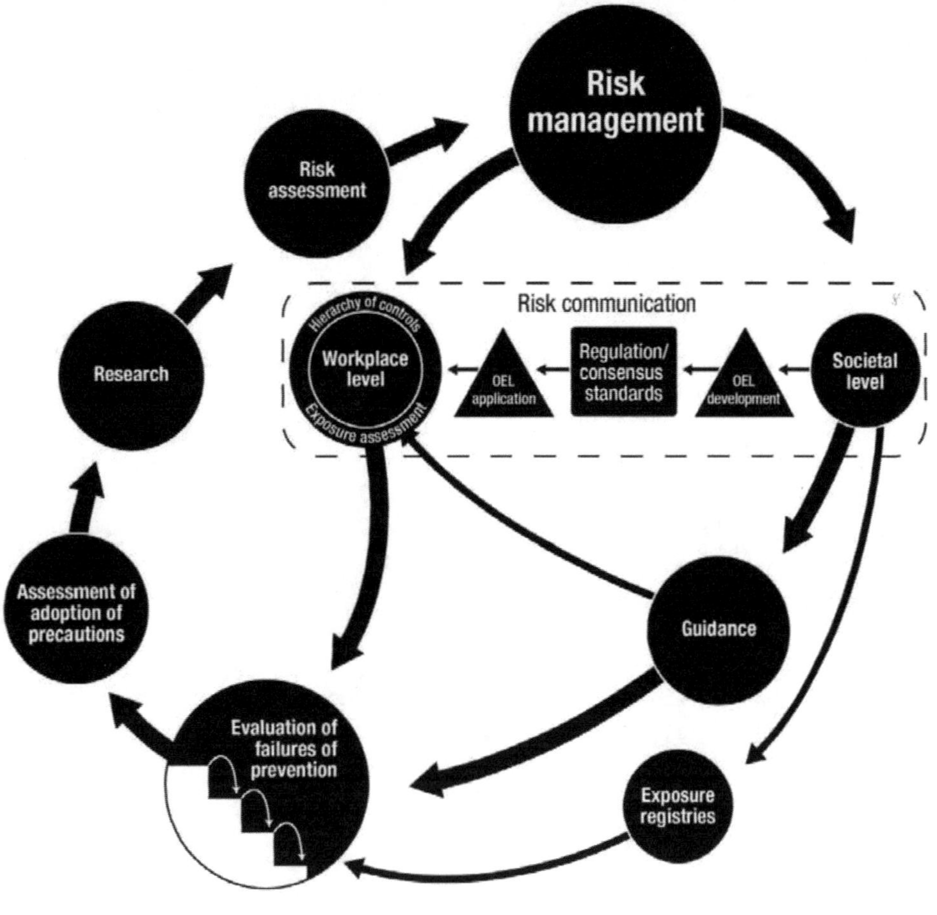

Figure 1. Naturalistic view of occupational safety and health risk maintenance for nanomaterials.

The critical issues to date are "what are the risks" and "what is safe". Responsible development of nanotechnology, in the face of uncertainty of the potential health risks (see [6-9] for review of the evidence), requires that a precautionary approach to risk management should be taken [4].

2. Societal level risk management

At the societal level, the initiators of risk management are laws, standards, regulations, guidance—soft and hard laws. These efforts need to be seen globally, as well as locally, since what happens in one country may affect what happens in another, and as Murashov et al. (2012) [10] noted: "Anticipatory international consensus standards established at the introductory stage of nanotechnology could also potentially hinder the "race to the bottom" where nations compete for jobs by lowering workplace safety standards." In addition to general control guidance, the critical feature of societal level risk management is the development of OELs. Two approaches to developing OELs have been identified, one where there is adequate health information to conduct a quantitative risk assessment to derive an OEL, and the other, where there is limited information requiring a pragmatic approach that relies on professional judgment (Figure 2).

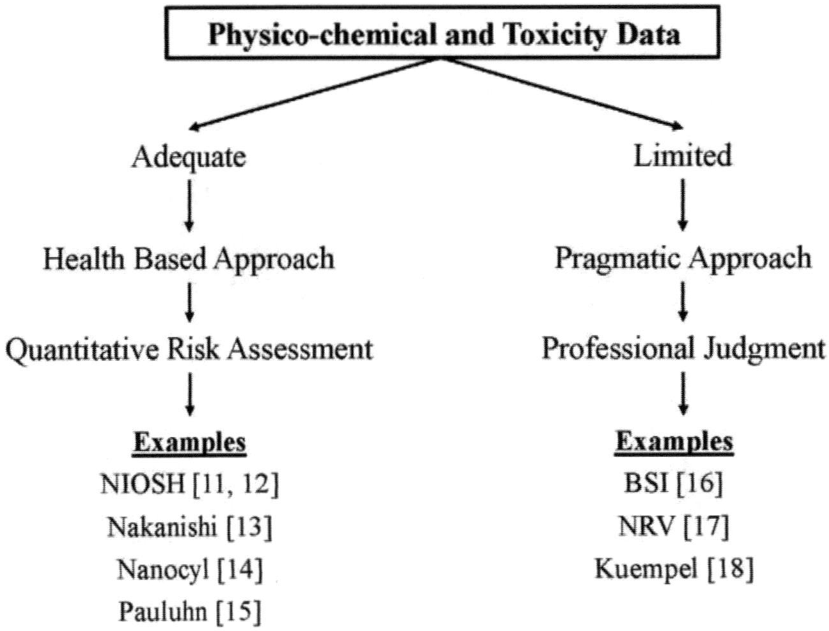

Figure 2. Approaches to developing Occupational Exposure Limits for engineered nanomaterials.

Examples of the first approach include the NIOSH recommended exposure limits (RELs) for titanium dioxide [11] and carbon nanotubes (CNT)/carbon nanofibers (CNF) [12]. Examples of the second approach include British Standards Institute (BSI) guidance [16], the provisional German/Dutch Nano Reference Values (NRV) [17] and is described by Kuempel et al [18]. Variations in non-regulatory OELs based on risk assessments have been conducted including those for carbon nanotubes, with proposed OELs ranging from 1-50 μg/m^3 [12-15,19]. These values may seem quite divergent but, in fact, this divergence generally reflects variability in the use of safety/uncertainty factors and the manner which the working life risk of exposure is assessed (up to 45 years). Moreover, there is greater divergence between proposed OELs for nanomaterials and the OELs for the same bulk material. This discrepancy is evident in the U.S. when comparing the proposed OELs for CNTs with the OEL for graphites. These bulk OELs are often hundreds of times higher than for their nanoscale variants. Where pragmatic OELs are developed, hazards and control bands may be used to identify what level of control to apply. The critical research need for hazard and control banding is to further improve the parameters that define the risk bands (i.e. determining the health risk). The banding strategies developed for nanomaterials have been recently reviewed [20], and approaches to link quantitative risk assessment with exposure control banding schemes (for categorical decision-making and validation) have been described [21].

3. Workplace risk management

The actual application of RM for ENM occurs in the workplace. There is a range of workplaces in which workers are potentially exposed to ENMs that include research laboratories, start-up and pilot operations, manufacturing, incorporating manufactured ENMs in products, and end of life activities including recycling (Figure 3).

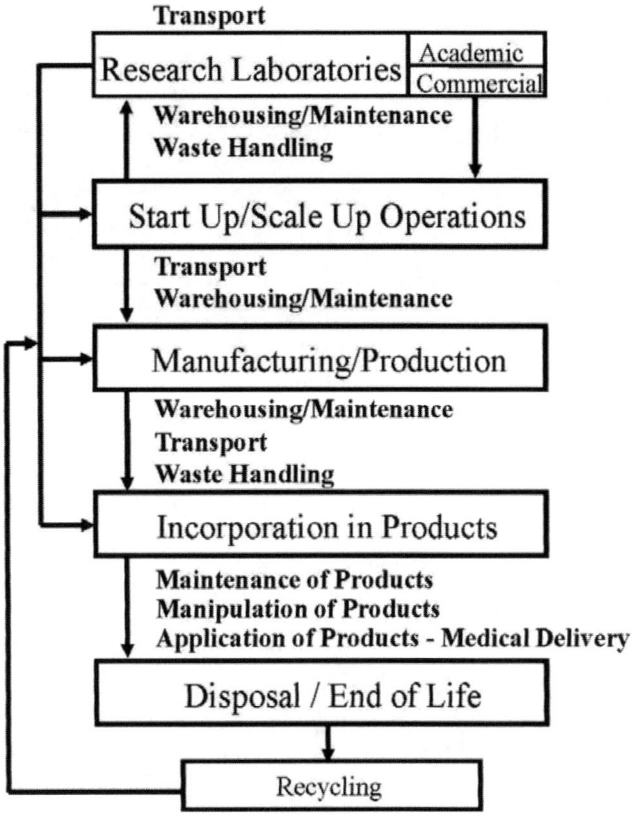

Figure 3. Workplaces where exposures to ENMs could occur (Adapted from [22]).

Also, exposure scenarios include maintenance activities in each of these types of workplaces, and transport between them. These various workplaces could require different risk management strategies.

The principal approach that is recommended for risk management of ENMs is the establishment of ENM-risk management program that is integrated into the overall company health and safety program or system (Figure 4) [22]. The core of the ENM risk management program, as with all occupational safety and health (OSH) risk management programs, is implementation of the well-established hierarchy of controls (Figure 4).

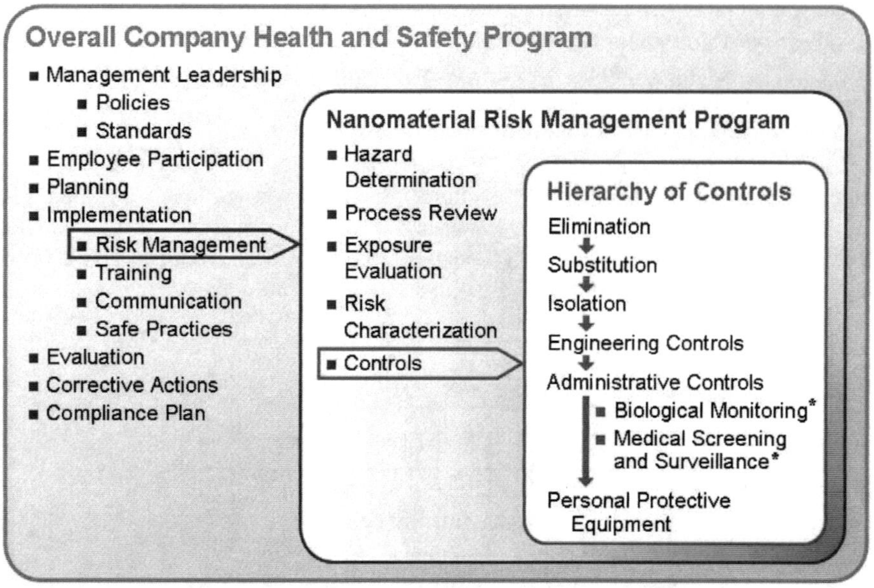

Figure 4. Nanomaterial risk management should be part of an overall company health and safety program [22]. The elements with an asterisk are not actually part of the classical hierarchy of controls, rather they help evaluate controls; however, they may be viewed as part of a hierarchy of prevention that include control and evaluative strategies. (Reprinted with permission of the journal.)

Experience for more than 100 years has shown the effectiveness of the hierarchy of controls (e.g. engineering and personal protective equipment [PPE]), to control fibers and dusts in general industry, biologically active powders in the pharmaceutical industry, and radioactive aerosols in the nuclear industry [22-25]. One limitation in employing an effective RM process is that workers and various employers downstream from the original manufacturer may not know whether materials they receive are nanomaterials or contain nanoparticles due to insufficient or inadequate information on safety data sheets (SDS). Recent studies have shown a large percentage of ENM SDS's were in need of improvement [26,27].

4. Effectiveness of risk management

The limited data that are available on exposure supports the need to periodically evaluate the application and effectiveness of the controls used to manage ENM risks. It is useful to consider the risk management process holistically as a cascade of interventions to prevent harm to workers (Figure 5). Each step represents opportunities to intervene to prevent adverse effects. Interventions at the higher levels can reduce the burden downstream. Working from a cascade framework brings together disparate elements into a cohesive system that is "greater than the sum of its parts [28]." An important element in the cascade is surveillance which provides feedback to upstream processes. The environmental monitoring, part of such surveillance, presents unique challenges since currently available analytical methods and instrumentation may not provide the specificity or sensitivity to cover the broad range of nanomaterials being produced or used. However, a combination of exposure assessment options may provide adequate data to appropriately support risk management needs [29-31].

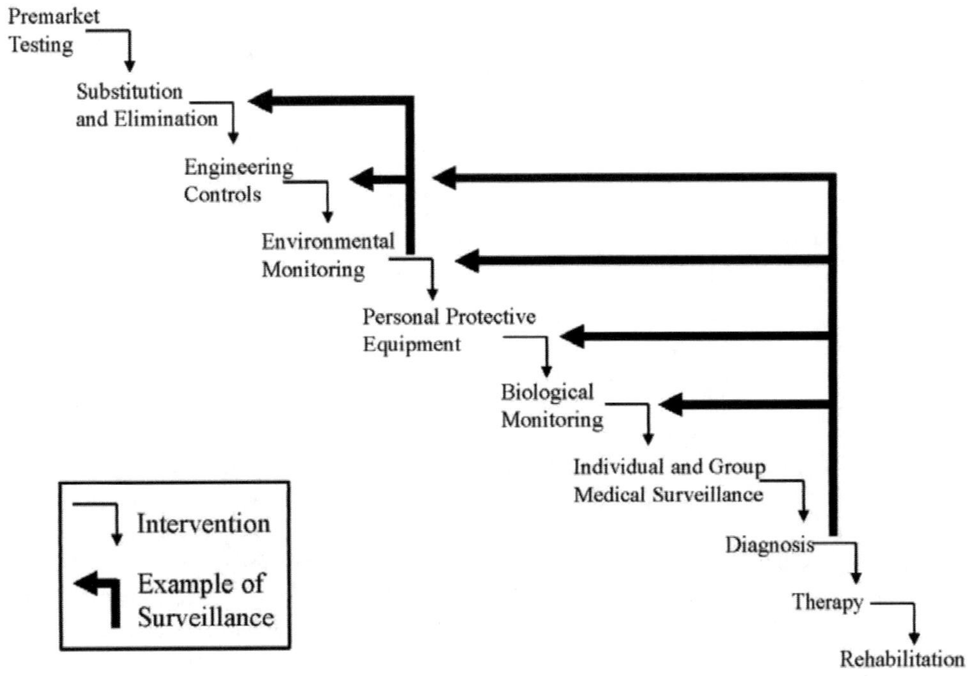

Figure 5. The cascade of occupational health is a model that illustrates the interacting roles of interventions and surveillance in risk management. (Adapted from [28]).

Ultimately, the effectiveness of risk management is based on the extent to which employers have adopted precautionary guidance to control exposures in the face of scientific uncertainty about the extent of the health risk. Numerous government agencies and other organizations have issued precautionary guidance for nanomaterials (e.g., [16,32-34]). Initial surveys of industries conducted in 2004 and 2010 on the use of effective risk management practices have shown that the implementation of precautionary guidance is not at the highest levels [35]. These surveys had relatively low response rates and could have been influenced by responder bias (i.e., more response by the more adherent employers). There is need for further national and global assessment of the adoption of precautionary guidance, and the development of targeted information campaigns for sectors and subsectors where adoption is low.

5. Future issues in RM for ENMs

5.1. Need to develop categorical OELS

It is unlikely that it will be possible to assess toxicologically (using animal studies) all the ENMs in, or entering, commerce [21,36,37]. Therefore, it may be necessary to screen materials individually or as categories and assign them to hazard and control categories (based on physical-chemical parameters). This screening could involve simple characterization of materials and matching them on structure-activity relationships (SAR), or other analogous factors where the hazard is known. For some materials, where high commercial volumes are anticipated, screening may also include a tiered approach that includes a literature assessment followed by *in vitro* testing and *in vivo* testing for those most likely to be of concern.

5.2. Designing out hazards

Hazards of ENMs can be mitigated by considering the hazard potential in the initial design phase of the ENM and in the design of production processes. A recent workshop in 2012 assembled material scientists, toxicologists, and occupational safety and health specialists to address these issues [37]. Such prevention through design (PtD) or "safety by design" efforts need to be supported by corporate leadership that is committed to protecting the workforce and the environment by adopting sustainable practices throughout the life cycle and value chain of an ENM. The challenge is to design a material that maintains the desired functionality while mitigating the potential toxicity. Similarly, production processes can be designed to reduce exposures and hence risk.

5.3. Monitoring the health of the workforce

Thus far, the health of the ENM workforce has been considered in an anticipatory way by conducting animal studies, extrapolating results to workers, and implementing precautionary controls. The nanomaterial workforce also should be continuously monitored and assessed. Funding agencies and employers' associations should commit resources to support significant workforce health surveillance and assessment. This will involve support for medical surveillance as well as for exposure registries, and epidemiological studies [38,39]. Monitoring of the workforce is critical to demonstrating that nanotechnology is being developed responsibly. A pioneering effort in this regard is the national surveillance program in France which initially will monitor workers exposed to carbon nanotubes and titanium dioxide [40]. Similar efforts are needed in other countries. There are sufficient preliminary findings in animals and from experience with incidental nanoparticles to suggest health endpoints to look for (e.g. various respiratory and cardiovascular effects) [6,8,9,41,42]. Animal studies have also yielded an informative array of candidate biomarkers of exposure and effect that could be assessed cross-sectionally and possibly prospectively [43]. There are many technical and logistical issues in developing and conducting epidemiologic studies of workers exposed to ENMs, but there should be a commitment of funds for international research to pursue the health of the current and future workforce [44-45].

5.4. Address advanced nanomaterials

The current generation of "passive" nanomaterials are predicted to be succeeded by generations of more advanced materials [10, 46]. These materials will be advanced in comparison to the current generation by being more active, more integrated and complex, and capable of being linked in systems. It is not known if these materials will be more hazardous than the current generation, but in some cases, they will be designed to be more interactive with biological systems, to change properties during functional operation in response to a stimulus or external signal, and be assembled from the "bottom up" from atoms or molecules [10]. For the most part, these materials appear to be still in laboratory development and not in commerce. But there is not a clear understanding of their ultimate use, location, and the current workforce handling them. More significantly, it is not known to what extent such materials are being targeted for toxicological tests, screening, or anticipatory hazard assessments. There is need to address all of these uncertainties before, like with the passive nanomaterials, they are introduced in commerce without any indications of their potential health risks.

6. Conclusions

The potential health risks of exposure from nanomaterials to workers need to be anticipated and managed if nanotechnology is to be developed responsibly. Absent such efforts, workers may be harmed and society deprived of the timely realization of benefits of the technology. Critical for effective management of risks of nanomaterials is the need for a naturalistic view—one that sees where science and society are in the development of the technology. Nanotechnology is new, early in its natural history, and it needs to be treated cautiously. By actively managing the risks to the workforce through the hierarchy of controls, and then confirming the effectiveness of those management efforts through medical screening, surveillance and epidemiologic studies, society can demonstrate a responsible approach to the development of nanotechnology.

Acknowledgements: The authors thank the following for comments on early drafts of this manuscript: Candace Tsai, John Peters, Michael Ellenbecker, and John Weaver.

Disclaimer: The findings and conclusions in this report are those of the author(s) and do not necessarily represent the views of the National Institute for Occupational Safety and Health.

References

[1] Invernizzi N 2011 Nanotechnology between the lab and the shop floor: what are the effects on labor? *J. Nanopart. Res.* **13**(6):2249-68.

[2] *Nanotechnology research directions for societal needs to 2020: Retrospective and outlook.* 2011 [cited 2012 November 20]; Available from: http://www.wtec.org/nano2/Nanotechnology_Research_Directions_to_2020/.

[3] *Consumer Products Inventory.* [cited 2012 November 20]; Available from: http://www.nanotechproject.org/inventories/consumer/.

[4] Schulte PA, Salamanca-Buentello, F 2007 Ethical and scientific issues of nanotechnology in the workplace *Environ. Health Perspect.* **115** 5-12.

[5] Schulte PA, Geraci CL, Hodson L, Zumwalde R, Castranova V, Kuempel ED, Methner MM, Hoover M, Murashov V 2010 Nanotechnologies and nanomaterials in the occupational setting *Ital. J. Occup. Environ. Hyg.* **1** 63-68.

[6] Castranova V 2011 Overview of Current Toxicological Knowledge of Engineered Nanoparticles *J. Occup. Environ. Med.* **53** S14-S17.

[7] Donaldson K, Borm, PJA, Oberdörster, G, Pinkerton, KE, Stone, V, Tran, CL 2008 Concordance between in vitro and in vivo dosimetry in the proinflammatory effects of low-toxicity, low-solubility particles: The key role of the proximal alveolar region *Inhal. Toxicol.* **20** 53-62.

[8] Duffin R, Tran, L, Brown, D, Stone, V, Donaldson, K 2007 Proinflammogenic effects of low-toxicity and metal nanoparticles in vivo and in vitro: Highlighting the role of particle surface area and surface reactivity *Inhal. Toxicol.* **19** 849-856.

[9] Oberdörster G, Oberdörster E, Oberdörster J 2005 Nanotoxicology: An emerging discipline evolving from studies of ultrafine particles *Environ. Health Perspect.* **113** 823-839.

[10] Murashov V, Schulte, P, Howard, J 2012 Progression of Occupational Risk Management with Advances in Nanomaterials *J. Occup. Environ. Hyg.* **9** D12-D22.

[11] NIOSH 2011 *Current Intelligence Bulletin 63: Occupational Exposure to Titanium Dioxide* NIOSH Publication No. 2011-160 (Cincinnati, Ohio: National Institute for Occupational Safety and Health).

[12] NIOSH 2010 *Current Intelligence Bulletin: Occupational Exposure to Carbon Nanotubes and Nanofibers* Draft for public comment. NIOSH Docket Number: NIOSH 161-1 (Cincinnati, OH: National Institute for Occupational Safety and Health).

[13] Nakanishi J 2012 *Risk assessment of manufactured nanomaterials: carbon nanotubes (CNTs).* in, Final report issued on August 12, 2011. NEDO project (P06041) "Research and development of nanoparticle characterisation methods".

[14] Nanocyl 2009 *Responsible care and nanomaterials case study Nanocyl* Presentation at European Responsible Care Conference (Brussels, Belgium: The European Chemical Industry Council (CEFIC), Prague).

[15] Pauluhn J 2010 Multi-walled carbon nanotubes (Baytubes (R)): Approach for derivation of

occupational exposure limit. *Regul. Toxicol. Pharmacol.* **57** 78-89.

[16] BSI 2007 Nanotechnologies Part 2. PD 6699-2:2007: guide to safe handling and disposal of manufactured nanomaterials (London: British Standards Institution).

[17] van Broekhuizen P, van Broekhuizen F, Cornelissen R, Reijnders L 2012 Workplace exposure to nanoparticles and the application of provisional nanoreference values in times of uncertain risks *J. Nanopart. Res.* **14** 770-795.

[18] Kuempel ED, Geraci, CL, Schulte, PA 2012 Risk Assessment and Risk Management of Nanomaterials in the Workplace: Translating Research to Practice *Ann. Occup. Hyg.* **56** 491-505.

[19] Aschberger K, Johnston, HJ, Stone, V, Aitken, RJ, Hankin, SM, Peters, SAK, Tran, CL, Christensen, FM 2010 Review of carbon nanotubes toxicity and exposure-Appraisal of human health risk assessment based on open literature *Crit. Rev. Toxicol.* **40** 759-790.

[20] Brouwer DH 2012 Control Banding Approaches for Nanomaterials *Ann. Occup. Hyg.* **56** 506-514.

[21] Kuempel ED, Castranova, V, Geraci, CL, Schulte, PA 2012 Development of risk-based nanomaterial groups for occupational exposure control *J. Nanopart. Res.* **14**. DOI: 10.1007/S11051-012-1029-8.

[22] Schulte P, Geraci C, Zumwalde R, Hoover M, Kuempel E 2008 Occupational risk management of engineered nanoparticles *J. Occup. Environ. Hyg.* **5** 239-249.

[23] Naumann BD, Sargent, EV, Starkman, BS, Fraser, WJ, Becker, GT, Kirk, GD 1996 Performance-based exposure control limits for pharmaceutical active ingredients *Am. Indust. Hyg. Assoc. J.* **57** 33-42.

[24] Maiello ML, Hoover MD (eds.) 2011 *Radioactive air sampling methods* (Boca Raton, FL: CRC Press).

[25] Peterson JE 1973 Principles for controlling the occupational environment. In: *The industrial environment—Its evaluation and control* (Cincinnati, OH: U.S. Department of Health and Human Services, Centers for Disease Control and Prevention, National Institute for Occupational Safety and Health) DHHS (NIOSH) Publication No. 74-117. Available: http://www.cdc.gov/niosh/pdfs/74-177.pdf.

[26] Eastlake A, Hodson L, Geraci CL, Crawford C 2012 A critical evaluation of material safety data sheets (MSDS) for engineered nanomaterials *J. Chem. Health and Safety*, **53** S108-S112.

[27] Lee JH, Kuk WK, Kwon M, Lee JH, Lee KS, Yu IJ 2012 Evaluation of information in nanomaterial safety data sheets and development of international standards for guidance on preparation of nanomaterial safety data sheets *Nanotoxicol.* **Early Online 1-8**.

[28] Kreider T, Halperin, W 2011 Engineered Nanomaterials Learning from the Past, Planning for the Future *J. Occup. Environ. Med.* **53** S108-S112.

[29] NIOSH 2009 *Approaches to safe nanotechnology: managing the health and safety concerns associated with engineered nanomaterials.* (Cincinnati, OH: U.S. Department of Health and Human Services, Centers for Disease Control and Prevention, National Institute for Occupational Safety and Health) DHHS (NIOSH) Publication No. 2009-2125.

[30] Dahm MM, Evans, DE, Schubauer-Berigan, MK, Birch, ME, Deddens JA 2012 Occupational Exposure Assessment in Carbon Nanotube and Nanofiber Primary and Secondary Manufacturers: Mobile Direct-Reading Sampling *Ann. Occup. Hyg.* online. http://annhyg.oxfordjournals.org/content/early/2012/10/25/annhyg.mes079.long

[31] Dahm MM, Evans, DE, Schubauer-Berigan, MK, Birch, ME, Fernback, JE 2012 Occupational Exposure Assessment in Carbon Nanotube and Nanofiber Primary and Secondary Manufacturers. *Ann. Occup. Hyg.* **56** 542-556.

[32] J NIOSH 2009 *Notification on precautionary measures for prevention of exposures to nanomaterials.* (Japan: Labour Standard Bureau, Ministry of Health, Labour and Welfare) Japan NIOSH Notification No. 0331013.

[33] Safe Work Australia 2009 *Engineered nanomaterials: evidence on the effectiveness of*

workplace controls to prevent exposure. (Commonwealth of Australia 2009) ISBN # 978-970-642-32884-32887.

[34] ANSES 2010 *Development of a specific control banding tool for nanomaterials.* (French Agency for Food Environmental and Occupational Health and Safety) No. 2008-SA-0407.

[35] Engeman CD, Baumgartner, L, Carr, BM, Fish, AM, Meyerhofer, JD, Satterfield, TA, Holden, PA, Harthorn, BH 2012 Governance implications of nanomaterials companies' inconsistent risk perceptions and safety practices *J. Nanopart. Res.* **14** 749.

[36] Schulte PA, Murashov, V, Zumwalde, R, Kuempel, ED, Geraci, CL 2010 Occupational exposure limits for nanomaterials: state of the art *J. Nanopart. Res.* **12** 1971-1987.

[37] National Institute for Occupational Safety and Health 2012 *Safe Nano Design workshop: molecule to manufacturing to market* [cited 2012 November 27]; Available from: http://cnse.albany.edu/Outreach/NIOSHPresentations.aspx

[38] Schulte PA, Trout, DB 2011 Nanomaterials and Worker Health Medical Surveillance, Exposure Registries, and Epidemiologic Research *J. Occup. Environ. Med.* **53** S3-S7.

[39] Schulte PA, Mundt, DJ, Nasterlack, M, Mulloy, KB, Mundt, KA 2011 Exposure Registries Overview and Utility for Nanomaterial Workers *J. Occup. Environ. Med.* **53** S42-S47.

[40] Boutou-Kempf O, Marchand, J-L, Radauceanu, A, Witschger, O, Imbernon, E, Hlth Risks, N 2011 Development of a French Epidemiological Surveillance System of Workers Producing or Handling Engineered Nanomaterials in the Workplace *J. Occup. Environ. Med.* **53** S103-S107.

[41] Kuempel ED, Smith, RJ, Dankovic, DA, Stayner, LT 2009 Rat- and Human-based Risk Estimates of Lung Cancer from Occupational Exposure to Poorly-Soluble Particles: A Quantitative Evaluation In: L. Kenny (Ed.) *Inhal. Part X.*

[42] Li N, Nel, AE 2011 Feasibility of Biomarker Studies for Engineered Nanoparticles What Can Be Learned From Air Pollution Research *J. Occup. Environ. Med.* **53** S74-S79.

[43] Erdely A, Liston, A, Salmen-Muniz, R, Hulderman, T, Young, S-H, Zeidler-Erdely, PC, Castranova, V, Simeonova, PP 2011 Identification of Systemic Markers from A Pulmonary Carbon Nanotube Exposure *J.Occup. Environ. Med.* **53** S80-S86.

[44] Schulte PA, Schubauer-Berigan, MK, Mayweather, C, Geraci, CL, Zumwalde, R, McKernan, JL 2009 Issues in the Development of Epidemiologic Studies of Workers Exposed to Engineered Nanoparticles *J. Occup. Environ. Med.* **51** 323-335.

[45] Riediker M, Schubauer-Berigan, MK, Brouwer DH, Nelissen I, Koppen G, Frijns E, Clark KA, Hoeck J, Liou S-H, Ho SF, Bergamaschi E, Gibson R 2012 A Road Map Toward a Globally Harmonized Approach for Occupational Health Surveillance and Epidemiology in Nanomaterial Workers *J. Occup. Environ. Med.* **54** 1214-1223.

[46] Subramanian V, Youtie J, Porter AL, Shapira P 2010 Is there a shift to "active nanostructures"? *J. Nanopart. Res.* **12** 1-10.

Journal of Physics: Conference Series **429** (2013) 012063 doi:10.1088/1742-6596/429/1/012063

Risk assessment of nanomaterials and nanoproducts – adaptation of traditional approaches

J Jahnel, T Fleischer, S B Seitz

Karlsruhe Institute of Technology, Institute for Technology Assessment and Systems Analysis (ITAS), Karlstrasse 11, 76133 Karlsruhe, Germany

E-mail: jutta.jahnel@kit.edu

Abstract. Different approaches have been adopted for assessing the potential risks of conventional chemicals and products for human health. In general, the traditional paradigm is a toxicological-driven chemical-by-chemical approach, focusing on single toxic endpoints. Scope and responsibilities for the development and implementation of a risk assessment concept vary across sectors and areas and depends on the specific regulatory environment and the specific protection goals. Thus, risk assessment implication is a complex task based not only on science based knowledge but also on the regulatory context involving different parties and stakeholders. Questions have been raised whether standard paradigms for conventional chemicals would be applicable and adequate for new materials, products and applications of nanotechnology. Most scientists and stakeholders assume that current standard methods are in principle applicable to nanomaterials, but specific aspects require further development. The paper presents additional technical improvements like the complementary use of the life cycle methodology and the support of risk-based classification systems. But also aspects improving the utility of risk assessment with regard to societal impacts on risk governance are discussed.

1. Introduction

Nanomaterials are substances with a high diversity commonly characterized by its nano-size. Their properties differ significantly from those of lager particles of the same material. This makes them suitable for new or improved applications in areas, such as in medicine, environment and energy production. However, these new properties are just the same as those which concern scientists, policy makers, a number of stakeholders and parts of the general public. These concerns have to be taken seriously because experiences of the past showed that new materials may be a source of new threat for human health and the environment [1]. Moreover there is a need for a responsible approach to ensure that potential safety issues are being addressed carefully because nanomaterials are already used in a number of commercial applications. The scientific community – in particular the toxicologists – is requested to answer the question whether nanomaterials pose environmental, health and safety (EHS) risks or not, and to provide policy makers with the appropriate knowledge to perform risk assessment (RA) as a prerequisite for science-based risk management. The current concept for assessing nano-specific risks is the conventional expert-based chemical RA procedure. This is a well-established process which is specified with the Regulation on Registration, Evaluation, Authorization and Restriction of Chemicals (REACH) [2].

A nanospecific RA goes beyond present standard strategies for conventional chemicals. EU committees like the Scientific Committee on Emerging and Newly Identified Health Risks

(SCENIHR) and the European Food Safety Authority (EFSA) evaluated the current RA procedures and concluded that the methods are in principle applicable to nanomaterials, but specific aspects require further development [3, 4]. SCENIHR proposed additional technical tools for the improvement of the standardized RA procedure like the use of the life cycle methodology [5] or the support of preliminary classification systems [6].

Moreover the traditional exposure-response paradigm arises from a narrow toxicological perspective and will not consider adequately societal impacts and societal needs for understanding risk in a broader sense than experts. Despite of serious methodological uncertainties the conventional RA approach is based on confidence in the knowledge used and the possibility of assessing and managing uncertainty as described by SCENIHR [7]. Accordingly, a wider concept is needed which allows for a plurality in perspectives, actors and different kinds of knowledge. In addition regulation based on quantitative RA is an inherently slow governance process which is not intended to facilitate decision-making. The paper highlights the challenges and strategies for improving RA of nanomaterials in the context of risk and innovation governance.

2. The traditional risk assessment paradigm

The classical RA is a well-established and formalized process intended to "calculate or estimate the risk to a given target organism, system or (sub)population, including the identification of uncertainties, following exposure to a particular agent, taking into account the inherent characteristics of the agent of concern as well as the characteristics of the specific target system" [8]. While the hazard of a substance is its potential to cause harm whereas risk is the likelihood of that harm occurring, taking into account wider considerations of exposure and uncertainty. The RA process consists of four steps: hazard identification, hazard characterization including dose-response assessment, exposure assessment, and risk characterization.

According to the Risk Commission [9], the scientific RA process primarily deals with consequences of the effects of noxious agents to human health quantifying the probability of a harmful effect to individuals or populations. The framework was developed for conventional chemicals as an information and decision supporting tool for regulations in the context of safety and sustainability. Associated uncertainty in the progress is managed by the application of safety factors. In general, the traditional approach is a chemical-by-chemical approach, focusing on a single substance, a single media and a single toxic endpoint. In cases where the risk of adverse effects at exposures below a safe level is expected to be practically zero, the Margin of Exposure can be calculated to describe the quotient between expected exposure and the No-Observed-Adverse-Effect-Level. Thus, the output of RA is based on a substance specific quotient produced on an absolute basis.

Triggers for chemical RA are in general the quantities of a chemical, the properties like persistence or bioaccumulation and the effects like acute and chronic toxicity, sensitization, genotoxicity, reproductive toxicity or carcinogenicity. Thus, information requirements include physicochemical properties, toxicological and ecotoxicological information. Chemical RA has been the standard approach for assessing the potential health and environmental risks of bulk chemicals and subsequently has been applied to nanomaterials in recent years [10-12]. In Europe this process is defined for chemicals by the EU chemicals policy REACH (Regulation on Registration, Evaluation, Authorization and Restriction of Chemicals) [2] and the detailed guidance documents provided by the European Chemicals Agency (ECHA).

As a decision support tool RA uses the weight of evidence approach: This process is based on the totality of data in a holistic manner and facilitates transparency and quantification in decision-making. The data provided by toxicology are ranked and scored considering the varying quality, relevance and consistency. This results in the so-called strength of evidence, which is inversely related to the degree of uncertainty. In addition individual data are clustered in evidence lines. Complementary and contradictory evidence of varying strengths and evidence lines are considered in the weight of evidence process. SCENIHR stated that this weighing of evidence is important as a guideline for preliminary RA and for determining the priority for further assessments [7].

Accordingly, classical RA privilege scientific knowledge that can be measured, weighted and monitored. But also values, ethics and tacit forms of knowledge are important factors. There are several inherent value judgments in RA and questions raised if value-based decisions should be left to political decision-makers [13].

RA for products considers the matrix in which the chemical or substance is delivered to the end-user. Scope and responsibilities for the development and implementation of RA for a concrete product and risk management context vary across sectors and areas and depends on the specific regulatory environment. For example RA of nanoproducts falls under the specific regulations for foods, novel food, cosmetics or biocidal products. With regard to RA purposes the assessment of the hazard potential of nanomaterials "as manufactured" has to be complemented by an assessment of nanomaterials "as consumed". As an example, the nanospecific RA for foods porposed by EFSA follows a comparative approach using the data of a relevant comparator like a non-novel material or the corresponding bulk material. With the aid of special criteria like solubility, persistence and translocation it will be decided whether a specific toxicological assessment is required. A tiered weight of evidence approach divided the toxicological assessment into an initial toxicological profiling and supportive studies. Finally, nanomaterials "as used" in its intended food matrix will be evaluated [4, 14].

3. Challenges for a nanospecific risk assessment

Nanotoxicology emerged from the classical toxicology and studies in particular the biological effects of engineered nanomaterials on living organisms and in ecosystems [15]. There is a consensus that the classical measures of toxicology are in principle applicable to nanomaterials, but standard procedures of RA have to be modified [10]. SCENIHR stated in their opinions on RA of nanomaterials and nanoproducts that the current methodologies are generally likely to be able to identify the hazards associated with the use of nanomaterials. However, they highlighted the need for modifications of the guidance on the assessment of risks [3, 11]. In its interim review the Working Party on Manufactured Nanomaterials of the OECD concluded as with traditional chemicals, it is clear that each nanomaterial may pose specific challenges, but in most instances, they can be addressed with existing methods. In some cases, it might be necessary to adapt methods and test guidelines. But it will not be necessary to develop completely new RA approaches [16]. The adaptation of regulations like REACH and its detailed guidance is an ongoing process [17]. However, serious doubts and concerns are expressed that a conventional paradigm may not be sufficient to capture all the dimensions of risk which may arise not only from its material toxicity but also from its interactions with complex biological and environmental systems [18]. The main limitations of current procedures for the RA of nanomaterials are summarized with the following points:

- The identification and definition of the term "nanomaterial" poses a challenge for framing a "substance class" with a high diversity [19].
- Equipment and methods for characterization and detection of nanomaterials are often not appropriate and need further optimization. It is still an analytical challenge to detect nanomaterials in biological matrixes. Furthermore there is still no agreement about a concept for the dose or concentration of nanomaterials in test systems [10, 18, 20].
- High quality exposure data is also largely missing. Many exposure related studies are published on occupational scenarios while much fewer studies are published on environmental and consumer exposure as well as about both acute and chronic exposures [12, 21].

- There is still a need of standardized toxicological methods as well as appropriate controls. Detailed experimental factors - such as surface functionalization, dispersing behavior in biological media or the use of solvents in case of non-dispersing nanoparticles (e.g. fullerenes) in aqueous media are not addressed sufficiently in many studies [12, 21].
- Studies that showed no significant (hazardous) effects are usually not published, even though they are crucial to relive nanomaterials from the suspicion of hazard [22].
- There is an ongoing debate on the significance of high-dose in vitro or in vivo studies conducted so far and whether or not the used methods are suitable for hazard characterization [23].
- For eco-toxicological studies it is in general difficult to simulate real environmental scenarios [12, 21].

Besides the described knowledge gaps and methodological uncertainties the most important challenge is the fact that nanomaterials share no common characteristics besides the nano-scale size. In the nanotoxicology community it is consensus that only a "case-by-case" assessment is responsible and sound. Thus, RA requires the full dataset for each and every kind of nanomaterial. This makes the progress of gathering the relevant data extremely slow. Therefore, a complete RA is only possible for a small selection of high abundant nanomaterials [12, 22].

4. Strategies for improving nanospecific risk assessment

4.1. Combining life cycle methodology with the risk assessment procedure

Life cycle assessment (LCA) is a systematic assessment of the potential environmental impacts of a defined good or service throughout all life cycle stages, contributing to the production of the product under investigation, its use and disposal [24]. LCA is a most extensively developed and standardized methodology. The assessment typically covers a broad range of environmental impacts, such as climate change, resource depletion, and toxicity on human health caused by releases of chemical agents. In contrast to RA the results of LCA are normally presented on a relative or comparative basis, such as the comparison of two alternative solutions for product development, to make informed decisions regarding the relative environmental sustainability of products. The LCA inventory is a mass-based accounting system that does not take into account spatial, temporal, dose–response, and threshold information. Hence, this assessment tool could be helpful to quantify potential impacts of nanoproducts across different life cycle stages. In excess of the traditional chemical RA the manufacturing aspect but also consumption and disposal aspects are addressed. However, nanospecific LCA has its own challenges and limitations. There is a lack of robust data regarding emissions and exposures related to the production, use and disposal of nanomaterials and products containing them and subsequently a need to establish a database which includes the most common production pathways. As discussed already for RA knowledge about the environmental fate, behaviour, and toxicological impacts of nanomaterials are generally limited. The unclear metrics used in these assessments, the inconsistencies in viable data sources and the challenge of proper identification and characterization of nanomaterials results in serious methodological uncertainties.

Formulating strategies to use LCA and RA together for the assessment of chemicals has been pursued in a range of research efforts for more than a decade, and remains a topic of on-going work. Some proposed approaches range from the use of the same data, the complementary use of the results as well as efforts to integrate these two tools into on [5]. The proposed integrative approaches could be allocated to a more traditional RA approach applied in a life cycle perspective ("LCA-based RA'') or to the more conventional LCA supplemented by RA in specific life cycle steps (''RA-complemented LCA'') [5]. However, the full integration into one tool is considered neither desirable nor meaningful as the two tools answer different questions and have different system boundaries. For example the two methods have differing scopes, aims, and end results: while LCA assesses a wide variety of environmental impacts of a product or system related to the functional unit, RA deals with the health and environmental risks of a single substance at a particular point in the chemical's life cycle. As these different systems have their own meanings of terms like "life cycle" or "characterization" it may be more beneficial to use these two technical tools in a complementary manner. The following approaches were proposed:

- The performance of RA at each or selected stages of a product's life cycle.
- The identification of potential health issues on a system-wide and hypothetical basis by LCA which is followed by a detailed RA.
- LCA as a strategic tool to prioritize data which is needed to complete by RA.

Obviously, both LCA and RA methods require substantial amounts of technical data and strong expert knowledge. Both methods have been criticised for not being able to effectively handle uncertainties or the lack of data adequately [5].

4.2. The support of risk-based classification systems
Despite of advances in frameworks and assessment methodologies, a comprehensive characterization of predictive properties for all nanomaterials is far from complete. Thus, some authors proposed a preliminary relative risk ranking or risk classification of nanomaterials. Zuin et al. [25] used physico-chemical and toxicity data for a relative ranking of the hazard potential of quantum dots, fullerenes, single-walled carbon nanotubes and carbon black. The classification system of Olson and Gurian [26] is based on the properties persistence and toxicity. They recommended that persistence should be characterized first because this determine the burden of proof for non-toxicity. Tervonen et al. [7] described a classification system for nanomaterials based on a set of performance metrics that measure both the toxicity and physico-chemical characteristics as well as the expected environmental impacts through the product life cycle. The parameters included properties of nanomaterials like agglomeration, aggregation, reactivity, charge, critical functional groups, size, bioavailability, bioaccumulation and the toxic potential. In addition a relative ranking of the exposure potential of selected nanoparticles were proposed using different transport and persistence properties in the aquatic environment [27].

The various nanomaterials will be clustered in different risk categories with the aid of the weight of evidence method, a formal decision analysis or expert judgments. The purpose of the proposed classification and categorization systems is not to select a single best alternative, but to preliminarily rank or group the alternatives through a structured process. The results are important for prioritizing further studies or management measures. A detailed RA may be reserved for nanomaterials found to be both persistent and toxic and for high exposure scenarios [7].

An example for a classification system was introduced by the German NanoKommission. This stakeholder commission on nanotechnologies was established by the German Federal Ministry for the Environment, Nature Conservation and Nuclear Safety (BMU) and presented a key dialogue panel within the German Federal Government's Nano-Initiative. A proposed preliminary assessment tool structures nanomaterials in three categories, the so-called 'levels of concerns' and combined them with voluntary measures for risk management in the following way:

1. Probably hazardous – concern level high:
   - Criteria: Exposure occurs; materials have high mobility, reactivity, persistence or toxicity of the materials.
   - Measure: A concept is required for measures to minimise exposure or to avoid certain applications.

2. Possibly hazardous – concern level medium:
   - Criteria: Exposure cannot be ruled out; materials have unknown agglomeration or deagglomeration behaviour; too little is known about materials' solubility and biodegradability; the possibilities for release not of nanoparticles from matrices have not yet been explored.
   - Measure: A concept is required for measures to reduce exposure of humans and the environment.

3. Probably hazardous – concern level low:
   - Criteria: Exposure can largely be ruled out; materials are soluble or biodegradable; materials are bound in matrices; materials form stable aggregates or agglomerates.
   - Measure: No measures in addition to those for "good work safety practice" (or "hygiene practice") are required.

The criteria for the different categories are based on current studies and need to be refined if further specific assessments in the area of health and environmental protection become available [20].

Another example is the Swiss 'Precautionary Matrix for Synthetic Nanomaterials'. This approach was introduced as a key element under the Swiss Action Plan for synthetic nanomaterials and was revised on the basis of users' experience at the beginning of 2010. The precautionary matrix "provides a structured method to assess the "nanospecific precautionary need" of workers, consumers and the environment arising from the production and use of synthetic nanomaterials. Upon entering a limited selection of nanomaterial-specific and application-specific parameters into an electronic form (size of the particles, reactivity and stability, their release potential, the amount of particles) the matrix provides a simple risk classification:

- Class A: risks specific to nanomaterials are low, no further clarification necessary.
- Class B: possible risks, further clarification and/or risk reduction needed.

The Precautionary Matrix may be regarded as an instrument that supports companies. It is to be used in the context of duty of care and industry self-supervision. The approach functions simultaneously as a differentiation aid, a detector of gaps in knowledge and an early warning system but should not in any way replace a quantitative RA process [20, 28].

4.3. The role of the social context: concern assessment

In order to consider societal impacts and societal needs for understanding risk in a broader sense than experts, the classical toxicological-driven RA paradigm should be widened for more plurality in actors and different kinds of knowledge. Actors in the RA process are the industry and manufacturers which have to ensure that substances brought on the European market do not adversely affect human health or the environment. On the other hand scientists like toxicologists, and legislators as well as scientific committees, European and national agencies and politicians are also engaged in assessment procedures. Classical RA however overemphasises the role of technical experts and scientific knowledge that can be measured and weighted. It is also important to consider the contextual information from the general public and the civil society organisations. The IRGC framework for the risk governance of nanotechnology [29] is a sophisticated risk management model involving a multitude of different actors in a dynamic process with various iterations and feedbacks. It acknowledges that risk governance decisions have to be taken in instances of complexity, uncertainty and ambiguity. Therefore, strategies should be based on a corrective and adaptive approach and take into account the level and extent of available knowledge and a societal balancing of the predicted risks and benefits. The framework includes two concepts for the governance of manufactured particulate nanomaterials:

It integrates a scientific risk-benefit assessment (including environment, health, and safety (EHS) and ethical, legal, and other social issues (ELSI)), with an assessment of risk perception and the societal context of risk, called "concern assessment". The IRGC framework is a cyclical process and consists of four phases:

1. "Pre-Assessment": trigger or initiator of the whole assessment and management process
2. "Risk Appraisal": comprises two elements: risk assessment and concern assessment.
3. "Tolerability and Acceptability Judgment": brings together the classic risk characterization and risk evaluation as a new element.
4. "Risk Management": has to react not only to new scientific results regarding a hazard or an exposure to it. It also reacts to changing societal or cultural factors like altering expectations on risk reduction procedures, new judgments about tolerability and acceptability of risks, developing value systems or shifting risk perceptions of different actors.

Some authors criticised this framework because public participation is still perceived as a factual input, as part of an expert-driven process, rather than a real empowerment of citizens. The risk appraisal stage is subdivided into the traditional RA and concern assessment without a concrete inclusive concept. This might only be tempting, "if not intended to fall into the trap of identifying the former as legitimate (i.e. scientific), while the latter could be viewed as 'political', hence of a different sort of legitimacy" [13].

In detail, Renn and Walker defined concern assessment as "a social science activity aimed at providing sound insights and a comprehensive diagnosis of concerns, expectations and perceptions that individuals, groups or different cultures may link to the hazard" [30]. The perception of technological risks depends on psychological factors such as perceived threat, familiarity, personal control options and positive risk-benefit ratio. But also political and cultural factors such as perceived equity and justice, visions about future developments and effects on personal interests and values are important. A systematic assessment of the concerns and preferences of the various actor groups and the public at large, a systematic feedback of its results to the related regulatory and legislative processes are necessary prerequisites to improve our understanding of the likely societal responses to the developments in nanomaterials and nanotechnology.

While chemical RA can build upon a long tradition of methodological development, concern assessment is still in its early stages. Among a set of well-established methods that social science uses to study perceptions of nanotechnology's benefits and risks within individuals, groups or the society as a whole are quantitative and qualitative methods. Each of them has its own pros and cons. Quantitative methods – including surveys which are designed to ascertain large and therefore representative datasets as well as experimental studies using non-probability samples – for example, allow for testing and revising existing hypothesis, and making statements about defined groups of people. Typical examples are large, standardized polls within a representative sample of a population. In contrast, qualitative methods are rather designed to gain insights into individual arguments, ideas or values and to explore new aspects of an issue. Thus, they are designed rather open (not standardized) to capture even unexpected facts. Beside in-depth interviews, focus groups are typical examples of qualitative methods. The results of focus group discussions can broaden the perspective of various actors involved in risk governance but a simple interpretation of the participants' positions voiced as stable attitudes would be misleading [20, 31, 32]. The 'translation' of societal values, concerns and perceptions into concrete measures for risk governance is still a subject of debate.

5. Recommendations for risk assessment in the context of risk governance

Although the traditional RA framework is a powerful approach for bulk chemicals, its use for estimating potential risks of nanomaterials in the near-term may be limited due to a kind of "paralysis by analysis" problem. An improvement of research in broader issues than chemical-based aspects such as decision-making, risk governance and a systematic evaluation of complementary RA tools is needed and was increasingly stipulated [33]. In general, RA should support timely decisions and it should be transparent and open to inspection by all. In addition iterative, tiered, flexible and adaptive procedures are recommended which also consider improved social and cultural aspects. According to the OECD "adaptive management" means that "the substance is produced and used under a certain set of conditions based on a preliminary assessment, while additional data are collected to periodically evaluate the initial assessment and to modify the conditions as needed to ensure health and safety" [16]. Incorporating product life cycle considerations into this tiered process may prioritize RA needs.

On the other hand the precautionary principle should support decisions in the absence of scientific certainty. The precautionary principle is embedded in the EU law and applies especially within EU product authorization procedures. The principle runs through legislation for example in the "no data, no market" idea of REACH or the pre-market authorization in special food regulations. This principle allows governmental bodies to intervene with risk management decisions like case-by-case decisions whenever necessary. In the general food law framework (EC 178/2002) [34] it is acknowledged that "scientific risk assessment alone cannot provide the full basis for risk management decisions"– leaving open that risk management decisions could be partly based on ethical principles or particular consumer interests. According to the idea of responsible innovation it is also necessary to promote the participation of stakeholders since the early stage of innovation through the involvement of the business and epistemic community.

The recently re-evaluated traditional RA framework by the National Research Council goes in this direction: additional steps were proposed to integrate risk and non-risk information and stakeholder involvement [16]. In addition, various elements like iterative RA and stakeholder engagement are currently implemented in REACH. Thus, adaptation of traditional RA is a necessary and ongoing process which should not only consider preliminary tools but also serve for facilitating regulatory measures. In this context the question remains whether the compatibility of RA with the precautionary principle will need a further paradigm shift.

References

[1] Oberdörster G, Oberdörster E and Oberdörster J 2005 *Environ Health Perspect*, **113(7)** 823-839
[2] Regulation EC/1907/2006 of the European Parliament and the Council of 18 December 2006 concerning the Registration, Evaluation, Authorization and Restriction of Chemicals (REACH). *OJ L* **396**/1, 30.12.2006
[3] SCENIHR 2007 *Opinion on the appropriateness of the risk assessment methodology in accordance with the technical guidance documents for new and existing substances for assessing the risk of nanomaterials* (Brussels: European Commission) http://ec.europa.eu/health/ph_risk/committees/04_scenihr/docs/scenihr_o_010.pdf
[4] EFSA 2011 *EFSA Journal* **9** 2140
[5] Grieger K D, Laurent A, Miseljic M, Christensen F, Baun A, Olsen S I 2012 *J Nanopart Res* **14** 958-981
[6] Tervonen I, Linkov J, Figueira J Steevens J, Chappell M and Merad M 2009 *J Nanopart Res* **11**

757-766
[7] SCENIHR 2012 *Opinion on the Memorandum on the use of the scientific literature for human health risk assessment purposes – weighing of evidence and expression of uncertainty.* (Brussels: European Commission) http://ec.europa.eu/health/scientific_committees/emerging/docs/scenihr_s_001.pdf
[8] OECD 2003 *OECD Series on Testing and Assessment* **44** *ENV/JM/MONO* **15** (Paris: OECD)
[9] Risk Commission 2003 *Ad hoc Commission on "Revision of Risk Analysis Procedures and Structures as well as of Standard Setting in the field of Environmental Health in the Federal Republic of Germany". Final Report* (Berlin)
[10] Rocks S, Pollard S, Dorey R, Levy L, Harrison P and Handy R 2008 *Comparison of Risk Assessment Approaches for Manufactured Nanomaterials. Final Report* (London: Defra)
[11] SCENIHR 2009 *Opinion on risk assessment of products of nanotechnologies.* (Brussels, European Commission) http://ec.europa.eu/health/ph_risk/committees/04_scenihr/docs/scenihr_o_023.pdf
[12] Aschberger K, Micheletti C, Sokull-Klüttgen B and Christensen F M 2011 *Environment International* **37(6)** 1143-1156
[13] Senjen R and Hansen S F 2011 *C. R. Physique* **12** 637-647
[14] Cockburn A, Bradford R, Buck N, Constable A, Edwards G, Haber B, Hepburn P, Howlett J, Kampers F, Klein C, Radomski, M, Stamm H, Wijnhoven S and Wildemann T 2012 *Food and Chemical Toxicology* **50** 2224-2242
[15] Oberdörster G 2010 *J Intern Med* **267(1)** 89-105
[16] OECD 2012 *OECD Series on the Safety of Manufactured Nanomaterials* **33**: ENV/JM/MONO **8** (Paris: OECD)
[17] ECHA 2012 *Guidance on information requirements and chemical safety assessment: Appendix R7-1, R7-2, R8-15, R10-2, R14-4 Recommendations for nanomaterials* (Helsinki: European Chemicals Agency)
[18] Maynard A D, Aitken R J, Butz T, Colvin V, Donaldson K, Oberdörster G, Philbert M A, Ryan J, Seaton A and Stone V 2006 *Nature* **444(7117)** 267-269
[19] Commission Recommendation of 18 October 201 on the definition of nanomaterial. *OJ L* **275**/38, 20.10.2011
[20] STOA 2012 Fleischer T, Jahnel J, Seitz S B *NanoSafety - risk governance of manufactured nanoparticles. Final report* IP/A/STOA/FWC/2008-096/LOT5/C1/SC3 (Brussels: European Parliament, Science and Technology Options Assessment)
[21] ENRHES 2010 *Engineered Nanoparticles: Review of Health and Environmental Safety. An International Collaborative Review involving Researchers from Edinburgh Napier University, the Institute of Occupational Medicine, the Technical University of Denmark, the Institute for Health and Consumer Protection of the European Commission's Joint Research Centre, and the Institute of Nanotechnology. Project Report.* http://www.nanowerk.com/nanotechnology/reports/reportpdf/report133.pdf
[22] Krug H F and Wick P 2011 *Angewandte Chemie International Edition* **50(6)** 1260-1278
[23] Oberdörster G 2010 Concepts of Nanotoxicology *NanoAgri 2010 Conference* http://www.nanoagri2010.com/fao_mini_papers_extra_files.pdf
[24] Hauschild M, Jeswiet J and Alting L 2005 *CIRP Annals – Manufacturing Technology* **54(2)** 1-21
[25] Zuin S, Micheletti C, Critto A, Pojana G, Johnston H, Stone V, Tran L and Marcomini A 2011 *Nanotoxicology* **5(3)** 445-458
[26] Olson M S and Gurian P L 2012 *J Nanopart Res* **14(4)** 1-7
[27] O'Brien N J and Cummins E J 2011 *Risk Analysis* **31(5)** 706-726
[28] Hodge G A, Bowman D M and Maynard A D 2010 *International Handbook on Regulating Nanotechnologies* (Cheltenham: Edward Elgar)
[29] IRGC 2006 *White Paper on Nanotechnology Risk Governance* (Geneva: International Risk

Governance Council)
[30] Renn O and Walker K 2008 *Global Risk Governance: Concept and Practice Using the IRGC Framework* (Dordrecht: Springer)
[31] Fleischer T and Quendt C 2007 *Wissenschaftliche Berichte FZKA* 7337.
[32] Fleischer T, Haslinger J, Jahnel J and Seitz S B 2012 *International Journal of Emerging Technologies and Society* **10** (2012) 79-95
[33] Grieger K D, Baun A and Olsen R 2010 *J Nanopart Res* **12** 383-392
[34] Regulation EC 178/2002 of the European Parliament and of the Council of 28 January 2002 laying down the gegeral principles and requirements of food law, establishing the Eurpean Food Safety Authority and laying down procedures in matters of food safety. *OJ L* **31**/1, 1.2.2002

Risk assessment method for the implementation of materials divided up to the nanometric scale

L Gridelet[1], P Delbecq[2], L Hervé[3], G Fayet[4], D Fleury[4], S Kowal[4], P Boissolle[5]

[1]SOPROREAL, 137 rue Jacques Duclos, F – 93600, Aulnay Sous Bois

[2]L'OREAL, 1 avenue Eugène Schueller, F - 93600, Aulnay Sous Bois

[3]SANOFI, 1 avenue Pierre Brossolette, F - 91385, Chilly Mazarin

[4]INERIS, Parc Technologique Alata, BP2, F – 60550, Verneuil-en-Halatte

[5]AIR FLUX CONCEPT, 53 Av du Maréchal Lyautey, F - 75016, Paris

Abstract. A new approach of assessing the risks inherent in the implementation of powders, including nanomaterials has been developed. This tool is based on the method of the OHB (Occupational Hazard Band) widely spread in the chemical industry. The European classification and CLP scales of toxicity have not been modified; only the control of exposure has been worked at. The method applies essentially to the prevention of the exposures by airborne materials, whatever their particle size. The skin exposure is not treated there specifically for the time being. The method considers exposure based on seven parameters to take into account the characteristics of the materials used, their emission potential, the conditions of use, as well as classic parameters of the characterization of the exposure as the duration and the frequency. The method stresses on a pragmatic exploitation of the current knowledge and of the available data, bearing in mind that a lot of them are not easily accessible to plant operators. The product of the reflection is then positioned on a hazard x exposure matrix from which 3 levels of priority of action are defined, as in the classical OHB method applied to pure chemical risk. This approach fills in a gap in term of risk assessment and avoids jeopardizing all that had been set up for years, while introducing newelements of reflection accessible to all operators.

1. Introduction

The evaluation of risks at the workplace is a must in the process of occupational risk control. It must be conducted under the responsibility of the employer. The fact that nanomaterials show more unknowns than givens enhances the importance of assessing the risks in order to better prevent their occurrence. There are many risk assessment methods, mainly associated to the chemical risk. Given their reactivity and their physical characters, the nanomaterials make necessary the adaption of existing tools to the new context in term of health and industrial hygiene. The main challenge is to provide professionals with appropriate information and tools so that they can better exchange with their partners when building an exposure case, and better prioritize the preventive actions to be taken. It is important that the method is dedicated to operational workers so that they can take active part to their own safety rather than being spectator. The safety and hygiene staff provides support and guidance, and insures the quality of the process of risk assessment in a continuous improvement approach. The following risk assessment method leans upon « control banding » in use in the industry for long, and easily accessible to operators. It combines toxicology data with the conclusions of exposure scenarios. However, given the current level of uncertainty of toxicological data of nanomaterials, the toxicity scale was deliberately kept unchanged.

2. Risk assessment method

2.1. Objectives and scope of the method

The present risk assessment method tackles the handling of solid materials, of powders, of solid aerosols, and beyond of solids divided up to the nanometric scale. This method is based on « control banding » which consists in distributing on several levels the toxicological hazard of materials on one side, and the potential exposure to these materials on the other side. With respect to hazard, this method leans upon the OHB (Occupational Hazard Band) toxicology scale commonly used in chemical risk management (1, 2). The assessment of exposure takes into account the conditions of use of materials, their characters, the quantities used, as well as their frequency and duration of use.
The risk assessment process can be summarized as follows:

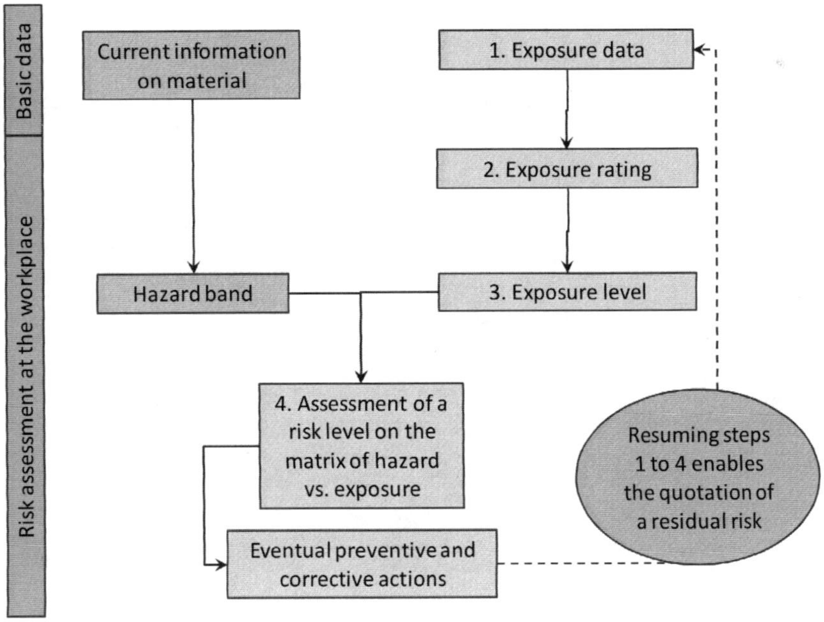

Figure 1: Risk assessment process scheme.

The application of this method results in a semi quantitative quotation of risks which must have its follow-up in the appropriate sizing of risk control measures. This risk assessment process has to be usefully supplemented with specific measurements of exposure at the workplace using appropriate techniques which will enable to check the adequacy of risk control means recommended with actual working situations.
This risk assessment tool can also be used to provide information in the risk assessment studies preliminary to operations like maintenance and cleaning, or even in the preparation of specific procedures intended to manage accidents. The PPE's (Personal Protective Equipment) are not considered in this risk assessment method. As a matter of fact, the risk has to be assessed in the absence of the PPE's that must be used as the last protection means, only once the collective preventive actions and the collective corrective actions have been taken to reduce the initial risk.

2.2. *Control banding*

The toxicology scale used in the present risk assessment method is the OHB scale, very commonly used in the industry, which encompasses 5 levels. This classification scale has been recently revised in 2010, in order to match the requirements of CLP. This toxicology scale is slightly different from most scales commonly used at this moment that tend to consider the CMR 1A & 1B materials in the higher toxicity band, but that are leaving at the same time the acute toxic 1 and 2 at a lower concern.
The CLP toxicology scale has not been adjusted for the specific risks inherent to nanomaterials for the following reasons. The transition period between the old labeling and the new one is not yet finished and it is preferable to let the industry finalize the implementation of CLP before moving to a new classification, even though a slight adjustment. Another reason is that the toxicological data related to nanomaterials are not yet finalized for most of them. Since the occupational exposure limits are currently missing, it is recommended to lean upon the data issued by the suppliers and to get as much information as possible before implementing any process, or before using any new material.
In the total absence of toxicological properties, nanomaterials shall be rated at OHB 4, which is a rule that applies for any unknown chemical. It is recommended that the materials that show most concerns like the bio persistent fibbers are rated at OHB 5 by application of the principle of precaution. However, from a practical stand point, rating at OHB 4 or 5 does not make a lot of difference to the extent that both OHB categories can result in similar recommendations of safety at the workplace.

2.3. *Exposure levels*

The exposure levels are calculated as follows:

$$Ie = H \times S \times E \times C \times Q \times F \times D$$

The parameters H, S, E, C, Q, F, D are based on a pragmatic approach of assessing risks in labs and plants. The hermetism parameter H takes into account that a lot of implementations are performed with a physical barrier between the material and the operator, hence with no contact at all during the whole task being analyzed which enables to put into practice the principle that no exposure means no risk in normal operating conditions.
The properties of the powders implemented and their behavior within the process applied are integrated in the characters of the solid S and the emission potential E. The expression of these parameters is very much impregnated with the observation of industrial processes within the cosmetics.
A novelty compared to the other risk assessment methods is the air containment C that is taken into account as an essential parameter that drives or prevents the exposure to airborne materials. Based on the current state of the art, a dynamic barrier like a top-down air flow generated at the ceiling combined with a specific extraction that enables a slight depression at the workplace is amongst the best ways to protect operators from the exposure by inhalation to materials. Given all the different

configurations that may be encountered in an occupational environment, it is reasonable to let operators to assess the risks associated to their tasks while taking into account the characters of their collective protection equipment.
As with other control banding methods, quantity Q, frequency F and duration D are used to qualify the exposure to materials during the task being observed. The quantity Q covers all scales from research up to manufacturing, while frequency F and duration D take into consideration classical job description parameters.

Evaluation of risk
The calculation of the exposure level Ie of a given task is next reported in a OHB vs exposure matrix that allows highlighting the levels of risk on 3 levels (high, medium and low) and their corresponding priorities of action.

2.4. Residual risk
For every task identified in the examined process, a level of priority of action must be concluded. All the tasks result in an action plan that must be taken and finalized before the process is started.
The risk assessment has to cover all the phases of implementation, in particular the reception, the weighing, the handling, the quality controls, the elimination of waste and of the effluents, the maintenance of equipments including the equipments of collective protection, as well as the cleaning of the workplace.
The actions of protection and/or prevention which ensue from evaluations of risks are decided by the employer and confirmed by operational staff in charge of applying them in logic of continuous improvement. These actions cover ideally:

- Keeping the operators away of the danger zones where they have nothing to do, or even the replacement of certain materials when it is possible
- The collective protection equipments
- The personal protection equipments (PPE) which must be taken as the last solution to be implemented only after having finalized all the possible actions of collective protection
- Detailed procedures, drafted in a practical and practicable language so that they are easily practiced
- Training, information and authorization of the staff, as specific and preliminary requisites in the process to be operated
- A possible medical follow-up in the situations requiring it

The EHS staff, as an expert of the risk assessment method and as the support of the operational staff realizing their evaluation of risks, has to validate the actions decided before their implementation in logic of continuous improvement.
Once the evaluation has been finalized, the EHS staff has to assist the operational staff in the quotation of the residual risk, resulting from the application of protective and / or preventive measures decided, so that no unwanted situation continue during the implementation of the processes that have been examined.

2.5. Documenting
The documentation of all the reflections of evaluation of the risks and the actions which ensued from it for all the examined processes is ideally managed by the site Management in due time by the regulations, or in defect during at least all the duration of implementation of the above mentioned processes.
Also, the traceability of the implementation of materials and of their maintenance as well as of processes must be kept by the site Management in the conditions and deadlines determined by the local regulations.

References

[1] Evaluation des risques au poste de travail - ERPT – Sanofi Aventis CRVA, revision 5, juin 2007
[2] Evaluation and control of chemical risk – Rhoditech, 2001

Journal of Physics: Conference Series **429** (2013) 012065 doi:10.1088/1742-6596/429/1/012065

Concerns related to Safety Management of Engineered Nanomaterials in research environment

A Groso, Th Meyer

Occupational Safety and Health, School of Basic Sciences, Ecole Polytechnique Fédérale de Lausanne, 1015 Lausanne, Switzerland

Contact e-mail: thierry.meyer@epfl.ch

Abstract. Since the rise of occupational safety and health research on nanomaterials a lot of progress has been made in generating health effects and exposure data. However, when detailed quantitative risk analysis is in question, more research is needed, especially quantitative measures of workers exposure and standards to categorize toxicity/hazardousness data. In the absence of dose-response relationships and quantitative exposure measurements, control banding (CB) has been widely adopted by OHS community as a pragmatic tool in implementing a risk management strategy based on a precautionary approach. Being in charge of health and safety in a Swiss university, where nanomaterials are largely used and produced, we are also faced with the challenge related to nanomaterials' occupational safety. In this work, we discuss the field application of an in-house risk management methodology similar to CB as well as some other methodologies. The challenges and issues related to the process will be discussed. Since exact data on nanomaterials hazardousness are missing for most of the situations, we deduce that the outcome of the analysis for a particular process is essentially the same with a simple methodology that determines only exposure potential and the one taking into account the hazardousness of ENPs. It is evident that when reliable data on hazardousness factors (as surface chemistry, solubility, carcinogenicity, toxicity etc.) will be available, more differentiation will be possible in determining the risk for different materials. On the protective measures side, all CB methodologies are inclined to overprotection side, only that some of them suggest comprehensive protective/preventive measures and others remain with basic advices. The implementation and control of protective measures in research environment will also be discussed.

1. Introduction

The increasing production and use of engineered nanomaterials (ENMs) contributed to an evolving concern in occupational risk assessment and management of exposure to these materials. Currently, the assessment of possible health risks as a result of exposure ENMs is associated with significant uncertainties [1]. This situation has given rise to initiatives of governmental authorities, policy makers, industrial organizations, and civil society organizations to advocate the application of the precautionary principle for risk management. Recommendations on working safely with nanomaterials have been developed in the past decade by government agencies and occupational health organizations [2-5]. Diverse good laboratory practices are proposed on the universities' side as well [6-8].

Control banding (CB) was developed in the pharmaceutical industry as a pragmatic tool to manage risks resulting from exposure to a wide variety of potentially hazardous substances in the absence of firm toxicological and exposure data [9]. In control banding approach Hazard and Exposure bands are estimated and combined into broad risk classes. Different levels of protection are suggested consequently to control the risk. In the context of uncertainty in the production and use of manufactured nanomaterials, CB approach can be very helpful in implementing a risk management strategy according to a precautionary approach [10]. In recent years a few teams published works [11-15] on qualitative risk estimate of ENPs based on this approach and it seems to be, at the moment, a largely accepted strategy to handle 'unknowns' in management of Nanomaterials safety.
However, the extent to which safety guidances are followed and recommended methodologies applied is not well known. The validation of the effectiveness of the exposure controls and measurements methods remain a key research needed [1].

Regarding universities' laboratories, Jesus Santamaria and his team [16] have conducted in 2010 an online survey to identify which safety practices researchers are following. The responses of the 240 participants shed some light on what was going on. The questions covered: details of the materials and processing methods used; safety measures; waste disposal procedure, and knowledge of legislation for handling nanomaterials. One of the most surprising results [17] is that nearly three quarters of respondents reported not having internal rules to follow regarding the handling of nanomaterials (approximately half of them didn't have rules and over a quarter were not aware of any internal regulations).

In this work we analyze our experience in implementing risk assessment and management of nanomaterials safety at EPFL in Switzerland.

2. EPFL method for Nanolaboratory classification

In collaboration with accidents insurance and State Secretariat for Economic Affairs, as well as nanomaterials users and producers, the occupational safety and health team at the Ecole Polytechnique Fédérale de Lausanne (EPFL) developed a procedure for managing the OHS risks relevant to research laboratories producing and using nanomaterials [18]. The methodology (will be called EPFL method here in) is similar to control banding approaches. It consists of a schematic decision tree allowing classifying laboratories handling or producing ENPs into three exposure potential classes (from Nano 3 - highest exposure potential to Nano 1 - lowest exposure potential). The laboratories are classified by taking into account physical state of the substances, quantities of substances as well as agglomeration state (see figure 2 in reference [18]). The idea behind making the methodology simple is to allow a rapid determination of the precautionary exposure potential class by users of nanomaterials themselves.

The first differentiation in the decision tree for exposure potential class determination regards the environment, whether the process is carried out in a closed (complete process confinement) or open system. In case the process is not fully enclosed (glove box or completely sealed environment), different types of activities with nanomaterials are correspondingly discussed:

- Activity with nanofibers
- Activity with nanoobjects in powder
- Activity with nanoobjects in suspension
- Activity with nanoobjects in solid matrix.

Inside these categories, hazard potential classification is based on the quantity of nanomaterial as well as on the aggregation/agglomeration state (for activities with nanopowders).

Journal of Physics: Conference Series **429** (2013) 012065 doi:10.1088/1742-6596/429/1/012065

For nanopowders we also distinguish *production* and *handling*. Very often, particles are supplied by other laboratories or external suppliers, where occupational safety and health team cannot control the process as well as for home-made particles. Furthermore, users manipulate such particles more often in confined spaces. Limits for exposure potential classes' in case of *handling* are therefore lower than those for *production*.

The hazards related to nanomaterials suspension are not only influenced by the nature of particles but also by the dispersant. The decision tree is organized accordingly: For manipulated quantities superior to 1 liter the nature of the used dispersant (flammable, toxic etc.) is considered.

The preparation of composites is either treated as "Activity with nanoobjects in suspension" or "Activity with nanoobjects in powder" when performed in solution or in dry conditions, respectively. The laboratory is treated as Nano 1 if material characterization and post-preparation processing activities do not include any mechanical or thermal treatment. If dust can be released during the manipulation or if composites are friable, laboratory is treated as "Activity with nanoobjects in powder".

Protective measures are subsequently recommended for each of three exposure potential classes, and they are divided in technical, organizational and personal, as custom in health and safety/ occupational hygiene practice when managing the risks. For example, laboratories with exposure potential Nano 3 will require rather extensive technical measures with capture at source, exiting air filtering with at least a F7 filter [EN 779 - European Standard for ventilation filters. F7 has 80-90 % average efficiency for 0.4 μm particles], and access restrictions using a security vestibule (double door). Examples of organizational protective measures are the following:

- Each laboratory must have a responsible person (nano-officer).
- An ordering/receiving procedure must be established with identified collecting points.
- Pregnant women are allowed to work with nanomaterials only with a special work authorization issued by an occupational physician.
- Lab safety audits are performed by occupational health and safety specialists.
- Permanent laboratory staff working in Nano 2 lab and every person working in Nano 3 are subject to medical surveillance.

As illustrated, the set of measures is quite extensive and gives detailed information on all elements when handling/producing nanomaterials, including delivery, storage, cleaning, medical surveillance etc. [18].

3. Implementation of the management of the nanomaterials safety in practice

The presented approach for management of nanomaterials is currently being implemented at EPFL for research and teaching labs dealing with nanomaterials. Using above mentioned decision tree, the researchers themselves determine preliminary exposure potential level for each process involving nanomaterials.

The results of inquiry (see table 1 for part of results) tell us that: there are about 30 processes that classify laboratory to Nano 3 type (more than 70 in Nano 1 type). The main types of activities are: use of different types of nanopowders for producing new materials, production and use of nanofibers and a multitude of processes including nanomaterials in suspensions. The most frequently used types of particles are: TiO_2, Au, SiO_2, Carbon nanotubes, C black, Fe_2O_3, Ag, RuO_2, Fe_3O_4, ZnO etc.

Table 1. Excerpt from the results of inquiry on production/use of nanomaterials at the EPFL.

Process	Number of labs concerned	Lab classification according to EPFL methodology
Manipulating nanoparticles in suspension, less than 1 litre	28	Nano 1
Production of nanopowders, less than100 mg per batch. Particles agglomerate.	10	Nano 1
Use of more than 10 mg of nanoparticles in powder per experience. Particles agglomerate.	6	Nano 2
Use of more than 1 mg and less than 10 mg of nanoparticles in powder. Unknown agglomeration.	7	Nano 2
Manipulation of dry nanofibers.	15	Nano 3
Use of more than 10 mg of nanopowders per experience. Particles do not agglomerate.	5	Nano 3
Production of more than 100 mg of nanopowders per batch. Unknown agglomeration properties.	5	Nano 3

As an illustration of methodology use, we will discuss a simple process of manipulation of C black nanoparticles. These particles (commercial name FW200) with average diameter of 13 nm and specific surface area of 550 m^2/g are received in 500 grams containers and first weighted in order to distribute the powder into smaller containers. Then 30 grams of carbon black is weighted from a small container and added to the previously prepared liquid polymer. The working surface is cleaned using tissue and a solvent. The approximate time needed for operation of weighting/cleaning is about 30 minutes. The prepared mixture is then stirred in magnetic stirrer in order to obtain paste for further studies. This process is done with 2-3 times per week frequency. Work is performed under a fume hood and with a FFP3 respiratory mask.

If a part of process in which powder is used is analyzed using EPFL methodology to obtain exposure potential, decision tree will rapidly give: Activity with powder form, Use of nanomaterial, Quantity more than 100 mg per experiment, unknown particles agglomeration state. This will lead to *Nano 3* as safety classification, requiring the most extensive risk mitigation measures; the process should be done in the restricted access room, under a fume hood and using overall with hood-Tyvek® style, overshoes, close fitting safety goggles, 2 pairs of gloves and FFP 3 mask. The contaminated material tissue used for wet cleaning etc. is disposed in a special bin dedicated to toxic waste.

After laboratories' nano levels are determined, a workplace visit for each process classified as Nano 2 or Nano 3 are performed in order to validate/complete the analysis by occupational safety and health specialists. For the above-mentioned process, as an example, we have applied three other CB methods in order to check which risk band they recommend. Summary of the analysis is given below.

ANSES CB method [11]: In order to determine hazard band, this method suggests using classification of the nano substance by authorized authority if the former exists. C black is classified as carcinogenic of category 2 by EU-GHS/CLP [19]; According to ANSES CB this corresponds to hazard band 4 (HB4). To determine Emission potential band (EP) we have considered that C black has high or moderate dustiness powder, giving final emission potential band EP4. Final risk control level (CL), obtained by combining HB4 and EP4 is CL5. For the case of CL5, the ANSES CB method advises full containment and review by a specialist required: seek expert advice.

Another method used is Control Banding Nanotool [20]. Based on the knowledge of the nanomaterial characteristics and review of the operation in the field, the CB Nanotool gave value 51.25 for severity (High severity) and 80 (Probable) for probability score (see table 2). As it can be seen in table, 5 out of 13 factors contributing to severity are unknown contributing largely to the high severity score.

Table 2. Results of risk evaluation using CB Nanotool

Severity factor	Points	Probability of exposure	Points
Surface chemistry, reactivity and capacity to induce free radicals	10	Quantity of NP used during the task	25
Particle shape	0	Dust making capacity	30
Particle diameter of NP	5	Number of employees occupying similar working place	5
Solubility of NP	10	Operation frequency	15
Carcinogenicity of NP	5.625	Time for procedure	5
Reproductive toxicity of NP	5.625		
Mutagenicity of NP	5.625		
Dermal toxicity of NP	5.625		
Toxicity of parent material	0		
Carcinogenicity of PM	3.75		
Reproductive toxicity of PM	0		
Mutagenicity of PM	0		
Dermal hazard of PM	0		
Total severity	51.25	Total probability of exposure	80

Combination of obtained severity and probability values for the analyzed activity indicated that the overall risk level (RL) is 4 (the highest risk level). As risk mitigation measure for RL4, the CB Nanotool suggests consulting an expert (general ventilation is proposed for RL 1, Fume hoods or local exhaust for RL 2 and containment for RL 3).

Application of Stoffenmanager Nano web tool [12, 21] for risk analysis of the same process gives Hazard band D (fourth on the scale of 5). When taken into account that the process is performed in a fume hood and with a FFP3 mask, one obtains exposure band 2 (on the scale of 4) so the obtained risk priority band is equal to II (medium risk priority). Since existing protection is already taken into account when doing risk analysis, different protective measures can be investigated when using the Stoffenmanager Nano tool and observe their effect on the risk priority band. For example, if it is assumed that the process is done with full containment, exposure band is reduced to 1 (lowest), but due to high hazard band (D), the risk priority band still remains the same (medium).

Summary of results of risk classification and control measures using different methods is given in table 3.

Table 3. Results of risk evaluation for the process of C black weighting, using four different tools

	EPFL	**ANSES tool**	**CB Nanotool**	**Stoffenmanager**
Risk classification of the process	Level 3 on the scale of 3	Level 5 on the scale of 5	Level 3 on the scale of 4	Level 2 on the scale of 3
Control measures	Technical, organizational and personal measures for Nano 3	Full containment and review by a specialist	Containment of the process	Measures included in the analysis process

In similar way, these risk analysis methods were applied to evaluate the process of dry cleaning of the quartz tube used for the synthesis of Multi Walled Carbon Nano tubes (MWCNT) by Chemical Vapor Deposition. MWCNT in question have 20 nm diameter and several microns length; 30 grams are synthesized per batch. After MWCNT synthesis, the tube is emptied into a beaker and then cleaned with a dry tissue for about 5-10 minutes (synthesis and cleaning are realized one time per week in average). FFP3 respiratory mask, lab coat and close fitting goggles are used for manipulation.

EPFL method gives the highest exposure potential classification (Nano 3). It is assumed that fibers in question are biopersistant; ANSES CB methodology therefore gives the highest risk band (full containment and review by specialist). The CB Nanotool gives RL3 risk level requiring containment of the process. Stoffenmanager gives hazard class E (highest) and time weighted exposure class 1 (lowest). The obtained risk score is the highest one on the scale of 3.

4. Discussion

4.1. Results of risk analysis using different tools

From the analysis described above as well as other analysis of processes involving nanomaterials in our research laboratories ($BiVO_4$, $LiNbO_3$ fibers, Al_2O_3, TiO_2 etc.), we could deduce that EPFL method is easy to use (very little information is required for doing analysis). Stoffenmanager Nano explicitly focuses on work environment. Its database contains certain number of particles type in a list, so the user is not asked to search for information on their properties for hazard band determination. This tool is mainly intended to be used by non-expert SME's employers and employees [10]. Among the CB tools, CB Nanotool requires the most extensive effort in information retrieval on nanoparticles properties (as surface chemistry, reactivity, toxicity etc.) This methodology was developed to support the front line occupational health and safety specialists in the first place [10] since nanomaterials users are less willing to do safety information search for the substances they are using if the former are not accessible in Material Safety Data Sheets (MSDS). In case of ANSES CB methodology, one of the key elements for its application is the use of MSDSs according to the GHS system. ANSES CB is intended to be used by persons adequately qualified in chemical risk prevention [10].

One of the main difficulties when using those tools remains the lack of nanomaterials' specific information and MSDSs in hazard band determination. Namely, the majority of MSDSs for ENMs, when available, provide most of their safety and health information based on the bulk parent material [15]. In a recent study [22] on 97 MSDSs of ENMs, it is found that most of the time, MSDS do not include sufficient information on safety of nanomaterials, such as their toxicity and physicochemical properties. As a consequence, severity factors often give value 'unknown' and the result of risk analysis for a particular process is essentially the same with a simplified methodology requiring very little input information (as EPFL methodology) and one that takes hazardousness/toxicity into account

(CB Nanotool and ANSES CB methodology). CB methodologies give the same or similar (high) hazard band resulting in high risk/ protection.

Since new research initiatives arise, the CB tools will certainly evolve [10]. New information will also become available on toxicity and exposure. Indeed, taking into account, when available, information on surface chemistry, solubility, carcinogenicity, toxicity etc., instead of 'unknowns' is crucial when hazard band is determined; this will allow more sophisticated CB tools to give more precise results of analysis. To create a consistent standard for the information provided on safety, health, and environment matters for manufactured nanomaterials-containing products, guidance for the preparation of nanomaterials-specific MSDS, including both nanomaterials and mixtures of nanomaterials with conventional non nano scale materials was recently initiated by ISO TC 229 [22].

Regarding exposure band determination, in order to give dustiness level of the product, ideally dustiness tests should be made [23]. Furthermore, information on the agglomeration state of the particles is also necessary. Namely, it is suggested that for the same mass concentration, non-agglomerated dry nano - particles should be given a higher dustiness/mistiness rating than agglomerated or liquid-suspended nano-particles [14, 18]. Without detailed analysis, it is often difficult to know whether particles agglomerate. Even if the answer is known, the potential of deagreggation either in aerosols form or after absorption by body fluids will often be a key issue [24].

4.2. Challenges in exposure studies (measurements) in research environment

At the moment, quantitative and sensitive measurements of workers' exposure to nanomaterials are difficult, requiring multitude of instruments for measuring different parameters (average concentration, particles size distribution, elemental analysis...), in process near and far field. At this stage, it would be interesting for OHS specialists in academia research environment to do at least survey measurement for sources of emission/ potential exposure for a particular processes using simple portable instruments as portable Condensation Particle Counter [25] or Miniature Diffusion size classifier [26]. According to our experience at the moment, due to small quantities of substances used and rapidity of the processes, emissions are difficult to detect.

4.3. The implementation and control of protective measures

As already discussed above, the applied qualitative methodologies to analyze the risk of different processes involving nanomaterials give in general high risk band demanding high protection. For the highest risk levels, consulting an expert/specialist is demanded in CB Nanotool and ANSES CB. According to ANSES CB, an expert in this context is person highly knowledgeable in toxicology and industrial hygiene [11]; this person or group of people has to make a full risk assessment which is not possible to do using CB methods. In practice, finding adequate expert to complete risk analysis might be difficult. The other difficulty is related to process containment that is suggested but often not possible to implement due to the size of equipment for example. One should not forget that taking materials out of contained equipment might also represent a risk.

A technical measure generally suggested is the use of enclosed ventilation, as ventilated booth, fume hood and similar. An important point recommended by EPFL method is filtering (HEPA F12 filter) of all extraction air (therefore from enclosed ventilation) coming from the labs Nano 3 before this air reaches the ventilation system of the building. Air contamination outside the research labs should be avoided in this way.

As far as organizational measures are concerned, EPFL method recommended restricted access to the labs; only authorized personnel can access the Nano 3 laboratories. Lab personnel also do cleaning. All nanomaterials contaminated waste is treated as toxic waste.

Regarding medical surveillance, the reports [4] indicate that level of knowledge today doesn't allow advising a specific medical survey, or indicators of exposure or effects. Still, certain consensus is obtained at international level [27] to recommend that potentially exposed workers should have periodical medical survey with 'conventional' exams, specific for potential target organ. On Swiss national accident insurance company's (SUVA) initiative, the potentially exposed workers (exposures

30 days or 200 hours per year) at the EPFL are subject to rather extensive medical tests comprising: lungs radiogram, blood and urine test, spirometry and ECG with 2 years frequency. The idea is to use results of these examinations as input for database to make epidemiological studies; although the data might be limited due to high turn-over of the staff and constantly changing processes in academia research [28].

5. Conclusions

Making rules and policies is one aspects but it is even more important to implement control, continuous improvement and adaptation of the mitigation measures. Monitoring, adaptation, control and review of measures in place have high importance. As research' processes and equipment evolve very fast, the use of simplified methods with dedicated correctives measures and actions is essential.
Due to lack of knowledge/ application of precautionary approach in all methodologies used at the moment, protective measures have an 'over-protection' tendency; there is high chance that these measures will be reduced as knowledge will increase. As far as academia research is concerned, being on 'over-protective' side is far better than having insufficient protection; one should not forget that this environment has an important role in educating people who will work in industry afterwards.

Acknowledgements

SUVA (non-profit insurance company under Swiss public law), SECO (State Secretariat for Economic Affairs), Medical services EPFL, A. Petri-Fink, A. Magrez and M. Riediker are gratefully acknowledged for their valuable contribution. This work is partially supported by The Swiss National Science Foundation through the projects numbers 200021-115900 and 205321-125299.

References

[1] Kuempel E D, Geraci C L and Schulte P A 2012 Risk Assessment and Risk Management of Nanomaterials in the Workplace: Translating Research to Practice *Annals of Occupational Hygiene* **56** 491-505
[2] Good Nano Guide http://www.goodnanoguide.org/Basic
[3] International Organization for Standardization 2008 ISO/TR 12885: Nanotechnologies — Health and safety practices in occupational settings relevant to nanotechnologies,
[4] AFSSET 2008 Avis de l'Agence française de sécurité sanitaire de l'environnement et du travail relatif aux nanomatériaux et à la sécurité au travail,
[5] Institut de recherche Robert-Sauvé en santé et en sécurité de travail (IRSST), Canada 2009 Best Practices Guide to Synthetic Nanoparticle Risk Management,
[6] Best Practices for Handling Nanomaterials in Laboratories (Boston: MIT Environment Health and Safety Office) http://ehs.mit.edu/site/sites/default/files/files/University_Best_Practices.pdf
[7] Nanotoolkit 2012 (California: California Nanosafety Consortium) http://www.cein.ucla.edu/PDFs/nano_toolkit_April_2012.pdf
[8] Nanomaterials safety program 2009 (New Hampshire, USA: University of New Hampshire) http://www4.dcu.ie/science_and_health/pdfs/UNH-nanomaterials-safety-program.pdf
[9] Maynard A D 2007 Nanotechnology: The Next Big Thing or Much Ado about Nothing? *Ann. Occup. Hyg.* **51** 1-12
[10] Brouwer D H 2012 Control Banding Approaches for Nanomaterials *Annals of Occupational Hygiene* **56** 506-14
[11] ANSES French Agency for Food, Environmental and Occupational Health and Safety 2010 Development of a specific control banding tool for Nanomaterials,
[12] Duuren-Stuurman B, Vink S, Brouwer D, Kroese D, Heussen H, Verbist K, Tielemans E, Van Niftrik M and Fransman W TNO 2011 Stoffenmanager Nano: Description of the conceptual control banding model,

[13] Precautionary Matrix for Synthetic Nanomaterials (Switzerland: Federal Office of Public Health) http://www.bag.admin.ch/themen/chemikalien/00228/00510/05626/index.html?lang=en
[14] Paik S Y, Zalk D M and Swuste P 2008 Application of a pilot control banding tool for risk level assessment and control of nanoparticle exposures *Ann. Occup. Hyg.* **52** 419-28
[15] Zalk D M, Paik S Y and Swuste P 2009 Evaluating the Control Banding Nanotool: a qualitative risk assessment method for controlling nanoparticle exposures *Journal of Nanoparticle Research* **11** 1685-704
[16] Balas F, Arruebo M, Urrutia J and Santamaria J 2010 Reported nanosafety practices in research laboratories worldwide *Nature Nanotechnology* **5** 93-6
[17] Questionable safety practices in nanotechnology labs around the world http://www.nanowerk.com/spotlight/spotid=14778.php
[18] Groso A, Petri-Fink A, Magrez A, Riediker M and Meyer T 2010 Management of nanomaterials safety in reserch environment *Particle and Fibre Toxicology* **7**
[19] United Nations 2009 Globally harmonized system of classification and labelling of chemicals (GHS),
[20] Paik S Y, Zalk D M and Swuste P 2008 Application of a pilot control banding tool for risk level assessment and control of nanoparticle exposures *Ann Occup Hyg* **52** 419-28
[21] Van Duuren-Stuurman B, Vink S R, Verbist K J M, Heussen H G A, Brouwer D H, Kroese D E D, Van Niftrik M F J, Tielemans E and Fransman W 2012 Stoffenmanager Nano Version 1.0: A Web-Based Tool for Risk Prioritization of Airborne Manufactured Nano Objects *Annals of Occupational Hygiene* **56** 525-41
[22] Lee J, Kuk W, Lee J, Lee K and Yu I 2012 Evaluation of information in nanomaterial safety data sheets and development of international standard for guidance on preparation of nanomaterial safety data sheets *Nanotoxicology* **Ahead of print**
[23] Nano Stoffenmenager 2012 (Netherlands: TNO) http://nano.stoffenmanager.nl
[24] Brouwer D, van Duuren-Stuurman B, Berges M, Jankowska E, Bard D and Mark D 2009 From workplace air measurement results toward estimates of exposure? Development of a strategy to assess exposure to manufactured nano-objects *Journal of Nanoparticle Research* **11** 1867-81
[25] CPC 3007 (USA: TSI) http://www.tsi.com/condensation-particle-counter-3007/
[26] miniDiSC 2012 (Switzerland: Matter Aerosol) http://www.matter-aerosol.ch/index.php/products/discmini
[27] Gaffet E Aspects sécurité des nanomatériaux et nanoparticules manufacturés 2009 (Editions Techniques de l'ingénieur)
[28] Marendaz J L, Friedrich K and Meyer T 2011 Safety Management and Risk Assessment in Chemical Laboratories *Chimia* **65** 734-7

French registry of workers handling engineered nanomaterials as an instrument of integrated system for surveillance and research

I Guseva Canu[1], O Boutou-Kempf[1], L Delabre[1], S Ducamp[1], Y Iwatsubo[1], JL Marchand[1], and E Imbernon[1]

[1]French Institute for Public Health Surveillance (InVS), Occupational health department (DST), 94415 St Maurice, France

E-mail: i.guseva-canu@invs.sante.fr

Abstract. Despite the lack of data on the human health potential risks related to the engineered nanomaterials (ENM) exposure, ENM handling spreads in industry. The French government officially charged the InVS to develop an epidemiological surveillance of workers occupationally exposed to ENM. An initial surveillance plan was proposed on the basis of literature review and discussions with national and international ENM and occupational safety and health (OSH) experts. In site investigations and technical visits were then carried out to build an adequate surveillance system and to assess its feasibility. The current plan consists of a multi-step methodology where exposure registry construction is paramount. Workers potentially exposed to carbon nanotubes (CNT) or nanometric titanium dioxide (TiO_2) will be identified using a 3-level approach: 1-identification and selection of companies concerned with ENM exposure (based on compulsory declaration and questionnaires), 2-in site exposure assessment and identification of the jobs/tasks with ENM exposure (based on job-expose matrix, further supplemented with measurements), and 3-identification of workers concerned. Data of interest will be collected by questionnaire. Companies and workers inclusion questionnaires are designed and currently under validation. This registration is at the moment planned for three years but could be extended and include other ENM. A prospective cohort study will be established from this registry, to pursue surveillance objectives and serve as an infrastructure for performing epidemiological and panel studies with specific research objectives.

1. Introduction

During the last decade important body of toxicological and exposure data was generated in the framework of occupational safety and health (OSH) research focusing engineered nanomaterials (ENM) [1,2]. However the lack of epidemiological and biomedical data on the human health potential risks related to the ENM exposure is a substantial gap for validating experimental results, risk-assessment modeling and developing consistent risk management guidelines if needed [3]. Considering that despite these uncertainties, ENM handling spreads in industry, the French Ministries of Health and of Labour have given the French Institute for Public Health Surveillance (*Institut de Veille Sanitaire, InVS*) responsibility for designing the protocol of an epidemiological surveillance system of workers likely to be exposed to engineered nanomaterials (ENM).

2. Exploratory feasibility study

An initial surveillance plan was proposed within a multidisciplinary working group held by the French Institute for Public Health Research (*Institut de recherché en santé publique, IReSP*) on the basis of literature review and discussions with national and international ENM and OSH experts [4]. An exploratory study was then conducted with the aim of assessing the feasibility of an epidemiological surveillance system of workers occupationally exposed to ENM in France. Ten companies producing or incorporating carbon nanotubes (CNT), carbon black, titanium dioxide (TiO2), or amorphous silica were contacted to be visited from 2008 to 2010. These four ENM were chosen considering available toxicological data, quantities manufactured in France and projected for production development, the choice of France in the framework of the sponsorship program for the testing of ENM sponsored by the Organisation for Economic Co-operation and Development (OECD) and social perception factors. In site investigations and technical visits of eight companies which had agreed to participate allowed to collect critical information such as the number of workers likely to be exposed to nanomaterials, conditions of exposure, medical follow-up, and collaboration issues. The main conclusions outlined by this exploratory study were as followed: 1-relatively good acceptance of cooperation in public research laboratories and R&D facilities, but some disagreement issues on the ENM definition in some ENM producing facilities; 2-variable likelihood of exposure (low on R&D facilities but probable on facilities producing precipitated silica, carbon black, TiO_2); 3-generally small number of workers likely to be exposed on each site (from 15 to 120 on private facilities and 250 on the public R&D laboratories), 4-absence of a specific medical follow-up for nanoworkers other than annual clinical exams including in some cases lung function tests, blood analysis (blood cell counts, creatinine, transaminase, C Reactive Protein), or chest radiography [4]. According to these conclusions, an operational plan of the French integrated surveillance system should account for possible cooperation and comprehension issues by supplying all necessary tools and conditions before its actual implementation on the national level.

3. French ENM integrated surveillance system development

3.1 Definition and objectives of a surveillance system

Public health surveillance is the ongoing, systematic collection, analysis, and interpretation of health-related data with the *a priori* purpose of preventing or controlling disease or injury [5]. Four major goals are commonly assigned to epidemiological surveillance: 1-to detect timely unusual health situations, 2-to assess the magnitude of a health problem to make decisions affecting public health policy and allocation of resources, 3-to contribute to further research, and 4-to evaluate the effects of prevention and intervention efforts [6]. Occupational health surveillance is a specific field of public health surveillance related to either occupational exposure (hazard surveillance) or adverse health outcomes (injuries, disorders or diseases). Actually, it might be intended to embrace both workers' health surveillance and work environmental surveillance and thus to introduce the concept of indicators of change that are either exposure indicators or effect indicators [7]. Surveillance information may be collected by a variety of means, such as administrative registers and statistics, expert assessment systems, workers survey or questionnaires, exposure databases registers of accidents and diseases, and multi-source policy-directed systems. Depending on the source of classification and the level of analysis for classifying systems, four types of surveillance systems could be distinguished: registers, workplace observations and occupational health and safety (OHS) service data, working conditions survey, and integrated systems [8]. The integrated surveillance systems look at multi-source information focused on both exposure and health outcome information that is combined into one system. Such systems are considered to be well responsive to workplace or organizational-level needs and promising for improving workplace health and safety [8,9]. Exposure data is the focus in the integrated surveillance system and existence of an exposure registry enables the development of such systems [8].

3.2 Exposure registry

While the term register is applied to the file of data that can be related to a population base, i.e. the actual document or database, the registry is the system of ongoing registrations [10]. An exposure registry is a system for collecting and maintaining in a structured record, comparable information on persons with known or suspected occupational or environmental exposure to a hazardous substance [11]. Because exposure registries may have unintended consequences and high costs to workers, employers and society, they are not warranted in every situation. A registry would be warranted if it meets established criteria or rationale. Scientists and OHS experts recognized that in the case of nanotechnology workers with unknown exposure level and unknown health hazards there might be good reasons to consider registration of workers with ENM exposure [12,13].

3.3 French multi-company ENM exposure registry

Since carbon black and amorphous silica have been produced at nanometric size for decades in France, for starting this registry a priority was given to two more recent ENM: CNT and TiO_2. The French registry is intended as an individual registration of the ENM handling workers likely to be exposed either to CNT or TiO_2 while their employment in the nanotechnology-related industrial or R&D companies in France. The workers registration is planned to be exhaustive in terms of the exposed workforce on a company-level. The development of the ENM registry is based on a multi-level and multi-step approach, as presented in Figure 1.

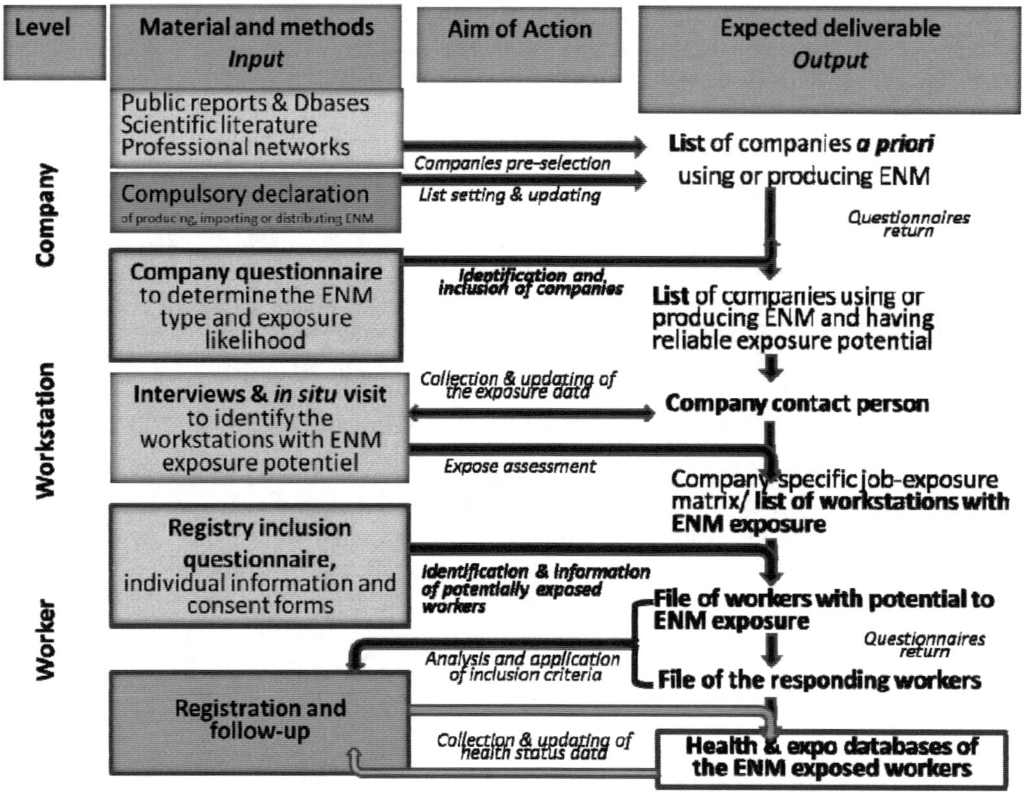

Figure 1. Multi-step development of the French ENM exposure registry.

One or more specific actions are associated to each level, resulting in a specific deliverable. At company level, the expected deliverable is a list of companies producing or using TiO_2 or CNT based on the detailed description of process/tasks and work conditions known to have a reliable potential of ENM emission and exposure. Various available sources including published data and reports as well as information exchange within the professional network might be useful for pre-selecting such companies. Furthermore, this preselected company list could be completed and systematically updated thanks to the compulsory registration of ENM. This registration is instituted by law entering in application on the first of January 2013 and requiring a compulsory declaration of types, quantities and use purposes of ENM produced, imported or distributed in France and downstream users. Founded on the European Community definition of the nanomaterials, this registration should resolve uncertainties in identification of the ENM of interest for the surveillance system and facilitate the relationship with companies. At the workstation level the expected deliverable is the list of workstation with exposure concern which will be abstracted for each company from the company-specific job-exposure matrix (JEM). Human resources department will then be asked to extract a file of workers who are affected to one or more workstations on the list. Finally, on individual level, workers likely to be exposed to ENM will be contacted and provided with registry-related documents, namely a detailed information letter, individual participation consent form and the registry inclusion questionnaire.

3.3.1 Methods and tools for exposure assessment

Considering the difficulties to measure ENM exposure in workplace in a scientifically and methodologically relevant way along with the heterogeneity and uncertainties of available industrial hygiene measurement data, the French ENM registry is firstly designed to contain qualitative and semi-quantitative exposure data. The methods used to gather these data may include expert exposure assessment, technical observation visits and review of company data on exposure and exposure management. Additional tools specifically developed for the ENM surveillance are questionnaires and company-specific ENM JEMs:

3.3.1.1 Company enrolment questionnaire is designed with aim to address the likelihood of ENM exposure by analysis of the process used to produce or transform ENM at companies pre-selected as of first attention and to establish contacts with company OSH manager and occupational physician for further collaboration (see Fig.1). This questionnaire was reviewed by inner and outer InVS exposure experts and cross-validated with other questionnaires on ENM exposure [14,15] to harmonize it for potential international collaborative studies.

3.3.1.2 Company-specific ENM JEMs. In contrast with generalist JEMs, where exposure assessment must take into account a large number of occupational categories to link the information on exposure to specific workplace hazard for use in large population-based studies, specific JEMs designed for a company or industrial sector have a more specific field of application and may be developed using a more detailed classification of occupations and better exposure information [16]. Within the specific ENM JEMs, workstations in each company will be analysed and classified by industrial hygienists in terms of exposure likelihood and exposure frequency and intensity level, resulting in qualitative or semi-quantitative indicators of exposure. Focusing the exposure assessment on a company should allow more precision in the JEM data and reduce misclassification issues. The semi-quantitative ENM JEMs could be by further consolidated or replaced with quantitative data from available company industrial hygiene measurement data [17] and/or from supplementary studies focused on exposure biomarkers [18] or exposure measurement which may be conducted in frame of collaborative scientific projects.

3.3.1.3 Worker registry inclusion questionnaire is designed with aim to further individualize occupational exposure data assessed through the specific ENM JEMs. This should enable us to compute individual cumulative exposure scores and to model the exposure with accounting for personal protective equipment compliance and work regimen (partial time/shift work, etc.). Moreover, this questionnaire encompasses items on past occupational history and associated exposure but also items on health status and anamnesis, life-style and habits such as smoking, alcohol consumption and physical activity. By gathering this information it will be possible to address most potential confounding factors in future epidemiological studies of potential health effects of the ENM. The questionnaire is designed in accordance with questionnaires from three other national population-based health surveys for further comparisons of health statistics.

3.3.2 Connection with existing national databases and registers

Intended as an individual registration of the ENM handling workers, the registry will enable InVS to use workers' identifying information for connecting the registry record with other public health databases and national registers. Actually, the national register of causes of death, registers of cancer, medical database of the standardized hospital discharge summaries from the national hospital information system (*Programme de médicalisation des systèmes d'information, PMSI*), medical insurance database of health care and drug consumption from the national health insurance information system (*Système national d'information inter-régimes de l'assurance maladie, SNIIRAM*) are some of relevant available sources of information on workers' health status and disease history. The abstracting of data from these sources is possible only within special inter-institutional agreements after approval of the Research ministry consultative commission on the treatment of

information in the field of health research (*Comité consultatif sur le traitement de l'information en matière de recherche dans le domaine de la santé, CCTIRS*) and autorisation of the French authority of privacy and individual rights protection (*Commission nationale de l'informatique et des libertés, CNIL*). The obtaining of CCTIRS and CNIL approvals is expected for early 2013.
On the other hand, workers' past and present occupational history encoded using the most recent editions of international classifications, such as the International standard industrial classification (ISIC 2008), and the international standard classification of occupations (ISCO 2008), should enable the InVS to extract relevant information on occupational exposure from other exposure databases available on national and international level. For instance, national population-based JEMs such as *MatGene* for exposure to occupational agents classified as carcinogenic, mutagenic or repro-toxic (CMR), *Ev@lutil* for asbestos and other fibers, *MatPUF* for particulate matter exposures might be useful for abstracting data on associated exposures. These databases are developed by InVS in collaboration with Universities or other research and expertise institutions. When necessary, the Eurostat correspondence tables would be used to recode occupations and activities into other ISIC or ISCO editions.

3.4 Layout of an integrated ENM surveillance system
The occupational surveillance of the ENM exposure and possible health effects may enter into operation with the ENM registry and general prospective follow-up of the ENM exposed workers

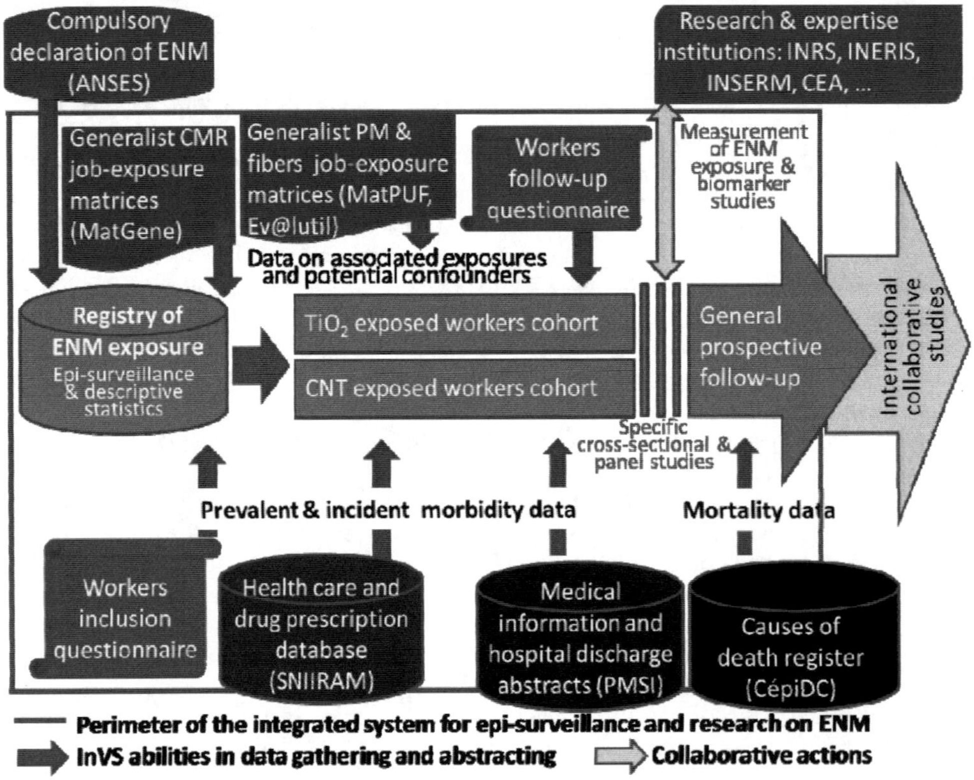

. **Figure 2.** Pattern of the French integrated ENM surveillance system.

The connexion with other health and exposure databases as well as workers follow-up questionnaires (intended on every two-year basis) should make it possible to gather and abstract all necessary information for meaningful statistical analyses within epidemiological cohorts and collaborative panel and cross-sectional studies on specific scientific questions (Figure 2). These collaborations are suitable and might be strengthened by specific collaboration agreements. A collaboration agreement has been signed with the French agency for food, environmental and occupational health safety (*Agence nationale de sécurité sanitaire de l'alimentation, de l' environnement et du travail, ANSES*) and two other agreements are about signature at the National Institute of Safety Research (*Institut national de recherche et de sécurité, INRS*) and the Occupational medical inspection department at the ministry of Labor.

The registration of first potentially exposed workers should start in 2013. The registration is at the moment planned for three years but can be extended and include other ENM. The ENM exposure registry should be considered as an instrument of integrated system for epidemiological surveillance and national and international OSH research in the field of ENM. This will be able to provide early data concerning human health potential risks associated with ENM in France.

References

[1] Kendall M, Holgate S. Health impact and toxicological effects of nanomaterials in the lung. *Respirology* 2012;17(5):743-758.
[2] Nel A, Xia T, Madler L, Li N. Toxic potential of materials at the nanolevel. *Science* 2006;311(5761):622-627.
[3] Aschberger K, Micheletti C, Sokull-Klettgen B, Christensen FM. Analysis of currently available data for characterising the risk of engineered nanomaterials to the environment and human health - Lessons learned from four case studies. *Environ Int* 2011;37(6):1143-1156.
[4] Boutou-Kempf O, Marchand JL, Radauceanu A, Witschger O, Imbernon E, André JC, et al. Development of a French epidemiological surveillance system of workers producing or handling engineered nanomaterials in the workplace. *J Occup Environ Med* 2011;53:S103-S107.
[5] Lee LM, Thacker SB. Public health surveillance and knowing about health in the context of growing sources of health data. *Am J Prev Med* 2011;41(6):636-640.
[6] Thacker SB, Stroup DF, Parrish G, Anderson HA. Surveillance in environmental public health: Issues, systems, and sources. *Am J Publ Health* 1996;86(5):633-638.
[7] Fine LJ. Surveillance and occupational health. *Int J Occup Environ Health* 1999;5(1):26-29.
[8] Brook D, Cowley S, Else D, Leggett S. International review of surveillance and control of workplace exposures. Wellington: NOHSAC Technical report No.5; 2006. 99 p.
[9] LaMontagne AD, Herrick RF, Van Dyke MV, Martyny JW, Ruttenber AJ. Exposure databases and exposure surveillance: Promise and practice. *Am Ind Hyg Assoc J* 2002;63(2):205-212.
[10] Last JM. A dictionary of epidemiology. International Epidemiological Association. 2nd ed. New York, Oxford, Toronto: Oxford University Press; 1988
[11] Schultz MG, Sapp II JH, Cusack CD, Fink JM. The national exposure registry: History and lessons learned. *J Environ Health* 2010;72(7):20-25.
[12] Riediker M, Schubauer-Berigan MK, Brouwer DH, Nelissen I, Koppen G, Frijns E, et al. A road map toward a globally harmonized approach for occupational health surveillance and epidemiology in nanomaterial workers. *J Occup Environ Med* 2012;54(10):1214-1223.
[13] Schulte PA, Mundt DJ, Nasterlack M, Mulloy KB, Mundt KA. Exposure registries: Overview and utility for nanomaterial workers. *J Occup Environ Med* 2011;53(6S):S42-S47.
[14] Schubauer-Berigan MK, Dahm MM, Yencken MS. Engineered carbonaceous nanomaterials manufacturers in the United States: Workforce size, characteristics, and feasibility of epidemiologic studies. *J Occup Environ Med* 2011;53(6S):S62-S67.
[15] Schmid K, Danuser B, Riediker M. Questionnaire on nanoparticle use in Swiss industry. Swiss nano-inventory. Lausanne, Switzerland: Institute for Work and Health; 2010. 22 p.
[16] Goldberg M, Kromhout H, Guenel P, Fletcher AC, Gerin M, Glass DC, et al. Job exposure matrices in industry. *Int J Epidemiol* 1993;22(S2):S10-S15.
[17] Guseva Canu I, Faust S, Knieczak E, Carles M, Samson E, Laurier D. Estimating historic exposures at the European Gaseous Diffusion plants. *Int J Hyg Environ Health* 2012 (in press);
[18] Guseva Canu I, Laurier D, Caer-Lorho S, Samson E, Timarche M, Auriol B, et al. Characterisation of protracted low-level exposure to uranium in the workplace: A comparison of two approaches. *Int J Hyg Environ Health* 2010;213(4):270-277.

Journal of Physics: Conference Series **429** (2013) 012067 doi:10.1088/1742-6596/429/1/012067

Legal issues of the environmental safety regulation in the sphere of nanotechnology in Russian Federation

Ekaterina A. Belokrylova

PhD in Law, Associate Professor, The Director of The Environmental Law and Policy Center
Institute of Law, Udmurt State University, Russia
Member of international legal association AiSDC (Lausanne, Switzerland)

E-mail: belokrylova.ekaterina@gmail.com

Abstract. Nowadays one of the principal innovative spheres in Russia is considered to be a rapid development of nanotechnology and nanomaterials and implementation its products into almost all fields of social and industrial life. Nevertheless, despite the pointed advantages an appearance of nano products has set not only a positive example of the scientific progress development and dynamics, but also a potential threat of possible risks for the environment and people's health in the case of its massive use when there is a lack of research about its impacts. In Russian Federation the problem of nanosafety regulation is extremely actual. A number of international documents have noted a lack of legal regulation in the sphere of nanotechnologies in Russian Federation. Thus, according to the strategic EU program, it's mentioned that in 2009 on an annual meeting of OECD Tour de Table Meeting in Paris (November, 2007) Russia took responsibility to develop a long term program of nanoindustry development by 2015 (Nanotechnology Action Plan for Russia-2015). The key role should be dedicated to the nanosafety aspects and potential risk assessment of nanomaterials for environment and humans' health. Unfortunately, until now Russia did not provide this document for discussion at the international level by the partners of nanoconsortcium. Indeed, from 2007 until nowadays in Russian Federation there are no a single federal act of legislation (federal'nyi zakon), establishing the state foundations of regulation of nanosafety. The only one Federeal Legal Act in the described sphere is the Federalniy Zakon from 19 July 2007 «About the Russian corporation of nanotechnologies», which has established the principles of organization, activites, functions and termination of Rosnanotech. However, there are no articles dedicated to the environmental safety provision in regards of nano products applications. Also there is no complex federal act consolidating legal status of action in the field of development and application of nanoproducts, inspite of fact, that these relationships have a significant financial segment. Thus, according to the Federal Act from December 13, 2010 «About federal budget for 2011 year and the period of 2012 and 2013 years" the expenses for realisation of program «The infrastructure development of nano industry in Russian Federation for 2008-2011» were 250 milliard of Russian rubbles. The profound analyses of the current legal systems has demonstrated that nowadays the main legal regulation in nano sphere consist of some range of *frame document*s (accentuated by the author). Thereby, one of the most dynamic and developed field of innovative activities in Russian Federation – nanotechnology - is left out of environmental and legal area of protection and that is might lead to the raising level of ecological risks at the stage of creation and application nano products to the environment and humans' health. During the analyses of annual norms of the Report OECD « Nano technologies: the Environment, health and safety» the conclusion is following –

Russia has an extremely low degree of conceptual realization of the program in the sphere of forming the policy connected with nanotechnology and its impact on the environment and also a lack of implementations of norms into the national legal systems in terms of the assessment criteria of nano safety EHS (Environmental, Health and Safety) and ELSI (Ethical, Legal and Social Issues). To the great regret, there is no independent and precise legal act about the ways to creat and apply nano products with the certain definitions and principles and, more importantly, with the level of legal obligations and responsibility. This gap is not possible to fill by just altering and editing the existed legal acts due to the lack of the state Russian regulation. Thus, one of the most dynamic fields of innovative activities – nano technologies – is practically out of the regulation. It might lead to an increase of ecological risks' level in the process of creation and application of nano products to the environment and health of people. As a result, implementation of international legal recommendations in the field of safety regulation of nanotechnology is quite crucial for Russian Federation.

Like any other state, Russian Federation has its own legal system. It consists of accurately consolidated hierarchy based upon a legal validity of documents.

Due to that in the theory of Russian Law, there are two principal kinds of legal documents: legislative acts and subordinate acts.

The foundation of legislative acts is the Constitution of Russian Federation, dated from 12 December, 1993. It regulates common provisions of all specific areas of law: it also includes the environmental law, where the main norms for guaranteeing of nanotechnology nanomaterial's safety had been set. The Constitution of Russian Federation has got the highest legal power and not a single legal act can violate its clauses. The key provision of the Constitution, which regulates issues about nanotechnology safety, is article 42. It vests a right of every citizen for a favorable environment, reliable information about its state and compensation for a harm, that have been caused to life, health and property of a citizen by any kinds of unsafe activities.

For more comprehensive and detailed implementation of Constitution's provisions in Russian Federation there is a legal possibility of creating federal constitutional acts and federal acts which will formulate a principal mechanism of state politics realization in the economic, social, political and other spheres. Particularly, for the basic regulation in the sphere of environmental safety realization federal acts are issued – The Federal Act "About Environmental safety", The Federal Act "About Atomic Energy", The Federal Act "About Waste", The Federal Act "About Environmental Impact Assessment", The Federal Act "About Regulation of Genetically Engineered Activities" and etc.

The Constitution of Russian Federation, the federal constitutional acts and federal acts is the sphere of the regulation, which is the principal because it forms the basic postulates of any activity in Russian Federation. The foundation of subordinate acts of the ordinances of the President of Russian Federation and decrees of the Government of Russian Federation, which are being issued following the federal legislation and also supervision and audit for the observance of federal acts.

A separate niche in the system of subordinate regulation occupy official acts of suitable departments, for instance, by the Department of public health and social development, its divisions, agents and services. An essential place in the system of official acts of departments in the sphere of nanosafety has the Federal service on customers' rights protection and human well-being surveillance (in Russian - Rospotrebnadzor), led by the Main Sanitary Inspector of

Journal of Physics: Conference Series **429** (2013) 012067 doi:10.1088/1742-6596/429/1/012067

Russian Federation. Precisely that organization guaranties the main regulation of relations in the sphere of development and application of nanotechnologies [1].

In Russian Federation it is compulsory to register new legal acts, connected with civil, political, socio-economic and ecological rights and responsibilities of citizens and also their implementation guaranties, secured in the Constitution of Russian Federation and other subordinate acts. It is considered to be a supplementary guaranty of fulfillment those kinds of acts. It is also legal evidence that these acts don't contradict Constitution of Russian Federation.

Acts, issued by Rospotrebnadzor don't have to be obligatory registered in the Ministry of Justice of Russian Federation and are approved by the Main Sanitary Inspector of Russian Federation. In the sphere of relations, which regulate the principles of nanosafety in Russian Federation Rospotrebnadzor issues resolutions, methodical instructions and letters. Traditionally, these acts belong to the system of normative-technical documents, which have a subsidiary and voluntary characteristic.

It was a little excursus into the theory of regulation in Russian Federation, conducted in order to orientate in the system of legal regulation of relations, connected with specifics of guaranteeing nanotechnology's environmental safety in Russian Federation.

A rise of considerable amount of subsidiary documents has become a first step on a way of politico-legislative creation in the sphere of development and implementation nanotechnologies.

The foundation of the process has been approved by the President of Russian Federation The basics of politics in Russian Federation in the area of science development and technologies for the period until 2010 and further prospect from 30 March 2002. According to that, the environmental priorities while developing and implementing of nanotechnologies should be the main tasks of Russian scientific policy. It is connected with the requirements of environmental legislation in Russian Federation in the sphere of realization of any kinds of industrial activity. Hence, the President of Russian Federation has announced that a development of legal acts in the sphere of safety provision ought to become a primary target.

Furthermore, a Program for coordination of activity in the sphere of nanotechnologies and nanomaterials in Russian Federation has been appeared, approved by the Direction of the Government of Russian Federation July 25, 2006 has been announced that one of the main tasks was an increase of a level for environmental safety and an improvement of habitat with a use of nanosystematic technology for the environmental monitoring.

An annual Message from the President V.V. Putin to the Federal Committee of Russian Federation from April 26, 2007 declared and first revealed an importance of regulation in the sphere of nanotechnology. On the basis of that document it had been offered to develop a complex of necessary documents. A rapid fulfillment of suitable activities in the sphere of nanoindustry development is one of strategic national priorities of Russian Federation.

It would have been logical to suppose that the mentioned initiative could find its objective expression in activities of appropriate authorities, which were willing to prepare a draft of legislative acts about a procedure of activities in nanotechnologies areas. Those authorities could develop a basis for preparation and approval of subordinate regulatory acts.

Up until now in Russian Federation there is no federal legislative act about procedure of nanotechnologies' development and application, which could have contained all necessary terminology; the major principles and kinds of activities in the sphere of nanotechnology and also a liability for violation of norms and requirements in the field of environmental safety. In addition, there are even no technical regulations while a practical activity in the current field has been actively in progress [2].

In the theory of Russian law this situation is called "legal gap", which could not have been resolved by altering already existed regulations because in Russian Federation there are no suitable acts with allied legal characteristic and an object of legal regulation.

It is worth noting highlighting, that in the significant majorities of industrial areas of Russian Federation such acts exist. For instance, the federal acts "About Atomic Energy", "About Waste", "About Regulation of Genetically Engineered Activities", "About Radioactive Waste", "About Destruction of Chemical Weapon" and so on. According to the mentioned acts a set of subordinate regularity acts have been issued, which provide a mechanism of their implementation (decrees, resolutions, directions and etc.).

Among the main subsidiary acts in the sphere of nanoregulation in Russian Federation two programs have to be mentioned particular – The Program of nanoindustry's development in Russian Federation until 2015 (January 17, 2008) and The Strategy of nanoindustry's development in Russian Federation (April 24, 2007). These documents regulate an activity for analysis potential threats related to a development and technical provision of nanoindustry. A special attention had been drawn to the following areas:

- ✓ A system of technical regulations;
- ✓ Guarantee of nanotechnologies' safety and its products;
- ✓ An insurance of risks in the sphere of nanotechnologies.

Nowadays the mentioned area of relations is on a stage of dynamic development.

A system of nanosafety regulation can be divided into two principle categories: 1) legislative sphere; 2) subordinate (by-law) sphere [3].

I. Legislative nanosafety regulation in Russian Federation.

Nevertheless, there are federal acts which, in any event, regulate a field of nanoactivity. One if the example is The Federal Act from July 19, 2007 "About Russian nanotechnologies' corporation" (Rosnanoteh) – the only act in the mentioned sphere. It has been set, that Rosnanoteh is *the main* specially authorized body in the established area and acts towards assistance of nanotechnologies politics and nanoindustry realization [4]. Nevertheless, in the text of this act there is a lack of regulation about guaranteeing of nanosafety; only organizational and financial support of nanoprojects has been mentioned. In addition, the article 3 of this law clarifies that Rosnanoteh has the principle authority in the sphere of state politics implementation of prospective nanotechnologies and nanoindustry. Among aims, tasks and functions of Rosnanoteh there is no mentioning about a participation in the nanosafety regulation, but the main subject of its activities is an organizational and financial support of nanoprojects [5].

Besides the above-listed law, in the sphere of nanotechnologies nowadays in Russian Federation there are legislative acts which have only an indirect connection to provide nanosafety, for instance:

- ✓ The Federal Act from 30 March 1999 N 52-ФЗ "About Sanitarian and Epidemiological Well-being of Inhabitants";
- ✓ The Federal Act from 26 June 2008 N 102-ФЗ "About Proving of Measurements Unity";

- ✓ The Federal Act from 18 July 1999 N 183-ФЗ "About an Export Control";
- ✓ The Federal Act from 27 December 2002 N 184-ФЗ "About Technical Regulation";
- ✓ The Federal Act from 10 January 2002 N 7-ФЗ "About Environmental Protection";
- ✓ The Federal Act from 21 November 2011 N 323-ФЗ "About Health Protection in Russian Federation" and etc [6].

  Due to that, it became evident that these acts have only a common connection to provide nanosafety in Russian Federation.

  Thus, relations in the sphere of nanotechnology have no independent *legislative* regulation.

II. Subordinate (by-law) nanosafety regulation in Russian Federation.

The existed legal vacuum has been actively filled with a lot of subordinate acts issued by suitable departments. Thus, according to the Order of Public Health Ministry of Russian Federation (August 5, 2009) a Working Group has been created in order to organise a development of subordinate and methodical documents covering questions of nanomaterials' and nanotechnologies' safety and methods of Nanosafety Risk Assessment.

The precise due dates for issuing and approving three categories of documents have been set. The covered areas were:

- ✓ medico-sanitary provision of nanoindustry safety;
- ✓ development of requirements applied to products of nanoindustry, its processes designing, production and its market turnover and utilisation;
- ✓ priorities and issue of recommendations to provide an environmental protection and health of employees engaged in nano technological factories and also inhabitants of areas close to those industries.

It is stated that the Working Group under the Public Health Ministry and Social Development in Russian Federation must prepare suggestions for the principle federal legislation and subsidiary acts in the nanosafety regulation. Unfortunately, in the context of this Order there are no exact dates and up until now not all mentioned categories of documents have been prepared.

Currently in Russian Federation all acts about nanosafety are being regulated by the means of normative technical acts, authorized in the form of resolutions, methodical instructions (in Russian - МУ) and methodical recommendations (in Russian - МР), approved by The Main Sanitarian Inspector of Russian Federation.

The main document in the field of nanosafety is The Concept of toxicological researches, methodology of risk assessment, methods of an identification of quantitative detection of nanomaterials (October 31, 2007). It is very important that the document has been registered by the Ministry of Justice of Russian Federation. As it was stated before, it indicates of high importance and highlights that this act is connected with the constitutional rights of citizens in the sphere of nanosafety and its guaranteeing.

Due to the fact that is it one of the most influences and essential documents in the field of nanosafety, it is worth listing its key parts:

- ✓ The characteristics new properties and behavior of nanomaterials in the environment and biological objects;
- ✓ The particularity of Risk Assessment on the stage of manufacturing and use of nanomaterials; the existed methodology of risk assessment nowadays cannot be completely applied to nanomaterials because it is based upon a full toxicological assessment of a specific substance or its combination and detection of "dose-effect'. In this regards it is necessary for every individual nanomaterial to be comprehensibly studied in the toxicological aspect with an identification of allowed twenty-four-hour dose or conditionally bearable (monthly) inflow. It is also necessary to create data resources for biosafety of nanomaterials. Until now those resources are not enough in Russian Federation.
- ✓ The analysis of data about safety of manufacturing and use of nanomaterials;
- ✓ The procedure for supervision and toxicological research of nanomaterials. A development and maintaining of nanoparticles and nanomaterials register (in accordance with the Federal Register of potential hazardous chemical and biological substances) has to be fulfilled by Rospotrebnadzor.

The next key acts issued by Rospotrebnadzor in the field of nanosafety is The Assessment on nanomaterials' safety (October 12, 2007), published in the form of methodical recommendations. The mentioned document establishes an Algorithm of identification of high level of a potential danger for nanomaterials towards a health of human being.

There are three levels of potential danger of nanomaterials:

1. Low level of potential danger (low priority). A suitable nanomaterial has to be assessed according existed indicators of its components in traditional form. An extra research of specific biological effect of nanomaterials is not needed.
2. Medium level of potential danger (medium priority). A suitable common toxicological assessment is needed and when it is necessary, some kinds of extra researches have to be conducted.
3. High level of potential danger (high priority). A comprehensive complex of researches have to be conducted on the subject of nano materials' penetration through biological membranes and tissues of organisms plus a common toxicological assessment which includes studies contained a test of genotoxicity.

The Assessment of Nanomaterials' Safety consists of a list of methods for studying the main physical, chemical and molecular-biological characteristics of nanomaterials and also approximate schemes for organizing of experiments to conduct a research about general toxic effects of nanomaterials.

For the purpose of The Assessment of Nanomaterials' Safety it recommendable to apply The Procedure of Medico-Biological Assessment of nanomaterials' impact engaging laboratory animals according to their morphological attributes and metabolic parameters (October 17, 2011).

Except all above-listed documents nowadays in Russian Federation there more than 30 approved methodical acts of Rospotrebnadzor in the sphere of nanosafety regulation, which have a departmental and not compulsory power. The vast majority of those acts have been issued within a timeframe of 2010-2011.

It is worth mentioning that there is a Policy of nanotechnologies' Classification and its products according to level of potential harm (December, 17, 2010).
Thereby, in Russian Federation the regulation of nanosafety has been fulfilled most purely on a subordinate (by-law) - "lowest" departmental level. It proves that the essential statutory provisions in the field of nanoregulation don't have an imperative legislative binding. Currently this is one of the key challenges in Russian Federation in the sphere of nanorelations' regulations. A situation might be resolved by a issuing a separate federal legislative act which will become a "catalyst" of other sub-laws creation.

References:

[1] See full source: State Internet portal of Russian Federal service on customers' rights protection and human well-being surveillance, http://rospotrebnadzor.ru/news [access 15/11/12].
[2] See full source: State Internet portal «Nanotechnologies and nanomaterials», http://www.portalnano.ru/toplevel/?id=2 [access 15/11/12].
[3] See full: Belokrylova E. A. The specifics of formulating a state politics in Russian Federation in the sphere of legal provision of environmental safety on the stage of nanotechnologies development and application. Law&Policy. №5. 2011. 10 p.
[4] See full source: Ekaterina A. Belokrylova. Some legal problems of environmental safety of nanotechnology and nanomaterials in the Russian Federation, Vestnik Udmurtskogo Universiteta, N4, 2012 http://vestnik.udsu.ru/2012/2012-024/vuu_12_024_14.pdf [access 15/11/12].
[5] See full source: Ekaterina A. Belokrylova. The Legal Problems of Nanotechnology Environmental Safety Provision in the Russian Federation: The Foreign Country's Experience, Nanotechnology Law&Business, http://www.nanolabweb.com/index.cfm/action/main.default.viewArticle/articleID/370/CFID/5073669/CFTOKEN/44254839/index.html [access 15/11/12].
[6] See Full source: Russian Legal Internet portal Garant, http://www.garant.ru/ [access 15/11/12].

REGULATION AND SAFETY IMPLEMENTATION OF NANOTECHNOLOGY FOR CHEMICAL ENTERPRISES IN THE CENTRAL EUROPE SPACE

A FALK[1], S HARTL[1] and F SINNER[1,2]
[1] Project Manager, BioNanoNet Forschungsgesellschaft mbH, AT
[2] Deputy Head of Institute, JOANNEUM RESEARCH Institute for Biomedicine and Health Sciences, AT
E-Mail: andreas.falk@bionanonet.at

Abstract. As result of the gradually increasing nanotechnology sector there is the necessity of a contemporary analysis of the present regulations used for nanomaterials, to outline the current situation of the nanotechnology sector, to promote international cooperation and research's coordination to overcome disciplinary boundaries, to fill the gap between more and less experienced regions and to turn investments in R&D in industrial innovations.
The general objective of the Central Europe project NANOFORCE, which is developed by national and regional chemistry associations and R&D Centres of the Central Europe area, is to foster the innovative nanotechnology-sector networks across Central Europe regions by bringing together public and private organizations to carry out collaborative and interdisciplinary researches on nanomaterials (in the frame of REACH Regulation) and to turn the most promising laboratory results into innovative industrial applications.
To build up a legal advisory board for chemical enterprises starting in nanotechnology, a state of the art report on existing safety procedures and nanotech related regulations was produced to give an overview on currently available regulations used by chemical industries and manufacturing companies within the European region to secure their products. The main emphasis was placed on REACH regulation to search for relevant sections concentrating on nanomaterials which are applicable for nanotechnology. In addition, all relevant directives and amendments of REACH were screened with regard to identify gaps where action is still needed and give possible recommendations for the European Commission. Beyond literature research a questionnaire for producers, users, researchers and financiers was developed with the goal to collect information about the nanotechnology sector in the CE region concerning development, financial status, and international cooperation within joint ventures, safety and nanotoxicology.

Introduction

Nanotechnology and nanosciences are said to be key technologies for future development of new products and materials and have experienced a significant upturn in the past decade within the European economy. The term "nano" is commonly used in research and development for particles and materials at a molecular or atomic scale from 1 to 100 nm, with the result of having specific functions or properties due to the small size. The importance of this new "horizontal" or "enabling" technology is seen in the ability to bring together different research areas benefiting from an interdisciplinary approach towards innovations [1].

According to a report published by the International Labour Organization in 2010 on *emerging risks and new patterns of prevention in a changing world of work* approximately 20% of worldwide produced goods will in some extent be based on nanotechnology [2].

A study has shown that the global volume for products derived from nanotechnology should be growing from €200 billion in 2009 to €2.5 trilion by 2015 [3]. Today's global market is estimated to hold 11 million tonnes of nanomaterials and employing more than 300 000 people in the nanomaterials sector in Europe [4].

The positive economic impact of nanotechnology is said to be the main force to drive fundamental changes within technology sectors leading to benefits and new application [5].

The EU science and research commissioner Janez Potocnik said that *"'Europe needs to invest in knowledge to maintain its competitive edge in the global economy – Nanotechnology is a key area where Europe is in the lead, and we must ensure that we stay there. Nanotechnology has enormous potential for European industry and for society in general, so a clear strategy and decisive action is needed for research in this area. At the same time, we must take into account any possible health, safety and environmental risks and address them as early as we can"* [6].

Many different focus areas can be discovered due to the complexity of broad term 'nanotechnology', such as

"cosmetic, food, food contact material, biocidal products, plant protection products, waste, energy, bio engineering, etc."

That has resulted in the need of clear regulations for the safe use and implementation of nanotechnology. Nanomaterials have been under current investigations on risk and benefit assessment in many research groups due to their specific effects resulting from shape, morphology, size including surface area, functionalization, atomic structure and particle chemistry.

Despite the advantages to be seen in the use of nanomaterials for industry and medicine, several concerns have occurred during the past years. Environmental risks as well as health and safety concerns [7] should be addressed by the use of specific risk assessment tools to be performed in the EU and coordinated by authorities. Drawing attention to exposure (risk, hazard) as well as standardization, this is known to be taken into account to be able to draw realistic scenarios about potential risks. The Scientific Committee on Emerging and Newly Identified Health Risks (SCENIHR) has pointed out that there is a significant uncertainty about the potential risks coming from nanomaterials, therefore a case-by-case risk evaluation has to be done with each single substance [8].
One of the major problems facing nanotechnology and the production of nanomaterials is said to be the lack of knowledge transfer on risks that might be associated with the implementation of new technologies and materials such as nanomaterials. Evidences on toxicity and risk assessments for nanomaterials have been available on a certain level, but are still missing a common approach. This has led to high quantities of disorganized research data, a lot of double work carried out and caused confusion on the consumer and users side. Together with the lack of current available regulations for the safe use of nanomaterials on the market these problems cause barriers for technology transfer of research to innovative applications. Recently, on national level are already started activities to consolidate and structure existing expertise, for example within the European Center for Nanotoxicology which is an Austrian initiative focused on nanotoxicology and nanomedicine [9].

Nanotechnology for Chemical Enterprises

Focusing on the development of innovative public-private partnerships in the nanotechnology sector and bringing together public and private organizations to turn the most promising laboratory results into innovative industrial applications is proposed by the Central Europe project NANOFORCE.
The project consortium has planned to set up an international service hub for nanotechnology including an expertise database to provide stakeholders with tangible sources for nanotechnology and nanotoxicology and to promote data transfer as well as communication and dissemination. Furthermore the implementation of nanotechnology responsible management is aimed together with setting up an information pool for governmental bodies to give recommendations on safe use of nanotechnology. A framework analysis on existing regulations for nanomaterials and their safe use and handling is performed to give possible recommendations on existing needs and gaps to the European Commission.

Journal of Physics: Conference Series **429** (2013) 012068 doi:10.1088/1742-6596/429/1/012068

Firstly presented within a Central Europe Programme are actively performed lab analysis on nanomaterials provided by companies using standardized OECD testing guidelines and evaluation methods with the output to provide safety data sheets and exposure scenarios.
Within the aim of NANOFORCE this paper is focused to introduce the work which has already been done in the project especially on the evaluation of the state of the art on existing safety procedures and nanotechnology related legislations.

Approach for nanomaterials on the EU market
Nanotechnology has been part of a major new technology development that are said to be able to have a significant impact on the economy but to be likely followed by various unexpected consequences [10]. The goal of policy makers and regulations to guarantee the safe manufacturing of products and to promote new technologies as well as to generate benefits while avoiding possible risks that may occur has to be transmitted at a regular basis for the insurance of a safe handling of nanomaterials [11].

When entering the European market each product and good needs to be subjected to a standardized authorization procedure. The EU directive on general product safety determines the security measures of products prior to market placement, to avoid harmful risks to consumers, described by the European Committee for Standardization (CEN) and the European Committee for Electrotechnical Standardization (CENELEC). Products that accomplish these requirements are signed with the 'CE' symbol and may circulate freely on the European market [12].

Nevertheless the precautionary principle should be enabled in any case facing possible dangerous effects on human health, animal or plant health and environmental damages [13]. For the implementation of nanomaterials on the European market, industry should be encouraged to follow the precautionary principle to 'do your best to avoid possible negative side effects of nanotechnologies'.

Definition of Nanomaterial
Furthermore, a point of discussion which has been addressed was the recommendation on a definition for nanomaterials launched in October 2011 by the European Commission, defining a nanomaterial as

"a natural, incidental or manufactured material containing particles, in an unbound state or as an aggregate or as an agglomerate and where, for 50 % or more of the particles in the number size distribution, one or more external dimensions is in the size range 1 nm-100 nm. In specific cases and where warranted by concerns for the environment, health, safety or competitiveness the number size distribution threshold of 50 % may be replaced by a threshold between 1 and 50 %. [...]" [14].

The definition has aimed at providing a standardized identification of nanomaterials and to open new perspectives on risk assessment tools and safety regulations. Due to the lack of knowledge of the specific characteristics and hazards of nanomaterials the new definition has been granted the ability to provide a new opportunity to monitor and assess the proceeding risks. Only risk assessment tools can determine whether nanomaterials are classified as hazardous goods and whether further action is warranted to minimize possible risks. Another fundamental objective of this definition has been to ensure that nanomaterials are handled cross-sectional. The 'Second Regulatory Review on Nanomaterials' published in October 2012 has handled the implementation of the recommendation to be used by Member States, European Union agencies and companies. Standardization measures are carried out to compile existing definitions for nanomaterials and the definition will be evaluated again in 2014 by the European Commission [15].

REACH Regulation for nanomaterials

The regulatory framework for the management of chemicals (REACH) is said to be the main legislation in nanotechnology for the overall controlling of production, product market placement and usage of substances in preparation or in articles. The key elements of REACH also applicable for nanomaterials can be found in registration of chemicals manufactured or imported in the EU, unless specifically exempted. A dossier on physiochemical data and (eco-)toxicological properties has to be included which is generated within collaborations of companies. To examine the potential hazards and risks for chemicals, additional toxicological tests with results to be included in 'chemical safety reports' (CSR) have to be performed. An evaluation of the substance and the data from the dossiers is carried out by the European Chemicals Agency (ECHA) together with the Commission and the Member States to evaluate potential risk on health and safety [16].

According to Bernadette Quinn, PhD from ECHA - substance identification and data sharing unit - within a panel discussion at SENN2012, Helsinki 2012, there have been three registered substances to be 'nano' being followed by four registration after the launch of the recommendation for definition on nanomaterials in October 2011.

Compared to approximately 4000 registered substances, nanomaterials have taken up a relatively small proportion on registered chemicals. Due to upcoming deadlines and the implementation of the definition on nanomaterials a change is expected [17]:

- *REACH: 1 Dec. 2010 phase-in substances produced or imported with volumes of over 1,000 tonnes/year.*
- *REACH: 1 Dec. 2013 phase-in substances with volumes of over 100 tonnes/year.*
- *REACH: 1 Dec. 2018 phase-in substances with volumes over 1 tonnes/year.*

The REACH regulation is determined to be the most appropriate framework for dealing with potential risks of nanomaterials, but specific adaptations on REACH for nanomaterials and the preparation of design modifications as well as amendments are expected to be necessary [15].

To study the ability of REACH to cover nanomaterials altogether a recent (unpublished) state of the art report on existing safety procedures and nanotechnology related regulations was produced within the NANOFORCE project, with several regulations with relevance for nanomaterials to be found for cosmetics, food, food contact material, food additives, biocidal products and restriction of hazardous substances in electrical and electronic equipment.

Cosmetics

At present, there is said to be an inadequate information on the risks associated with nanomaterials; still the need to ensure a high level of protection to human health for every cosmetic product that contains nanomaterials is assessed. Additionally cosmetic products containing nanomaterials have to be notified to the Commission by the responsible person six months prior to market placement, except where they have already been placed on the market by the same responsible person before 11 January 2013. To guarantee the safety of consumers and to follow information demands "a catalogue of all nanomaterials used in cosmetic products placed on the market, including colorants, preservatives and UV filters" (Art. 16. (10.)(a)), will be published by the Commission with additional labelling obligations to "list the relevant materials with the word 'nano' in brackets" (Art. 19. (1).) [18]

Food

Regulations for the presence of engineered nanomaterials in food should ensure the information of consumers by clearly indicating the nanomaterial in the list of ingredients and labelling such ingredient with the word 'nano' in brackets [19].

Food contact material and food additives

Nanoparticles are defined as engineered substances with significant changes in size and physical properties which might be leading to changes in the toxicity. Those have to be addressed on a case-by-case basis for risk assessment. Nanoparticles which are used in food contact materials should therefore only be used prior to an explicit authorization [20]. Food additives which are produced by new or significantly changed methods like nanotechnology leading to alterations in particle size need to be considered as different agents requiring a new entry in the community list or change in the specification prior to market placement [21].

Biocidal products

Nanomaterials used in biocidal products are managed differently than any active substance except where explicitly mentioned. Biocidal products containing nanomaterials need to be labeled 'nano' in brackets next to the material itself. Beginning in September 2015 Member States have to report to the Commission the correct implementation of the regulation providing information on the use of nanomaterials in biocidal products including risks. Nevertheless the Commission is encouraged to review the scientific progress of the provision on nanomaterials on a regular basis [22].

Restriction of hazardous substances in electrical and electronic equipment

Following the precautionary principle, electrical and electronic equipment produced with or using nanotechnology and nanomaterials have to be notified to the Commission. Hazardous or possible hazardous substances are said to be substituted by environmentally friendly alternatives to ensure the safe use as well as human and environmental health over their life cycle [23].

Conclusion

Based on current knowledge, it can be assumed that nanomaterials show similarities in toxicity to customary chemical substances. Therefore risk assessment has to be tailored to the individual case and on the information available (case-by-case) for the material. The definition on nanomaterials which was presented by the European Commission in 2011 will be integrated into EU regulations keeping in mind that the overall coverage of nanomaterials is managed by REACH which is discovered to be the most appropriate framework for dealing with potential risks of nanomaterials. One of the problems has been addressed for nanomaterials to have entered a very late point of discussion in REACH which will lead to additional changes and comments and an essential control of further developments to be monitored carefully by the Commission including a new report to be presented in three years.

Within REACH there are still issues to be addressed for nanomaterials to proceed with the safe implementation of nanotechnology on the European market:

- registration for imported and manufactured nanomaterials <1 tonne/year
- harmonized labeling of nanomaterials
- safety reports with exposure assessment for registered nanomaterials
- reporting requirements for all nanomaterials on the market
- adaption of the precautionary principle at all levels of the production cycle

Not only regulations but several guidelines with focus on needs of the good implementation of nanomaterials in products including a safe market placement are placed at disposal for consumers and producers to help ensuring the correct use of the new technology. Safety research and standardization - addressed via bottom-up approach by consolidation of national expertise (ref. to www.EURO-NanoTox.eu) - shall assist the safe implementation of nanotechnological applications and products. However, findings of the project NANOFORCE have shown significant needs in the standardization of the regulatory body for nanomaterials, a better implementation and development of simple and standardized testing methods, a uniform data collection and harmonization tools for a better knowledge dissemination and transparency for data to be available for stakeholders.

Acknowledgment
The authors like to thank the NANOFORCE project consortium for their valuable input to the reports.

References

[1] COMMUNICATION FROM THE COMMISSION Towards a European strategy for nanotechnology, COM(2004) 338 final, 12.05.2004, Access: http://ec.europa.eu/nanotechnology/pdf/nano_com_en.pdf

[2] International Labour Organization: 'Emerging risks and new patterns of prevention in a changing world of work' (2010) Access: http://www.ilo.org/public/portugue/region/eurpro/lisbon/pdf/28abril_10_en.pdf

[3] Forfás: 'Ireland's Nanotechnology Commercialisation Framework 2010-2014', August 2010, Access: http://www.forfas.ie/media/forfas310810-nanotech_commercialisation_framework_2010-2014.pdf

[4] EUROPEAN COMMISSION: 'High-Level Expert Group on Key Enabling Technologies - Final Report', June 2011, Access: http://ec.europa.eu/enterprise/sectors/ict/files/kets/hlg_report_final_en.pdf

[5] Working Paper 2010.05: 'The EU approach to regulating nanotechnology', etui - European Trade Union Institute, Aida Ponce (2010), Access: http://www.etui.org/Publications2/Working-Papers/The-EU-approach-to-regulating-nanotechnology

[6] Commission submits plan for keeping Europe at the forefront of nanotechnology, Date: 14.06.2005, accessed on 02.11.2012 at: http://cordis.europa.eu/fetch?CALLER=EN_NEWS&ACTION=D&SESSION=&RCN=23973

[7] Roblegg E, Sinner F, Zimmer A 2009 Health Risks of Nanotechnology *EURO-NanoTox-Letters* ISSN: 2074-8515, **No 001**, p. 1-18

[8] Scientific Committee on Emerging and Newly Identified Health Risks (SCENIHR): 'Risk Assessment of Products of Nanotechnologies' (2009), Access: http://ec.europa.eu/health/ph_risk/committees/04_scenihr/docs/scenihr_o_023.pdf

[9] Roblegg E, Falk A, Froehlich E, Zimmer A, Sinner F 2011 European Center for Nanotoxicology EURO-NanoTox – An Austrian Activity *Academy of Science Malaysia Journal*, **5 (1)**

[10] Scientific Committee on Emerging and Newly Identified Health Risks (SCENIHR) -modified opinion (after publication) on: 'The appropriateness of existing methodologies to assess the potential risk associated with engineered and adventitious products of nanotechnologies' (2006) SCENIHR/002/05, Access: http://ec.europa.eu/health/ph_risk/committees/04_scenihr/docs/scenihr_o_003b.pdf

[11] Seaton A, Tran L, Aitken R, Donaldson K 2010 Nanoparticles, human health hazard and regulation *Journal of the Royal Society Interface* **7**, S 119 - S 129

[12] EUROPEAN COMMISSION: 'Guide to the implementation of directives based on the New Approach and the Global Approach' (2000) Access: http://ec.europa.eu/enterprise/policies/single-market-goods/files/blue-guide/guidepublic_en.pdf

[13] COMMUNICATION FROM THE COMMISSION on the precautionary principle, COM (2000) 1 final, 2.2.2000, Access: http://eur-lex.europa.eu/LexUriServ/LexUriServ.do?uri=COM:2000:0001:FIN:EN:PDF

[14] COMMISSION RECOMMENDATION of 18 October 2011 on the definition of nanomaterials, (2011/696/EU), 20.10.2011, Access: http://eur-lex.europa.eu/LexUriServ/LexUriServ.do?uri=OJ:L:2011:275:0038:0040:EN:PDF

[15] COMMUNICATION FROM THE COMMISSION TO THE EUROPEAN PARLIAMENT, COUNCIL AND THE EUROPEAN ECONOMIC AND SOCIAL COMMITTEE - Second Regulatory Review on Nanomaterials, {SWD(2012) 288 final} COM(2012) 572 final, 3.10.2012, Access: http://eur-lex.europa.eu/LexUriServ/LexUriServ.do?uri=COM:2012:0572:FIN:en:PDF

[16] Spencer W, Panko J, Paustenbach D 2009 The European Union's REACH regulation - a review of its history and requirements *Critical Reviews in Toxicology* **Vol. 39**, No. 7, pp. 553 - 575

[17] ECHA - European Chemicals Agency : 'REACH 2013', accessed on 05.11.2012 at http://echa.europa.eu/reach-2013

[18] Regulation (EC) No. 1223/2009 of the European Parliament and of the Council on cosmetic products of 30. Nov. 2009, in : OJ of the European Union L 342/59 (22.12.2009), Access: http://eur-lex.europa.eu/LexUriServ/LexUriServ.do?uri=OJ:L:2009:342:0059:0209:en:PDF

[19] Regulation (EU) No. 1169/2011 of the European Parliament and the Council on the provision of food information to consumers [...] of 25 Oct. 2011, (awaiting publication in OJ) in : OJ of the European Union L 304/18 (22.11.2011),

[20] Commission Regulation (EU) No. 10/2011 on plastics materials and articles intended to come into contact with food of 14. Jan, 2011, in : OJ of the European Union L 12/1 (15.1.2011), Access: http://eur-lex.europa.eu/LexUriServ/LexUriServ.do?uri=OJ:L:2011:012:0001:0089:EN:PDF

[21] Regulation (EC) No. 1333/2008 of the European Parliament and of the Council on Food additives of 16. December 2008, in : OJ of the European Union L 354/16 (31.12.2008), Access: http://eur-lex.europa.eu/LexUriServ/LexUriServ.do?uri=OJ:L:2008:354:0016:0033:EN:PDF

[22] Regulation (EU) No. 528/2012 of the European Parliament and of the Council concerning the making available on the market and use of biocidal products of 22. May 2012, In : OJ of the European Union L 167/1 (27.06.2012), Access: http://eur-lex.europa.eu/LexUriServ/LexUriServ.do?uri=OJ:L:2012:167:0001:0123:EN:PDF

[23] Directive 2001/65/EU of the European Parliament and of the Council on the restriction of the use of ... hazardous substances in electrical & electronic equipment of 8. June 2011, in : OJ of the European Union L 174/88 (1.7.2011), Access: http://eur-lex.europa.eu/LexUriServ/LexUriServ.do?uri=OJ:L:2011:174:0088:0110:EN:PDF

Journal of Physics: Conference Series **429** (2013) 012069 doi:10.1088/1742-6596/429/1/012069

Defining Occupational and Consumer Exposure Limits for Nanomaterials – First Experiences from REACH Registrations[4]

K Aschberger[1,5], Z Klöslova[2], G Falck[2] and F M Christensen[1,3]

[1]Nanobiosciences Unit, Institute for Health and Consumer Protection, European Commission-Joint Research Centre, 21020 Ispra, Italy

[2]European Chemicals Agency, Annankatu 18, 00121 Helsinki, Finland

[3]Current affiliation: COWI, Parallelvej 2, 2800 Kongens Lynghy, Denmark

Email: Karin.aschberger@ec.europa.eu

Abstract. By 1 December 2010 substances manufactured or imported in the EU $\geq$ 1000 t (as well as certain other substances) had to be registered under the REACH Regulation 1907/2006. The Joint Research Centre (JRC) in close cooperation with the European Chemicals Agency (ECHA) carried out an analysis and assessment of what type of information on nanomaterials was provided in the received registrations. The aim of the assessment was to develop options for an adaptation of the REACH regulation to ensure proper information generation and reporting and an appropriate risk/safety assessment of nanomaterials (Nano Support project). It should be noted that this analysis and assessment was not a compliance check of the dossiers. From 26000 submitted registration dossiers covering 4700 substances finally 25 dossiers (19 substances) were identified to cover nanomaterials or nanoforms of a substance. It is possible that other dossiers are considered to cover nanomaterials or nanoforms by the registrants, however such dossiers could not be identified to address nanoforms given the information contained in those dossiers. The identified 25 dossiers were subject to a detailed analysis and assessment of information provided for all endpoints including substance identity, physico-chemical properties, human health, environmental fate & behaviour, ecotoxicity, PBT[6] assessment, Classification and Labelling as well as the attached Chemical Safety Report documenting the Chemical Risk/Safety Assessment. In order to evaluate how the safety of workers and consumers was ensured, it was appropriate to check how the "Derived No (Minimum) Effect Levels" (DN(M)ELs) were established for substances, covering nanomaterials or nanoforms. DNELs were established mainly for long term inhalation exposure of workers. Half of the assessed dossiers included an oral long term DNEL for the general population. DNELs were usually not specific for nanosized forms and, in the few cases where they were calculated for nanosized materials, they were not derived from hazard data for the nanoform. Different methods for deriving the DNELs were applied and few dossiers

[4] The opinions expressed in this publication are those of the authors and not necessarily those of the European Commission

[5] To whom any correspondence should be addressed

[6] Persistent, Bioaccumulative, Toxic

derived DNELs by applying the default assessment factors in the REACH guidance.. Several DNELs were based on available Occupational Exposure Limits (OELs) for inhalable and respirable dust or the nuisance dust levels, which have not been established for nanosized materials. In general lower (i.e. less strict) assessment factors were applied with different types of justification. All DNELs were expressed in the mass metrics. It is important to note that submission, identification and selection of the dossiers addressed in this study was done before the adoption of the EC recommendation (2011/696/EU) on a definition of nanomaterial and before the publication of the revised ECHA guidance documents that include recommendations for nanomaterials.

1. Introduction/Background

Nanomaterials are covered by the "substance definition" of Regulation 1907/2006 on Registration, Evaluation, Authorisation and Restriction of chemicals (REACH) [1] as REACH addresses chemicals in whatever size, shape or physical state. By 1 December 2010 substances produced and imported ≥ 1000 t in the EU had to be registered under REACH which potentially also included nanomaterials. In addition a number of very toxic and a number of "new" (under REACH "non phase-in") substances were registered. As part of an Administrative Arrangement with Directorate General Environment of the European Commission (DG ENV), the Joint Research Centre (JRC) in close cooperation with the European Chemicals Agency (ECHA) made an analysis and assessment of what type of information on nanomaterials was available in received registration dossiers (the Nano Support Project[7]). For dossiers that included information on nanomaterials or nanoforms, it was assessed whether the information submitted was adequate for a risk assessment of the nanomaterials/nanoforms and based on that developed options for specific nanomaterial provisions building on the current REACH framework [2]. An assessment of the consequences of these options for industry, consumers, human health and the environment is currently ongoing.

45 registration dossiers possibly addressing nanoforms/nanomaterials were identified (see figure 1) based on a combination of:

- automated searches of the IUCLID[8] database (IT tool used for submitting Registration under REACH) with the 26000 submitted registration dossiers covering 4700 substances
- focused searches on substances selected for assessment in the OECD WPNM (Working Party on Manufactured Nanomaterials), and
- other substances considered to include nanoform from a technical perspective.

Following an examination of the information on substance identity, state, granulometry and other relevant endpoints, 20 dossiers were excluded from further assessment as based on the information provided it could not be concluded whether they were actually covering nanomaterials or nanoforms. The remaining 25 dossiers, covering 19 different substances were identified as likely to cover nanoforms/nanomaterials. These dossiers included three registrations that had explicitly selected "nanomaterial" as the form of the substance in an optional IUCLID field. The further assessment identified three categories of registration dossiers, where

I) the registrants recognised that nanoforms or particle size < 100 nm were inside the scope of the dossier (8 dossiers /5 substances);

II) substances considered to exist only as nanomaterial without a bulk form (12/9), and

[7] Administrative Arrangement between JRC and DG ENV No. 07.0307/2010/581080/AA/D3:Scientific technical support on assessment of nanomaterials in REACH registration dossiers and adequacy of available information. The support from all colleagues contributing to this project is acknowledged: ECHA: Bernadette Quinn, Abdelquadar Sumrein, Julia Fabrega Climent; JRC: Stefania Gottardo, Nanna Hartmann, Christian Micheletti, Hubert Rauscher, Juan Riego Sintes, Birgit Sokull-Klüttgen, Stefania Vegro.

[8] International Uniform Chemical Information Database

Journal of Physics: Conference Series **429** (2013) 012069 doi:10.1088/1742-6596/429/1/012069

III) the assessors identified nanomaterials on the basis of the presence of a 'nanotail' in the provided particle size distributions (5/5) (see figure 1).

It is recognised that since December 2010 some additional nanomaterials have been registered as non phase-in substances, but these were not within the scope of the Nano Support project.

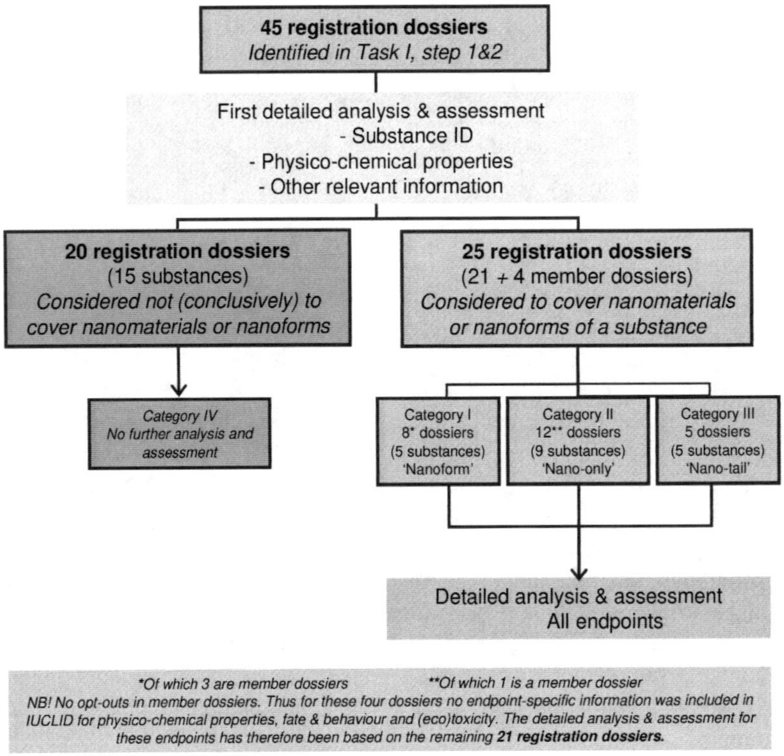

Figure 1: Overview of dossier identification, categorisation, analysis and assessment

2. Assessment of information on nanomaterials/nanoforms in REACH registration dossiers

The identified REACH registration dossiers concluded to be likely to address nanomaterials or nanoforms were subject to a detailed analysis and assessment of all endpoints including substance identity, physico-chemical properties, human health, environmental fate & behaviour, ecotoxicity, PBT assessment, classification and labelling, as well as the attached Chemical Safety Report documenting the Chemical Risk/Safety Assessment. The results and conclusions of this analysis and assessment are briefly summarised below with focus on issues of relevance for this paper, namely to present and discuss how "derived no effect levels" (DNELs) for workers and consumers had been addressed in the dossiers investigated.

It should be stressed that the assessment was not a compliance check of individual registration dossiers but focused on how the REACH regulation could be adapted/clarified to ensure a proper information generation and risk/safety assessment of nanomaterials. Substance names cannot be revealed as this information was considered confidential in relation to dissemination from this project.

2.1. *Substance Identity and physico-chemical properties*

The examination of substance identity and key physico-chemical properties (granulometry, surface area and other physico-chemical properties) revealed that the identification of substances was mainly based on the chemical composition and the information provided in the substance identity sections of the dossiers alone did generally not enable the assessors to identify the form(s) of the substance or to characterise it as a nanomaterial or nanoform of a substance. When the particle size was reported, it was generally not specified, whether it referred to primary particles, aggregates, agglomerates or other and usually detailed information on the method and the sample preparation was not provided. For granulometry usually only one data set per substance was submitted even for joint registration dossiers, irrespective of the fact that these data depend on the manufacturing process of individual registrants. The methods for particle size measurements were often incapable of measuring particles with a size below 100 nm. Measurements of the surface area were provided in a few cases. Surface treatment/modification was described in general terms in a few dossiers, but was generally not provided. Generally one set of physicochemical data or waiving arguments were provided without specifically addressing whether the data would also apply to the nanoform.

In conclusion: the information provided in the dossiers was in general not sufficient to identify a nanomaterial or nanoform and would not also not be sufficient in relation to the EC definition recommendation 2011/696/EU [3]. Based on that, the proposed options for an amendment of REACH suggests that nanomaterials/nanoforms should be explicitly described in the scope of the registration dossier, including more detailed characterisation of the nanomaterial/nanoforms.

2.2. General observations for human health and environment

The general observations for human health and environmental endpoints were that the test material was usually not well characterised and sometimes only trade names were reported. There was no or little distinction between different forms (bulk, micro/nanoform) in the test material description and particle sizes were very rarely reported. For inhalation studies usually only the mass median aerodynamic diameter (MMAD) was reported. When read across was applied to other related substances, this was based on the chemical composition (e.g. Metal-ion) or physico-chemical properties (e.g. (in)solubility), however a size dependency or particle specific effects were not addressed. As some of these substances exist in various forms with different particle sizes and aggregation/agglomeration status, it was difficult to assess the appropriateness of the provided test data in relation to the form(s) registered without a better and more detailed characterisation of the test material. This applies also to substances of category II, which are considered to exist only as nanomaterial without a bulkform.

Based on that, the developed options require that nanoforms within the scope of a registration dossier should be explicitly addressed in the endpoint sections and a detailed description of the test material and sample preparation should be provided.

2.3. Specific observations for Human Health

2.3.1. Biokientics. Knowledge of the biokinetics of nanomaterials provides an important basis for the estimation of systemic exposure and systemic toxicity. In addition, it could provide a basis for the decision on the validity of extrapolating available data from the bulk material to the nanomaterial or between different nanoforms of the same nanomaterial, and the justification for read across to other related substances and grouping. It should be noted that REACH (Annex VIII, point 8.8) does not require a specific toxicokinetic study but an assessment of the toxicokinetic behaviour of the substance to the extent that can be derived from the relevant available information. Information on absorption, distribution, metabolism and excretion via different application routes was provided in most registration dossiers. The conclusions were often made either based on data from toxicity studies (often not with nanoform) or read across to chemically related substances, without giving special attention to the particular nature of the nanoform. Dermal penetration was tested specifically for 2 substances, where dermal exposure to the nanoform is significant, suggesting that these nanoparticles were not absorbed. Altogether, studies with different forms of substances were presented, however the fact that size and other characterisers may alter the toxicokinetics of a substance/particle was not specifically addressed.

2.3.2. Acute and repeated dose toxicity/reproductive toxicity. Most studies were performed using the oral or inhalation route with either bulk forms or no specific test materials information was provided that would allow identifying nanoforms. If read across was applied it was based on the leading ion, not taking into consideration the particle nature of the material. Dermal toxicity tests were usually waived, mostly referring to the insolubility of the substance. In repeated dose inhalation studies, some effects of different particle sizes have been observed, with the smaller particle size being more potent in inducing toxic effects. Information on particle size specific potency was not further elaborated in the dossiers, rather observed effects in the lung were not used for deriving nano- or size-specific conclusions. All materials tested showed low acute toxicity via all exposure routes and all but 2 substances were not classified for repeated dose toxicity.

Little information was provided for reproductive and developmental effects and studies were waived for the arguments that no systemic toxicity is anticipated or that long term studies or developmental screening studies did not indicate a reproductive toxicity hazard. In addition, for some substances, the natural abundance and high tolerance of the elements of the substance, and the absence of effects with the leading ion or structural analogues, were used as justification. One dossier provided a study (OECD 422; Combined repeated dose toxicity study with the reproduction/developmental toxicity screening test) with a nanoform.

2.3.3. *Irritation and Sensitisation.* Results specific for nanomaterials/nanoforms were reported from one *in vitro* skin and one eye irritation test, respectively. Some dossiers reported unspecific effects, either to the eye or respiratory tract, which could be the consequence of deposition of particles, but was not considered as inherent property of the substance and to warrant classification. Few sensitisation studies were provided and most dossiers concluded on non-sensitising effects based on physico-chemical properties or on the absence of effects in humans over long exposure periods (however without exact knowledge on the particle size exposed to). All substances were concluded to be non-irritating to skin, eye and respiratory tract and one substance was classified for skin sensitisation. In general, the nanoform of the substance was not explicitly addressed.

2.3.4. *Genotoxicity and Carcinogenicity.* All dossiers addressed the endpoint genotoxicity either by presenting data with the registered substance, by reading across from related substances or by concluding based on a weight of evidence approach. Some substances were explicitly tested in the nanoform for mutagenicity *in vitro* and this mostly with bacterial mutagenicity tests. The suitability of this test for (nano)particles was not specifically discussed, although it is known that insoluble particle may not penetrate the bacterial cell wall and this test has not a good predictivity for mutagenic effects in mammalian cells/organisms. Few nanoforms were tested *in vivo* and results with the bulkform were usually taken to draw conclusions. Three dossiers provided data for carcinogenicity including inhalation and oral feeding studies, which were (except on study) however not specific for nanomaterials. Carcinogenic effects observed in rats for one nanoform were attributed to rat lung overload and therefore considered not relevant for humans. Several dossiers concluded 'non-carcinogenic' by applying a weight of evidence approach based on data from repeated dose toxicity studies, read across, mutagenicity and/or epidemiological data. One substance was classified as carcinogenic (II B) in one of the three submitted dossiers. Another substance was classified as mutagenic and carcinogenic, but this was based on a category approach based on the leading ion rather that effects attributable to particle nature.

3. Human health exposure limits – DN(M)ELs

3.1. Derived No(Minimal)-Effect Levels - DN(M)ELs

For substances produced or imported ≥ 10 tons, derivation of no-effect levels is part of a Chemical Safety Assessment (CSA) if required. Thus, the registrant based on the available and generated hazard information has to define the relevant threshold levels for exposure below which risks for human health (and for the environment) are considered to be controlled [4, 5]. The Derived No-Effect Level (DNEL) is the level of exposure to the substance above which humans should not be exposed.

The DNEL may vary depending on the *exposure pattern* to the substance which is usually defined by a combination of the *population* likely to be exposed to the chemical, i.e. workers, consumers or humans exposed through the environment; the *frequency and duration* of the exposure, e.g. single/acute exposure or continuous exposure for eight hours, and the *route* of exposure: dermal, inhalation or oral.

The DNELs are normally calculated by dividing the dose descriptor for no/lowest health effects from a toxicity study (e.g. no(lowest) observed adverse effect level = N(L)OAEL) by an assessment factor (AF). Since those dose descriptors are usually obtained from experimental animal data, assessment factors are required to allow for extrapolation to real human exposure situations.

From all health effects, the lowest DNEL for each exposure pattern is documented in the Chemical Safety Report (CSR) and in the safety data sheet, where required. In case no safe threshold level can be obtained, like for example, for genotoxic substances, a semi-quantitative value, known as the DMEL or Derived Minimal Effect level may be developed if data allow. The DMEL values represent exposure levels where the likelihood that the identified adverse effect occurs in a population is sufficiently low to be of tolerable concern.

3.2. DNELs in registration dossiers covering nanomaterials

The 25 dossiers likely to cover nanomaterials/nanoforms were analysed, whether and how nanomaterial/nanoform specific DNELs were derived and on which data they were based.

One dossier (Category II, nano only) derived DNELs for all exposure routes and for workers, consumers and the general public. For all other dossiers, DNELs were established mainly for long term inhalation exposure of workers, either for systemic or local effects in the lung. In 11 dossiers, in addition to the DNEL for workers, a long term oral DNEL for the general population was provided. Two dossiers did not present any DNELs for consumers.

Justifications for not providing DNELs for various routes (mostly the acute DNELs) included

- Absence of acute toxic and/or local effects,
- No systemic availability and effects anticipated,
- Absence of or negligible exposure either due to route of exposure (oral, dermal), use (consumer) or particle nature (hardly respirable as aggregated/agglomerated),
- Absence of classification for hazardous effects or
- The long term DNEL was considered sufficient to protect also for short term effects.

Different methods were used for deriving the DNEL in the 19 dossiers:

1. Use of animal data and application of AFs for inter-/intraspecies differences, or duration using REACH default values for AFs
2. Applying AFs different from REACH default values, as they were considered too conservative, e.g. in comparison to existing limit values (OEL, MAK[9] values) for similar compounds
3. Use of general dust limit or OEL for inhalable (10 mg/m^3 or 4 mg/m^3) and respirable dust (3 mg/m^3)

[9] Maximale Arbeitsplatzkonzentration (Maximum workplace concentration)

4. Use of NIOSH Recommended Exposure Levels (REL) (divided by 2 for the general public)
5. Use of human data: large multi-centre studies, NOAEL = 3/5 of concentration with prevalences of symptoms (chronic bronchitis) without any further AF (DNEL = NOAEL in humans) or internal values from human data without further consideration of additional AFs
6. Route to route extrapolation from oral data to inhalation exposure e.g. using a specific method or by a specific absorption rate for the different exposure routes (= evaluation based on ion and not on particle nature)
7. DNEL derived from structural analogue
8. Recommendations for oral intake of the specific ion

All DNELs were expressed in the mass metric. In the analysed and assessed dossiers covering several forms, the DNELs were not derived or generally not indicated to be specific for nanosized forms, nor were they derived from hazard data for the nanoform. Even the dossiers that had used the IUCLID picklist for the nanoform did not derive a nanospecific DNEL from the available hazard data. For a substance that is used in high tonnages for both, bulk and nanoform, the DNEL was derived from hazard data with the bulk form, and considered by the registrant to be also valid for the nanoform. An available chronic inhalation study with the nanoform of this substance, which was included in the dossier was not taken into consideration in relation to the DNEL derivation.

For inhalation DNELs, the assessment factors for interspecies variability were usually lower than the default values in the REACH guidance or no interspecies factors were applied, as the rat was considered more sensitive than humans. In general no assessment factors for extrapolation of duration (e.g. from subchronic to chronic) was applied, for example in one case with the justification that the NOAEC for effects seen in the lung would not depend on time, due to constantly operating clearance mechanisms (while recognising that the severity of effects may increase with increased exposure duration).

Some dossiers discussed why NOAECs from animal studies would not be appropriate as a starting point for a DNEL – as the exposure patterns and particle characteristics do not mimic the conditions in the occupational environment, where particles form aggregates/agglomerates that are hardly respirable, – as prolonged exposure does not give the animals the normal recovery period for lung clearance and – as rats (more than other rodents or humans) are specifically sensitive to develop lung tumours from lung overload to poorly soluble particles due to ineffective clearance.

Two dossiers derived the DNEL from human data which in principle should be considered the most relevant source of information on human toxicity. Several issues addressing the reliability should however be considered when using human data, e.g. valid method for observing effects, confounding factors, statistical power, sufficient length of follow up, causal association etc. [6]. One major weakness of the use of human data is the difficulty of exposure characterisation. Quantification of exposure especially over longer periods as well as characterisation of particle sizes to which people were exposed are a challenge, especially when production processes may have changed. The absence of effects with bulk materials should therefore not be used for deriving DNELs for the nanoform and the use of human data should in general be critically discussed.

3.3. Discussion on different methods to derive exposure limits

There are in general different approaches to determine maximum level of exposure to nanomaterials and some of them have been discussed and published in the recent years [7-12]. The applicability of the REACH methodology for deriving DNELs for nanomaterials was assessed in the REACH Implementation project on Nanomaterials (RIPoN [13]) which aimed at providing the basis for the updated guidance on the information requirements and safety assessment of nanomaterials under

REACH. The report concluded that for most part, the current guidance provided sufficient flexibility to address areas of uncertainty, data gaps and if justified, deviations from the default approach/AFs. Based on risk assessment approaches, case study results and the wider particle toxicology literature, an alternative approach for extrapolating from experimental animals to humans for inhalation exposure was suggested for consideration and development in relation to its suitability for possible future incorporation into guidance. Some of the conclusions of the RIP-On projects were implemented into the updated REACH guidance [4]. There it is recognised that in deviation from the default assessment factor during the derivation of a DN(M)EL for (nano)particles, a calculation of the actual lung dose could be performed. However as there are considerable differences in ventilation rates, deposition patterns, and clearance rates between humans and animals, all of these factors should be taken into account. It is recognised that a higher factor maybe appropriate if there is the potential for accumulation of particles. This may be especially prudent for such exposure routes as the lung and certain particle morphologies associated with reduce clearance, e.g. long straight, biopersistent fibres which are likely to be retained if deposited in the non-ciliated airways leading accumulation of particles following repeated exposure.

Other approaches for deriving human exposure limits from animal studies are usually based on differences between humans and animals in particle deposition, ventilation rate and lung retention biokinetics in the lung [9-11]. The US National Institute for Occupational Safety and Health (NIOSH) [9] found that for carbon nanotubes a working lifetime exposure 0.2 – 2 µg/ m^3 (8-h time-weighted average, for deposited and retained lung doses respectively) would be sufficient to protect from early-stage pulmonary effects. For the Recommended Exposure level (REL) however they had to consider the upper limit of quantification (LOQ) of the currently recommended analytical method for measuring airborne CNT (7 µg/m^3), which might not assure an appropriate protection of the worker.

Current exposure limits are predominantly mass based approaches. It is however recognised that for nanomaterials mass might not be the best predictor of certain lung diseases. Based on current knowledge toxicological evidence is primarily relating inflammation to particle surface area [13] and therefore also surface area and/or particle number should be taken into account. NIOSH recognised that the particle surface area of nano-TiO_2 seems to be the dominating factor in toxicity and therefore lower exposure limits for the nanosized (ultrafine) TiO_2 [8] were recommended. The suggested REL for ultrafine/nano TiO2 is thus lower than for the micro form, however still based on the mass metric due to acknowledged differences with measuring other metrics.

A general problem of hazard based exposure limits is the currently large deficiencies in hazard data. This was also a conclusion of our previous assessment of nanomaterials risks and appraisals to derive safe exposure levels (DNELs) for specific nanomaterials such as carbon nanotubes, fullerenes, silver and titanium dioxide [11, 14-17]. In the absence of sufficient hazard data and a commonly agreed methodology to derive hazard based exposure limits, precautionary nano reference values could be used as a comprehensive and useful risk management tool [18]. It is therefore strongly recommend that research efforts focus on the development of exposure assessment strategies and tools to improve the effectiveness of methods for controlling occupational and consumer exposure to nanomaterials.

4. Summary and conclusions

From the huge amount of dossier submitted for the first registration deadline, 25 dossiers covering 19 substances could be identified to likely address nanomaterials or nanoforms. An analysis and assessment of these dossiers showed that the information provided in the substance identity sections did generally not provide sufficient information to enable identification and characterisation of nanomaterials or nanoforms of the registered substances. Few toxicity studies provided in the dossiers were performed with nanoforms of a material. Very often the provided test material descriptions for toxicity studies did not allow identifying what form had been tested. DNELs were derived predominantly for long term inhalation exposure and rarely for other exposure routes or short term exposure. DNELs were not specific for nanoforms, even in dossiers which were flagged as nanomaterials and in dossiers clearly addressing different forms, including nanoforms. Different methods were used for establishing DNELs. Besides following the REACH guidance, different calculations were used or already established exposure limits for respirable and inhalable dust were used.

Therefore it is concluded that from the analysed and assessed registration dossiers, very little information and experience on nanospecific exposure limits can be retrieved. In that context it should be noted that most of the high tonnage substances exist in different forms/sizes and that the submission, identification and selection of dossiers was done prior to the adoption of the EC recommendation (2011/696/EU)[3] on the definition of nanomaterial. It was therefore up to the registrant to decide how to address different form(s) within the dossier. In addition, the dossiers had to be prepared prior to the publication of ECHA guidance documents including recommendations for nanomaterials.

Registration deadlines for substances of lower tonnages are pending. It can be assumed that due to the definition recommendation, the available guidance and the increased awareness and experience, the thoroughness of the dossiers with respect to information specific for nanomaterials will be improved. Nano Support and results from similar projects might also trigger an amendment of the legal requirements to more explicitly require information on nanomaterials.

References

[1] European Parliament and Council. Regulation (EC) No 1907/2006 of the European Parliament and of the Council of 18 December 2006 concerning the Registration, Evaluation, Authorisation and Restriction of Chemicals (REACH), establishing a European Chemicals Agency, amending Directive 1999/45/EC and repealing Council Regulation (EEC) No 793/93 and Commission Regulation (EC) No 1488/94 as well as Council Directive 76/769/EEC and Commission Directives 91/155/EEC, 93/67/EEC, 93/105/EC and 2000/21/EC, OJ(2006).

[2] European Commission. Nanosupport Project - Scientific technical support on assessment of nanomaterials in REACH registration dossiers and adequacy of available information. Brussels: JRC, ECHA2012 March 2012 Contract No.: AA 07.0307/2010/581080/AA/D3.

[3] European Commission. Commission Recommendation of 18 October 2011 on the definition of nanomaterial, 2011/696/EU (2011).

[4] ECHA. Updated Guidance on Information Requirements and Chemical Safety Assessment for nanomaterials. 2012 [updated 18/10/2012]; Available from: http://echa.europa.eu/en/view-article/-/journal_content/3df5b7b9-a36d-4e74-811b-3aeee23366f8.

[5] ECHA. Guidance in a Nutshell - Chemical Safety Assessment. Helsinki: ECHA; 2009 [updated 15/10/2012]; Available from: http://echa.europa.eu/documents/10162/13632/nutshell_guidance_csa_en.pdf.

[6] ECHA. Guidance on Derivation of DNEL/DMEL from Human Data. Helsinki: ECHA2010.

[7] van Broekhuizen P, van Veelen W, Streekstra W-H, Schulte P, Reinjnders L. Exposure Limits for Nanoparticles: Report of an International Workshop on Nano Reference Values. *Annals Of Occupational Hygiene*. 2012 July 2012;56(5):**515-24**.

[8] NIOSH. Current Intelligence Bulletin; Occupational Exposure to Titanium Dioxide April 2011 ed: Department of Health and Human Services; Centers for Disease Control and Prevention; National Institute for Occupational Safety and Health; 2011. p. **140**.

[9] NIOSH. Current Intelligence Bulletin: Occupational Exposure to Carbon Nanotubes and Nanofibers 2010.

[10] Pauluhn J. Multi-walled carbon nanotubes (Baytubes®): Approach for derivation of

occupational exposure limit. *Regulatory Toxicology and Pharmacology*. 2010;57(1):**78-89**.

[11] Aschberger K, Christensen FM. Approaches for establishing human health no effect levels for engineered nanomaterials. *Journal of Physics: Conference Series*. 2010;304:8.

[12] Pauluhn J. Poorly soluble particulates: Searching for a unifying denominator of nanoparticles and fine particles for DNEL estimation. *Toxicology*. 2011;279(1–3):**176-88.**

[13] Aitken RA, Bassan A, Friedrichs S, Hankin SM, Hansen SF, Holmqvist J, et al. Specific Advice on Exposure Assessment and Hazard/Risk Characterisation for Nanomaterials under REACH (RIP-oN 3) - Final Project: REACH Nano Consultation 07 July 2011.

[14] Christensen FM, Johnston HJ, Stone V, Aitken RJ, Hankin S, Peters S, et al. Nano-TiO2 – feasibility and challenges for human health risk assessment based on open literature. *Nanotoxicology*. 2011;5(2):**110-24**.

[15] Aschberger K, Johnston HJ, Stone V, Aitken RJ, Hankin SM, Peters SA, et al. Review of carbon nanotubes toxicity and exposure--appraisal of human health risk assessment based on open literature. *Crit Rev Toxicol*. 2010 Oct;40(9):**759-90**.

[16] Aschberger K, Johnston HJ, Stone V, Aitken RJ, Tran CL, Hankin SM, et al. Review of fullerene toxicity and exposure--appraisal of a human health risk assessment, based on open literature. *Regul Toxicol Pharmacol*. 2010 Dec;58(3):**455-73**.

[17] Christensen FM, Johnston HJ, Stone V, Aitken RJ, Hankin S, Peters S, et al. Nano-silver - feasibility and challenges for human health risk assessment based on open literature. *Nanotoxicology*. 2010 Sep;4(3):**284-95**.

[18] van Broekhuizen P, van Broekhuizen F, Cornelissen R, Reijnders L. Workplace exposure to nanoparticles and the application of provisional nanoreference values in times of uncertain risks. *J Nanopart Res*. 2012;14::25.

Development of an Integrative Program of Nanosafety: Promote the Coordination Between Industries and Risk Assessor

Claude Emond[1,2], Serge Kouassi[2], and Frédéric Schuster[3]

(1) BioSimulation Consulting Inc., 2915 Ogletown, #1182, Newark, DE, 19713, USA
(2) Université of Montréal, Département de santé environnementale et santé au travail (DSEST), P.O. Box 6128, Main Station, Montréal, Québec, Canada
(3) Commissariat de l'Énergie Atomique (CEA), Saclay, France

E-mail: claude.emond@umontreal.ca

Abstract. Nanomaterials are widely present in many industrial sectors (e.g., chemical, biomedical, environment), and their application is expected to significantly expand in the coming years. However, nanomaterial use raises many questions about the potential risks to human health and the environment and, more specifically, to occupational health. The available literature supports the ability of the lung, gastrointestinal tract, and skin to act as significant barriers against systemic exposure to many nanomaterials. However, because a potential risk issue exists about the toxicity of nanomaterials to the biological material, tools need to be developed for improving the risk management of the regulators. *The goal is to develop a tool that examines the current knowledge base regarding the health risks posed by engineered nanoparticles to improve nanotechnology safety prior to the marketing phase.* The approach proposed during this work was to establish a safety assessment constructed on a decision-control pathway regarding nanomaterial production and consumer's product to integrate different aspects. These aspects include: (1) primarily research and identification of the nanomaterial base of physicochemical properties, toxicity, and application; (2) the occupational exposure risk during the manufacturing process; (3) and the engineered nanomaterial upon the consumer product. This approach provides important parameters to reduce the uncertainty related to the production of nanomaterials prior their commercialization, reduce the reluctance from the industry, and provide a certification tool of sanitary control for the regulators. This work provides a better understanding of a critical issue of nanomaterials and consumer safety.

Acknowledgement. Special thanks to the Réseau de recherche sur les aspects éthiques, économiques, légaux, sociaux et environnementaux des nanotechnologies (NE3LS), Nanotechnology Ethical, Environmental, Economic, Legal, and Social Issues); the Institut de recherche Robert-Sauvé en santé et en sécurité du travail (IRSST), and BioSimulation Consulting Inc. for funding this work.

1. Introduction

Nanotechnology is growing in an exponential cadence around the world [1], and there is much anticipation in industrial and economical fields of using this technology. Today, the use of nanotechnology in aerospace, construction, chemistry, foods, cosmetics, and textiles and by vehicle manufacturers is undeniable. Based on the data acquired from Woodrow Wilson Center since 2005, we estimate that 1,750 products were commercially available in January 2013, representing an annual revenue of $750 billion US dollars, which is three times more than the $224 billion in revenue from the same source in 2009 [2]. This increase is expected to reach $1 trillion in 2015, representing 2 million jobs [3].

Although nanotechnology has generated a lot of revenue, several hundreds of workers are potentially exposed during the development of nanoparticles in the laboratory and during production. Unfortunately, research from the health and safety organization (HSO) working to improve the evaluation of risks to workers is lagging far behind. Multiple *in vivo* studies in animal and syntheses of the literature helped to highlight that manipulation or production of nanoparticles in both research and development. Several studies demonstrated that nanoparticles can affect different organs (i.e., the heart, lungs, kidneys, and the reproductive system) [4; 5], and nanoparticles can cause toxic effects, which have been addressed as granulomas, fibroses, and tumor responses in the lungs, and genotoxic and cytotoxic effects [6]. These issues are a major concern in the international scientific community, especially in terms of Nanosafety. Moreover, it appears that, in general, the "nano" size of a substance is oftentimes more toxic than the larger size of the same substance.

Despite all of these preliminary efforts to understand the toxicity of nanoparticles and the concerns raised by the HSO about these issues, there is not enough information available in literature to determine specific exposure limits. There is also not enough information to establish a unique methodology for evaluating occupational exposures to nanomaterials; therefore, a precautionary approach is suggested by the experts.

For this work, we developed a methodology to anticipate the potential risk of nanomaterials, and then used this information to develop a better safety assessment. The objective of this study was to develop an assessment table matrix to allow safe handling of nanomaterials prior the commercialisation. More specifically, the study aims included: *(1) conducting a literature review of the health issues related to nanotechnology, (2) developing a health assessment table matrix to identify the determinants of occupational exposure to nanomaterials, and (3) validating the assessment table matrix by experts in safety assessment, in collaboration with companies using nanotechnology.*

2. Methodology

The first step in our approach was to conduct an analysis of the scientific literature by using different databases such as PubMed, MEDLINE, ScienceDirect, and TOXNET, while taking into account the importance of the physicochemical and toxicity identifications of nanomaterials.

In addition, we reviewed reports and reference documents from recognized organizations (e.g., Institut de recherche Robert-Sauvé en santé et en sécurité du travail [IRSST, The Research Institute Robert-Sauvé Occupational Health and Safety]; Agence nationale de sécurité sanitaire de l'alimentation, de l'environnement et du travail [Anses, the French Agency for Food, Environmental, and Occupational Health Safety]; Institut National de la Recherche et de la Sécurité [INRS, the French National Research and Safety Institute] Institut national de l'environnement industriel et des risques [INERIS, the French National Institute for Industrial Environment and Risks]). We also reviewed information from the search engine, Google, and from SAFENANO's Web site (safenano.org), the Organisation for Economic Co-operation and Development's site (www.oecd.org), The Project on Emerging Nanotechnologies' site (www.nanotechproject.org), and NanoImpactNet's site (www.nanoImpactnet.eu).

2.1. Data Synthesis

Nanoparticles have specific properties that make it difficult to characterize their physicochemical properties. However, the literature suggests that many parameters (i.e., size, chemical composition, surface reactivity, solubility, and the formation of aggregates/agglomerates) influence exposures, which are essential for better characterizations of the nanomaterials [7-10]. These parameters, which do not comprise the final list of factors, influence the risk of exposure and the potential health effects on exposure.

Exposure pathways are the concerns that profoundly influence the exposure assessment of nanomaterials. Because of their "nano" size, the main pathway for nanoparticles is via the airway route. So far, the available data show that nanoparticles can be distributed to different regions of the respiratory tree according to their size [4]. After inhalation, nanomaterials can also be found in the gastrointestinal tract via swallowing. The absorption of nanomaterials through the skin is still under investigation; however, it has been shown that professional tasks such as handling and maintenance of nanomaterials can result in dermal exposure to workers [11]. At the same time, the real nature of nanomaterials (i.e., powder and liquid suspension), the synthesis methods used, the degree of confinement, the quantity, the protection in the workplace, and the ability of the products to reach the air or the work surface are all factors that influence the exposure assessment [12; 13].

Appropriate toxicological tests to determine the effects of nanomaterials on health require specific tools that are yet to be harmonized and developed. This data collection will depend heavily upon the cooperation of industry and research [11; 14; 15].

2.2. Methodologies for assessment of nanomaterials

Several methods on the exposure assessment of nanomaterials are proposed in the literature (e.g., Lux Research Inc., DuPont and Environment Defense; Federal Office of Public Health Switzerland). These methods are mostly based on the standard model of risk assessment of chemicals using a strategy of quantitative or qualitative treatment of the data. An approach that is more qualitative and corresponds to the Control Banding model is very popular for this assessment [16]. The major methods used for assessment are discussed in the following paragraphs:

The Lux Research method is a quick and easy tool for qualitatively assessing the risk of specific nano-finished products to humans and the environment throughout a life-cycle approach. However, several important criteria are not used in the assessment criteria for this tool (e.g., criteria quotations for risk assessment of toxicity, non-differentiation between risks to human and environmental risks, how to handle products containing several types of nanomaterials) [17].

The DuPont and Environment Defense method is used to assess the nano-specific risks to humans and the environment from products containing manufactured nanomaterials. These risks are evaluated over the nanomaterials' entire life cycle. This method contains elements to aid management so they can base their decisions on the risk assessment (e.g., evaluate the effectiveness of risk management, methodological framework of actions to be taken based on an analysis of the situation). In comparison to the Lux Research method, the number of criteria is insufficient in DuPont and Environment Defense method, and the quotes tool does not take into account the toxicity and ecotoxicity at each stage of the life cycle [17].

The Federal Office of Public Health Switzerland has proposed a method of a precautionary matrix for synthetic nanomaterials to estimate and distinguish nano-specific risks of a product containing nanomaterials for workers, consumers, and the environment [14]. The method is to combine the scores associated with each of the criteria considered with a formula to derive a risk score. Different risks (i.e., employee, worker, and environment) are then characterized in terms of scores. However, this method is limited because it estimates the potential effects of nanomaterials. The score, given the level of these potential effects, is based only on two criteria for the entire life cycle: redox activity level and/or catalytic nanomaterial and stability in the relevant media (physiological and environmental conditions). This method considers that the toxicity remains constant during entire life cycle of the product and that the mechanism of toxicity is restricted only based on the formation of reactive species of oxygen.

The Control Banding model proposed by Paik et al. (2008) is an approach based on using a limited number of factors to assess the level of risk to reduce the complexity and increase the applicability [18]. This model uses three levels, or bands, of engineering controls based on the fundamentals of hygiene, plus a fourth level of control for most at-risk situations. In this approach, regular updating of new information is needed to ensure continuous and effective control of exposure [13].

2.3. Development of the assessment table matrix

As previously stated, the current data on the toxicity of nanomaterials remain fragmented and do not provide quantitative risk assessment of exposures. Therefore, using a methodology for determining safety assessment (qualitative) exposures to nanomaterials is an alternative to help reduce the risk of exposure to nanomaterials. Among the recent methods of qualitative assessment proposed in the literature, two have captured our attention: the model control banding specific case of nanomaterials proposed by the National Agency Food, Environment and Occupational Health Safety [19] and the methodological guide for characterizing emissions and occupational exposures to nanomaterials developed jointly by the Commissariat à l'énergie atomique et aux énergies alternatives (CEA, Atomic Energy and Alternative Energies Commission), INERIS and INRS [19; 20]. We designed our evaluation matrix based on these two methods.

The motivation of this new approach is based on the literature, which report the existence of a particular biological reactivity of nanoparticles in relation to their sizes, compared to the observed particles of the same composition, but in larger sizes. This reactivity of nanoparticles on cell and tissue biomaterials can represent a hazard to humans if they are exposed. However, continued development of this new technology is sufficiently important to our economy, but with respect to biological integrity.

From this complete review, we are proposing a new table matrix based on the published approaches, as well as the risk assessment methods used in toxicology since they were developed by the National Research Council in 1983 and used a default approach in environmental risk assessment [16]. This new approach contains the different phases such preliminary assessment of exposure, exposure dose/response characterization, and risk characterization (See Tables 1A and 1B in the appendix at the end of this paper). Each of these phases is associated with techniques and methodologies proposed to characterize the nanomaterial. Each phase drives the assessors to different plateau; thus, each plateau of the assessment can be stopped during the evaluation if the risk is considered to be negligible.

3. Results

The strategy of this table matrix is based on recommendations concerning the characterization of potential exposure to nanomaterials that are currently documented in the literature. This strategy relies on a step-by-step approach and takes into account proposals for the techniques and methods used to better characterize nanomaterials.

The Table Matrix developed in this works can be used by all employees responsible for health workers and consumer safety, stakeholders, and users of nanotechnology. The purposed of this table matrix is to contribute to a better understanding of nanomaterials to anticipate their potential effects. The analytical framework of this table matrix consists of the following four main steps [21]:

1. Primary evaluation of the exposure
2. Characterization of exposure and dose/ response
3. Risk Characterization
4. Restitution of the results.

This table matrix operates on a decision tree that describes each step for evaluation (Figure 1).

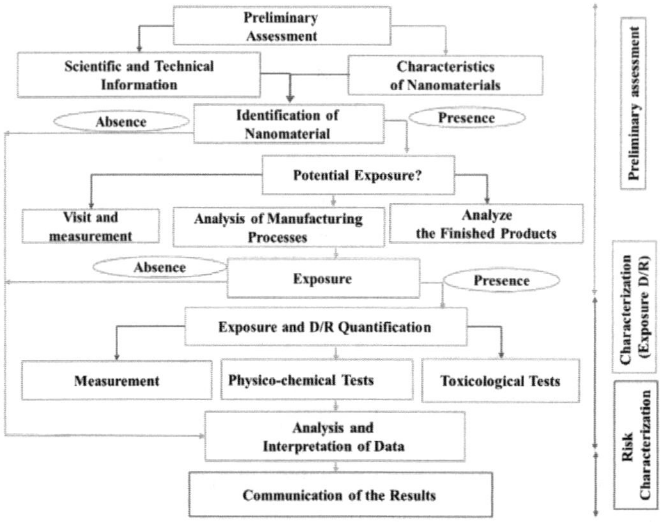

Figure 1. The decision tree for a safety assessment of nanomaterials.

4. Preliminary Assessment of Exposure

The preliminary assessment is one of the predominant steps in the evaluation of nanomaterials. In this step, technical equipment are used to identify nanomaterials (e.g. Condensation particles counter, Optical particles counter, SMPS,ELPI…).

4.1. Characteristics of nanomaterials

This step is based on the criteria used identify nanomaterials. Materials that do not meet the definition of nanomaterial will not be considered in the evaluation table matrix.

4.2. Scientific and technical information

These data refer to all available technical and scientific documentation about the nanomaterials, such as the following:

- Name of the material or nanomaterials, the unique registration number [9], and the main commercial uses
- Structural formula and the molecular and basic morphology of the chemical
- Composition of the nanomaterials used, including the degrees of purity and impurities and additives, and descriptions of surface chemistry (i.e., coating or modification)
- Catalytic activity.

4.3. Characteristics of nanomaterials

In this step, we document public or confidential information available about the potential nanomaterial study. This part includes a list of what we know about the nanoparticles, including the following:

- Raw materials, intermediate products, and final products
- Characteristics of the materials (e.g., fine powders, granular, colloids)
- Chemical composition of metals, oxides, or polymers
- Shape (e.g., spherical, cuboidal, aggregates/agglomerates)

- Size (e.g., number)
- Physicochemical properties (e.g., specific air, surface reactivity, solubility)
- Type of nanomaterials (e.g., carbon nanotubes, quantum dot, dendrimers)
- Dispersion medium (e.g., aerosols, gels, colloids).

At this step, the facts available should give us enough information to decide whether the particle is nanomaterial. If we have determined that it is not a nanomaterial, then we stop the process here.

4.4. Potential exposure to nanomaterials

During this step, we will investigate the potential exposure to nanomaterials to precisely target our analysis of the various manufacturing processes and working conditions. This step include three options that complement one another: the site visit, an analysis of the manufacturing processes, and an analysis of the finish product. For the last option, there is no question of analyzing each finished product; instead, there is a question about whether there is instability in nano-finished products that can induce toxicity.

4.4.1. Site visit

A visit to the production site allows investigators to take note of all working processes, including the delivery of raw materials, packaging and storage of final products, and waste disposal. The analysis must also take into account the following:

- Potential emission sources
- Routes of exposure (e.g., inhalation, dermal, ingestion)
- Magnitude and duration of exposure
- Ability of atmospheric contamination
- Configuration of the workplace
- General and local ventilation (system efficiency)
- Personal protective equipment (individual or collective)
- Preventive and safety measures.

During nanomaterials manufacturing, many techniques are designed to control the characteristic parameters such as size, shape, composition, and the degree of agglomeration of the nanoparticles to disperse systems and grains for bulk materials. These processes can affect the potential risk of exposure and the exposure routes. Therefore, an analysis of the manufacturing processes is necessary for the preliminary assessment. The instability of nanomaterials in the finished product is possible, and can subsequently cause a health risk to the consumer, especially to children. A thorough evaluation of the product before it is placed on the market would reduce this risk.

4.5. Characteristics of nanomaterials

At this step, we know that nanomaterials are present in the workplace and that there is a potential source of exposure. This next step of the process will provide information about quantitative exposure and the dose/response relevant interaction. The quantitative exposure is subdivided to contain the measurement and the physicochemical tests steps.

4.5.1. Characterization of exposure

The measurement of the sample uses different techniques, which will allow us to identify and characterize the target nano-aerosol at potential emission sources. Nano-aerosol can be measured at the source, in the breathing zone, in an area near or far from a work station. After identifying the likely nanoparticles at different emission sources, many types of additional measures may be considered to determine the number concentration (x/cm^3) or surface ($\mu m^2/m^3$), particle size distribution, chemical composition, specific surface area and surface reactivity, the shape of these particles, their crystalline structure, and solubility and their ability to form aggregates or agglomerates. The choice of instrument will be based on the type of study to be performed.

The physicochemical analysis is one of the critical steps in the risk assessment of aerosol (or nano-aerosol) because it is used to identify different chemical components such as ions, metals, and organic compounds. This analysis also provides information on the shape and size, chemical nature, the aggregates or agglomerates, and the type of aerosol. Moreover, the capacity of transition metals, such as copper, chromium, and vanadium, to catalyze the production of reactive species should not be underestimated [22]. All of these features can better assess the risk of potential exposure.

The exposure quantification provides no information about the toxicity of nanoparticles. This is essential to mention that toxicity of nanoparticles is not yet proven before this step. Thus, care must be taken to protect the health and safety of workers and consumers. In this regard, a better understanding of the physicochemical characteristics of nanomaterials and the biomonitoring or biological indicators of the exposure will increase the safety of workers. The next step will provide information about and quantification of the toxicity resulting from the exposure. Specifically, this step will help address the question: is there a reason to be aware of this exposure or not?

4.5.2. Toxicology assessment (dose/response quantification)

As with any chemical, the potential toxicity associated with the nanoparticles depends, in particular, on their physicochemical properties (i.e., size, surface reactivity, solubility in biological fluids). At this step, we need a test that provides rapid and low-cost information about the toxicity. So the in vitro tests represent a priori the better approaches. Tests to determine viability or proliferation, membrane integrity, oxidative stress and inflammation, or metabolic activities need to be prioritized as a first step to characterize toxicity.

However, knowledge of different exposure routes appears to be an essential prerequisite to the toxicological evaluation. Plus, nanoparticles are capable of producing reactive oxygen species [15], resulting in phenomena of oxidative stress and inflammatory responses, which can potentially induce DNA damage and lead to genotoxicity [23]. During the toxicological assessment of nanoparticles, one should pay particular attention to the target organs (i.e., liver, kidneys, lungs, Central Nervous System, and spleen). Toxicological tests may also be referred to the potential immunologic and reproductive toxicant nanoparticles.

4.6. Characterization of the risk of exposure

An analysis of the results allows for data to be produced on the basis of the previous tests. An analysis will provide the relevant information based on the samples collected, the techniques used to make the samples, the instruments and analytical methods employed, and the comparison of the results with the bibliographic data.

5. Discussion

This new approach presented in this paper represents the first intention to develop an integrated strategy to assess nanomaterials during the synthesis, far before the commercial integration in the finished nano-products. This approach is based on the default environmental risk assessment approach developed and published in the red book in 1983[16], but adapted for the nanotechnology reality and limitation. Nevertheless, it is more than that reason why our group wanted to make a significant effort in finding a balance between protection of biological integrity and economic capacity for manufacturers to incorporate these analyses into the Risk Assessment for Health. The objective of this work was to identify and maximize the analysis with the concept of integrity of a Risk Assessment for Health expert, but also based on an economic realistic perspective. This approach should allow manufacturers to identify the limitation of utilization with regard to their nano-products.

6. Conclusion

This work combines different and complementary steps, including, but not limited to, considering the method and preparing for analysis, noting the observations, conducting the analyses, quantifying the exposure and dose response, following a general process for evaluation, and comparing the observation and measurements to scientific literature. This new table matrix was also validated with European and Canadian company generated nanomaterial. The Table matrix we develop here provides a better understanding about the critical issue of nanomaterials with regard to worker and consumer safety.

7. Reference List

[1] L. Goldman, C. Coussens. Implication of nanotechnology for environmental health research. 1-71. 2005. The national academies press.
[2] B. Jurron. Nanotech's Evolving Environmental, Health, and Safety Landscape. Conference présentée au congrès Nanosafe. Conference présentée au congrès Nanosafe . 2010. Grenoble.
[3] M. C. Roco, *J. Nanopart Res* 13, (2011) 427-445.
[4] C. Emond, *Journal of Physics: Conference Series* 304, (2011) 1-8.
[5] V. Castranova, V. Vallyathan, *Environ Health Perspect* 108 Suppl 4, (2000) 675-684.
[6] W. Lin, Y. w. Huang, X. D. Zhou, Y. Ma, *Toxicology and Applied Pharmacology* 217, (2006) 252-259.
[7] A. Helland, M. Scheringer, M. Siegrist, H. G. Kastenholz, A. Wiek, R. W. Scholz, *Environ science & technol* 42, (2008) 640-646.
[8] S. J. Klaine, P. J. Alvarez, G. E. Batley, T. F. Fernandes, R. D. Handy, D. Y. Lyon, S. Mahendra, M. J. McLaughlin, J. R. Lead, *Environ Toxicol Chem.* 27, (2008) 1825-1851.
[9] G. Oberdorster, A. Maynard, K. Donaldson, V. Castranova, J. Fitzpatrick, K. Ausman, J. Carter, B. Karn, W. Kreyling, D. Lai, S. Olin, N. Monteiro-Riviere, D. Warheit, H. Yang, *Part Fibre. Toxicol* 2, (2005) 8.
[10] V. Stone, B. Nowack, A. Baun, N. van den Brink, F. Kammer, M. Dusinska, R. Handy, S. Hankin, M. Hassellov, E. Joner, T. F. Fernandes, *Sci Total Environ* 408, (2010) 1745-1754.
[11] INRS, *INRS* ND 2286, (2008).
[12] C. Ostiguy, Échantillonnage et analyse de surface (Ch 24) in Manuel d'hygiène du travail, Du diagnostic à la maîtrise des facteurs de risque, Modulo-Griffon, Qc 2004,.
[13] C. Ostiguy, B. Roberge, L. Ménard, C.-A. Endo. Guide de bonnes pratiques favorisant la gestion des risques reliés aux nanoparticules de synthèse. R-596. 2008. IRSST.
[14] S. Chevillard. Risque toxicologique des nanoparticules « Effets sur la santé ? 2010.
[15] D. M. Brown, K. Donaldson, P. J. Borm, R. P. Schins, M. Dehnhardt, P. Gilmour, L. A. Jimenez, V. Stone, *Am J Physiol Lung Cell Mol Physiol* 286, (2004) L344-L353.
[16] NRC, *Risk Assessment in the Federal Government: Managing the PRocess,* National Academiy Press, Washington, 1983.
[17] AFSSET. Évaluation des risques liés aux nanomatériaux pour la population générale et dans l'environnement. AFSSET. 1-223. 2010.
[18] S. Y. Paik, D. M. Zalk, P. Swuste, *Ann Occup. Hyg.* 52, (2008) 419-428.
[19] ANSES. Développement d'un outil de gestion graduée des risques spécifique au cas des nanomatériaux. Saisine N°2008-SA-0407 . 2011.
[20] INRS. Recommandations en matière de caractérisation des potentiels d'émission et d'exposition professionnelle aux aérosols lors d'opérations mettant en œuvre des nanomatériaux. Guide_méthologique_CEA_INERIS_INRS_V14, 32P. INRS. 1-32. 2011.
[21] S. Fiorito, A. Serafino, F. Andreola, A. Togna, G. Togna, *J. Nanosci. Nanotechnol.* 6, (2006) 591-599.
[22] D. M. Brown, V. Stone, P. Findlay, W. MacNee, K. Donaldson, *Occup Environ Med* 57, (2002) 685-691.

[23] N. Singh, B. Manshian, G. J. Jenkins, S. M. Griffiths, P. M. Williams, T. G. Maffeis, C. J. Wright, S. H. Doak, *Biomaterials* 30, (2009) 3891-3914.

8. Appendix

Table 1a. The table matrix for the preliminary assessment

Steps	Objectives	Methodology	Techniques and Materials
Preliminary assessment of exposure	***Identification of nanomaterials***	**General Technique and Scientific Information** (Material Safety Data Sheets and technical information fact sheets provided by the producer; articles or summary documents from the scientific literature)	Scientific literature
		• Characteristics of nanomaterials (e.g., powders, fiber or colloid, shape, and length; aggregates or agglomerates) • Mass volume, specific area; specific surface volume • Applications of nanomaterials (foods, medical biology, environment) • Toxicological properties (e.g., toxicology data sheet	
		Analysis of the Manufacturing Process	
		• Synthesis process, confinement degree in every step of the synthesis • Raw materials used, intermediate products and final products • Quantities of nanomaterials handled; application of the nanomaterials • Ability of the nanomaterials to be suspended in the air **Site visit:** • Potential emission sources of aerosol and dust levels; background, number, and importance • Ability of nanomaterials to contaminate working atmospheres • Information about any exposure or co-exposure likely increasing the risk • Technical activities at work (close to the target source, time, habit) • Routes of exposure (e.g., inhalation, dermal, ingestion) • Importance and duration of the exposure • Configuring the workplace • General ventilation and local (system efficiency) • Actual performance of prevention measures (e.g., ventilation, filtration, personal protective equipment) for the products handled • Fire and explosion risks • Measurement, prevention, and safety procedures	

Table 1b. The table matrix for the exposure, dose response assessment and for the risk characterization risk.

Steps	Objectives	Methodology	Techniques and Materials
Exposure and dose/response characterization	***Quantification of the exposure***	**Characterization of Materials** • Measurement of target aerosol and the background aerosol • Physical and chemical properties of nanomaterials (e.g., size, morphology, surface area, surface reactivity) • Particle number concentration (particles/cm^3), surface geometry ($\mu g^2/m^3$), particle mass concentration ($\mu g/cm^3$) • Morphological properties (e.g., size distribution, shape) • Particle size distribution (e.g., number, mass) • Chemical composition (e.g., inorganic, ionic, organic compounds) • Solubility of liquid materials	Condensation particle counter (**CPC**) Optical particle counter (**OPC**) Scanning mobility particle sizer (**SMPS**) and fast mobility particle sizer (**FMPS**) Electrical low-pressure impactor (**ELPI**) Electronic microscopy (**EM**) Cascade impactor (Sioutas, micro-orifice uniform deposit impactor [**MOUDI**]) Chromatography analysis (inductively coupled plasma-mass spectrometry [**ICP-MS**] or gas chromatography-mass spectrometry [**GC-MS**])
	Quantification of the dose/response	**Toxicological tests (in vitro or in vivo)** • Acute toxicity, cytotoxicity • Altered membrane • Blood parameters • Oxidative stress • Inflammatory response • Genetic modification	Enzyme-linked immunosorbent assay (**ELISA**) High-performance liquid chromatography (**HPLC**) Western blot Real-time polymerase chain reaction (**RT-PCR**) Flow cytometry
Risk characterization	***Analysis and interpretation of the results***	• Method and preparation of analysis • Terms of observations and analyses • Exposure response • General process of evaluation (e.g., exposure, aerosol) • Comparison of the results with scientific literature	Technical and scientific expertise; scientific literature
Retrieval of results			

Nanosafe 2012: International Conferences on Safe Production and Use of Nanomaterials IOP Publishing
Journal of Physics: Conference Series **429** (2013) 012071 doi:10.1088/1742-6596/429/1/012071

Managing Operating Procedures in Distributed Collaborative Projects

A Hool[1], M-G Beuzelin Ollivier[2], F Roubert[3]

[1]MatSearch Consulting Hofmann, Chemin Jean Pavillard 14, 1009 Pully, Switzerland
[2]Powder Technology Laboratory, Ecole Polytechnique Fédérale de Lausanne-MXD 333, Station 12, 1015 Lausanne, Switzerland
[3]Business Information Systems Department, University of Westminster, 115 New Cavendish Street, London, W1W 6UW, United Kingdom

E-mail: publications@samplebook.eu

Abstract. In recent years, large distributed collaborative projects have become very prominent in scientific research, allowing exchanges between laboratories located in different institutions and countries and between various domains of competence. Particularly the work on nanotoxicity – a field which has only been under investigation for a few years and is still lacking regulatory framework – highlighted the need for well-controlled methods, as well as rules for the handling and disposal of used materials. To obtain comparable and reproducible results of experiments conducted in a distributed context, the standardisation and proper documentation of the applied methods is crucial. The European project NanoDiaRA, whose aim is to develop nanoparticles and biomarkers for the early diagnosis of inflammatory disease, faces this situation as it involves 15 European partners and brings together different scientific cultures and professional backgrounds. Protocols especially developed for Superparamagnetic Iron Oxide Nanoparticles and a management system were designed and implemented within the NanoDiaRA project to fulfil those needs. The main goals were the establishment of standardised Standard Operating Procedures assuring transparency and reproducibility and the provision of access to these protocols to every project partner, as well as their clear allocation to carry out precise measurements and production steps.

1. Introduction

Large distributed scientific research projects, involving a large number of scientists from different institutions and scientific disciplines, have become increasingly prevalent in the recent past. In Europe, this is mainly the result of the European Commission's efforts to implement Research Framework Programmes (FP) aiming to foster collaborative links between different research institutions. In addition to the usual Good Laboratory Practice (GLP) and Good Manufacturing Practice (GMP) requirements which underlie any research project, large distributed research projects have brought along additional challenges. These can arise from the diversity of geographical and scientific backgrounds, differences in scientific cultures and methods or divergences in institutional approaches and management practices. To this effect it is essential to ensure a flawless exchange of data, methods and research samples between all the partners involved in such a collaborative project. Particularly it is essential for the following requirements to be fulfilled:

- Manufacturing steps, experimental measurements and production procedures have to be reproducible for every partner.
- Exchange of experimental samples, clinical probes and scientific data must be properly recorded.
- Sensitive samples have to fulfil specific requirements like correct deposition of potentially hazardous materials and appropriate storing. Strict rules have to be followed in line with the regulations of the partner providing the samples, those of the partner receiving the samples as well as the regulations of the project itself.
- Ethical standards have to be adhered to with regard to scientific data and, even more, clinical samples. Again, regulations of all partners involved have to be met.

Another major challenge of distributed collaboration is the effective use and integration of existing information located at the various involved sites. This can be a particularly sensitive problem in the area of clinical studies for which patients are recruited for sample collection and documentation of cases. It requires efficient data management measures to ensure the effective recording, investigating and communicating of the collected information [1]. In addition, the use of human biological specimens in scientific research is also a major issue with regards to bioethics. Collection, handling and subsequent storage of specimens must abide by very strict ethical rules. This is even more complex in the context of collaborative projects where different regulation systems exist. One of the primary concerns is to ensure that the safe and ethical handling of the samples and data security is guaranteed, as well as a proper use and safe disposal of these samples [2].

In the past few years, several institutions have been reconsidering the risk problems of nanomaterials. Many papers have been published regarding their toxicity, but a closer look at the approach showed a lack of proper data and methodology [3][4]. This makes it even more urgent than in other fields to ensure the adequate establishment and communication of methods. In the EU-funded FP7 project NanoDiaRA, partners come from different institutions across Europe and a wide area of diverse scientific disciplines. The project aims at developing nanoparticles and biomarkers to be used for early *in vivo* diagnosis of inflammatory diseases. This requires overall alignment and reproducibility of the various scientific and technological practices in the different research fields. Superparamagnetic Iron Oxide Nanoparticles (SPIONs) are produced by the Laboratory of Powder Technology (LTP) at the Ecole Polytechnique Fédérale de Lausanne (EPFL) and distributed to the other European partners who work in the fields of biology, biotechnology, medicine, physics and microtechnology. The production process of these particles and the exchanges with the other partners in relation to the particle use and the associated methods and results served as a basis to develop clear research protocols and an electronic management system that ensures a proper utilization and documentation of these protocols as well as of the related research samples. The methods which were used to establish the research protocols are described in section 2, followed by the standards put in place for the development and implementation of an electronic management system which are discussed in section 3.

2. Protocols for distributed research projects

2.1. Distributed scientific research

Conducting science within a collaborative context can fertilize research by bringing together people with different expertise, but can also raise several challenges that add to the usual scientific practices. This section explores the context of distributed projects in more depth to justify the need for the setting up of effective research protocols to support collaborations.

Cummings and Kiesler [5] conducted a very comprehensive study of 62 different collaborations – multidisciplinary, cross-institutional and both. They acknowledge that multidisciplinary collaborations are fertile grounds for innovation, i.e., the successful execution of original and inventive ideas and actions. However, they emphasize that these types of collaborations pose major challenges related to coordination, i.e., the combination of different parts of a project to achieve a common goal. They posit that these challenges are even intensified when in addition the collaboration is cross-organisational. Principal Investigators (PIs) find it difficult to synchronize all the efforts of a consortium of scientific partners who are geographically dispersed and who have different ways of operating. They believe that tighter coordination mechanisms driven by information technologies (IT) have a great potential but that the monitoring of their efficiency is crucial.

Hoekman et al. [6] emphasize distributed research within the European context. They highlight the tension between the benefits of easier cross-border collaborations, which allow the scientists to work together with partners from different countries, and the advantages of cultivating close collaborations which appear to result in more effective research. Jackson et al. [7] go beyond exploring the effect of geographical and institutional dispersion on collaborative research and focus on the temporal dimension. They contend that collaborative scientific work needs to be structured around four temporal rhythms – organizational, infrastructural, biographical and phenomenal – and that in order for the collaboration to operate successfully, additional efforts need to be invested in temporally coordinating the practices of the various partners to align these rhythms.

2.2. Working with research protocols

There have been many in-depth explorations of the use of research protocols particularly in the medical field, as medicine is very practice-based but also heavily knowledge- and research-driven. Timmermans and Berg [8] have investigated how the use of very detailed research protocols is embedded within the researchers' working practice. Its development often is a dynamic process that requires multiple interchanges between partners and a string of revisions before establishing the final usable product. Mørk et al. [9] provide a detailed insight of the utilisation of research protocols in various research projects. They thoroughly examine the role of protocols and associated Standard Operating Procedures (SOPs) and identify examples where the production of a research protocol is an individual activity used to carefully track down all the details of the technical operations necessary to conduct an experiment and faithfully document the related technical settings. The emphasis is here on the reproducibility of the scientific operations and on the quality of the experimental work. They also analyse situations in which the development of a research protocol is more of a collective and iterative process which requires in-team and across-team negotiations and the implementation of many adjustments before obtaining a workable document. Finally, they look at these contexts in which the research protocol is part of a much larger structured set of protocols and emphasise the importance of the interlinking of these protocols to support a broader experimental process. The main roles of research protocols in the context of distributed research projects are explored. They can be used to concentrate and bring together the focuses of the researchers, as vehicles for scientific communication and coordination between the various partners, as aide-memoire to pin down the details of the operational procedures to be undertaken during the experimental process, and, finally, as a driver for standardization, interoperability and ultimately quality assurance.

2.3. SOPs in the NanoDiaRA project

When NanoDiaRA started in early 2010, operating procedures to synthesise, characterise and manipulate nanoparticles *in vitro* and *in vivo* had already been in use by the different scientists involved in the project. Each contributing research group had its own portfolio with their laboratory requirements in relation to their own competences and domain of expertise.

Similar methods were applied at different locations across the NanoDiaRA consortium, and results were expected to be comparable (e.g., characterisation of nanoparticles, interaction of nanoparticles with living cells etc.). Certain partners provided the nanoparticles or developed new techniques, while other partners had to implement new processes to improve or manipulate these nanoparticles. The same characterisation methods had to be repeated at various time points in the development chain and at different sites to ensure reliability. Therefore, the various results obtained at different stages of this process had to be recorded, particularly when new methods were implemented. To meet these challenges, identically structured SOPs were established for the project in order to harmonise the different practices across the collaborative project.

A SOP is a protected document which should be shared among partners in different laboratories in order to ensure the reproduction of similar functions, i.e., conducting an experiment, prepare samples, exploit data, prepare administrative documents and send samples [10]. Those *modi operandi* fulfil GLP and GMP requirements to comply with the quality standards expected by the project. This can be done by following basic rules such as shared efficiency of the method and uniformity of the result in compliance with given regulations [11]. The following sections discuss the development process of these SOPs and focus on the particular requirements which have to be fulfilled when producing them.

2.3.1. Participation. The procedure to be implemented for the sharing of SOPs should be established in agreement with all the users involved [12]. In complex research projects, people with different technical backgrounds and scientific cultures may need to make use of the document. Therefore, it has to be understandable by all the collaborators. Any comment made by any members of the consortium who may potentially utilise the protocol has to be taken into account as it may help making it more explicit.

2.3.2. Concision. The content of the SOP should provide sufficient information to be understandable by all the consortium members but at the same time has to be concise enough to provide a maximum of reading clarity [12][13]. Long sentences should be avoided and meaningful examples can be included. The terminology used has to be clear and precise to avoid any unnecessary complexities and thus to allow any reader to get the essence of the protocol without the necessity of further inquiry.

2.3.3. Customisation. Each protocol should be produced in a way that it is possible to tailor it to specific needs and adapt it to different situations. Every individual protocol may have been produced in a certain context but may be reusable in a slightly different one; a certain range of options should be on offer, e.g., "for urine proceed..."; "for organs proceed…".

2.3.4. Identification. It is essential for SOPs to be unique; only one version must exist at a given time [12][14]. It must not be allowed for different researchers to have separate SOPs for similar but slightly different methods. The aim is reproducibility, and even the smallest variations in the procedure need to be recorded in a leading document, as they may have an impact on the result.

2.3.5. Version management. For traceability reasons, there should be no ambiguity on which version of the document is used for which experiment and at what time point. Each new updated version of a previous protocol should be clearly distinguishable from the previous one to avoid any confusion [12][14].

2.3.6. Qualification/Restriction. For measurements and characterisation methods, limits and ranges must be set: All users should know what specifications are to be expected as a result of this, for example nanoparticles of 40 nm size; no toxicity on cells… [11][15].

2.3.7. Availability/Accessibility. Even if the versions are tightly controlled and regularly updated, all SOPs should be made accessible to all the project partners across the consortium [12][16][17]. This allows the traceability of all experiments, avoids variation among the consortium members and promotes constructive discussions between the collaborators.

2.4. Practicalities of the use of SOPs

One of the NanoDiaRA project partners, the LTP at EPFL, already had SOPs in place when the project started. To extend their utilisation to the rest of the project, they used their expertise, for example for setting limits of measurement and their previous experience in SOP management to modify the existing documents in order to satisfy the global project needs. New protocols were also implemented to fulfil the consortium's requirements. To meet the new challenges of an extended collaboration, it was essential to develop very clear standardised templates for the protocols as well as a proper file organisation system to effectively manage their use.

2.4.1. Templates. Standardised templates were introduced to ensure that all future generated protocol documents would follow the same pattern to foster concision, consistency and reproducibility. A full reorganisation of the supporting system was then implemented for several purposes: SOP identification and segregation, document version management and access management.

Templates were created in order to facilitate the production of new or already existing procedures so that laboratory technicians and scientists are strongly encouraged to standardise their protocols in a simple and sustainable way. The SOP documents include practical information such as the title, the author's name and the person responsible for its control as well as the date when the SOP becomes effective. A summary provides a quick description of the procedure, i.e., a clear and concise explanation of the aim of this procedure. For a laboratory methodology, a list of materials and machines required for conducting the experiment or making the measurements should also feature.

As part of the technical description of the procedures, all the steps were numbered to allow better readability but also to enable internal referencing – e.g., "8. Repeat steps 1 to 7 twice" – and referencing to laboratory notebook (lab book) entries. These sections have to be filled in with great care and precision and remain concise so that the method is understandable to everyone. The language used is English and the terminology should be carefully selected to describe accurately how to proceed through each of the steps. The use of subsections is highly recommended in order to provide a well structured layout and to clearly separate each of the phases of the documented protocol or method, e.g., 1. Preparation of the sample; 2. Assay; 3. Analysis etc.

An appendix section was also included to provide space for recording complementary information that could be helpful while following the SOP, for example images illustrating a technique, graphs for back calculations, etc. [12]. All pages are numbered and include a reference to the name and version of the SOP.

2.4.2. File organisation. It is crucial to support the development of standardised protocols with a robust and flexible management system to avoid misleading, overlaps, duplications or handling errors [11]. This section discusses the SOP management system which was introduced to accompany the introduction of the unified project operating procedures. A new electronic system was thus designed at LTP to include the following areas: draft version, corrected version, final version and PDF forms to produce a step by step follow-up of the SOP preparation.

When a new SOP form is created, a strict naming convention has to be followed to include the type of SOP – i.e., cell-related; nanoparticles synthesis; human resources etc. –, the SOP number, the version – with a "D" for Draft – and the title, for example "HR 1-2D Arrival of new collaborator". As soon as the form is completed by the author, it is forwarded to a document control manager who will carry out a first review of this document. At the end of this process, the form is printed out and reviewed by the concerned users who are strongly advised to add comments, as those are essential to improve the quality of the document [11]. All comments and edits, if accepted, are then saved in the updated file and a second revision process takes place until the final version is accepted by everyone; the "D" is subsequently removed from the naming, and the final version is saved. Finally, the operating procedure document is printed out and signed by the author, the document control manager and the supervisor. This paper-based copy is then stored in a ring binder located in the laboratory for general accessibility. This master copy is also scanned and converted into a PDF file to safely keep an electronic version which can be passed subsequently onto other partners. The new version replaces the older one, if existing. The electronic copy of the older version is archived in the "old SOPs" section of the system, while the old paper version is removed from the laboratory. For monitoring purposes, old versions should be kept in order to be able to track down, for instance, which batch of nanoparticles was produced or which version of the method was used; important changes may have occurred in the SOP which could have affected the result. In addition, to avoid any mix-ups with older versions while new ones are effective, old versions should not be made available to anyone except to the document control manager [12]. Finally, the critical question of the dissemination of the SOP among the consortium partners is considered. It needs to be carefully managed because of the complexity of the collaboration with a high number of dispersed partners in a distributed research context. Once the SOP document is finalised, a satisfactory system with an easy approach should be found to allow anyone to access it and to upgrade the document whenever needed, but still under close supervision.

3. Information Technologies within distributed research projects

3.1. Collaboratories

To overcome the problem of different people at different locations, there have been many attempts to design information technology-driven collaborative platforms. Those are focused on group productivity [18] and on the reduction of the geographical and time-based constraints imposed by large multidisciplinary and cross-institutional collaborative setups. One main area of study has been the conceptual design and sometimes practical development of technology-driven collaboratories, i.e., the idea of an interconnected ever-accessible laboratory in which scientists can interact with each other and access instruments and data irrespective of location and time.

Finholt [19] puts forward a very detailed account of the historical evolution of this concept and provides an in-depth analysis of the opportunities and challenges that the model of collaboratories offers. He contends that even if collaboratories will not change the way science is done, they still represent a very relevant interconnection between digital technologies and resulting changing scientific practices. On one hand, collaboratories and other technological-driven collaboration platforms can enable an increase of influential research partners from various fields, a more efficient sharing of the resources and a more effective combination of observational data with theoretical visualisations. Thus, collaboratories can allow the bridging of the gap between various disciplines and also between theoreticians and experimentalists, shaping scientific practices towards a more multi-dimensional view of the research problem under study. On the other hand, the inherent constraints of the use of a computerised platform can lead to a reduction of the sense of a common global vision and a loss of common ground. Hence, the challenge for the developers is to provide the teams with tools which can counterbalance the diminution resulting from the lack of co-location [20] particularly for tightly coordinating experimental tasks which require frequent interactions and feedback between partners.

3.2. Management of the experimental process

In recent years, many attempts have been made to develop IT tools to support the experimental process at the heart of the work of the scientists. These range from Electronic Laboratory Notebooks (ELNs) which aim at extending the functionality of lab books, to Laboratory Information Management Systems (LIMS) which attempt to integrate the overall documentation of the experimental process [18][21].

Scientists involved in the experimentation as part of a scientific project need to record the experimental practices they undertake. They also need to thoroughly document all the activities which support the experimental process. Mackay et al. [22] offer a detailed account of what the lab book and its potential digital extension represents for the various categories of contributors to a scientific project. For the experimenter, the lab book should be distinguished from the simple production of scratch notes and is used to methodically record and track the experimentation procedures alongside the related observations, analyses and results. Their primary concern is that the lab book should be a universal platform embedded within their scientific practices which can seamlessly support the implementation of experimental operations. In their view, it is a simple and flexible tool which allows the capturing of meaningful information and its effortless retrieval. It is structured in a chronological manner to provide an accurate temporal account of the "live" tasks which were executed, but can also include some reflective elements when edited at a slightly later stage. In the practice of researchers, it often goes beyond the simple recording of scientific facts – it can assemble a wide range of materials for a variety of purposes such as experimental techniques, spontaneous observations, snippets of data, partial results, external documents and images. For the project management team, the lab book provides the opportunity to monitor the progress of the various members of the project. It is also seen as a centralised and durable repository of essential project information which can be used as evidence for discovery and as the basis for further exploitation.

There has been a tendency to consider IT to support the documentation of the experimental process and thus to implement a set of digital platforms. The latter provide scientific research teams with ubiquitous tools to be used in and out of the laboratory to manage the research-based operations, but also to help coordinate these in a distributed environment. Myers [18] distinguishes between these electronic lab books which are designed to cater for individual needs, and those developed to foster collaboration and support coordination between team and consortium members. Tabard et al. [23] propose an analysis of the functionalities offered by various electronic notebooks and experimental management systems. Schraefel et al. [21][24] make attempts at designing and developing tools which explicitly retain the specificities of the paper media, but extend it to provide additional benefits such as ease-of-use, flexibility and ease-of-share. Tabard et al. [23] stress that a new generation of tools should be considered to provide better support of increasingly complex multi-disciplinary and cross-institutional projects. In their view, the digital laboratory platform should be a central storage space of information, designed to support interactions between the various stakeholders of the scientific projects and networked collaborations between the various members of the research group. It should offer sufficient flexibility to allow the various teams to adopt an organisational strategy which is suitable to their specific needs and practices, and provide a simple means for the involved collaborators to share knowledge, know-how and specific techniques. Finally it should offer mechanisms to keep track of a wide range of heterogeneous physical and digital data such as internal and external written communications, research protocols, ongoing results and experimental samples.

3.3. Development of an Electronic Sample Book for NanoDiaRA

3.3.1. Driving Principles. For NanoDiaRA, several strategies were adopted with the aim to meet the high standards required by such an interdisciplinary project and to ensure the proper management of scientific data and samples. In a first stage, the partners from the different disciplines developed SOPs as described in section 2. An electronic system was initially developed to track the exchange of samples within the project. In a second stage, an internet-driven Electronic Sample Book (ESB) was developed to merge these two areas of functionalities. The idea was to provide all the partners with a documentation system to handle all the samples, methods and protocols used within the project. This should be conducted with a view to ultimately ensure the traceability and reproducibilty of each step of the production and each produced result. The goal was to design and develop a tool that is easy to use and that can in principle support any collaborative project, with a heavy focus on fulfilling GMP and GLP requirements as well as ethical regulations. Another important idea was to avoid the necessity of installing the system at the partner institutions and the need of an extensive training of the users, as this would have taken too much time and effort for a time-limited project like NanoDiaRA.

The data produced as part of the project is stored in a structured manner in the ESB by defining the samples as the basic unit of research. Technically, each sample is represented by a database record. Each sample record consists of different fields depending on its type. The sample types, their properties and associated processes for measuring and producing such a sample can be defined by the various involved institutions.

For NanoDiaRA, nanoparticles, MRI pictures, biological samples etc. can be created as records in the ESB. All necessary information is recorded by the users who fill in the essential details for each specific sample. As the users enter the required information for their specific samples, it is possible to create types for every item that can be recorded and exchanged within a project. Furthermore, one type can be integrated into another (e.g., an MRI picture of an animal, or a biofluid containing particles), thus allowing all the properties of the merged samples to still be traceable. Then any created records can be sent to other groups. To track the shipment of a sample, the *state* property of the record is changed to "on its way", and the group who is meant to receive it is notified that the sample has been sent. The ESB allows the monitoring of all scientific actions and of their results, helps harmonizing research protocols and prevents loss of samples.

For the purpose of the NanoDiaRA project, particle suspensions were synthesised at LTP and sent to other partners. The group working at EPFL was used as the basis to develop the first prototype of the system and establish proper and useable documentation and sending practices. Each new batch of particles was recorded into the ESB in order to track it and provide the related information. Ranges of acceptance were defined and implemented, thus allowing anyone consulting the database to determine the potential acceptance of the samples. Each sample sent to another location needs to be tracked and followed accurately throughout the process: i.e., the quantity, the properties, its usage and the obtained results. This enables for levels of the various consumables and for the needs of each of the partners to be constantly under control. Finally, essential information regarding samples and other items can be transferred from the databases located at the various partners' locations and recorded in the ESB.

3.3.2. Fulfilment of GMP, GLP and Intellectual Property requirements. All partners using the ESB can provide their established SOPs and other experimental process descriptions, details of the outputs of the experimental processes, i.e., "sample properties", an entry for every sample and clinical probe they are sending to or receiving from other project partners, and a description of the production steps, tests or other scientific work that are relevant for the project.

Similarly to a paper lab book, the ESB should ensure that after a result has been filled in once, no further manipulation of this entry is possible. If the properties of a sample need to be edited, the updates have to be entered as the record of a new result. This guarantees that the information provided at a certain time by the person responsible for the sample is accurately documented and is subsequently made available for any potential search in the case of Intellectual Property (IP) related questions. However, in the case of an accidental typography error, there is an option to mark a result as faulty for a certain period of time after it has been filled in. To ensure that everything is recorded, the faulty result is not discarded but appears as crossed out. For all other more significant changes, the administrator has to be contacted and can decide whether a manipulation should be modified or rolled back. Logging every data submission of all users registered in the ESB guarantees the effective monitoring of every action. Manipulations made by the administrator can be hidden to avoid confusion but are still recorded in the system. An independent server provider stores all the data as long as necessary; this is valid beyond the end of the project. Therefore, it ensures proper long term documentation of all information which may become crucial for IP questions that could occur during the project or as a result of the achievements of the project. Data monitoring and security is provided in NanoDiaRA by a subgroup in which academic and industrial partners are informed constantly. A connection with the clinical database at Charité Berlin was established, transferring selected project-relevant parameters as read-only data to the partners. The connection and selection of parameters was conducted in accordance with Charité's data safety regulation; therefore it serves as a prototype for the integration of internal partner data thus meeting the high standards of a clinical environment.

4. Conclusion and Recommendations

The management of large distributed collaborative projects is a complex process which requires thoughtful planning and organisation. The main focus must be to generate a valuable output of the results gained through the research project. This becomes especially relevant with regards to relatively new fields like nanoparticles and their toxicity. In this new environment, the establishment of methods and protocols are an evolutive and ongoing task, and the significance and impact of results is still under consideration [25][26].

Modern technologies nowadays have the potential to ease workflows between partners and provide the basis for a complete documentation accessible to everyone at all times. Different electronic systems can be put in place and adopted by all partners at various locations, but this takes time and capacities, raises the risk of safety leaks and jeopardises data protection. In addition, one of the main challenges specific to the NanoDiaRA project was to put in place specific techniques to fit the peculiarity of the materials in use and meet the challenges of a research area which is still lacking standards, as well as to develop clear procedures to handle and eliminate these materials in use in the project. Moreover, a fine line had to be found as it was important not to impose changes which were too radical for the partners involved.

The afore-mentioned challenges can be met by a two-fold approach:

- First, a clear methodology has to be implemented to attempt to unify the methods in place in each collaborating institution. Obtaining standardisation and complete reproducibility of the methods used by every partner can increase the value of the produced results and enhance the overall scientific impact of the project. To enable this harmonisation of methods in one unified methodology, relevant aspects for development and use of SOPs have to be defined.
- Second, flexible electronic information solutions ought to be designed and used to fulfil the specific needs of a distributed collaborative project, in accordance with the standards adopted by all partners involved, as well as GMP and GLP requirements. The focus hence should be on the ease of setting up and using of the system, a clear navigational structure, as well as on the possibility to operate rapid and easy comparisons of the produced results.

Both of these steps are crucial for the establishment of a proper workflow and the generation of comparable and valuable results across the distributed collaborative project. In NanoDiaRA, we made a first attempt to implement them in a structured and explicit way. We hope that the standards that we established will increase the scientific value of the results obtained throughout the lifecycle of NanoDiaRA, and that they can be used as standards for other projects.

Acknowledgments

The authors are supported by the FP7 project NanoDiaRA. Thanks to Margarethe Hofmann-Amtenbrink and Heinrich Hofmann for their valuable input and to Margret Pauels, Katharina Mader and Raphael Scholl for proof-reading.

References

[1] Mager S R, Oomen M H A, Morente M M, Ratcliffe C, Knox K, Kerr D J, Pezzella F, Riegman P H J 2007 Standard operating procedure for the collection of fresh frozen tissue samples *European Journal of Cancer* **43** 828–834

[2] Ravid R 2008 Standard operating procedures, ethical and legal regulations in BTB (brain/tissue/bio) banking: what is still missing? *Cell and tissue banking* **9** 151–67

[3] Hristozov D R, Gottardo S, Critto A and Marcomini A 2012 Risk assessment of engineered nanomaterials: a review of available data and approaches from a regulatory perspective. *Nanotoxicology* 1–19

[4] Krug H F and Wick P 2011 Nanotoxicology: an interdisciplinary challenge. *Angewandte Chemie (International ed. in English)* **50** 1260–78

[5] Cummings J and Kiesler S 2005 Collaborative research across disciplinary and organizational boundaries *Social Studies of Science* **35** 703–22

[6] Hoekman J, Frenken K and Tijssen R J W 2010 Research collaboration at a distance: changing spatial patterns of scientific collaboration within Europe *Research Policy* **39** 662–73

[7] Jackson S, Ribes D, Buyuktur A G and Bowker G C B 2011 Collaborative rhythm: temporal dissonance and alignment in collaborative scientific work *Proceedings of the ACM 2011 conference on Computer Supported Cooperative Work - CSCW '11* pp 245–54

[8] Timmermans S and Berg M 2003 *The gold standard: the challenge of evidence-based medicine and standardization in health care* (Philadelphia: Temple University Press)

[9] Mørk B E, Aenestad M and Ellingsen G 2004 Research protocols as a vehicle for the (re)production of knowledge *5th European Conference on Organisational Knowledge, Learning and Capabilities, Innsbruck, Austria, 2.-3. April 2004*

[10] Gough J and Hamrell M 2010 Standard operating procedures (SOPs): how companies can determine which documents they must put in place *Drug Information Journal* **44** 49–54

[11] Hattemer-Apostel R 2001 Standard operating procedures - a novel perspective *The Quality Assurance Journal* **5** 207–19

[12] Lindgren C 2009 Standard operating procedures *Cell Therapy: cGMP Facilities and Manufacturing* pp 109–20

[13] Gough J and Hamrell M 2010 Standard operating procedures (SOPs): how to write them to be effective tools *Drug Information Journal* **44** 463–8

[14] Gough J and Hamrell M 2009 Standard operating procedures (SOPs): why companies must have them, and why they need them *Drug Information Journal* **43** 69–74

[15] Hastings R 2010 Quality control in FISH as part of a laboratory's quality management system *Fluorescence in situ Hybridization (FISH)* **659** pp 249–59

[16] Grounds M D, Radley H G, Lynch G S, Nagaraju K and De Luca A 2008 Towards developing standard operating procedures for pre-clinical testing in the mdx mouse model of Duchenne muscular dystrophy *Neurobiology of disease* **31** 1–19

[17] Minei J P, Nathens A B, West M, Harbrecht B G, Moore E E, Shapiro M B, Bankey P E, Johnson J L, Freeman B, McKinley B, Moore F and Maier R V 2006 Inflammation and the host response to injury, a large-scale collaborative project: patient-oriented research core standard operating procedures for clinical care II. Guidelines for prevention, diagnosis and treatment of ventilator-associated pneumonia *The Journal of trauma* **60** 1106–13

[18] Myers J 2003 Collaborative electronic notebooks as electronic records: design issues for the secure electronic laboratory notebook (ELN) *Collaborative Technologies Symposium (CTS'03)*

[19] Finholt T A 2003 Collaboratories as a new form of scientific organization *Economics of Innovation and New Technology* **12** 5–25

[20] Olson G and Olson J 2000 Distance matters *Human-computer interaction* **15** 139–78

[21] Schraefel M C, Hughes G, Mills H and Smith G 2004 Breaking the book: translating the chemistry lab book into a pervasive computing lab environment *Proceedings of the 2004 ACM annual conference on Human Factors in Computing Systems - CHI '04* pp 25–32

[22] Mackay W, Pothier G, Letondal C, Bøegh K and Sørensen H E S 2002 The missing link: augmenting biology laboratory notebooks *Proceedings of UIST 02* **4** pp 41–50

[23] Tabard A, Mackay W and Eastmond E 2008 From individual to collaborative: the evolution of prism, a hybrid laboratory notebook *Proceedings of the ACM 2008 conference on Computer Supported Cooperative Work - CSCW '08* pp 569–78

[24] Schraefel M C and Dix A 2009 Within bounds and between domains: reflecting on making tea within the context of design elicitation methods *International Journal of Human-Computer Studies* **67** 313–23

[25] Clift M J D, Gehr P and Rothen-Rutishauser B 2011 Nanotoxicology: a perspective and discussion of whether or not in vitro testing is a valid alternative. *Archives of toxicology* **85** 723–31

[26] Oostingh G J, Casals E, Italiani P, Colognato R, Stritzinger R, Ponti J, Pfaller T, Kohl Y, Ooms D, Favilli F, Leppens H, Lucchesi D, Rossi F, Nelissen I, Thielecke H, Puntes V F, Duschl A and Boraschi D 2011 Problems and challenges in the development and validation of human cell-based assays to determine nanoparticle-induced immunomodulatory effects *Particle and Fibre Toxicology* **8**

AUTHOR INDEX

AUTHOR INDEX

AUTHOR INDEX

AUTHOR INDEX